Alberta's Petroleum Industry and
the Conservation Board

Alberta's Petroleum Industry and the Conservation Board

DAVID H. BREEN

THE UNIVERSITY OF ALBERTA PRESS

First published by
The University of Alberta Press
Athabasca Hall
Edmonton, Alberta
Canada T6G 2E8

Copyright © The University of Alberta Press 1993

ISBN 0-88864-245-8 cloth

CANADIAN CATALOGUING IN PUBLICATION DATA

Breen, David H.
    Alberta's petroleum industry and the Conservation Board

    Includes bibliographical references and index.
    ISBN 0-88864-245-8
    1. Petroleum industry and trade—Government policy—Alberta.
2. Gas industry—Government policy—Alberta. 3. Alberta. Energy
Resources Conservation Board. I. Title.
    HD9574.C23A43 1993   338.2728097123   C92-091719-4

Typesetting by
The Typeworks,
Vancouver, British
Columbia, Canada

Printed on acid-free
paper.

Printed by
John Deyell
Company Ltd.,
Willowdale, Ontario,
Canada.

To the early resource conservationists,
who fought for the conscientious management
of Alberta's natural endowment

CONTENTS

ILLUSTRATIONS

FIGURES

TABLES

FOREWORD

THE PETROLEUM AND NATURAL GAS CONSERVATION BOARD was created in 1938. Its responsibilities have since been expanded to all types of energy resources, and it is now known as the Energy Resources Conservation Board (ERCB). The occasion of the organization's fiftieth anniversary caused many of us to think about the significant role that the Board has played in the development of the province's energy industry. We also recognized that, even though detailed files on Board business were in existence, little had been done to ensure that the information on the Board's role was complete, properly preserved, and pieced together in a manner understandable to those who were not present as the events had unfolded. This led to the history project, an undertaking that involved many current and former Board staff members. The project included not only written material but also recorded interviews with many of the capable and colourful people who had figured prominently in the development of the organization that we know today.

This book is the showpiece of the history project. It details the Board's early years, describes events, explains why they happened, and introduces us to many of the personalities who made them happen. Dr. Breen has employed his exacting standards for detailed research and analysis in preparing this work, and he has written it in a style that informs while it retains our interest. For this, the Board is thankful. Special thanks are also due to those individuals, primarily former Board staff members, who devoted much time and attention to the collection and synthesis of the mass of information. This book deals with the first two decades of the Board's history. The efforts of Dr. Breen and his research assistants leave a record that will facilitate additional detailed analysis and serious written works on the Board's later history. It is my hope that this will occur in the future.

There is little doubt that this work will take its place as an important part of the written history of the development of public policy in Alberta and of the oil and gas industry in North America. Those interested in these subjects will find the book especially valuable, as will those who worked for the Board throughout the years.

The book has the potential to play another role for current employees of the Board, particularly those who have joined the organization in recent years. The awareness of the important role played by the Board in the early development of oil and gas policy and of the problems it faced will undoubtedly increase their understanding of the organization and their pride in its accomplishments. I expect that they will note the shift in emphasis for the organization from that of a regulator and policymaker in the early years to the primary role of regulator that we see today. This change is largely because of the shift in society whereby an impartial arbiter, at arms length from government, is required to make regulatory decisions regarding energy developments.

Notwithstanding the somewhat modified role of the organization, employees will see in Dr. Breen's assessment of its early history the emergence of the same basic principles that guide the Board today. They are the twin objectives of serving the public interest of the province and of doing so in an efficient manner based on the premise of working co-operatively with other involved parties. The problems and stakeholders have changed over the years, both in complexity and in number, but these early principles remain totally relevant. The efforts of working towards these objectives, through technical excellence and innovative approaches to regulation, have resulted in a highly respected regulatory agency.

This book chronicles the important role played by one organization in the development of energy policy in the province of Alberta. Many of us take pride in that organization. We can experience that same sense of pride in this book, which not only describes our beginning but also improves our understanding of why we exist.

G.J. DeSorcy
October 1992

PREFACE

ALTHOUGH IT IS a provincial agency, Alberta's Energy Resources Conservation Board (ERCB) is one of the most important regulatory bodies in Canada. Decisions of this agency have had profound consequences, often beyond Alberta. The significance of the ERCB, known from 1938 to 1957 as the Petroleum and Natural Gas Conservation Board (PNGCB) and from April 1957 to 1971 as the Oil and Gas Conservation Board (OGCB), stems directly from the key role that it has played in the Canadian energy sector. Curiously, the Board remains little known, and even in its home province the true nature of the Board's responsibilities is not well understood outside industry circles.

The ERCB's importance and its prescribed role are largely defined by the unique characteristics of the resources that it has regulated since 1938. Petroleum, including its associated by-products, was quickly recognized as essential to the well-being of modern communities, and access to assured sources of supply has remained a cornerstone of the foreign policies of industrial nations since World War I. Even the United States, the first industrial nation blessed with an abundant local supply, soon realized that it was in the national interest to ensure that production was managed so as to ensure long-term maximum recovery.[1] The idea that oil and natural gas resources warranted special attention came not only from the recognition that petroleum supply had become a matter of strategic importance but also from the growing awareness that the peculiar properties of this resource rendered traditional frameworks of exploitation and development inappropriate. Oil and natural gas are migratory resources, which means that the reserves underlying one tract of land can often be produced from a well located on another. Therefore, when a number of operators are producing independently from the same oil pool without any restrictions or regulations, the rational response of each producer will be to drill and produce as rapidly as possible to avoid being drained by his neighbours. Such frantic production to gain short-run advantage usually has a negative long-run effect. Most oil reservoirs are production-rate sensitive, and the total ultimate recovery of oil from the reservoir will be reduced if it is produced at too fast a rate.

In recognition of the fundamental importance of petroleum and the difficulties inherent in its exploitation, North American governments began about 1915 to introduce legislation to protect the public interest and to bring some order to the chaotic relations existing among contending oilfield interests. Alberta's Petroleum and Natural Gas Conservation Board was established in 1938, and like its counterparts in most oil-producing U.S. states, it set out to address two essential concerns. Empowered to prevent the wasteful exploitation of the province's oil and natural gas resources, the first item on the new Board's agenda was, to the extent practical, to eliminate the flaring or burning of unmarketable natural gas from the Turner Valley oilfield. Another problem to be faced immediately was that of devising some appropriate formula that would permit the sharing of production fairly among all those who possessed the legal right to produce. In the new Board's arsenal of powers to deal with the twin problems of conservation and equity, the most important was its authority to set an allowable production rate from each producing well. The Board's influence was enhanced further when it was later assigned a critical role in determining the quantity of natural gas that could be removed from the province.

The essential core of the Board's story, therefore, is about decisions. Board decisions, especially in the cumulative sense, have played a key part in shaping both the operating framework and the character of the petroleum industry as it has evolved in Alberta. Regulations formulated and enforced by the Board have affected production costs and discovery and development rates, as well as the relationship between producers. Even though it is a provincial agency, the ERCB's decisions and regulations affect the price and supply of a resource that comprises a vital component of the national economy.

Although the significance of the ERCB's regulatory role in the oil and gas sector of the Canadian economy is apparent, its influence in the narrower sphere of individual corporations should not be overlooked. Depending on the ranking criteria, six to eight of Canada's 20 largest corporations are oil and gas companies.[2] Both the operating environment and the financial well-being of these large and powerful corporations are affected directly by decisions emanating from the ERCB headquarters in Calgary.

Although the regulatory decisions of the ERCB and its predecessors, the Oil and Gas Conservation Board and, before that, the Petroleum and Natural Gas Conservation Board, have had a far-

reaching local and national importance, this has been almost ignored in the literature on the petroleum industry and energy policy.
Among the few to present a general history of Alberta's petroleum
industry, Eric Hanson is one of an even smaller number who have
made more than passing reference to the Conservation Board.[3] His
study offers a useful summary of how the Board operated during
the 1950s, but there is no attempt to understand the Board within
the context of the political and economic forces that brought about
its creation and shaped its evolution. Although there is no discussion of either alternative approaches or analysis of the consequences of Board actions, it must be acknowledged that Hanson
was writing in the midst of the boom that he was describing and
lacked perspective. His preoccupation was with the dramatic statistics of growth and development.

Perhaps the best of the more recent studies is *Prairie Capitalism:
Power and Influence in the New West* by John Richards and Larry
Pratt.[4] They take advantage of their greater time perspective to venture well beyond the more descriptive commentaries of earlier
writers and offer an interpretive assessment of Alberta's postwar oil
boom. Although the focus of their study is not specifically upon Alberta's petroleum industry, they do acknowledge the significance of
the ERCB and draw certain conclusions regarding its regulatory role.
As they see it, the Board was the centrepiece of a regulatory framework that was adopted uncritically from the practice in the U.S. oil-
producing states that they believe fostered the interests of the large
oil companies. Based more on apparent similarity of legislative
form than careful research, it is a conclusion that nonetheless has
found favour with subsequent commentators.[5]

In most popular accounts and nearly all company histories, the
Conservation Board's profile is almost nonexistent. In their haste to
get on with the "real" story, most authors have quickly passed over
the seemingly tedious landscape of the regulators and fixed their
gaze upon an apparently more colourful topography peopled by
roughnecks and entrepreneurs.[6] Such accounts of the petroleum industry might have colour and drama, but they present a partial picture only. Along with the oil barons and politicians, the various
chairmen of the Conservation Board must be considered among the
principal players in the history of the Alberta oil industry. With
little or no attention focused upon the regulatory environment or
upon the regulators, the written history of the petroleum industry
and our understanding of it remain incomplete.

Given the singular importance of the Board's role, what explains
its nearly nonexistent profile, even within the more serious and

analytical literature on the petroleum industry or regulatory bodies? In part, this might have to do with the remarkable political stability that has characterized postwar Alberta. Since World War II, there has been but one change of government, and this essentially was a change from one conservative administration to another. Moreover, opposition parties in the Alberta Legislature have seldom mustered more than a handful of members. This has meant that debate about development and regulatory policy has been muted, perhaps with the exception of the gas export question of the early 1950s, and that continuity rather than change has typified both policy and personnel in government and in the senior levels of the provincial civil service. With a history unmarked by dramatic change, with no great controversies debated in the public arena, with no great scandals or outrageous personalities, there has been little about the ERCB to attract the casual observer or the writer in search of a "good" story.

But what of the more serious professional observers? Energy policy has been an area of often embittered national debate for nearly two decades. Although a few scholars have been moved to address topics in this area, without exception the focus has centred upon policies, politics and institutions at the national level.[7] In these studies, Alberta's ERCB has usually received but passing mention, even though it is the initial point of regulation in the oil and natural gas sector and has played a principal role in shaping the regulatory environment surrounding this vital resource area.[8] This oversight is unfortunate but understandable. Although it is one of the most important regulatory agencies in Canada, its decisions have emanated from a city far removed from the familiar path of national policy-oriented social scientists preoccupied with institutions in Ottawa.

That a full and serious study of Alberta's Conservation Board is long overdue is readily apparent, but not just to derive a more balanced understanding of national energy policies and the dynamics of the Alberta and federal relations from which these policies emerged. Such a study, as has been suggested earlier, is also one of the critical elements of any comprehensive history of the Alberta petroleum industry. The organization, pattern and pace of the industry's development in Alberta over the past half century have been influenced by the policies of this Board. Moreover, the impact of Board decisions has extended well beyond the immediate confines of the petroleum industry. It is clear that this agency is one of the most important institutions to have shaped Alberta's postwar development. A careful look at the political environment and process that created and subsequently refined the legislation that governed the Board's activities, along with an examination of the forces that influenced the Board in its key decisions, would seem therefore to offer promising rewards.

The history of the ERCB is also worthy of study in a more theoretical sense. Much effort, particularly in the United States, has been expended studying the behaviour of national regulatory bodies. The consensus that emerges from the more recent literature is that regulatory agencies have not been particularly effective in protecting the public interest in the long run. Not long after its creation, it seems that a typical agency is inevitably co-opted by the industry it was designed to regulate.[9] Professional regulators soon come to see and interpret the world through eyes similar to those of the regulated industry. Differences come to be more of detail than substance. Does the history of the ERCB lend confirmation to this recognized pattern? It must be asked how successfully the Board has fared on behalf of Albertans in the face of the variety of private interests that frequently clamour for special consideration. In the name of resource conservation, has this regulatory agency functioned ultimately, as one writer recently asserted, as a client of the major oil companies to establish "a producers' cartel that eliminated price competition to the detriment of Canadian consumers."[10] The purpose of this study therefore is to examine the ERCB with particular attention to the legislation that defined the Board's role, the important issues that the Board has had to address, the process by which it has come to important policy decisions, and the leadership role of Board chairmen. Policy decisions rest upon the judgements of key individuals, and to clarify the factors that shaped and motivated these judgements the Board and its activities must be considered within the political and economic setting in which it was created and functioned. Advantageous perspective on the Board and its activities is also to be gained through reference to an outside measure, and for this reason some comparison with similar U.S. agencies will also find a place in this study. Finally, it must be stressed that this is a historical study and not an economic analysis. The long-term economic implications of certain policies or decisions must remain an indeterminate question, perhaps to be pursued more appropriately in a specific or specialized examination.

This study of the Conservation Board begins with a review of essential geological, engineering and legal concepts, followed by an examination of the U.S. experience from which early petroleum and natural gas conservation emerged. The subsequent discussion is divided into two naturally defined periods. Part I reviews the discovery phase of petroleum development in western Canada, with particular attention to Turner Valley and the problems there that gave rise to the conservation movement in Alberta. Close examination is given to the debate on petroleum and natural gas conservation and to the 1938 legislation that established the Petroleum and Natural Gas Conservation Board.

Part II begins with a close look at the regulatory foundation established during the Board's formative years before the Leduc oil discovery in 1947. This discussion is followed by an examination of the Conservation Board's role during the critical first decade of Alberta's post-Leduc oil boom, when the Board moved through youthful trial and adjustment to become a mature technically competent conservation authority. Special consideration is given to the natural gas export decisions of the 1950s, the surplus oil production capacity question, the development and enforcement of Board regulations, and the changing technology of oil and gas conservation. A central theme here is how the Board, confronted by individuals and an industry often preoccupied by short-run perspectives emanating from constant focus on the yearly balance sheet, managed to promote long-range conservation policies. In 1959, Conservation Board Chairman Ian McKinnon left Alberta for Ottawa to chair Canada's new National Energy Board, and petroleum and natural gas regulation at both the provincial and federal levels entered a new phase.

All relevant documentary sources for this period at the Board's Calgary headquarters were made available for this study. These include Board minutes from 1938, letters to oil and gas field operators from 1932, minutes of meetings with industry groups, internal correspondence and numerous reports on topics of timely concern. The transcripts of evidence presented at Board hearings, which, along with the Board's "decision Reports," are part of the public record, have been used extensively. While the Board's court-of-record status has ensured the preservation of most records of historical importance, the great wealth of information at the Board nonetheless is compromised by the regrettable loss of its chairmen's correspondence for the 1940s and most of the 1950s and 1960s. To some extent, this deficiency can be made up through consultation of the premiers' papers, especially those of E.C. Manning, available at the Provincial Archives of Alberta in Edmonton. This collection contains most, and perhaps all, of the written communication between the various Conservation Board chairmen and Premiers Aberhart and Manning. Other correspondence between the premiers and fellow politicians, at both the provincial and federal levels, private citizens and oil company executives, on matters relating to the Conservation Board and its activities, offers helpful background in many areas. Deputy ministers' correspondence and other relevant documentation from the Alberta Department of Lands and Mines, later the Department of Mines and Minerals, as well as from the Department of Intergovernmental Affairs at the Provincial Archives, also allow important insights into the formulation of policies governing the administration of Alberta's petroleum resources. The extensive tape-recorded commentary prepared by Ernest Manning at the University of Alberta Archives was similarly helpful. At the Public Archives of Canada in Ottawa, the C.D. Howe Papers and various files from the collection of the Department of the Interior and its successors, as well as the Department of Munitions and Supply and the National Energy Board, contributed important elements of the national context. The transcripts of certain commissions, particularly Alberta's McGillivray and Dinning commissions, as well as Canada's Borden Commission, provided another rich source upon which this study relied. Tape-recorded interviews undertaken by the Glenbow Archives in Calgary as part of the "Petroleum Industry Oral History Project," plus the numerous interviews with former Board members and employees, and industry officials, carried out expressly for the preparation of this history, have added a further important dimension to the foundation of sources upon which this study rests. The "Scrapbook Hansard" record of debate in the Alberta Legislature; the popular press, particularly the Edmonton dailies, the *Bulletin* and *Journal,* and Calgary's *Herald* and *Albertan;* and the main trade journals, the *Western Oil Examiner,* the *Daily Oil Bulletin* and *Oil in Canada,* have been used to shed light on the political, economic and technical environment in which Alberta's oil and natural gas legislation evolved and which the province's Conservation Board reflected.

ACKNOWLEDGEMENTS

THIS BOOK IS ADDRESSED to all those who have an interest in or curiosity about the petroleum industry and the manner in which it developed in Alberta. It represents a huge undertaking that would have been impossible without the generous assistance of many institutions, organizations and individuals. Standing first among these varied groups is the Energy Resources Conservation Board. Without the initiative, research funding and unrestricted access to records that the Board provided, this study would have been impossible. The faith of former Board chairman Vern Millard and current chairman G.J. DeSorcy in this project, and their desire for an open and comprehensive approach to the study of petroleum regulation in Alberta, was crucial. The completion of this study extended well beyond the date originally estimated, and I am particularly grateful for the patience and unstinting support of deputy Board chairmen Norman Strom and Frank Mink. Board librarian Liz Johnson and her staff consistently offered helpful and congenial assistance. I am also pleased to acknowledge the important contributions of Barry Scott, Grant King and others in the Board's drafting department, particularly Joan Richardson, who prepared most of the maps and figures for this book. Among the many former and current Board staff whose interests and special efforts have enriched this study are Ian Cook, Rod Edgecombe, Joan Evans, Eileen Flegg, Don Hannah, Simone Marler, Al Mayer, Anita Miltimore, Merv Mumby, Alyn Olive, Olga Potter, Bob Pow and Kelly Schieman, and the late Red Goodall and Ken Fuller.

Others outside the ERCB who have made important contributions are William Epstein, a native Calgarian and distinguished United Nations disarmament adviser, who provided invaluable information on the drafting of Alberta's first Conservation Act; Hubert Somerville, former deputy minister of the Department of Mines and Minerals, who helped to clarify various aspects of the legislation and regulations of the period; and Mark Frawley and Ronald McKinnon, who furnished helpful background on their fathers—two of the more important individuals who figure in this study. I also wish to acknowledge the help and encouragement that I received from geologist-historian Aubrey Kerr. My research would have been seriously deficient had not two former Alberta premiers, the Honourable E.C. Manning and the Honourable Peter Lougheed, permitted me to consult their papers. I am sincerely grateful for their consideration.

Archivists of the National Archives in Ottawa, the Provincial Archives in Edmonton and the University of Alberta Archives have been essential contributors to my research for the project. In particular, I would like to thank Doug Cass at the Glenbow Archives in Calgary. I also wish to acknowledge the generosity of the *Nickle Oil Bulletin,* the Canadian Petroleum Association and Nova Corporation, who allowed me access to important records in their possession.

This study has gained substantially from the observations and suggestions made by the readers of earlier drafts. In this regard, I wish to thank Gordon Connell and a former Board member, the late Doug Craig. George Govier and Frank Manyluk read the entire manuscript, and I am grateful for the benefit of their insightful comments. Valuable observations were also received from the unnamed referees chosen by the University of Alberta Press. The final product also owes a great deal to Mary Mahoney-Robson's editorial guidance, the proficient copy-editing of David Evans, Kerry Watt's design and the comprehensive index prepared by Eve Gardner. In addition, I must extend my sincere thanks to Pat Schulze, whose commitment and good humour remained unfailing through the typing of numerous manuscript revisions.

The publisher wishes to acknowledge the Book Publishers Support Programme of Alberta Culture and Multiculturalism for a grant to prepare the manuscript for publication.

Last, I wish to acknowledge my singular obligation to retired Conservation Board engineers Murray Blackadar and George Warne. Their assistance began with the start of the project in September 1985 and their contribution stands apart from all others. My debt to them for their painstaking research, criticism, constant encouragement and friendship can never be repaid.

That for which I am beholden to my wife, Patricia, and family, Derek, Stuart and Patrick, is of a different order. They offered remarkably patient and essential support to an often absent and distracted husband and father. To them I shall ever be in debt.

What merits this book might possess rest heavily upon the contributions of those named; what deficiencies of fact or interpretation exist rest solely with the author.

Conservation, Reservoir Characteristics and the United States Experience

BEFORE EXAMINING THE ORIGIN and evolution of Alberta's Conservation Board, it is helpful to clarify certain principles of petroleum geology and reservoir engineering. Similarly, some knowledge of the peculiarities of oil and gas property law and the structure of petroleum industry in North America is essential to understand what conservation measures are intended to achieve. A logical place to begin is with the term "conservation."

MEANINGS OF CONSERVATION

Government departments or specialized agencies, typically called "conservation boards" or "commissions," regulate the petroleum industry at the provincial or state level, and their entire range of activities is commonly described by the phrase "conservation regulation." The statutes that define the powers and responsibilities of such bodies are generally known as "conservation acts or laws." Such general usage renders the term "conservation" almost incapable of specific definition. Having absorbed many related ideas to assume a general meaning, "conservation" has also retained a more precise usage. The confusion that surrounds the purpose and activities of petroleum conservation agencies stems in good part from the different meanings ascribed to the word "conservation" by the public, the petroleum industry and economists.

The popular notion of conservation is rooted in the traditional credo that emerged at the turn of the century. Thoughtful Americans realized that three centuries of frontier experience had finally run their course. As scholars contemplated what impact the closing of the frontier might have upon U.S. society, it became apparent that the notion of the United States as a country of limitless lands and resources was no longer appropriate. The idea that present use should be curtailed so that the needs of future generations might be met gained wide support. Conservation came to be seen as a restraining force that had to be fixed upon a free enterprise economy to protect the public interest against rapacious private greed. From the latter part of the nineteenth century in the United States, and mainly after the turn of the century in Canada, so-called "conservation" measures were applied to natural resource areas, including forest preservation, fisheries and wildlife, parklands, streams and rivers, as well as mineral exploitation. Although such measures were controversial and inevitably touched upon conflicting private and public interests, they have generally been sustained by the common belief that long-range social and economic benefits would be foregone without public regulation.

For early conservationists, preservation for future use was the essence of their concern, but the related idea of efficient production to eliminate waste emerged more slowly. It was not until the importance of exhaustible and nonrenewable mineral resources to modern communities became more apparent that the idea of promoting economically efficient methods of production to assure the maximum ultimate recovery of the resource was seen as the appropriate complement of conservation. For traditional conservationists the emphasis nonetheless remained on preservation—efficiency could not be equated with conservation. The reason, as Erich Zimmerman has explained, is that efficiency promised to lower costs which, if reflected in lower prices, might stimulate demand and thus cause "an accelerated exhaustion of earth materials, which to [the traditionalists] appeared as the very antithesis of conservation."[1] This primary focus on preservation has remained central to the popular notion of conservation.

Economists have their own notion of conservation. They argue that conservation involves maximizing the present value of a resource: in other words, getting the most petroleum in the present for the least investment. Their concept of conservation is centred upon the necessity of efficiency. Inefficiency or "misallocation" in the use of resources is one of the discipline's classic concerns, and a considerable literature on the theory of conservation has developed. The central concept centres upon the maximization of social benefits over time through correct distribution of resource use. Based upon the fundamental doctrine of maximum efficiency being the essential measure of social benefit, an economic definition of conservation might be summarized "as action designed to achieve and maintain the optimum time distribution of use of natural resources."[2] The critical concept in the definition is "optimum time distribution." It does not denote, however, the concern for future supply that agitated traditional conservationists. As between present and future use, the concept is neutral. As Lovejoy and Homan, two economists noted for their work on petroleum conservation, have explained, optimum time distribution offers "a way of defining an economic test of efficiency to be applied to various sets of circumstances."[3] In other words, the economic problem facing the resource owner is to determine which production alternative will yield, in present value terms, the maximum return over the life of the resource. Similarly, society wants ideally to obtain the maximum net benefits from a resource over time. To this end, economists have generally been critical of the petroleum industry and regulatory bodies and have structured elaborate models as a guide to move government, industry and individuals in the direction of what they argue would represent more rational policy formulation. At the same time, economists have argued for the acceptance of a more rigorous and precise definition of conservation that would conform to economic logic and for the removal of what they consider are inappropriate regulations and restraints that compromise the economic criterion of efficiency.[4] Adoption of the economists' definition of conservation has definite implications, the most important being the development of reservoirs as a unit.

The petroleum industry and its regulators use the term "conservation" in a much more general and usually less consistent way. Industry publications and the comments of conservation officials denote a blend of traditional conservation philosophy, insights from practical engineering experience, and economic theory.[5] In 1950, early in his career as a member of Alberta's Conservation Board, Dr. George Govier explained to a Vancouver audience in an address on the subject of oil and gas conservation that the realization by western Canadians that their resources were not inexhaustible accounted for the broad acceptance of conservation principles. He went on to explain that

> conservation involves the efficient use of natural resources, the development of these resources in such a way as to protect the interests of future generations, and the elimination of all economically avoidable waste. It may be defined as "The preservation of natural resources for economical use." The concept of the elimination of waste is paramount.[6]

The traditional idea of preservation for future use is seen to be part of Govier's concept of conservation, and his concern about the prevention of waste touches upon one of the two fundamental objectives that the petroleum industry has come to associate with "conservation" programs. To prevent oil and gas waste, regulators focused on ensuring maximum recovery from underground reservoirs. Significantly, critical economists have often observed that this concern about waste is focused by statute and regulation almost entirely upon the production and storage of petroleum and only to a limited extent on refining and end use. The second major purpose associated with conservation programs is the protection of "correlative rights." This takes the form of measures designed to ensure that production is allocated so that each owner of oil and gas rights in a field has an opportunity to obtain his equitable share.

Closely allied with these primary objectives, the prevention of waste and the protection of property rights, is an important—though implied—goal that has influenced the thinking of regulators, or at least those who drafted the conservation acts. Underlying the regulatory apparatus is the compelling desire to maintain a healthy domestic producing industry to sustain the local or regional economy. The industry's role as a critical generator of employment and government revenues has never been far from the minds of provincial or state legislators. Considerations of national security have also influenced the manner in which the industry has been regulated in the United States and, to a lesser degree, in Canada.

It is apparent that the term conservation has a diverse lineage. This, with the broad-ranging and not always consistent mandate of so-called conservation agencies, renders precise definition unlikely. It is doubly important therefore to understand the strands of thought, diverse and inconsistent as they might be, that are to be found within the concepts that have guided industry and petroleum conservation bodies.

If conservation practice does not exhibit the tidy parameters or theoretical purity that some would prefer, it is hardly surprising. Knowing the necessary difference between what is theoretically desirable and what is politically acceptable is one of the first laws of practical and successful administration. Petroleum conservation regulations did not emerge full-blown from a solidly constructed theoretical foundation. Rather, they developed incrementally in response to specific problems, in light of field experience, in the wake of changing technology, or, in other words, in the face of shifting historical forces. A proper understanding of the role and activities of Alberta's Conservation Board demands therefore a careful analysis of these historical forces. The historical forces set in motion by the discovery and increasing importance of petroleum were shaped in the first instance, however, by the mineral's peculiar physical properties. It is necessary therefore to consider first the factors that determine the occurrence of petroleum and attend its behaviour, so that implications that follow from these characteristics will be clearly understood.

ELEMENTS OF GEOLOGY AND RESERVOIR MECHANICS

Petroleum and Natural Gas

Petroleum is a mixture of naturally occurring hydrocarbons that can exist in a solid, liquid or gaseous state, depending upon the temperature and pressure to which it is subjected. Almost all petroleum is produced in liquid or gaseous form, and these constituents are referred to as either crude oil or natural gas, depending upon the hydrocarbon mixture.[7] Whether the substance is crude oil or natural gas is largely determined by the number of carbon atoms in its molecular structure. For both oil and gas, the greater the number of carbon atoms the higher the specific gravity. Crude oil might also contain impurities such as sulphur and minor amounts of certain metals that affect the later processing at the refinery stage.

Natural gas is found with oil, and frequently on its own. Within the reservoir, gas might exist in solution with crude oil, but upon being produced to normal atmospheric conditions, it separates in the gaseous form. Gas produced from either an oil or gas reservoir may be "wet," that is containing heavier hydrocarbons that can separate out in liquid form under surface conditions. These liquids are variously known as condensate, distillates or natural gas liquids, and they differ from crude oil in their molecular composition. In certain reservoirs, gas can be found that does not contain liquids. This gas is simply referred to as being dry. Natural gas can also contain impurities, especially hydrogen sulphide or carbon dioxide.

Although natural gas is important as a fuel in its own right, it can also be of substantial importance in the production of oil. This relationship between oil and gas is central to an understanding of the concerns that motivated the search for effective petroleum conservation measures.

Geology of Western Canada

The Western Canada Sedimentary Basin is the chief geological feature relating to the occurrence of oil and gas. Stretching from the Northwest Territories to South Dakota, it is the largest basin within the vast multibasin freeway that extends from the Arctic islands through the Mackenzie Delta and great central plains in varying widths to the Gulf of Mexico. The Canadian portion of the Western Canada Sedimentary Basin, shown in Map INT.1, occupies an area

MAP INT.1   Western Canada Sedimentary Basin

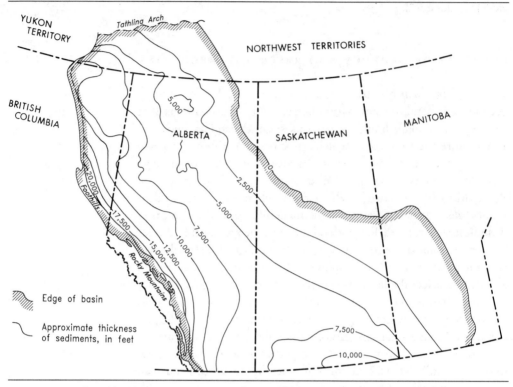

SOURCE: ERCB, adapted by J.R. Pow.

of some 540,000 square miles. It is bounded on the north by a geo-logical feature known as the Tathlina Arch, located in the south-western corner of the Northwest Territories; on the east by the sedi-mentary section's erosional edge at the Canadian Shield; on the south by the international boundary; and on the west by the Rocky Mountains. The Basin consists of undeformed strata, except in the narrow foothills belt where they were severely distorted when the Rocky Mountains were formed, between about 75 million and 27 million years ago.[8]

The Alberta portion of the Basin covers some 92% of the prov-ince's 255,000-square mile area. Undeformed sedimentary strata occupy an area of approximately 223,000 square miles beneath the plains, lowlands and uplands of the province and, as shown in Fig-ure INT.1, thicken gradually from zero at their northeastern edge to over 19,000 feet at the eastern boundary of the foothills. The se-verely deformed strata of the foothills belt extend over some 11,500 square miles and are believed to average between 15–20,000 feet in depth. The cross-section presented in Figure INT.1 shows the rela-

tionship and relative thickness of the principal geological formations. Most of the oil and gas reserves in the gently dipping, relatively predictable sedimentary strata that characterize the plains
portion of the Basin have been found in Cretaceous and Devonian
formations. In the more geologically complex western edge of the
Basin, in the Rocky Mountain foothills, oil and gas reserves occur
mainly in the folded and faulted formations of the Mississippian
age. The heavier crude oil and drier gas reserves are found in the
eastern areas, whereas the lighter crude oil and wet and sour gas
(containing hydrogen sulphide gas) are found mainly in the western
portion of the Basin. Coal occurs mainly in the Cretaceous sediments, with the higher heating value coals being encountered in the
mountain and foothills areas.

Reservoirs

Oil and gas reservoirs are commonly referred to as pools. A general
area underlain by one or more pools, often in strata found at different depths, is known as an oil or gas field. Rather than existing as
a vast cavern filled with oil or gas, as is often pictured in the minds
of laymen, reservoirs actually consist of sections of porous rock or
sand with oil and gas in the pore spaces. As one author has explained, "In geologic time, oil and gas, and associated salt water
migrated through the pores in rocks (usually called sand), and when
a barrier or impermeable material was encountered, the migration
stopped and the oil and gas became trapped. The result was an accumulation of oil and gas, or a pool."[9] The pore space in reservoir
rocks can vary from less than 5% to over 35%. For limestones a
range of 5 to 12% is common, and for sandstones, 15 to 35%.[10]
From 5 to 70% of the total pore space can be occupied by water
(connate water), thus reducing the space available for oil or gas.
Oil- and gas-bearing rocks vary not only in terms of their "porosity" but also in their "permeability," or the ease with which they
permit the migration of fluids. In the case of oil, flow within the reservoir is also a function of its viscosity. A low viscosity or light oil
moves more readily than a sticky, highly viscous heavy oil. Remaining variables that define the quality of a reservoir are the depth or
thickness of the petroleum-bearing formation and its areal extent.
So far, the largest single oilfield discovered in Alberta is the Pembina field, which contains the large Cardium pool, comprising an
accumulation in Cretaceous sediments. The largest gas accumulation by area is known as the Southeastern Alberta Gas system. The
several characteristic "structures" known to geologists as likely
sources of trapped oil and gas in Alberta are shown in Figure INT.2.

FIGURE INT.1   Schematic Cross-section through the Alberta Portion of the Western Canada
Sedimentary Basin

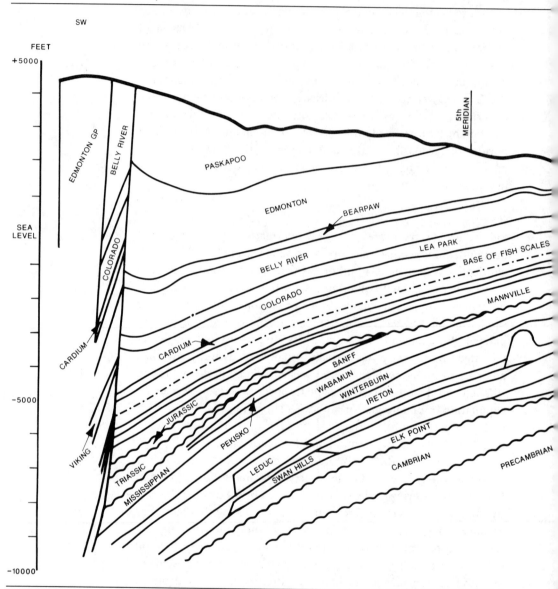

In each case, the seal is formed by a nonporous stratum, which pre-
vents further upward migration of fluids. Often the oil or gas zone
is underlain by extensive water-bearing porous sections, which
could be part of the same aquifer underlying other oil or gas pools.

On its own, oil possesses little energy to expel itself from the

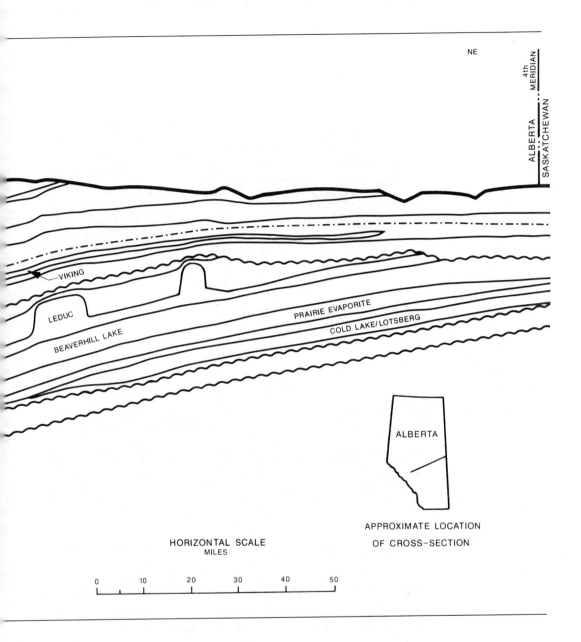

NE

4th MERIDIAN

ALBERTA

SASKATCHEWAN

VIKING

LEDUC

PRAIRIE EVAPORITE

COLD LAKE/LOTSBERG

BEAVERHILL LAKE

ALBERTA

APPROXIMATE LOCATION
OF CROSS-SECTION

HORIZONTAL SCALE
MILES

0   10   20   30   40   50

pores of rock and flow to the surface. Either water encroachment or gas expansion provides the natural driving force that allows production to occur. These forces become operative only with the release of pressure that takes place when a hole is bored into the reservoir rock to provide a connection between the high-pressure oil

FIGURE INT.2   Examples of Several Entrapment Conditions for Oil
and Gas

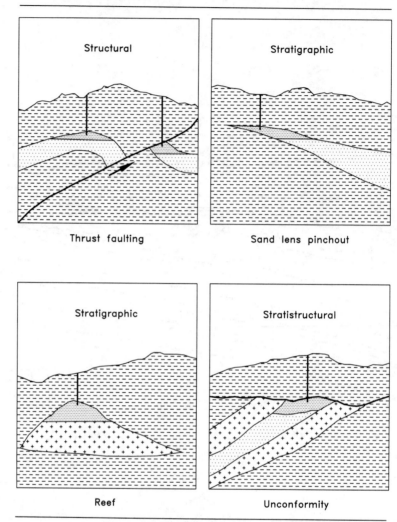

reservoir and the much lower pressure existing at the surface. Without sufficient pressure differential between the reservoir and the
surface, oil will not rise to the surface, hence the importance of
maintaining reservoir pressure. There are three main mechanisms,
or "reservoir drives" in the parlance of the petroleum industry, by
which oil can be displaced and driven to the wellbore: water-drive,
gas-cap drive, and dissolved-gas or solution-gas drive. The first period in the producing life of a reservoir, when the oil is displaced by
one or more of these natural reservoir drives, is known as the "pri-

FIGURE INT.3   Water-drive Reservoir

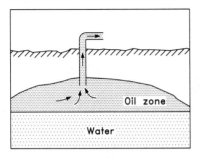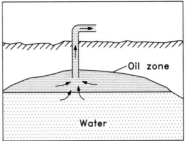

Original conditions   Partly depleted

mary production" or "primary recovery" phase. In the early development of the industry, the natural drives were allowed to function
until they were depleted or exhausted, and then "secondary" or
"improved recovery" was attempted through the introduction of
more costly artificial drive mechanisms. In modern petroleum reservoir management, the artificial mechanisms are introduced at the
optimum time, which is usually long before the natural mechanisms
are depleted.

Water-drive Reservoir

The water in most water-bearing formations has a fluid pressure
that increases with depth beneath the surface. When the reservoir
rock is penetrated by the wellbore, the compressed water expands
and moves towards the region of pressure release, driving the oil in
front of it, as shown in Figure INT.3. Water, the displacing agent,
typically encroaches from below the oil ("bottom-water drive") or
from the edges of the reservoir ("edge-water drive"). Although the
recovery efficiency of a water-drive is normally superior to other
natural drive mechanisms, with recoveries as high as 70% of the
oil-in-place being reported, high efficiency is achieved only when
the reservoir characteristics are extremely favourable and the advance of the water front is uniform and slow. It is important therefore to produce from a water-drive reservoir at a rate in keeping
with the rate of advance of the water front.[11]

Solution-gas-drive Reservoir

Even where water exists in the reservoir, if the crude oil contains
gas in solution, a solution-gas drive could develop, supplementing

FIGURE INT.4   Solution-gas-drive Reservoir

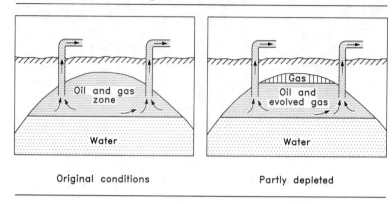

Original conditions   Partly depleted

any water drive. In such reservoirs, the lighter hydrocarbon components remain dissolved in the liquids at formation pressures in excess of the saturation pressure—the pressure at which gas normally begins to come out of solution. When the wellbore taps the reservoir and production occurs, the reservoir pressure declines. After the saturation pressure is reached, the dissolved gas comes out of solution and expands, thus displacing the oil and driving it to the surface, as shown in Figure INT.4. As the drop in pressure below the saturation pressure extends away from the wellbore and increases, more gas comes out of solution to displace the oil. Solution gas is the most common natural energy source for oil production, but being a depletion-type mechanism it is inherently inefficient. Recoveries ranging from as low as 5% to as much as 30% of the oil-in-place are typically expected. Pressure declines quickly and continuously in solution-gas reservoirs, thus requiring the assistance of well pumps and making some form of artificial pressure maintenance desirable at an early stage. Field production practices that enhance this relatively low rate of recovery are therefore of particular concern to conservation authorities.

Gas-cap-drive Reservoir

Gas-cap drive is the third type of displacement process, and its recovery efficiency typically ranges between that of the solution-gas and water-drive mechanisms. A gas-cap drive occurs where gas under pressure is trapped above the oil as shown in Figure INT.5. Displacement of the oil occurs through the downward expansion of the overlying gas cap that takes place when the penetrating wellbore allows the release of reservoir pressure. With production, the reser-

FIGURE INT.5   Gas-cap-drive Reservoir

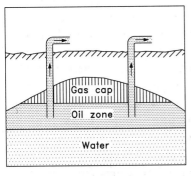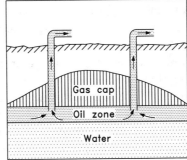

Original conditions   Partly depleted

voir pressure declines, and some of the gas-cap gas will be produced with the oil if production rates are too high. Therefore, as for water drive, the gas-cap mechanism is production-rate sensitive, and the downward rate of movement of the gas-oil interface should be kept low. Preservation of the gas cap that provides the driving energy is essential, and this accounts for the efforts of conservation authorities to reduce and normally prohibit the significant production of gas-cap gas until most of the recoverable oil has been produced.

Gravity Drainage and Combination Drive

In reservoirs, the force of gravity is always at work, and gravity effects can sometimes play an important role in recovery as original gas or water drives decline, especially in thick or steeply tilted, highly permeable reservoirs. In such situations, oil moves downwards to producing wells and gas coming out of solution migrates to the crest of the reservoir, forming a gas cap. This gas cap expands as pressure declines further, sometimes providing an effective driving mechanism.

Additionally, there are certain reservoir situations where some of these drives can function in combination. A gas-cap drive, for example, might be combined with a water drive or a gas-cap drive with gravity drainage. Similarly, a solution-gas drive can be augmented by bottom-water encroachment.

The reservoir characteristics described so far are critical factors in determining the number and location of wells that ideally should be drilled into an oil-bearing stratum to assure efficient and maximum recovery. Porosity and permeability of rock, as well as oil viscosity, define the portion of a reservoir that can be drained by a single well in a given time. Also, wells need to be located where they can gain the greatest advantage of whatever natural drive is available. It is apparent therefore that the most efficient production of petroleum usually requires the development and working of a reservoir as a single unit. This can be achieved readily where the field is developed and operated by a single owner, as is often the case in the Middle East. In North America, however, where ownership is typically divided, the operations of separate producers should be co-ordinated into a single plan. Such co-ordination might require "unitization" to ensure that the reservoir is produced in a manner that is acceptable to society and to maintain equity among the various competing producers. Conservation authorities have been established in many producing regions to accelerate the achievement of optimized production operations.

OIL AND GAS LAW

Operation of the reservoir as a unit is not only the most effective measure to ensure efficient production but also a means of protecting property or correlative rights. The problem is attributable to the unique "fugitive" or "fugacious" nature of petroleum. A surface property or lease boundary line is no barrier to the migration of oil or gas in the reservoir. One owner or well operator, through certain operating practices, is able to draw oil away from neighbouring properties. In the absence of regulation, as was characteristic in the early years of the petroleum industry's development in North America, the courts were soon clogged with litigants contesting one another's "rights;" however, the judges seeking to resolve conflicting claims knew even less about oil and the nature of reservoir mechanics than did the oil operators. Without precedent to go by, they looked in desperation for analogies that would help them deal with the increasing number of court actions relating to this strange new resource. Finally, in 1889, the Supreme Court of Pennsylvania found what was to become a guiding precedent for the following half century. In *Westmoreland Natural Gas Company v. DeWitt*, oil was likened to wild game, which English game law decreed could be captured by right by whoever found, or could lure, game on his land.[12] This recognition of title to oil and gas produced, even though it might have been drained from the property of others, came to be known as the "Law of Capture" or the "Rule of Capture." This concept also carried within it the old English common-law tradition that applied to underground water. It was accepted that anyone might take as much water from a well as they wanted.

In Canada, the Law of Capture was understood to apply from the earliest days of petroleum exploration and it gained explicit authority in *Borys v. Canadian Pacific Railway and Imperial Oil Limited, 1953*. In addressing a case that had progressed through the Canadian courts to the Judicial Committee of the Privy Council in London, what was then the final court of appeal, Lord Porter observed:

> [Water, oil or gas] are fugacious and are not stable within the [reservoir] although they cannot escape from it. If any of the three substances is withdrawn from a portion of the property which does not belong to the appellant but lies within the same [reservoir] and any oil or gas situated in his property thereby filters from it to the surrounding lands, admittedly he has no remedy. So, also, if any substance is withdrawn from his property, thereby causing any fugacious matter to enter his land, the surrounding owners have no remedy against him. The only safeguard is to be the first to get to work, in which case those who make the recovery become owners of the material which they withdraw from any well which is situated on their property or from which they have authority to draw.[13]

As Lord Porter explained, the logical solution open to an adjacent landowner was to drill an offset well and commence production as quickly as possible before more of "his" oil or gas was "captured" by others.

The right to drill and produce as one pleased, reinforced by the Rule of Capture, had predictable results. There was a vast proliferation of wells and massive overproduction, as owners of land in producing areas drilled frantically to beat or counter the efforts of neighbours and then produce to capacity. The tremendous waste resulting from unrestricted operations was soon apparent and became a matter of public concern. In the early legislation that tried to curb some of the wasteful excesses, a second doctrine of lasting significance emerged. When the State of Indiana tried to compel the Ohio Oil Company to abide by newly enacted statutes designed to prevent waste, the company strenuously resisted and the case eventually reached the Supreme Court of the United States. There, in 1900, Chief Justice White proclaimed in favour of Indiana. In his judgement supporting Indiana's waste-prevention statute, White accepted the state's right to legislate in the public interest and challenged the validity of the wild-game analogy. He pointed out that all landowners able to draw from a common supply source of oil and gas have an equal privilege to take, and that the unrestrained exercise of such privilege by one

> may result in an undue proportion being attributed to one of the possessors of the right to the detriment of others, or by waste of one or more, to the annihilation of the rights of the remainder.[14]

The owners of a common source of oil and gas possessed "correlative" rights. The concept of correlative rights was slow to gain acceptance, especially in some western states where the Rule of Capture doctrine in the popular mind was judged to be more in keeping with frontier traditions, and the two concepts co-existed in uneasy tension.

An Outline of Regulatory Problems

Without regulation, and governed only by the Law of Capture, a host of troubling problems quickly materialized. An ever-expanding pattern of offset drilling was characteristic. A density of five to 10 wells per acre was not uncommon in many areas, with the extreme of 27 wells in a single acre being reached in the Kilgore field in East Texas. The imperative of producing the oil before someone else did led to prodigious waste. Misuse of reservoir energies lowered ultimate recovery, and copious production stored in primitive facilities contributed to dramatic waste and pollution on the surface. Salt water produced with the oil flowed freely into streams and contaminated local water supplies. Wooden derricks, with attendant sheds almost side-by-side, and scattered debris, combined with the presence of highly volatile substances, made fire a fearsome and constant hazard. At the same time, small producers found themselves in an increasingly desperate situation. Unregulated production meant unstable and usually depressed prices, with the advantage in the hands of the large refining interests. Wasted resources and extremely low prices were also seen to have a deleterious influence on the local governments' short- and long-run revenue prospects.

Both public and private groups began campaigns that eventually led to regulations that sought to govern the method and rate of production. The regulatory approach came in three broad areas: limitations were placed upon the right to drill, restrictions were placed on the location of wells, and restraints were placed on production.[15] In turn, as is typically the case, the regulations created a series of new problems. There were complex technical and engineering questions to be faced, not to mention the entrenched legal rights of private property supported by the powerful North American tradition of the individual's right to the undisturbed pursuit of opportunities for economic gain.

Motivation to limit the right to drill came from the desire to protect reservoir energy in the interest of achieving the maximum ultimate recovery, to minimize recovery costs by eliminating the drilling of unnecessary wells, and to protect correlative rights. But even as late as the early 1940s, such motivation was challenged by the widely held opinion that more wells brought forth more oil.[16] More difficult to contend with was the persistent opposition, particularly in the United States, of numerous owners of tiny tracts who insisted on their right to drill at least one well. Also, the large oil and gas field development and service industry was not enthusiastic about restricted drilling or the shutting in of unnecessary wells. Typically, this sector has been quick to remind government of the large numbers it employs in small communities and rural areas.

The principle of restricting the rate of production so that each owner had the right to produce or receive a "fair" share of the recoverable gas or oil in the reservoir was accepted by the majority of producers almost from the first. Although the attraction of rate restriction perhaps had more to do with the possibility of bringing about price recovery and stability than with conservation, the great stumbling block has always been finding a mechanism or formula acceptable to all parties—small independent companies, large integrated multinationals and the public. Protection of the broad public interest was usually assigned to a legislatively created board or conservation agency.

Deciding who should produce how much, taking into account reservoir conditions and marketing factors, is a complex matter. All producers seek to maximize production, and the means used to measure each producer's fair share is therefore of vital concern and the subject of constant debate. The allocating or prorating of production in Alberta has led first to the assignment of an allowable to each pool, and then to the division of the pool allowable among the various wells in the pool. Ultimately, this means setting production quotas for hundreds of pools and thousands of wells. The problem is complicated in that no two oil pools have exactly the same physical characteristics and they produce a variety of crude oils. Market demand is not for crude oil per se but for specific crude oils for different product blends. Oil from all pools is therefore not interchangeable or equally marketable at a given time. For example, recognition that the special characteristics of heavy crude oil do not make it interchangeable in the market place with light and medium crude oil has led Alberta to exclude heavy oil from the proration system. Beyond this difficulty, quota allocation must not only serve the protection of property rights (production based on under-ground reserves) but also be consistent with the effective elimina-tion of waste. The latter requires that no pool be produced at a rate in excess of its Maximum Efficient Rate (MER), that is, beyond the rate at which a deterioration of the efficiency of use of reservoir en-ergy will occur. This means that each pool's allowable production should not be determined simply with reference to the total provin-cial allocation and the total recoverable reserve for each reservoir. Rather, permitted production rates must have regard for the MERs at which various reservoirs can be produced to avoid underground waste. Different pools might require different rates of production. It will take longer therefore to produce the reserves from some pools than from others.[17]

Related to the principle of waste reduction is the problem of stripper or marginal wells. These are wells capable of only low-volume production, in many cases defined as less than 10 barrels a day, and are typically approaching the end of their life cycles. Such wells are usually exempted from production restrictions on the grounds of preventing physical waste, that is, leading to premature abandonment and the loss of recoverable oil underground. The cu-mulative production from stripper wells is substantial; moreover, it is higher cost oil, which in a restricted market is produced at the ex-pense of available cheaper oil. In other words, the prevention of physical waste may give precedence over near-term economic waste, a point to which economists have been quick to draw atten-tion.

Devising a set of rules or formulas that offers an acceptable com-promise between competing interests is complex. It is rendered even more difficult when, as is often the case, restriction of production from existing wells exists paradoxically alongside policies designed to encourage exploration for new fields and to enhance the produc-tive capacity of old ones. If the latter policies are successful, this means that production quotas assigned to new and enhanced wells must come at the expense of production from old wells, unless there is a corresponding expansion of the market or decline in production capacity from old reservoirs.

Other elements that complicate the regulatory picture include the potential conflict or contradictory results that can follow from the differing approaches or emphases of other government agencies and other government jurisdictions. The vast heavy-oil deposits of east-central Alberta extend well into Saskatchewan. Similarly, the immense natural gas fields of the Peace River area are divided in some cases by the British Columbia-Alberta border. On Indian lands within the provinces, ultimate authority over the development and production of oil and gas rests at the federal level. The significant impact of federal authority also comes in the form of the National Energy Board recommendations regarding natural gas or oil export applications and other national policies that shape domestic market demand and price, as well as measures that shape the pace and character of petroleum exploration in the Arctic or off Canada's East Coast. Seldom have measures applied within or across jurisdictional boundaries functioned in harmony. It follows that problems with implications across jurisdictional boundaries have often been difficult to resolve. It is apparent therefore that regulatory problems in the "conservation" area stem in good part from the remarkably complex and sometimes volatile environment in which conservation agencies function, in addition to the diverse and sometimes conflicting nature of their responsibilities.

CONSERVATION TECHNIQUES

Prorationing of Oil

When prorationing was first introduced, the common practice was to prescribe a flat per well allowable. This approach simply allocated pool allowables in proportion to the number of wells on the property. Equally common was allocation related to each well's potential, that is, the amount of oil that it could produce in a given time. Both approaches were eventually shown to be unacceptable, and in recognition of the complexities discussed above, contemporary conservation authorities take a number of factors into account in their attempts to fashion acceptable allocation formulas and procedures. These include acreage, or a unit that corresponds to the reasonable drainage area of a well; the nature and efficiency of the reservoir drive mechanisms; the thickness of the oil-bearing section; the viscosity of the oil; the estimated recoverable oil initially in place; and the extent to which it has been produced.

Rateable-take Regulations

Major integrated oil companies are naturally inclined to favour the purchase of their own oil to meet refinery needs; this incentive is even greater when there is surplus capacity and prorated or restricted production. To ensure that nonintegrated, and especially smaller, independent producers have an equal opportunity to market their allowable production, some governments have enacted legislation that declares each buyer a "common purchaser." Such statutes require the purchaser to buy rateably without discrimination from all producers in the pool who are making reasonable offers to sell. Many producers are also potentially handicapped, because to market their product they require access to pipeline systems that are owned by others. Pipeline companies might therefore be designated as "common carriers" and be required to receive and transport oil or gas rateably without discrimination among sellers. The protection of correlative rights is the basis for common purchaser and common carrier provisions.

Well Spacing

Well-spacing regulations were one of the earliest regulatory measures and are common to nearly all petroleum-producing regions. Regulation of well spacing with production control lies at the heart of conservation regulation. The first regulatory order of Alberta's newly formed Petroleum and Natural Gas Conservation Board in 1938 decreed that henceforth the number of wells to be operated in the Turner Valley field would be restricted to one well for every 40 acres, and without Board authorization no well was to be closer than 660 feet from an adjoining property boundary.[18] Production control, through restriction of well density, was presented first as a means of protecting correlative rights. It was soon realized that, to a degree, wider well spacing also offered more economic longer-term production. Dense drilling and high initial rates of production were shown in most cases to compromise ultimate recovery. In addition, excessive drilling was recognized to add unnecessarily to reservoir development costs and therefore represent an economic waste. Thus, all regulatory bodies specify minimum distances between wells and assign a minimum acreage to each well, depending upon fluid and reservoir characteristics. Well-spacing regulations are enforced by requiring "drilling permits" that set out exact locations of proposed wells. Conservation authorities are further empowered to require the pooling of smaller parcels of land to form "drilling units" that conform to the minimum specified acreage size. If property owners cannot reach voluntary pooling agreements, sharing arrangements might be imposed.

The historical trend in the period of restricted markets has been towards larger well-spacing units. In this progression, conservation authorities have typically been supported by major oil companies, which usually have large investments on extensive parcels of land, and resisted by smaller operators and mineral owners. In recent years, with near-capacity markets, there has been a movement to somewhat closer spacing in some areas for improving productivity and reservoir drainage in some complex reservoirs or for tertiary recovery processes. Regulation of spacing remains the essential method of controlling production in most eastern and midwest states, including Illinois, Indiana, Ohio, Pennsylvania and Kentucky.[19]

Unitization

Today, it is almost universally recognized that the most efficient way to develop most reservoirs is to treat them as single producing units. The ultimate in conservation practice is often represented in unitized fields, and economists have frequently argued that unitization should be made compulsory.[20] Reservoir unitization often offers the most promising means of effecting the essential objectives of conservation, namely, effective reservoir pressure maintenance, production rates that do not exceed the reservoir's MER, avoidance of unnecessary development costs, the most effective basis for secondary recovery programs, and equitable division of benefits. As one proponent has argued, unitization of reservoirs

> seeks to eliminate the multiplicity of competing interests for purposes of operating a common reservoir, while retaining separate interests for purposes of sharing equitably in common costs and benefits. It creates a consolidated private interest which coincides with the public interest in conservation.[21]

Despite general consensus regarding the obvious merits of unitization, no U.S. petroleum-producing state has been prepared to go beyond encouraging voluntary unitization, and producers have been remarkably slow to come to agreements on their own. In Alberta, the Conservation Board encourages, but does not compel, unit operation.[22] Compulsory unitization is provided for in Saskatchewan legislation.

Improved Recovery

Even with the best natural production methods, a significant percentage of the oil is left behind. Traditionally, conservation commissions have tried to create policy environments that encourage introduction of improved recovery methods where the natural drive mechanism is inefficient. Consistent with their objective of reducing waste, conservation regulators in many jurisdictions ensure that important technical information is gathered and retained and made generally available; promote unitization as the basis from which improved recovery schemes can be launched; and, most importantly, offer economic incentives, such as higher production allowables, to further the cause of increasing total recovery. Whether or not to undertake improved recovery programs is usually as much an economic problem as it is an engineering one. The increased operating costs of an improved recovery project must be set against the revenues anticipated from projected added production. Generally, it costs more to produce secondary than primary oil, and the industry has always been quick to draw this to the attention of regulatory agencies and government so that the appropriate "allowances" might be made. In the end, the pace, nature and pattern of improved recovery programs are considerably influenced by conservation policy.

The essential technical objective of improved recovery is maintenance or restoration of formation pressure and flow capacity in a reservoir and improved displacement of oil from the reservoir rock to enable the recovery of as much of the remaining oil as possible. There are four main methods of improved recovery: water flooding, gas injection, chemical flooding and thermal recovery. Water injection or water flooding is the most commonly used method, and is often begun during the primary production phase. Water, including water produced with the oil, is frequently put back into the formation, not only to maintain pressure but also to improve the sweep of the oil towards producing wells. The selection of where to place injection wells and how fast to inject the water is therefore of critical importance. Gas is also used to maintain or replace pressure in reservoirs. If gas is injected into the reservoir under conditions where it does not mix with the oil to sustain or increase reservoir pressure, it is known as "immiscible gas injection."

Other processes are usually referred to as enhanced recovery methods, because they not only increase the formation pressure but also improve displacement and overcome restrictive reservoir forces. In "miscible gas injection," injected gas acts as a solvent in displacing the oil at appropriate pressures, depending on the composition of the gas used and the oil in rock pores. Injected gases include carbon dioxide, butanes, propane, nitrogen, methane, and methane enriched with other light hydrocarbons. Detergent-like agents are used in "chemical flooding." Injected into the reservoir with water, they decrease surface tension and help to break the oil into droplets and thus ease its removal from the rock pores. Thermal methods are sometimes part of the initial recovery process applied in Alberta heavy oil and oil-sands reservoirs. They utilize the injection of heat to reduce the viscosity of the reservoir oil and to constitute an improved displacement mechanism.

Gas Flaring

The tremendous glow generated from flared gas was sufficient in the 1930s to keep the Turner Valley oilfield almost in a state of perpetual daylight. It was this waste, combined with the realization that the excessive gas production was having a damaging effect on the reservoir, that brought the question of oil and gas conservation to public debate and eventually brought forth legislation designed to curb such a practice. Regulations restricting gas flaring or venting were initially intended to prevent the squandering of gaseous hydrocarbons produced incidentally with oil or natural gas liquids. Normally, provision is made for some flaring if operators have no economical alternative. Regulations are sometimes necessary to encourage operators, or groups of operators, to construct a gathering system so that the gas can be processed and marketed, or to implement a program of gas injection for reservoir pressure maintenance.

Gas-oil Ratio Penalties

Nearly all jurisdictions in oil-producing regions require conservation of gas in oil reservoirs. Typically, excess gas production, except when it is returned to the reservoir, results in penalties being applied to oil production in proportion to the excess. Conservation agencies determine gas-oil ratio penalty schedules from careful reservoir analysis. Gas-oil ratio limitations are intended to prevent overly rapid and inefficient depletion of the natural reservoir energy.

Salt Water Disposal

Salt water is often produced unavoidably with oil or gas, especially in the later stages of the production of a pool. Sterile, salt-encrusted soil is not an uncommon sight in some of the United States' oldest oilfields. The purpose of salt water disposal regulations is to prevent pollution of surface soil and fresh water, as well as shallow potable water zones. Discharge into fresh surface waters or evaporation in unlined open pits is prohibited in favour of injection which, in addition to effective disposal, could have the supplementary advantage of helping to maintain reservoir pressure. Enforcement is normally handled through mandatory water disposal permits and by field inspection. Water-oil ratio penalty regulations, like gas-oil ratio penalty regulations, normally provide for credits where the fluid is returned to the zone of origin.

Casing and Well Abandonment

All oil- and gas-producing regions in North America require the casing of oil and gas wells and proper abandonment procedures. These rules describe the manner in which wellbores are to be lined and, when wells are abandoned, how the borehole is to be plugged and the site rehabilitated. The regulations are intended to prevent damage to reservoirs and contamination of fresh groundwater, surface water and soil. Worker and public safety is also a consideration. Conservation boards, or an appropriate government department, are usually given the responsibility for creating and enforcing such requirements, and do so through regulations, permits and field inspections.

Natural Gas Conservation

Wastage of natural gas can contribute to wastage of oil. It is apparent from the foregoing discussion that oil conservation and gas conservation are intimately bound together, and that many of the laws and regulations are common to both gas and oil. There are nonetheless several unique aspects of gas conservation worth special mention.

Natural gas migrates much more readily than oil through producing formations. The spacing pattern for gas wells is therefore wider, typically one well per square mile. Given the physical characteristics of natural gas, the need for improved recovery procedures is rare; however, where the gas is rich in natural gas liquids, hydrogen sulphide or helium, displacement of the rich gas with another gas could be warranted. Rate of production is not usually a factor, and most gas wells can be produced at high rates without concern about reservoir damage. Where the gas overlies water in the reservoir, high rates of production can lead to premature water production, well damage and loss of recovery. In such cases, maximum rates of production are often prescribed.

Generally, waste has been associated with production beyond market demand. Initial concern about gas waste came on two fronts. Communities near gas fields were quick to adopt relatively inexpensive natural gas for heating and commercial use. Citizen and business groups were quick to seek assurance of the long-term supply of this remarkable fuel. Similarly, companies supplying the natural gas, and their bankers, wanted assurance of substantial reserves before extensive pipeline delivery networks were put in place. Gas produced with oil tended to be more expensive to gather and market than that produced from high-capacity gas wells, and its availability was closely related to the rates at which oil was produced. In addition to concern about reserves and availability, and once the delivery systems were in place, consumers began to challenge the pricing practices of the natural gas companies. Usually, consumers did not have access to competing supply systems. The response of most governments was to regulate the gas transmission and distribution industry as a public utility. Conservation agencies became involved with the measuring and monitoring of gas reserves to ensure that waste or uncontrolled removal to distant markets did not jeopardize long-term local supply. By the 1960s, the definition of wasteful use in many areas had broadened to include end use. Hence, in many states, and in Alberta, the manufacture of carbon black was prohibited or restricted on the premise that it constituted a wasteful use of the fuel. Alberta later restricted the use of natural gas for electric energy generation, except for accommodating peak loads, and encouraged the use of the more plentiful coal for this purpose. The responsibility of protecting correlative rights in the field by careful policing of rateable-take regulations also fitted naturally into the mandates of conservation authorities. In some jurisdictions, the institutionalized pricing system for natural gas, based on long-term contracts, brought the further obligation of fixing minimum wellhead prices to ensure conservation and exploration to expand reserves to meet long-term requirements.

THE PETROLEUM INDUSTRY IN NORTH AMERICA: ITS OPERATING FRAMEWORK

The origin and evolution of petroleum conservation was influenced significantly by the unique structure of the petroleum industry in North America, and it is necessary therefore to be aware of certain features of the industry's operating framework. This is a subject that has been closely examined by M. Adelman, J. Blair, R. Engler and others;[23] thus, only a brief overview is required here.

The most striking and commonly noted characteristic of the petroleum industry in North America is its domination by a handful of corporate giants that include some of the world's most profitable corporations. Known as the "majors," these are companies that integrate all four of the basic industry activities—producing, transporting, refining and retail selling. The integrated corporation usually has co-ordinated control over every stage of the oil flow and a potential profit centre in each of the four sectors. In contrast to the cost of entering the production sector, the capital cost of acquiring a presence in the refining and pipeline areas is extremely high. John D. Rockefeller, the first of the great oil barons, was quick to realize the advantage to be gained through the control of refining and transporting in the maintenance of industry stability and profit.

Although their dominance is readily apparent, this is not the feature that distinguishes the North American petroleum industry from the offshore industry. Rather, it is the existence of thousands of large and small independent producers that makes the North American setting unique. The presence of the "independents" indicates not only that integration in the petroleum industry is decidedly less than that elsewhere but also that it is less integrated than the other basic mineral industries within North America. Although it is usual for most mineral companies to process their own ore, this is not the case in the petroleum industry, where most producers exist without the capacity to refine their own product. In addition, it is important to note that the independents are a sufficiently diverse group as to make generalization difficult. Although some are engaged in as many as three of the industry's main functions and have an economic perspective that differs little from that of the majors, most nonetheless are small companies whose interests are concentrated in the production sector, with activities limited to a few fields or even a single pool. Even though as individual companies their economic power has usually been limited, their number and their situation have had a great effect upon the formulation of regulatory policy.

The industry perspective of the independent companies was conditioned mainly by the peculiar physical properties of their resource and the market system in which they operated. Traditionally, oil was marketed in the field at prices posted by the buyers, usually refinery companies or their affiliates. Since the demand for crude oil was derived from the aggregate of demands for various refined products, the buyers' posted prices reflected the unit value of refined products expected from a projected refinery product mix. Posted prices set by the refiners also took into account variations in crude oil characteristics from field to field as well as distance to the nearest refinery. The purchased oil was collected at the field by a pipeline firm that often was affiliated with the refinery company. Consequently, in many fields, integrated companies were at once the buyers, sellers and transporters of oil. In such situations, where other nonintegrated producers had no alternative buyer-transporters nearby, a major company had the power to discriminate with respect to price and purchased volume in favour of its own production within and between reservoirs. This particular buyer-seller relationship was a primary cause of waste and inequity, and thus is of direct relevance to the problem of petroleum conservation.

THE HISTORICAL EXPERIENCE IN THE UNITED STATES

Regulatory approaches seldom emerge to be put in place as pure theoretical constructs; rather, they are the product of historical forces in conflict. While local and national forces are most important, the Canadian, and more particularly the Alberta, approach to oil and gas conservation was largely conditioned by what had happened, and what was happening, in the United States. The province's first attempt to implement a full-fledged conservation scheme in 1932 was based heavily upon the experience and technical advice of F.P. Fisher, a highly respected consulting engineer from Mount Vernon, Ohio.

The Alberta government in 1938 selected W.F. Knode, a petroleum engineer who had experience with the equivalent Texas agency, the Texas Railroad Commission, as chairman for its newly formed Conservation Board. Knode was succeeded after a short interval by R.E. Allen, another American petroleum engineer. Allen's advanced thoughts on petroleum conservation had been published in various technical journals,[24] and he brought with him senior-level administrative experience gained in the service of the producer-operated conservation authority in California. Almost immediately after setting up its Petroleum and Natural Gas Conservation Board, the Government of Alberta appointed a royal commission, popularly known as the McGillivray Commission, after its chairman the Honourable Mr. Justice A.A. McGillivray, to inquire broadly into all matters relating to petroleum and petroleum products. The Commission heard evidence through 1939 and early 1940. Among the numerous witnesses brought forth, two of the most prominent were Dr. George G. Brown, head of the petroleum engineering program at the University of Michigan and a highly regarded consultant in industry circles, and Dr. John W. Frey, a senior Washington bureaucrat who was the assistant director of the Petroleum Conservation Division under the secretary of the interior and who had long been involved in federal efforts to promote petroleum conservation in the United States. Both spoke extensively on the necessity of proper conservation practice and, identified as experts, were quoted at great length in the Commission's eventual report to the provincial government.[25] Through the 1940s and early 1950s, technical advice was frequently sought from respected American academics, institutions and lawmakers. Dr. Brown was brought back to Alberta in 1941 to study the Turner Valley oilfield and to recommend a rate of production that was consistent with the best engineering practice of the day. In 1943, Dr. Thomas Weymouth came to Alberta, on loan from the U.S. Petroleum Administration for War, to help devise a plan for the conservation of gas produced by Turner Valley oil wells. Consultation with the University of Michigan, particularly with Brown's colleague, Dr. Donald Katz, continued after the war.

The University of Michigan's influence upon Alberta policy was also reinforced on another level. Studying under Dr. Katz was a gifted Albertan, George Govier. On receiving his Sc.D., Govier returned to the Faculty of Engineering at the University of Alberta. In short order, Govier became head of the Department of Chemical Engineering and, given his background and field of research, was a logical appointment to the three-man Alberta Petroleum and Natural Gas Conservation Board in 1948. Govier remained with the Board until 1978, having served as chairman from 1962.

Alberta drew not just upon U.S. technical advice and experience. The province also sought the guidance of Fort Worth lawyer Robert E. Hardwicke. Perhaps the pre-eminent authority on oil and gas law of his generation, Hardwicke played a major part in drafting conservation statutes in Texas and other oil-producing states. His trip to Edmonton in 1949 to perform such a role north of the border underscores the nature of the continental forces that so profoundly shaped the character of the petroleum industry's development in the Canadian West.

The legal and technical advice lent by U.S. experts rested upon three-quarters of a century's experience that had commenced in 1859 along Oil Creek in Pennsylvania.[26] (Although it did not develop as extensively, the industry in Canada began almost a decade earlier at Oil Springs in southwestern Ontario.) In August 1859, at nearby Titusville, Edwin L. Drake, a drifter and former railroad conductor, struck oil at the remarkable depth of 69.5 feet. Soon, jets of black oil were shooting past the derrick tops of scores of wells, many spaced only a few feet apart, along a seven-mile stretch of the creek. No one knew how to handle the liquid that emerged, and it was held in hastily built containers of every description and in open pits. What was not lost by fire or seepage into the creek was loaded in leaky barrels and hauled away by wagon. Oil that had

sold for $20 a barrel in 1859 had plunged to 50¢ in 1861. The se-
quence of events at Titusville, where the discovery was followed by
frantic speculation, side-by-side drilling, prolific production accompanied by heedless waste, unchecked pollution and eventually price
collapse, foreshadowed a pattern that would be repeated again and
again.

The first conservation regulations began to appear in the late
1870s and early 1880s and were almost entirely concerned with
external damage and pollution caused by oil production. Such legislation, often spurred by the urging of government technical men,
defined wellbore casing requirements intended to prevent the interchange of fluids among strata that fouled groundwater and in some
areas damaged coalmines.[27] Among the earliest regulations were
those intended to curb surface waste and pollution, particularly by
calling for the proper plugging of abandoned wells. Rude storage
guidelines gradually emerged. Measures to reduce fire hazard, as
well as laws to prevent flaring gas from gas wells and burning gas in
flambeau lights, were common by the turn of the century.

From 1890, oil exploration and production began to move from
the Appalachian region westward to the mid-continent, the Gulf
area and California. But, even as late as 1899, and despite the opening of new fields in Kansas by the Standard Oil Company, and after
the big strikes in the Bakersfield, Kern County, Santa Barbara and
Coalinga regions made California a leading oil producer, Appalachian oilfields still supplied 93% of total U.S. oil production. In the
first years of the new century, the balance of production shifted
dramatically. By 1910, Appalachian fields accounted for less than
10% of production. Anthony Lucas and Patillo Higgens struck
Spindletop in 1901. It proved to be one of the richest gushers in the
world. Yielding more than six million barrels within six months of
its discovery, it marked the beginning of full-scale petroleum exploration throughout Texas. Rich oil finds followed at Sour Lake, Sar-
atoga and Humble to establish the foundation upon which the contemporary petroleum industry in Texas was built. In Oklahoma, a
series of major discoveries, including Bartlesville and the Glenn
pool near Tulsa in 1905, followed by strikes at Cushing in 1912
and Healdton in 1913, turned the state into the United States's largest oil producer, a position it retained until 1928.

The legislative record follows the expansion of the industry out
of Pennsylvania and the Upper Appalachian region into the north-
central states, then into the south, mid-continent and Gulf states.
Prolific discoveries characterized by enormous waste and falling
prices led Oklahoma to break new ground and enact conservation

laws that were more comprehensive than those of any other state. Drilling regulations were adopted in 1905, and by 1909 were supported by an inspection system to ensure enforcement. Appreciation of expanding petroleum production in the state was recognized by the creation of the Oklahoma Corporation Commission in 1907, which was charged with responsibility for certain oil and gas matters.

Development of the vast Glenn pool immediately put the state and its new Commission to the test. This pool covered a large area and its wells were big producers of high-grade oil. The oil produced was soon far in excess of what the market could absorb or what could be transported by existing facilities. In this situation, great advantage lay with the integrated companies. Having their own pipelines, storage facilities and refineries, they could absorb the production from their own properties. The nonintegrated operators were helpless; they had to take what price was offered—if they could get an offer and if they had access to a pipeline. At the same time, if they did not produce, they had to stand by and watch their properties being drained by the producing companies that owned pipelines. The incentive therefore was to produce and store production. Storage, however, was a great problem: some erected steel tanks, and some put up wooden structures, but most pumped oil into prepared earthen pits. Huge volumes were lost through seepage, evaporation, fire and washouts. The gravity of the situation is underscored by a saying of the time: "More oil has run down the creeks from Glenn pool than was ever produced in Illinois."[28]

Oklahoma responded to the petitions of crude oil producers with a series of precedent-setting enactments between 1909 and 1915. To remedy the desperate situation of the independent operators, the state enacted an anti-trust bill classifying pipelines as "common carriers," obligating them to "rateable taking" from areas within the pipelines' reasonable reach.[29] Having pronounced in 1909 that a purchaser and carrier of crude oil must not discriminate in anyone's favour,[30] the Oklahoma Legislature went on in 1913 to compel rateable production from common reservoirs of natural gas. At the same time, the Oklahoma Corporation Commission was holding hearings concerning problems of flush production and ruinous crude oil prices. The Commission eventually concluded that, although common purchaser and common carrier provisions did much to improve the position of small producers, it was nonetheless an inadequate remedy in face of the appalling waste that was accompanying increasing surpluses. In its view, oil was best stored in the ground until the market was ready to accept it. The answer seemed to lie in limitation through some scheme of prorationing of production. To this end, the Commission issued an administrative order stating that producers could not take more oil from their wells than the market would accept at 65¢ per barrel. Restraining production by fixing a minimum price proved an unsatisfactory mechanism, and though the order was ignored by most producers it nonetheless foreshadowed a new level of state intervention.

Still the problem continued to escalate. In 1913, the Healdton field was discovered, and as daily production in this field approached 100,000 barrels, a vast new pool was struck near Cushing. Oil prices plummeted, and wasteful and excessive production now threatened the stability of the state economy. In desperation, Oklahoma legislators passed the United States's first statewide proration law in 1915. The measure gave the state the power to restrict the production of private operators whenever excessive waste was evident in a particular field, whenever market demand for existing oil was inadequate, or when prices dropped below production costs. The standard for proration was well potential: each well would be allowed to produce that percentage of its potential that the total pool allowable was of the total of all the potentials of all the wells in the pool.[31] Enforcement was left to the Corporation Commission.

In February 1917, Oklahoma greatly expanded its system of regulation to include virtually every major activity in the oil-producing industry. Henceforth, all natural gas produced in Oklahoma was to be metered, written permission had to be secured before any well could be drilled, deepened or plugged, and a sequence of approved conservation practices had to be followed for each of these activities. Oil and gas companies were required to make monthly reports on their drilling operations and on their production. Before a pipeline company could connect any well, its owner had to produce a certificate showing that he had complied with state conservation laws. Daily drilling records had to be kept and could be inspected at any time. The task of monitoring adherence to the regulations was assigned to the Corporation Commission, and in recognition of its much expanded supervisory role a separate oil and gas department was created within the Commission.

Many of Oklahoma's pioneering precedents were gradually adopted by other oil-producing states. In 1917, pipeline companies in Texas were declared common carriers. Since pipelines carried petroleum and were therefore clearly a means of transportation, authority to oversee their operations was given to the Texas Railroad Commission. In Texas too, the ill effects of a drill-as-you-please environment were becoming increasingly apparent. Within a year of its discovery, the Spindletop field produced the incredible volume of more than 17 million barrels; within three years this had dropped to less than two million barrels.[32] To check the pattern of discovery, followed by frantic development, flush production and great waste, then sharp decline, Texas passed a general waste prevention statute in 1919 that gave wide regulatory powers to the Railroad Commission and became the basis of subsequent statewide oil and gas regulation.

The next phase of regulatory development was greatly shaped by the ideas and advocacy of Henry L. Doherty. Described by one writer as "a keen, independent thinker" who "enjoyed his role as the intellectual of the petroleum industry," Doherty is recognized as one of the fathers of modern conservation practice.[33] Doherty was actively interested in research related to reservoir behaviour, and he became increasingly agitated over the immensely wasteful production methods characteristic of most U.S. oilfields. He became convinced that the only satisfactory remedy lay in compulsory unit operation of petroleum reservoirs enforced by federal law.[34] A director of the American Petroleum Institute (API), he decided first to take his reform ideas before his fellow directors in the hope that this important organization might be persuaded by his ideas. The API reception was less than enthusiastic: some of the directors strenuously denied that there was a serious problem, and all abhorred the idea of federally enforced compulsion. Unable to convince the directors of the Institute, Doherty called upon his friend Calvin Coolidge at the White House in the fall of 1924. Also urging the president to give special attention to the conservation of petroleum resources was Secretary of the Interior Hubert Work, who was convinced of the need to establish a federal oil commission to study and promote the conservation of oil deposits.[35] Prompted by Doherty, Work and his own concern about national security needs, Coolidge created the Federal Oil Conservation Board (FOCB) in December 1924. This action by the president had the extremely important effect of identifying petroleum conservation as a matter of national concern, and it greatly stimulated discussion and research in petroleum problems.

The FOCB was composed of the secretaries of war, the navy, the interior and commerce. Functioning through a technical and advisory committee made up of representatives of the four departments, the Board immediately set out to obtain information from which facts could be obtained and issues clarified. Then, early in 1926, the FOCB began open hearings, inviting both professional experts and practical oilmen. Division of opinion at these hearings was clearly drawn. Industry executives were mainly unwilling to admit that there was a serious waste problem, and even those who were prepared to acknowledge shortcomings in the industry's approach to production were not of the view that legislative "interference" was the solution, especially federal legislation. By contrast, professional experts lined up one after the other to decry the deplorable waste, with many arguing in favour of unit operation as the fundamental solution.[36]

Between 1926 and 1933, the FOCB issued a series of reports to the president urging the promotion of uniform state conservation regulations consistent with the principles of reservoir mechanics. Unit operation was presented as the ideal. In later reports, increased emphasis was given to the control of supply through import restrictions and domestic production quotas. In its final report in 1932, the Board presented a specific proposal for an interstate compact that would permit a state-federal agency to forecast demand and allocate production quotas to producing states. Although it would still take several years for a consensus to solidify, and even though the critical problem of finding a way to get around the anti-trust laws to permit co-operative agreements among oil producers remained to be addressed, the FOCB had nonetheless played an extremely important role. It had served to focus discussion on national oil policy, and its studies were an important factor in shaping the opinions of public officials and business leaders.

By the time of the FOCB's final report, the situation facing the oil producers and the oil-producing states was desperate. A series of major new fields were discovered through the late 1920s, culminating in "Dad" Joiner's incredible East Texas find in the fall of 1930. A dramatic new surge of production commenced just as the impact of the Great Depression and declining demand had begun to be felt. Demoralization and confusion were compounded by the actions of state and federal courts. Conflicting interpretations on whether regulation to restrict production to market demand was an acceptable method of preventing physical waste or an unacceptable attempt to fix prices had been brought forth.[37] In Texas, where the market had collapsed, sending oil prices down to as little as 10¢ a barrel, Governor Sterling had been compelled in August 1931 to declare martial law and send the National Guard into the East Texas oilfield to protect life and property as producing oil wells were shut down. A similar situation saw troops sent into the Oklahoma City field.

In this desperate environment, prejudice against federal participation eased, and the prospect of a joint federal-state conservation board along the lines suggested by the FOCB appeared bright. With the election of Franklin D. Roosevelt in November 1932, however, the FOCB became defunct, the compact idea was dropped, and the problems of the petroleum industry were approached in a different manner. Roosevelt was not nearly so reticent about using federal powers to bring order to a big sector of the national economy. The *National Industrial Recovery Act (NIRA)*, passed in the summer of 1933, permitted the establishment of an industry code regulatory system. This was extended to include the oil and gas industry. Under the Oil Code, adopted later in 1933, a far-reaching and unprecedented system of import restriction, minimum prices, administration approval of new reservoir development plans, limitation of domestic production to total demand less imports and allocation of allowed domestic consumption among states was established. Secretary of the Interior Harold Ickes was anxious to go even further, and to have the oil industry declared a public utility and regulated accordingly. Roosevelt's "Code" approach, while offering federal structure and sanction, left administrative responsibility to personnel recruited from the industry and the internal processes of proration in each state to state regulatory authorities. Having closely involved industry in drafting the Code, the president then cleverly balanced his creation by appointing Ickes as administrator of the Oil Code.

By the end of 1933, many oilmen rated the Oil Code a great success: production had been cut back, prices had firmed and begun to move upward, and the industry showed signs of stabilizing. Despite production cutbacks, one problem still loomed. Although Ickes's administrative hand was strengthened through section 9(c) of the NIRA, which prohibited interstate shipment of oil produced in violation of state law or regulation, there remained nonetheless a significant volume of "hot" or "bootleg" oil being moved, mainly by small producers, across state boundaries. Early in 1935, this problem took on a new dimension that threatened to bring the whole program crashing down. With its famous "Panama" decision of January 7, the U.S. Supreme Court declared that federal officials did not have the power to enforce production quotas determined by state agencies and that section 9(c) of the NIRA was therefore invalid.[38] Anticipating the court decision, Ickes readied a remedial bill, which Thomas Gore and Tom Connally introduced in the Senate within a day of the hostile verdict. Known as the Connally "Hot Oil" Act, the new measure was rushed through both houses of Congress with the industry's eager anticipation.

The success of the Connally Act in bolstering the efforts of state conservation agencies was no sooner confirmed when the Court struck the existing regulatory structure a second crippling blow. In June 1935, the Supreme Court declared the entire *NIRA* unconstitutional, which meant the end of the Oil Code. Despite Secretary Ickes's plea for a renewed and strengthened federal regulatory role, Roosevelt waited cautiously for consensus. The benefits of enforcing conservation measures and restricting production had been clearly demonstrated, but while many in the industry, particularly the larger integrated companies, were well disposed to the idea of self-regulation with federal involvement, much along the line of the now defunct *NIRA*'s Oil Code, it was also apparent that there was strong opposition to any direct federal regulatory role among a large segment of the industry, especially in Texas. Federal control of production and distribution of petroleum promised to be a Herculean task fraught with political hazards.

Roosevelt was not inclined therefore to challenge the initiative already taken by several of the large oil-producing states. Governors E.W. Marland of Oklahoma, James Allred of Texas and Alfred Landon of Kansas had been in active discussion about the possibility of organizing an interstate oil compact for some time. By the spring of 1935, two clearly different draft proposals had emerged. Governor Marland, supported by a majority of the producing states, proposed a compact whose prime purpose would be the economic stabilization of the industry. This would require an agency to determine market demand and to recommend to each participating state the relative part of the total demand for domestic production that could be produced without waste. Administration of such a program would be by a joint state-federal committee. Governor Allred advocated a different approach. He was unalterably opposed to the principle of conservation regulation for the purpose of price control. Allred would consider only the matter of preventing physical waste, and this he believed was solely the business of the State of Texas. Prepared to concede to federal regulation of some form of hot oil law and restriction of foreign oil imports, he would not countenance any further participation. Allred proposed therefore an interstate compact that would promote the standardization of state conservation laws.[39]

If Marland had the majority of delegates inclined to his view, Texas controlled by far the greater amount of oil. No agreement on curtailment of oil production would be workable without the state that accounted for half the country's oil supply, but the Texas delegation was unyielding. Thus, the draft to which delegates ultimately assented was patterned essentially on the Texas proposal. Kansas, New Mexico, Illinois, Colorado, Oklahoma and Texas informed Washington of their accord, and Congress formally ratified the agreement, establishing the "Interstate Compact to Conserve Oil and Gas" on 27 August 1935.

The created agency, the Interstate Oil Compact Commission, possessed no power to determine and distribute petroleum quotas among the states. Article II of the Act stated plainly, "The purpose of this Compact is to conserve oil and gas by the prevention of physical waste thereof from any cause." Article III required the compacting states to enact or continue to enforce laws to prevent a series of specified physical wastes.[40] The terms of the Compact limited it to the functions of collecting and disseminating technical and legal information to the member states, of serving as a forum for the discussion of common problems at semi-annual meetings, and of promoting the enactment of comprehensive conservation statutes in each of the producing states. At the time of approval, only Oklahoma, Texas, Kansas, New Mexico and Louisiana had such comprehensive legislation.

Within each state, regulators still had to struggle with the problems of operating proration systems that divided production among reservoirs and wells, of regulating purchasers and pipeline companies, of formulating rules of reservoir development, and of devising means to regulate drilling, completion and operation of individual wells closely, in an environment of rapidly changing technology, aggressively competing economic interests, partisan politics and public scrutiny. Although the Compact Commission had no direct role to play in these matters, its semi-annual meetings offered politicians, regulators and industry personnel the opportunity to discuss shared problems and to consider experience and insights in a manner that allowed the member states to progress in a generally complementary manner.

Within this larger continental setting, and with reference to the operating structure of the North American petroleum industry, and partly within the context of this historical experience, Alberta began to address the issue of oil and gas conservation in the early 1930s.

PART I

The Conservation Movement in Alberta
1908–1938

Oil well and camp at Oil City, Alberta, in today's
Waterton Lakes National Park, at the site of 1902
oil discovery. This Rocky Mountain Development
Company operation reflects the "dig, burrow, blast
and drill at will" attitude of the day. Glenbow Archives NA-3465-3.

ONE

Turner Valley "Waste" Gas and the Early Conservation Movement

THE INITIAL REGULATORY FRAMEWORK

What has been done in the United States will be done in Canada. There were people who would not buy Standard Oil Stock at a dollar a share twenty-five years ago, but there are no careful investors today, with the history of the oil industry before them who will hesitate to buy Rocky Mountain Co. stock at $1.00 a share. It is not a question of "Will it pay, but how much." [Prospectus: Rocky Mountain Development Company, 1902].

Such is the advice offered in the first prospectus issued by a petroleum company in western Canada.[1] The descriptive pamphlet produced by the company mirrors the spirit of the time. In their request for investment capital, the promoters of the Rocky Mountain Development Co. presented the public a compelling lesson in speculative mathematics. Over the preceding five years, the average price of a barrel of oil in the United States had been $1.15, whereas the cost of production had averaged about 5¢. A well producing a mere five barrels daily would earn 150% annually on its cost; one producing 10 barrels, 300%, and so on. To excite speculative appetites still further, readers were reminded that "Oil is worth today in Winnipeg, about $10 a barrel wholesale, or 25 cents a gallon." "Figure," the brochure urged, "300 barrels a day per well at $10 a barrel, 90,000%." But this was only the beginning of the fantastic generation of profit possible in the oil business. "Figure how many wells 3,840 acres of land will support, allowing one well per acre." Beyond motivating the would-be investor to rush out and buy company stock, these figures enabled one to "easily figure how Rock-afeller [*sic*] became a billionaire."[2]

The oil strike and limited production at "Oil City" along Cameron Creek near Waterton Lakes in 1902 was enough to launch Alberta's first oil boom but insufficient to sustain interest beyond 1905. Although would-be oil millionaires would have to wait a decade before Alberta's next and more substantial oil boom brought a second chance, the flurry of activity that generated the Rocky Mountain Development Co. prospectus is illustrative of the laissez-faire environment that attended resource exploitation in the immediate post-frontier period. Just as there were no restraints upon the excesses of stock promoters, only the most rudimentary guidelines governed the removal of minerals. The essential requirement was that one had to obtain the consent of the mineral rights holder, then one might dig, burrow, blast and drill at will.

In the case of western Canada, this usually meant obtaining rights from the Crown, as this is the most significant factor that distinguishes the Canadian from the U.S. experience, where such rights were usually held by the surface owner. When title to Rupert's Land, which included the area that today comprises Manitoba, most of Saskatchewan, and parts of Alberta and the Northwest Territories, was surrendered by the Hudson's Bay Company to Canada on 15 July 1870, ownership of lands and minerals passed to the federal government in the name of the Crown.[3] The acquired lands and mineral resources were administered from Ottawa by the Department of the Interior from 1873 to 1930, when authority within the region bounded by the prairie provinces was transferred to the appropriate provincial governments.[4]

The legislation that guided the federal government in its administration of the natural resources of the western interior was embodied in the *Dominion Lands Act* of 1872.[5] In detail, the Act first defined how Manitoba and the North-West Territories would be surveyed, and then laid the foundation for a system of land classification based on use. Under separate headings, Hudson's Bay Company lands, educational endowment lands, military bounty lands, homestead lands, grazing lands, haylands, mining lands, coal lands and timberlands were all identified and discussed. General provisions of the Act gave the governor-in-council power to withdraw particular categories of land from some or all of the Act's provisions and, as is characteristic in the legislation of parliamentary governments, authority to provide for the disposal of the various types of land under separate regulations.[6]

Separate regulations gradually emerged for each classification of land. Federal officials first directed their attention to the formulation of regulations governing homesteads, licences for timber operations, and the terms under which grazing leases could be obtained for ranching activities. Regulations concerning mining and coal lands evolved more slowly. Patent or title extended under the 1872 Act did not separate surface and mineral rights, although specifically designated coal lands were withdrawn from the operation of the Act and the minister possessed the authority to remove land from sale and institute a system of leasehold in the case of certain areas that might in the future prove "to be rich in minerals."[7] As geological knowledge of the western interior expanded and economic activity increased, regulations governing the Crown's disposition of natural resources became more elaborate. By 1881, orders-in-council had established a framework governing access to coal lands, and regulations governing the disposition of minerals other than coal were introduced in 1884. Then, in a fundamental redirection of policy, the federal government proclaimed that, from 31 October 1887, for all lands west of the third meridian only surface rights would be surrendered and "all mines and minerals which [might] be found to exist within, upon or under the lands" were reserved to Her Majesty.[8] Henceforth, no lands were alienated without a clause in the patent specifically reserving mines and minerals to the Crown. This was a change of far-reaching consequence. At a stroke, Canada set upon a course of resource exploitation that differed significantly from that in the United States where typically land titles combined surface and subsurface rights. Nowhere is the longer-term consequence of this difference more apparent than in the development of the petroleum industry in western Canada. The region west of the third meridian (a line that divides Saskatchewan in east-west halves) was settled only slightly in 1887, and the change meant that mineral title in this area would not be widely held outside of the Hudson's Bay Company, the Canadian Pacific Railway Company, and the Calgary and Edmonton Railway Company, all of whom had received land grants before the policy change.

As economic activity in the western territories began to expand and diversify following the completion of the Canadian Pacific Railway, individual minerals other than coal became the focus of attention, and specific regulations were fashioned from the initial regulations of general application. In 1890, the 1884 Regulations relating to lands containing mines and minerals other than coal were amended as regards petroleum.[9] The general 1884 Regulations stipulated that "no mining location or mining claim shall be granted until the discovery has been made of the vein, lode or deposit of mineral or metal within the limits of the location or claim."[10] In recognition that this provision could retard the development of lands that might contain petroleum, the amendment provided that an applicant could obtain entry for a location upon swearing an affidavit that "from indications he verily believes that petroleum exists on the location applied for." The Department of the Interior would then reserve the location for up to five years, and would only sell the property to the applicant upon proof that there was at least one oil well in operation yielding petroleum in paying quantities. This amendment stands as evidence of the emerging awareness that the unique physical properties of petroleum required a distinct regulatory approach. The amendment therefore marks the humble but true beginning of petroleum regulation in western Canada.

The next development in the nascent administrative framework governing petroleum exploration came in 1898 after the minister of the interior had received a number of requests from individuals desiring to prospect for petroleum on lands in southern Alberta. Anticipating the first independent petroleum regulations, the order-in-council of 6 August 1898 gave the minister authority to reserve an area not exceeding 640 acres for up to six months for any prospector. If oil was found in paying quantities, the reservation could be purchased at $1.00 per acre, subject to a royalty of 2½% on sales.[11] These regulations, which applied only to the area south of the Canadian Pacific Railway line in the territorial district of Alberta, were extended on 31 May 1901 to all unappropriated dominion lands.[12] Also added was the requirement to submit sworn monthly production statements, along with the warning that any attempt to defraud the Crown would be met with cancellation of land title. In 1904, the area that could be reserved for an individual or company was increased to 1,920 acres, along with the proviso that the length of the tract could not exceed three times its breadth.[13] The price charged for the first 640 acres remained as before, the rate for the remainder of the reserved area, however, was set at $3.00 per acre, and henceforth the patents transferring title would specifically convey surface and petroleum rights rather than undifferentiated mineral rights as had been the case. In 1906, a further amendment extended the application of the petroleum regulations to include "the reservation and sale of lands for natural gas purposes."[14]

On 2 May 1910 all previous orders-in-council were rescinded and the first comprehensive set of "Regulations for the disposal of petroleum and natural gas rights" came into effect.[15] The new regulations represented a fundamentally different approach to the disposal of petroleum and natural gas title. From this point, the Crown was prepared to lease rather than to sell such rights. Leasehold of up to 1,920 acres was made available for 21 years (renewable for a further 21) at 25¢ per acre for the first year and 50¢ per acre payable in advance yearly thereafter. Anxious to encourage development, the federal government also introduced into the regulations the important principle of accepting specified development expenditures as a credit against the lease rental.[16] To lend further encouragement to the search for oil in the West, Ottawa also inserted into the regulations the provision that no royalty would be levied on the sale of "petroleum" until after 1 January 1930. In the case of natural gas, however, there was no such commitment. It was left to the minister of the interior to levy such a royalty rate as he might think appropriate. To retain the lease, the holder, in addition to meeting the Crown's rental, had to have appropriate machinery and equipment on the leasehold within one year, and boring operations had to be commenced within 15 months. Later, in 1910, following a nudge from the British Admiralty outlining the growing strategic importance of petroleum, the Canadian government rounded out the regulations with an amendment giving the Crown at any time the right to expropriate all crude oil or its products "for the use of His Majesty's Canadian Navy."[17] Agreed to more in a spirit of deference to the mother country than out of insight about the role oil was destined to play in warfare and in evolving industrial economies, the amendment was nonetheless a harbinger of the larger international forces that would always be a part of the critical background in which domestic petroleum policy, knowingly or otherwise, would have to be framed.

Under the petroleum lands sales policy that had existed up to the inauguration of the lease system in 1910, only 12 parcels, covering 16,028 acres, had been alienated in Alberta.[18] Four of the sales were in the Waterton Lakes area, another was in the vicinity of Bow Island, southwest of Medicine Hat, and the remaining seven, covering more than two-thirds of the total acreage, were in the north along the Athabasca River near Fort McMurray. The acres that were to launch Alberta into its first real oil boom, however, were those acquired by William Stewart Herron in 1911 as leasehold under the new system.

PETROLEUM EXPLORATION AND DEVELOPMENT IN
ALBERTA TO 1921

Alexander Mackenzie was the first to comment on the presence of petroleum in western Canada. He wrote:

> At about twenty-four miles from the Fork, are some bitumenous [sic] fountains, into which a pole of twenty feet long may be inserted without the least resistance. The bitumen is in a fluid state and when mixed with gum, or the resinous substance collected from the spruce fir, serves to gum the canoes. In its heated state, it emits a smell like that of sea coal. The banks of the river, which are there very elevated, discover [sic] veins of the same bitumenous quality.[19]

Subsequent travellers commented on the bituminous sands along the Athabasca, and following the acquisition of the western territory by Canada in 1869 the area was frequented by a series of expeditions sent forth by the Geological Survey, whose task was to investigate the potential resources of central Canada's vastly extended hinterland. In 1872, the first of these expeditions was led by John Macoun, the noted botanist and western explorer. Reporting later for the Geological Survey, Macoun wrote of tar oozing from the earth:

> When we landed, the ooze from the bank had flowed down the slope into the water and formed a tarred surface extending along the beach over one hundred yards and as hard as iron; but in bright sunshine the surface is quite soft and the men tracking along the shore often sink into it up to their ankles.[20]

The most important party to venture next into the Athabasca Basin was that led by the renowned geologist Robert Bell in 1882. Also employed by the Geological Survey, Bell examined the geology of the region much more thoroughly and concluded that the oilsands rested on a vast reservoir of petroleum.[21] Quickly gaining currency, for decades the theory continued to lure geologists, engineers and entrepreneurs to the oil sands. But even though the Geological Survey, technical people in other branches of the federal civil service, and a few others would remain captivated by this potential Eldorado in the northern wilderness, the area remained too remote to attract much more than scientific interest, especially when there seemed to be promising possibilities in the more accessible south.[22]

The first indication that there existed a potential for petroleum discovery in southern Alberta came in 1883. Laying track across the western prairie in 1882 and 1883, railway engineers experienced great difficulty obtaining adequate water supplies. To address this problem, drilling was commenced on a series of deep wells along the CPR right-of-way. The well at Langevin Station,[23] about 25 miles along the main line west of Medicine Hat, failed to find water; however, at the 1,155-foot mark, the drillers hit natural gas and, as though nature intended to lend drama to the first significant discovery of natural gas in western Canada, the gas ignited to consume the drilling rig in a great ball of fire.[24] Eventually managing to extinguish the raging fire, the drilling crew started a second well just a few feet away, and in the fall of 1884 made a similar though, in this instance, fireless discovery of natural gas.[25] Near Cassils, 38 miles farther northwest along the main line track, the search for water was no more successful. Here, also in 1884, drillers discovered natural gas. The ubiquitous Geological Survey of Canada was not long in investigating and offering informed comment. In a paper, entitled "On Certain Borings in Manitoba and the Northwest Territory," presented to the Royal Society of Canada in May 1886, Dr. George M. Dawson explained that, although the wells at Langevin did not yield a sufficient quantity of good water, they did demonstrate

the very important fact that a large supply of natural combustible gas exists in this district at depths of 900 feet and over, in the sandy layers of the "Lower Dark shales." In consequence of the generally horizontal position and widespread uniformity in the character of the rocks, it is probable that a similar supply will be met with over a great area of this part of the Northwest, and that it may become in the near future a factor of economic importance.[26]

In 1900, when gas was struck again near the railway's South Saskatchewan River crossing, CPR President Sir William Van Horne agreed to a request from several of the leading citizens of the nearby community of Medicine Hat for the loan of a drilling rig so the well could be deepened. The result was the discovery of an immense, uncontrollable gas flow that launched the community on a search for the effective exploitation of this new resource and, eventually, to the formation in 1904 of a municipally owned utility to supply fuel for home and commercial use.

Medicine Hat's success, combined with the recognition that oil

Gas well near Elizabeth Street School, Medicine Hat, Alberta, 1907. The earlier discovery of a large natural gas reservoir had launched the community's municipally-owned utility in 1904. Medicine Hat Museum and Art Gallery Archives PC525.134.

could be expected to exist in combination with natural gas, prompted the CPR to reconsider its initial conclusion that natural gas was mainly a nuisance. In February 1906, the company hired Eugene Coste, one of Canada's most experienced geologists, to search for oil under its lands. As an incentive, the CPR promised Coste $25,000 above his expenses and salary if a significant discovery resulted.[27] Coste was a logical choice. Earlier, as an employee of the Geological Survey of Canada, he had become familiar with the geology of the southern prairies. Complementing the decision to hire Coste, and further reflecting the CPR's changed attitude towards the resource potential of its vast prairie land holdings, was the company decree that henceforth petroleum rights would not be surrendered to the purchasers of railway land.[28]

While Coste was busy investigating CPR lands in southeastern Alberta, a group of Calgarians, also inspired by events in Medicine Hat, incorporated the Calgary Natural Gas Company and began searching for natural gas. This company, drilling on the Colonel Walker estate in east Calgary in late October 1908, struck an encouraging gas flow at the 800-foot level. J.S. Dennis, senior CPR official in Calgary, immediately informed his superiors in Montreal. He noted that the well was close to the railway's main line and also near land that the CPR was developing as a new industrial subdivision.[29]

As Dennis and his senior colleagues waited for the Calgary well

Eugene Coste, CPR geologist in 1906 and founding president of the Calgary-based Canadian Western Natural Gas, Light, Heat and Power Company in 1911. Canadian Western Natural Gas Company Limited.

to be drilled to its planned depth, news came in February 1909 of an important natural gas strike by Coste at Bow Island, southwest of Medicine Hat. The CPR moved immediately to consolidate its land position by acquiring all available mineral leasehold on crown lands adjoining railway property in the discovery area. Although confirming Coste's geological speculations, the southern gas discovery was still far from Calgary, the region's most important centre. The prospect of an important gas find virtually on CPR property

"Old Glory," the Bow Island Field's 1909 discovery well. Note unburnt gas blowing to atmosphere. Canadian Western Natural Gas Company Limited.

in the city therefore led company officers to discount Coste's opposition to assisting the Calgary company and to agree to help finance completion of the more convenient well. The CPR's investment of $5,000 was rewarded in May 1909 by a gas strike in east Calgary.[30]

While A.W. Dingman, manager of the Calgary Natural Gas Company, began organizing the connection of industrial and residential customers to the newly discovered natural gas supply, Coste commenced drilling more wells near Bow Island to determine the extent of the southern gas field. Mindful of the earliest gas finds along the CPR main line between Medicine Hat and Calgary, Coste also extended his search northward to the Brooks and Bassano areas, and by the summer of 1910 had a producing gas well near each of these communities. Coste had been remarkably successful. Under his direction, six wells had been drilled—all of them successful gas producers.[31] He had not, however, found the oil in which the railway was particularly interested, and the most productive Bow

Island gas wells were 150 miles from Calgary, the region's only sub-stantial market.

Contemplating what might be done with the natural gas found on its land, the CPR at this juncture began to think in favour of some arrangement with the Dingman group, who possessed the critically important exclusive franchise to supply Calgary with natural gas.[32] Coste, who was not about to let the opportunity promised by his discovery slip from his grasp, quickly organized the financial back-ing of a number of prominent eastern capitalists, the most impor-tant of whom was the Honourable Sir Clifford Sifton. His strategy was to make such an attractive offer for the assets of the Calgary Natural Gas Company that the Calgarians could not refuse.[33] For holdings valued by Calgary accountants at $69,943, Coste offered $200,000 in bonds and $300,000 worth of common shares in a successor company capitalized at eight million dollars. Gaining ac-ceptance, and with the prized franchise in his possession, Coste be-

Eugene Coste (centre) and his remarkably successful drilling crew at Old Glory No. 1: (l-r) driller W.R. "Frosty" Martin and roughnecks H. Gloyd, G.W. Green and A.P. Philips. Canadian Western Natural Gas Company Limited.

The labour-intensive job of constructing this Bow Island to Calgary gas pipeline was finished in just 86 days in the spring of 1912. Canadian Western Natural Gas Company Limited.

gan negotiations with the CPR. The agreement ultimately concluded stipulated that Coste's Prairie Fuel Gas Company would operate two drilling rigs on railway lands, until at least 12 wells were completed; lay a gas pipeline to Lethbridge, and then to Calgary; and purchase at least 50% of all the gas required to supply the gas company's markets from CPR wells. Payment to the CPR would be 6% of the gross monthly receipts gained from the sale of gas originating on railway land. If Coste happened to discover oil on railway land, it was further agreed that it would be turned over to the CPR at cost. In addition, Prairie Fuel would transfer $200,000 in fully paid capital stock, giving the railway a minority interest. For its part, the railway promised to transfer all drilling equipment, tools and pipe at the well sites; to give Coste the exclusive right to market gas from the Bow Island, Brooks, Bassano and other selected CPR lands; and to allow the gas company to lay pipeline through and along the railway right-of-way.[34] The memorandum agreement signed on 15 February 1911 gave Coste one year to finish organizing his company and until 1 November 1912 to have the Lethbridge portion of the pipeline completed.

The gas company's organization was completed on 19 July 1911 when it was reincorporated and renamed the Canadian Western

Natural Gas, Light, Heat and Power Company. The blowing in of a tremendous new gas well at Bow Island in September 1911 confirmed the success of the new company's drilling program. Construction of the gas transmission line started on 22 April 1912. Eighty-six days later the line was complete, not just to Lethbridge but also to Calgary, and twelve thousand citizens assembled to witness Coste's triumph as the gas was turned on in Calgary on 17 July 1912. The development that Dr. G.M. Dawson had prematurely anticipated in his address to the Royal Society of Canada a generation earlier had at last materialized.

During the spring of 1911, as Coste was concluding his negotiations with the CPR and preparing the second phase of his drilling program, William Stewart Herron was investigating the gas seepages along Sheep Creek, a short distance southwest of Calgary. Lacking Coste's formal training, Herron was nonetheless a keen student of geology. Noticing the telltale curve of an anticline on a rocky outcrop in the vicinity of the gas seepage, he concluded at once that the gas was a petroleum derivative. When this was confirmed from samples sent to the Universities of Pennsylvania and California, Herron directed his attention to the acquisition of mineral rights to as much of the nearby land as possible. Finally, in July 1913, Herron was able to combine his land, some 3,200 acres of surface and subsurface rights in the immediate proximity of the gas-seep, with Calgary capital to form the Calgary Petroleum Products Ltd.

The new company began drilling in January 1913. Frequent delays imposed by a complex geology, the limitations of cable-tool drilling technology, and the repeated need for further infusions of capital led to local predictions of failure as 1913 passed into the spring of 1914. Then, about 5:00 p.m. on May 14, the company drilling crew pierced an oil-bearing stratum at 2,718 feet. The pressure released by the wellbore sent oil gushing 15 to 20 feet above the derrick floor. The remarkable straw-coloured liquid (naphtha) was of such a light gravity that it could be pumped directly into the gas tanks of the unsophisticated vehicles of the day and was described by Calgary newspapers as "practically pure gasoline." Christened by the press as the "Dingman Discovery," after A.W. Dingman, the manager of Calgary Petroleum Products, the well launched a speculative frenzy. From makeshift brokerage offices, frantic promoters hawked the "attractions" of dozens of new oil companies. City police had to be called to supervise and keep order among the unruly crowds lined up to purchase shares in the likes of X-Ray Oils, Lucky Mary, Lucky 7, Bonanza King, Utopia, Hermit

Calgary Petroleum Products Co., Ltd.'s 1914 Turner Valley discovery well (known as Dingman No. 1 and later as Royalite No. 1). Glenbow Archives NB-16-604.

Oil, Planet Oil, Big Dipper, Prosperity Oil, and even the Sherlock Holmes Oil and Gas Company.[35] Within three months of the discovery, more than 400 companies with an authorized capital of well over $300 million had been granted provincial charters.[36]

By the late summer of 1914, the first Calgary oil boom had collapsed in the face of disappointing results in the field; embittered actions alleging theft, swindle and stock fraud in the courts; and the preoccupation of international capital with the shadow of war in Europe. The excesses of May, June and July 1914 would take almost a generation to erase, and for the next decade legitimate local petroleum companies starved for capital were able to proceed only intermittently.

Even the discovery company, Calgary Petroleum Products Ltd. (CPPL), was near the end of its financial tether by the fall of 1915, and it was only after securing $5,000 from the CPR that it was able to finish drilling the discovery well to its planned depth. By the summer of 1916, CPPL was again in dire straits. Wells No. 1 and 2 were both producing several million cubic feet of wet or gasoline-producing gas a day, but both required further development work to become significant producers. To accomplish this and begin well No. 3, managing director Dingman again sought CPR assistance.

The railway remained one of the few bodies interested in what, perhaps prematurely, was being called the Turner Valley oilfield—

and with good reason. As P.L. Naismith, the manager of the railway's Resources Department, explained to one of the company's senior officials, CPPL "are drilling in the heart of a district which is very largely owned by the CPR and who would be materially benefited by any discovery in that field."[37] As the CPR considered Dingman's most recent request, a new and much stronger player arrived on the scene to capture the railway's attention. The newly interested party was Imperial Oil, a subsidiary of one of the world's most powerful oil conglomerates, Standard Oil of New Jersey. For the CPR, the prospect of working with a company of Imperial's stature was immensely attractive. Here was an organization with vast capital resources that could evaluate and develop the petroleum potential of railway lands to a degree that local western capitalists, dependent on the CPR for supporting investment, could never achieve.

From Standard Oil's perspective, some kind of arrangement with CPR also had definite attractions. The situation in Alberta had been closely monitored since the 1914 discovery. Commenting to his executive colleague C.O. Stillman in the midst of the Calgary oil boom, President Walter Teagle of Imperial Oil explained:

> you realize fully the importance of the discovery of a productive oil field in Canada and how such a discovery would affect our business. For this reason it seems to me that we should be very much alive to the situation and if there is any likelihood of paying production being developed in Canada, we should, if possible, try and arrange to be in on the ground floor with leases of our own, so that from the very outset we might occupy an important position as producers of oil in Canada as we now occupy as refiners and distributors.[38]

The two geologists who were subsequently dispatched to Turner Valley were not impressed, and Jersey Standard's chief geologist reported that there was little to commend the region for more serious consideration, but some leases might be taken as a precautionary measure. This indifferent assessment was seconded by the reports of the two experienced oilmen from West Virginia who Standard hired to go secretly to Alberta in the early summer of 1915.[39] The acquisition of petroleum leasehold as a defensive measure, however, posed something of a problem, given that section 40 of Canada's "Petroleum and Natural Gas Regulations" denied crown leasehold to any company "directly or indirectly controlled by foreigners or by a foreign corporation."[40] Imperial was restricted therefore to the search for petroleum rights on privately held lands that, unlike crown lands, were not burdened with the restrictions of the "Petroleum and Natural Gas Regulations." To this end, discussions were initiated in the summer of 1916 with the CPR, holder of mineral title to hundreds of thousands of acres scattered across the Canadian prairie.[41]

Talks commenced in an informal manner, and the CPR was surprised to learn that it was the Viking-Wainwright area of east-central Alberta, rather than the foothills region southwest of Calgary, that had attracted Imperial's geologists.[42] In February 1917, Teagle presented the railway with a concrete proposal. The oil company requested that the railway withhold from leasing to others the oil and gas rights on CPR lands within a vast block of more than ten million acres centred around Wainwright and straddling the Alberta-Saskatchewan border until 31 October 1917.[43] During this period, Imperial geologists would select, from different parts of the area set aside, up to fifty thousand acres of mineral rights to be leased from the railway. Afterwards, but before December 1919, Teagle promised, Imperial would spend not less than $100,000 on drilling operations.

The railway negotiators were not impressed. Their calculations revealed that, in the proposed reservation area, the CPR held the oil and gas rights to approximately 2.15 million acres. A.M. Nanton, a key member of CPR's Resource Department Advisory Committee, warned, "We must not endorse an agreement that will have the effect of tying up a large area to the oil people, without having the power to cancel the arrangement unless there is full and continuous development."[44] This was the principle that had guided Sir Thomas Shaughnessy in his negotiations with Eugene Coste half a decade before. The railway strategy that would shape negotiations with Imperial Oil through the decade from 1917 was to keep the amount of land reserved to a minimum and the commitment to exploration and development to a maximum. The reverse strategy guided the oil company; it endeavoured to have the maximum acreage set aside but wanted to keep development obligations to a minimum. Both corporations understood the other's stance and bargained from strong positions.

As the CPR contemplated Imperial's proposal through the spring of 1917, its negotiating stand was immensely strengthened by the activity of another of the major competitors in the global quest for oil. In April 1917, the Shell Transport Company Ltd. of London, Jersey Standard's arch rival, approached Prime Minister Sir Robert Borden to express a general interest in the "exploration of territories in Western Canada."[45] Shell was thinking of a Middle East-sized

MAP 1.1 Shell and Imperial Concession Requests, 1917

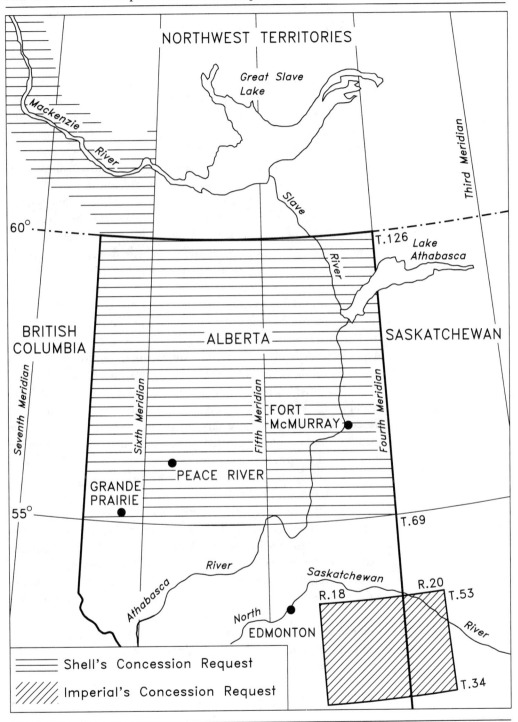

exploration concession and asked for "the exclusive oil and natural gas rights" to all of Alberta north of 55°, plus that portion of the Mackenzie River Valley south of 65°, some 328,000 square miles in all.[46] For a government used to offering exploration leasehold in units of 1,920 acres, or three square miles, Shell's request seemed excessive, and in stages the London company responded by reminding that 90% of the concession territory would be returned in five years and that perhaps there could be some direct involvement by the Canadian government and a sharing of profits.[47] Drawing attention to the pattern of development in Pennsylvania, Texas and California, where oilfields were divided into numerous smallholdings, Shell described the tremendous waste of capital and the "almost criminal" waste of petroleum that was characteristic. By contrast, Shell proclaimed that "scientific" development by a single company would eliminate the stock fraud or swindles that commonly attended North American oil booms and would ensure the largest possible percentage of production for human use.[48]

Knowing that section 40 of the "Petroleum and Natural Gas Regulations" did not define British-owned companies as "foreign" and that the Shell Transport Company was receiving a receptive ear in Ottawa, Imperial pulled out all the stops to prevent the "Yellow Peril," as Shell was uncharitably known in Standard circles, from gaining such a significant beachhead in Imperial's backyard. Imperial knew that it had to get an exploration program under way without delay. As one company official put it, "Action and lots of it is deemed absolutely necessary to answer the situation that the pending application has created." Therefore, an agreement with the CPR took on a note of some urgency.

While trying to speed negotiations with the CPR to a successful close, W.J. Hanna, Imperial's new president, was also attempting to persuade the Canadian government to recognize the "menace to the country" contained in the Shell proposition. He argued that the Shell concession would absorb "all the potential oil land of Canada with only some very negligible exceptions." He also assured that Imperial had budgeted an exploration expenditure for the coming year of nearly half a million dollars, "a large part of which it [was] intended to devote to the territory for which this application [was] being made." Finally, Hanna offered the grave warning "that the granting of such a concession would be very nearly equivalent to a destruction of the entire present investment of the Canadian oil industry."[49]

In the end, Imperial was successful. Agreement was reached with the CPR, but the acreage to be held, from which Imperial could select petroleum leaseholds, was sharply reduced and a strictly binding schedule of development was imposed.[50] More importantly, for the moment at least, the "Yellow" tide had been stemmed. For this, Imperial could thank the timely coincidence of planned negotiations between Ottawa and the prairie provinces concerning the surrender of federal control of lands and resources and in particular Alberta's emphatic opposition to the Shell proposal.[51]

In keeping with the sense of urgency that had brought Imperial to define and protect its interests in western and northern Canada, the company launched an energetic exploration and drilling program along the Mackenzie River, near Fort Norman in the Northwest Territories, and in east-central Alberta. Success came first in the North with a significant strike at Fort Norman in August 1920. To the south, Imperial did not do so well, the 11 drilling crews active on the Alberta prairie had nothing to show for their efforts. At the same time, the pressure to establish the most promising land position and achieve success had not diminished. This was the product of escalating concern about a predicted world oil shortage that gripped Washington, D.C., and European capitals in 1920. Shell had withdrawn from the scene, only to be replaced by two new companies. One was the Northwest Petroleum Exploration Company, organized by a group of wealthy New York capitalists and directed by Ralph Arnold, a respected American consulting geologist and petroleum engineer;[52] the other was Whitehall Petroleum, a London company that was amply capitalized and almost as well connected as Shell.[53] Both were interested in the vast acreages of mineral title held by the CPR.

The lack of success on the prairie and the presence of substantial international capital assessing prospects in Alberta compelled Imperial to look more closely at an area its geologists had earlier dismissed, the foothills region immediately south of Calgary. Although the CPR had extensive acreage in the area, the most immediately attractive lands in Turner Valley were those held by the discovery company, Calgary Petroleum Products Ltd. (CPPL). If Jersey Standard planned to play a major or dominant role in what yet might become western Canada's first producing oilfield, some arrangement had to be reached with the owners of CPPL. The scenario was not an unfamiliar one to the executives, who supervised Jersey Standard's empire from 26 Broadway in New York. In this instance, however, circumstance contrived to ease their Canadian subsidiary's task. In late October 1920, a fire destroyed the Calgary company's absorption plant. Deprived of their refinery facility, the dispirited directors of the near-bankrupt company were in a receptive mood. In late December, the shareholders of CPPL agreed to Imperial Oil's proposal for the organization of a new company to be known as the Royalite Oil Company Ltd.[54] The terms called for the surrender of assets to the new Imperial-controlled enterprise in return for a 25% equity, plus a commitment to complete well no. 3, to drill a fourth well and to rebuild the gasoline extraction plant. As a shareholder and creditor of the Calgary company, the CPR was asked for and gave its blessing to the arrangement. The vulnerability of the capital-starved independent oil operator and the pattern that would characterize the development of western Canada's petroleum industry was foreshadowed by the demise of the first Turner Valley oil company. Development of the petroleum industry in Alberta now entered a new phase.

Drawn to Turner Valley by the industry's belief in an impending world oil shortage and the determination of the parent company to be directly involved in every promising oil play around the globe,[55] combined with several years of indifferent results on the Alberta prairie, Imperial arrived with a corporate mentality that displayed more of a defensive character than enthusiastic endorsement of the region's potential. The situation that confronted Imperial in Turner Valley offered little to promote a change of attitude. There were no successful companies to inspire confidence and challenge competitive instincts, only a handful of semi-active Canadian concerns ever in a desperate search for capital to complete the drilling of wells that often had commenced years before. Most of the dozen or so "producing" wells were lucky to average more than five barrels per day, for a total field production of 11,032 barrels in 1920.[56]

As newly incorporated Royalite Oil Ltd. began operations in the spring of 1921, it confronted a regulatory environment that had undergone no essential alteration since the 1914 discovery. The regulations remained focused upon the three essential elements that had been brought together for the first time in the consolidated "Petroleum and Natural Gas Regulations" approved on 19 January 1914.[57] First, Ottawa was anxious to promote the rapid and orderly development of this highly valued resource. Second, in recognition of the special strategic importance of this sector of the resource economy, was the desire to ensure that ultimate control remained in federal hands. Third, there was the concern that development should occur in a manner that would prevent inappropriate waste. Development was encouraged by a system of leasehold land tenure that permitted the cost of holding land to be diminished by allowing normal lease rentals to be reduced by the amount spent on certain aspects of exploration and development. Refusal to surrender crown leasehold to all but British and Canadian citizens or companies and the right to reassume possession of all land, works and production was the product of Ottawa's recognition of petroleum as a vital resource. In 1914, conservation, in the form of prohibition against the unnecessary waste of natural gas, the injurious access of water to oil-bearing formations and measures to ensure the proper closing of wells, was added to the former concerns of development and national security.

The appearance of a conservation clause in the regulations came in response to observations of development problems in southern Alberta gas fields by Mines Branch officials of the Department of the Interior. Section 29 decreed that, should a well be abandoned for any cause, "the lessee shall be at liberty to withdraw the casing from the said well, but in order to prevent [surface] water gaining access to the oil-bearing formation, the lessee shall immediately close the well by filling it with sand, clay or other material."[58] The lessee was further enjoined to "take all reasonable and proper precautions to prevent the waste of such natural gas" as might be discovered, and "immediately upon discovery, to control and prevent the escape of such gas." Measures to prevent salt water from gaining access to oil-producing formations were also required. At this juncture, however, the federal government was content merely to identify remedies, and for want of serious enforcement the new conservation measures remained for the following decade, more a statement of desired rather than required practice. This new section in the regulations did nonetheless establish conservation of the resource as a legitimate public concern, and most importantly section 29 set the stage for subsequent intervention by giving the minister of the interior the right

> from time to time [to] make such additional regulations as [might] appear to be necessary or expedient governing the manner in which boring operations shall be conducted, and the manner in which the wells shall be operated.[59]

This enabling provision proved remarkably enduring. It remained throughout the period of federal administration, and the similar clause to be found in all Alberta's subsequent oil and gas conservation acts is the direct descendant of section 29.[60]

Except for the years spanning World War I, Ottawa's overriding concern was with development, rather than conservation or considerations of national security. This is apparent in the alteration of contentious section 40, which made Canadian or British citizenship

Looking east, Turner Valley townsite, circa 1918. Under federal jurisdiction, oil development in the WWI years reflected greater concern with rapid development than with conservation. Glenbow Archives NA-2674-40.

a prerequisite to obtain petroleum leasehold from the Crown. Bending to Alberta protests that this restriction prevented the inflow of U.S. capital essential to the industry's development[61] and the expressed hostility of the U.S. Congress,[62] the Canadian government approved a less restrictive measure in January 1920 that simply required that "Any company acquiring by assignment, or otherwise, a lease under the provisions of these regulations, shall be a company registered or licensed within His Majesty's Dominions."[63] A further measure of the government's development emphasis can be seen in the pattern of regulation amendments to ever widen the base of expenditures that could be credited against lease rentals. In May 1918, the regulations were amended to permit the acceptance of expenditure in the satisfaction of the rental beyond the third year to the fourth and fifth years of the term of a lease.[64] In the interest of promoting development, the Department of the Interior also backed away from any substantial departure from its initial decision that no royalty would be charged upon sales of petroleum up to 1 January 1930. In December 1919, Ottawa proclaimed that, for petroleum leasehold granted henceforth, the Crown would retain the right to set a production royalty as might from "time to time be fixed by the Governor in Council."[65] A month later, the annual per acre lease rental was doubled.[66] Industry protest was immediate. It was argued that exploration costs were already excessive and that one could not expect a company to incur large expenditures without knowing in advance what tax or royalty would have to be paid. An appropriate amendment was soon forthcoming. It stated that for a five-year period after discovery no royalty would be applied; for the next five-year period, the Crown would set a rate of not less than 2.5% and not more than 5% of the sales or output of the well; and for the third five-year period, the rate would be between 5 and 10%, and thereafter 10%.

In such amendments, Ottawa was consistent, and the regulations desired were those that, within reason, favoured development by private capital. Summing up federal petroleum policy after control had been surrendered to Alberta in 1930, a senior bureaucrat in the Department of the Interior explained:

> The extensive area over which the search for oil in Alberta has been made, and the important fact of abundant supply of cheap oil in the United States makes it impossible to say that Alberta has yet been properly tested for oil. Encouragement of operators willing to risk capital in what may take years to bring a return or may become lost, has been the keynote of Dominion petroleum legislation.[67]

Even though the industry was struggling and still in its infancy, there were two other changes to the "Petroleum and Natural Gas Regulations" before 1921 that set important precedents.

First was the decision in October 1920 to open dominion forest reserves to petroleum exploration.[68] The Rocky Mountain Forest Reserve bounded the entire western flank of the Turner Valley exploration area, and for nearly a decade oil companies had pressed for access only to meet Ottawa's flat refusal. Unbeknown to the applicants, the Rocky Mountain Forest Reserve, in addition to protecting an important drainage area on the eastern slopes of the Rocky Mountains, had assumed an equally important though de facto function. In the summer of 1913, the British Admiralty had advised the Canadian government that, should any of the extensive areas under investigation in Alberta come forward for actual development by petroleum interests, it would be prudent to ensure that a portion of these lands be withheld as a naval or crown reserve.[69] Later, when the Canadian government decided that the idea had merit, it became apparent that the creation of such a reserve might be realized unofficially through the medium of another kind of crown reserve already in existence. Therefore Ottawa used authority conferred by the *Dominion Forest Reserve and Parks Act* of 1911 to deny access to petroleum companies.[70] Thus, when retreating in favour of development and opening forest reserves to petro-leum exploration in 1920, Ottawa was nonetheless determined to maintain the crown reserve principle. The amendment to the regu-lation provided therefore that leaseholds up to 3,840 acres could be obtained in forest reserves; this was double the maximum permit-ted elsewhere. The area leased then had to be divided into two rec-tangular parts of equal area, one of which would be reserved for the Crown.[71] This principle was an extremely important feature of sub-sequent Alberta legislation in the 1940s and 1950s.

The second amendment to the regulations to establish a prece-dent of lasting importance came just a few months later. Having ex-perienced the clamour and conflict among individuals and groups competing for the rights to cancelled petroleum leasehold, the De-partment of the Interior began a new procedure in March 1921. Henceforth, when a lease was cancelled, usually because the holder had been unable to generate enough capital to begin drilling and was deeply in rental arrears, the petroleum rights would not again become available for applications under the regulations but would be reserved to the Crown for disposal by public tender.[72] The auc-tion or bonus-bidding approach to the disposal of petroleum and natural gas rights was also to reappear to form part of the critical legislation that was to shape the post-World War II development of the petroleum industry in Alberta.

The evolution of regulations governing the exploitation of natu-ral resources on dominion lands necessarily brought forth changes in the field. More complex regulations demanded a more sophisti-cated approach to inspection. At first, the Department of the Inte-rior simply made use of its network of homestead and timber in-spectors, and would call upon the nearest inspector to visit and report on any resource-related activity in which the Department was interested. It was the extensive development of coal deposits that gradually pushed Ottawa to implement a separate inspection system. The government's decision in 1901 to levy a royalty on the production of coal, and in 1907 to lease rather than to sell coal lands,[73] added the need for the technical supervision of mineral leases on dominion lands. In 1909, therefore, the position of in-specting engineer was created in the Mining Lands and Yukon Branch of the Department of the Interior, and O.S. Finnie, a McGill-trained engineer, was selected to assume the responsibilities of the new post.

Beginning in 1909 and continuing through 1911, and with the loaned assistance of D.A. Macaulay, chief engineer of the Interna-tional Coal and Coke Company of Coleman, Alberta, Finnie inspected and reported on all the coalmines in western Canada. The primary objective of the survey was to establish revenue; it being necessary to compute the amount of coal that had been mined. For many mines, this was unknown, and often no royalty payments had been made. Secondarily, Finnie was interested in conservation, and his reports allowed an evaluation of the economy of the mining methods employed and whether or not proper attention was being paid to achieving as high a percentage of coal recovery as was possible.[74]

The emerging concern within the Mining Lands and Yukon Branch regarding the wise and proper exploitation of natural resources was reflective of a more broad-based current of opinion that had grown in North America through the first decade of the new century. With the active support of President Theodore Roosevelt in the White House, leading figures across the continent, particularly from universities and the senior levels of government service, warned that North America was not a continent of unlimited resources and that the wasteful excesses of the pioneer period could no longer be tolerated. Alarmed by the increasing rate at which coal and iron supplies were being exhausted, as well as the wastage and reckless destruction of timber, water and soil resources, in May 1908 Roosevelt called all the state governors, members of his Cabinet, justices of the Supreme Court, and the heads of the various scientific bureaus in Washington, together with other noted citizens, to examine the question of the conservation of natural resources. The conference led to the appointment of a National Commission of Conservation. This was followed by the decision to hold a North American conservation conference in Washington in February 1909. The governments of Canada, Newfoundland and Mexico were invited to send delegates. Impressed by the report of the Canadian representatives who attended, Ottawa also decided to create a Commission of Conservation.[75]

Chaired by Sir Clifford Sifton, the Commission was composed of 20 appointed members and a number of ex-officio delegates. Of the former, at least one member appointed from each province had to be a member of a university faculty, and the latter included the federal ministers of the interior, mines and agriculture, plus the minister responsible for the administration of natural resources from each of the provinces. The Commission's mandate was to:

> take into consideration all questions which may be brought to its notice relating to the conservation and better utilization of the natural resources of Canada, to make such inventories, collect

and disseminate such information, conduct such investigations inside and outside of Canada and frame such recommendations as seem conducive to the accomplishment of that end.[76]

Although the Commission had no executive or administrative powers and was purely an advisory body, it played an important and public role in promoting conservation. This was because the Commission had strong bipartisan support in both Houses of Parliament. It had a remarkably competent membership that combined a high level of scientific knowledge with administrative and political experience, and it had in Clifford Sifton, former Liberal minister of the interior, an energetic, capable and highly committed chairman.[77]

Obliged to lay a report before both Houses of Parliament each year and having the power to authorize special studies, the Commission divided its work among seven standing committees.[78] One of the most active of these was the Committee on Minerals, chaired at the time by Dr. Frank D. Adams, dean of the Faculty of Applied Science at McGill University. Adams was an ardent and influential spokesman for the conservation cause. Conservation of the "national domain" was the subject Adams chose for his presidential address before the Royal Society of Canada in 1914, and his comments reflect the sentiments that guided the conservation movement in Canada. Adams acknowledged that "Canada [had] been blessed with great natural resources," but he warned his audience that "each and all of these, however, already show signs of serious depletion."[79] Although Adams accepted the notion that the present generation held the national domain in trust for future generations, he was anxious to dispel the common idea that this meant hoarding and restricted development.

> Our forests, our lands and our fisheries will, if properly worked, not only yield this generation a larger profit, but they will be handed on to our successors in a more highly productive condition than that in which we received them. We are prosperous now but we must not forget that it is just as important that our descendants should be prosperous in their turn. Each generation is entitled to the interest on the natural capital, but the principle should be handed on unimpaired.[80]

Adams lamented "the subordination of the consideration of the welfare of the nation to the pursuit of personal wealth," which seemed to him to be so widespread. He saw this as "destructive of all true national life and to the development of a strong and happy people."[81] Adams urged Canadians to awaken to the need for conservation policies and turn away from the "practice of personal aggrandizement and selfish waste."[82]

As chairman of the Commission of Conservation's Committee on Minerals, Adams moved forward aggressively. He sought first to survey mining practices in Canada. For this and related work, the Commission hired W.J. Dick as mining engineer of the Commission. Dick spent the summer of 1911 inspecting coal mining operations in western Canada, and in his section of the Commission's report, submitted to Parliament in 1912, he commented at some length on wasteful mining practices. His report also drew to parliamentarians' attention the "enormous quantities of natural gas [that] have been wasted both in Eastern and Western Canada."[83] To underline and lend visual impact to the statement, a picture was included of the flaring Pelican Portage gas well, which had been burning out of control for years. In the Commission's view, such carelessness and ignorance was a legitimate public concern and "a proper field for legislative control."[84] Recommended were regulations requiring the plugging of abandoned gas wells and the filing of the records of all holes drilled as well as the levying of a tax on wasted gas.

The concern about waste gas remained a constant theme in subsequent reports by the Commission of Conservation. In 1914, the Commission devoted an entire section of its report to the "Importance of Bore Hole Records and the Capping of Gas Wells" that included extracts of appropriate U.S. legislation to serve as a model for desired Canadian regulations.[85]

The following year, in its report, the Commission included a specially prepared chapter by Adams entitled "Our Mineral Resources and the Problem of Their Conservation." A special section was devoted to the problem of natural gas, which Adams identified as "the most perfect fuel with which we are furnished by nature."[86] The problem, he explained, was the common belief that the supply of natural gas was so great as to be practically inexhaustible. In his attempt to undermine this perception, Adams pointed to the irrefutable evidence provided by the experience in the United States where gas fields had been operated for more than 30 years. In one field after another in which the supply was supposed to be inexhaustible, production had declined to a fraction of the original volume. The decline in yield, Adams argued, was in large part the product of the enormous waste that was allowed to occur in the period immediately after discovery when gas appeared abundant and it was next to impossible to persuade people that it would not last forever. But even in the Ohio gas fields, where "the operators tried to believe that the gas was being formed within the earth as fast as it was being allowed to escape," decline came nonetheless.

Adams asked Canadian politicians to bear in mind the experience in the United States, to realize that there was every prospect that productive new gas fields would soon be discovered in the West, and to ensure that appropriate preventative measures were taken. They could begin by ending the "inexcusable" waste at Pelican Portage, Alberta, where for the previous 17 years escaping gas had been burning continually like an immense torch.[87] Reminding that only Ontario required that unused gas wells be plugged, he urged that the remaining provinces and the federal government follow Ontario's lead in checking the waste of natural gas by exacting a tax of 2¢ per thousand cubic feet, with a rebate of 90% for the gas that was actually used. This was an idea that Ottawa would pursue at various times over the subsequent 15 years, on each occasion to face the unyielding opposition of western producers.

Adams was also anxious that the public be educated to realize that they must take responsibility themselves to end certain practices, since it was impossible to legislate against every wasteful activity. Among such undesirable practices, he singled out the custom of selling natural gas at so much per burner per month instead of at so much per 1,000 cubic feet.

> The inevitable tendency of this, as seen in Medicine Hat at the present time, is to allow the gas in the street lamps to burn all day, seeing that it costs no more to do so, while at the same time it is easier to let it burn than to turn it out, and the spectacle of gas blazing throughout the day conveys a general suggestion of the abundance of a product which one can afford to waste so lavishly.[88]

This concern was consistent with the principles that underlay Adams's conservation philosophy. In his view, a country's mineral resources required special attention. Renewable resources, such as agriculture, forests and fisheries, if properly managed, could yield an annual interest like well-invested money, while the capital remained unimpaired or even increased in value. But nonrenewable resources were "like a sum of money or treasure hidden in the ground. It does not renew itself, and every amount abstracted leaves just so much less for future use."[89]

The conservation message promoted by Adams and the Commission of Conservation found a receptive ear, especially among the professional and technical staff in the Department of the Interior charged with administering the Dominion's lands and resources in the West, including Adams's former student, O.S. Finnie.

Although the much greater volume of production dictated that coal remain the centre of conservationists' concern in the mineral sector, the remote and modest attempts at petroleum extraction were nonetheless subjected to Ottawa's scrutiny. Finnie was sent to inspect a number of oil wells drilled near Fort McMurray in north-eastern Alberta in 1912. As this assignment was being completed, two companies began drilling in the foothills near Calgary. It was soon apparent that additional technical assistance was needed, and another of Adams's students, C.C. Ross, was engaged as senior mining inspector. Concern about illegal mining led the Mining Lands and Yukon Branch to open an office and to station a permanent mine inspector in Calgary in 1916.[90] To keep abreast of the dramatic burst of petroleum development that had occurred since 1914, and particularly, in keeping with section 29 of the 1914 regulations, to see what measures operators were taking to ensure that wells were being properly protected against the damaging intrusion of surface water, all foothills oil and gas wells were inspected in 1917. This inspection was extended in 1918 to the Medicine Hat and Bow Island wells, the well at Pelican Rapids on the Athabasca River that had been burning out of control since 1897, and other wells on the Athabasca and Peace rivers. The inspection confirmed in the Department's mind the need to monitor the emergent petroleum industry more closely. To ensure that adequate supervision could also be carried on during the winter as well as the summer, C.C. Ross was transferred permanently to the Calgary office in 1918.

In preparation for a much more active and continual presence in the field, early in 1919 the minister of the interior, under the authority of section 29 of the "Petroleum and Natural Gas Regulations," announced more rigorous boring and operating regulations.[91] In recognition that mere cancellation of the lease might not in some cases be regarded as a sufficient penalty for failure to take proper precautions to prevent the waste of natural gas and the injurious access of water, the minister would henceforth have the right to:

> take such effective means as may appear to him to be necessary or expedient in the public interest to control and prevent the escape of the natural gas. . . . or to prevent water from gaining ac-

cess to or escaping from such well, and to recover from the lessee
of the location upon which the well was bored, all costs and expenses incurred by the Crown in stopping the escape of natural
gas, or the ingress or egress of water to or from the well.

The revised drilling and operating regulations also required that,
before commencing boring operations, the lessee notify the minister
of his intention and provide information regarding the location,
size and weight of the casing to be used and the kind of drilling rig
to be employed. Before abandoning a well and removing the rig or
pulling the casing, the lessee had to have the permission of the Department and use approved methods of closure.

Once established in Calgary, the Department moved ahead confidently to promote resource conservation. First, to put its own
house in order, a crew under the guidance of Department technical
staff successfully extinguished the Pelican Rapids well, and later in
1919 a third McGill graduate, S.E. Slipper, was hired and appointed petroleum engineer. Slipper's specific duties were to inspect
all drilling operations, to report on the waste of gas and improperly
abandoned wells, to cap wells that were out of control, and generally to see that the technical provision of the regulations under section 29 were properly carried out.[92]

Slipper's first tour of inspection revealed the need for further
modification of the section 29 regulations. If petroleum conservation was to be promoted effectively, then it was important to improve the regularity and quality of information submitted to the
Department. Writing from Medicine Hat in the late autumn of
1919, Slipper urged that drillers be required to keep a daily log and
make it available always for the inspection of Department engineers, that drillers be prepared to submit core samples if requested,
that the pressure of all producing natural gas wells be tested annually in the spring or autumn and forwarded to the Department, and
that operators be compelled to give notice immediately if their wells
showed evidence of water encroachment.[93] Slipper was simply requesting what had come to be considered enlightened engineering
practice and the kind of measures that the Commission of Conservation had been advocating for some years.[94] The endeavour to create an effective administrative framework that would ensure the
constant and orderly expansion of knowledge about changing field
and reservoir conditions would remain a somewhat elusive objective for decades.

Slipper and his colleagues often found it difficult to get the level
of co-operation they desired from small western companies. First,

they were confronted with a general public hostility towards continued federal control of lands and resources. The strength of western, and particularly Alberta, feeling was made clear from the first.
On the eve of provincehood, the *Calgary Herald* spoke for many
southern Albertans when it was learned that the new prairie provinces would be denied the rights accorded their older sisters in Con-
federation. In an editorial headed "Shall Alberta Eat Dirt," the paper screamed:

> Alberta has been deprived of her public lands and the natural re-
> sources. She has been robbed of the freedom of self-government
> enjoyed by Manitoba and British Columbia.
>
> She is made a political Cinderella among the sisterhood of Cana-
> dian provinces.
>
> Albertans are treated by parliament as inferior to residents of
> older provinces and as unfit to be given the usual controls over
> their own affairs.[95]

The feeling would lose none of its intensity until federal control
was eventually surrendered in 1930.[96] To those of such temper, mining inspectors and other Department of the Interior officials situated throughout the prairie region were seen as agents of an increasingly resented federal and even "foreign" authority. In
addition, the university-educated engineers of the Mines Branch
had to contend with certain powerful and deeply entrenched biases
that had grown out of two centuries of North American frontier experience. There was the frontiersmen's usual hostility to regulations
of any kind. Regulations always seemed to be made by those in distant centres of power who wished to deny local opportunity. The
world of the frontier settler was typically one of "abundant" natural resources, and this made the conservation ideal harder to accept,
especially when, as was so often the case, the message was being
promoted by someone from outside the region. Finally, pioneer
communities gave pride of place to the doer rather than to the
thinker, to the man of practical experience rather than to the man
of books. Frontier men were "practical" men, and it was no accident that westerners engaged in the search for oil often referred to
themselves as "practical" oilmen. Schooled by experience rather
than at university, they had their own geological theories, their own
ideas about reservoir dynamics, and their own thoughts about the
need for conservation. Moreover, such ideas and experience were

often rewarded by success in the field, and so it was with some hauteur that they chose to label themselves as "practical" oilmen.

The powerful frontier inclination that favoured the practical man over the professional man was reinforced in this instance by the structural division that characterized the petroleum industry in North America. Self-educated businessmen-promoters were the driving force behind most of the small independent producing companies of the interwar period. These men lived by their wits in a high-risk environment that constantly pitted them against the large major companies. Among the features that increasingly distinguished the majors was their corps of university-trained professionals. It did not go unnoticed among the independent operators that this also characterized the new regulatory authorities. Similarly, it was not hard for some independents to imagine an alliance of professionals employed by the majors and in government service, the main objective of which was to rein in the independent sector. Such tensions lasted well into the 1950s and stand as one manifestation of an industry in transition from pioneer to a more technically advanced enterprise.

Although they had the weight of enforceable federal legislation behind them, Mining Lands and Yukon Branch engineers still had to work with great tact to achieve their objectives. As one engineer explained, they quickly learned "the best way to get cooperation, especially from the practical man is to see he has something in return."[97] This came in the form of technical advice, which in the early years was usually sought after local approaches or remedies had been tried and had failed. Slipper's success in helping kill a wild well near Medicine Hat in the summer of 1920 enhanced the Department's credibility, as did help with water shut-off problems, advice in setting casing, sharing geological information and analysis of oil and gas samples.[98] The decision in September 1920 to hire C.W. Dingman as assistant mining inspector also helped the Department gain the approval of local oilmen. Well-known and respected in Calgary petroleum circles, Charles Dingman's oilfield experience had begun in 1913 with his uncle's company, Calgary Petroleum Products Ltd. Before joining the federal civil service, he had managed the Medicine Hat gas wells of the Canada Cement Company. Dingman's intimate knowledge of field operations and personnel in Alberta contributed greatly to the subsequent success of the Calgary Office of the Mining Lands and Yukon Branch.

With Dingman's addition to the technical staff of the Calgary office, the Department completed its inspection of Bow Island gas wells, finished supervising the repair of several Medicine Hat wells and in 1921 tried unsuccessfully to cap a Peace River well discharging gas and salt water. In 1922 the efforts of the now well-established Calgary office culminated with the publication of the region's first *Field Manual for Operators*. The publication marks something of a turning point in the history of federal supervision of western Canada's oil and gas industry. From this point, the regulations did not undergo any fundamental change. The essential policy objectives and basic features of field practice that were to characterize the regulatory environment until World War II were set in place. It was no small achievement. As one federal engineer recalled, it had

> taken years of education on the part of the Department to get the drillers to keep proper logs of their wells, to collect samples of the formations drilled through, advise when oil, gas or water are struck and to give facilities for their sampling. This information is desired from all wells, and equally from those without scientific management and often with indifferent drillers. It was only by close personal contact that such good results were achieved and the foundation was laid in the early days of the Calgary office.[99]

Although the small operators occupied the greater part of the time and attention of Mining Lands and Yukon Branch officers in Calgary, after 1921 the party that exercised the greatest influence upon the regulatory framework, and whose support was crucial to the successful implementation of any measure, was Imperial Oil. Upon arrival in Turner Valley, Imperial efficiently and immediately set about consolidating its position. The "Petroleum and Natural Gas Regulations" were reviewed, and discussion commenced with the federal government about certain changes that would be to Imperial's advantage. With the objective of strengthening further its land base in Turner Valley, negotiations with the CPR were extended, and to insure for itself a key role in the transportation of petroleum products from Turner Valley, Imperial sponsored a pipeline bill in the Alberta Legislature.

With the Alberta Legislature assembled for its traditional spring sitting, the pipeline proposal necessarily came first on Imperial's agenda. The proposal to incorporate the Imperial Pipe Line Company was presented as a private member's bill in early March 1921, thus launching Alberta's and Canada's first important pipeline debate. At the outset, it was argued that pipelines should be declared common carriers. The idea drew bipartisan support as well as bipartisan opposition, and at once the debate became heated and prolonged. Opponents of the bill repeatedly cited experiences in various U.S. states where large companies had used their control of pipelines to beat down the small independent operators and where the fight to compel pipelines to act as common carriers had gone on for years.[100] George Hoadley, the Independent member for Okotoks, whose constituency included Turner Valley, argued that the most thorough investigation should be made into experiences of other oilfields, so that the House would have some knowledge of the eventual ramifications before granting a franchise to "the child of the United States Standard Oil octopus which would enable the immense monopoly to practically take possession of the oil output of Alberta."[101] On the other side, it was argued that a common carrier provision would "promote indiscriminate drilling which would mean that fields would be drained of oil much sooner than they should be," and the example of certain Texas fields was presented as a case in point.[102] Others expressed the concern that to deny Imperial's request would seriously jeopardize continued expenditure by one of the few companies with sufficient capital to undertake serious development. Attorney General J.R. Boyle wondered if the honourable members opposing the incorporation were trying to tell Imperial Oil that the company was "free to spend vast sums in exploration work but if oil were found, they were not to pipe it out."[103]

With certain members alleging that they were being threatened outside the house,[104] with filibusters inside, and with the Speaker struggling to keep order, Premier Stewart sought a compromise. He proposed that, rather than an unlimited franchise, Imperial be allowed to construct pipelines up to 150 miles in length until such time as the government could prepare a general pipelines act. When it became clear that its incorporation proposal was unlikely to pass, especially in its original form, Imperial asked its sponsor, William Rae, the member from Peace River, to withdraw the bill. In doing so, Rae read a statement from the company saying that it wanted no favours, only to be treated fairly, and that, if it could not invest its own capital to build its own pipeline, the government must undertake to build necessary pipelines at public expense.[105]

Rebuffed by the Alberta Legislature, Imperial had better luck with the federal government. Still with little to show for its steadily rising expenditure, Imperial sought an important change in the "Petroleum and Natural Gas Regulations." In November 1921, Imperial bluntly informed Minister of the Interior Sir James Lougheed, who happened to be a major Royalite shareholder, that:

We have been seriously considering the advisability of abandoning for the present our operations in Western Canada, but we realize that our abandonment of the field would have a serious
effect upon the operations and plans of others, and upon the
country generally. We have been urged by public bodies and representative Canadians, both in the East and in the West, to continue our explorations for at least another season.[106]

What Imperial wanted was a relaxation of the Regulations that
would permit the company to carry forward, as a credit against the
yearly rentals charged to leaseholders of crown land, 40% of the total exploration expenditure rather than the 25% currently possible.
Ottawa was faced with a situation in which it was even more vulnerable than that warned of by Imperial in its earlier opposition to
the proposed Shell concession. If it wanted significant exploration
to continue in Alberta, it would have to give careful consideration
to the interests of the one company that possessed the land, capital
and expertise to achieve the most promising results. In late December, the Regulations were amended to give the minister of the interior power to allow whatever portion of the expenditure incurred
by a leaseholder to be credited against rentals as he might decide
advisable.[107] Imperial got its 40%, but as 1921 drew to a close there
was little else to cheer about. There were still no encouraging results from the field and the recently elected United Farmers of Alberta (UFA) government in Edmonton could be counted on for little
sympathy.

ALBERTA'S FIRST NATURAL GAS CRISIS

Not only were the managers of dominion lands and resources in
Ottawa confronted with the threatened collapse of significant oil
exploration in the West but also a dark shadow was suddenly cast
over what had been the one bright spot on the oil and gas horizon—
the development of the natural gas fields near Medicine Hat. In late
1919, rumours had begun to circulate that the Medicine Hat and
Bow Island gas fields were in serious decline. The reality of the situation was clouded by a simmering conflict that had developed between the City of Calgary and the Canadian Western Natural Gas,
Light, Heat and Power Company over the selling price of natural
gas. By late 1920, inspecting engineers from the Mining Lands and
Yukon Branch Calgary office confirmed that the fields were in

trouble.[108] At the same time, Calgary and the gas company remained at an impasse. The company insisted that it was absolutely impossible for them to obtain the necessary revenue for further exploratory work to find new supplies of natural gas without a rate increase. Canadian Western wished therefore to be released from its contract with the city.[109] Remembering that Calgary ratepayers had voted in a December 1920 plebiscite five-to-one in favour of holding the gas company to the existing price of gas as stipulated in its franchise, and anticipating how voters would react to a proposed rate schedule that was nearly double the existing levy, city aldermen steadfastly refused the gas company's demands. The threat of a serious gas shortage brought both senior levels of government into the fray: the provincial government to effect a compromise between the city and Canadian Western, and the Department of the Interior to address the larger question of natural gas conservation.

Increasingly desperate to resolve the stalemate in time to allow some chance to increase natural gas supplies before the 1921–22 winter heating season, and knowing that required hearings on any rate change before the Alberta Board of Public Utility Commissioners could be immensely time-consuming, Calgary and the gas company agreed to convene a formal conference to see if they could agree on a joint document to be presented to the Commissioners. It was agreed that the findings of the conference would be binding and would be presented as the final deposition of the two parties to the Board. If it worked, the plan promised to save months of deliberation.

A committee of technical experts was appointed to hammer out an agreement. The two city representatives were Roswell H. Johnson, a consulting geologist and professor of oil and gas production at the University of Pittsburgh, and Frank P. Fisher, a consulting engineer and former general manager of the Mid-Continent Natural Gas system of the Cities Service Company. Appearing for Canadian Western were Thomas R. Weymouth, consulting engineer and chief engineer of the United Natural Gas Company of Oil City, Pennsylvania, and Cecil Elmes, middlewestern manager of Sanderson and Porter, Consulting Engineers. A.I. Payne, technical adviser to the Board of Public Utility Commissioners, was the fifth member of the conference and its chairman.[110]

After reviewing the technical data and coming to an agreement on existing field conditions and productive capacities, the most promising sources of additional gas supplies, and the value of company property and its operating expenses, the committee set forth its recommendations on 15 August 1921. The conference decided that "consideration of public policy make it imperative that the small domestic consumer receives favoured treatment and when shortage conditions occur he should be the last to suffer."[111] With this in view, two separate rate schedules were unanimously agreed upon. The domestic rate would be increased from 35 to 50¢ per thousand cubic feet of gas, and large consumers would be charged 40¢ per thousand on the first 150,000 cubic feet per month, 35¢ on the next 850,000 and 30¢ on anything over 1,000,000 cubic feet per month. The better rate given large users was balanced with the provision that their supplies would be reduced in favour of the small consumers in the event of shortage. Canadian Western was directed to build a pipeline to Turner Valley and to purchase natural gas from Royalite Oil Co. on terms set out by the Commission; to drill a test well immediately in the Foremost field and, as soon as adequate supplies were proved, to construct a connecting pipeline to the Bow Island system (not less than two development wells were to be drilled in the Chin Coulee field); and, finally, to purchase from the Great West Natural Gas Corporation not less than one billion cubic feet of gas per year from the Redcliff field.[112] The Red-cliff field offered the only substantial amount of untapped gas immediately available and was essential therefore to the immediate solution of the impending winter gas shortage.

When the Board of Public Utility Commissioners signified their agreement, almost everyone else joined in a chorus of dissent. Calgary ratepayers protested that they had voted overwhelmingly against an increased price for natural gas.[113] Residents of the Medicine Hat and Redcliff areas were furious that "their" gas, and with it the future of their region, was to be surrendered to Calgary.[114] From the Winnipeg office of Wood Gundy and Company investment brokers came the stern warning to the Alberta premier that the proposition to divert gas from the Redcliff area to Calgary was causing a "growing feeling of anxiety on the part of people who [had] invested their money in the city of Medicine Hat." Wood Gundy explained that investors who had bought Medicine Hat securities did so believing that the city's future was assured because of an adequate supply of cheap fuel. Diversion of gas could materially shorten the life of the field. If this happened, Medicine Hat "would no doubt go into default," and Wood Gundy reminded:

> if an Alberta city the size of Medicine Hat is ruined as a result of the proposed scheme, the effect on the credit of every municipality in Alberta as well as the effect on the Province itself will be disastrous.[115]

Medicine Hat Petro-
leum Co., Ltd. cable
tool rig near Bulls-
head, Alberta, circa
1923. Medicine Hat
Museum and Art Gal-
lery Archives P532.1.

Redcliff, Alberta, residents enjoyed ready-made
Victoria Day fireworks at this gas well flare in the
early 1920s. Medicine Hat Museum and Art Gallery Archives P56.22.

An eerie sight. Locals view this ice-encrusted derrick in the Cypress Hills, flowing gas at 20,000
mcf/d plus water, on 26 December 1928. Medicine
Hat Museum and Art Gallery Archives P76.11.

Having enacted special legislation in its 1921 session that conferred upon the Public Utilities Board the power to require Canadian Western to augment its supply of gas and thus solve the Calgary problem, the Alberta government now found itself involved in a controversy that had broadened to engulf the entire southern third of the province. At this juncture, the minister of public works refused to approve the location of the necessary pipeline for Canadian Western to gain access to the Redcliff field. Compelled by one government body to collect Redcliff's gas and denied access by another, the bewildered company pleaded with the premier to intervene. Company President H.B. Pearson cautioned that

As these drillers installed drive pipe on the Bullshead well, western oil operators tried unsuccessfully to reduce federal customs duties on drilling tools and pipe from the United States. Medicine Hat Museum and Art Gallery Archives P532.2.

if the Government yields to the pressure now being exercised by Medicine Hat and Redcliff, it is going to establish an exceedingly dangerous precedent, which will give rise to many difficult and embarrassing situations in the future. Whenever large supplies of natural gas are developed, the nearest ambitious town, aspiring to become an industrial centre through the possession of cheap natural gas, will demand the same protection as Medicine Hat and Redcliff.[116]

With matters once more at an impasse, the debate intensified to take up also the issue of natural gas conservation. Admittedly with a special interest to serve, the mayor of Calgary informed the premier of the City Council's belief that "as a general principle, gas should not be used for industrial purposes except in certain special preferred industries requiring specialized treatment."[117] This was a question that weighed increasingly on the petroleum section of the federal Mining Lands and Yukon Branch for most of the natural gas in question was being produced on dominion lands.

Department engineers had been watching the southern gas fields closely for two years and had reliable data on the magnitude of decline. In late December 1920, O.S. Finnie informed the head of the Mining Lands and Yukon Branch that, from the pressure curve histories of other fields, it was evident that the Bow Island and Medicine Hat fields had reached a stage of rapid decline, and he wondered whether it would not be advisable to place some provision in the regulations prohibiting or at least restricting the use of natural gas for industrial purposes. In his view, Medicine Hat furnished a "conspicuous example of the prodigal use of natural gas." Finnie argued that there might be some justification if no other fuel were available:

> but with coal at their very doors, it does not seem fair to the householder that the gas should be used in such lavish quantities by a few companies or industries. The result is exactly as expected the field will be depleted in a few years and there will be nothing left either for industrial or domestic purposes.[118]

Finnie concluded that, in western Ontario and elsewhere, governments did not allow the use of natural gas for industrial purposes.[119]

By June, the idea had made considerable headway among senior Department of Interior staff in Ottawa. The concept was also gaining converts in the United States.[120] Before coming to a decision, however, the Department sought the opinions of its field engineers in Calgary. In two convincing reports, S.E. Slipper put the notion to rest. He pointed out that the great consumption of gas was not by industrial users but by domestic use in the cold months of the year; that the high peak-load was far harder on the fields than the regular industrial load; and that, as it was extremely difficult to maintain equipment in a field that only served periodic high peak-load consumption, the steady industrial load had a balancing effect, serving to maintain the wells and equipment in proper working order. Finally came the clinching argument, Slipper asserted that if the use of natural gas was restricted to domestic purposes it "would put a stop to all development of the natural gas resources, and would hinder, to some extent, the exploration for oil."[121] In his following memorandum, Slipper admitted that there was no doubt that public opinion was in favour of restricting natural gas to domestic use, but he felt it essential that people realize that the "great crime" in the industrial consumption of natural gas was the low rate at which gas was sold. In his view, the best thing to do would be to place a royalty on industrial gas. A flat rate royalty of 5¢ per 1,000 cubic feet, he thought, would be stiff enough to "prohibit the use of the gas for brick making and other rough industries, without restricting it from more efficient utilization."[122]

Adopted by the Mining Branch, the royalty solution was presented to the deputy minister with the notice that the CPR charged a royalty of 10% of the market value of gas produced from wells on its lands.[123] In the end, the idea was not taken up, it seems that there was some fear among politicians that even this measure would put development at risk.[124] Enthusiasm for such a royalty remained within the Department nonetheless and was on the verge of implementation in 1929 only to be set aside once more when the impending transfer of resources to the prairie provinces made such a departure impractical.[125]

The most important part of the answer to the natural gas supply crisis was seen to be the promotion of exploration and development rather than conservation. This approach was certainly more acceptable to Premier Greenfield; it averted a vicious feud between the two largest southern Alberta communities, a contest from which there could be no political winner. The strenuous insistence by Alberta municipalities of their right to an assured long-term natural gas supply for domestic use and for industrial development brought forward during Alberta's first natural gas crisis stood as a warning to future premiers of the political hazards that lay in any scheme of gas removal from the province. The sensitive issue of natural gas supply and the correlative issue of gas waste would remain near the forefront of Alberta politics for the next 35 years.

More interested in encouraging development than conservation, the deputy minister of the interior took the unprecedented step of journeying to Edmonton and Calgary in late February 1922, so that he might meet western oilmen, hear their increasing complaints, and seek their recommendations for improvements to the Regulations.[126] At Edmonton, those choosing to meet W.W. Cory were mainly individuals with interests in the McMurray oil-sands. It was largely oilmen associated with smaller Turner Valley companies who assembled to greet Cory at the Mining Lands Office in Calgary. Separate private meetings were held with Dillon Coste of Canadian Western and with a small group led by R.B. Bennett.[127] The most important of the recommendations accepted included one that attempted to protect the subsurface rights of leaseholders by prescribing that no well could be drilled closer than 200 feet from a lease boundary. Another endeavoured to prevent surface owners causing unreasonable delay by giving the Department the right to grant permission in certain cases to enter upon a small portion of the surface during the period of arbitration. To help small companies raise capital, the Department also decided to allow leaseholders the right to divide and to assign to others parcels as small as five acres, knowing full well that such a policy raised the danger of uncontrollable speculation.[128]

In their attempt to reduce federal levies, western oilmen were unsuccessful. Cory refused to reduce lease rentals and could only express his sympathy for the requests of Alberta producers for a bounty for their oil similar to that paid to Ontario producers[129] and that they be relieved of the customs duties on drilling tools and casing imported from the United States. The deputy minister of the interior could only promise to pass these requests on to the Departments of Trade and Commerce and Customs and Excise, under whose jurisdiction such matters fell.

If Calgary oilmen had been listening carefully to Cory's general comments about the evolution of the "Petroleum and Natural Gas Regulations," however, they could not have been too sanguine about their prospects. At one point, he had remarked:

> When the oil business started back here some years ago, we had regulations we thought were all right. Nobody kicked, because nobody was digging for oil. We have got up to this point now, and I think we will have to go a little longer on some of these things before we can convince the 23 or 24. . . . other men in the council, [Cabinet] who are not interested in oil, and who do not know much about it, why provisions should be made that are going to cut the Government's revenue.[130]

Such sentiments only strengthened the growing determination among Albertans to put an end to federal control of provincial lands and resources.

ROYALITE NO. 4 AND THE SECOND TURNER VALLEY
OIL BOOM

In keeping with the agreement under which it acquired the assets of
Calgary Petroleum Products, Royalite worked through 1921 to
complete drilling of the third Dingman well, while Imperial, the
parent company, carefully monitored the drilling results of the six
rigs it had working elsewhere on the prairies. It was the continued
poor results on the plains, the modest encouragement that came
with the limited success of the third well on the new Turner Valley
property, and especially its obligation under the agreement to commence a fourth well that finally moved Imperial to initiate discussion with the CPR once more.

The preferred drilling site for the fourth well was on a leasehold
that Calgary Petroleum Products had acquired from the railway,
but Imperial did not like the lease it had inherited. The terms were
more onerous than those the oil company had obtained earlier on
CPR leaseholds in eastern Alberta. Imperial asked for lease conformity as a minimum concession, and its western solicitor, R.B. Bennett, went so far as to suggest that the company might elect to drill
on government land where, if oil was discovered, no royalty would
be due until 1931.[131]

Conscious that any alteration would be taken as a precedent, the
CPR took until late August 1922 to reach an understanding. The
railway finally agreed to write new leases for the two 320-acre parcels Imperial had acquired in Turner Valley, but not simply to make
them conform with the earlier petroleum leases held by Imperial.
The new lease provided for renewal in perpetuity as the oil company requested, instead of for 10 years as in the old form. Also, the
railway surrendered some say on decisions about subsequent development wells, but the company remained firm on the question of
royalties. Not only was the 10% petroleum royalty retained but
also, in anticipation of a naphtha gas discovery, the railway added
an additional 5% royalty on the market value of any gasoline that
might be produced as a natural gas by-product.[132]

Within a month of the new lease agreement, a Royalite drilling
crew was at work on the site,[133] and in the early fall of 1923, at a
depth of 2,890 feet, a flow of some 7,000,000 cubic feet of wet gas
per day was secured. Now known officially as Royalite No. 4, it
seemed that this fourth well on the original CPPL leasehold might be
the big producer everyone had been waiting for. Royalite decided to
suspend drilling operations during the difficult cold season and to
market the already discovered natural gas flow in Calgary over the
winter. In the spring, drilling resumed, and by the late sum-

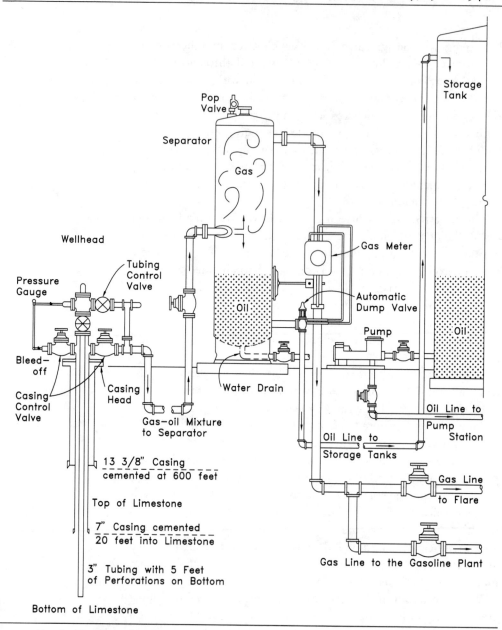

Pop Valve

Separator

Gas

Wellhead

Tubing Control Valve

Pressure Gauge

Gas Meter

Oil

Automatic Dump Valve

Bleed–off

Pump

Oil

Casing Control Valve

Casing Head

Gas–oil Mixture to Separator

Water Drain

Storage Tank

Oil Line to Pump Station

Oil Line to Storage Tanks

13 3/8" Casing cemented at 600 feet

Top of Limestone

7" Casing cemented 20 feet into Limestone

3" Tubing with 5 Feet of Perforations on Bottom

Bottom of Limestone

Gas Line to Flare

Gas Line to the Gasoline Plant

SOURCE: Based on Royalite Oil Publication.

mer, as the wellbore penetrated beyond the traditional producing horizons of the Dakota and Kootenay formations into the Dolomitic limestone, the hopes of the previous fall evaporated. On October 24 the company was on the verge of abandoning further drilling when suddenly, at the 3,740-foot level, some 310 feet into the limestone, a tremendous gas flow was struck. Estimating the flow at 21,500,000 cubic feet of wet gas per day, the drilling crew commenced closing in the well. The moments that followed were dramatic.

After being closed, "the gauge registered 200 lbs. in the first minute, 400 lbs. in the second minute, 600 lbs. in the third minute."[134] With growing alarm, the drilling crew watched the pressure gauge until it reached 1,150 lbs. and then ran for cover. Watching from a safe distance, they saw the rising pressure lift 94 tons of casing and drilling tools to the top of the derrick, then the gas burst free, allowing the casing to subside into the borehole. Now the drillers were confronted with a massive wet gas well blowing wild. The situation was made even more difficult when, as often happens in the endeavour to control escaping gas, a stray spark set the well ablaze on November 9. Royalite's crew was finally able to smother the fire and bring the well under partial control. Owing to what for the technology of the time was phenomenal pressure, along with the damaged condition of the hole, it was judged necessary to let the well flow continuously.[135] Most importantly, however, the company was able to install a separator and to recover from the gas nearly 500 barrels of naphtha a day[136] (see Figure 1.1). Naphtha sold for $6.00 a barrel, making Royalite No. 4 the outstanding commercial success that had inspired the efforts of the pioneer oilmen who had organized the Calgary Petroleum Products Company, identified the site, and acquired the petroleum rights to the land upon which No. 4 well was situated.

In light of its success, Royalite decided to lay a 4-inch welded pipeline to the newly constructed Imperial refinery in Calgary some 35 miles distant. In the interval, oil was hauled from the Valley in five tank trucks.[137] The discovery also hastened the conclusion of discussions that had been under way for some time with Canadian Western. Royalite agreed to erect a scrubbing plant to remove the toxic and odorous hydrogen sulphide from the gas and Canadian Western commenced construction of a 10-inch pipeline to collect Royalite natural gas in the Valley (see Map 1.2).

During 1925, Royalite No. 4 produced 166,500 barrels of light gravity oil, which exceeded by several thousand barrels the combined production of all other Canadian wells in 1924.[138] The impact

MAP 1.2   Turner Valley Natural Gas Pipelines, 1921–1928

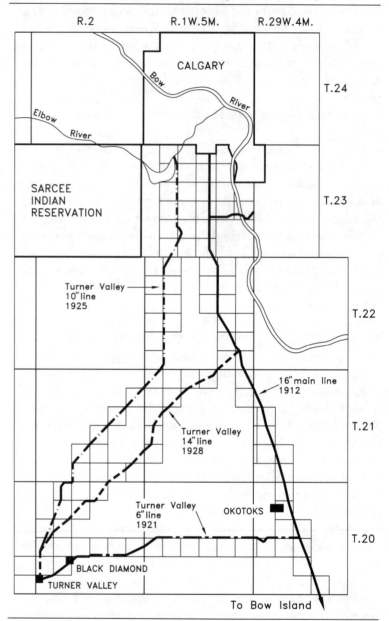

SOURCE: Data provided by Canadian Western Natural Gas Company Limited.

was immense. Royalite No. 4 seemed to prove that better producing horizons existed in lower formations. New wells heading for the limestone below the Kootenay formation were soon under way, and old companies rejuvenated with fresh capital began to consider the prospects of deepening wells abandoned years before. The Royalite success also allowed the more modest achievements elsewhere in the province to be interpreted in a better light. Discovered in 1914, the Viking gas field was finally connected to markets in Edmonton by an 80-mile pipeline in late 1923. This sparked the drilling of additional wells to ensure an adequate long-term supply. In the Wainwright area, farther southeast of Edmonton, the British Petroleums Limited of Vancouver had obtained enough oil production from several wells drilled in 1923–24 to excite investors in the Alberta capital.[139] The Medicine Hat gas field, in the southeast corner of the province, also continued to develop. Since 1904, the city had drilled 17 wells, and various industrial concerns had put down a further 20 successful wells in the Medicine Hat-Redcliff area. Although the Bow Island field appeared close to its end, the newly developed Foremost field, a little farther south, had functioned successfully as the major source of Calgary's gas supply since 1923. Still farther south, along the Alberta-Montana border, another oil and gas play was generating a good deal of excitement. Apart from the distinction of having the largest gas well yet to be found in Canada, what especially attracted interest was that the discovery seemed to establish a northern extension to the Kevin-Sunburst field of northern Montana, which was producing both oil and gas. Scattered as they were over such a broad area of the province, these activities generated a buoyant optimism that launched Alberta into its second significant oil boom.

Imperial moved ahead aggressively to expand company landholdings. An arrangement was quickly concluded with the CPR that essentially gave the oil company the right of first refusal for the unassigned petroleum rights on railway lands in the Turner Valley area.[140] By early summer 1926, Imperial's subsidiaries had five producing wells, 17 rigs that were drilling, and 36 additional drilling locations had been selected.[141] As the year ended, there were 34 different companies drilling for oil, and Albertans waited expectantly for the Texas- or Oklahoma-sized discovery that had eluded them for a decade.

The now buoyant atmosphere, however, was not entirely without blemish. Increasing production of naphtha, for which there was a ready market, brought increasing production of associated gas, for which there was insufficient demand, and this was beginning to

Main Street, Turner
Valley, circa 1918.
Glenbow Archives
NA-67-51.

generate critical comment. One of the parties drawn by the criticism to review what was happening in Turner Valley was the provincial government. Operators found this just a little unsettling, for
the United Farmers of Alberta (UFA) had been mainly critical of the
petroleum industry and how it operated in Alberta. As an opposition party in the Alberta Legislature, they had been instrumental in
the defeat of the Imperial Oil pipeline bill. After being elected by Albertans to form the government later in 1921, they began to act
upon a number of the complaints registered by their supporters.
One of the more important of such initiatives by the farmers' government had been the gasoline price inquiry. Launched in the fall of
1923, this inquiry had precipitated, or at least coincided with, the
decision by Imperial Oil, the supplier of nearly 90% of gasoline in
the province, to reduce prices by 4¢ per gallon.[142] The Greenfield
government, true to its Progressive roots, stood squarely behind the
notion that the province's resources belonged ultimately to the
people, and on this ground had also sought, by passage of a mineral
tax act, to tax certain areas of mineral rights that had been alienated from the Crown.[143] Although the CPR, arguing that the Alberta
legislation was repressive and unjust, eventually succeeded in persuading the federal government to disallow it, the provincial government's attitude towards the petroleum industry was nonetheless
made clear.[144]

It was with some apprehension therefore that the industry
viewed the UFA government's increasing interest in the flaring of
Turner Valley gas. Attention to what was happening in Turner Val-

ley was stimulated partly by the issue of natural gas exports to Montana. When it was learned that certain interests in Great Falls were anxious to purchase natural gas from the extremely productive Imperial–Rogers well just inside the Alberta border, Premier Greenfield wrote to inform the federal minister of the interior of Alberta's opposition. Taking a stance that would characterize the essential position of all Alberta governments before 1987, Greenfield explained that the province would not "consent to the exportation of natural gas from the Lethbridge or any other field until all the requirements of Alberta are fully protected."[145] Its opposition to the export of natural gas put the government in an awkward position. If natural gas was such a limited and valued resource that export could not be contemplated, how could the burning of so-called waste gas at Turner Valley be countenanced at the same time?

While the government pondered the dilemma, the situation in Turner Valley began to attract more comment, and not only from southern Albertans. In a stinging editorial, the *New York World* rebuked Albertans for their imprudent behaviour.

> Inhabitants of the Turner Valley oilfield up in Alberta, seem to be rather proud of the fact that the huge flaming jets of natural gas from one of their wells, burning day and night, have so warmed the land in the neighborhood that many wild flowers have bloomed weeks and weeks ahead of the right seasonal time. Instead, they ought to be ashamed to confess the waste of a precious commodity, and one that is even more precious in that region than in others where cheap fuel is less needed. They are only doing, however, what has been done again and again here in the United States—and now is bitterly regretted wherever the supply of natural gas is exhausted or failing fast.[146]

Reporting the New York paper's assessment for its readers, the *Calgary Herald* acknowledged that in the reproof there was certainly "food for present reflection." The *Herald* also astutely observed that little action had been taken to this point for fear that "it might result in the checking of valuable development operations in search of oil," thus underscoring the Alberta government's further dilemma.

In Edmonton there was the will to act, but great uncertainty about when and how. The question of how to deal with the situation was difficult, not only because of the elusive nature of a formula that would protect the public interest in Turner Valley and yet not stifle development but also because Alberta possessed only lim-

Flaring at Turner Valley Okalta Well, 1929. In 1931, an American consultant alleged that Turner Valley "constituted by far the greatest waste of natural gas taking place on the continent." Glenbow Archives NB-16-601.

ited powers to deal with resource matters. In September 1925, Greenfield received the signal for which he had been waiting. Charles Stewart, the Liberal minister of the interior and the man Greenfield had replaced as premier of Alberta, wrote to the provincial government to urge the preparation of conservation legislation. The problem, as Stewart explained, was that, although his department had the power to prevent the waste of gas from lands acquired under lease from the federal government, in those cases where the petroleum rights had been earlier conveyed without restriction, the federal government was powerless to act. He added that his senior official in Calgary, C.C. Ross, was convinced that the development of the industry had reached the point where such legislation was necessary and that the Department's technical staff would be happy to advise about complementary legislation if it would be helpful.[147] The result was the passage of the *Oil and Gas Wells Act* in the next sitting of the Alberta Legislature.[148]

Presented by new Alberta Premier John Edward Brownlee, the Act provided for the regulation of oil and gas wells in all areas that did not come under the scope of existing federal regulations, namely, CPR and Calgary and Edmonton Railway lands, Hudson's Bay Company lands, and homestead lands taken before 1887. Proclaiming that waste was unacceptable to the public and operator alike, the new legislation gave the government power by order-in-council to establish regulations governing every phase of oil and gas development. The Act anticipated general regulations to prevent the waste of oil and gas and to prevent production that threatened in any way the premature exhaustion of common reservoirs, as well as specific regulations that would permit the government to deal with such matters as the location of wells, restricting production of gas wells to a certain percentage of open flow capacity, prescribing proportional production whenever supply was in excess of market demand, and the takeover of dangerous wells. Giving notice of the government's intention to bring the actual regulations before the next session of the Legislature, the Act also offered the prospect of consultation with technical experts and operators "in order that the regulations may be as effective as possible and impose no further hardship upon individuals than is necessitated by the prime importance of conserving oil and gas."[149] Finally, the Act contemplated the possible formation of a "Board of Reference," consisting of operators, technical experts and others who the government might appoint and charge with the responsibility of administering the regulations.[150]

For an industry inclined to be apprehensive about the UFA's intentions, the *Oil and Gas Wells Act,* had a stabilizing effect. It provided the first clear statement on how the provincial government would go about regulating the industry and, most important, it seemed to assure that the oilmen would have a role to play in the formulation of the actual regulations. Given the impending transfer of control of lands and resources to the province, this was a message of particular importance. In its consultative approach, the Alberta government was following precedent set by federal politicians and civil servants connected with the Mines Branch of the Department of the Interior. In both approach and essential content, the new legislation also set the pattern for the future.

While the months passed and Brownlee contemplated the complex technical and political foundation upon which suitable regulations might be drawn, his government continued under pressure to allow the export of natural gas from the border area. Explaining government thinking on the matters of gas export and conservation in the summer of 1926 to an applicant whose request was being denied, the premier wrote:

Your request was mainly supported by reference to the condition resulting from the oil development in Turner Valley which apparently has resulted in a great surplus. The Government feels however, that this development, which can hardly be stopped may lead to very rapid depreciation of this field.... Again, if there is to be a policy of conservation in this Province, it would appear we will have to consider as one phase the saving of dry gas so as to draw first on the wet gas resulting from oil development.[151]

Unable to determine any consensus among local operators, other than a general desire to be left alone, Brownlee looked outside the province for independent technical advice on conservation policy. He first called upon F.P. Fisher, the American consulting engineer who had drawn the attention of Albertans at the Calgary Natural Gas Conference hearings in 1921. Basically, there were two ways of approaching the gas flaring problem. The first and most desirable was to find a market for the product, the second was to restrict production for future use. Fisher came to Edmonton to confer directly with the premier and warned against trying to deal with the surplus problem by artificially expanding the local industrial market. "It is unsound economics," Fisher informed Brownlee, "to attempt by means of gas sold far below its intrinsic value to induce the inauguration of an industrial enterprise that can only exist with the aid of what amounts to a fuel subsidy."[152] He pointed out that Alberta's great handicap of distance and transportation costs precluded extensive development of manufacturing for outside markets, and that industries attracted by fuel subsidies seldom survived the exhaustion of the subsidized natural gas supply. Hence, such a policy put a premium on the wasteful and early exhaustion of gas fields and brought only temporary industrial benefit. In Fisher's view, the industrial development of Medicine Hat and Redcliff rested upon a questionable foundation. A more suitable market approach to the surplus problem would be to look beyond Alberta's borders. In the meantime, Fisher pointed out that there was the exceptional opportunity of preserving Turner Valley gas for future use by piping it to be stored in the now depleted Bow Island reservoir.[153]

Brownlee, however, was enough of a political realist to know that, for the moment at least, he dare not contemplate the removal of gas from the province. While the premier weighed Fisher's advice, and while Minister of Public Works Alex Ross and Attorney General J.F. Lymburn reviewed the oil and gas regulations of Oklahoma, the *Calgary Herald* warned of the great danger of hasty legislation and of the mere copying of the regulations of a state where the industry was in a highly developed stage and where there was enormous oil production. The paper urged legislators to be wary lest they cripple the Alberta industry while still in its pioneer state.[154]

The *Herald* hardly needed to have warned the Brownlee government to be wary. Continued study of the matter meant that 1927 passed without the regulations being presented to the Legislature. Delay, however, failed to bring insight, the problem only grew more intractable. More wells reached the limestone naphtha horizon. This brought increased naphtha production, but also greater volumes of gas to be flared. At the same time, oil prices were beginning to decline, making smaller producers even more resistant to suggestions that flaring and, hence, their production be restricted.

Beset on all sides, in desperation the premier ordered yet another study, this one to be done "quietly" by A.A. Carpenter, chairman of the Board of Public Utility Commissioners. In a telegram asking for the co-operation of federal technical staff, Brownlee explained to Stewart that he had in mind an "enquiry not by public commission but rather a quiet study situation, having in mind [the] necessity of avoiding any action which may hinder oil development."[155] The minister responded, to say not only that his department would be happy to lend one of their most experienced petroleum engineers to work with Carpenter but also that they had recently been conferring with Imperial Oil about the waste gas problem and Alberta could expect to find the company co-operative. To both Albertans, Stewart and Brownlee, this was reassuring. In their view, the ideal solution would be a conservation program worked out among the companies themselves.[156] This was a goal that, despite numerous setbacks, would remain constant for the next decade.

In A.A. Carpenter, the premier had identified perhaps the best qualified Albertan to examine the waste gas problem in the Turner Valley field. Few knew the industry as well. Carpenter had a thorough knowledge of operations in the field, and from his vantage point as chairman of the Board of Public Utility Commissioners he had acquired an unrivalled knowledge of the complex world of corporate accounting and rate structure formulation. Carpenter outlined clearly the nature of the problem. Unencumbered with the side issues and pleadings of vested interests that clouded debate in the press and in the Legislature, he evaluated concisely the more plausible solutions advanced to date and identified a course of action the government might pursue. Although later studies would offer much more extensive comment on the complex geology and reservoir dynamics of the Turner Valley field, conclusions did not alter.

Faced with a situation where there were almost as many figures as there were commentators, Carpenter set out first to define the magnitude of the problem. After assessing the relevant data, he determined that total yearly consumption in southern Alberta was currently about 12 billion cubic feet and total production 37 billion cubic feet, leaving an annual wastage of some 25 billion cubic feet.[157] But this was only the preliminary stage of a much worse situation to come. Carpenter pointed out that 33 wells were in the active drilling stage, and one of these, Okalta No. 1, had just come in with a tremendous estimated production second only to that of Royalite No. 4.

Given the vast overproduction, the next question that had to be addressed was a technical one, could the wells be shut in? As Carpenter put it, if the wells could be shut in, "the limitation of output and the consequent conservation of natural gas becomes merely a matter of policy."[158] There was one feature about the wells in Turner Valley that distinguished them from those of any other field discovered up to that time, and this was the extremely high well pressures. Observers had been astounded by the 3,000 pounds estimated pressure required to lift the casing to the top of the derrick as Royalite No. 4 blew in on 24 October 1924. Since that date, the well had never been completely shut off, and this was also the case with all the major producing wells that followed. The reason, as Carpenter explained to the premier, was the alleged concern of "the Imperial Oil Company, that any attempt at shutting in would possibly not only endanger the wells, but also the lives of those taking part in the attempt."[159] Noting that, in Wyoming, wells with pressures greater than 1,700 pounds had been capped, Carpenter argued that even though Turner Valley pressures were known to be higher, there was no reason why under proper supervision an effort should not be made to find out whether or not it was possible to shut in the wells. The Utility Board chairman knew this to be essential before any rational marketing or conservation program could be attempted. The only way to establish the field's condition and to determine the rate of decline was to shut the wells in long enough to get a proper measurement of field pressures and to follow this with periodic tests. Determination of how the field was standing up, and projecting from that how long the gas supply would last, was an essential first step for any group that might contemplate investing in a pipeline to deliver the gas to distant markets.

An endeavour was necessary not only to determine the potential extent and longevity of the gas supply but also to understand more precisely what was happening in the Turner Valley reservoir. Carpenter informed the premier:

> it is now an accepted fact that by conserving the natural gas in an oilfield, the production of oil is prolonged and greater quantities of oil can be eventually recovered from the oil sands by holding back the flow of gas in the proper degree than if the gas were allowed to flow without restriction.[160]

Although the conservation of natural gas was a legitimate concern in its own right, conservation of gas produced in connection with the production of petroleum (casing-head gas) was doubly important for it had a direct bearing on the amount of oil that could be produced ultimately from the reservoir.

What was to be done? Carpenter pointed to California, where it seemed the greatest progress had been made in conserving natural gas, and he saw the steps taken in that state as being particularly instructive. In contrast to Alberta's meagre production, some 10,716 oil wells in 32 California oilfields produced 3,643,345 barrels of oil and more than one billion cubic feet of gas daily. Of the latter, about 24% was flared.[161] Even though California possessed a vast reserve of dry gas, efforts were being made to reduce the waste of casing-head gas even further. Carpenter drew particular attention to the considerable success achieved as "the result of an agreement between the operators themselves, although the State initiated the proceedings by the appointment of a Committee of Investigation."[162] The approach of California operators to the "waste" gas problem followed along three lines. First, in certain fields where the gas-oil ratio was high, such as the Ventura Avenue field, production was cut back. In this regard, it was drawn to Brownlee's attention that California gas-oil ratios (the amount of gas produced with one barrel of oil) were much superior to those in Turner Valley. With the exception of several tiny fields, the greatest amount of gas produced with a barrel of oil was about 4,000 cubic feet. Royalite No. 4 produced 40,000 cubic feet per barrel of naphtha, and even greater quantities were produced in most of the other Turner Valley wells.[163] Second, California operators found merit in storing surplus gas in older depleted fields, and Carpenter outlined several of the more ambitious examples. In this regard, the report noted that, although there had been discussion about sending Turner Valley gas to Bow

Island, this had still not happened. Third, oil producers in California made special efforts to extend existing markets for natural gas. Assessing such possibilities in Alberta, Carpenter echoed the earlier remarks of F.P. Fisher. The manufacture of carbon black from natural gas was dismissed as "merely a limited form of waste," and the prospect of building a community of industrial users beyond those intending to supply a limited local market could not, in Carpenter's view, be seriously entertained. Supplying the larger prairie cities to the east seemed to offer the best prospect of substantially expanding the market for Turner Valley gas, but investors would want to be "satisfied that the field was of sufficiently long life to warrant the investment" before the expense of a pipeline could be justified. Having reviewed the methods being adopted in California, Carpenter advised, "There does not appear to be any reason why somewhat similar methods should not be adopted in Turner Valley."[164]

Although Premier Brownlee could gain little comfort from the knowledge that the surplus gas problem was certain to get worse, Carpenter's "quiet" study helped reinforce the idea in the premier's mind that the operators must find their own solution, and, most importantly, the California model suggested the approach the Alberta government might use to achieve this end.

Brownlee directed his attention first to the matter of gas exports. He proceeded with caution, knowing that to permit the export of natural gas from the province as a partial solution to the Turner Valley problem involved great political risk. Alberta communities remained adamantly opposed to any scheme of gas removal. Moreover, Calgary, the largest city in the province, had joined Medicine Hat and Redcliff as another community dedicated to the promotion of industrial development by offering the attraction of cheap and abundant supplies of natural gas. When the mayor of Calgary learned that consideration was once more being given to the export of gas to Montana, he hurried off a letter to the minister of the interior expressing the city's opposition. "Our domestic and industrial needs are growing," the mayor wrote; he went on to say that Calgary's "greatest industrial attraction" was the presence of a huge gas supply virtually at the city's gates.

During the year 1928 our favoured position in this regard brought to this city the Manitoba Rolling Mills, The Dominion Bridge Company Ltd., and the Dominion Wheel and Foundries Ltd., all of whom are erecting plants this year and will be large consumers of gas. Remove the attraction only temporarily and

our future must be affected adversely. Our citizens would be very loath if any industry in the United States were led to rely upon a fuel supply which is our main industrial attraction and may at any moment be necessary for Southern Alberta consumers.[165]

In May, the minister of the interior journeyed to Calgary to meet with Turner Valley operators and city officials to discuss the matter of gas exports and to try personally to promote some kind of a consensus on how to approach the Turner Valley gas problem. In acknowledgement that the question of gas exports was intimately tied to the wastage of gas at Turner Valley, it was decided to establish a joint federal-provincial commission to study "the gas situation in the Turner Valley field."[166] The commission was to be chaired by Charles Camsell, federal deputy minister of mines, with C.C. Ross and A.W.G. Wilson, a chemical and industrial engineer also from

Central part of Turner Valley Field, showing increased drilling activity after the prolific Royalite No. 4 oil discovery ushered in the field's second oil boom in 1924. Glenbow Archives ND-8-421.

the Department of Mines, contributing the technical expertise on the federal side. Dr. R.C. Wallace, president of the University of Alberta, and A.A. Carpenter were named the provincial representatives. Faced with the task of formulating a comprehensive resource policy and establishing a new government department to take over from federal administrators now that the lands and resources transfer agreement was on the verge of being signed, Brownlee was happy to let his federal colleague take the initiative in dealing with this particularly thorny issue.[167]

The committee immediately set about its task. Several members were assigned specific areas to investigate: Carpenter was to examine the possibility of markets in Winnipeg, Regina and Saskatoon, and Ross was sent to visit gas fields in the southwestern United States to examine methods used to control high-pressure gas wells. Professionals outside the committee were also contracted for special assignments. Dr. George Hume, of the Canadian Geological Survey, was asked to investigate the possibilities of underground storage of waste gas in the vicinity of Calgary; two American engineers, L.C. Jones and Dr. B.T. Brooks, were asked to examine the possible manufacture of chemical products from Turner Valley gas; and Professor W.G. Worcester, of the University of Saskatchewan, was employed to see if clays in the Turner Valley area were suitable for ceramic industries that would in turn use natural gas as fuel.[168]

Although the committee went over much the same ground as earlier studies had, it did contribute more substantial evidence to support already advanced conclusions. In an interim report of 25 October 1929, the joint committee dismissed the prospect of developing a substantial industrial market for Alberta gas and advised that under any circumstance, except complete closure of the wells, "there will be a very large surplus of gas for which no market can be found in this territory." The conclusion was that, "if there is an opportunity of selling the surplus gas to a foreign consumer, this should be done."[169] The committee's final report was presented to both governments on 8 January 1930, and the seriousness of the situation in Turner Valley was illustrated dramatically with reference to California. In August 1929, the average wastage in 37 California fields was marginally over 1,000 cubic feet of gas for each barrel of oil produced; in Turner Valley for June 1929, the loss was 59,000 cubic feet per barrel, or 590 times greater.[170] It was time, the committee insisted, to recognize the Turner Valley field for what it was; no amount of wishful thinking would make it an oilfield when it was truly a wet gas field. The committee warned that "present operations are rapidly and wastefully dissipating a great natural resource."

The problem, according to the committee, was that the 40 or so producing companies were primarily interested in producing as much naphtha as they could in as short a time as possible. Each operator, the committee observed, was surrounded by others.

Unless he develops his property rapidly, particularly along the boundaries, the supplies of gas and naphtha along the edges of his property can be drawn off by his adjacent neighbors. Offset drilling thus becomes a necessity to the operators and corresponding increases of wastage follow.[171]

What the committee members were seeing in Turner Valley was the classic and lamentable phenomenon that had been repeated in countless oilfields across North America over the previous half century.

Despite the obvious damage to the field and the hardship that attended in the unregulated wake of the rule of capture, Turner Valley operators could not agree on a policy of controlled production involving the introduction of conservation measures. The committee saw little chance that attitudes would change, and decided therefore that "the only solution" that would make conservation possible and protect the public interest would be the imposition of rigid regulations. They recommended that such regulations be "administered uniformly throughout the area by one body representing both federal and provincial parliaments." The committee was convinced that proper and efficiently administered regulations would best serve the interests of the public and the producers engaged in the development of what was seen as a national asset. This, they insisted, was "the first and most important step to be taken; *other measures can only be palliative*."[172] In his covering letter, submitted to Premier Brownlee with the report, Committee Chairman Charles Camsell offered the Alberta leader a rebuttal to the opponents of regulation who could be expected to respond in the local press with the traditional argument that any restrictions on production would mean an end to further exploration and development. Camsell argued the reverse, that regulation and production controls would bring investment, because industries could then be assured of an adequate supply of gas over a period long enough to warrant substantial capital investments.[173]

Camsell's letter also informed the premier of the committee's desire to point out that, under any scheme of regulated output and consequent commercial development, there would still be a large supply of natural gas surplus to Alberta's needs and therefore available for export. Having waited in anticipation of such credible backing on the natural gas question, Brownlee moved ahead quickly. After circulating the joint committee's report, and gaining the cautious approval of Calgary's mayor, he informed the minister of the interior of Alberta's approval for limited gas export to Montana.[174] The premier was nonetheless careful to set out the terms of Alberta's consent. The applicant had to agree to submit to the decision of the provincial Public Utilities Board regarding the maximum amount of gas that could be exported in any season, post with the Board the price paid for gas in the field, pay the province a "fair and reasonable tax as mutually agreed or fixed by the Tax Assessment Commission," and understand that no subsequent claim for other than waste gas would be entertained.[175] A one-year export licence was subsequently granted by the federal Department of Trade and Commerce.[176] The position taken by Alberta in the negotiations regarding this first removal of gas from the province, particularly the premier's firm insistence that a provincial agency have absolute jurisdiction over the price and quantity of natural gas to be exported, set an important precedent that established both the principle and the approach that would guide Alberta in the natural gas removal[177] debates of the early 1950s.

Brownlee now proceeded with the urgent matter of oil and gas regulations. Urgency stemmed not only from the growing dimension of the waste problem but also because it had now been established that control of crown lands and resources would pass from Ottawa to the Province of Alberta on 1 October 1930.[178] The province had not only to prepare regulations to govern activities in the petroleum sector but also an entirely new government department and administrative apparatus had to be organized.

PROVINCIAL CONTROL OF RESOURCES AND THE
STRUGGLE TO FIND A REGULATORY FRAMEWORK

Heeding the advice of the joint committee on conservation, Premier Brownlee secured the services of C.C. Ross, the senior and most experienced petroleum engineer among the federal government's Calgary staff, and J.A. Allan, professor of geology at the University of Alberta, to direct the preparation of Alberta's first petroleum and natural gas regulations. Ross and Allan proceeded in what had become the expected manner. They prepared a draft, which was submitted to, and then discussed clause by clause with, a committee of operators appointed by the Oil and Gas Association of Alberta. Thus, when the final draft of the proposed regulations were submitted to the premier in mid-March 1930, the authors could claim that the document was acceptable to the industry.[179] The suggested regulations largely followed the existing federal approach in well-spacing, reporting, casing, control of water, and abandonment. They differed, however, in the substantial authority assigned to a government-appointed "director." This official could take whatever means he judged necessary in the public interest to control and prevent the waste of natural gas and oil and then recover from the offending operator all costs incurred in the action. Also included was the provision that, before the commencement of a well, all operators be required to furnish a bond of up to $10,000 to be held by the government for the life of the well as a guarantee that operations would be conducted in accordance with the regulations.[180] Having in hand draft regulations that were apparently acceptable to the Turner Valley operators, the Brownlee government waited for October 1 and the official transfer of resources in anticipation that five years of public agitation over the flaring of "waste" Turner Valley gas could be laid to rest at last.

The long-awaited surrender of federal control over lands and resources in Alberta was greeted with great fanfare, and was followed on October 2 by the appointment of the Honourable R.G. Reid as minister responsible for the newly created Department of Lands and Mines. Administrative control in the petroleum section, however, was delayed while Edmonton and Ottawa settled differences regarding the transfer of records. In the interim, federal legislation remained in force and under the supervision of Mines Branch personnel in Calgary. Finally, when arrangements were completed on 1 February 1931, almost all the professional staff at the Calgary office of the Petroleum and Natural Gas section of the federal Department of Mines transferred to provincial service, and Alberta took over the supervision of all oil and gas operations.[181] Accepting positions in the Petroleum and Natural Gas Division of Alberta's new Department of Lands and Mines were petroleum engineers William Calder, who was named director of the Division, Charles Dingman, G.R. Elliott and Vernon Taylor; geologists J.G. Spratt and R.M.S. Owen; and statistician F.K. Beach. Dingman, the senior member of the group, had spent his entire career in the Alberta oil and gas fields, and had been with the federal Mines Branch since 1920. Calder had joined in 1926 after considerable field experience, especially in the area of pressure control and well shut-off. Spratt also began work in 1926 to help strengthen the capacity of the Calgary office to interpret the increasing number of rock and core samples that were being collected. The desire to manage better the mounting

TABLE 1.1 Alberta Lands, 30 September 1930

Land Category	Acres (Approx.)
Total area of province	163,382,400
Water area of province	4,150,400
Land area of province	159,232,000
Lands patented as at 30 September 1930	42,930,000
Lands alienated as at 30 September 1930	5,314,000
Area of unalienated provincial lands as at 30 September 1930	84,467,170
Total area within forest reserves	12,436,500
Water area within forest reserves	43,000
Land area within forest reserves	12,393,500
Total area within dominion parks	13,434,240
Water area within dominion parks	635,000
Land area within dominion parks	12,799,240
Indian reserves	1,328,090
Area of surveyed land, including water areas but excluding dominion parks and Indian reservations	70,007,000

SOURCE: Alberta Department of Lands and Mines, *Annual Report,* for year ended 31 March 1931, p. 9.

body of statistical data being collected on petroleum and natural gas activities had brought Beach from the Irrigation to the Mines section of the Department of the Interior's Calgary establishment in 1927. Elliott arrived a year later to contribute his lengthy California field experience to the considerable expertise already assembled by the federal government in Calgary. An increasing workload had brought the addition of another engineer, Taylor, in 1929, and the following year a petroleum geologist, Owen.[182] This meant not only that Alberta gained a body of experienced and highly competent individuals but also that the transfer of the oil and gas industry from federal to provincial administration was remarkably smooth. Indeed, little had changed. Although they now reported to a minister in Edmonton rather than in Ottawa, the group of individuals charged with overseeing the development of Alberta's oil and gas industry remained almost identical. The apparent continuity of personnel, policy and approach was reassuring to an industry that was already suffering the effects of the Depression and a declining market.

TABLE 1.2 Petroleum and Natural Gas Rights Alienated from
the Crown

CPR	9,821,972 acres
Calgary and Edmonton Railway	1,534,048 acres
Hudson's Bay Company	1,796,301 acres
Homestead lands	139,656 acres

SOURCE: NAC, RG15A, vol. 2000, statistics gathered for Dysart Commission.

TABLE 1.3 Annual Production of Natural Gas and of Naphtha in the
Turner Valley Field, 1925–1931*

	Gas (Mcf)	Naphtha (Barrels)	Gas/Oil Ratio
1925	8,760,000	165,717	52.9
1926	8,800,000	211,008	41.8
1927	13,870,000	290,270	47.8
1928	21,320,000	410,623	51.9
1929	61,450,000	908,741	67.6
1930	114,080,000	1,314,039	86.8
1931	169,380,000	1,345,689	126.0

SOURCE: Alberta Department of Lands and Mines, *Annual Report*, 1933.
* These figures differ slightly from Board estimates shown in Appendix II.

Alberta's new Department of Lands and Mines inherited administrative responsibility for a vast area of nearly 150 million acres.
The different categories into which these lands fell, including dominion parks and Indian reserves where federal control was retained, is shown in Table 1.1. Separate from, but returned to the
province along with the remaining surface area, were the unalien-
ated subsurface rights. This included not only the mineral rights to
nearly all the remaining public lands (nearly 50% of the province)
but also those to virtually all the alienated homestead lands, thus
putting the petroleum and natural gas rights to slightly more than
90% of the province under Alberta's control. Such rights beyond
Alberta's direct authority were held mainly by the Canadian Pacific
Railway, the Calgary and Edmonton Railway Company, and the
Hudson's Bay Company, as indicated in Table 1.2.

Only a relatively small portion of the petroleum and natural gas
rights to Alberta lands had been permanently alienated; however, a

further 2,263,076 acres were set aside under lease. This area was divided into 17,219 leaseholds and assigned to the responsibility of Calder and his colleagues in the Petroleum and Natural Gas Division of the Department of Lands and Mines.[183] Also assigned to Calder's supervisory care were the 70 producing gas wells and 56 producing oil wells situated on the transferred leasehold.[184]

The primary concern of the professionals assembled in the Alberta government's new Petroleum and Natural Gas Division was Turner Valley waste gas. Carpenter's 1928 warning to Premier Brownlee that matters could only get worse had proved only too true. New wells were being brought into production, gas-oil ratios were on the increase, and increased naphtha production was being achieved therefore through the ever greater production of gas that had to be flared for want of a market.

Appropriate regulations were urgently required, and Calder was hardly established in his new Edmonton office before he submitted to Premier Brownlee a report on the Turner Valley situation, thereby adding his personal contribution to the collection of reports already in the premier's possession.[185] The new director of the Petroleum and Natural Gas Division was harsh in his criticism. "This enormous waste," he charged,

> has solely been due to the operators who have intentionally avoided the development of the shallow crude oil horizons and concentrated all efforts in drilling deep and expensive wells to exploit the wet gas horizon in the limestone. The crude oil horizons being nearer the surface could have been developed at approximately one half the cost per well as compared with the expenditure necessary for the deep wells drilled to date. Operators, therefore, having created an uncalled for hazard are solely to blame for the present waste.[186]

Calder wanted all Turner Valley wells shut in at once for a 24-hour pressure test, and the operators of wells found to be defective and unable to hold the pressure be required to make such wells pressure-tight immediately. He explained to the premier that all efforts of the Calgary branch of the Department of the Interior to make such pressure tests had been met by the refusal of the operators to co-operate on the grounds that their wells would be damaged. Drawing from his own specialized experience in the area of well shut-offs, Calder assured the premier that the mechanical difficulties of shutting in wells were greatly exaggerated by the operators and that original field pressures were nowhere near as high as conventional wisdom held. He went on to urge that further drilling into the limestone be disallowed and gas withdrawal from producing wells be limited to 25% of their open flow volume. For this to be achieved, it was essential that an equitable prorationing scheme be established and that the government give special encouragement to those seeking to pipe gas to more distant markets.

To underline the necessity of immediate action, Calder drew Brownlee's attention to the most recent field production data. During December 1930, the gross production of wet gas in the Turner Valley field amounted to 537 million cubic feet per day. After being stripped of the naphtha content, 51 million cubic feet of gas per day was used for field requirements and to meet the needs of the Calgary gas company, leaving the balance of 486 million cubic feet per day to be destroyed by burning. Using the current market value of 5¢ per 1,000 cubic feet, Calder calculated the value of the burnt gas at $24,300 a day and set this against the $18,718 earned by the naphtha production of 5,348 barrels per day. He emphasized that the difference of $5,582 represented a waste expenditure of $1.04 for every barrel of naphtha produced.[187]

Calder acknowledged that any plan to cut back production would face the strenuous objection of the small independent operators, and perhaps even legal challenge. To confront this, he recommended that the government work through an advisory committee composed of representatives from the owners of mineral rights and from various oil and gas operators associations. Brownlee remained nonetheless as cautious as ever. He had in hand the draft regulations proposed by Ross and Allan, and the authority under the 1926 Oil and Gas Wells Act to put these measures in place, yet he delayed. In good part, this stemmed from the greatly deteriorated economic situation in the province generally, and in Turner Valley specifically, from the previous spring when Ross and Allan had submitted their report. Demand for petroleum products was dropping and prices were in decline. The phenomenal strike by "Dad" Joiner on the Daisy Bradford farm in East Texas in October 1930 assured that the emerging glut of crude oil would be no passing phase. In Turner Valley, development declined sharply and many wells in the earlier drilling stage were abandoned. Exploration in several of the more promising areas elsewhere in the province had stopped completely. In this situation, small Turner Valley producers became even more resistant to any talk of curtailing production.

Brownlee's response was to call yet another conference. The UFA premier remained fixed upon the California approach, and he persisted in his efforts to persuade the Turner Valley operators to work out an acceptable conservation program managed by their own organization. In the Alberta Legislature, Brownlee tried to reassure his colleagues that, if the conference went as planned, it might not be necessary to introduce any legislation dealing with gas conservation.[188] The meeting was held in Edmonton on March 10 while the spring session of the Legislature was under way. Gathered to meet with the UFA Cabinet and its advisers, Calder and University of Alberta President Dr. Wallace, were various small producers and their legal advisers, representatives from the major leaseholders, Royalite, Hudson's Bay Company and the Calgary and Edmonton Oil Corporation, the gas companies, as well as Calgary's city solicitor, and Calgary Alderman J.H. Ross. In his introductory remarks, Premier Brownlee informed his audience that the government thought it only fair to consult with those involved in oil and gas production before introducing legislation to try to solve the waste gas problem. He then proceeded quickly to the point. What would be the operators' reaction to cutting back the production of wells to 40% of their open flow capacity, as provided for under the old federal regulations, or to 25% as was being considered by the provincial government?[189] "Impossible," was the reply of the first speaker to respond. Another wondered whether, in limiting production, "it would not to some extent be confiscating property." The representative of McDougall-Segur Oils contended that "any cutting down of production should be deferred until such time as companies that had just completed expensive drilling programs had been given an opportunity to get on their feet." One operator expressed the opinion that cutting back on any arbitrary percentage basis was the wrong approach. "Every well," he stated, "behaves differently, and you will kill production unless you make a thorough study and then deal with wells individually." The premier's alternative proposal to divide the field into areas and alternate production seemed to draw a marginally more acceptable response. Brownlee also sought guidance on the matter of well-testing. He noted that, to this point, efforts to take tests had met the definite refusal of some companies and asked whether the operators would have "any objection to the government's taking the widest powers possible to have wells in Turner Valley tested?" Drawing no stated objection from the operators on this matter, the premier announced that such testing was among the government's first objectives. Co-operation was hoped for, but the government would exercise its full power against those who resisted.

While discussion was continued in a confidential meeting, consisting of a smaller representative group, the essential forces in contention behind the closed doors were made abundantly clear in the open part of the conference. When the trend of discussion seemed to be going in the direction of recommending another study, Calgary's City Solicitor L.W. Brockington spoke for those who were growing weary of continued delay. He was joined by Calgary Alderman Ross, who noted that, if the suggestion to hoist conservation measures for six months while the field situation was studied were implemented, "72 billion cubic feet of gas, or enough to supply the requirements of Calgary for the next ten years," would be wasted.[190] On the other side, a lawyer representing one of the smaller companies pointedly asked the premier, "Have you considered whether the province has a right to restrict where perpetual rights—that is, where the right of gas production goes with the soil—are held?"[191]

When the meetings concluded, Brownlee announced that his government favoured the appointment of an advisory committee to examine the problem thoroughly. Since this would take some months, some "tentative action," including a provision to restrict production to 40% of potential well capacity, might be necessary.[192]

To reinforce its authority to formulate regulations, particularly in the area of well-testing, which the government had identified as a priority concern, and to demonstrate to Albertans that action was being taken, a revised *Oil and Gas Wells Act* was piloted through the Legislature in late March.[193] In addition to a new provision, specifically mentioning the right to take closed pressure and open flow measurements,[194] the new bill also required that a bond of up to $10,000 be posted before drilling commenced on any well to insure that drilling operations were conducted in accordance with the regulations, thus following the earlier recommendation of Allan and Ross. The remaining new departure was intended to benefit the UFA's farm constituency; it permitted the government to set the "maximum price at which fuel oil (casinghead oil), produced at any well shall be sold at the well to purchasers taking delivery thereof at the well."[195] The 1931 Act also extended several of the measures common to the 1926 legislation. The general waste prevention clause, which allowed the formulation of regulations "generally to conserve gas and oil, or to prevent waste or improvident disposition," was altered to permit the government to require the cleaning out or deepening of wells. In other regards, the new *Oil and Gas Wells Act* was much like its predecessor. By order-in-council, the government could fashion regulations dealing with drilling licences; drilling locations; offset drilling; cementation of well casings; the keeping of logs, records and geological samples; the restriction of the production of oil and gas to any percentage of open flow capacity; and the takeover of any wells that were judged a menace to good conservation practice.

The first order-in-council under the new Act came on May 6. It ordered the "flow of gas, or gas and oil" from every well in the province restricted to 40% of its potential capacity.[196] It seems, however, that most of the operators chose to ignore the order.

While the engineers from the Petroleum and Natural Gas Division continued their well-testing program and Brownlee tried to find the elusive formula that would allow the formation of an advisory committee whose proposals would stand a reasonable chance of acceptance, Calgary natural gas consumers grew impatient. With millions of cubic feet of gas being burned as waste, almost on their doorstep, Calgarians began to question why their natural gas utility bills should not be reduced. In May 1931, the City of Calgary accordingly brought action before the Board of Public Utility Commissioners, seeking a reduction in the rate paid for gas in Calgary. The resultant hearings brought the Turner Valley waste gas problem once more to public attention, and this time into much sharper focus. In the preparation of their cases, the gas company and the city both retained the services of specialist engineers to study the Turner Valley situation. From their studies came evidence that was dramatic and alarming. The waste occurring at Turner Valley was more excessive and the consequences more severe than anyone had imagined. Albertans learned that a fuel equivalent of 25,000 tons of coal a day was being flared, that a year's supply for the entire pipeline system in southern Alberta was being lost in each 11- or 12-day period.[197] Putting this in grim perspective, a respected American consulting engineer was drawn to comment that what was happening in Turner Valley "constituted by far the greatest waste of natural gas taking place on the continent."[198] Alberta operators were unable therefore to diminish the immensity of their flagrant waste, as they had in the past, by pointing to even more excessive waste elsewhere.

Calgarians were stunned to learn that, if the present rate of loss continued, the field would be exhausted within two to five years. Testimony at the hearings further revealed that no new source was available to replace Turner Valley gas, except at substantial cost and consequently higher prices for the consumer. The attendant public outcry made it impossible for Premier Brownlee to procrastinate further. On June 13, and under duress, he at last named the long-awaited Advisory Committee. Appointed were Judge A.A. Carpenter, chairman of the Board of Public Utility Commissioners; William Calder, director of the Petroleum and Natural Gas Division; Eric Harvie, of Okalta Oils and Regal Refineries; B.L.

Thorne, CPR, Natural Resources Division; Stanley J. Davies, consulting geologist; R. Van A. Mills, Hudson's Bay Oil and Gas Company; and R.O. Armstrong, field manager for the Royalite Oil Company. At their first meeting on June 26, it became apparent that a technical subcommittee was required to address technical complexities and conservation alternatives on a full-time basis. They recommended that the government engage a respected petroleum engineer from outside Alberta to chair the committee, and that the independent producers and the Imperial Oil interests each appoint an engineer to assist. Agreeing to the request, Brownlee called once more upon F.P. Fisher of Mount Vernon, Ohio; the independents chose Stanley J. Davies of Calgary, and Imperial selected R.O. Armstrong.

The government's new resolve was also reflected in the action taken on another long-awaited matter. On 10 July 1931, the comprehensive regulations respecting drilling and production operations of oil and gas wells that had been contemplated in the 1926 and 1931 *Oil and Gas Wells Acts* were finally presented.[199] While the new regulations greatly strengthened the ability of the engineers in the Petroleum and Natural Gas Division to achieve their supervisory obligations, they also helped to set a tone that assisted the task of the two committees. Most of the operators sensed that this time the exercise was in earnest.

Charged with the straightforward but immensely complex task of devising the most equitable and acceptable method of conserving Turner Valley's gas resources, the committee commenced collecting technical data. The context of their work was defined further on July 28 when, in response to the mayor of Calgary's appeal for a clear statement of government conservation policy, Premier Brownlee made a public announcement that Turner Valley gas production would be limited to 100 million cubic feet per day.[200] The committee was left alone to devise whatever scheme or approach was required to gain the voluntary support of Turner Valley producers, but the field production figure could not exceed 100 million cubic feet per day. Turner Valley producers, especially the smaller independents, were horrified; existing production was well over 500 million cubic feet per day. Having been given 100 million as an acceptable figure by Fisher, the premier, with wide public support, stood his ground. The province's production target still called for a yearly production that was more than three times what the market could absorb.[201]

His familiarity with the Turner Valley field updated with the most recent tests completed by the Petroleum and Natural Gas Division engineers, Fisher commenced a series of intense negotiations.

Even though there was wide agreement that the solution lay in the pooling of Turner Valley production, there was wide disagreement on a formula to determine how the production should be shared. Throughout the summer of 1931, the committee worked its way through a series of drafts for a voluntary pooling agreement. Finally, in early September, with the fifth draft, Fisher felt he had wide enough general acceptance to submit the document to the industry, the public, and the Advisory Committee for general scrutiny and comment.[202]

The *Calgary Herald* published the complete text of the proposed agreement, asked all Calgary gas consumers to study it carefully, and extended editorial support with the observation that it was regrettable "conservation measures were not put into effect years ago. The longer conservation is delayed, the more blame will attach to those responsible."[203] With the proposal's public presentation, it immediately became apparent that there was also strong opposition on the part of some of the independent producers. Speaking for this group, W.S. Herron, the foremost pioneer developer of the Turner Valley field, expressed his vigorous objection to the confiscation of his property. He warned that "most assuredly, if the wells and production can be expropriated, the pipelines and scrubbing plant can be expropriated also." Herron protested with the angry defiance of the traditional frontier entrepreneur against the collectivist philosophy of the United Farmers of Alberta.

> I was one of a number of delegates who went to Edmonton a few months ago to confer with Premier Brownlee and other Ministers concerning the Petroleum regulations that were then being drafted. During our discussion, I was struck very forcibly by a remark that Mr. Brownlee made. He said that he and his fellow Ministers thought that "the natural resources belonged to the people," meaning that the oil and gas rights, although they might be alienated by title or lease granted from the Crown, and regardless of the fact that the owner or lessee may have spent a large sum of money on development, nevertheless their claim was at most, only secondary.

> If our class government enforces these principles, I would think they would have reason to fear that it might act as a boomerang on their own supporters, thousands of whom have comfortable farm homes and title to many thousands of acres of land in Alberta, and they might some day find themselves in the same circumstances as the Kulacks in Russia who had to give up their

lands and all stock and farm implements to the Bolsheviks who dispossessed them.

It is then you would hear those same farmers squeal who are getting assistance from all classes of tax payers in the form of rebated tax on gasoline, a bonus of 5¢ a bushel on wheat, guarantees to the banks on their loan to the wheat pool, and dairy pool, seed grain advances, feed, free transportation for stock, and every assistance of which their imaginations can conceive.

But after having passed legislation to shut in the Turner Valley wells and expropriate the property of those who invested in the oil industry, they will have no cause for complaint if the communists demand that they give up their farms for the benefit of those who have none.[204]

Herron's polemic reflected the strongly held conviction of the quintessential self-made man of the frontier. It was Herron who first recognized Turner Valley's potential, obtained the first petroleum and natural gas leasehold in the valley, and organized the discovery company—Calgary Petroleum Products. Herron mortgaged all his assets and poured his life savings into the development of the Turner Valley field. By dint of his ingenuity and unrelenting hard work, first in the timber trade, then in ranching and cartage, and ultimately in the petroleum industry, he had begun to prosper after nearly 20 years of perseverance. It was a bitter pill to see the government threaten to limit his production just when years of hardship were beginning to pay off. Herron was convinced, and it was a view shared by many of the smaller operators, that the main beneficiaries would be the big companies—Imperial Oil and Canadian Western Natural Gas, Light, Heat and Power.[205] This made conservation measures doubly hard to accept.

For three generations, the Herron family had survived on the remote edge of the frontier, using the resources that nature seemed to provide in abundance. When resources and opportunities in one area diminished, they moved on to where resources and opportunities seemed more plentiful. Herron had grown up with the view that natural resources were there to be used by those who had the initiative and capital to exploit them. Beyond the basic protection of property rights, Herron believed that human progress and wellbeing were best advanced with as few restraints and as limited government intervention as possible. It was a philosophy deeply entrenched in the West. It was reinforced by the structural division within the petroleum industry and it was a factor that the proponents of petroleum conservation would have to contend with for decades to come.

Opposition to the draft agreement also came from the City of Calgary. Although quick to assert that they favoured the plan in principle, civic leaders thought, like some of the smaller operators, that the proposal was far too generous to Canadian Western and Imperial Oil. The problem was that the pooling scheme devised by Fisher and his committee was going to be financed partly by Calgary natural gas consumers through a rate increase of 2¢ per thousand cubic feet. Their rationale was that, since the companies were being asked to sacrifice in some cases as much as two-thirds of their recent rate of earning, it was only fair that those who would profit by a prolonged gas supply—the gas company and the consumers—should offer some compensation. In an open letter to the premier, Mayor Andrew Davison expressed vigorous opposition to "the fixation of rates upon a series of precedents largely invented in the United States and elsewhere by ingenious experts with a company experience and a company mentality," and he urged that Brownlee look to government revenues, Imperial Oil and the gas company's considerable surpluses to finance the agreement.[206]

Comments, criticisms and suggestions collected, Fisher commenced another round of meetings with Brownlee in Edmonton, Mayor Davison and his advisers, the president and counsel of the gas company, a combined gathering of Oil and Gas Association delegates and government leaders, as well as numerous private conferences with individual representatives of operating companies. Finally, on 16 October 1931, Fisher submitted his committee's report to Minister of Lands and Mines R.G. Reid. The report outlined what in the committee's view was "the most logical method" of bringing a Turner Valley conservation plan into effect. Included also was a draft agreement, copies of which the government was to send to the operating companies with an official request to execute the agreement.[207] Of the several changes differentiating the final from the September draft, one of the more important was an increase in the field price of natural gas of 4¢ rather than 5¢ per 1,000 cubic feet, with Calgary domestic consumers picking up 1¢ rather than 2¢ of the increase.

By October 21, companies with more than 50% of the independent production had indicated their willingness to sign the agreement.[208] Others in the independent group protested that they could not sign. They argued that since Stanley J. Davies, the so-called representative of the independent producers on the technical committee, did not speak for them and had not consulted with them, they had insufficient technical information to be able to determine what was in their best interest. It seemed to them also that the situation in the Turner Valley field was not nearly as grave as had been made out, and that action on the proposed agreement should be postponed for a year while further tests were made.[209] Then, the Canadian Western Natural Gas, Light, Heat and Power Company notified the government that, "if certain concessions were made in rate-making principles, they were prepared to assume a portion of the four-cent increase in recognition of the value to them of gas conservation."[210] On November 16, Vice-president A.M. McQueen of Imperial Oil announced that the agreement was unfair to their interests and therefore unacceptable.

In late November, a conference between Fisher and Imperial representatives resulted in certain revisions to suit the oil company, but an early December conference with Calgary and gas company officials failed to find a common basis for sharing the additional cost of gas. At this juncture, the Calgary and Edmonton Corporation appeared to say that, while they wished to co-operate, there would have to be certain modifications. Soon after, the Calgary and Edmonton Corporation presented its own modified proposal and obtained signatures of agreement from a number of the companies, although some of the acceptances came with a covering statement that "they repudiated the necessity for conservation."[211] Finally, in late February 1932, a meeting of Imperial, Calgary and Edmonton Corporation, and the larger independents agreed on the Calgary and Edmonton Corporation proposal for the unitized operation of the Turner Valley field, but with the proviso that control of the pool be by a board of five directors: two to be elected by the companies associated with Imperial, two by the independent companies, and one by the royalty holders. The smaller independents immediately protested that this would result in complete control by the Imperial group and would therefore be unacceptable.

Fisher had striven for eight months to find a conservation agreement acceptable to all parties. His efforts stand as testimony both to his own remarkable perseverance and to the widely divergent interests he sought to accommodate. It was now abundantly clear that a voluntary agreement would never be reached, and Fisher advised the minister of lands and mines that legislative action would be required.[212] This was a bitter blow for political, perhaps more than philosophical, reasons. Brownlee was convinced that a voluntary approach of the kind that existed in California was desirable, and for nearly six years, often in the face of criticism within his own party, he had promoted this goal.

Although Brownlee was under no illusion as to how difficult it was going to be, even with legislation, to impose conservation upon the smaller producing companies, the prospect of effective administration in the oil and gas sector of the provincial economy was not entirely bleak. If the federal government, with its primary emphasis on the development of the region's petroleum resources, had left the province to deal with the conservation conundrum, it had also left Alberta with a nucleus of highly competent professional staff committed in the public interest to the elimination of wasteful production practices. Almost from the beginning of its development, the Turner Valley field was monitored by public servants whose knowledge of North America's conservation practice was current and whose technical expertise was close to the best available to the dominant companies. It was a beginning that boded well for the future; it would be political will rather than technical competence that would be put to the test.

With enabling legislation in place and the issues of conservation, after long debate, laid out clearly before the public, the time for the Brownlee government's testing was at hand.

TWO

The Turner Valley
Gas Conservation Board and the Failure
of Alberta's First Proration Program

THE SITUATION FACING Premier J.E. Brownlee in the winter of 1932 was hardly one to be envied. His government had made little headway in remedying the effects of the still deepening depression. His fractious UFA followers in the Alberta Legislature were in a more rebellious mood than ever and the signs pointed to an increasingly restive electorate. It was into this inhospitable political environment that Brownlee was compelled to bring forward for consideration the long-standing and controversial question of Turner Valley waste gas. The Speech from the Throne on 10 February 1932 announced the government's intention to present conservation legislation during the current session. In the wearisome debate that ensued, nothing new emerged, only the more insistent statements of established positions.

Making an eleventh-hour appeal for a voluntary solution, Brownlee declared in the throne speech debate that there was still time for the various interests to come together and avoid a government-imposed solution.[1] In the days that followed, he tried to persuade his fellow politicians of the folly of presenting arbitrary legislation that would likely draw challenge in court, lengthy litigation, and thus delay in achieving the conservation objective.[2] Even though Brownlee's sense of the situation would prove prescient, his efforts had failed to bring results for so long that his ideas had lost credibility and he was unable to play other than a defensive role. First, in a heated exchange, Brownlee had to deny the allegation of F.C. Moyer, leader of the Independents, that he had been paid $1,000,000 to stall on the Turner Valley gas question.[3] Then the premier had to face down a Labour motion, supported by six members of his own party, calling for public ownership and development of Alberta's natural resources.[4] Outside the Legislature, Brownlee had to counter a newspaper and radio campaign, sponsored by some of the smaller independent oil companies, that was telling Albertans that the entire conservation scheme was designed to promote the interests of the Imperial Oil Company.[5]

With Brownlee trying vainly to defend his government's record, debate moved from the throne speech to formal hearings before the Agriculture Committee of the Legislature on March 16. A parade of witnesses went over the old ground once more. Technical experts, including Frank Fisher and J.A. Allan, professor of geology at the University of Alberta, argued that the need for conservation measures was urgent. Clarence Snyder, who had been a driller in the field since the Dingman discovery, spoke for some of the smaller operators, saying that reports of declining field pressures had been much exaggerated, and R. Shouldice warned the government against "confiscating the investors' rights."[6] Presentations continued in the same vein through the third day of hearings. Dr. Robert C. Wallace, a geologist and president of the University of Alberta, urged immediate action to curb waste and prevent the early decline of the field. W.S. Herron, who more than any other was responsible for the discovery of oil in Turner Valley, countered that "it was ridiculous to say the field would be exhausted in two or three years."[7] The secretary of the Calgary Stock Exchange, Alderman Harold Riley, presented petitions signed by 7,200 Calgarians asking that the government defer any action in Turner Valley until economic conditions improved, and A. Davison, the mayor of Calgary, read a resolution from Riley's fellow councillors expressing emphatic opposition to the continuing waste. To add to the welter of uncompromising positions, Calgary City Solicitor L.W. Brockington later reminded the Legislature that although the resolution passed by the city council deplored the economic waste in Turner Valley, Calgary had never agreed to any plan of compensation for restricting production under which the added cost would be divided between city consumers and the gas company.[8] Before the session's midnight adjournment, Brownlee also had to endure the stinging criticism of his old colleague Herbert Greenfield. The former UFA premier, now representing 13 of the independent companies, castigated the government on all counts: for delay and the lack of a definite policy, for trying to push a conservation program that would give a monopoly to one interest in the field, and for making alarmist statements about the short life of the field unless rigid controls were adopted.[9]

The next day's hearings continued to produce more heat than light. In desperation, the premier named President Wallace to chair a committee that included Arthur A. Carpenter, chairman of the Board of Public Utilities Commissioners, and Frank Fisher, and then dispatched them to Calgary in a final bid to reach a compromise solution. The resulting round-table conference in Calgary on March 22 seems to have floundered over the composition of the board to oversee conservation measures. Proposed was a board consisting of two representatives appointed by Imperial Oil, two independent operator representatives, and a fifth member to be appointed by the government. Convinced that Turner Valley operations should be carried on under private rather than government control, and that a government-appointed member would impair the board's impartiality, Imperial rejected the proposal.[10]

The only strategy left untried was direct intervention by the premier. Accordingly, Brownlee called representatives of the contending groups to meet with him in Edmonton on March 30. His efforts gained only a further heated exchange of views, climaxed by a fistfight in the corridor outside the meeting room.[11]

On April 2, the beleaguered Brownlee government put its draft Turner Valley conservation bill before the Legislature. Although passage was swift, sharp debate attended the bill at each stage. Calgary members were particularly vigorous in their opposition to the compensation clause which promised higher gas rates to southern Alberta consumers. In the end, Brownlee was not prepared to face the wrath of the southern Alberta voters mobilized by the Calgary City Council, and the controversial clause was dropped from the measure that passed the Alberta Legislature on August 6.

THE *TURNER VALLEY GAS CONSERVATION ACT* AND THE SPOONER CASE

With its new *Turner Valley Gas Conservation Act,* the Alberta government hoped to bring substantial order to the developing provincial oil and gas industry. The Act provided for the creation of a cabinet-appointed three-man board to be known as the Turner Valley Gas Conservation Board (TVGCB). Charged with promoting general conservation practice in the Turner Valley area, the board was also instructed specifically to achieve what for professional petroleum engineers had become the ultimate objective of sound reservoir management. Under the new legislation, the TVGCB was directed to "endeavour by means of negotiation to bring about the unified operation of the wells in the area by means of pooling or otherwise."[12] If consensus could not be reached by 1 October 1932, the Board was to submit its own scheme for cabinet approval. The Act also fixed natural gas production at not more than 200 million cubic feet per day and directed the Board to prescribe a daily rate of production for each well. So that it could proceed with prorated production, the new agency was given authority to take whatever measures were required to undertake a proper well-testing program. After it had ascertained the productive capacity of every well, and ordered production at what it considered appropriate rates, the Board was then to gather additional technical information, including the probable amount of gas in the Turner Valley field; the estimated immediate and prospective requirements for domestic and industrial gas consumers; the likely productive life of the field; and the methods that might be used to bring about efficient conservation in the production area. Finally, to encourage the co-operation of the smaller producing companies was the provision that the TVGCB hold a public inquiry before 1 October 1932 to review progress and to determine if the aggregate allowable production of natural gas needed to be adjusted and if there were any to whom payment ought to be paid on account of any detrimental consequences that might have happened as a result of Board activities under the Act.[13]

To accomplish these assigned responsibilities, the TVGCB was given the authority to hire a professional staff, the right to take whatever measures were deemed necessary to implement a thorough well-testing program, and the power to take evidence under oath. Most important was the additional right to formulate orders stating the total daily production allowed each well. The penalty upon conviction for noncompliance with Board orders was modestly set at not more than $500 and costs, with $100 for each additional day in default.[14]

In keeping with the UFA's penchant for open government, the TVGCB was directed to keep a "full and complete" record of its activities and to ensure that such records were available to the public.[15] Also, to accommodate the UFA's lean treasury, the cost of operating the TVGCB was to be defrayed by a levy on each 1,000 cubic feet of gas and each gallon of naphtha sold during each quarter.

His legislation having gained assent, Brownlee announced that, although the government had practically decided on two of the prospective Board members, he would be travelling to Ottawa to discuss with the federal Department of Mines who from outside Alberta might make a suitable third appointee.[16] Once pressed to action, the premier now moved with dispatch. On April 23, the Alberta government named Dr. Wallace, president of the University of Alberta, and Arthur A. Carpenter, chairman of the Board of Public Utilities Commissioners, to the new Turner Valley Gas Conservation Board,[17] just as the press had been predicting. The third member, John McLeish, was the "outsider" who Brownlee had sought in the East. McLeish was the director of Mines Branch of the Department of Mines in Ottawa, and like Wallace and Carpenter, he simply added the Board responsibilities to his existing commitments.

The Wallace appointment is noteworthy for it marks the University of Alberta's and particularly Henry Marshall Tory's, contribution to the Progressive's and the United Farmers of Alberta's ideal of regulation in the public interest. Tory, the university's founding president and Wallace's immediate predecessor, was dedicated to the vision of his university being actively involved in provincial affairs, including the search for practical solutions to problems confronting Albertans.[18] For two decades, he had worked to build an institution able to contribute professional knowledge, particularly in the area of applied science. Through his participation in the waste gas debate and by taking his administrative and scientific expertise to the TVGCB, Wallace was acting out Tory's ideal. He began a tradition that would see University of Alberta scientist-administrators exercise their talents in aid of oil and gas conservation.

The TVGCB held its first meeting six days later in the Legislature in Edmonton. After it was decided that Carpenter would be the chairman, the Board began the task of selecting a field staff. This was a critical matter, for the individuals chosen had to be men whose technical competence and knowledge of the industry in Alberta were without question. They had to be men who as individuals, if not as agents of a regulatory authority, enjoyed the respect of the segment of the industry that was opposed to government interference. The obvious primary candidate was Charles Dingman, an Albertan who had been closely associated with the industry for many years and who was well liked even by the "practical" oilmen who were inclined to disparage the theories and ideas of the professionally trained technical people. For proven technical competence, the Board also selected one of Dingman's colleagues in the Petroleum and Natural Gas Division of the Alberta Department of Lands and Mines, G.R. Elliott. Geologist J.O.G. Sanderson and engineers R.V. Johnson and F.W. Shelton were also appointed to work in the field under Dingman's direction. Further, it was decided that the Board's

Installing meter at Royalite No. 18 in the Turner Valley Field to accurately measure gas production. ERCB, Elliot Collection.

These photos—part of a series taken circa 1932 by TVGCB engineer, G.R. Elliott—are the only known visual account of early well-testing activities.

Collecting drip from spray on blowdown of British Dominion No. 2 in the Turner Valley Field. ERCB, Elliot Collection.

Board official, J.O.G. Sanderson, checks readings on a wellsite recording gauge (right) against that of a master gauge (left) at Royalite No. 7. ERCB, Elliot Collection.

F.W. Shelton changes meter chart at Royalite No. 9 well. ERCB, Elliot Collection.

Board engineer, R. Johnson (left), Mercury Oil's Mr. Cameron (centre) and Royalite's Mr. Phelps (right) gauge naphtha at the Mercury No. 1 well. ERCB, Elliot Collection.

TABLE 2.1 TVGCB Gas Flow Allowance for Naphtha-producing Wells
of Imperial Subsidiary Companies, Imperial Contract
Companies and Independent Companies

	No. of Wells	Gas Allowance (Mcf)	Assumed Corresponding Daily Naphtha Production (barrels)
Imperial subsidiary companies	32	79,181	655.6
Imperial contract companies	14	45,691	440.3
Independent companies	37	71,545	704.7
Total	83	196,417	1800.6

SOURCE: TVGCB, Minutes, Schedule to Order No. 1

office should be located in Calgary and that a stenographer and statistician would be required.[19]

At its next meeting on May 4 in Calgary, the Board issued its soon-to-be-famous "General Order No. 1," thus marking the first significant attempt to impose conservation upon the reluctant producers in the Turner Valley field. The order, to be effective from May 9, set a maximum field production of 200 million cubic feet per day, and then specified an allowable daily rate for each of the 83 naphtha-producing wells in the valley.[20] Resulting field production was divided between the Imperial subsidiary companies, the companies with which Imperial held production contracts, and the independent producers, as shown in Table 2.1. Lacking facilities for continual measurement of the gas flow from each well, the Board used as the controlling factor the gas-naphtha ratio determined for each well by the Alberta Department of Lands and Mines during February and March 1932. Set out on the order schedule, opposite the description of each well, therefore was the "Gas Allowance," the "Estimated February-March Gas Naphtha Ratio" in Mcf per barrel, and the "Assumed Corresponding Daily Naphtha Production," as shown in the examples in Table 2.2. The accompanying notice to operators informed them that the permitted daily production was subject to revision after the Board had the opportunity to conduct further testing.

As announced in the letter and contemplated in the *Turner Valley Gas Conservation Act,* the Board field staff immediately set about organizing a comprehensive well-testing program. For this

TABLE 2.2 Partial Schedule to TVGCB Order No. 1

Company	Well No.	Gas Allowance (Mcf)	Estimated February-March Gas Naphtha Ratio in Mcf per Barrel	Assumed Corresponding Daily Naphtha Production
Model Oils Ltd.	1	949	22	43.1
Okalta Oils Ltd.	1	1352	272	5
Okalta Oils Ltd.	*2&3	3855	299	12.9
Spooner Oils Ltd.	1	Bad condition	—	—
	2	Bad condition	—	—
	4	771	163	4.7
Royalite Oil Company Ltd.	6	6445	159	40.5
(Incl.#12)	7	1784	261	6.8
(Incl.#11−18−20)	8	12343	441	28
	9	762	261	3
	11	—	—	−incl. 8
	12	—	—	−incl. 7
	13	707	261	2.7
	14	5035	350	14.4
	15	—	—	—
	16	633	441	1.4
	17	2877	176	16.3
	18	—	—	−incl. 8
	19	4948	176	28.1
	20	—	—	−incl. 8
	21	1640	261	6.3?
	22	—	—	—
	23	6042	149	40.5
	24	1438	550	2.8
	25	834	157	5.3

SOURCE: TVGCB, Minutes, Schedule to Order No. 1
* If the owners wished it, two or more wells could be operated as a unit.

purpose, Frank Fisher was persuaded to remain in Alberta for another month to advise Board field staff. The nature and method of
the tests had been discussed at the conference held by Dr. Wallace
with the operators in Calgary just before the enactment of the conservation legislation, and a formula had been agreed upon by the
larger companies. Intended to yield data upon which an equitable
scheme of good production practice and conservation could be
implemented, these tests involved the metering of all wells for
a certain interval, the taking of their closed-in pressures during a
72-hour shut-in period, the taking of a 10-day naphtha production

test with the wells operating at two-thirds of the field closed-in pressure, and a test of gas flow taken at 500 lbs. pressure.[21] Before testing could proceed, however, one fundamental problem had to be solved. There were nearly 100 wells in the field, but none were metered; thus, the Board had to find 100 meters. In the end, only 25 could be obtained, which meant that it was necessary to divide the field into four areas, with tests to be undertaken in each successively. With testing ready to begin at last and Frank Fisher due to return to Ohio, the Board members, in anticipation of the reluctance of some of the smaller producers to co-operate, decided it would be prudent to engage the advisory services of an engineer who had wide experience in gas production, operating and pooling problems but no previous connection with Turner Valley. Frank McC. Brewster, the man chosen, was president of Belmond Quadrangle Drilling Corporation, Bradford, Pennsylvania, and came with high recommendations from both the Department of Mines in Ottawa and the head of the U.S. Bureau of Mines in Washington.

The idea that Brewster's presence might give the launching of the Board's testing program greater sanction was dashed almost immediately. Objection came first from an unexpected quarter. The Association of Professional Engineers of Alberta asked Board Chairman Carpenter to explain why a qualified Canadian engineer had not been hired.[22] Much more serious was the challenge of Spooner Oils Ltd. It registered its objection to "General Order No. 1" in court, and while the question of whether or not certain aspects of the *Turner Valley Gas Conservation Act* were within Alberta's constitutional competence was being debated, an interim court injunction put the Board's order and the well-testing program on hold.

The action of Spooner Oils bears witness to its desperate situation. Order No. 1 had the effect of reducing Spooner's normal naphtha production from approximately 90 to 4.7 barrels per day, thus threatening the company's ruin and the consequent loss of nearly $400,000 of the shareholders' invested capital.[23] Having succeeded with others for so long in delaying conservation measures in Turner Valley, the company—with its back to the wall—now sent its lawyers to attack the constitutional validity of the Conservation Act with a view to preventing the exercise by the Board of the powers conferred by the Act. A dozen equally threatened independent companies watched anxiously from the sidelines.[24]

On 24 June 1932, the Alberta court presented its judgement in favour of the TVGCB, thereby dissolving the injunction and permitting the Board to proceed with the enforcement of production allowances and field testing. Since Spooner Oils immediately appealed the lower court's decision to the Appellate Division of the

Alberta Supreme Court, however, great uncertainty continued to surround Board activities and initiatives. This accentuated the many difficulties that Dingman and his colleagues had to face in the field. It meant that the host of technical problems had to be addressed in a largely unco-operative and even hostile environment. Model Oils proclaimed that their No. 1 well was "an efficient well and [was] not wasting natural gas and should be exempted from conservation." The company warned that the well could not be shut in for testing without serious damage, for which the Board would be held entirely responsible.[25] Mercury, Miracle, Homestead Oils and others informed the Board that they would not comply with Order No. 1. This left little choice but the threat of legal action. Board Chairman Carpenter cautioned the McLeod Oil Company that, if the Board's order was not complied with at once, court action would commence and that "in view of the Company's attitude with regard to the Board's order, the maximum penalty, provided for by the Act, will have to be pressed for."[26] Summoned to police court for noncompliance, New McDougall-Segur Oil Company Ltd. agreed to submit to Order No. 1 "as far as it is practically possible," but added the proviso that it was doing so "without prejudice" to its right to recover damages for the one-third loss of production that the conservation order would entail.[27] Although not overtly hostile and obstructionist, the larger independent companies were convinced that their views were not receiving sufficient consideration and they warned Carpenter that their co-operation might be withdrawn.[28] Hoping to ease the mounting pressure, on July 19 the TVGCB abandoned its controversial gas-naphtha ratio, as the critical factor in determining allowable production, in favour of a simple percentage of each well's gas flow potential, calculated to ensure that aggregate production did not exceed the 200 million cubic feet per day provided for in the Act.[29] Opposition to Board measures continued nonetheless and Imperial Oil added its name to the ever-growing list of companies complaining of grievous injury and seeking special consideration. In early August, Imperial informed the Board that, since Model Oil's No. 1 well was continuing to produce at full capacity in defiance of Order No. 1 and was rapidly draining Imperial's larger adjacent property, it would operate its nearby Foothills No. 1 well in the same manner as its competitor.[30] Moreover, to make up for the loss that it claimed to have suffered, Imperial stated its intention to continue producing Foothills No. 1 at its maximum rate for at least three months beyond whatever time the Model well might be brought under the Board's conservation order. Imperial persisted in its course despite Carpenter's admonition.

As TVGCB engineers struggled in the field to bring conservation to Turner Valley, lawyers from the Attorney General's department battled on the Board's behalf before the Supreme Court of Alberta. The argument before Justices Harvey, Clarke, Mitchell, Lunney and McGillivray centred on five questions: whether or not the Act as a whole was outside the constitutional powers assigned to provincial governments; whether or not parts of the Act encroached upon the exclusive right of the federal government to regulate trade and commerce; whether or not the Alberta Legislature had improperly undertaken to clothe the lieutenant-governor-in-council with the authority to appoint judges, to assign them judicial duties, and to endow them with judicial capacity and power; whether or not the levy charged to meet Board expenses was really a form of indirect taxation, and thus beyond the powers granted to provincial governments; and whether or not the Act was a violation of the terms of the statutory agreement respecting the transfer of natural resources from the Dominion to the province of Alberta, and thus not applicable to the land held by Spooner Oils.[31] Of special concern to J.J. Frawley, counsel representing the Attorney General, was Spooner Oil's subsidiary allegation that Board Order No. 1 was inappropriate because it presumed wrongly that the ratio of gas production to naphtha production was constant at all pressures.[32] Evidence in support of this assertion was submitted in the form of an affidavit from Clarence Snyder, who was recognized as perhaps the most experienced driller in Turner Valley. Frawley anticipated that this was intended as the basis of an argument claiming Board Order No. 1 could cause permanent injury to Spooner's wells. It is probable that this was one of the factors that led the Board to abandon its much questioned gas-naphtha ratio as the critical allocation factor even before the case went to trial. When the judges' decision was presented on November 14, the Brownlee government was relieved to learn that its beleaguered conservation agency had survived the test. The Appeal Court found in favour of the TVGCB on all counts but one, it held that the provision of the Act imposing a levy upon operators to meet the expenses of the Board was in fact an indirect tax and therefore *ultra vires*. Since no charge had yet been exacted, this restriction caused the Board no great inconvenience.

The feeling of relief, however, was short-lived. Almost immediately, Spooner Oils appealed the decision of the Alberta court to the Supreme Court of Canada. With the matter not finally settled, at least in the eyes of the independents, it meant continued resistance to Board activities in the field and continued pressure on the government to withdraw the offending legislation. This pressure increased through the autumn, as the TVGCB, in keeping with its obligation under the Act, went about organizing a public inquiry to be held in Calgary from January 10 to 13.

At the inquiry, much attention was devoted to the question of gas reserves. Dr. Oliver B. Hopkins of Imperial Oil estimated that, to 400 psi abandonment pressure, the field would ultimately produce 600 billion cubic feet; B.F. Hake, representing the largest group of independent operators, estimated 723 billion cubic feet; Stanley J. Davies, speaking for another group of independents, projected 476 billion cubic feet; whereas S.E. Slipper of the Canadian Western Natural Gas Co. presented 470 billion cubic feet as the likely reserve. The Board also heard various proposals for a more relaxed approach to conservation. Slipper was one of the few to speak in favour of the Board's policy of restricting the production of low-pressure wells, capable of producing only negligible quantities of naphtha, in favour of higher-pressure wells that achieved substantially greater and thus more efficient production. Also presented for the Board's attention was a resolution containing the names of 24 companies asking for the removal of the regulation restricting the drilling of wells into the limestone.[33]

In light of the evidence presented at its public inquiry in Calgary, the information acquired through its testing program, and its field experience over the preceding months, the Board prepared its report. Submitted to the government on 22 February 1933, it offered little that was new. The Board estimated the amount of gas available in Turner Valley to 400 psi abandonment pressure or above to be a minimum of 512 billion cubic feet, which at the current rate of production meant a seven-year supply.[34] This came with the caution that, until the entire field was put on a metered basis, no really good estimate would be forthcoming. With Calgary's requirements estimated at approximately 10 billion cubic feet per year, it was apparent that there was only an adequate long-term supply if waste burning was sharply curtailed.

In its recommendation of what was to be done, the Board also went over the same wearisome ground. One of the responsibilities assigned to Carpenter and his colleagues had been to promote voluntary pooling so that there could be unitized operation of the field. In this, they had failed. As the report explained, there were two basic problems: essential reservoir data was deficient and the operators remained unable to reach agreement on any plan. Since the Board was not prepared to recommend that unitized operation be brought about by compulsion, "conservation by pooling" was impossible. Instead, the Board recommended "conservation by proration," and presented a plan that it felt would make sure that the assigned gas quota would be used to ensure the greatest amount of naphtha production under prevailing conditions and to promote the efficient operation of the field.[35] The proposal called for 70% of the allowable to be apportioned according to acreage and available pressure (that is, potential production from the acreage) and 30% according to naphtha production. Essential to the proper working of this, or any plan of field management, was the accurate monitoring of each well. The government was advised therefore to insist that each well be equipped with the necessary meters and pressure gauges and to enforce strictly the taking of measurements in a systematic fashion.

From the outset of the conservation debate, the provincial government had tried to overcome the resistance of small producers with the assurance that it was prepared to work out some kind of compensation scheme for those who might be affected adversely. As with most other aspects of conservation, this turned out to be a complex issue, and in the end it was left as one of the questions to be studied by the TVGCB. In their report, Carpenter and his fellow board members also left the matter unresolved, stating that the matter could not be discussed intelligently until a conservation plan had been in effect for at least a year. They recommended that compensation be left for the consideration of another body selected specifically for that purpose.

Unable to ease the uncertainty on the compensation question, the Board in its report did offer relief on an even more contentious issue. As part of its program to control excess gas production, the Board had put a ban upon the drilling of new wells into the Turner Valley limestone formation. Knowing that the production of additional gas would only complicate the problem and result in the lowering of production quotas to existing producers, this seemed a sensible restriction to the Board. The Board's position was based also on the knowledge that crude oil had been found in three different horizons at depths much shallower than where the naphtha-carrying gas was met in the limestone. The Board's attempt to direct the search for oil to these shallower horizons, however, was seen by the smaller producers as a "dictatorial" measure, and they had protested vigorously. To win over the small independent companies, the TVGCB was prepared to retreat on this issue, and its report recommended that the drilling restriction be ended, noting that the drilling of additional wells would not "materially affect the total allotment of gas for the field" under the plan submitted.[36]

The report submitted by the TVGCB was very much a reflection of the environment in which it was produced. Since its creation in April 1932, the Board had made limited headway. The industry remained divided, with a core of smaller independent companies adamantly opposed to any kind of regulatory interference. Uniform cooperation was never achieved, and the Board found it next to impossible to enforce compliance.[37] Only in its testing program was the TVGCB largely able to achieve its objective. In the face of great difficulty, Dingman and his colleagues in the field were able to offer the government more authoritative information on reservoir conditions and on available reserves of natural gas. With the submission of the Board's report in late February 1933, the UFA government was once again made aware of the need for conservation in Turner Valley and of the kind of practical adjustments necessary to implement a workable conservation program. The question, once again, was whether or not Brownlee and his cabinet had the political will to move ahead. Like the report itself, the response of the Alberta government was shaped by the near desperate environment in which Brownlee found himself in the spring of 1933. The provincial economy hovered on the brink of collapse, and nowhere was this more apparent than in the oil and gas sector. Market demand for Alberta natural gas, which had dropped by 30% since 1930, was still declining.[38] Naphtha production was also in retreat, and had dropped a staggering 40% to 810,000 barrels during the fiscal year 1932–33. Although this was a consequence of TVGCB conservation measures, as well as the declining productivity of Turner Valley wells, and was more than a million barrels short of Alberta's yearly consumption, the growing glut of cheap oil at the Alberta-Montana border kept prices depressed. The light and heavy crude oil situation was equally dismal. Heavy crude-producing wells at Ribstone were abandoned by their owners, and operations in the Wainwright heavy oil area had come to a complete standstill.[39] A real measure of the lack of confidence and the gloom hanging over the industry was apparent in the decline of drilling activity. The total well footage drilled in Alberta in 1930 was 295,752 feet, in 1931 the total was 127,243 feet, and in 1932 it was a mere 40,492 feet.[40] Development in the industry had almost come to a stop, and this bore heavily upon the Alberta government as it brought the TVGCB report before the Legislature's Agriculture Committee for hearing.

In early March, just as the hearing was about to begin, a group of independent Turner Valley operators issued a statement that added the final critical elements to the stage being set for the debate in the

Legislature. The statement claimed that, if the Board's plan were put into effect, 13 of the small independent companies would have to cease operations and another 10 would follow shortly after. Arguing for the primacy of the "law of capture," the independents charged that:

> Even if the Board could determine the specific amount of gas under a certain area, it has erred in allowing to each leaseholder a specific portion of that gas, as it has overlooked the right of capture. No leaseholder has rights to production from his lease in Turner Valley or elsewhere until he has drilled that lease.[41]

When TVGCB Chairman A.A. Carpenter appeared before the Agriculture Committee to give a summary of his Board's report on March 7, he was immediately asked if the liquidation of some Turner Valley oil companies would result if the proposed gas conservation plan went ahead. He was compelled to admit that this was likely.[42] Despite the counter that without conservation many more companies would face early liquidation, Carpenter, Wallace, Brewster and other proponents of efficient production were unable to escape from the defensive position in which they had been caught. A familiar series of speakers and criticisms followed. Opponents charged that, if the scheme went ahead, it would be a serious blow to the provincial economy. Resurrecting an old bogey, Harvey Price, of East Crest Oils, warned that the Board plan would give "Imperial Oil Company a chance to kill out the smaller companies"[43] Others, not prepared to dismiss the idea of conservation completely, spoke in favour of a simpler, if less technically sound, approach. Their proposal was that every well in the field should be allowed to flow at 40% of its open flow capacity, which meant that wells with inferior gas-oil ratios would not be penalized with reduced quotas. This idea was given an immense boost when J.H. Ross, the Calgary alderman who had long led the fight to curb the flaring of "waste" gas, announced his approval of the percentage of open flow approach.[44]

The government's response to the TVGCB report and proposal came in the form of a motion presented by the premier to the Legislature on April 6. Brownlee stated his government's conviction that the work of conservation begun by the Board in Turner Valley had to be continued, but his accepted motion declared that this would be carried on under the direction of the Petroleum and Natural Gas Division of the Department of Lands and Mines. The motion provided further that the total gas production of all wells in the field would be restricted "as nearly as possible" to 240,000,000 cubic feet per day. In addition, all gas wells were to be equipped with meters to measure production, and the regulations prohibiting the drilling of wells into the dolomite limestone were rescinded.[45] Being particularly sensitive to the fragile state of the provincial economy and worn down by the aggressive unrelenting opposition of the independents, Brownlee had gone most of the way to meet their demands. If, however, the result was considerably short of what the Board and their technical experts—Fisher, Brewster, Calder and Dingman—would have held desirable, the essential principle of conservation had been established and the metering of all gas wells, the critical first essential of any effective system of field supervision, had also been achieved.

As those Turner Valley producers with less efficient wells began to increase their gas production to 40% of capacity from the roughly 25% to which they had been restricted formerly,[46] and others planned new wells into the limestone, Dingman and his colleagues tried to re-establish the direct authority of the Petroleum and Natural Gas Division of the Department of Lands and Mines in the field. Their efforts to implement the government's latest attempt at conservation in Turner Valley, however, were barely under way when a new setback had to be faced.

On 3 October 1933, the Supreme Court of Canada finally delivered its decision on the Spooner appeal. Although agreeing with the lower court that the *Turner Valley Gas Conservation Act* did not deal with matters beyond the scope of powers assigned to the provinces, to the great dismay of the conservationists, the Supreme Court held that the Alberta Act was invalid insofar as it related to lands held by Spooner Oils.[47] The seeming paradox in the court's decision—that the province possessed the constitutional authority to enact conservation legislation but not the right to impose conservation measures upon Spooner Oils—arose from the particular manner in which Arthur Spooner had acquired his petroleum leasehold. Spooner was the holder of a lease from the dominion government granted under the Department of the Interior's petroleum and natural gas regulations of 1910 and 1911.[48] Section 2 of the agreement that transferred the control of public lands and resources from Ottawa to each of the prairie provinces in 1930 provided that the provinces would honour the terms of contracts relating to lands, mines and minerals drawn earlier with the federal government and would not "affect or alter any term of any contract to purchase, lease or other arrangement by legislation or otherwise."[49] The chief justice of the Supreme Court concluded that the regulations created under the authority of the *Turner Valley Gas Conservation Act* concerning the production of natural gas and naphtha did "affect" the terms of Spooner's lease and, as such, were invalid.

The implications of the Supreme Court decision were far-reaching. Not only were the conservation regulations struck down insofar as they related to the Spooner leasehold, but also they could not be made to apply to any of the old petroleum and natural gas leaseholds acquired under the 1910 and 1911 federal regulations. This meant that a significant portion of the acreage comprising the Turner Valley oilfield, much of which was held by the smaller independent companies, lay beyond the control of the provincial government, at least as far as imposing any kind of production control was concerned.

Minister of Lands and Mines R.G. Reid attempted to play down the impact of the Supreme Court decision by insisting that the government's gas conservation policy would not be affected. He argued that the ruling was with reference to the *Turner Valley Gas Conservation Act* and was not against the principle of gas conservation. "We are now proceeding," Reid announced, "under the general laws of the province and by the *Oil and Gas Wells Act*, we get around the point raised in the supreme court decision."[50] It was plain to almost everyone else that there would be no "getting around" the decision. The *Calgary Herald* observed that the government could be faced with numerous civil actions for loss of revenue if it tried to enforce production restrictions, and therefore, insofar as Turner Valley was concerned, gas conservation was "a dead issue."[51] The assessment was correct, but for reasons that were more complex than the paper might have imagined. The UFA government was preoccupied and tired. What energy it could muster to sustain its vision of regulation had shifted to focus almost entirely upon broad issues of economic planning intended to counter the Depression.

INDECISION AND DRIFT

With the Spooner decision, Calder and Dingman and their colleagues found themselves thrust back almost to where they had been two years previously. Their situation was not unlike that faced by others trying to impose basic conservation measures upon the petroleum industry and was symptomatic of a struggle that was being waged through much of the continental interior between Calgary, Alberta, and the Texas Gulf Coast. Contending interests, both private and public, which were constantly before the courts in Texas, Oklahoma, Kansas, California and Washington, as well as in Alberta and Ottawa, were being drawn into a process that was beginning to define the balance between private rights and the public interest in the production of oil and gas. The complexity of the issues and the tenacity with which the opponents of regulatory interference resisted before the courts and in the political arena meant that the decisions of the judges and the legislation of the politicians were often contradictory and that, above all, the process would be slow. In April 1931, the Texas Railroad Commission (TRC) issued its first proration order, only to have it struck down four months later by the federal court on the grounds that the real intent of the measure was not to prevent waste but to fix prices. It would take two more years of chaos and embittered struggle in the courts before the Railroad Commission's legal authority to set production quotas was confirmed. It was not until 1936, in a series of judicial decisions known collectively as the Clymore case, that Texas courts backed TRC orders to stop flaring, and not before an estimated billion cubic feet of gas was being burned or dissipated into the atmosphere each day.[52] Success came a little quicker in Oklahoma. In 1932, in a landmark decision, the U.S. Supreme Court held in the Champlin case that production quotas based upon market demand was "a reasonable method of preventing physical wastes, and that the effect on price, if any, was incidental."[53] Observing the outcome, Kansas amended its statute, using the same language as Oklahoma to define prohibited wastes, thus putting proration programs on solid legal footing in Texas, Oklahoma and Kansas. With these decisions, the courts also gave authority to a new concept of the oil pool as consisting of three resources: oil, gas and reservoir energy. Whereas the first two were considered as belonging to the owners of the mineral rights, reservoir energy had to be considered as a common attribute of the pool. The focus of private rights remained centred upon gas and oil, and common or public rights increasingly, though not exclusively, became associated with reservoir energy.

In Alberta, the prospect of a renewed effort to find some means of addressing the "waste" gas problem had all but vanished. Premier Brownlee, whose approach to Turner Valley conservation matters had been cautious at the best of times, was overtaken by a serious personal scandal, which politically neutralized him and forced his resignation in July 1934. Just a few days before the Spooner decision, Alberta newspapers headlined the sensational story of a legal action launched against Premier Brownlee by Vivian

MacMillan and her father. The young woman accused Brownlee of seduction and claimed civil damages.[54] Preoccupied with the revelations of the Brownlee trial and the mounting political threat posed by the sudden emergence of the Social Credit movement, the UFA government put the Turner Valley waste gas problem near the bottom of its list of political priorities.

Left essentially on his own, William Calder, director of the Petroleum and Natural Gas Division, did his best to promote sound production and development practices through the medium of the "Drilling and Production Regulations" established under the authority of the *Oil and Gas Wells Act* of 1931. His difficulties, frustration and commitment to the conservation cause are all apparent in his first annual report following the disbandment of the Turner Valley Gas Conservation Board. Calder explained to the members of the Alberta Legislature that "owing to a depleted staff and limitations of expenditure, the usual annual inspections of gas wells by divisional staff had to be omitted" and "the only information available respecting the condition of wells [was] that submitted by the operators."[55] Such information, Calder noted, was unlikely to include essential details on defective wells. Although acknowledging that there had been a welcome reduction in the daily waste of gas at Turner Valley, Calder was not prepared to let the matter rest. He insisted that there was "still ample evidence that further curtailment was imperative."[56] According to the report, a majority of the Turner Valley operators were exceeding their gas quota withdrawals. In addition, Calder warned that the newly completed Royalite absorption plant would mean only a slight increase in the efficiency of naphtha production, for the independents were still dissipating enormous quantities of recoverable naphtha into the air with the accompanying gas. Given the situation, the director could not resist editorial comment.

> Inasmuch as the present wastage of gas entails a considerable loss in naphtha to the independents, it is strange that a collective agreement is not arrived at whereby all operators would agree to tube their wells and voluntarily demand a reduction of the allowable gas withdrawals to 25 percent. In the event of operators failing to make such arrangements the only alternative available would be the full enforcement of Section 28 of the Drilling and Production Regulations so that all wells would be tubed to reduce the gas/oil ratio.[57]

The roundabout appeal to the government for direction to enforce the regulations fully did not bring a response.

Success had been achieved in one area. Calder was able to state that all Turner Valley operators had complied with the requirement to install production meters. This had not been achieved, however, until the seriousness of the department's intent had been revealed by shutting in and sealing two wells for noncompliance. But even this modest success was not without its cost. It revealed the director of the Petroleum and Natural Gas Division to be the most important remaining obstacle to unfettered activity by the most recalcitrant independents. The resulting campaign to get rid of this "conservation enthusiast" eventually surfaced in the Alberta Legislature in early 1935. Labelling Calder as "the most tactless man I ever knew," Calgary Conservative MLA Hugh Farthing warned that it was no time to harass the industry in the current depressed climate and that the "autocratic rule of William Calder" should be ended.[58] It was not the UFA government, however, that was destined to decide Calder's future.

On the 22 August 1935, Brownlee, R.G. Reid (his successor as premier) and every other UFA member of the Legislature were thrust from office. Offended by the Brownlee scandal and frustrated with the failure of the United Farmers of Alberta to deal effectively with the hardships caused by the Depression, Albertans turned en masse to William Aberhart, giving his Social Credit Party 56 of the 63 seats in the Legislature.

Despite the remarkable majority, action to address Alberta's problems was slow in coming. Admittedly, the problems were not ones that could be solved readily, but beyond this the Social Credit party had not developed detailed proposals on how it intended to address most of the difficulties faced by the province. Rather, the focus of Social Credit energies had been upon educating the Alberta public on the principles of social credit monetary theory. Aberhart had told his audiences that they had only to appreciate the fundamental truth revealed in the theory, and then trust in an elected Social Credit government to find the experts to put the theory to practice. As Aberhart explained to his audiences,

> You don't have to know all about Social Credit before you vote for it; you don't have to understand electricity to use it, for you know that experts have put the system in, and all you have to do is push the button and you get the light. So all you have to do about Social Credit is to cast your vote for it, and we will get experts to put the system in.[59]

In consequence, Social Credit had no particular plans about what to do in Turner Valley. Beyond this, since every Social Credit member was new to the Legislature, there was no one who carried with them knowledge of the complex waste gas issue gained in previous legislative debate or in hearings before the House Agriculture Committee. In addition, before Social Credit could get organized to deal with this and other issues, the party had to organize a by-election to get their leader elected to the Legislature. Although Aberhart had organized and led Social Credit to an astonishing victory, he had sought to be seen as being above the political fray and therefore did not run for a seat himself.

The drift that attended the party's delay in formulating a petroleum policy in all probability would have lasted even longer but for an earlier chance meeting between William Aberhart and Charles C. Ross on a train to Vancouver in late July 1935. Aberhart discovered that Ross was a mining engineer who, before the transfer of natural resources to Alberta's control, had been the senior federal officer in the Calgary office of the Department of the Interior's Petroleum and Natural Gas Division. During the course of the journey, the two became acquainted, and Aberhart later persuaded Ross to stand for office.[60] There was, however, one hurdle to be surmounted. Before departing for his Vancouver holiday, Aberhart and his advisory board had already selected the province-wide slate of Social Credit candidates. Since these individuals were now in the midst of campaigning, Aberhart waited until after the election to name one of his hand-picked followers to step down and thus allow Ross to be named to cabinet pending confirmation in a by-election. Ross was therefore among the group sworn in as the first Aberhart cabinet on 3 September 1935. By-elections held subsequently in the constituencies of Okotoks-High River and Athabasca on 14 November 1935 returned Aberhart and Ross respectively by acclamation.

Ross began his career as Social Credit minister of lands and mines with an important advantage, at least in the eyes of Calgary oilmen directly concerned with Ross's portfolio. He had earned the respect of the industry as a federal administrator, and as a latecomer to the Social Credit party he was not identified as one of the enthusiasts who had stumped the countryside in support of Aberhart and monetary reform. It was public knowledge that Ross had agreed to join cabinet on the understanding that he did not fully accept social credit monetary principles.[61] Thus, while the new Social Credit government was looked upon with great apprehension, Ross was the one individual in the Aberhart cabinet who enjoyed some credibility among Calgary oilmen. Perhaps wishing to build upon this base of goodwill and to fix himself firmly in control, Ross im-

mediately sought the resignation of William Calder, his old col-
league from Department of Interior days and now director of the
Petroleum and Natural Gas Division, and replaced him with
Charles Dingman.[62]

As Calder's successor, Charles Dingman was left to carry on the
conservation struggle with Calgary oilmen while Ross endeavoured
to make some headway on petroleum matters with his fellow cabinet members. This proved an almost impossible task. Through the
rest of 1935 and early 1936, the government was almost totally
preoccupied with steering the province clear of bankruptcy and
with an involved debate about how social credit measures might be
introduced. The essential nature of Ross's difficulty was made omi-
nously apparent in late November 1935. After weeks of careful
preparation in industry and government circles, Ross advised
Turner Valley operators that, as of December 1, a schedule of reduced gas production quotas would take effect. The reaction of the
small opposing group of independent producers was swift and predictable. They were able to launch a successful "twelfth-hour" intervention, the nature of which was later revealed to J. Harvie, deputy minister of lands and mines, by the still confused and resentful
Charles Dingman. Explaining the curious sequence of events between 9:30 a.m. and 9:15 p.m. on November 28, Dingman wrote:

> On the morning of the 28th ultimo at about 9:30, the premier
> telephoned me and I visited his office. Mr. Becker who was there
> previous to my arrival stated that he had erected a plant in
> Turner Valley capable of processing five million cubic feet of gas
> per day, which was dependent entirely upon the gas from the
> Anaconda No. 2 well, the revised quota of which was 2,685,000
> cubic feet per day and that this quota would put his plant out of
> business. It might be noted in this case that his previous quota
> was 4,188,000 cubic feet and that if he were at all familiar with
> the field he would know that the wells, particularly in this loca-
> tion, are becoming less productive regardless of any quota issued
> by the Department.[63]

Dingman explained that he went on to inform the premier and Bec-
ker that the revised quota was based upon the maximum production that this department's engineers felt could be allowed if the
field ever was to have any degree of protection from "extremely
wasteful overproduction." After hearing the explanation, Aberhart
directed Dingman to telephone Ross in Ottawa and ask the minister
if he thought it advisable to postpone putting the revised quotas

into effect. Ross's response was that he had been in communication with a number of Turner Valley operators and had received no adverse criticism. He did not consider it necessary therefore to postpone going ahead with the new quotas. When informed of Ross's reply, Aberhart expressed his concurrence. In the afternoon, Dingman and Aberhart received a telegraph message from the *Western Oil Examiner* in Calgary stating that there was much adverse criticism regarding the planned revision. The editor of the *Examiner* warned that "small independent companies which have not been forced into selling out to Imperial will in some instances likely be compelled to do so under restrictions of production effective December 1," and appealing to Aberhart's partisanship went on to say that the independent operators felt that "they were discriminated against by [the] old government through ignorance or worse" but hoped the Social Credit government would give more considerate treatment to an industry that was creating wealth from the province's natural resources.[64] Moreover, the telegram assured that the heavy gas flow from the latest Royalite discovery set at rest any danger of a gas shortage. When Ross called Dingman in the evening, the telegram was discussed, and once more the minister expressed his belief that the criticism was limited to one or two individuals. Next, as Dingman related in his letter to Harvie, "the premier telephoned me stating that he had been in communication with Mr. Ross, giving me a message from Mr. Ross to postpone the application of the revised quotas until his return from Ottawa."[65]

The protesting operators organized a write-in campaign, and in the interval before Ross returned Aberhart received numerous letters from royalty holders, saying that conservation would mean the end of the small income upon which they were dependent during these difficult times, and from workers who feared for their jobs.[66] One man advised, "now is the time we need to use wisdom, and the talents God gave us; to help ourselves to the Riches that is [sic] given us in the resources of this wonderful province of ours. Should we preserve it in the earth and fold our hands, and let ourselves and families Perish [sic]. Why in the name of all that is sane should we preserve unlimited Riches."[67] On his return, Ross found Aberhart unwilling to proceed with gas conservation, and gas flow in the field continued nearly wide open.[68]

Ross and Dingman soldiered bravely on. Dingman began once again the familiar and now tedious ritual of trying to persuade the independents that conservation would mean little overall reduction in naphtha production in the short run and much greater production in the long run. In a detailed report sent to all operators and

MLAs in January 1936, he reviewed Turner Valley production history from 1928, calling particular attention to field experience during 1932. With the aid of a graph, Dingman showed that when the field was heavily produced in April 1932, 436 million cubic feet of gas per day produced 2,609 barrels of naphtha (see Figure 2.1). In contrast, during the last five months of 1932, when the short-lived Turner Valley Gas Conservation Board was restricting gas production, an average daily withdrawal of 224 million cubic feet produced 2,073 barrels of naphtha per day. In other words, the recovery rate in barrels per million cubic feet of gas increased by about 59% while bringing a drop in naphtha production of only 20%.[69]

Dingman's report received wide attention but changed nothing; the balance of forces remained the same. Although the majority of informed observers and most of the large producers found the evidence convincing, this was a group that had been convinced for years. On the other hand, many of the smaller, so-called "practical" oilmen, for a variety of reasons, many of which had little to do with the essential principle of conservation, remained opposed, and their numbers had been bolstered as the Depression wore on by oilfield workers who feared for their jobs or sought employment.

Through the spring, pressure to address the problem began to mount inside and beyond Alberta's borders. According to an editorial in the financial pages of the Saskatoon *Star-Phoenix*, the situation defied common sense. It noted that Alberta's director of the Petroleum and Natural Gas Division had presented an "unanswerable argument that controlled flow increases the naphtha yield and conserves the gas pressure and the life of the valley," and concluded the indictment with the question, "When is 'Charlie' Ross going to act?"[70]

It was not until June that Ross felt ready to try to tackle the conservation question once more. On June 19, both Aberhart and Ross spoke to a large audience assembled at Black Diamond in the heart of Turner Valley oil country. The premier spoke to reassure Albertans that their government would persevere despite the opposition of the bankers and the federal government. Ross courageously took the opportunity to speak of the urgent need for conservation. He decried the continuing waste, pointing out, that since 1912, "$104 million worth of gas was wasted to produce $23 million worth of oil."[71] With the premier sitting at his side, Ross noted that, "as last year's quotas were not put into effect," the field had been operated wide open, pressures had dramatically declined, and recovery had dropped to 212 gallons of naphtha per million cubic feet of gas, which was only one-third the recovery obtained when

FIGURE 2.1 Turner Valley Production History, 1927–1936

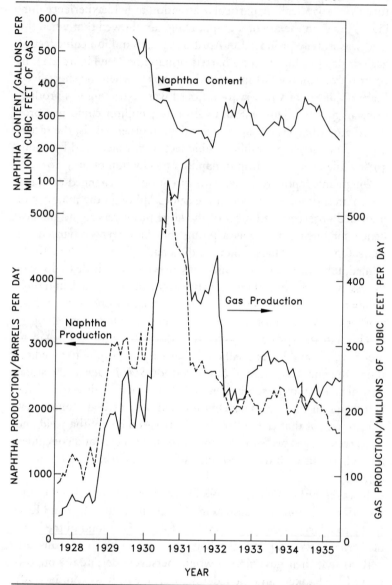

SOURCE: Adapted from Provincial Archives of Alberta, Premiers' Papers, 74.101, 550a, January 1936.

the field was new. He reminded his audience that, when Turner Valley gas was no longer available, "the consumers of gas would be called upon to pay in cash for the wastage." The minister concluded by saying that he was happy to acknowledge the encouraging new discovery made at the southern end of the valley, but nonetheless he wanted "to issue a warning that it [did] not change the picture as far as the Turner Valley gas supply is concerned to any great extent."[72]

The encouraging new discovery to which Ross referred was the Turner Valley Royalties No. 1 well. Having blown in on June 16, the well produced nearly 3,000 barrels of crude by the time of the Ross speech, and production was continuing at 900 to 1,000 barrels per day. On June 20, Ross pronounced the strike to be "the most important development in Turner Valley since Royalite No. 4 penetrated the limestone in 1924."[73] The Royalties well was the nearest thing to a gusher yet drilled in Alberta, and the excitement soon spread across the country, moving the *Financial Post* to proclaim "one may begin to think of Turner Valley as something a little more than a gas field."[74] Drilled into a deeper part of the Rundle formation on the southwest flank of the Turner Valley oilfield, the new well gave an immediate and significant surge to oil exploration. By September, daily crude production in the valley reached 10,000 barrels, well beyond the capacity of the local market. As production continued to expand, crude prices began to fall, and independent Calgary oilmen approached the federal government to help find a way to open markets for Alberta oil beyond provincial boundaries and to examine the continued importation of Montana crude oil into Alberta.[75]

Throughout the summer months, Ross had continued to preach the message of conservation, but the problem had become more complex. To the vexed issue of natural gas waste was now added an emerging oil glut. The classic oilfield problem of overdrilling and overproduction was beginning to unfold. When oil was discovered under one property, those with petroleum rights on adjoining acres immediately began drilling to prevent "their" oil from being "captured" or drained away. Incentive to drill new wells existed, even though the glut promised to get worse and crude oil prices were declining. Despite this environment, the smaller independent producers remained mainly opposed to production controls. Most could not afford to delay return on their invested capital, and in the case of the small Alberta oilmen, they remained ever optimistic that new markets could be found. This point of view was also supported by certain groups of investors outside the province, and like the locals they aggressively pressed their message upon the Alberta government. Such a group, visiting Calgary to assess Turner Valley prospects in the autumn of 1936, warned that government interference would halt further development.[76] In Toronto, the *Star* cautioned that "any action resulting in restricting the flow of oil in Turner Valley would be a national mistake of the first order."[77] Such comment, however, had little impact upon the minister of mines and minerals. Ross remained convinced that conservation in the form of production controls had to remain the priority concern. The premier, on the other hand, was becoming ever more concerned that conservation might be a threat to his first priority—development and jobs.

It was not that Ross was opposed to development. Under his initiative, a number of measures had been taken to encourage exploration. Companies were offered larger leaseholds at more nominal rents.[78] In defined "wildcat" districts, the driller of a discovery well was offered a rebate on production royalties, and as a special concession to small operators with limited capital the minimum lease area was reduced from 160 to 40 acres.[79] Ross, however, did see conservation as an essential component of any orderly program of oil and gas field development. This conviction rested upon nearly 20 years of professional experience in the Turner Valley area. It was under Ross's initiative that the federal government had tried to push ahead with conservation measures in 1925. Ross had also been a key member of the team that had advised the Alberta government on suitable petroleum and natural gas regulations on the eve of the resource transfer. Having been in the front ranks of those fighting for conservation in Turner Valley for so long, neither his personal nor his professional integrity could now allow compromise.

In contrast, Aberhart looked upon the matter as a politician. Activity in the Turner Valley oilfield represented almost the only bright spot in an otherwise bleak provincial economy. As one who had solemnly pledged to end "poverty in the midst of plenty," Aberhart was not prepared to chance any measure that could put development at risk. His inclination in this direction was strengthened by his bias in favour of the "little man." The cry of the small producers, that quotas would benefit the big operators, especially Imperial Oil, at their expense, seems to have weighed heavily. Aberhart saw the big eastern banks holding Alberta in bondage, and it was not hard for him to see the Toronto-based Imperial Oil Company in the same light.[80]

Differences between the premier and his minister of lands and mines remained irreconcilable. When he learned in late December that Ross was planning to shut in certain Turner Valley gas wells, Aberhart asked for and received his resignation.[81] There was still a want of political will in the premier's office. Faced with what he perceived to be a choice between development and conservation, Aberhart chose to let matters drift, thereby following the pattern of expediency established by his predecessors Greenfield and Brownlee. As always with conservation issues, it narrowed down to whether short- or long-term perspectives gained dominance.

THREE

Founding the Petroleum and Natural Gas Conservation Board, 1937–38

NATHAN ELDON TANNER and his family returned to Cardston from Edmonton in the autumn of 1936 after the conclusion of the second session of the Alberta Legislature. The months in the capital had been trying ones for Tanner, not only had there been little success in turning social credit theory into practical legislation but also growing tensions within the Social Credit party had made his task as novice Speaker of the assembly even more arduous. Tanner's well-earned rest, however, was short-lived. Late one evening, just before Christmas, Tanner was awakened by a telephone call from the premier. Aberhart wanted to know if Tanner would like to accept an unidentified cabinet position. The conversation was followed several days later by a wire asking Tanner to come to Edmonton "as expeditiously as possible" to assume the position of minister of lands and mines.[1]

The new minister inherited a difficult situation. Although the more substantial elements of the business community and petroleum industry were uneasy about Aberhart and his followers, the presence of Charles Ross in the cabinet had been moderately reassuring. With Ross's departure, the bigger oil companies feared the worst, and the statement of the new minister, just minutes after he was sworn in on 6 January 1937, did little to enhance his credibility in the eyes of the industry. Tanner's introductory announcement that his policy would be "to develop the natural resources and industries within the province for the benefit of the people as a whole," seemed to reflect Aberhart's suspicion of big business.[2] The *Edmonton Bulletin* immediately drew a connection between Tanner's remark and the sentiments expressed previously at the Edmonton district Social Credit convention. Delegates had censured

C.C. Ross, the former minister of lands and mines, and passed a resolution declaring that "outside capital was not required" and "should not be allowed to participate" in the development of Alberta's resources.[3]

Even though the small producers could relax, knowing that they had been successful once again in forestalling the immediate "threat" of conservation, the larger companies remained apprehensive about what further consequences might result from Ross's resignation. The new minister was from southern Alberta, but he had no previous connection with the petroleum industry or, for that matter, any other business enterprise. From the perspective of the business community, he was a complete unknown.

N.E. Tanner, like 24 of the other elected Social Creditors, came to Edmonton with a teaching background. Born in Salt Lake City, Utah, in 1898, Tanner grew up in the staunchly Mormon community of Cardston, just 17 miles north of the Alberta-Montana border. Upon completion of high school, he took a six-week teacher-training course and in 1918, aged 20, commenced his teaching career as principal of a three-room school at Hill Spring, a village near Cardston. In 1927, Tanner agreed to become principal of Cardston High School, the largest and most important school in the Mormon community. Identified as an especially able educator, dedicated youth leader and energetic churchman, Tanner was soon acknowledged as a community leader. Recognition of the esteem in which he was held came in 1933 with his call to become a bishop in the Church of Jesus Christ of Latter-day Saints and his election to the Cardston Town Council. It is not surprising therefore that when Tanner endorsed the Social Credit message being presented by William Aberhart, fellow school principal and evangelist, in the spring of 1935, the local community approved of him as their Social Credit candidate to run in the coming election.[4]

Although there was nothing in Tanner's vocational experience to match that of Charles Ross, and upon which the petroleum industry might draw confidence, he nonetheless began his term as minister of lands and mines with one advantage that Ross had never been able to claim. This was the complete confidence of both the premier and Ernest Manning, his most trusted disciple. Tanner was also inheriting a department that had the singular distinction of not being in utter destitution and totally dependent upon the near-bankrupt provincial treasury.

The novice minister of lands and mines had barely begun acquainting himself with his new responsibilities and their related issues of concern when he received the department's *Annual Report* for the fiscal year ending 31 March 1937. It announced a revenue surplus of nearly one million dollars.[5] Natural gas consumption was noted to have moved marginally upwards to nearly 19 billion cubic feet, still well short of the peak of a little over 23 billion cubic feet recorded in 1930. Oil production, however, had reached a record 1,447,661 barrels, up approximately 200,000 barrels over the previous year.[6] Moreover, the rate of production was in the midst of a steep climb. As the report stated, "the outstanding feature of the year was the bringing in of the Turner Valley Royalties No. 1 well on 16 June 1936." In the nine months that had elapsed between the discovery and the preparation of the department's report, five more wells within a mile and a half of the TVR well had been brought into production and additional drilling was under way. In its enthusiastic emphasis upon increased production and exploration activity, the department's report left unstated that production had already passed local market requirements and that crude oil prices were slipping.

The coming in of new wells, the announcement of new drilling ventures, and the evidence of the growing interest of eastern Canadian capital remained the focus of press, public and government attention throughout the winter and spring of 1937. At a banquet organized by the Oil and Gas Association of Alberta to introduce Tanner to the industry, the minister assured his audience that it was "going to be one of the greatest development years in the history of the province," that he did not "favour numerous investigations into the oil industry," and that the policy of his department "would be to protect investors as far as possible." Further, he promised that "he would do his best to decide problems affecting the industry with the aid of the able staff of the Department of Lands and Mines."[7] The assembled oilmen were reassured, especially by Tanner's implied message that he would be guided in his actions by the professional staff in his department rather than by some of the more extreme anti-business elements within the Social Credit caucus. Dr. Link, Imperial Oil's chief geologist, was quick to second Tanner's wise decision to discuss oil problems with such able members of his staff as "'Charlie' Dingman, Grant Spratt and Vernon Taylor," and he pronounced the minister "OK."[8] Dr. R.J. Manion, a former federal minister of railways, and engineer-businessman Major-General G.B. Hughes, both of whom had travelled from Toronto so that they could use the meeting to announce a half-million dollar six-well drilling program, declared themselves favourably impressed with Tanner. They advised members of the oil industry "to co-operate, regardless of political opinions, with a man who was evidently sincere and determined to do his best."[9] In keeping with the positive tone, and for the minister's benefit, Hughes also declared that Alberta need not fear for the lack of a market for its oil production. He pointed to the huge requirements of the British navy, and asserted that building a pipeline over the mountains to the coast would be "a small matter in admiralty estimates."[10]

Caught up in the excitement of Alberta's first significant crude oil discovery, investors such as Hughes chose to ignore geographic realities. They preferred instead to focus upon the prospect of replacing Canada's substantial crude imports with newfound Alberta oil and went ahead with additional drilling programs. These wider-ranging exploration activities led the Alberta government in turn to make an important change in its petroleum regulations. On 3 March 1937, the government commenced a policy of setting aside certain areas to be known as crown reserves.[11] Included in such reserves at first were all lands in unsurveyed territory and lands within Alberta forest reserves. Companies and individuals wanting land inside the reserves were required to divide the areas in which they were interested into "equal parts of equal value from surface indications, for the discovery of oil."[12] They could then apply to lease one of the parts, the other being reserved to the Crown. A variant of the crown reserves idea, first pressed upon the Canadian government by the British Admiralty in June 1913 as a means of ensuring a long-term oil supply for the navy and applied in 1920 by the federal government to petroleum leasehold taken in forest reserves, the concept of protecting the public interest through the reservation of certain tracts remained a central feature of Alberta petroleum policy for decades.[13]

The anticipated rush to acquire exploration leasehold within the new crown reserves failed to materialize in the end. It was apparent by the spring of 1937 that the latest Alberta oil boom rested upon a precarious foundation. The first warning came from a familiar quarter and raised an all too familiar issue. In its annual report issued on 5 April 1937, the Canadian Western Natural Gas, Light, Heat and Power Company expressed its "grave concern" about the continuing waste of natural gas and cautioned that "the period during which we have been able to supply our customers with remarkably cheap gas from Turner Valley is rapidly drawing to a close."[14] Reflecting upon the gas company's report and the prospect of higher fuel costs, the *Calgary Herald* editorialized that the continuing waste was excusable only if it could be shown that there were commensurate benefits to the public. "If," the *Herald* reminded, "it should be proven here that, as in so many other instances, the public has been gypped, it would be well to remember that responsibility rests with governments, elected to office by the owners of all natural resources."[15]

The *Herald*'s message passed unnoticed. The Alberta government was totally preoccupied with other matters. In a Sunday broadcast the week previous to the release of the gas company's report, Premier Aberhart had been compelled to inform Albertans that he would be unable to keep his pledge to have social credit policies working in Alberta within 18 months of taking office. This announcement foreshadowed an orthodox budget and brought forth ringing protest from the party and the public. In the furious budget debate that followed, the long-simmering feud between the Douglas and Aberhart factions within the Social Credit party burst into the open. Twenty-five Social Credit back-benchers demanded the implementation of true social credit measures in keeping with the principles of the founding theorist, Major C.H. Douglas, and threatened to remove Aberhart from party leadership unless results were immediately forthcoming. Struggling to maintain his hold on the party, Aberhart bowed to the demands of the insurgents and agreed to the creation of a Social Credit Board, which would translate, with the aid of "experts" sent from England by Douglas, social credit theory into effective legislation.

As spring turned to summer, the situation in Alberta remained chaotic. With most of the important pieces of Social Credit legislation stalled before the courts, and given the still-raging battle within the party threatening the government's collapse, cabinet ministers had little time to devote to the concerns of their respective departments. The petroleum industry was therefore largely left to manage its own affairs. At Turner Valley, the wasteful flaring of natural gas continued unabated, and the press continued to chronicle the successful completion of additional oil wells. Increasingly, Turner Valley appeared as an oasis of prosperity in the otherwise bleak economic landscape.

Albertans and Turner Valley operators were jolted into reality in early September when Imperial Oil and the British American Oil Company announced simultaneously a sharp reduction in the price their refineries would pay for crude. The price for naphtha was dropped by 24¢ to $2.36 per barrel and crude oil prices were reduced by 19 to 39¢ per barrel depending on the gravity. The new crude oil price scale ranged from $1.36 to $1.84 per barrel.[16] The price decrease underlined dramatically that the petroleum industry in Alberta had entered a new phase. Production had moved well beyond what the local market could absorb, and Alberta passed from the status of importer to exporter of oil.

It was the rapidly expanding oil glut that once again drew government attention to the situation in Turner Valley. On September 11, the minister of lands and mines and Charles Dingman met in Calgary with producers, representatives of the distribution companies, and the Alberta Oil and Gas Association. The result was the announcement of a voluntary prorationing agreement limiting the production of each well to 65% of its capacity and the promise of a study to look into the situation.[17] If the minister believed that he could now look forward to some reduction in the cumulative 200,000 barrel surplus that filled almost every available storage tank in Calgary and in the Valley, he was immediately awakened to the reality of oilfield politics and the implacable division that separated large and small producers. Although officials of Imperial Oil and British American, the major refiners of Turner Valley crude oil, proclaimed that prorationing was now in effect, independent oil producers countered with a statement saying that they had not agreed to prorationing at the meeting with Tanner. According to the independents, "what producers were called together for was to listen to an ultimatum from the major purchasing company," which had done little to extend distributing facilities or to seek broadened markets for Alberta's increasing production.[18] The answer to the problem, the small producers argued, was not prorationing, but rather action to get the federal government involved in the development of a national oil policy. They concluded:

> The absurdity of proration being necessary with a daily production of ten thousand barrels should be apparent when it is realized that this amount is but a fraction of the daily Canadian requirements, the balance being imported and a considerable amount coming from Montana.[19]

Still preoccupied with the ongoing feud within the party, and between the party and the federal government, Tanner now faced the additional burden of having to pick his course carefully through a minefield of issues that had divided the industry for more than a decade. His response was not unlike that of his UFA predecessors. He was quick to seize upon the independents' panacea of expanded markets as the solution to the difficulties facing oil and gas producers. The call to look to the federal government to help solve the market problem was particularly attractive, for not only was Ottawa's support essential to any scheme of shipping Alberta oil east but also, if discussions proved barren, Edmonton could lay the blame at Ottawa's door and claim federal indifference to Alberta's interest. This strategy fitted easily with the current practice of presenting Ottawa as being responsible for the Aberhart government's inability to enact social credit measures intended to ease the burden of the Depression. The much more difficult and contentious issue of oil and gas conservation was allowed to drift. Later in a prepared statement to Canadians outside Alberta, the Aberhart government explained that its hands had been tied by the Natural Resources Transfer Agreement. "Gas wastage in the Turner Valley has seriously depleted the gas reserves in Alberta, and this can be attributed *solely* to the restrictions imposed upon the Province at the time of the transfer of the natural resources to provincial control, which has prevented the Alberta Government inaugurating any scheme of conservation."[20]

Given the frigid relations that existed between Edmonton and Ottawa, it was the initiative of the independent oil producers rather than that of the Alberta government that brought the federal government to review carefully the oil situation in Alberta. On 18 September 1937, the Petroleum Producers Association of Calgary approached federal Minister of Mines T.A. Crerar with the request that the Dominion Tariff Board, or another appropriate agency, be instructed to make a full investigation into the production, refining, transportation and marketing of Canadian oil.[21] In their memorandum to Crerar, the Albertans argued that, since 90% of Canada's petroleum requirements were imported, making oil and oil products the largest single imported commodity, it was in the national interest that Alberta's expanding oil production gain a larger share of the Canadian market. It seemed to Alberta's producers that the obstacles to such development lay with the refiners and the railways, thus their request for an investigation. In the short term, while the study was under way, the Petroleum Producers Association requested that the importation of foreign oil into Manitoba, Saskatchewan and Alberta be prohibited. The minister was reminded that Calgary was hundreds of miles closer to Regina and Winnipeg than the Oklahoma oilfields from which oil was currently being shipped.[22] Especially galling was the continued shipment of Montana oil into southern Alberta, and Calgary oilmen left no doubt about the strength of their feelings on this. "Shall Canadian oil be stored in the ground," they asked, "while foreign oil is allowed to be imported and bootlegged across the Montana border? If so, it is simply legalized plunder."[23]

This plea by the independent oil producers began the first chapter in the decades-long struggle of Alberta oilmen to gain access to eastern markets. They sought simply what in the eyes of most westerners was seen to be the fair application of Canada's long-standing National Policy. As the Alberta oil producers put it:

This Commoil No. 1 well in Turner Valley illustrates flaring practises in the 1930s. Glenbow Archives NA-67-126.

It was the feeling of all interested in the industry that we would be able to ship our raw products into Eastern Canada and receive their finished products in return, providing the exchange was equitable.[24]

In Thomas Crerar, the Calgary oilmen found a sympathetic ear. Although Crerar had returned to the Liberal fold following a short period as leader of the national Progressive party in the early 1920s, he remained a powerful champion of western interests.

Before taking the independents' request for a Tariff Board review to his cabinet colleagues, the federal minister of mines and resources sought comment from the senior officials within his own department. Among these civil servants was a small core of individuals who were well informed about the development of the petroleum industry in Alberta. Included in this group were John McLeish, the director of the Bureau of Mines and a former member of the Turner Valley Gas Conservation Board, and T.G. Madgwick, a petroleum engineer who had been a senior member of the professional staff who had watched over the industry in Alberta before 1930, when responsibility for western lands and resources resided with the Department of the Interior. Madgwick's assessment was astute. First, alerting the minister to the political reality, he pointed out that there was some ambiguity about just who was represented by the Petroleum Producers Association of Calgary. According to Madgwick, if any quantity of oil was to be removed from Alberta, a pipeline would have to be laid to the Great Lakes. The problem was that such a pipeline was a costly undertaking and unlikely to be built until it became clear that the Turner Valley oilfield contained sufficient reserves to sustain large long-term production. In Madgwick's view, there was good reason to doubt that Turner Valley would prove to be of sufficient size. The key to the eastward movement of any Alberta oil in the short term therefore was freight rates, and until this situation was clarified Madgwick urged that oil be kept in the ground. He reminded that "the waste of gas in order that immediate revenue be gained by the producers of Turner Valley was so colossal as to become the wonder of the industry," and expressed his hope that "better councils would prevail in regard to the current over production of crude oil."[25]

As the federal Department of Mines and Resources contemplated how it might respond to the growing oil glut and the producers' appeal, others also began to draw attention to the parallel problem of waste. By mid-autumn, there was a growing sentiment that the Aberhart government had had sufficient time to formulate a comprehensive petroleum policy and that further delay was intolerable. In a blistering editorial on November 25, the *Calgary Herald* called attention to Social Credit's lamentable procrastination. "The Hon. N.E. Tanner," the paper charged, "is now an enthusiast on the subject of gas conservation in Turner Valley, but as a matter of record the first move in the direction of lessening the appalling waste of this valuable natural product was made by the UFA government."[26] Readers were reminded that for several years before Social Credit came to power the total production of gas in the Valley was limited to 200 million cubic feet per day, and that all new wells brought in were permitted to produce only their share of the established quota. Once in power, Aberhart had cancelled the existing conservation order and subsequently thwarted the efforts of Minister of Lands and Mines C.C. Ross. In consequence, there had been no government regulation of Turner Valley gas production since Social Credit took office, and gas continued to be wasted at the rate of approximately 250 million cubic feet per day. "It would seem," the *Herald* concluded, "high time that Mr. Tanner intervened more effectively than he has done so far."[27]

As the clamour for action began to mount once more, the government started to listen with greater interest. In good part, this was because a powerful new voice had been added to the conservation lobby. What was happening was that the small producers were beginning to separate into two distinct groups whose interests were leading them towards opposing positions on the conservation question. Although a core of companies remained who were concerned mainly with naphtha production and opposed to conservation, another group of companies had emerged recently who were concerned primarily with crude oil production and looked more favourably upon production controls. The latter group saw the wasteful burning of natural gas as a threat not only to the amount of oil that they might produce from their individual holdings but also, by threatening to shorten the life of the entire field, the uncontrolled production of Turner Valley gas would make the building of a pipeline less likely and in turn preclude obtaining vital access to eastern markets. This was the message that the Alberta government began to hear increasingly through late 1937 and early 1938. For years, the small independent naphtha producers had warned the government that gas conservation in Turner Valley would cause their ruin and kill further investment; but now, from the larger group of independent crude oil producers, came the reverse argument that further investment and development of the Turner Valley field was dependent upon the implementation of effective conservation measures. Anxious to maintain the pace of development in Turner Valley above all, the government became more receptive to the crude oil-producing segment of the industry.

It was the "New Year's message" of J.H. McLeod, president of Royalite Oil Company, and the resultant howl from Turner Valley crude oil producers, that finally jolted the Social Credit government to act. On 4 January 1938, McLeod announced that the price of Montana crude had collapsed, that there was now no recognized posted price, and that distress prices were in effect. "Consequently," McLeod declared, "certain steps must be taken to assure Turner Valley of its market on the prairies."[28] The most important of these steps was a cut of 22¢ a barrel in the wellhead price offered for Turner Valley crude, which brought the average price down to $1.26 a barrel. The next day, the *Calgary Herald* reported that one U.S. operator was offering Montana crude at 65¢ an imperial barrel and that an all-out "crude war" threatened.[29]

Ten days later, N.E. Tanner announced that conservation legislation would be introduced at the forthcoming session of the Alberta Legislature. The minister admitted that some of the major oil companies had not yet given their consent to the plan, but he assured that the government would "endeavour right to the last to reach an amicable agreement with these companies before legislation is drafted."[30] In his desire to consult closely with the operators, Tanner was following the practice established from the first by the Department of the Interior's Petroleum and Natural Gas Division and later followed by the United Farmers of Alberta. Once adopted by Tanner, the emphasis on consultation with the industry on conservation matters also remained characteristic of Tanner's Social Credit successors.

The Alberta government's belated interest in addressing the conservation problem was stimulated further by the Canadian Tariff Board hearings in Calgary on January 27 and in Ottawa from February 1 to 5. Held in response to the earlier request from the Alberta Producers Association, the hearings revealed that the prospect of placing a duty on imported Montana oil to assist the access of Turner Valley crude to eastern and even southern Alberta markets was remote. Representatives from farm organizations in the Lethbridge-Foremost district, supported by several Saskatchewan groups, protested that a duty on crude oil would raise prices and inflict a hardship on farmers.[31] In Ottawa, the focus of the hearings shifted to the marketing of Turner Valley crude, and most of the time was spent discussing the relative values of Turner Valley, Cut Bank, Montana, and mid-continent crudes. In a brief on behalf of the independent producers, Professor George Granger Brown, from the University of Michigan, reported that Turner Valley crude produced more gasoline per barrel than mid-continent crude, and when processed at Sarnia it would offer a price difference in its favour of 22¢ per barrel.[32] The real problem—apparent to all—was the cost of transporting the oil to Sarnia. Only if the reserve of crude oil in Turner Valley seemed sufficient could the merits of a pipeline be entertained seriously. On the question of the probable amount of recoverable oil in the Turner Valley field, however, there was diversity of opinion. In the end, it seems that most of those present were prepared to agree with the assertion of R.V. LeSueur, vice-president of Imperial Oil, that a reliable estimate of oil reserves in Turner Valley was impossible given the current level of knowledge.[33] In the near term, therefore, the only realistic option was movement by rail, and W.S. Campbell, president of the Alberta Petroleum Producers Association, pointed out that the recent freight rate reductions were more than offset by the latest drop in the wellhead price for crude.[34] Representatives attending for the railways would give no opinion of whether or not further reductions might be possible, and the hearing was adjourned until April 4 to give the railways additional time to consider.

N.E. Tanner and Ernest Manning, provincial secretary and minister of trade and industry, had sat in on the Calgary portion of the hearings, and for them the message that emerged from the federal Tariff Board's exercise was clear. In the short term, nothing was going to happen, but the oil glut was going to get worse. There would be no tariff remedy to help Alberta oil gain access to eastern markets. The only solution was longer term, and it was tied to establishing the extent of Turner Valley reserves and a pipeline. The pressing need for the moment was to see that the limited local market was divided equitably among producers, so that the investments of smaller operators would be protected and the waste characteristic of unregulated production prevented. It was necessary also, if the longer-term solution were to be achieved, to protect the already proven Turner Valley crude reserve and to encourage further exploration by putting an end to the wasteful dissipation of natural gas—the essential reservoir energy required to drive the oil to the surface. The conservation legislation already under discussion with Turner Valley operators promised to deal with both these short- and longer-term concerns.

As Dingman and his colleagues in the Petroleum and Natural Gas Division of the Department of Lands and Mines continued the interminable and seemingly impossible task of drafting conserva-

Scrubbing Plant near Turner Valley townsite in 1930s with conventional drilling rigs in background. Glenbow Archives ND-8-436.

tion legislation that would satisfy their technical objectives and, at the same time, find favour with Turner Valley producers, Tanner began to think about who could head the board contemplated in the impending legislation. His decision rested on now familiar logic. Given the intensity of feeling and the sharp divisions that divided Turner Valley operators, Tanner decided to look for an outsider, for someone not identified with the government or with any of the operator factions. The minister wanted someone with experience, personal force and professional stature who could be acquired on temporary assignment to pilot the board through its crucial first year. On March 17, he telegraphed J.W. Finch, director of the Bureau of Mines in the Department of the Interior in Washington, to say that Alberta was enacting legislation to deal with the conservation of oil and gas resources and to ask the director if he would recommend the names of competent petroleum engineers experienced in this area who might be prepared to come to Alberta.[35] From the list of names submitted, Tanner selected the firm "Parker, Foran, Knode and Boatright," a group of consulting engineers in Austin, Texas, who had been influential, Finch explained, in the formulation of conservation legislation in Texas.[36] He telegraphed to ask R.D. Parker if he would be able to come to Alberta for a short consulting assignment. Tanner could not have been directed to a more qualified source of assistance. Trained as a civil engineer, Parker

was perhaps the most experienced petroleum conservation regulator in private practice in the United States. Before establishing his
own consulting firm, Parker had headed the Texas Railroad Commission's Oil and Gas Division. He served the Commission for 26
years, until his celebrated resignation in June 1934.[37] Unavailable
himself, Parker recommended his colleague, W.F. Knode.[38] With the
Alberta Legislature due to begin debate on the government's proposed conservation legislation in just over a week, time was critical.
Tanner immediately wired back requesting that Knode come to Edmonton for consultation as soon as possible.

In addition to getting some last-minute advice from Knode, the
Alberta government's other pressing concern was Ottawa. The Su-
preme Court decision in *Spooner Oils Ltd. v. Turner Valley Gas
Conservation Board* of 3 October 1933 had put an end to Alberta's
first attempt to implement a comprehensive conservation program.
If a new conservation authority was not to suffer the fate of the
Turner Valley Gas Conservation Board, it was necessary that the
Canadian Parliament pass an appropriate amendment to the 1929
Natural Resources Transfer Agreement. This was to ensure that Al-
berta's proposed conservation legislation could apply to all lands
outside national parks and Indian reserves, including petroleum
and natural gas leasehold granted under federal authority before
the transfer of lands and resources to provincial control in 1930.
Thus, on the day that Tanner's telegram went out to W.F. Knode in
Corpus Christi, Texas, another telegram was on its way to Ottawa
from Tanner's deputy minister. This second telegram expressed the
Aberhart government's anxiety that the Transfer Agreement
amendment was still before Parliament, even though the current
session of the Alberta Legislature was nearing its end.[39] Prime Minister Mackenzie King, however, was little inclined to take any action
for the convenience of the Alberta premier.

THE OIL AND GAS CONSERVATION ACT, 1938

William F. Knode arrived in Edmonton on 29 March 1938. Expected to advise in the drafting of Alberta's impending oil and gas
conservation legislation, Knode brought an appropriate background to the task. He grew up in the area around Manington,
Fairview and Waynesburg, West Virginia, as the oil industry was
developing in the region. His father owned a torpedo company that
"shot wells" to improve oil flow, and Knode got his first oilfield job
at age eight. Intent upon an oilfield career, Knode enrolled at the

University of West Virginia in 1917. Since petroleum engineering had not yet gained the status of a separate program, at this or most other universities in the United States, Knode took the next closest thing—mining engineering. In his studies, he concentrated on geology and hydraulics, the flow of gas and fluid through porous media. After graduation in 1921, he went to the oilfields of northern Oklahoma, and through the following decade he worked for various companies in Kansas, New Mexico and Venezuela. Employed by Shell Oil in west Texas in the late 1920s, Knode was on hand to observe the formation of the Central Proration Committee by independent operators in that region. The Committee hired several engineers and set up field committees to study and compare drilling and production data and the behaviour of fields under different flowing conditions, so that they might better understand reservoir performance. Knowledge of reservoir dynamics was still in its infancy, and Knode was attracted to the work being done. This led him to take a position with the Central Proration Committee in 1930 and, a year later, to join the engineering staff of the Texas Railroad Commission, apparently on a consulting basis.

These were exciting formative years in the history of North American petroleum conservation. The Texas Railroad Commission's first proration order, limiting each well in the East Texas field to about a thousand barrels a day, was issued in April 1931. Four months later, in the famous MacMillan case, a federal court struck down the order on the grounds that it bore no relation to physical waste but was really an attempt to fix prices. The battle to bring proration and conservation to the Texas oilfields raged for the next four years. Knode was a direct participant during the critical period in the struggle to establish oil and gas conservation in the most important oil-producing state. In 1934, with newly acquired expertise, he left the Commission, along with his supervisor, R.D. Parker, to establish the consulting firm of Parker, Foran, Knode and Boatright in Austin, Texas. Under Parker's continued tutelage, Knode continued to develop his background in the petroleum conservation field, and it was this expertise that the Alberta government hoped to use to advantage in the spring of 1938.[40]

What Tanner seems to have had in mind was that Knode would come to Alberta to review the legislation that had been already drafted by Dingman and his colleagues in the Department of Lands and Mines, to offer constructive comment based upon his Texas experience, and to lend his authority to the presentation of the legislation before the Turner Valley operators. Rather than being the essential creator, Knode was to be the essential defender of Alberta's new conservation legislation. Even the latter role was considered to be only short term. Both parties were cautious, and the possibility of Knode serving as chairman of the soon-to-be-created Petroleum and Natural Gas Conservation Board was not mentioned until after Knode had been in Edmonton for a week, and then only for a six-month term.[41] For his part, Knode was not sure that he wanted to be away from Texas for even that long. It was agreed therefore that Knode would serve as chairman for an initial period of three months, after which he would be permitted to return to the United States for "at least six weeks." Then he would return to Alberta for a second three-month period, and if, after this second visit, it was deemed unnecessary for Knode to remain in Alberta continually as an active chairman, he would be held on a retainership basis by the Board for a further year and a half. As late as July, Knode was still contemplating the management of the Board's affairs on a part-time basis, commuting back and forth from Texas so that he might continue his consulting business there.

Secure in the knowledge that the government's intended conservation legislation had the approval of a recognized outside expert and that the services of a suitable Conservation Board chairman had been secured, Tanner turned his attention to Ottawa and the progress of the necessary federal enabling legislation. Initially, it had taken Tanner three trips to Ottawa just to persuade the prime minister and his cabinet to amend the Natural Resources Transfer Agreement to remove restriction on the scope of intended provincial conservation legislation. When agreement had been reached on March 5, the Alberta government had anticipated that parliamentary assent would follow quickly, thus permitting the presentation of the desired conservation bill before the Alberta Legislature was prorogued in early April. As March passed, however, Tanner's position grew increasingly awkward. He was committed to presenting the government's long-awaited conservation legislation; yet, in Ottawa there was still no sign of the special bill confirming the amendment. In desperation, just a few days before the Legislature was scheduled to rise, Tanner was compelled to proceed with an Act that spelled out the government's precise intent, but could not be applied to meet the growing crisis in the Turner Valley field. The *Oil and Gas Conservation Act* assented to 8 April 1938 was necessarily restricted by a clause stating that the Act would only come into force on a day proclaimed by the lieutenant-governor-in-council after Parliament had ratified the agreement reached on March 5.

The bill that Tanner presented for the Legislature's consideration was essentially a refined version of the 1932 *Turner Valley Gas Conservation Act*. Like its predecessor, the new conservation measure represented an approach to regulation dear to the hearts of agrarian governments in North America. The idea of regulation by an independent commission was a product of the dramatic industrial development that occurred in the last half of the nineteenth century, particularly in its centrepiece, the railroad. Organized farmers in the midwest states had been quick to challenge what they saw to be abusive freight rates set by railroad monopolies. In their search for means to control the railroads, farmers' governments gradually brought about a momentous shift in the character of business regulations in the United States. At first, individuals and governments had relied on the judicial process for the enforcement of laws intended to protect against discriminatory railroad practices. It was soon apparent, however, that the courts were not able to deal effectively with the complex economic world of freight rates and corporate management. This resulted in the search for a new administrative mechanism "which would be more flexible than the legislature and more competent than the judiciary in dealing with complicated economic matters."[42] A solution was found in the creation of a new agency independent from both the legislature and the courts and vested with certain of the powers traditionally held by each. Created by statute and generally known as "boards" or "commissions," those hybrid agencies were typically formed to regulate the development and conditions for carrying out an activity, such as transportation, communication and the production and sale of energy. Possessing the power to investigate, hear and adjudicate simultaneously, such bodies draw together within one decision-making process the three methods by which the state acts—rule-making, adjudication and administrative enforcement. They are distinguished from "administrative tribunals" in that they generally possess broad powers of regulation and because "their process of hearing and individual adjudication is based upon norms which are usually subjective."[43] The norm of "public interest" is often presented in the enabling statute as the basis upon which a board or commission must base its decisions.

Faith was placed in the new commission or board idea by those who believed that executive departments of government could provide neither the high level of specialized expertise that was required for effective regulation in some areas nor the desirable neutral environment, free from partisan political consideration, necessary to promote regulation in the public interest. The commission, it was argued, not only had the advantage of political independence but also offered great advantage over common recourse to the courts. Judicial procedures were slow and costly. Courts could not initiate actions on their own and were confined to cases brought before them. Beyond this, they were seriously deficient in their ability to gather relevant information. Having no independent means of investigation, they depended solely upon the litigants to develop the facts in any controversy. In contrast, boards could initiate actions on their own and could carry out extensive investigations to gather information to suit their own purpose. It was perceived, moreover, that the commission, as an expert body, could gather and evaluate complex masses of information and, through accumulated experience, function as a critically important source of advice to legislators drafting regulatory policy.

Those who supported the commission idea were also inclined to point out that such an agency could serve the interests of the regulated, especially the smaller enterprise. The commission could act as a tribunal to adjudicate conflicting interests and, in certain situations, save regulated companies from the effects of cutthroat competition to the benefit of private and public interests alike.

The commission idea was not without its critics, and more will be said of this later. In Great Britain, where the commission approach has an even older history, criticism was intense from the beginning. While the railroad was still in its infancy, the vexatious question of carrying charges led to the creation of a panel of Railway Commissioners in 1846. This perceived trend towards government commissions led Joshua T. Smith, in a book published in 1849, to denounce all commissions as "the chosen instruments of schemers and enemies of public liberty."[44] In this more inhospitable climate, the idea of the independent regulatory commission did not at first flourish and, it was from the U.S. experience therefore that Canadian, and particularly prairie, governments drew their inspiration.[45]

Canadian legislators, especially in western Canada, observed the activities of state boards and commissions and the popular agitation to bolster state regulators that led in 1887 to the passage of the *Interstate Commerce Act,* which created the first national commission in the United States concerned specifically with the regulation of economic matters. In Canada, similar concerns, expectations and agitation led to the creation of the Board of Railway Commissioners in 1899, and by the turn of the century the commission idea was solidly entrenched as a preferred vehicle of regulation. Alberta's first regulatory board was created soon after provincehood was achieved in 1905. The appointed five-member Provincial Board of Health established by the *Public Health Act* of 1907 was given wide power to issue and enforce rules, orders and regulations to protect and promote the health of Albertans.[46] A more directly relevant part of the province's regulatory experience, from which the Conservation Board ultimately emerged, was the *Public Utilities Act* of 1915.[47] This Act called for the appointment of a three-member Board of Public Utility Commissioners to regulate tram and street railways in the public interest. In its delineation of the structure, jurisdiction and powers assigned to the Board, the Act manifests the classic thinking underlying the commission approach to regulation. To ensure neutrality and independence, the Act provided that commissioners' terms of office would be for 10 years and that they could not own any kind of interest in a public utility.[48] The commissioners were given authority to examine "all questions relating to the transportation of goods or passengers," and specifically to investigate any situation when there was "reason to believe that the tolls demanded by any public utility exceed what is just and reasonable." The Board had the power to investigate upon its own initiative and after hearing to fix by order "just and reasonable individual rates, joint rates, tolls, charges or schedules." To ensure that it possessed the technical competence to meet its prescribed mandate, the Board was given the power to hire "expert" advice. Other responsibilities assigned to the commissioners included the obligation "to impose and enforce regulations for the safety and protection of employees of any public utility" and, where it was deemed to be in the public interest, "to decree the joint usership of the means of distribution, such as poles, conduits and other equipment." Having made the decision to use the commission form of regulation in the public utilities sector, Alberta legislators were not reticent about granting the Board sufficient power to effect its purpose. The Board was given the right to summon witnesses, to issue commissions, and to take evidence outside the province, and in its hearings it was to be governed by its own rules of conduct rather than being bound by the technical rules of legal evidence. A decision of the Board upon any question of fact or law within its jurisdiction was binding; no Board order, decision or proceeding could be appealed.[49] Finally, no officer or employee of the Board was personally liable for actions carried out under the authority of the Act. The panoply of powers vested upon the commissioners, especially the denial of right of appeal, seems at first glance to stand in contradiction to the democratic impulses traditionally ascribed to agrarian reformers. In truth, however, the assignment of such power represents not so much a contradiction as it does a testimony of the great faith Alberta legislators were prepared to place in the integrity of "neutral" boards, which they held were more insulated from improper influence than the courts. The place of the board as the preferred regulatory mechanism gained even greater sanction after 1921 when the United Farmers of Alberta assumed the mantle of government.[50]

The broad base of popular acceptance of boards, and more specifically of "experts," was strengthened further with the election of Social Credit in 1935. Social credit theory gave a central place to the role of experts in the creation of an economic order designed to "end poverty in the midst of plenty." Social Credit's contribution, however, was of a different order than that of the progressive ideology espoused by the UFA. It came in the form of a general predisposition among the elected Social Credit members of the Alberta Legislature towards the idea of administration by experts. Although it took the Aberhart government several years to establish the Conservation Board, this is explained mainly by the internal conflict that preoccupied the party through most of the period and not by any apparent division in cabinet regarding the appropriate type of administrative authority required in Turner Valley. In fact, once the decision was made, the cast of mind dominant within the Social Credit caucus was such as to allow the creation of a board with stronger powers than its UFA predecessor might have found acceptable. Although Social Credit would emerge later as the ardent champion of laissez-faire free enterprise, it never lost the strong authoritarian streak that characterized the party's founders and its first years in office.

The Oil and Gas Conservation Act presented by Tanner to the Alberta Legislature in the spring of 1938 was in essence a direct legacy of the continent's radical agrarian heritage. In its general form, the Act was cast in the mould created by the Granger movement in midwestern states, such as Iowa, Illinois and Wisconsin, in the 1870s and 1880s[51] and adopted by Albertans after the turn of the century. In its specific form, it drew heavily upon the laws defining the scope and activities of conservation commissions set up in the western oil states of Texas and Oklahoma in the 1920s and 1930s. It was this continental regulatory experience, combined with the particular situation in Turner Valley, that produced the 1938 conservation legislation—the essential detail of which has remained at the heart of Alberta's conservation legislation to the present.

The central feature of Tanner's bill was its provision for the creation of a board of not more than three persons to be known as "The Petroleum and Natural Gas Conservation Board."[52] Board members were to be appointed by the lieutenant-governor-in-council; the first chairman was to hold office for one year, and at the pleasure of the government thereafter. Every subsequent chairman and every other member of the Board was to be appointed for a five-year term, after which they would serve as long as the government saw fit. Removal of any member was possible only by vote of the Legislative Assembly. In deciding to have an appointed Board, Alberta was following its own tradition rather than that of the Texas Railroad Commission, which was the inspirational source for other aspects of the province's conservation legislation and program. This was a decision of particular consequence. In Texas, commissioners were elected for six-year terms by statewide ballot. It was thought that appointed commissioners were more vulnerable to the influence of regulated groups, whereas elected commissioners remained subject to public control. This meant in practice, however, that commission leadership also had an overt political dimension. It meant that partisan politics became a part of the agency culture. In practical terms, for example, each of the Texas Railroad commissioners had the de facto right to appoint one-third of the Commission staff.[53] By contrast, the appointed Alberta Board was structurally removed from the cut and thrust of provincial politics, and this was a key factor in shaping a different agency culture. Unlike the Texas Railroad Commission, political cronyism was not a cross that the Petroleum and Natural Gas Conservation Board would have to bear.[54]

According to the Act, the Board's general purpose was to effect the conservation of Alberta's petroleum and natural gas resources by:

(a) preventing the exhaustion from a producing petroleum area of the energy necessary to produce petroleum by methods shown to be uneconomic in that such method of production allows this exhaustion without proportionate recovery of petroleum to the end that the maximum ultimate recovery of petroleum can be attained;

(b) prorationing the production of petroleum or natural gas from the wells in any area to the economic markets available in such manner that an uneconomic reduction of price is not brought about and in such a manner that an equitable share of the available markets for petroleum or natural gas is available to each producing well.[55]

In addition, the Board was also specifically charged with the duty of enforcing any regulations made under the provisions of the *Oil and*

Gas Wells Act, 1931.[56] The new legislation was to apply to every producing well, regardless of whether the well was on leasehold granted by the railway, the Hudson's Bay Company, the Crown in the right of Canada before 1930, or afterwards by the Crown in the right of Alberta.[57] Universal application was essential for any effective conservation program, and for this reason, as the Spooner case demonstrated, Ottawa's ratification of the amendment to the Natural Resources Transfer Agreement was crucial.

For the purpose of preventing the "wasteful or uneconomic" production of the province's petroleum resources, the Petroleum and Natural Gas Conservation Board was accorded wide powers. These included the right to acquire such employees and professional staff as it considered necessary to carry out its obligations; the right, subject to approval of the lieutenant-governor-in-council, to make regulations or orders prescribing the conditions under which production could be permitted and the amount of production permitted from any well; the right to determine well spacing; and the specific right to prohibit the operation of any well unless the gas produced in excess of market demand was replaced in such geological horizon and in such a manner as the Board might approve. Further, the Act assigned the explicit authority to prorate the production of petroleum "to the available economic markets in such a manner that in so far as it is possible the maximum ultimate recovery of petroleum may be obtained and equity between operators in a common pool may be maintained."[58] In support of such authority, the Board was also given the power to determine the amount of crude required to supply prevailing market demand.

So that it might have access to the information necessary to discharge its duties at all times, and to undertake whatever investigation it might see fit, the Board was given the coercive power of inquiry common to a court of law. Each member of the Board, pursuant to the *Public Inquiries Act,* was bestowed with the power to take evidence under oath and to compel the attendance of witnesses and the production of documents. For its part, the Board was not obliged to hold hearings or to give reasons for its decisions.

As if to demonstrate the strength of the government's determination that its new Board should not in anyway be thwarted in the pursuit of its assigned objectives, the section of the Act conferring powers with respect to conservation was concluded with a clause conveying to the Board a remarkably wide-ranging authority

> to prescribe rules and regulations as to the production, transportation, distribution, or use of all or any petroleum products, and

the uses which may be made thereof or the amount which may be produced transported or used, either generally or in any area at any specified well or wells and for any specified purpose.[59]

A separate section provided that every order or regulation that the Board might make would have the same force as if it had been an integral part of the Conservation Act. Failure to comply with any such order of the Board was punishable on summary conviction by a fine of up to $2,000 and costs, plus $500 for each day after the first day during which the default continued.

Finally, to ensure that Board members could go about their tasks with authority and without interference, the Act provided legal immunity. No officer or employee of the Board could be held personally liable for actions carried out in the pursuance of any of the provisions of the Act or any regulation or order made under the authority of the Act.[60]

Having decided to remove the most important elements of petroleum industry regulation from the Department of Lands and Mines to an independent board beyond the immediate reach of government, the Social Credit government found it easy to decide that the cost of its operation should be raised outside also. It was provided therefore that all expenses incurred in the administration of the Act would be obtained through an annual assessment upon every producing well in the province.[61]

In all, Alberta's new conservation legislation was impressive. The bill's provisions placed it clearly on the leading edge of what the technical experts of the day considered ideal.[62]

The bill gained passage in the Alberta Legislature with limited debate. Most concern was generated from within Social Credit ranks, and the minister of lands and mines was compelled to emphasize to his colleagues that the "establishment of the three-member commission would not cost the taxpayers anything," and to assure that conservation in Turner Valley would not be followed by higher gasoline prices to farmers.[63]

Outside the Legislature, the *Calgary Herald* gave its blessing to the legislation in an editorial headed "Conservation at Last."[64] Although this was merely consistent with the paper's long advocacy of the conservation cause in Turner Valley, given its stature as the most important newspaper in southern Alberta and as the voice most stridently opposed to social credit in the province, it was a significant endorsement. In the *Herald*'s view, the minister in charge of bringing in this long overdue legislation had gone about things in the right way; industry had been consulted in advance and its ideas taken into account.

The *Herald*'s prediction of satisfactory results, however, was premature. Ottawa remained undisposed to move things forward on Alberta's behalf, and there were a collection of irritants that continued to sour relations between the two governments. While the foundation statutes of Social Credit monetary reform were being finally and decisively beaten down by the federal government in the courts through the spring and early summer of 1938, Alberta was demonstrating its displeasure by refusing to continue customary payments for the lieutenant-governor's upkeep, thus forcing him to vacate Government House. The Aberhart government also retaliated by rejecting Prime Minister Mackenzie King's request for Alberta's permission to amend the *British North America Act* so that Ottawa would have authority to institute a national unemployment insurance scheme.[65]

Embittered federal-provincial relations was not the only obstacle. The anticipated resistance of some of the smaller Turner Valley operators was even stronger and more widespread than expected. The hostile naphtha producers, whose position was most immediately threatened by conservation, were joined by the most vocal independent refiner Leon L. Plotkins, the owner of the Lion Oils Limited, as well as many of the smaller crude oil producers, all charging that their activities were going to be curtailed mainly for the benefit of Imperial Oil.[66] At the same time, other operators were pressing the minister of lands and mines to name the Conservation Board members without further delay and to get on with the urgent task ahead.

For his part, Tanner found himself in an increasingly awkward situation. It did not seem appropriate to appoint members to a board that could not formally exist until Ottawa passed the enabling legislation that would render Alberta's Conservation Act operative. Under mounting pressure and in desperation, Tanner finally accepted the expedient of naming members to a board, the existence of which was still just anticipated, and whose appointments would receive formal confirmation only after the federal government had ratified the March agreement. On May 16, Tanner announced that the chairman of Alberta's Petroleum and Natural Gas Conservation Board would be W.F. Knode, assisted by C.W. Dingman and Fred C. Cottle. He went on to identify Knode as a prominent consulting engineer from Austin, Texas, and to emphasize that he came "highly recommended by the United States Bureau of Mines."[67] Dingman's appointment was expected. Having served on the staff of the short-lived Turner Valley Gas Conservation Board and as director of the Petroleum and Natural Gas Division of the Department of Lands and Mines, he was the most experienced and best informed person on the petroleum industry in government service. Cottle, auditor for the Public Utilities Commission, was also a well-known career civil servant. Tanner had planned his announcement to coincide with Knode's return to Alberta from Texas, and the next day, accompanied by Deputy Minister John Harvie, Knode and Dingman, he met with a delegation of the Alberta Petroleum Producers' Association to review various features of the *Oil and Gas Conservation Act*.[68]

While Knode continued the task of trying to persuade reluctant producers and trying to lay the groundwork for conservation measures, conditions in the Turner Valley field continued to deteriorate. Several large new crude wells came on stream and precipitated a market crisis. In early June, major local refiners notified producers that, in the face of current market demand and increasing production, they were compelled to reduce their purchase of crude oil from 40% to 30% of each well's capacity.[69] Also early in June, Calgarians were jolted to learn that gas pressure in the Turner Valley field had declined to such an extent that, for the first time since the Royalite discovery in 1924, gas would likely have to be pumped to the city to meet winter fuel needs.[70] Under steadily mounting pressure to do something about gas waste and crude oil markets, the beleaguered Tanner and Knode pleaded that "their hands were tied." Agitation then shifted to Ottawa. The Calgary Board of Trade sent a telegram to the prime minister, stating that the situation in Turner Valley required the "immediate attention of the Oil and Gas Conservation Board" and calling attention to the urgency of the enabling legislation.[71] As mid-June passed without a federal response, Premier Aberhart was forced to appeal humbly and directly to Prime Minister Mackenzie King—the man he had constantly vilified. In a telegram to King, Aberhart pleaded:

> Conditions acute in Turner Valley through bringing in of additional big producing wells and imperative that our Oil Conservation Board function without delay. Legislation approving Transfer Agreement delays action until first of month following assent. Would appreciate your cooperation to have Canada's legislation assented to before June 30th otherwise our Conservation Board will be unable to act for another month.[72]

Perhaps thinking that this latest demonstration of federal power had been played to its maximum advantage, King replied on June 24 to say that the much anticipated legislation had been given royal assent that afternoon. Tanner then announced that Alberta's new Petroleum and Natural Gas Conservation Board would start to function officially on July 1.[73]

Knode could at last get down to his assignment. He was the right man for the task, for he was experienced, absolutely convinced of the necessity of enforcing proper production practices to conserve reservoir energy, and possessed of a personality at ease in the rough environment of the frontier oilfield. He was not verbally bound, as was the minister of mines and minerals, by a strict Mormon rectitude that, when provoked, would permit only the oath, "Jupiter."[74] Knode was intimate to all the inventive and colourful malediction of the oil patch and relished its boisterous use. He would not be cowed by the raucous protests of the Turner Valley naphtha producers, who must have seemed an almost insignificant group when compared to the array of conservation opponents in Texas.

The product of professional commitment and personality, Knode's determination was undoubtedly strengthened by the knowledge that he was the most highly paid individual in government service. At $1,000 per month, plus $8 per day subsistence, for as long as he was absent from Corpus Christi, Texas, Knode's salary was more than three times that of fellow Board member Charles Dingman, two-and-a-half times that of Deputy Minister of Lands and Mines J. Harvie, and twice that received by Tanner and other members of the Alberta cabinet. By contrast, the first professional engineers hired by the Board earned from $75 to $150 per month.[75]

The announcement that the Conservation Board would commence operations as of July 1 was followed quickly by a series of orders-in-council naming Calgary as the site of the Board office, defining the boundaries of the Turner Valley oilfield, and giving the Board authority to borrow up to $25,000 to pay for its operations until the industry levy of so much per producing well could be determined and collected.[76] Of these, the decision to locate the new Board's headquarters in Calgary was of the greatest importance. It was, of course, a logical decision in the immediate sense, by far the most significant oil and gas production in the province was near Calgary, and almost all the companies had their headquarters in the city. The location of the Board office in the southern centre helped to emphasize the conservation agency's complete separation from the Petroleum and Natural Gas Division of the Department of

The Petroleum and
Natural Gas Conser-
vation Board's first
head office in the
Alberta Government
Telephones Building in
downtown Calgary.
Alberta Government
Telephones.

Lands and Mines in Edmonton, even though the Board's new Calgary headquarters was in a government building that also housed
the division's branch office. The decision was also of major significance for Calgary in the long term; it helped to shape the city's future. The move reaffirmed Calgary's place as the centre of western
Canada's emerging petroleum industry, a status first conferred in
1918 by the federal Department of the Interior when inspecting engineers were permanently established in a Calgary office. Once
established, the Board's presence helped to support Calgary's position as the financial and administrative centre of the petroleum industry after the industry's exploration focus and field operations
shifted mainly to the Edmonton area in the 1950s and 1960s.

W.F. Knode chaired the first meeting of the Petroleum and Natural Gas Conservation Board on July 4. His central objective at this
meeting was to name a support staff to begin the collection and
preparation of oil and gas well data upon which the foundation of
the Board's conservation program would rest. For this purpose,
four engineers were engaged, M.D. Kemp, H.G. Bagnall, A.W. Lees
and L.D. Publicover, the first named being the senior man. Also
hired was an accountant and statistician, J.W. Kraft, and two ste-

nographers, G.L.A. Doherty and P.S.L. Falk.[77] Five weeks later, a fifth man, G.A. Connell, was added to the engineering group.

If the initial step of engaging qualified professional field staff seems obvious, it was nevertheless of significance. In following what had been Department of Interior tradition in this regard, Knode's approach to appointments also was likely conditioned by the principled stand taken on hiring practice by R.D. Parker, his Texas colleague. Parker, a 26-year veteran of the Texas Railroad Commission, had resigned just as he was about to be fired for resisting the Commission's frequent practice of giving preference to political connection over professional experience in the selection of field personnel. On his highly publicized departure, Parker declared:

One of the first objectives of the 1938 Petroleum and Natural Gas Conservation Board was to hire an engineering team to staff the Black Diamond Field Office. (l-r) Gordon Connell, Andy Lees and Lloyd Publicover. Herb Bagnall is absent. ERCB, Corey Collection.

> Apparently I made the fatal mistake of firing political favorites of all three commissioners.... Every time I fired one, some one of the three commissioners would put him back at work, with resultant weakening of the morale of the entire organization.[78]

The pattern of staffing established at the outset was important. Alberta's Conservation Board began, and would remain, outside the network of political patronage.

While the Board's newly acquired engineering staff went about determining the potential daily production of each producing oil well as the necessary prelude to prorated production, Knode began

The Board's blunt message to operators in 1938 was clear: wastage of gas and oil like this large earlier flare at Royalite No. 4 would be tolerated no longer. Glenbow Archives NA-701-9.

a series of meetings with Turner Valley operators.[79] His message was blunt, Alberta fields would produce as much oil and gas as was necessary to meet a legitimate market and no more. Naphtha would be considered a by-product of gas wells, and these wells would be allowed to produce gas prorated to actual market demand. "Wastage of gas and oil would be prevented."[80]

The Board's first explicit conservation measure came on August 4 in the form of an order to prevent excessive drilling. After an examination of the spacing of existing wells and wells currently being drilled in the Turner Valley field, and considering the acreage held by various leaseholders for future drilling, the Board concluded what informed observers had known for a long time, that there had been excessive drilling. Acting upon what they perceived to be in the best interest of the future development of the field, the Board issued an order restricting subsequent drilling to not more than one well for every 40 acres.[81] The order specified further that no well could be located closer than 660 feet from the boundary of the minimum area or 1,320 feet from another well. This order was important not only as the first direct step in the long-awaited conservation program for Turner Valley but also for establishing an important precedent in administrative process. The well-spacing order offered the prospect of some flexibility, by permitting an application for a special drilling permit that would allow for some variation from the 40-acre minimum. By stating explicitly that any decision regarding such an application would be preceded by a *hearing* of interested parties, the Board's first order initiated a process that had not been prescribed in either the *Oil and Gas Wells Act, 1931* or the *Oil and Gas Conservation Act, 1938* but would become the central feature of the Board's regulatory approach.[82]

While the collection of data upon which to base an acceptable proration order frantically continued, Board activities were overtaken by the fortuitous circumstance of a rising crude oil market. This allowed the Board a few more weeks grace, and seemed to herald a more hospitable environment for the introduction of what was certain to be a controversial order. In the meantime, the Board had to work with the existing system of production control that Imperial had imposed upon the Turner Valley field in September 1937.[83] At that time, as production surpassed both market demand and the capacity of the pipeline system connecting the field to the Imperial refinery in Calgary, Imperial announced that it would accept only a certain percentage of each well's open flow capacity. It was generally accepted by all parties that such an approach was inequitable and deficient on technical grounds. Crude oil producers anxiously awaited release from what was essentially a buyers' proration plan in favour of the Board's neutral and more technically sophisticated approach to proration. Their wait was eased as the Board, after consultation with Imperial, announced on August 13 that the allowable quota for producing wells would be raised from 37% to 48% of capacity and total field production to 25,000 barrels per day.[84] On August 19, the quota was raised to 51% and on August 26 to 56%.[85]

The Board's long anticipated proration schedules were presented on August 31. They came immediately after a conference with members of the Alberta Petroleum Association and in the form of two Board orders: one for the production from oil wells, the other for production from gas wells.[86] The oil schedule contemplated a total field production of 28,363 barrels per day divided among 52 producing wells in such a manner as would best promote a uniform rate of withdrawal from the common reservoir, a requirement that engineers had recently come to understand was fundamental to achieving maximum recovery. Each well's allowable was arrived at through the application of a formula that took into account the well's gas-oil ratio, bottom-hole pressure, acreage, and rate of flow through a two-inch nipple. The Board was convinced that each well in the field would be allowed "to produce its fair share of the available economic market" and that "the new daily allowables result-

ing from the adoption of this formula would permit the field to supply the present market demand with greater ease and with considerably less waste than resulted from the system of buyers' proration,
presently in effect."[87] The new production schedule was to come
into effect on September 2.

The Board's gas schedule was based upon an estimated "total
monthly economic market demand" of 1,200 million cubic feet or
40 million cubic feet per day, well down from the estimated 150 to
200 million cubic feet then being produced daily. This projected demand had to be shared among 101 producing gas wells, the latter
being defined as wells that "produced natural gas with a gas-oil ratio in excess of thirty-one thousand cubic feet for every barrel of liquid recoverable" under normal field pressures and temperatures.[88]
To establish each well's proper share, the Board was guided by a
formula that took into account three field factors: assigned
acreage,[89] rate of flow at two-thirds closed pressure, and bottom-
hole pressure, plus two overriding external factors. None of the
aforementioned potential factors could account for more than 25%
of a well's allowable production, and the maximum production of
any well could not exceed 40% of its open flow potential. In recognition that compliance with its gas proration order would require
some rearrangement of production and transportation facilities in
the field, the Board delayed the order's date of effectiveness until
September 14.

Reaction to the proration orders came quickly and was predictable. Anticipating the end of what had been a nearly 20-year struggle, the *Calgary Herald* declared its satisfaction with the headline
"Calgary's Midnight Sun Will Disappear Next Week When Gas
Waste Halted."[90] Gas well operators hinted darkly of a possible
court challenge, whereas crude oil producers expressed general satisfaction.[91] For the latter, one of the blessings, in addition to the
promise of a fairer distribution of market share, was the end of the
oil contracts, which the refineries had compelled their suppliers to
sign, forbidding the sale of any surplus oil to any other refiner. In a
last effort to gain willing compliance, the Board accepted the advice
of the Alberta Petroleum Association to postpone the implementation of its gas order until October 15, a month beyond the original
target date.[92] This was not enough for the Alberta Petroleum Association's radical fringe. On September 16, Miracle Oils Ltd., Mercury
Oils Ltd. and Gas and Oil Products Ltd., the company operating
the absorption plant in the south Turner Valley field, took a
certiorari application before the Supreme Court of Alberta in the
hope of persuading the court to quash, or at least set aside, the new

Board's first gas well production order.[93] Mercury Oils alleged that the Board order would mean the shutting in of the company's four wells and the loss of approximately $11,000 per month. Miracle claimed that it would lose $7,000 per month, whereas Gas and Oil Products held that they would no longer receive sufficient gas to operate, thus putting at risk the $500,000 that had been invested in its plant. The dissident companies sought by injunction to have the offending Board order set aside while the court heard their contention that both the *Oil and Gas Conservation Act*, which created the Board, and the subsequent amendment to the natural resources agreement between the province and the federal government were *ultra vires*.

Knode welcomed the challenge. William Epstein, a young lawyer working for the Board at that time, remembers going with the Board chairman to Turner Valley to confront Mercury Oils' president, Albert Mayland, who had promised to keep his company's wells producing. Mayland was a prominent Calgary businessman and one of the most successful independent Turner Valley operators.[94] Standing nose-to-nose with Mayland, Knode informed him that the Board would have seals put on the valves of Mercury's wells, and "if he broke those seals he would be thrown in jail."[95] Used to the litigation that characterized the working environment of the Texas Railroad Commission, Knode looked forward to having Alberta's new conservation act tested or, as he preferred to say, "validated" in court. The Board's legal council, A.L. Smith, and his assistant, Epstein, were less enthused. They believed that defending the Act would be a difficult uphill fight.[96]

The Supreme Court of Alberta agreed to hear the motion and on October 11 before Mr. Justice W.C. Ives, the companies' lawyers commenced their argument to demonstrate that the Board was an illegally constituted body whose orders could not be put into effect. In presenting their case, S.J. Helman and J.C. Mahaffy went over much of the same ground covered earlier in the Spooner case. They held that the Act was *ultra vires* because it interfered with the regulation of trade and commerce, which was a federal responsibility; because it provided for the taking of property from the plaintiffs for the benefit of others without compensation; and because it altered the terms of the leases of the plaintiffs without their consent. For good measure, they contended that, far from bringing conservation, the Board's measures were doing quite the reverse. The sworn statements of several well-known local engineers were presented, including that of Stanley J. Davies, who declared, "I have no hesitation in saying that since the coming into force of the orders with relation to the so-called oil or crude wells the amount of wastage of gas has been increased tremendously."[97] Critics such as Davies focused their complaint upon what they perceived to be a basic injustice. On the one hand, the Board sought to restrict gas wells to a volume of approximately 40 million cubic feet per day; on the other, the Board was permitting the flaring of about 90 million cubic feet of gas per day that was jointly produced with crude oil and for which there was no market. Helman and Mahaffy concluded that the basis for prorationing had been arrived at by "non scientific and inequitable methods," and that the combined effect of the Board's activity was the creation of an unlawful combine to stifle competition.[98]

Fearful that the Conservation Act might not survive the test of Helman and Mahaffy's forceful argument, Smith shifted his defence strategy. Rather than focusing on the legal argument, he took a different line, saying that any decision made regarding this matter was far too important to the future of the industry and to Alberta to be decided before a judge in chambers where solicitors could only debate questions of law. Smith insisted that this was an area where matters of law could not be decided properly without reference to highly technical facts that demanded the presence of qualified witnesses.[99]

Mr. Justice Ives agreed. On October 15, he announced that he was dismissing the application.[100] Though A.L. Smith, counsel for the Board, and J.J. Frawley, for the Attorney General's department, had been successful in having the application thrown out, the Board had nonetheless suffered a severe battering. The characterization of the Board as "dictatorial" had made headlines, and the decision of Smith and Frawley not to argue the validity of the Conservation Act, but instead to base their defence upon the plaintiff's "improper proceedings," did little to enhance the Board's credibility.[101] In open defiance, the embittered naphtha producers at the south end of the Turner Valley field simply refused to recognize the Board's gas proration order. Meanwhile, another company took up the challenge. Charging that "the gas conservation board is ruining the Alberta oil industry," C. Fisher, the managing director of Model Oils, declared that the Model-Spooner-Reward Nos. 1 and 2 wells "would continue to be operated at maximum possible efficiency, despite orders of the Board," and he defied the Board to prosecute.[102] Fisher's action was commended immediately by the president of Davies Petroleum, who stated that he might be compelled to follow a similar course.[103] Meanwhile, Miracle and Mercury Oils directed their lawyers to appeal the Ives decision.[104] It was clear that the conservation legislation was going to have to be revised.[105]

Just when the Board needed all the friends in the industry that it could muster to stave off the expected, but serious, challenge from the naphtha sector, it was dismayed to learn that support among the crude oil producers was eroding rapidly. The problem here centred upon an unanticipated decline in the crude oil market. At the end of August, the Board had optimistically based its first crude oil proration order upon a projected field demand of 28,363 barrels per day, and it had been widely held that, since producers were no longer contractually bound to a single refinery, additional markets would be found elsewhere within the prairie region. On September 21, when it was necessary for the Board to adjust the field allocation down to 22,000 barrels per day to help reduce the volume of oil that had accumulated in storage, there seemed little reason for alarm; however, the third proration order, presented on October 18, and calling for a reduction in field production to 14,500 barrels, was greeted with shock. The spreading gloom became evident when, eight days later, a new Board order called for a reduction to 11,500 barrels per day.[106] Not only had prorated production been cut back by nearly two-thirds in the space of three months, but also five new producing wells had come on stream to compete for a share of the sharply diminished allowable field production. Most threatened were the smaller independent crude oil producers, some of whom now began to question the confining regulations of the Petroleum and Natural Gas Conservation Board and to agitate to be left alone to produce any volume for which they as individual companies could find a market.[107]

The naphtha group, in combination with a growing number of disgruntled independents from the crude oil sector of the industry, now presented a more formidable obstacle to the Alberta government's oil and gas conservation program. As the remaining support for the Board shrank to a core of larger producers, the government became more vulnerable to the charge that conservation was really a device to promote the big companies and foreign capital at the expense of the local men whose pioneering efforts had established the industry.[108] By mid-October, the Alberta government realized that it had a full-fledged crisis on its hands. The Board that it had confidently created in the spring had been undermined by a collapsing market and was now floundering as its orders went widely unheeded. With an emerging oil glut, investment was already drying up, and there was beginning to be talk of winter unemployment in the oilfield.[109] At the same time, the depressed market brought forth a renewed attack on the refiners. Two groups became highly critical of the spread between the purchase price of crude oil and the retail price of gasoline. Turner Valley crude oil producers sought an explanation as to why the price they received was so low, and consumers were asking why the price they paid for gasoline was so high relative to the declining price of crude oil.

The Social Credit government's response was twofold. First, Provincial Secretary E.C. Manning announced that Mr. Justice A.A. McGillivray of the Supreme Court of Alberta had been appointed to chair a royal commission that would conduct a thorough investigation of the production, refining, transportation and marketing of petroleum and petroleum products in Alberta. Beyond its sweeping general mandate, the commission was specifically charged to find "the fair and equitable field price which should be paid for crude petroleum in the Province of Alberta," and was also to find "what the fair and equitable price and/or cost of petroleum products sold to consumers in [Alberta] should be."[110] Second, Premier Aberhart announced that his government intended to hold a special fall sitting of the Alberta Legislature, so that the *Oil and Gas Conservation Act* could be reconsidered.[111]

As part of the reconsideration process, the government stated its intent to have Turner Valley operators come to Edmonton to give members of the Legislature the direct benefit of their views. Although the government was prepared to go once more through the ritual of seeking guidance, it is clear that its central objective was fixed. In an editorial a few days after Aberhart's announcement, the *Calgary Herald* correctly interpreted that "the chief purpose of the special session of the legislature is to present amendments to the Act which will make it more courtproof."[112]

The dissenting independents attempted, as they had in the past, to prepare the ground by circulating to each member of the Legislature a memorandum outlining the "inequities" of government-imposed conservation.[113] In their document, the independents argued that Board policy was misconceived, and they endeavoured to establish in the minds of the members of the Legislature the idea that the principle directing the Board and its supporters was proration and not conservation. The prorationing of crude oil production, they maintained, was "not necessary or proper to prevent the waste of gas," and the only substantial effect would be to hand the field over to the major companies whose record in the search for new oil, they alleged, was notoriously poor.[114] Although it is apparent that the arguments presented by this group of independents did carry some weight with individual MLAs, they had little impact upon those who were actually drafting the government's new conservation bill.

While hostile operators organized, Frawley, Smith and Epstein, with special assistance from John Weir, dean of the Faculty of Law at the University of Alberta, worked to put the final touches on the government's second attempt to produce comprehensive and effective conservation legislation.[115] The precise nature of how the Social Credit government intended to proceed was revealed with the release of a draft copy of the proposed legislation a week before the scheduled fall session of the Legislature. Section 44 of the new Act proclaimed that, in those areas designated by the Act, the Board's decisions and orders were "final and conclusive" and not open to question or review in any court, and also that any proceeding of, or by or before, the Board could not be "restrained by injunction, prohibition or other process" or be "removable by certiorari or otherwise into any court."[116] The worst fears of the Board's critics were realized, and their sentiments were reflected in the response of the *Calgary Herald* to the Speech from the Throne opening the Legislature. Stretched across the front page was the headline "Gov't Plans Drastic Valley Control, Even Own 'Police' to Enforce Orders."[117] The paper predicted that the control measures would generate a storm of protest in the special session just under way in Edmonton. The next day, *Herald* columnist Fred Kennedy wrote about the "dictatorial powers" to be given to the Conservation Board. He reported that not only had the government vested the Board with wide new powers to enforce its orders but also it had decided not to permit Board orders to be appealed.[118]

When the contending parties assembled on November 17 before the Agriculture Committee, the Alberta Legislature's equivalent of the Committee of the Whole House, it was to continue a debate that had persisted almost without let-up for more than a decade. It was to travel over the same wearisome ground once more; yet, the intensity of the confrontation between the opposing factions had not diminished, for there was the overwhelming sense that this was the final round. W.F. Knode was the first to be called as a witness before the gathered MLAs and industry representatives. It was the beginning of a gruelling ordeal. Over the three-day duration of the sitting, Knode faced a barrage of hostile questions that occasionally took on a personal edge when delivered by one of those who had been put off earlier by Knode's unyielding and sometimes brusque manner. The new Board chairman held his own nonetheless. What he lacked in formal debating style, he made up for in commitment and bulldog-like tenacity. Knode, with assistance from the Board's lawyer, A.L. Smith, and Ministers Tanner and Manning, was compelled to address four main questions: the impact of conservation

policy upon the small independent producer, the question of compensation for those injuriously affected by conservation measures,
the matter of the right to appeal Board orders, and the question of
the relationship between proration and conservation.

The attack commenced the moment Knode finished his short introductory address outlining the development of conservation policy in Texas and the essential features that characterized his approach to conservation in Turner Valley. Immediately, he had to
fend off a challenge to his credibility by a questioner who charged
that the advice Knode had just given the Alberta Legislature stood
in direct contradiction to the advice he had given a year earlier to
the Montana Conservation Board.[119] His next inquisitor wanted to
know if it would "still be the administrative policy of the Board
to . . . prorate the naphtha and gasoline to a point where it amounts
to confiscation of the interests of the small industries."[120] The operator went on to plead that men such as himself had drilled wells in
good faith, had done everything under permit and with approval of
the Department of Lands and Mines, and had brought in producing
wells that returned several thousand dollars per month. Now they
were not going to be allowed to produce, and Alberta investors
would not be allowed to profit from their enterprise or perhaps
even to recover the capital that they had invested. Was this not confiscation, did they not have a clear case for full compensation?
Knode's answer was, "The Board must . . . shut off that waste. The
effect on the particular producer who has been benefiting by a condition of waste I think shouldn't enter into the Board's consideration." His interrogator persisted," . . . [a] profitable well having
been secured according to the usages of the field at that time then
you come along and say 'Well, you are in a position where you have
a nice producing well—a profitable well, I am going to wipe that investment out.' Is that the policy of the Board?" Knode was not to be
moved. He responded, "If that well is creating waste, we are going
to stop that if we possibly can. I think that is definite enough."[121] Although it seems out of keeping with his Texas background and that
state's traditional emphasis on individual rights, it might nonetheless have been the difficulties that Knode experienced trying to initiate effective conservation measures in such an environment that underlay his philosophy on conservation. In his view, individual
rights, as manifest in the rule of capture, had to yield to the larger
interest of conserving reservoir energy for the benefit of all producers of a common pool and in the public interest to maximize recovery. Knode had no sympathy for those who engaged in the wasteful

production of oil and natural gas, even if they claimed to have invested in "good faith."

Similarly, Knode was adamant on the matter of prorationing. To him, conservation and prorationed production were inseparably linked. It was necessary to proration production not only to ensure equity but also to ensure that each operator producing from a common pool received his fair share of the resource. It was also necessary because, as Knode explained, maximum recovery required a uniform withdrawal from a common reservoir.[122] This was achieved by allocating production among individual wells, taking into account each well's particular characteristics. The Board chairman was prepared to concede that prorationed production in certain circumstances could be inequitable. It was pointed out by several of the smaller producers that the Royalite Oil Company (Imperial) was drilling a large number of new wells, and since it was impossible for them to keep up with Royalite's drilling pace, the allowable production from their wells would be cut back in favour of the new wells as long as the total allowable field production was not increased.[123] Problems would have to be worked out and adjustments would have to be made, but Knode was not prepared to yield on the proration principle.

On the third day of the hearings, Knode was called to give his summation. After having his knowledge of the Turner Valley situation called into question[124] and after listening to hours of criticism, typically veiled with mild obeisance to the principle of conservation, the edge of Knode's noted temper began to show. He was not going to accept the allegation, stated or implied, that he and his Board did not know what they were doing. "I am sorry," he began,

> that I haven't had 24 years, or how long that this field has been going, experience in the Turner Valley. I like the country up here—they are nice people, but unfortunately (maybe not unfortunately) I was born in the United States . . . and followed the oil business all my life, and was a 'roustabout' for the Standard Oil at $2.00 a day when I was about 16.

He stated that he and his colleagues knew that the question of Turner Valley gas proration would be a "dandy," and he explained that they had taken special care,

> knowing these fellows had only had 14 years gas production under conditions they knew were wasteful, to consider it, and be-

fore we entered the order we gave them still more time to consider, and after we entered the order we changed the effective date on that order by four weeks to give them time to restudy the problem which they had not given much consideration to for 14 years.[125]

Getting warmed up to his topic, Knode continued.

I begin to feel rather intensely about this thing, since all this hearing I have been impressed with one particular thing. In all the discussions the independent has presented his view, and very well presented it—the major companies presented their view last night and well enough, I guess, but there has been nobody presented the view of the man who has the basic ownership in that field and these companies... basically the province of Alberta is the basic owner in that field.... Then I say that this Legislature is justified in making any move that it sees fit, making any law that might be as stringent as possible to see that conservation is effected down there, because already the people of the Province have lost a lot of money in that structure.[126]

Knode wanted a strong Act that would allow the Board to get on with what he saw to be an urgent task in the public interest. To the shock and horror of some of those assembled, he warned that

I am ready now to recommend to this Government that if you cannot by legislation force conservation in the Turner Valley Field, in what I think is the most important piece of property that this Government is interested in in this Province; if you cannot bring that about by an Act and enforcement by a Board, then it would be my firm recommendation that this Government take over control of that field.[127]

In addition to his passionate conviction in the necessity of conservation, more than the representatives of the small independent companies in his audience realized at the time, Knode's working experience had left him with an emotional bias in favour of the independent companies.[128] There is a hint of this in his final appeal.

We have heard a lot of discussion about the various interests and how they are affected, but, gentlemen, who is going to benefit the most—by conservation, proration, or by the law of capture? I can tell you from experience that the major companies will certainly do well by the law of capture because, in various cases, what they were built under was the law of capture.[129]

Knode's suggestion that government control of the Turner Valley field might have to be considered certainly stimulated incentive to come to an agreement. Finally, in a late-evening session on 21 November 1938, a compromise was hammered out and consented to by all but two of the companies or interests represented. Despite Knode's view that compensation for those adversely affected by conservation orders was unnecessary, it was apparent to Tanner and Manning that this was the critical factor required to bring at least grudging acceptance from the majority of the fractious independents. Accordingly, a new clause was written into the draft bill, stating that, within six months of the legislation coming into force, the Board would prepare a definite compensation scheme for cabinet approval.[130] The existing Act did contain a section allowing a limited form of compensation if the Board deemed proper, whereas the new bill contemplated actual commitment to a specific scheme. In one other area, there was also a modest retreat. Tanner was persuaded to make an amendment that would allow court arbitration, in cases where the Board and an operator could not agree upon an assessment, to determine the amount to be contributed to meet the Board's operating costs. This was as far as the government was prepared to go in allowing a role for the courts. Although section 44, which denied appeal to the court regarding any "action, decision, and order of the Board,"[131] was almost universally condemned, Tanner and Manning held firm. During the early stage of the debate, Board lawyer A.L. Smith had attempted to still criticism by pointing out that Alberta's approach was not without precedent—decisions of the federal Board of Railway Commissioners were not open to appeal.[132] Even though he was no doubt aware, Smith chose not to mention the precedent that existed closer to home in the case of decisions made by the Board of Public Utilities Commissioners. Manning attempted later to deflect objections by reminding critics that operators truly did have the right to appeal. They could approach the cabinet, which had power under the Act to revoke, suspend or change any order made by the Board. He ventured the further suggestion that this offered the possibility of quicker redress than the courts.[133]

With almost the full consent of the Turner Valley producers secured, Tanner returned to the Legislature later in the day with his draft bill for third reading and assent. There he fought off last ditch attempts by Liberal and Conservative members of the Opposition to have the bill delayed, until after the findings of the McGillivray Commission were reported, and to have an appeal clause inserted.[134] On 22 November 1938, the *Oil and Gas Resources Conservation Act* became law. Apart from the fundamental change that came through denial of the right of appeal, and the substantial reshaping of the compensation section, the other changes were minor. A provision was added for the appointment of a deputy chairman to act in the place of the chairman if the latter were unavailable to carry on his duties. Owners or managers of refineries were required henceforth to keep and make available to the Board a record of all petroleum received—from whom, price paid, and the disposition of all refined products. The revised legislation also provided that government would pay the costs incurred by the Board enforcing the provisions of the *Oil and Gas Wells Act, 1931*. Most of the "new" enforcement measures were ones that the Board had access to previously through the *Oil and Gas Wells Act*. Now, the authority to shut wells down, to take possession of property, and to dispose of petroleum produced by wells that the Board might possess was to be found in the Conservation Act itself. Also conferred upon the Board was the right to purchase any well or wells that might be required for field repressuring or any other conservation purpose.

Opinion that the new Alberta legislation went to extreme lengths was not confined only to the smaller Turner Valley producers and certain members of the legal fraternity. T.G. Madgwick, the senior technical adviser to the federal Department of Mines and Resources, when called upon to review the newest Alberta Conservation Act, observed that section Nos. 44 and 46, which seemed "to place the acts of the Board beyond the Courts and yet empower it to expropriate property, temporarily at least, and to conscript the employees of the concern," were "little short of dictatorial." In his view, however, there appeared no reason for undue alarm.

> Provided [such powers] are to be used only as a club to bludgeon unruly or obstructionist operators into that cooperation without which the problems ahead of Turner Valley will never be overcome, no harm can come. The present Board is most unlikely to abuse its powers.[135]

For William Knode, it must have seemed a long time since March 29 when he first arrived in Edmonton from Texas. But both he and Nathan Tanner had survived their baptism of fire. More importantly, they had achieved their central objective, Alberta's Petroleum and Natural Gas Conservation Board was now effectively shielded from the possibility of legal challenge. In managing this, they provided a foundation of authority that progenitor agencies in the United States had never possessed. Given such a power basis, the essential requirement from this point would be the will and the abilities of the three men assigned the responsibility of translating the legislation into effective measures in the field.

PART II

Petroleum and
Natural Gas Conservation in Alberta
1938–1959

FOUR

Establishing a Regulatory Foundation,
1938–1947

THE INTENT, PURPOSE AND OBJECT of this Act is to effect
the conservation of oil resources and gas resources or
both in the Province by the control or regulation of the
production of oil or gas or both, whether by restriction or prohibition and whether generally or with respect to any specified
area or any specified well or wells or by repressuring of any oil
field, gas field or oil gas field and, incidentally thereto, providing
for the compulsory purchase of any well or wells. (*Oil and Gas
Resources Conservation Act,* 1938 Second Session)

Its mandate defined in a reinforced statute, the Petroleum and Natural Gas Conservation Board (PNGCB) once more set about the task
of bringing acceptable production practice to the Turner Valley oil
and gas field. The embittered 10-year struggle that preceded the
emergence of the Board predetermined that it would not be easy.
Although a small pocket of emphatically hostile companies—led by
Mayland Oil Co. Ltd.—promised to carry on the fight against provincial conservation legislation,[1] the majority of Turner Valley operators supported the idea of conservation; however, there was a
great range of opinion on how conservation could be equitably
achieved. Some were highly sceptical that the interests of the
smaller producers would receive "fair" treatment. The attitude of
many of the operators was one of "wait and see;" thus, despite its
immense coercive power, the Board's crucial first objective was to
establish credibility. It needed to put in place a field policy that
would not only move quickly towards acceptable conservation
standards but would also win the confidence of a diverse industry
group. The government and the Board realized that acceptance

Board Chairmen 1938–1962

W.F. Knode, 1938–1939. ERCB.

R.E. Allen, 1940–1941. ERCB.

J.J. Frawley, 1942–1943. ERCB.

Dr. E.H. Boomer, 1943–1945. ERCB.

A.G. Bailey, 1946–1947. ERCB.

D.P. Goodall, 1947–1948. ERCB.

I.N. McKinnon, 1948–1962 (absent on leave 1959–1962). ERCB.

G.W. Govier, Deputy Chairman, 1959–1962 (Chairman, 1962–1978). ERCB.

These wells at Little Chicago in the Turner Valley Field were typical of the 294 wells in Alberta in 1938 when the Board began to establish its regulatory authority. Glenbow Archives NA-2335-8.

would be achieved most easily if the push for conservation could be accompanied by an expansion of oil and gas markets. The PNGCB immediately acquired therefore a new responsibility not contemplated in the Act. It was not only to implement conservation in the field but also to play a major role in the search for new markets beyond Alberta. This additional obligation was undertaken willingly in recognition that an expanding market would greatly facilitate the acceptance of conservation practices.

The industry over which the Petroleum and Natural Gas Conservation Board hoped to establish its regulatory authority in the late autumn of 1938 was composed of approximately 60 producing companies, five with refining operations.[2] By far the dominant company was Imperial Oil Limited. It was the major refiner, and its subsidiary, Royalite Oil Company, with 28 wells, was the largest oil and natural gas producer. The British American Oil Company, Gas and Oil Products Limited and Lion Oils were the other significant refiners. Ranked well below Royalite were about a dozen middle-sized companies, each with at least four producing wells: Anglo-Canadian Oil Company, Brown Oils, Davies Petroleum, Home Oil, McLeod Oil, Mayland Oil, Mercury Oil, Miracle Oil, Model Oils, Okalta Oils, Spooner Oils and Sterling Pacific.[3] In addition, there were three natural gas companies. The Canadian Western Natural Gas, Light, Heat and Power Company Ltd. supplied Calgary, Lethbridge and a number of smaller southern Alberta communities. Northwestern Utilities Ltd. supplied Edmonton, and

TABLE 4.1 Oil Production in Alberta by Field for the Fiscal Year
1938–1939

Location	No. of Wells	Barrels
Turner Valley oil wells	70	6,011,112
Turner Valley gas-oil wells (naphtha)	103	579,518
Turner Valley shallow crude wells	4	9,038
Red Coulee	7	13,777
Wainwright	6	12,145
Other	5	25,731
Total	195	6,651,321

SOURCE: Alberta Department of Lands and Mines, *Annual Report,* to 31 March
1939, p. 68.

TABLE 4.2 Wells in Alberta Capable of Producing Natural Gas for
the Fiscal Year 1938–1939

Location	No. of Wells
Turner Valley	103
Medicine Hat/Redcliff	46
Bow Island	11
Foremost	6
Milk River	3
Viking	21
Kinsella	3
Battleview	1
Fabyan	2
Brooks	6
Total	212

SOURCE: Alberta Department of Lands and Mines, *Annual Report,* to 31 March
1939, p. 69.

the municipally owned Medicine Hat Natural Gas Company
looked after the needs of Medicine Hat. In all, the oil, gas and petroleum products industry directly employed 2,573 in 1938, and by
value petroleum production was already an important component
of Alberta's gross provincial product.[4] Still tiny by continental standards, the petroleum industry nonetheless represented the only significant growth sector in an otherwise struggling provincial econ-

TABLE 4.3 Gas Consumption in Alberta for the Fiscal Year
1938–1939

		Thousands of Cubic Feet
From Turner Valley		
Gas company	6,260,916	
Refinery	1,362,415	
Storage and compressor fuel	1,314,357	
Bow Island town	51,678	
Field use (largely estimated)	10,045,129	
Total		19,034,495
Brooks		49,319
From Wainwright—to town supply	98,269	
field use (estimated)	19,000	
Total		117,269
From Viking to Edmonton and towns		3,388,401
Medicine Hat		1,979,484
Redcliff		706,000
Range—exported to Montana		307,652
Red Coulee—field use, including export to Montana		39,918
Foremost		4,362
Total		25,626,900

SOURCE: Alberta Department of Lands and Mines, *Annual Report,* to 31 March 1939, p. 70.

omy. Moreover, oilfield workers' annual earnings averaged $1,531 compared to $431 for farm labourers. By the end of 1938, 37 limestone wells had been completed in Turner Valley, and elsewhere in the province 39 had either been completed or were still in the drilling stage. Oil production had climbed from 3,680,099 barrels in 1937 to 6,651,321 barrels, with a sales value of $8,432,258.[5]

Of immediate concern to the new Board were the province's 294 producing wells: 99 of which were gas wells, 92 produced crude oil, and the remaining 103 were mainly naphtha wells, producing a combination of gas and light oil. Nearly two-thirds of all the wells were situated in Turner Valley and accounted for virtually all the

TABLE 4.4 Petroleum and Natural Gas Conservation Board Staff: 31 December 1938–
31 December 1939

Name	Date of Employment	Date of Departure	Function or Position	Approximate Annual Salary
W.F. Knode	July 1938	July 1939	Board Chairman-Engineer	12,000
C.W. Dingman	July 1938	—	Deputy Chairman-Engineer	4,250
F.G. Cottle	July 1938	—	Board Member-Accountant	3,600
J.W. Kraft	July 1938	—	Statistician and Accountant	2,220
M.D. Kemp	July 1938	—	Office Engineer, Calgary	2,220
D.P. Goodall	Nov 1938	—	Chief Inspection Engineer	1,800
G.A. Connell	July 1938	—	Engineer in Charge, Turner Valley	1,800
H. Bagnall	July 1938	Nov 1939	Field Engineer, Turner Valley	1,200
A. Lees	July 1938	—	Field Engineer, Turner Valley	1,200
L.D. Publicover	July 1938	—	Field Engineer, Turner Valley	1,200
B.H. Corey	Nov 1939	—	Field Engineer, Turner Valley	1,044
S.S. Cosburn	Sept 1939	—	Field Engineer, Turner Valley (Geologist)	1,044
G. Doherty	July 1938	July 1939	Stenographer	1,080
P. Falk	July 1938	—	Stenographer	1,080
B.M. Smith	June 1939	—	Stenographer	1,080

SOURCE: ERCB, Minutes, 30 November 1938.

oil produced and about 74% cent of the natural gas *consumed* in Alberta[6] (see Table 4.3). The areal extent of the field upon which the new board concentrated its attention was relatively small. Rarely more than a mile and a half in width, it extended nearly 18 miles along a NW-SE axis (see Maps 4.2 and 4.4). On these few acres, however, rested the province's great hope for the revival of its stricken economy. The Board's initiatives in the Turner Valley field were awaited therefore with intense interest by both the private and public sectors.

The Petroleum and Natural Gas Conservation Board met for the first time under the revised Act on 30 November 1938. Assembled at their office in the "New Telephone Building" at 115–6th Avenue sw, first W.F. Knode and his colleagues confirmed the appointments of the eight Board employees hired in July, and then added to their professional body D.P. Goodall, who like most of the others came from the Petroleum and Natural Gas Division of the Depart-

FIGURE 4.1   Petroleum and Natural Gas Conservation Board
Organization Chart, 1939

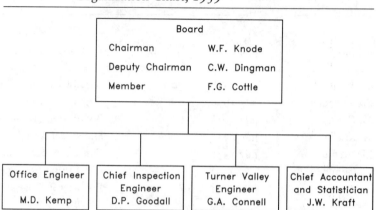

ment of Lands and Mines. This gave the Board a field staff of five engineers. Goodall was assigned supervision of drilling and production outside Turner Valley and G.A. Connell, with three assistants, was put in charge of activities in Turner Valley.[7] In the Calgary office was the senior engineer, M.D. Kemp, an accountant and two stenographers. Floyd K. Beach, a statistician employed by the Department of Lands and Mines, which had an office in the same building, also did some work for the Board.[8] (See Figure 4.1.)

The Board began its second attempt to control the production of oil and gas in Turner Valley. Board Order No. 1 called for a total field production of 12,500 barrels of oil per day and listed each well's allowable share.[9] The Board's order limiting natural gas production to slightly less than 40 million cubic feet per day was anticipated. There was, however, a modest compromise intended by the minister of lands and mines to placate the naphtha producers and to win at least their grudging support. Where the Board's abortive earlier order permitted the production of natural gas sufficient only to meet the market demand for fuel gas, the new schedule set aside a portion of the 40 million cubic feet per day field production allowance for the admittedly wasteful extraction of natural gasoline. This allowance was to continue until 15 January 1939, when the Board promised to restrict further the production of gas from all wells to the amount required "for lighting, heating and the operation of power, together with any further amount which is returned to the earth."[10]

The temporary concession to the absorption plant operators manifests the direct part being played in these critical first days in the Board's history by Minister of Lands and Mines N.E. Tanner. Board Chairman Knode saw no special reason to appease those who, in his view, had already benefited by 14 years of wasteful production and who had been unrelenting opponents of conservation; whereas Tanner, at yet another meeting with Turner Valley operators on the eve of his Board's renewed attempt to impose conservation, found reason for a more lenient and more political response.[11] Nonetheless, as 1938 drew to a close, conservation advanced at last from being merely an issue of debate to a concrete reality in Turner Valley. It had come none too soon. In the interval, an estimated 1,038,798,000 thousand cubic feet had been drawn from the Turner Valley reservoir, of which just 168,395,000 thousand cubic feet, or a mere 16% had been used, and much of this in an indifferent and wasteful manner.[12]

The imposition of production controls was only one aspect of the program to end this colossal waste. On 11 January 1939, the Department of Lands and Mines turned over the responsibility for the supervision of all drilling operations to Board engineers, and to assist the new inspectors to put an end to wasteful drilling and production practices, the government issued a more stringent and comprehensive body of drilling and production regulations on the same date.[13]

The regulations were based upon a comprehensive definition of "waste." In addition to its ordinary meaning, waste was declared to include the drowning with water of any stratum capable of producing petroleum or natural gas; permitting any natural gas well to burn wastefully; permitting the escape of natural gas into the open air, from a well producing both oil and gas or either, in excess of the amount that was necessary in the efficient drilling or operation of a well; the production of natural gas or crude petroleum in excess of transportation or market facilities or reasonable market demand; the operation of any well with an inefficient gas-oil ratio; waste resulting from the unnecessary, inefficient or improper use of reservoir energy, including the gas energy or water drive in any well or common reservoir; physical waste or loss resulting from drilling, equipping, locating or operating any well in a manner that would tend to reduce the ultimate recovery of any petroleum from any common reservoir; surface waste, including the storage of petroleum or any petroleum product in open pits or tankage or any receptacle declared to be inadequate by the Board; and the creation of fire hazards.[14] Finally, lest there be any doubt regarding the government's intention to provide comprehensive authority, the regulations provided that waste in "the production, transportation, distribution or use of natural gas or petroleum" could be defined by the Board as it saw fit.[15]

The regulatory approach to the elimination of waste began with a licensing system. No person could begin the drilling of any well unless he held a licence issued by the minister of lands and mines on recommendation of the Board under the authority of the regulations. Applications made to the Board for a licence to drill had to be accompanied by a plan, certified by the surveyor or engineer who made the survey, showing the exact location of the proposed well site and demonstrating that it was not within 440 yards (402 metres) of any producing well or any other well being drilled. Such well spacing was based upon the belief that the ideal or most efficient drainage area for each well was approximately 40 acres. The regulations did permit the Board to recommend the issuance of a licence for the drilling of a well nearer than 440 yards in certain defined instances. Such flexibility was a standard feature of oil and gas regulations throughout the continent and a factor of great potential significance. The setting aside of technical considerations in favour of other factors to allow closer spacing rested upon a rationale that was the product of North America's frontier heritage. Alberta's new regulations provided for the special situation where the applicant held the petroleum rights to an area of less than one legal subdivision (40 acres) but had no rights to any contiguous or adjacent land.

> Where the requirement to drill at a point at least 440 yards from a producing well would prevent the applicant from exercising his *right* to drill one well in respect to each legal subdivision in respect of which he has the right to search for, win and get petroleum and withdraw gas;[16]

the Board could issue a licence to drill a well at any designated point. The tradition that the continent's resource-rich frontier existed as a region where the "small man" was free to find his fortune predisposed legislators towards assuring access to lesser, and especially local, entrepreneurs. In Texas, oilfield development was characterized therefore by numerous exceptions granted to ensure that small mineral rights holders could drill at least one well. This made a mockery of well-spacing regulations and seriously compromised conservation objectives.[17] In Alberta's case, exemption from the minimum well-spacing standard remained more a possibility on paper than a principle in practice, except for heavy oil pools where spacing as close as 10 acres was permitted, beginning at about this time. Although the Alberta Board always seems to have been inclined to resist granting exceptions, it never had to face the same pressure as that put upon its U.S. counterparts by thousands of small farmers and ranchers who owned mineral as well as surface title. In Alberta, most settlers had not received subsurface rights with their homestead patents, such rights remained vested with the Crown. This distinction was of fundamental importance, for it gave the Alberta government and its Conservation Board a singular advantage in the implementation of regulatory policy.

Once a licence had been obtained, drilling could begin subject to the guidelines embodied in the regulations. While drilling, the operator was required to keep a daily record of operations, to ensure that the borehole was properly cased with approved casing, and to make certain by regular measurement that the drill hole did not deviate more than 4° from the vertical.[18] If it was determined that the bottom of the hole had wandered too far from the vertical, so as to infringe upon the drainage rights of adjoining properties, the Board could suspend further drilling operations or deny production until the well was redrilled. For example, in 1942, the owners of Foothills No. 10 in the Turner Valley field were compelled to redrill the last 1,600 feet of their well when it was found to have deviated outside the prescribed area. The regulations also restricted drilling beyond any "petroleum, natural gas or water stratum" without sealing off such stratum with mud-laden fluid, casing or cement. Before drilling operations could be suspended, the Board had to be notified so that it could make sure that this was done without damage to the reservoir. Similarly, the regulations established guidelines for well abandonment. Every effort was to be made to prevent water from entering producing horizons, and casing could not be removed without the Board's permission. Protection of the surface environment was also a matter of concern, and the regulations directed operators after derrick removal to fill up all excavations that had been made during drilling and production operations. Another of the more important measures required that the operator take drill-cutting samples at 10-foot intervals and that these be sent washed and "accurately tabulated" to the Board. When directed by the Board, the operator was also compelled to take and send core samples. How the samples were to be taken and cared for was carefully laid out in the regulations, thereby reflecting the Board's desire to promote and acquire a more sophisticated understanding of the characteristics of Turner Valley and other reservoirs. This commitment, sustained over the subsequent decades, would lead ultimately to the development of one of the world's outstanding core archives and research centres and an unrivalled knowledge of the Alberta portion of the interior sedimentary basin.

The new regulations also addressed production activities. Operators had to equip their wells with proper chokes, so that production could be restricted properly to the volumes allowed. Each natural gas well had to be equipped with a gas meter, and daily production records had to be kept for each well. Casing-head gas was to be conserved. The use of vacuum pumps, or other devices used to lift crude oil or natural gas, was prohibited except by special permission of the Board, as was the shooting, perforation or injection of acid in any attempt to improve a well's performance. Oil was not to be stored in earth pits or any other receptacle judged by the Board to be liable to cause waste or hazard, and storage tanks or batteries of tanks had to be surrounded by a dike or ditch. No salt water or drilling fluid was to be drained onto the land surface. To reduce the risk of fire, rubbish and debris were to be removed from the vicinity of wells and tanks, and smoking was prohibited within 75 feet of any derrick, or oil or gas well.

Infraction of any of these or other new regulations could result in the drilling licence being suspended or cancelled. If drilling had been completed and production already begun, the Board was authorized to affix seals on well, pipeline and tank valves to stop the further withdrawal or movement of gas or oil until the difficulty had been resolved to the Board's satisfaction.

The regulations presented in early 1939 differed little in purpose and substance from the first "Oil and Gas Regulations" presented by the Department of the Interior in 1922.[19] They did reflect, however, a more sophisticated understanding of reservoir dynamics that was the product of research and experience gained over the intervening period. In one sense, this mattered little; what was ultimately crucial was the willingness and ability of those charged with the responsibility to enforce the regulations. Although only time would reveal the commitment and the measure of success of the Alberta Board, its initial efforts to bring acceptable field practices to Turner Valley were auspicious. The Alberta agency began its work with a field office situated in the village of Black Diamond in Turner Valley and a sufficient number of field engineers to initiate an effective inspection system. This stood in contrast to the situation in most U.S. states where the low ratio of field engineers to drilling sites, producing wells and storage facilities made thorough inspection next to impossible. The effectiveness of the Turner Valley field staff's inspection efforts and the determination of Knode and his colleagues was soon confirmed. At its fifth meeting, on 11 April 1939, the Board took note that both Mercury Royalties No. 1 and National Petroleum No. 1 wells had produced "considerably in excess" of their assigned allowables and ordered both wells shut in until such time as the foregone daily allowables equalled the amount of the excess production.[20] The operators were notified further that they would be charged whatever costs the Board might incur in shutting off the wells. There continued to be instances where operators tried to operate outside the regulations, and ingenious methods were employed to avoid detection. At times, the Board's field staff found it necessary to perform inspections after dark.

While Board staff began the task of establishing the agency's presence in the field and gaining recognition and acceptance of regulatory policy, Deputy Board Chairman C.W. Dingman began work on the compensation scheme promised disgruntled naphtha producers in the *Oil and Gas Resources Conservation Act.*[21] The plan eventually devised and submitted to the Alberta cabinet on 18 May 1939 stands as further evidence of Dingman's and Knode's unwillingness to compromise on conservation principles. Although the naphtha producers no doubt hoped otherwise, there would be no compensation for foregone "wasteful" production. Dingman devised an elaborate formula to determine each well's "equitable share" of the actual aggregate amount of gas produced for "non-wasteful" purposes from all gas wells for the year. According to Dingman's plan, when a well was prevented from producing its full "equitable share" as a result of a Board order, then the operator would have a legitimate claim for compensation. Along with his compensation scheme, the Board deputy chairman forwarded the recommendation that since, in his judgement, it could not be said now that "any person [was] being injured by the gas conservation orders of the Board," there was no requirement for compensation; therefore, rather than bringing the scheme legally into force, it should simply be held in readiness for when it might be required, at which moment appropriate adjustments could be made.[22] Cabinet concurred, and it seems that improving market conditions temporarily muted the protests of traditional opponents.

Sensing by the spring of 1939 that it was finally well on the way to bringing Turner Valley under effective regulatory control, the Alberta government shifted its energies to the search for an enlarged market for the province's growing oil production. It was apparent that not only was the pursuit of a larger market essential to the continued development of the petroleum industry in Alberta and the improved health of the province's economy but also the foundation for acceptance of conservation regulation could be strengthened most readily in a buoyant economic environment. In the quest for markets, the new Conservation Board chairman was assigned a key role. It was, moreover, a role that presaged a primary function that Board chairmen would continue to play in the future. Quite beyond being the figure ultimately responsible for the effective management of the province's oil and gas conservation program, in April 1939 the Board chairman took the first step along a path that within a decade would lead to his confirmation as the government's pre-eminent expert on oil and gas matters. This meant that the Board, and especially the Board chairman, would play a doubly important role in evolution of the petroleum industry in Alberta.

In combination with the efforts of the Alberta Petroleum Producers Association (APPA), the minister of mines and minerals, with the assistance of William Knode, set about trying to resolve what had emerged as a daunting paradox. Turner Valley was capable of producing far more oil than the prairie market could absorb, and since wells were being produced at much below their capacity the incentive for continued investment was threatened. At the same time, unless the field could be developed further and production increased greatly, there was little hope of being able to break into the big markets of central Canada and the Pacific coast. One delegation, headed by Herbert Greenfield, president of the APPA, travelled to eastern Canada to initiate discussions with the federal government, the railways and the refineries. Their unachieved objective was to obtain lower freight rates, so that the daily production equivalent of 6,000 barrels of Turner Valley crude could be delivered to British American Oil Company's refinery in Toronto and Imperial Oil's refinery in Sarnia.[23]

LONDON CAPITAL AND A PIPELINE TO THE LAKEHEAD

In February 1939, N.E. Tanner also journeyed east. His purpose was to gain federal support for another delegation, which he hoped might lead to a more substantial solution to Turner Valley's market problem. From late 1937, the financial pages of London newspapers began to report on Turner Valley developments, and various British parliamentarians, including Prime Minister Neville Chamberlain, began to receive letters advising that, given the importance and vulnerability of the nation's oil supply, Great Britain should be looking closely at an attractive new source of supply developing "within the Empire" at Turner Valley.[24] In August 1938, Air Marshal Sir Edward Ellington visited the Turner Valley oilfield and spoke of Britain's need for increased supplies of aviation fuel. Interest quickened even further after a press report—originating in Berlin in mid-September—claimed that German capitalists were offering to build a pipeline from Alberta and would accept crude oil in payment.[25] Within days, an official from the Admiralty's contract department called upon the Dominions Office to announce that the "Fourth Sea Lord and the Naval Staff were showing increasing interest in [Turner Valley] from the point of view of developing a possible Empire source of oil supply . . . ," but that it might be better if some other department "approach the Canadian Government to ascertain in a preliminary way further facts as regards the situation in the Turner Valley field."[26] By early October, as the momentum of interest continued to build, various London financial houses began to show enthusiasm for a Turner Valley pipeline project.[27] This emerging interest was more than idle vision. Enough time had elapsed since the completion in 1931 of the world's first long-distance pipeline—the 1,000-mile, 24-inch pipeline from the Texas Panhandle gas field to Chicago—to refine and demonstrate the capability of new pipeline technology. The really crucial question on everyone's mind was what was the extent of Turner Valley's proven oil reserves. It seemed appropriate therefore to send to London an official delegation that could offer authoritative information.[28] Canada House in London had kept the Alberta Department of Lands and Mines informed about the situation in Britain, and it was this information that eventually brought Tanner to Ottawa to consult with his federal counterpart to ensure that any delegation sent to London from Alberta had, and could be seen to have, Ottawa's support. Federal Minister of Mines and Resources Thomas Crerar agreed that Dr. G.S. Hume, head of the Canadian Geological Survey, could accompany Tanner and Knode to the United Kingdom.[29] W.S. Campbell, past president of the Alberta Petroleum Producers Association, was later named as the fourth member of the group.

Chasing the old dream of attracting British investment capital, Tanner and his party departed for the United Kingdom on 2 April 1939. Just before the Canadian delegation was due to arrive, representatives from the Admiralty, the Air Ministry and the Petroleum Department met with colleagues in the Dominions Office to coordinate the position of the British government.[30] The chairman of the meeting, W. Bankes-Amery of the Dominions Office, set the tone of the discussion with his opening remarks. He observed that any consideration of the Alberta oilfields as a possible source of emergency supply had to be tempered with the realization that "any addition to United Kingdom purchases from Canada at such a time would raise almost insuperable exchange difficulties."[31] The well-informed director of the Petroleum Department pointed out that the Canadian delegation was not coming primarily to obtain backing from the British government, but rather to interest the financial community in the industry and to obtain capital there. But even this was not acceptable. Bankes-Amery pointed out that "in the present circumstances the Treasury would not want to see more sterling going to Canada" and that strong objections would likely be raised. It was concluded therefore that even though individual departments might offer technical help and information regarding the kinds of products they were prepared to buy, "no hope should be held out to them of government financial assistance for development and that, so far as possible, they should be encouraged rather to seek capital in Canada than in the City."[32] The obstacles to be overcome were more formidable than Tanner could have imagined.

The Alberta minister of lands and mines arrived with his colleagues at Canada House on 18 April 1939. There they were briefed by Canada's Chief Trade Commissioner Frederic Hudd, who was responsible for arranging the delegation's appointments and engagements. There followed a vigorous, four-week schedule of meetings with officials from government ministries, parliamentarians, oil companies, financial houses and private investors. Discussion almost always centred upon two questions: what was a reasonable estimate of the potential production of the Turner Valley oilfield and, assuming the potential was sufficient to justify a pipeline, in what direction and under what terms might a pipeline be built? In answer to the first question, Knode informed the Admiralty that while the field produced just 7,000,000 barrels in 1938, a reasonable estimate of potential annual production was 29,000,000 barrels, or 400 wells each producing 200 barrels per day.[33] At the Petroleum Department of the Ministry of Mines, Knode said that it was generally conceded that the minimum recoverable reserve of crude oil in the Turner Valley field was about 200 million barrels, or 25 million tons at 7.8 barrels to the ton. The delegation explained further that when the field was fully developed they saw it efficiently producing 60,000 barrels a day or 2,750,000 tons a year. It was noted that, with only 53 wells producing, the field for a short time in the autumn of 1938 had produced 28,500 barrels a day before being cut back for want of markets.[34] In addition, the Canadian delegation was always quick to add that, in focusing on Turner Valley, it was necessary not to forget that there were other promising areas of Alberta where some production had already been gained and where active exploration was under way. The importance of the role being played by Alberta's new Conservation Board to ensure that the province's oilfields would achieve maximum recovery was also pointed out. It was stressed that the powers of the Alberta Board were "a good deal stronger than the average in the United States." Stating that "the Mounted Police have already been out several times in order to close down wells of producers who were not observing the Act," the delegation sought to emphasize that these were powers that the Board would not hesitate to use.[35] British capital, it was implied, would be safe in Turner Valley.

On the matter of a pipeline, the Admiralty let it be known that it favoured a line east to Port Arthur [Thunder Bay, Ontario]. The Canadian delegation also expressed preference for this route, and observed that for the winter months, when the Great Lakes were frozen over, the pipeline could be used "for conveying the natural gas to towns along the route."[36] Knode and Tanner spoke of an "Empire pipeline," financed and operated as a common carrier independent of the oil companies and under the authority of a Board appointed by the federal government. They also mentioned the possibility of the pipeline company having an "exclusive franchise to transport Alberta oil" as a safeguard against the subsequent construction of a rival pipeline.[37]

The rounds of meetings with various investors, financial houses and oil companies, including Anglo-Iranian, continued until May 17. The most significant discussions seem to have been those with Robert Benson and Company, investment bankers, and their subsidiary, Kentern Trust. The Benson group had experience elsewhere with pipelines, particularly in Iraq, and were anxious to explore opportunities in Canada.[38] They met a number of times with the Canadian delegation, collected as much information as possible, and decided that Major Rex Benson would go to Ottawa for further discussions with the federal government.[39]

The Alberta delegation returned to Canada well pleased, believing that they had accomplished their objectives.[40] They were conscious nonetheless that conflicting estimates of Turner Valley's "proven" reserves remained a crucial obstacle. This was the question that Thomas Crerar immediately asked George Hume to address on his return from London. Hume informed the minister that he and W.F. Knode believed that in the "proven and semi-proven" areas of the Turner Valley oilfield there were approximately

280,000,000 barrels of recoverable oil, and an additional, but unproven, productive area that might contain a further 100,000,000 barrels. Eventual sustained production of 60,000 barrels a day seemed realistic, but Hume noted that stimulating the amount of drilling to reach that objective was unlikely "unless transportation facilities to allow the oil to reach a wider market are provided."[41] With this information in hand, Crerar waited to hear if British interest would materialize.

In late June, Major Benson and two colleagues arrived in Ottawa to meet with Crerar and his fellow members of the cabinet Fuel Committee: Minister of Transport C.D. Howe and Minister of Trade and Commerce W.D. Euler. The Benson group were interested in proceeding with the pipeline proposal "if they could have some assurance of a supply of oil in the field and a market for the oil in the East."[42] Benson was aware that the Ontario refiners had demonstrated no interest in a pipeline from Alberta. "The refiners," it was noted, "had a definite interest in other sources of supply," and "perhaps the Government might have a little influence with them in using Canadian oil in preference to imported oil."[43] The London bankers were informed that although it would be difficult for the Canadian government to give assurances about oil supply or markets, "the refiners might be called in conference to secure their views and their assurance that they would buy Turner Valley [oil] if it could meet competition from other sources."[44]

Benson and his advisers returned to London to draft some concrete suggestions for financial organization, and the Canadian government was left to contemplate what encouragement, if any, should be given to the pipeline scheme. In its review of the London and Ottawa meetings, the London *Financial Times* reported that "from a technical point of view construction of the pipeline is thought to present no difficulties" and the estimated cost of £6,000,000 to £7,000,000 for the proposed 1,200-mile pipeline and storage facilities at Port Arthur seemed reasonable. The remaining problem, according to the *Financial Times,* was convincing Ottawa that there was sufficient oil in Alberta and that it could be delivered to central Canada at a cost competitive with U.S. oil. The paper concluded, "Whether or not the scheme becomes a reality depends at the moment on the decision of the Ottawa government."[45] To evaluate the conflicting views he was getting from the interested parties within Canada, Crerar ordered his department to study and report on the pipeline question. Deputy Minister Dr. C. Camsell personally directed an investigation into the marketing of Alberta crude in Ontario, and he sent his department's petroleum engineer,

T.D. Madgwick, to Alberta to make a firsthand assessment of Turner Valley's potential. Camsell informed Crerar in early August that he was finding the attitude of refiners particularly difficult.

> I ran across a preconceived opinion that the whole project was visionary and was hardly worth investigating. I do not know yet what pressure can be brought to bear on the refineries of Ontario to give us their whole hearted [sic] co-operation on this problem.[46]

Camsell told Crerar that the real problem of getting Alberta oil into Ontario lay "in persuading the refineries of Ontario to take a sufficient quantity of oil to make pipeline transport feasible, even if that oil can be laid down at the refineries at competitive prices."[47]

Given the strength of the opposition, Madgwick's evaluation was crucial. His detailed report concluded that Turner Valley alone could not maintain production at the rate of 60,000 barrels per day for the 12-year period needed for amortization of a pipeline to the Great Lakes.[48] The game was over. Tanner and western producers continued to press Ottawa and provide updated information, but to no avail.

It was a bitter disappointment for the Alberta government to be informed that not only were Turner Valley reserves insufficient to warrant construction of a pipeline but also one of the great influences retarding further development was the "excessive" royalties levied by the province.[49] Ottawa was making sure that the fault for the retarded development of Alberta's petroleum industry was not going to be left entirely in its lap. This was but the first skirmish in a decades-long struggle to persuade Ontario refiners and consumers to accept Alberta oil. Tension between the two levels of government over the pipeline issue soon submerged, however, as Canadians became preoccupied with organizing the country's war effort. Moreover, it seemed to Albertans that the war could only enhance Turner Valley's prospects.

THE MCGILLIVRAY COMMISSION

The wartime situation put an entirely new perspective upon the production of oil and gas in Alberta, and thereby formed a new and crucially important backdrop to the other important petroleum sector initiative taken by the Alberta government in the autumn of 1938. This was the royal commission "to inquire into matters con-

Justice of the Supreme
Court of Alberta, A.A.
McGillivray, headed a
1939 royal commis-
sion into petroleum
pricing problems that
brought the new con-
servation board's poli-
cies under scrutiny.
Glenbow Archives
NA-2982-2.

nected with Petroleum and Petroleum Products." Known generally
as the McGillivray Commission, after its chairman the Honourable
Justice A.A. McGillivray, the Commission held a long-running series of hearings that continued through 1939 and formed the basis
of a report submitted in April 1940.[50] McGillivray was well-
acquainted with the petroleum industry in Alberta. He had prac-
tised law in the province since 1907, and from his vantage point on
the bench of the Alberta Supreme Court after 1931, he had heard
many of the more important cases relating to the industry, includ-

ing the famous *Spooner Oils Ltd. v. Turner Valley Gas Conservation Board*. He was also well known in political circles, having been leader of the Conservative party in the Alberta Legislature from 1926 to 1929. McGillivray's fellow commissioner was Major L.R. Lipsett, a farmer from Ardley, Alberta. A commission counsel, J.J. Frawley, from the Attorney General's department, and a commission accountant, F.G. Cottle, were appointed to assist. Like the commissioners, both men had a professional knowledge of the industry they were charged to investigate. Cottle was a well-known chartered accountant and member of the new Petroleum and Natural Gas Conservation Board, and Frawley had often been called to defend government legislation and regulations in the petroleum sector.

To bolster their own expertise and to ensure that they had the best available technical opinion to set against the varied evidence that they expected to receive from industry, Frawley was sent to the United States to find "men entirely independent of the industry, who yet could speak with a voice of authority concerning the industry, men whose reputations as petroleum experts [was] international and whose integrity [could not] be called into question."[51] The three individuals who Frawley persuaded to come to Alberta to testify greatly influenced the shape of the Commission's eventual report. The most important was Dr. John W. Frey, an associate director of the Conservation Division of the Department of the Interior in Washington, D.C. Dr. Frey was a graduate of the University of Chicago and the London School of Economics, his fields were geology, economics and world politics, and apart from a brief period as a faculty member of the University of Wisconsin, he had been a senior level adviser and administrator in Washington since 1928.[52] In addition to his general responsibilities as economic adviser to the secretary of the interior on petroleum matters, Frey was engaged in the enforcement of the controversial Connolly or "hot oil law" at the time of his departure for Alberta.[53] The second person was Dr. George Granger Brown, a well-known academic and consulting engineer. A professor of chemical engineering at the University of Michigan, Brown directed all petroleum-related research at that university and had to his credit a long list of consultancies for various oil companies, engineering firms and associations. He also had the advantage of being familiar with the petroleum industry in Alberta, having been hired previously by the Alberta Petroleum Producers Association to do a comparative analysis of Turner Valley and mid-continent crudes.[54] Also agreeing to appear before the Commission was Knode's former colleague, Dr. Byron B. Boatright of the Houston consulting firm of Parker, Foran, Knode and Boatright. After graduating as a mining engineer, Boatright worked for various oil companies and the United States Bureau of Mines before returning to the Colorado School of Mines to complete a doctorate in petroleum geology. From 1928 until he left in 1937, he was the head of the school's Department of Petroleum Engineering, and during the interval he developed a particular expertise in estimating oil and gas reserves.[55]

With the assistance of Frey, Brown and Boatright, McGillivray and his colleagues hoped to be able to gather and interpret sufficient evidence to determine whether or not Turner Valley crude oil producers were receiving a fair price for their product and whether or not consumers were paying too much for gasoline. It was the price difference between the $3\frac{1}{3}$¢ per gallon paid by refiners for Turner Valley crude oil and the wholesale price of $16\frac{1}{2}$¢ for standard gasoline that had prompted the unanimously carried resolution of the Alberta Legislature in favour of an investigation and brought forth the Commission.[56] This necessarily involved a critical examination of the philosophy of conservation and the impact of conservation practice upon oil and gas production in Turner Valley. After less than a year's operation, the Petroleum and Natural Gas Conservation Board, the legislation that gave birth to it, and the regulations that it was called upon to administer were to be put to rigorous scrutiny. The Commission's review and findings regarding conservation practice in Alberta are worthy of particular consideration, not only because they were central to the evaluation of production and pricing with which the commission was specifically charged but also because they emerged from the only substantive provincial review of the Conservation Board that has ever been undertaken.

The Commission's review can be seen as the final chapter in the almost unrelenting conservation controversy that had begun with Alberta's first natural gas crisis of 1920–21. The commission hearings allowed the independent producers one more opportunity to present their case—this time before a high-profile neutral authority. Once again, the old arguments were brought forth, bolstered in some cases by the admittedly limited, but nonetheless firsthand, experience under the supervision of the new Board.

Properly described by the *Calgary Herald* as a "staunch advocate of conservation," Dr. Frey laid out the principles of oil and gas conservation in an eloquent introduction, and then at length in subsequent testimony defended the legitimacy and the necessity of government intervention to protect the public interest.[57] Although it might have had an even stronger impact before a U.S. tribunal or commission, Frey began by establishing that in the United States the conservation of the nation's natural resources had had the sanction of America's founding fathers. He noted that one of the first actions of the Mayflower colonists after they landed at Plymouth Rock in 1620 was to pass an ordinance prohibiting the firing of trees within a certain boundary near the settlement.[58] With this oblique but powerful reference to the long-recognized legitimacy of the state acting to enforce conservation in the public interest, he went on to explain why the government's regulatory presence was essential in the production of oil and gas. Frey started from what he believed was a self-evident proposition that

> all oil fields and every oil well in the field should be produced in such manner as to produce the largest practical quantity of oil from that well and field, and that every engineering effort should be made to prevent the waste of oil.[59]

The greatest obstacle standing in the way of this worthy objective was the migratory character of oil and gas and the Rule of Capture, which legally ascribes ownership only when oil and gas are reduced to possession. The fundamental consequence was that production from a well always compelled an adjacent owner with the alternative of either losing oil under his lands or going to the expense of drilling an offset well to produce oil regardless of market conditions. Waste and the resultant ruin of once productive oilfields marked the short history of petroleum production in the United States. "Uncontrolled production, that is working under the Rule of Capture only," Frey insisted, "is destructive to the welfare of the state and the industry."[60] In contrast to the chaos imposed by the Rule of Capture in fields of multiple ownership, Frey presented the ideal of unit operation. The advantages gained where control resided with a single operator included the time to complete exploratory work before being compelled to formulate a plan for exploitation; being able to adopt a plan of well spacing that appeared best suited to the particular field; the incentive to use modern methods of pressure maintenance for the benefit of the entire field; the option of drilling up the field as rapidly or as slowly as the market demand for the field's quality of oil warranted; and the greater opportunity to adjust and readjust the plan of operations in accordance with constantly increasing knowledge of field characteristics and conditions.[61]

Although the economic and social advantages of unit operation were readily apparent, historical experience, including that of the Turner Valley field, demonstrated that it was extremely difficult to achieve where there was divided ownership. The forces mitigating against internal voluntary agreements were not only the consequence of wary operators striving to protect or enhance their positions but also of the fear of infringing laws against combination in restraint of trade, such as the anti-trust legislation in the United States and section 498 of the Criminal Code in Canada.[62] For this reason, Frey explained, conservation by government intervention was necessary to ameliorate the effects of the Rule of Capture and to secure the nearest equitable approximation to unit operation.[63] This meant in turn that an agency of government had to be created and charged with the responsibility.

Frey proceeded to outline the features that he believed were vital to the proper functioning of any conservation agency. The agency should be given "broad powers," but he cautioned that this had to be balanced with a clear-cut declaration of policy that included definitions of what constituted waste. It was necessary not only that the conservation agency be directed what to look for and what to act on but also how to act. This was crucial if the agency were to operate in an informed and equitable fashion. To this end, there had to be "provision for the taking of evidence, in other words hearings and on the basis of those hearings there should be findings and on the findings there should be a ruling."[64] Beyond this, Frey insisted that the possibility of a ruling being wrong had to be recognized, and for this reason there should be the provision for recourse to the courts. If the Conservation Act set out clearly the procedures, the circumstances, and the court to which an aggrieved party could take his case, Frey felt that the conservation agency would be sufficiently protected against frivolous and delaying litigation.[65]

Other than a conservation authority, Frey and his distinguished colleague, Dr. George G. Brown, saw no need for further government intervention in the petroleum industry. For both men, this was a position held first as a matter of principle and second as a consequence of their assessment of the situation in Alberta. They did not find convincing evidence that Imperial Oil held a monopoly position, that the field price of Turner Valley crude was inappropriate, that refiners' profits were inflated, or that the price paid by gasoline consumers was excessive.[66] Although many of the independent producers saw Imperial Oil's dominant position in a much more negative light and although none of the independents were inclined to accept that they were getting enough for their crude, they too remained largely unsympathetic to the idea of prices being fixed by the government. This had always been a solution favoured more by gasoline consumers, especially those in the farm sector.

Equally devoted to the idea of nongovernment interference, independent Turner Valley producers took special note of what Frey was saying about conservation. The gas-cap producers thought that the approach to conservation that Frey seemed to be suggesting offered some clear advantages over that being administered by Alberta's Conservation Board. J.C. Mahaffy, the lawyer representing Gas and Oil Products Ltd., skilfully used the fundamental conservation principles articulated by Frey as stepping-stones for a direct assault upon Alberta's conservation legislation, the procedure and policies of the Alberta Petroleum and Natural Gas Conservation Board, and particularly upon what he interpreted to be the powerful and misguided influence of the ex-Board chairman and current technical adviser, W.F. Knode. Mahaffy's questioning of the various witnesses over the long course of commission hearings and his presentations seem to have greatly influenced the Honourable Mr. Justice McGillivray, and in doing so played an important role in shaping the Commission's final report.[67]

This was a debate that Mahaffy had mastered. He had been round the course with it many times before, including the two sessions before the Agriculture Committee of the Alberta Legislature. As Frey was nearing the completion of his appearance before the Commission, Mahaffy began by seeking elaboration upon some of the important points that the Washington administrator had raised. "Now I understand you to say," Mahaffy asked, "that a good deal of horse sense has to be mixed with theory in working out these conservation schemes?" Frey responded, saying, "Yes, I do believe that it is not possible to work out a conservation scheme purely on a formula; that is, that any scheme has to be tinctured with a practical consideration that results from divided ownership."[68] Factors that Frey agreed could legitimately cause variation from a "theoretically perfect scheme" included the situation in a field where a considerable amount of development had been done before any conservation scheme was contemplated, the requirements of national welfare, and the potential impact upon competition. Frey explained that he thought

it would be highly undesirable to disregard the fact that you might throw all of the business into the hands of one or two companies. I think that should be given very careful consideration in the development of a plan.[69]

When pressed, Frey would not say whether it was a good idea "to leave the establishment of a conservation scheme for an oil field to the judgement of one engineer." But he did say, "I would want to be sure of my man,"[70] and he acknowledged further that conservation plans emanating from his department in Washington were the product of collective or team decisions.[71]

Having drawn forth these further observations from the most authoritative witness brought before the Commission, Mahaffy put them aside to be woven into his final argument later. Eventually, on 8 December 1939, he began by selectively identifying specific Turner Valley situations, which he then set in the light of Frey's general principles. First, he established, on the basis of evidence presented before the Commission, that natural gasoline—the product produced by his clients from the gas cap—was much valued as a gasoline blending agent by refineries in Winnipeg and Regina that otherwise had to be supplied at prohibitive cost from Oklahoma and Kansas.[72] Hoping to have established the need, even the dependence upon the product, of the gas-cap producers, and by implication that continued production was in the national interest, Mahaffy then launched his attack directly upon the Conservation Board. He was careful, however, to emphasize at the outset that this was not to be interpreted to mean "that we are opposing conservation in any way, shape or form."[73] Mahaffy drew the commissioners' attention to the situation faced by Gas and Oil Products Limited. The company had built its Turner Valley absorption plant in 1934–35. From that point, the plant had operated on a throughput of 31 million cubic feet of natural gas per day until mid-1938 when, under the order of the new Conservation Board, the flow was reduced to 12 million cubic feet per day. Now, Mahaffy explained, Turner Valley operators had been informed by a Board circular that allowables would again be drastically reduced and that the proposed schedule would allow the 12 wells upon which the absorption plant relied to supply only 3,208,000 cubic feet of gas per day.[74] It was claimed that such low allowables would not even permit the recovery of operating costs and that the wells would be shut in. Moreover, even if the operators managed to keep the wells going, Gas and Oil Products could not carry on its absorption operation with such a low volume of gas. According to Mahaffy, his clients were to be put out of business, whereas their principal competitor, Imperial Oil, would be permitted to continue producing natural gasoline by virtue of its subsidiary's exclusive contract to supply Canadian Western Natural Gas, Light, Heat and Power Company with natural gas. On top of this, he claimed that his company could not get a proper formal hearing.[75] As Mahaffy presented it, this all seemed to go against the grain of equity and practical balance that Dr. Frey held to be central to any conservation plan.

Mahaffy persisted. He argued that the fault lay with the Conservation Act, which did not compel the Conservation Board to hold hearings, to allow the appeal of Board decisions, or to contain explicit compensation provisions. The lack of such basic rights, he argued, was compounded by the inordinate and pernicious influence of one man, whose concept of conservation was narrow and whose technical understanding of the Turner Valley reservoir was at odds with predominant expert opinion. In support of his contention, Mahaffy brought forward several of the responses that Board Chairman Knode had made during questioning before the Alberta Legislature. For example, when asked whether the Board needed to consider its effects on individual producers in seeking to bring about conservation, Knode had answered, "It is my opinion that no Conservation Board can consider anything except the manner, or mere fact of conservation."[76] Mahaffy concluded.

> I make those quotations, Sir, to indicate . . . , that Mr. Knode felt that he only had one duty and that was to put in the narrow, stricter type of conservation as we sometimes describe it; that he had no consideration for the individuals, companies or equities in the field. Personally I know of nothing that has happened since, that causes me to think he has changed his mind and I submit that is not the type of conservation which Dr. Frey, for example, would recommend should be put into effect.[77]

Mahaffy did acknowledge that Knode's position had changed from that of chairman to technical adviser, but he maintained that the former chairman remained the "moving spirit" behind the Board's conservation policy.[78]

Even more serious was Mahaffy's contention that the gas-cap producers were the unfortunate victims of Knode's flawed understanding of the Turner Valley reservoir, and he observed that there were two theories regarding the Turner Valley field. The one held by Knode, and thus the theory upon which the Conservation Board's gas-cap policy was based, was that the field comprised a common reservoir in which there was free movement of gas and oil. The other theory, the one that Mahaffy noted was supported by most of the expert witnesses brought before the Commission, was that even though the field had a common reservoir the low permeability of the limestone and the presence of many dense or tight areas meant that there was little movement of gas and oil in the formation.[79] If the latter theory was correct, then the withdrawal of gas from the gas cap would not adversely affect the ultimate recovery of oil from the crude oil area. Mahaffy asked:

Should our business be junked and should the production of a valuable product, namely natural gasoline, be eliminated because the technical advisor to the Board holds an opposite view to that which I have now given expression? After all, it is only one man's opinion.[80]

Not content to let his argument rest, Mahaffy asked the commissioners to assume for the moment that the first theory was in fact correct, that there was free movement of gas and oil in the structure. He explained that

the gas under our property is bound to do one of two things. First, it will be withdrawn from the formation through the wells located on the northern and central portion of the gas-cap owned by the Royalite Oil Company and it will be sold by the Royalite Oil Company to the Calgary Gas Company; or second, if it does not move in that direction then it will be used by the producers of crude oil for the purposes of lifting their oil to the surface on the flank of the field and there this gas will be burned, because in the view of the Board it has performed its function.[81]

Yet the Board had taken no steps to provide compensation to the operators on the gas cap or to the operator of the absorption plant, which was dependent upon the gas produced at those wells, even though other operators benefited directly at their expense. This situation, Mahaffy maintained, hardly fitted the principles of equity and reason upon which "Dr Frey laid such stress."[82]

Mahaffy concluded his presentation with the recommendation that the Commission advise the Alberta Government to instruct the Conservation Board to make no further change "in the allowable production of wells located on the gas-cap of the Turner Valley field"[83] and to revise the *Oil and Gas Resources Conservation Act* so as to provide effective compensation to those who suffer loss as a result of conservation orders. He also urged that section 44 be eliminated or amended to allow appeals to the court by those who considered themselves aggrieved and to "comply with the perfectly reasonable attitude which Dr. Frey has said is followed in most of the Legislatures in the United States."[84]

Mr. Justice McGillivray fashioned his report from the nearly 16,000 pages of transcript testimony and more than 747 exhibits and presented it to Alberta's Lieutenant-Governor J.C. Bowen on 17 April 1940. Addressing the central issue, McGillivray wrote that the evidence compelled him to conclude that there is very real competition in this province; that prices are not out of line with prices in other places in which competition is keen; that the cost performance is reasonable and that the profits are not excessive. In such circumstances there is not the slightest occasion for the government to exercise government control for the protection of the public. On the contrary it would seem that the public in Alberta is adequately protected by the play of contending forces prompted by desire for gain.[85]

Although McGillivray found no reason to recommend direct government intervention in the petroleum industry in Alberta, he felt it necessary to elaborate upon the proper role of the state in business affairs. It was well known that many in both government and opposition ranks in the Alberta Legislature favoured government intervention in the petroleum sector. Indeed, the Commission had been specifically directed to consider whether the province should take over the wholesale and retail distribution of petroleum products. Moreover, the Commission's counsel, J.J. Frawley, and much to the distress of the Alberta Petroleum Producers Association, in his assigned role as guardian of the public interest had put forward for Mr. Justice McGillivray's consideration a specific proposal for the creation of a powerful government board with authority to set prices, not unlike the Board of Transport Commissioners from which railway companies had to gain approval for their freight rate schedules.[86]

McGillivray began by drawing a distinction between government ownership and government control. His position was that the public interest could often be properly served by the latter but seldom by the former. In support, he advanced the classic argument that government enterprise came at greater cost to the consumer. "This is so," he argued, "for the simple reason that those who carry it on have not the spur of self-interest to reduce cost in order that they may extend the profit to themselves."[87] As a general principle, he held that the government should not be in business competition with its own citizens. At the same time, however, McGillivray emphasized that he could not agree with those who shared the view of Dr. Brown, who had argued before the Commission that, in the long run, free competition would always serve to safeguard the public interest. "In our opinion," McGillivray wrote,

competition may be found to be very active and to adequately protect the public and again it may be found to be a mere cloak for collusive price raising behind which the public purse is co-operatively emptied and so government intervention may or may not be indicated according to the circumstances under consideration.[88]

In short, government intervention in the public interest was legitimate and sometimes necessary. It all depended upon the particular circumstance. After exhaustive investigation, McGillivray could find no reason to justify the creation of yet another and more powerful government board; however, and in keeping with his philosophy, he was quick to add that there was no certainty that this situation would continue, and that,

> as a general proposition, it is not sound for a government to concern itself about an industry's performance only when its abuse of privilege has caused such harm to the public as to make it notorious, but rather it is the duty of a government to hold a watching brief for the public, through some government agency that is both competent and just.

McGillivray therefore recommended that this responsibility of watching and reporting on "*all* of the activities of the petroleum industry" should be charged to the Petroleum and Natural Gas Conservation Board.[89]

Although he was prepared to advise that its responsibilities be broadened, McGillivray was not without harsh criticism of the form and practice of the existing Conservation Board. He accepted that such an agency was essential to the public interest given the unique character of the oil and gas resource, but he also believed, in light of the evidence brought before the Commission, that there was need for substantial improvement both in the legislation that established the Board's authority and in the procedures that the Board had devised to achieve its assigned objectives. McGillivray was of the opinion that, "since the Petroleum and Natural Gas Conservation Board as presently constituted [did] not enjoy the confidence of all the producers in Turner Valley," a new board should be created.[90]

The Commission's reported ideal was a five-person board, with one member who could be said to represent Alberta consumers; one member from industry who could bring an operator's perspective; a petroleum engineer who would have a proper understanding of engineering problems; a chartered accountant who could deal with the complex accounting problems that often followed from certain Board recommendations; and, finally, one member who had legal training. In McGillivray's view, this last individual was necessary to ensure that the Board had "a true appreciation of the equities which are involved in any worthwhile Conservation and Proration scheme," as well as to help the Board effectively counter inevitable challenges to its orders.[91]

On the matter of legislation, McGillivray was emphatic. "It is with some confidence," he reported, "that we make the recommendation that . . . The Oil and Gas Wells Act, 1931, and The Oil and Gas Resources Conservation Act be repealed and that new legislation be substituted therefor."[92] In McGillivray's view, the proper drafting of such legislation could be accomplished only if certain fundamentals were recognized at the outset. First was that the ideal in conservation could only be achieved under unit operation. Second, the evidence demonstrated clearly that it was now too late to implement unit operation in the Turner Valley field either by voluntary or compulsory means. This meant that the compromise measure of conservation and proration legislation had to be accepted. The great test was working out the details of such a compromise. McGillivray believed that success depended largely upon the recognition of one basic principle: "that in a field of divided ownership Conservation cannot as a practical matter be disassociated from the idea of economic equilibrium and the idea of equity."[93] The theme of equity was one that McGillivray turned repeatedly to in his report, and it underlines more than just the cast of mind characteristic of those who sit on the bench. It marks also the great impact made upon the commission chairman by the testimony of the small independent producers as well as that of Dr. J.W. Frey.

For the guidance of the Alberta government, McGillivray provided a list of specific measures which he felt had to form the core of acceptable conservation legislation. He recommended that, rather than leaving the Conservation Board with "unlimited power," a revised conservation and proration statute should contain a "clear cut understandable declaration of policy" and that the "specific powers which it is intended that the Board should enjoy should be specifically declared by the Legislature." Among the other measures judged essential was the necessity to ensure that anyone who might be prejudicially affected by an order of the Board had an opportunity of being heard before that order was made, and that there be a formal provision for such hearings, for the taking of evidence at such hearings, for findings by the Board on evidence, and for rulings to be made thereon. The commission report held further that such rulings should be open to appeal to the courts whenever a question of the proper application of the rules of equity or a question of the Board's jurisdiction was involved. Alberta legislators were reminded also that "all modern legislative schemes for Conservation and Proration take into account considerations of equity as well as those of engineering" and any ruling intended to prevent waste should be arrived at only after proper attention to the principles of equity. McGillivray did envisage one situation where the principles of equity and good engineering practice might have to receive secondary consideration. He advised that no conservation scheme should be enacted that would permit a monopoly to result from its effect.

McGillivray then advised how the Board's responsibilities should be expanded to protect the public interest. Convinced by his commission experience that the formulation of sound petroleum policy to serve the complementary interests of the public and the industry best depended greatly upon the range and quality of the information available to decision-makers, he proposed that the Conservation Board also become a sophisticated information gathering agency. This was to be accomplished by requiring the Board to monitor comprehensively all aspects of the petroleum industry worldwide. More than most Alberta politicians at the time, McGillivray recognized that Alberta's petroleum industry could only be understood within a global context.[94] Included among the Commission's specific recommendation were the following:

1. that this Board preserve for the benefit of the government, the industry and the public, the information gathered by this Commission with respect to the petroleum industry.

2. that from this starting point the Board should accumulate preserve and produce on request, any data as to the Turner Valley oil field and as to any other part of the Province, which can reasonably be expected to be of interest to those directly or indirectly concerned with the petroleum industry.

3. that the Board should be required to be at all times fully informed as to every branch of the petroleum industry, including exploration, production, refining, and wholesale and retail marketing.

4. that the Board be required to at all times be informed as to the world picture in respect to both crude and refined products and as to prices which obtain elsewhere in respect of these.

5. that it is important not only from the standpoint of the producers but from the standpoint of there being an incentive for drilling, that the Board be required to be informed as to whether or not the producers are obtaining proper field prices.

6. that the Board be required to be familiar with the cost and profit performance and the price spreads in respect of all branches of the industry.

7. that the Board be required to be informed about and concern itself with the possible enlargement of the economic market and to that end become thoroughly familiar with all relevant matters, such as the freight rate and pipeline situation.

8. that the Board should be required to make a full report to the Lieutenant-Governor in Council [every three months], not only upon its own activities but upon the activities of all branches of the petroleum industry in Alberta, ... and upon all other matters of importance which in anywise relate to the petroleum industry.

9. that in connection with such reports, the Board be authorized to add such recommendations as it may see fit to make.[95]

By establishing these measures in a new conservation act, McGillivray hoped to refine and formalize what was beginning to happen in practice. Already, the government had begun to seek the Board's expertise on petroleum matters outside the immediate conservation field. Industry, the smaller independent companies especially, also had begun to rely on the Board for certain types of field data and engineering advice.

Within the Commission's list of recommendations was one further important way in which McGillivray believed the Board might serve the public interest. He felt that the Board should have as one of its "primary objects" the responsibility to watch whether "competition of all kinds" remained a reality in Alberta. McGillivray was not unmindful, however, that unrestrained competition in certain sectors of the petroleum industry was not always in the public interest. Therefore, he qualified this recommendation with the advice "that the Board be required" to meet with industry to see what agreements might be reached that the Board might approve. "This suggestion," McGillivray explained, "is made to remove the fear of unjustifiable prosecution under the provisions of the Criminal Code and the *Combines Investigation Act*."[96] Underlying this proposal was the desire to promote the unitized operation of oilfields by removing what was perceived by some to be a major legal impediment.

Like most others before him who recognized the unique physical properties of oil and gas and who had closely studied the history of the petroleum industry in North America, McGillivray identified production of an oilfield as a single unit as the only truly effective way of eliminating the evils of the Rule of Capture. He dissented from the view presented before the Commission by some that, where there was a proper pattern of well spacing combined with rateable takings based on engineering findings, the Rule of Capture automatically disappeared. Although prepared to accept that conservation and proration had done much to eliminate the worst evils of unregulated production, McGillivray did not believe that any of the schemes brought to the Commission's attention could be said to have completely defeated the workings of the Rule of Capture. Under proration, it was still possible to withdraw oil from under the lands of its rightful owners. In his report, McGillivray explained that, under present prorationing methods, a person owning a well in a field of divided ownership was "limited in the amount of oil he could produce from his well, according to the number of wells in the field and the market demand at the time of production (of course having due regard to the other factors mentioned in any formula)."[97] As additional wells were drilled, and if there was not increased market demand to absorb this new production, then existing producers' "allowable" production was further reduced even to the extent that operating expenses might not be covered. This could put the small producer in an intolerable position. Although the law restrained production, and so deprived the operator "of the opportunity of making all the money which he could make by the efficient maximum operation of his well," it did not protect the operator "against being forced to make further capital expenditure in offset wells when other people have drilled adjacent to his properties quite regardless of the lack of need for further oil to satisfy the market or provide for depletion."[98] McGillivray pointed out further that the obvious counter of strict limitation on new drilling not necessary to meet market demand was no solution. This would give "a monopoly to those who have already drilled at the time of the limiting order and perpetuates the evil effects of the Rule of Capture in that those who have already drilled at this time are able to drain all of the lands from which oil will permeate to their wells and render it valueless to the adjacent owners."[99]

This assessment led McGillivray to two important related recommendations. Since the value of unit operation was self-evident, if and when new pools were discovered, the government should see to it that they were produced on a unit basis. Where voluntary unit operation in fields of divided ownership could not be achieved, where "lack of vision and greed are the only obstacles to the ideal plan of production, government intervention might well serve, even as it has done in connection with conservation and proration, where these same obstacles prevented the application of these measures."[100] Government intervention to force unit operation in Turner Valley, however, was not judged appropriate. McGillivray believed that development had progressed so far as to create insurmountable obstacles. Given that an ideal situation could never exist in Turner Valley, it was necessary to approach valley production schedules with a special sensitivity to ensure that basic matters of equity were not obscured by concentration on engineering or technical factors. McGillivray offered some special advice for Alberta's conservation engineers.

> The work of the engineer has to do with the unchangeable laws of nature; he is concerned with trying to understand those laws and to give effect to them in any engineering undertaking, but it is not as we understand it any part of the engineer's training to orient his work with the changeable but nonetheless ever-present social laws which hold men together in society. The conservation engineer is concerned with engineering efficiency in petroleum production which is a technical concept; it takes into account the objectively measurable causal factors and calculable effects in any process but there are limiting factors to the introduction and utilization of all that is efficient; these are social and economic. For example, from the economic standpoint it would be wasteful to eliminate petroleum loss which cost more to eliminate than to endure, even though in the opinion of the engineer it is quite possible to avoid that petroleum loss. Equally from the social standpoint it might be preferable to suffer a petroleum loss which the engineer says is avoidable, than to disregard the principles of equity and fair play on which our social system is founded.[101]

Motivating McGillivray's instruction to Conservation Board engineers was the testimony that had been heard before the Commission regarding the extent and manner by which production from the Turner Valley gas cap had been cut back. The controversial question of gas-cap production set the contending principles of engineering efficiency and equity in high relief. After carefully emphasizing that the Commission was making no findings as to what the Conservation Board had or had not done, McGillivray stated that the commissioners felt it their "duty" to point out that the most recent order proposed by the Board to restrict allowable natural gas production from the gas cap would have the effect of "rendering useless two absorption plants of great value without compensation, [and] that this would be considered a serious matter in any part of the Empire...."[102] He suggested that the Conservation Board should consider deferring its impending gas-cap production order until it had made an inquiry, set up in such a way as to satisfy everyone that it had not acted without proper regard for equity and without carefully weighing the opinions of engineers other than those of its own consulting engineer, William Knode.[103]

When presented to Albertans in April 1940, the report of the McGillivray Commission attracted little more than passing attention. Much of what the Commission had to say had been anticipated. The report immediately shielded the Social Credit government from the one area where there was an immediate political concern. By finding no evidence of an unreasonable price spread, the Commission gave the government the ammunition that it needed to still the widely held feeling that had been largely responsible for initiating the inquiry—that consumers were paying too much for gasoline. Although more cynical Albertans might have wondered about the coincidence, the retail price of gasoline dropped during the course of the inquiry and eased the pressure upon the Aberhart government.[104] Interest had also largely dissipated in other matters raised by McGillivray. Although Albertans had closely followed the early stages of the commission inquiry, they had become preoccupied with events in Europe by the spring of 1940. The German occupation of Denmark, the invasion of Norway, and the dramatic thrust of the German army through Holland, Belgium and France pushed all but concern for Canada's domestic war effort far to the background. Along with the new focus on the war came a new attitude regarding the prospects for Turner Valley crude oil. Already, markets had begun to improve, and well production allowables were starting to move upward.

The Alberta government now had great latitude in how it chose to respond to commission recommendations. Despite the almost universal condemnation of conservation legislation that denied the right of court appeal to an aggrieved party, the government remained unmoved and was not prepared to consider a restructured and more representative Board! Although not prepared to make any significant change in its conservation legislation, the government did realize that the dissatisfaction of so many Turner Valley operators could not be ignored completely. It hastened the search for a new Conservation Board chairman, who it hoped might have more success in winning the confidence of the small producers, and Board administrative procedure was more formally structured along the lines suggested by Dr. Frey. Henceforth, the Board would make a greater effort to ensure that its orders did not seem arbitrary. Although not anchored by legislation, as recommended by McGillivray, the Board's procedure came to rest upon a formal sequence of hearings at which evidence was gathered, followed by findings drawn from the collected evidence, which was succeeded by rulings or orders based upon the findings.

For all its efforts, and despite its thorough assessment of the petroleum industry in the province, the impact of the McGillivray Commission was modest. Events, at first beyond and then within Alberta's borders, conspired to submerge its influence during the following decade. McGillivray's untimely death in 1940 at 56 removed the one individual who might have reminded Albertans of the Commission's findings. Events that were stimulating the growth of the industry were also hastening the removal of the pioneer oilmen. Their last important battle was fought before the McGillivray Commission.

CONSERVATION AND WORLD WAR TWO

Through the spring and summer of 1940, the intentions of the Alberta government regarding oil and gas policy remained unclear. Possession of the McGillivray Commission's exhaustive report did little to arrest the indecision and drift in Edmonton. Aberhart and his cabinet colleagues were preoccupied with more pressing matters. They had just emerged badly battered from an embittered provincial election campaign in which the seriously divided Social Credit party had seen its once unassailable majority reduced by 20 seats.[105] Even in his home town, and the birthplace of the Social Credit movement, Premier Aberhart ran second to his old critic Mayor Andy Davison in the multimember Calgary constituency. This warning to Aberhart and his colleagues was reinforced just a week later in the general federal election where Social Credit candidates also fared poorly. The vain prospect of a significant party breakthrough outside Alberta to fight for Social Credit legislation at the federal level had proven illusory. When Aberhart met the new Legislature for the first time in May, he faced not only the continued hostility of the federal government but also a revived and raucous opposition that manifested a restive electorate, not to mention the divisions that continued to exist within the ranks of his own party.

The government's embattled and preoccupied state was reflected in the unsettled state at the Conservation Board in Calgary. For nearly a year, the Board had been without a chairman.[106] In the early summer of 1939, for reasons that remain unclear, the minister of lands and mines chose not to renew Knode's contract. Charles Dingman was left to run the Board as deputy chairman and Knode was appointed the Board's consulting engineer.[107] Knode's position with the Board was eroded further during the course of the McGillivray Commission hearings. Not only was Knode the focus of much hostility from the independent producers but also his technical competence was brought into question. It is not surprising that in the autumn of 1939, when the Alberta government sought an authoritative estimate of Turner Valley crude oil reserves and advice on production allocation, it went to a University of Michigan professor and consulting chemical engineer, Dr. George Granger Brown.[108]

Not prepared to continue with Knode as chairman, neither was the government prepared to promote Charles Dingman, the man who above all others had devoted himself to the cause of petroleum conservation in Alberta for 20 years and who, in all but formal title, had managed the Board since February 1939 when Knode went to England.[109] Unable to decide how its new conservation legislation should be imposed upon Turner Valley, and faced with mounting crude oil production allocation problems, the minister of lands and mines once again asked the petroleum division of the Department of the Interior in Washington, D.C., to recommend a highly regarded petroleum engineer who might be interested in coming to chair Alberta's new Conservation Board.[110] Tanner was still hoping that he might find an "outside authority" whose prestige and administrative skills would command widespread voluntary compliance with Board measures. Finally, in late August 1940, Dingman was advised that the government had made arrangements to appoint a "reputable engineer as Chairman of the Board and that, in these circumstances, the services of Mr. W.F. Knode, as consultant to the Board, need not be continued."[111] A few days later, Dingman learned that the "reputable engineer" who Tanner had found was Robert E. Allen, a consulting engineer from San Gabriel, California. Dingman promptly submitted his resignation. Having been passed over once before in favour of an outsider unfamiliar with the Alberta scene, Dingman was not prepared to accept the government's lack of confidence in his ability a second time. The surprise and widespread regret of Turner Valley producers was immediately conveyed, and Tanner was compelled to meet with the Alberta Petroleum Association (formerly the Alberta Petroleum Producers Association) to explain why the Alberta government had considered the appointment of a "high-cost" American engineer necessary.[112]

The government's announcement that J.J. Frawley was also being appointed to the Board for a five-year term was greeted with
even greater apprehension. Frawley was remembered particularly
for his recent submission to the McGillivray Commission recommending the creation of a single all-powerful board under which all
existing regulatory powers relating to the petroleum industry
would be centralized. Articulating the opinion of many oilmen, the
Western Oil Examiner declared that Frawley's contributions to the
oil industry had not been "particularly constructive" and expressed
the hope that his reponsibilities in the attorney general's department would occupy his attention.[113]

Robert Allen came with impressive credentials. Like his predecessor, W.F. Knode, he was born, raised and trained in the oilfields
of West Virginia; the two had briefly been classmates at the University of West Virginia. After graduating with a degree in petroleum
geology and engineering, Allen left for California in 1921. His arrival and first years on the West Coast coincided with the discovery
and development of the prolific Huntington Beach, Long Beach and
Santa Fe Springs oilfields. Allen gained his initial experience as a junior engineer and geologist with three of the more important independent oil companies. Stints as a private consultant, drilling contractor and field superintendent for the Continental Oil Company
rounded out his background up to the late spring of 1929 when he
became the supervisory engineer for the first attempt to institute
prorationing and conservation in California.[114]

To promote conservation, the California State Legislature passed
a gas conservation law that restricted production from wells with
an excessive ratio of gas to oil. Essentially, this was the limit of state
or public involvement. By early 1930, most fields in the state were
under curtailment supervised by the umpires, who were responsible
to the Conservation Committee of California Oil Producers. Participation nonetheless remained wholly voluntary. Some, particularly
smaller producers, refused any co-operation, whereas others, if they
chose to co-operate at all, accepted the pool and well allowables
recommended by the Conservation Committee of California Oil
Producers as only a rough guide.[115] It was in this environment, conditioned by the fierce determination of the independent operators
to avoid state regulation, that Robert Allen had gained his reputation as a conservation specialist.[116] The year before he left for Alberta, Allen had watched the independent operators pour millions
of dollars into a campaign mobilizing public opinion to defeat the
Atkinson Bill, which would have established a state-appointed conservation authority.[117]

By the time Allen left for Alberta in the summer of 1940, he was a recognized authority in his field and was especially well equipped to deal with Turner Valley's independent operators. Once in Alberta, Allen's considerable skills were immediately put to the test. The essential problem, as he chaired his first meeting of the Petroleum and Natural Gas Conservation Board on 21 September 1940, was the continued excessive production of Turner Valley gas. To this point, the Board after careful field study had issued three successive natural gas production schedules, all of which had drawn strong objection and even noncompliance by some operators. Allen arrived just in time to witness the Board's first attempt to use the court to enforce its will upon defiant operators. In an action against Mercury Oils Ltd. and the National Petroleum Corporation, the Board sought in accordance with section 46 of the *Oil and Gas Resources Conservation Act* to recover the costs incurred when it had seized and shut in a well belonging to each company for allegedly producing beyond their allotted quotas. This legal confrontation represented not just a demonstration of the Board's determination, it was also a test of the Alberta government's revised conservation Act. The independent companies continued to anchor their resistance on the old argument that the Act under which the Conservation Board functioned was ultra vires of provincial authority and, moreover, that the production quotas assigned were not for the purpose of regulating the wells in the interests of conservation, but rather for the purpose of regulating production in accordance with market demand.[118]

Allen began by outlining his background and views on oil and gas conservation before the Alberta Petroleum Association. In the company of the minister of lands and mines, he also undertook familiarization tours of Alberta's oil and gas fields, but these were compromised on two counts. Charles Dingman was unavailable for consultation and William Knode, the former technical consultant to the Board, refused Allen's request that he use the three-month interval specified in his termination notice to prepare a formal report on a pressure maintenance program for Turner Valley.[119] Therefore, it was in the area of administrative procedure that the new chairman's influence was most felt.

The most important of the administrative changes was the initiation of a much more regularized hearings policy.[120] This addressed one of the more pressing criticisms levied against the Board before the McGillivray Commission. Rounding out the new chairman's efforts to tighten the Board's procedures and administrative authority was the formal merger of the Calgary office of the provincial Department of Lands and Mines headed by Floyd K. Beach with the

Petroleum and Natural Gas Conservation Board. Allen announced that the merger, effective October 15, was made in the interests of efficiency and would eliminate the duplication of reports issued by the two offices.[121]

As Allen hurried to evaluate the available data on the Turner Valley field, he was confronted with a new and overriding factor. Another and even more powerful regulator came to Calgary to assess the situation in Turner Valley firsthand.[122] He was George R. Cottrelle, dominion oil controller and, like Allen, was interested in achieving maximum crude oil production from Turner Valley, but his interest was focused on the short term. Cottrelle's presence in Calgary marked the federal government's sudden preoccupation with the production and efficient distribution of strategic resources.

In the spring of 1940, the realization had finally dawned upon Ottawa politicians and military leaders that the war was going to be long and that the Allies were much more dependent upon Canadian material resources than initially forecast. The consequence was a further reorganization of Ottawa's wartime administrative apparatus. The War Supply Board, headed by Minister of Transport C.D. Howe, was disbanded, and the responsibility to mobilize Canada's defence production efficiently went with Howe to the new Ministry of Munitions and Supply. In addition to assuring the uninterrupted flow of certain crucial foreign goods and resources, Howe had to ensure that strategic resources at home were channelled efficiently into priority areas. To this end, on 24 June 1940, he established the Wartime Industries Control Board. The Board consisted of "Controllers," each with sweeping powers to regulate one of the essential economic resource sectors, timber, steel, oil and metals.[123]

George Cottrelle, a former Toronto banker and the chairman of Union Gas was appointed oil controller. His potential power was awesome. He could "take possession of and utilize any land, plant, refinery, storage tank, factory or building, used or capable of being used for mining, drilling for, producing, processing, refining or storing of oil," fix prices at which oil could be offered for sale and prohibit any person or corporation from dealing in oil without a licence issued by his office.[124] So supported, his task was to manage the nation's fuel supply. This meant ensuring that the escalating demands of wartime industry and the military were efficiently met while domestic consumer demand was effectively reduced to the bare minimum.[125]

Cottrelle was in Calgary to determine why, at various times since mid-July 1940, Turner Valley crude oil production could not meet regional demand.[126] Allen's job was going to be even more difficult than he could have imagined. In late November, after the normal monthly review of reservoir conditions, Allen announced that daily production in the Turner Valley field for December would be reduced from 25,400 barrels daily to 22,000 barrels daily.[127] On 12 December 1940, he informed all Turner Valley producers that the Board would be undertaking a general review of production allocation procedures, and he invited all to submit their "full and frank opinion of the present method," so that with the co-operation and assistance of all "a really constructive effort for the good of the Turner Valley field" could be achieved.[128] A day later, before one of the largest gatherings of oilmen ever to assemble in Calgary, Allen explained the reasons behind the recent reduction of Turner Valley allowables. He said that his own experience and the latest scientific information on efficient production demonstrated that "the rate of production had a very definite bearing on ultimate recovery." It was imperative that the Turner Valley field not be produced too fast. He went on to give a ringing defence of conservation, which he defined as "the efficient production of oil for beneficial use without avoidable waste." It was not, he stressed, "either hoarding or saving for the future."[129] Allen argued the case for conservation with examples from his experience in California.

Allen tried to establish the point that conservation was not only necessary to ensure maximum recovery but also the key to continued exploration and development. He pointed out that banks were more interested in well-managed oil properties, and where strict control was exercised, they were more willing to consider the oil beneath the ground as security for loans. Conservation and control would permit companies to borrow at better rates on their oil reserves and get on with development. And in this task, Allen assured his listeners, he had no doubt about the outcome. Commenting on his tour of exploration areas outside of Turner Valley, he remarked, "My opinion is so favourable, that it would embarrass me to tell you how optimistic I am." Discoveries would not happen without effort and cost, but Allen was emphatic. "I can assure you that the oil is here," he insisted.[130]

This seems to have been the setting in which Allen functioned best. Whereas Knode had spoken the language of the drilling crew, and by personality had exuded the rough-and-ready character of the oilfield, Allen was much more in his element in the boardroom, in the minister's office, or as an after-dinner speaker. He was a polished and persuasive spokesperson for the cause. Nonetheless, the extent to which Allen had been able to reduce the pressure to increase production in Turner Valley was almost immediately undermined. In early January 1941, the oil controller informed Alberta that he wanted more oil.

When releasing the January production allocations for each of Turner Valley's 130 producing oil wells, Allen had announced that, as a conservation agency, the Board could not in good conscience authorize a field production in excess of 22,000 barrels daily.[131] On January 15, the oil controller directed that the combined allotments of the oil wells in the Turner Valley field be increased to 25,000 barrels per day.[132] Once more in the national interest, the final say in the management of Alberta's oil and gas resources had reverted to Ottawa. In responding to Cottrelle's "request," Allen was careful, as he prepared the Board's new allocation for each well, to separate the allotment into distinguishable components: the "conservation allotment" and the "extra war emergency allotment."[133] He intended not only that Turner Valley operators would immediately make the distinction between production levels based upon sound engineering principles and the amount of excessive or "wasteful" production allowed to meet an urgent need but also, given the expectation that the life of the field would be greatly shortened, to leave for posterity a record of the Board's commitment to conservation. Allen's sense of how the situation would unfold and his resultant caution proved well founded. The directive from Ottawa to increase Turner Valley crude oil production was followed in just two weeks by a second "request" to expand output by an additional 1,000 barrels per day.[134] Through the spring, the gap between the optimum rate of production and what the field was directed to produce continued to widen.

The response of the Conservation Board to this serious and deteriorating situation was threefold. First, Allen sought to refine the allocation mechanism in such a way as to make it more acceptable to the smaller operators, and by simplifying procedures in light of a better understanding of Turner Valley reservoir behaviour. Second, Allen advised the Alberta government on various measures intended to encourage petroleum exploration. Also, the Board renewed its efforts to solve the fundamental problem in Turner Valley—the flaring of "waste" gas.

After reviewing industry recommendations for improving the existing method of production allocation, the Board announced several much desired changes in late January, along with the comment that in one respect it found the operators' response "somewhat disappointing." This was "the comparatively slight attention given to the subject of gas conservation without which," Allen reminded, "there can be no great measure of oil conservation."[135] In determining February production, Allen stated that three rather than four factors would be taken into account. The acreage factor was dropped, leaving gas-oil ratio, reservoir pressure and well productivity as the remaining variables determining allocation. For March production allocations, the formula was simplified even further. Henceforth, assigned production would be based upon a single factor that combined both well pressure and productivity components.[136] The Board also accepted the concept of a minimum well allocation or "minimum living wage." Although this measure was cheered by the operators of low-production marginal wells, it was also perceived by the Board as being consistent with its conservation objective. By assuring the operator enough production to meet operating costs and a modest profit, the premature abandonment of marginal wells with low gas-oil ratios could be prevented. This provision meant, for example, that the operator of a 5,000-foot well in Turner Valley could count on a minimum of 22½ barrels daily.[137] Allen also made another change that helped to appease several of the Board's severest critics. He agreed that the Miracle 2, Model-Spooner 1, Sterling Pacific 3 and Model 3 wells had been identified improperly as gas wells and should be reclassified and included in the oil well allocation schedule.[138]

Buoyed by the recent success of the Standard Oil Company of British Columbia's Princess 2 well and by the Conservation Board's official designation of the area in the vicinity of the well as the Princess oilfield, the Alberta government next introduced a series of expedients intended to encourage petroleum exploration that might lead to new oil discoveries. First, the existing regulations governing drilling and production of oil and gas wells were revised in the areas of well spacing, the coring and initial testing of producing formations, and the promotion of the unitized production of oil and gas pools.[139] Having strengthened the Conservation Board's hand in these areas, the government cancelled existing reservations to over one half million acres where little or no exploratory work had been undertaken, and then issued new regulations governing the reservation of petroleum and natural gas rights that made it impossible to hold lands for speculative purposes for a long period by merely paying the annual rental.[140] Moreover, money spent on exploratory work could now be credited against the rental. As a further incentive, Tanner announced that discovery wells would be exempt from certain royalty payments.[141]

To enhance the likelihood that its efforts to promote exploration would achieve results, the Alberta government encouraged individuals and companies to concentrate their efforts in selected districts.

MAP 4.1   Provincial Reserves, 1941

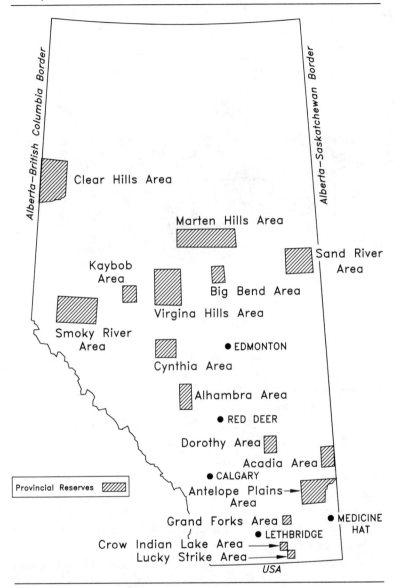

SOURCE: *Alberta Gazette*, 31 March 1941.

Early in 1941, Allen helped the minister of lands and mines identify those areas of the province that showed the best evidence of being potential oil lands. Of the 30 regions selected, the 15 most remote from transportation facilities and markets were withdrawn from possible reservation under the petroleum and natural gas regulations and set aside as provincial reserves.[142] (See Map 4.1.)

While the federal government pushed for greater wartime oil production, two of the Board's staff, Jack Storey (l) and Alex Essery (r), took a rare moment to relax on the steps of the Black Diamond Office in 1942. ERCB, Essery Collection.

While the government waited in the hope that the industry would respond quickly to its initiatives and expand the search for new oilfields, Allen returned his attention to the old and intractable problem of Turner Valley waste gas. In a letter to the producers in late March, Allen warned that the excessive withdrawals of gas and oil from the Turner Valley field were creating a serious situation, and that it was unlikely that the field would be able to maintain current production of 26,000 barrels daily for the duration of the war unless something was done to restore or maintain reservoir energy.[143] A few days later, the oil controller arrived in Calgary to say that the waste of gas in Turner Valley had to be stopped. "Something has got to be done at once and that is the job for the Conservation Board. Not only is energy being wasted but the gas burned contains gasoline," Cottrelle informed.[144] If Allen found it just a little galling to be told to get on with gas conservation by a senior federal administrator whose insistence on excessive crude oil production was a significant part of the problem, he did not say. Alberta oil producers, however, were not reticent about reminding Cottrelle that the more oil produced the greater the gas wastage was going to be. Cottrelle's answer was that the gas should be forced back into the field. This was also the preference of the Alberta government, and the vehicle through which it hoped to achieve this end, the *Oil and Gas Fields Public Service Utilities Act,* was currently under debate in the Legislative Assembly. The bill gave the government the power to grant exclusive franchises to corporations for operation of public service utilities in oil and gas fields. In particular, Tanner was anxious to grant such a franchise to a corporation, or group of corporations, to establish a pipeline to gather waste gas and return it to the producing structure in the field.[145] He explained that oil producers would have to pay the cost of pumping the gas into the field, but would receive back the value of the natural gasoline extracted from the gas before it was returned to the reservoir, and emphasized that the government would derive no revenue from the project.[146] Oilmen were sceptical of the plan; they believed that the cost of injecting the gas would likely be greater than the value of the natural gasoline extracted. Their question was, who would bear the cost? They were also uneasy about what appeared to be the Alberta government's interest in playing a more active role in the private sector.[147]

While the oilmen and the government debated about how best to deal with the gas produced by Turner Valley oil wells, Allen renewed his efforts to reduce gas well production. In an April letter, he informed absorption plant owners that gas production over the preceding eight months averaged more than 146 million cubic feet daily, of which an average of 92 million cubic feet or 63% had been wasted.[148] Allen cut allowable April production by nearly one-third and warned plant operators to expect further monthly cuts.[149] At the same time, Board engineers redoubled their efforts to monitor the performance of individual wells. To emphasize its determination to preserve reservoir energy, the Board identified the Model No. 3 well as a grossly inefficient producer and ordered it shut in.[150]

Despite all efforts, the situation continued to deteriorate. The much-hoped-for expansion of exploration activity failed to materialize and the rapidly declining production from the Princess discovery well dashed any hope that an important new crude supply would emerge from that quarter.[151] Without the contribution of supplementary supply, continued reliance on the Turner Valley field to meet the wartime demand of 26,000 barrels daily became serious and threatened the ruination of the field. By June, the Conservation Board faced a crisis. Allen informed Albertans that continued overproduction was causing the field to become steadily "more inefficient in its utilization of reservoir gas energy."[152] In April 1940, it took Turner Valley oil wells an average 3,030 cubic feet of gas to produce a barrel of oil; in April 1941, it required 4,185 cubic feet, an increase of 38%, and the rate of change was accelerating. He also noted that of the 145 crude oil-producing wells, the 32 wells with the highest gas-oil ratios produced 33.6% of the gas but only 7.6% of the oil. He announced that, commencing July 1, the Board planned to restrict sharply production from these wells.[153] This did not mean, however, that the other producing wells could simply make up the production loss. Total field production would have to be reduced.

On June 7, Allen sent the minister of lands and mines a lengthy letter. Deeply concerned, he lamented that "with a full realization of its actions this Board has been a party to sacrificing the future of the field as a contribution to the winning of the war. . . ."[154] He informed Tanner that the Board believed that the optimum rate of production to achieve maximum economic recovery was at maximum 11,000 barrels daily. In successive monthly steps, the Board had reduced the conservation quota of the field to 16,500 barrels daily, which was, Allen noted,

> still 50 percent greater than the optimum rate and we feel that this 50 percent tolerance over the optimum rate is as large a war contribution as we can justifiably authorize as a conservation agency that is deeply concerned about the post-war petroleum future of Alberta.[155]

Allen acknowledged that nothing was more important than winning the war, but he astutely reminded Tanner that the excessive production demanded of Turner Valley as a wartime levy needed to be examined closely. The letter indicated that only a small portion of Turner Valley oil production was being used for war purposes, such as training aircraft and local military transportation. "The great bulk of it continues to be used in the normal way as fuel for commercial and pleasure automobiles and for farm power." Allen also challenged Ottawa's position that the principal war contribution of Turner Valley crude oil production was the saving of foreign exchange for the purchase of aeroplanes and military supplies.[156] This, Allen argued, was but a pretence. It was understandable and acceptable only if the oil to replace a portion of Turner Valley crude production actually had to be imported, and this was not the case. Ottawa had made no attempt to implement even the mildest form of gasoline rationing to curb the pleasure use of vehicles. The Board chairman concluded, "It would seem then that a substantial part of the sacrifice of Turner Valley is not so much for direct war effort as to avoid the rationing of motor fuel in the Prairie Provinces." He warned Tanner that "the effect of near capacity production is little short of disastrous as witnessed by steadily declining pressures and the increasingly disappointing performance of new wells."[157]

Whatever the Alberta government might decide about the volume of crude oil production, a much more serious effort had to be made to maintain reservoir pressure. The Conservation Board's proposal to conserve the gas cap was to reduce withdrawals to market demand and to distribute the revenue derived from the sale of the gas on a rateable basis to all producers, thereby compensating producers who did not have an outlet for their gas except in the production of naphtha and natural gasoline. This would require building a gas trunk pipeline to tie in the producers at the south end of the field with the distribution point in the centre of the field. The hardest but essential part of this plan, and the aspect that Allen acknowledged would be "vigorously opposed," was the need to break the principal gas producer's exclusive contract to supply Canadian Western Natural Gas, Light, Heat and Power. Allen estimated that the concession that Imperial Oil's subsidiary, Royalite, would have to make was about one-fifth of its natural gas market and income.[158] The Board chairman concluded his appeal by urging the minister of lands and mines to use the authority conferred in the existing legislation to deal with the problem.

Consistent with, and issued the same day as the Tanner letter, was a unique Board order. Entitled "The Conservation of Petroleum Resources," Order No. 63 formally presented the Board's conservation philosophy to all Turner Valley operators.[159] The order identified seven fundamental principles with which the development and production of Alberta's petroleum and natural gas resources would have to conform. (See Appendix IV.) Having clearly articulated his concern about the situation in Turner Valley to the Alberta government and having issued his conservation commandments, Allen left the province for Washington to advise the U.S. government on growing oil supply and distribution problems. At first, it was considered that Allen was on loan to the U.S. government, but it soon became apparent that he would not be returning to Alberta.[160]

Allen left Edmonton with a clear statement of the problem and a program of remedies consistent with sensible conservation practice. Once more, resolution of the Turner Valley problem boiled down to a question of political will, and once again the politicians were found wanting. Allen had noted correctly that Ottawa's demand on Turner Valley was not being matched by an equally rigorous attempt to curb nonessential gasoline use. Full-fledged gasoline rationing did not come until April 1942, and even then it was the declining availability of a crude oil supply beyond Canada's borders rather than the price that was the critical factor. Alberta did not try to challenge the oil controller's requisition on Turner Valley. Manning and Tanner might have anticipated that the province's position would begin to receive more sympathetic attention among federal officials, since Allen's dire warning was coupled coincidentally with the resignation of the Conservation Board's deputy chairman, F.G. Cottle, and his departure to join the oil controller's senior staff in Ottawa. Perhaps the lack of will also had something to do with the Social Credit government's total lack of success in earlier challenges to federal power. Moreover, the tardy manner that characterized Alberta's efforts to address the Turner Valley waste gas problem undermined the possibility of taking the high moral ground in any protest that Ottawa's demands were ruining the field and wasting one of Alberta's most precious natural assets.

While the politicians temporized, the Board struggled to maintain its authority and presence in the field. Not only did the Board have to function with the leadership of Professor Karl A. Clark, a temporary deputy chairman, appointed from outside, but also the wartime economy was beginning to undermine the Board's presence in the field. Just two weeks before Allen's departure, the Board had been confronted with the resignation of half its field engineers, including that of the Board's senior man in Turner Valley, G.A. Connell.[161] This signalled the emergence of a problem that would have to be dealt with in the future, but for now the Board was compelled to recognize the value of remaining senior technical personnel by an immediate increase in salary. One of these, D.P. Goodall, was sent to take temporary charge in Turner Valley until another suitable field engineer could be trained.[162]

In Turner Valley, the persistent struggle against gas waste continued, and the Board shut in the wells of gas producers found guilty of producing beyond their quota. Crude oil producers found to be producing beyond their allocations had the excess production deducted from their subsequent allocations.[163] On the more fundamental question of completely disallowing production from the gas cap, as Allen and others before had recommended, no directions were forthcoming. By late July, as exploration efforts dried up, and Turner Valley crude oil production failed to meet the 26,000-barrel-a-day quota for the first time, both levels of government faced a new round of criticism for their inaction.[164] This brought a delegation of senior federal officials, including G.B. Webster, the assistant oil controller, and Dr. G.S. Hume, head of the Geological Survey of Canada, to Calgary to confer with Tanner.[165]

As discussions dragged on between federal and provincial politicians, the *Calgary Herald* lamented that after 15 years of intensive operation in the Turner Valley field the greater part of daily gas production was still burned in waste gas flares or blown to the air.[166] The problem was that Edmonton and Ottawa had become deadlocked over the rate at which the Turner Valley field should be produced. Observing the clash of opinion, the industry's local journal explained:

> Oil controller Cottrelle is believed among those who favour an all-out run for the wells, while the ACB [Alberta Conservation Board] holds that control must be exercised if the field is to be preserved longer than a few years.[167]

Finally, Cottrelle and Tanner accepted the suggestion of J.J. Frawley, deputy chairman of the Conservation Board, that a third party, Dr. George Granger Brown at the University of Michigan, be hired to give his opinion as to the correct rate at which the Turner Valley field should be produced.[168]

THE BROWN PLAN

Frawley's suggestion to call Dr. Brown was not surprising. Not only was Brown's reputation as a consultant firmly established in the United States but also he was already familiar with the Turner Valley situation, having been brought by Frawley to appear before the McGillivray Commission and subsequently hired in August 1939 to do a study of Turner Valley crude oil reserves. Senior Board Engineer M.D. Kemp forwarded to Brown the most recent Turner Valley well data information, and Cottrelle and Frawley met with Brown in early December at the University of Michigan to discuss the professor's preliminary findings. Brown presented the penultimate draft of his conservation plan for Turner Valley on 6 January 1942, and on January 9 he came to Calgary to discuss his plan with Calgary oilmen at a Conservation Board hearing.[169]

Brown's proposal was that the Board adopt the reservoir-fluid-volume method of regulating oil and gas production. The initial premise of this approach was based upon the principle of correlative rights. Each leaseholder has a right to the fluid—gas, oil or both—in the formation underlying his lease. To treat lease owners in an equitable manner on the basis of the reservoir fluid under their individual leases, it is necessary to limit each lease to the same maximum rate of withdrawal of reservoir fluid per acre per day. Under this method of operation, no well is allowed to produce at such a rate as to drain fluid from under adjacent leases. To ensure that each leaseholder had the opportunity to maximize his output from a common reservoir, it was necessary to make sure that the entire reservoir was produced at a proper uniform rate of with-drawal. In addition to achieving equity among leaseholders, deter-mining the proper rate of withdrawal served the public interest of promoting maximum recovery, and thereby met the fundamental principle of conservation. Exceeding a certain maximum rate of withdrawal for any appreciable length of time causes the pressure in the formation to fall so rapidly that the gas comes out of solution and, on release at the earth's surface, leaves much of the oil behind. Careful examination of the history of each well's performance in the Turner Valley field led Brown to conclude that the maximum rate of withdrawal to prevent excessive loss of energy and to achieve maximum recovery was 25 barrels of reservoir fluid per acre per day.[170] All leaseholders were treated alike for the maximum rate of withdrawal. Individual well production rates, however, were modified by an adjustment factor that decreased allowable production as the well's gas-oil ratio increased or as its 24-hour shut-in pressure decreased. If followed, Brown estimated that his plan would bring daily field production to 22,500 barrels of oil and 33.1 million cubic feet of gas.[171] He concluded his proposal with the declaration that whatever gas was not used should be returned to the reservoir (see Map 4.2).

Through February, the Board met with the Alberta Petroleum Association and individual operators to address industry concerns and objections. Essentially, the Board was confronted by individu-als and groups seeking special consideration, arguing that their part of the field or their well deserved special consideration, and, partic-ularly, why the gas-oil ratio penalty factor was too severe in their cases. Brown urged the Board to stick to its objective. Although he recommended that there be no compromise on treating the entire field on the same basis or on relaxing the gas-oil penalty factor, he did accept the idea of a "reasonably" gradual transition from the present allowable system to the new approach. Above all, Brown warned that, in view of the critical conditions in the southern part of the field, "no time should be lost if the Conservation Board de-sires to avoid future criticism."[172] He was confident that his plan, as well or better than any other that might be devised, took into con-sideration the practical difficulties that one could expect to encoun-

MAP 4.2   Turner Valley, Brown Plan, 1942

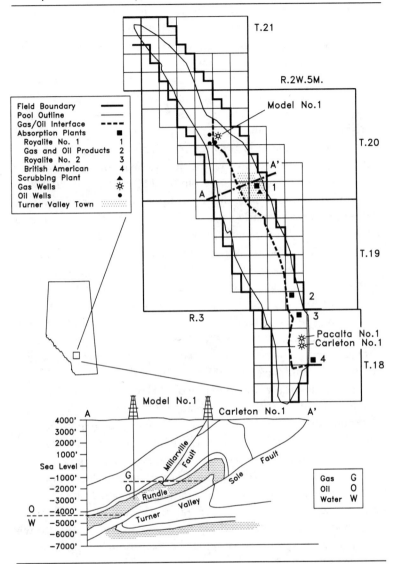

ter when "inaugurating a program of conservation in a field which has been allowed to run almost wide open for such a long period as the Turner Valley field."[173]

In the Brown plan, the minister of lands and mines had a proposal from an outside expert that was almost ideal. It met the standard of approved engineering practice and it offered a daily production from Turner Valley that was attractively situated almost midway between the 26,000 barrels demanded by the oil controller and the 16,500-barrel limit wanted by the Conservation Board. Tanner was cautious, however, and there was certainly no precedent of rushing decisions pertaining to conservation matters. The minister remained anxious about the degree of industry support. He knew that investment in new wells was lagging, and he was sensitive to charges that it was largely the misguided policies of the Alberta government that were retarding oil development.[174] His caution about the Brown plan might also have resulted from his earlier decision to give the industry a more structured role in generating and applying Conservation Board policy. In January, Tanner had been persuaded by a delegation from the Alberta Petroleum Association (APA) to allow the industry to form an advisory committee that would meet regularly with the Conservation Board to discuss matters affecting the industry. The new committee focused its attention on the proposed Brown plan and, having gained the minister's sanction, was determined to make its influence felt. A few months later, Tanner advised the APA that it was "essential that the Committee to work with the Board should be elected by a vote of all operators in the field."[175] The minister favoured and was prepared to give more weight to the Association's viewpoint, but he wanted to be certain that there "be no question but that such a Committee would have the confidence of all the producers."[176]

Introduction of the Brown plan would be gradual and with close consultation of industry at each stage. On February 25, Deputy Chairman J.J. Frawley announced the Board's acceptance of the Brown plan and explained that the plan would be implemented over a five-to-six-month period.[177] The first stage commenced on March 1 with a modified application of the Brown formula to determine well quotas yielding a total daily field production of 26,000 barrels. The next important stage was revealed on July 3 when the Conservation Board declared that full application of the Brown plan would be applied at once to the 35 producing wells in the northern portion of the Turner Valley field and in September for the remaining and larger part of the field south of Township 20.[178] (See Map 4.2.) This approach was contrary to Brown's insistence that to be effective his plan had to be implemented across the entire field at the same time, but it was the only approach that was politically acceptable. The division between the southern and new northern part of the field represented a political and economic reality. The majority of the oilmen whose operations were centred south of Township 20 had interests that varied from those whose business was based on crude oil-producing wells situated at the north end of

Turner Valley. Hostility towards the Board and its activities had always been centred among the small independent operators in south Turner Valley.

The task of bringing conservation to the southern part of the Turner Valley field was assigned to the new chairman of the Petroleum and Natural Gas Conservation Board, J.J. Frawley. Neither an engineer nor an American with experience in one of the oil states, as was the case with the other two who had held the title of chairman, Frawley was nevertheless especially well qualified for the task. Born in Sudbury, Ontario, in 1893, James Joseph Frawley graduated from University College, University of Toronto, in 1915, attended Osgoode Hall in Toronto, and journeyed west to join the legal department of the Canadian Pacific Railway immediately after being called to the Ontario Bar in 1919. He was called to the Alberta Bar in 1921, and in 1924 left the CPR to work for the Alberta attorney general's department, thus beginning a remarkable career in government service that eventually was to span nearly 50 years.

It was as assistant to the senior crown counsel in the famous Solloway and Mills stock fraud case in 1930 that Frawley's considerable ability first gained wide notice. There was little surprise when the United Farmers of Alberta government subsequently appointed Frawley registrar under their new *Security Frauds Prevention Act*. His reputation as a particularly able and dedicated civil servant not only survived the transition to Social Credit rule in 1935 but also enabled him to be entrusted in 1938 to manage Alberta's appeal before the Judicial Committee of the Privy Council in London to sustain the validity of three Social Credit bills that had been disallowed by the Supreme Court of Canada. Also in 1938, Frawley was assigned to supervise a small group charged with drafting one of the most important bills on the Alberta government's legislative agenda for that year, the *Oil and Gas Conservation Act*.[179] The insight that Frawley brought to bear in drafting this Act was informed partly by his experience representing the Alberta attorney general in *Spooner Oils v. Turner Valley Gas Conservation Board* (1933), the outcome of which was the collapse of the Alberta government's first attempt to impose conservation upon Turner Valley oilmen. No sooner was his work completed on the 1938 Conservation Act than he was appointed counsel to the McGillivray Commission. For the following year and a half, Frawley guided the proceedings of an exhaustive investigation covering every facet of the petroleum industry in Alberta. As an experienced and trusted civil servant, and with an excellent knowledge of the petroleum industry, Frawley's appointment to the Conservation Board on 18 September 1940 could hardly have been unanticipated. As Robert Allen departed for Washington in June 1941, Frawley was named deputy chairman.

It was not just in the oil and gas sector where Frawley's talents continued to be utilized. He was the key adviser to the cabinet during Alberta-initiated negotiations surrounding the various *Natural Resources Transfer Agreement Amendment* acts. Frawley was also influential in the development of the first true provincial bank. Established as consumer deposit and loan centres in hundreds of communities, the network of Alberta Treasury Branches rested upon the only significant piece of original Social Credit monetary legislation to escape federal disallowance. In managing this escape, Frawley's role in and outside the courtroom was a critical factor. By the early 1940s, Frawley had become an indispensable presence to the Social Credit government.[180]

By the time Frawley assumed the Board chairmanship in May 1942, the position had been vacant for nearly a year, and the press had been complaining for months that the industry remained "without the presence of a third star in the controlled firmament."[181] It is probable that Frawley's elevation to chairman might have come sooner had he been willing, but Frawley, above all, saw himself as a solicitor and was committed to a career in the Department of the Attorney General. In accepting the chairmanship, he insisted that, as a condition of his appointment, he retain his status as a solicitor in the Department of the Attorney General and that he continue to receive his salary from that source.[182] Frawley, it seems, had been persuaded to take on the chairmanship of the Petroleum and Natural Gas Conservation Board for a short term, to implement the Brown plan and to complete the task that had bedevilled all his predecessors—bringing full-fledged conservation to the southern part of the Turner Valley field.

As Frawley began the task, he was under no illusion that the Board's efforts would be without opposition. He had watched the independent gas-cap producers fight to stave off regulatory interference for nearly 20 years. In this final chapter in the conservation struggle, however, the real opposition to the Board's efforts came from quite another quarter, and this neither Frawley nor his political masters in Edmonton likely anticipated. On September 1, the Brown plan was duly applied to the entire field, and later in the month Dr. Brown arrived to interview operators and to review progress. To Brown's and the Board's relief, perhaps even surprise, and despite some grumbling about individual well allotments, the plan seemed to be meeting the general acceptance of Turner Valley oilmen. Yet the outward calm was deceptive. Background forces were at work that would soon bring the old problem and the old debate to the forefront once more.

On the international front, the oil supply picture began to take a serious turn. The invasion of North Africa in 1942 had sharply escalated Allied demands on U.S. supply. Mechanized warfare in the desert demonstrated dramatically the importance of oil—17,000 gallons of gasoline were needed to move a typical armoured battalion one hundred miles. With U.S.-supplied petroleum meeting 70% of Allied war needs, and with German submarine attacks and the Japanese occupation of Pacific oilfields threatening to make the Allies even more dependent on the United States's dwindling domestic sources, Washington became gravely concerned. The alarm in official circles spread gradually to the popular press.[183] By January 1943, this concern translated, among other responses, into an emphatic directive from the U.S. State Department to Canada to improve oil production and to get on with a much more vigorous exploration program.[184] The Americans were not impressed with Canada's efforts to develop promising oil exploration areas in western Canada or the apparently relaxed attitude of some central Canadian politicians and business leaders that access to U.S. supply could be relied upon. Washington's blunt misgivings sharply accentuated the dominion oil controller's concern about Turner Valley production. The problem was that, as Cottrelle's concern was increasing, Turner Valley's productive capacity had peaked and was now decreasing.

On the natural gas side, the war was also having a telling effect. The market for natural gas had been greatly extended. Stimulated by the general economic impact of Canada's war effort, Calgary began to grow, causing an accompanying rise in household and industrial demand for natural gas. Elsewhere in southern Alberta, several air force bases erected for the British Commonwealth Air Training Plan and prisoner-of-war camps were tied into the natural gas distribution network. Much more important than the expansion of the traditional heating market was the appearance of two large new industrial users. On the direction of Minister of Munitions and Supply C.D. Howe, a crown agency, the Allied War Supplies Corporation, had been set up to build and supervise the operation of two plants on the outskirts of Calgary. The first was an alkylate plant that took isobutanes, a product of the existing Turner Valley absorption plants, to produce the blending agent alkylate necessary in the making of high-octane aviation gasoline. To operate this plant at minimum capacity, required 329 barrels of isobutane daily, which in turn required the daily production of 85 million cubic feet of wet gas from the Turner Valley field. The second plant, Alberta Nitrogen Products Limited, produced ammonia and ammonium nitrate used in the manufacture of chemical products for munitions. Natural gas was the fundamental raw product, and the plant required nearly 10 million cubic feet daily, a volume amounting to more than one-third the Calgary utility requirement.

By late 1942, it became apparent that if wartime requirements were to be met the Turner Valley field would have to be used in a more rational manner. Wells connected to the existing natural gas pipeline gathering system were on the verge of being produced to excess to meet increased market demand, while a large volume of gas was still being flared at wells beyond the pipeline network. In concluding his report, Brown had reminded the Board that until all wells had been tied into the gas system and surplus gas returned to the formation, the conservation ideal would remain unachieved.[185] Intent on moving further towards that ideal, Frawley sought the advice of Robert Allen in Washington. Allen, now the assistant deputy coordinator for oil in the office of the Petroleum Administration for War, was greatly concerned about the situation in Turner Valley, not only because of his former role but also because of Washington's growing anxiety about the larger North American petroleum supply picture. Allen immediately sent from his staff, Holley Poe, director of the Natural Gas and Natural Gasoline Division. Poe journeyed to Calgary, reviewed production practice in Turner Valley, and was appalled at "the flagrant waste of natural gas."[186] He offered to send to Calgary his division's senior consulting engineer, Thomas R. Weymouth, a well-known figure in the U.S. natural gas industry with some knowledge of the Alberta situation. He had been in Alberta in 1921 as a technical adviser to Canadian Western Natural Gas, Light, Heat and Power Company.

Weymouth arrived in Calgary on 8 February 1943. He faced a formidable challenge that had largely eluded the best efforts of the train of experts who had preceded him. During the previous year, the year of the largest useful consumption of gas in the history of the field, more than 50% of all gas produced had still been wasted.[187] The knot that had hitherto prevented the untangling of the problem was the exclusive contract held by the Royalite Oil Company to supply the gas requirements of the Canadian Western Natural Gas, Light, Heat and Power Company. In the final analysis, conservation in Turner Valley hung on shared access to the available natural gas market.

Weymouth set out two foundation principles as he began discussion with the various Turner Valley factions. His eventual recommendation would be founded upon "the continued application of the Brown plan allowables and the maintenance of the status quo of all existing contracts."[188] He then demonstrated that a comprehensive program of conservation would add 12½ years to the life of the field.[189] Having established this point, Weymouth calculated what 12½ years of added supply would mean in revenue to be earned by each party involved and set this against the costs to be shared by each according to the degree of benefit gained. The projected revenue cost picture was sufficiently attractive that by March 3 Weymouth had an initial agreement. Essential features of the plan were that Royalite would move its No. 2 absorption plant from the Hartell area to the site of its No. 1 plant on Sheep Creek (see Map 4.3) and upgrade the efficiency of the consolidated plant. Royalite would also extend its gathering lines into the north end of the field for the purpose of collecting gas currently being flared by crude oil producers. This would leave British American to gather and process gas from the southern part of the field at its absorption plant and Gas and Oil Products Ltd. to collect wet gas from the south central section of the field at its plant at Hartell. The three companies would extend their gathering systems to ensure that all wet gas was delivered to the three absorption plants whose collective capacity was just enough to produce the isobutane required by Allied War Supplies Corporation. After being stripped of its natural gasoline, all residue gas was to be conserved rather than flared. This entailed building a pipeline to transmit residue gas from the British American plant to a compressor station and dehydration plant to be constructed near the Gas and Oil Products Ltd. facility. From here, the residue gas, including that from the Gas and Oil Products Ltd. plant, would be conveyed by a new pipeline to the site of the Royalite No. 1 plant at the southern end of the Calgary gas utility system. At this point, the gas was to be turned over to Royalite, which would add the gas to that being sold to the Calgary gas company or return it to storage in the gas cap, depending on market demand. Royalite would pay for the gas at a differential rate, depending on whether the gas was stored or sold. Included in the March 3 draft agreement was the commitment of all the involved parties to meet in Toronto on March 16 to work out final details on an acceptable allocation of gas supplies for the entire Turner Valley field and for the distribution of costs.[190]

The composition of the group assembled on March 16 and 17 in the boardrooms of the Canadian Bank of Commerce in Toronto underlines the importance that was attached to the proposed agreement. It was an interest that extended well beyond Alberta's bor-

MAP 4.3   Turner Valley, Weymouth Plan, 1943

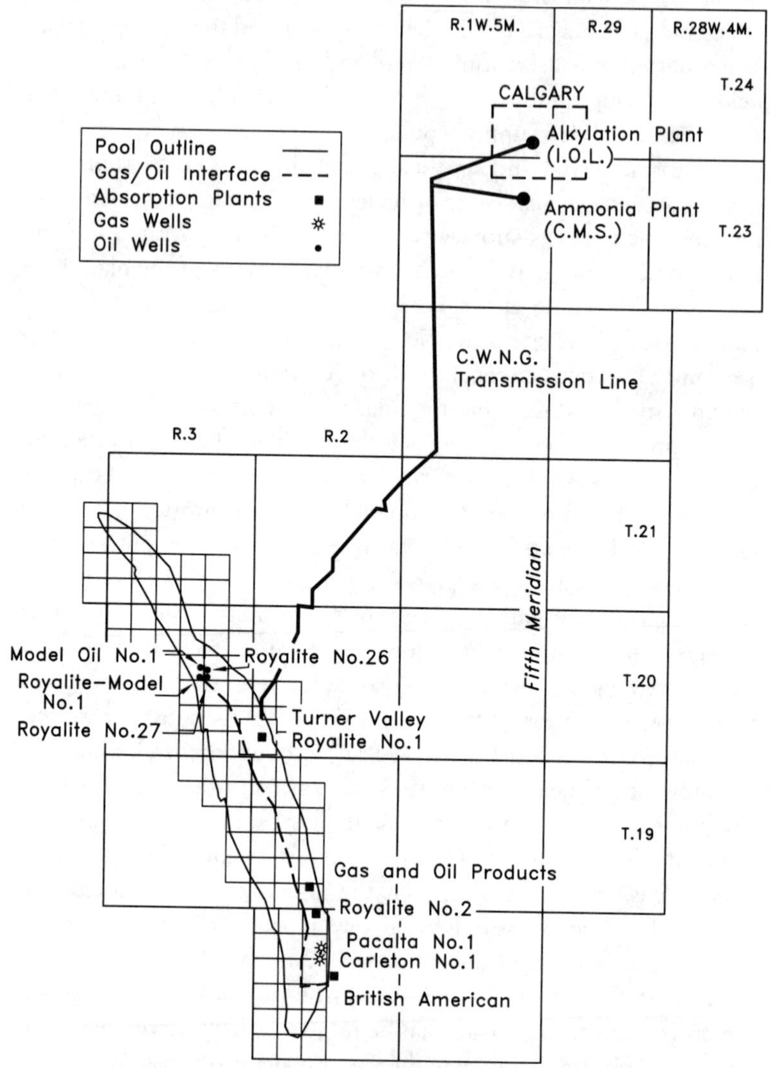

ders. In all, 16 individuals assembled to try to agree on a plan to bring the last and most important phase of conservation to Turner Valley. Representing the PNGCB was Chairman J.J. Frawley and M.D. Kemp; from the office of the Oil Controller was G.R. Cottrelle and three of his senior staff; Dr. D.M. Morison represented the Allied War Supplies Corporation; H.R. Milner the Calgary Gas Company; from Imperial Oil were Dr. O.B. Hopkins and D.J. Vandermeer; and from its subsidiary, Royalite, was R.E. Trammell;

the British American Oil Company was represented by W.K. Whiteford and J.A. McCutchin; Gas and Oil Products by A.H. Mayland and W.H. Jones; and from the U.S. Petroleum Administration for War, E.H. Poe and Thomas Weymouth.[191] After the meetings on the evening of March 17, Weymouth returned to his home in Greenwich, Connecticut, believing he had an agreement between the three companies—Royalite, British American and Gas and Oil Products—on an exchange of gas, and that he would return to Calgary within two weeks to put the plan before Turner Valley operators. His satisfaction, however, proved premature. On March 18, the vice-president of Imperial Oil, R.V. LeSueur, informed Frawley that his company did not feel that Royalite should proceed with the Weymouth proposal, especially the purchasing of gas to be stored and used years later when Royalite's present reserves were exhausted.[192] LeSueur's objection led to most of the group assembling for additional meetings in New York on March 25 and 30. These resulted in substantial agreement on the distribution of the financial burden, particularly that to be assumed by the Calgary Gas Company. Excluding the still to be determined cost of British American and Gas and Oil Products's extended gas gathering systems, the total estimated cost was $3,103,000. Since 19.24 billion or 16% of the 120 billion cubic feet of gas to be conserved by the program was a consequence of the excess production necessary to meet the isobutane requirements of the war effort, Allied War Supplies Corporation was asked to contribute $500,000 as its share of the cost, leaving $2,603,000 to be provided by Royalite and the Calgary Gas Company.[193] On May 25, at a conference in Montreal, Allied War Supplies Corporation announced that it would be unable to make a contribution on the ground that the Weymouth program was self-liquidating, and it would be improper therefore for the federal government to be involved.[194] After months of negotiation that had succeeded in bringing all the traditionally intractable elements together, it was a bitter disappointment. The Alberta government and the Conservation Board continued in the desperate hope that the oil controller and the minister of munitions and supply could be persuaded to allow crown assistance to break the impasse.

Within weeks, Alberta's wait upon Ottawa took on a sharply heightened intensity as a new crisis threatened not only the Weymouth modification but also the entire Brown plan and conservation in Turner Valley. On June 23, the Board's old adversary, Model Oils Ltd., representing a group of independent well operators, went to court seeking a judicial writ to command the Conservation Board to enforce its own allocation order, particularly as it applied to two Royalite wells. In his statement before the court, Model Oils' counsel, D.P. McDonald, pointed out that Board Order No. 139 had allotted the Royalite gas wells a total production of 9,004,795 thousand cubic feet for a one-year period commencing 1 November 1942. By 31 May 1943, McDonald charged that Royalite wells had already produced 122.2% of the amount allowed and that two of the wells, Royalite No. 26 and No. 27, which were offset wells to Model No. 1 and Royalite-Model No. 1, had overproduced to the extent of 156.7% and 150.7% respectively.[195] (See Map 4.3.) McDonald called attention to this violation of the Brown plan and the Board's order. He argued that, in failing to enforce its order, the Board had allowed Royalite to take gas from under Model Oils' adjoining acreage, thereby depleting the oil and gas that his company might produce in the future. Model Oils asked for compensation as provided for in section 17 of the Conservation Act. The vulnerability of the Board's position was implicit in Judge Ives's remark. "It does not look very well to me to have a public body of an administrative character dealing one way with one subject and another way with another subject." He adjourned hearings until July 21.[196]

The Board's position was more than embarrassing, because the credibility that it had fought so hard to win over the preceding five years hung in the balance. McDonald's disclosure that Royalite was overproducing was anything but a revelation; the monthly production figures were available to anyone interested. Of the 27 Royalite-operated wells in the gas cap, 21 were overproduced.[197] Earlier, Turner Valley operators had watched the National Petroleum No. 1 well shut in by the Conservation Board as soon as it had produced beyond its quota. Now the independents watched to see what would happen to Royalite, a company whose dominant influence, some believed, had always allowed it to have things its own way. Model Oils' letter to the minister of lands and mines regarding the company's application to the court articulated the sentiment of many. McDonald wrote that the principle of impartial enforcement was crucial if companies such as his were to commit themselves to conservation and the Brown plan, and that steps would have to be taken so that

> it will not be possible for the shareholders of this Company to feel that there is one law in this Province for the rich and one law for the poor with the choice of which law is to be enforced in the hands of the rich.[198]

For independent gas producers such as Model Oils, the conservation issue was tied directly to the question of access to the Calgary natural gas market. For them, the attraction of the Brown plan's approach to conservation was its implicit promise of such access. It was this essential aspect of the Brown plan that the Board, with the guidance of Thomas Weymouth, had been trying to achieve through the winter and spring of 1943. The Board was fully aware that Royalite's production quota under the Brown plan would be used up by late April and was counting on winning acceptance of Weymouth's scheme to overcome this final obstacle to comprehensive conservation in Turner Valley.[199] Failing the financial support of the federal government, Imperial Oil backed away from the Weymouth proposal, and the Board was left out on a limb. Rather than purchase gas from other producers, as was contemplated under the Weymouth scheme, Royalite simply abandoned the Brown quotas on its gas wells and increased production to meet the growing demand of its exclusive market. At the same time, the federal government, feeling it inappropriate to contribute $500,000 towards the cost of implementing the Weymouth scheme, insisted that it would "view with alarm any measure taken to reduce the supply of isobutane."[200]

Frawley's advice to the minister of lands and mines was to take a tough line and allow conservation objectives to prevail over war needs. "I have the definite feeling," he wrote, "that [Cottrelle] will not supersede us but will allow the iso-butane supply to go short."[201] Tanner was more cautious; the Board was directed to take no action until July 31. The minister's caution was in part a function of circumstance, because the crisis had come at a time when the political leadership in Edmonton was preoccupied. Throughout the spring session of the Alberta Legislature, Premier Aberhart's health began to deteriorate. In early June, the premier went to Vancouver for his customary holiday with his family. Within a few days of his arrival, he was admitted to hospital and died on June 23. The Alberta cabinet journeyed to the coast for Aberhart's funeral and then returned to Edmonton to convene a special caucus to elect a successor. Assembled on July 3, the caucus selected 34-year-old Ernest Manning to be the new premier of Alberta. Frawley waited as long as he dared and again wrote to Tanner to say that the Board had been placed "in a completely impossible position." As a result of the unauthorized production by Royalite, one independent operator had already commenced production beyond his well's quota, and Frawley advised that 15 to 20 wells could soon be operating in defiance of the Board. "The present situation cannot continue," he insisted. "The overproduced Royalite gas wells must be closed down unless the federal authority directs us that the violation of our Order must continue in the interests of the war effort."[202]

The Alberta politicians chose the federal option. Premier Manning sent a lengthy document to both Minister of Munitions and Supply C.D. Howe and Prime Minister Mackenzie King. Manning reviewed the obstacles that his government faced in trying to implement conservation and meet wartime demands. He informed Howe that the Alberta Conservation Board intended to enforce its order impartially on all producers, and then added the qualification,

> if the Oil Controller considers that the continued violation of the Board's order is necessary in the interests of the war effort or if he wishes the Board to make a new order which, though a negation of the conservation principles, he nevertheless considers necessary to meet the war emergency he need only direct the Board or the Government accordingly and his wishes will be respected and his direction readily implemented.[203]

Ottawa was not prepared to help the Alberta government off the hook. Cottrelle warned Howe, "There is no doubt in my mind as to that which the Alberta Government is trying to do, namely force us into support of the Weymouth plan." The oil controller insisted that the federal government had no responsibility for gas conservation. "We are not concerned about the supply of gas to Calgary twenty years from now—our first concern, in any event, is to win the war."[204] Clearly, the presence of former Deputy Board Chairman Cottle in the Controller's Office had had little effect in advancing Alberta's viewpoint. In his response to Manning, Howe followed the line of argument provided by his appointed oil controller. He pointed out that his department had "consistently held to the policy that conservation should be the responsibility of the Petroleum and Natural Gas Conservation Board" and hence any directives to operators should come from the Board. Howe went on to say that, although the Weymouth plan was "commendable," the current impasse would likely not have occurred if the Conservation Board had given precedence to the consolidation of the Royalite absorption plants rather than trying to impose the "larger scheme of conservation." He concluded, "I do not think it likely that we shall be in a position to help you beyond stating the war requirements at this Department, which we have already done in writing."[205] Howe and Cottrelle could or would not see that the absorption plant consolidation was but an interrelated part of a more fundamental problem accentuated by wartime demand, the precise character of which was determined in the oil controller's office. The intent of Weymouth's plan of conservation was to enable wartime requirements set by the oil controller to be met without overproducing any of the Turner Valley wells and to put an end to the continuing waste of gas. Ottawa's narrow focus and feeling on this issue is perhaps best revealed in Cottrelle's statement that Calgary's gas supply 20 years hence was not their concern. Quite apart from Calgary's future gas supply, whether federal politicians and bureaucrats were prepared to recognize it or not, Turner Valley's contribution to the war effort was not without cost to ultimate crude oil recovery. The federally dictated oil production rates, up to 60% in excess of recommended Board levels in some months, caused a further production of 10 billion cubic feet of solution gas out of the original 835 billion cubic feet in place. This excess gas production is estimated to have resulted in about a 1% or 1.2 million-barrel drop in the amount of oil ultimately recoverable from Turner Valley under primary production.[206]

As it had since 1930, the conservation buck stopped in Edmonton. Whether the continued ruination of the Turner Valley oilfield was to be stemmed would depend upon political will in the provincial capital.

On 2 August 1943, the Alberta Supreme Court issued the anticipated writ directing the Conservation Board to enforce Order No. 139 as it applied to Royalite. From this point, events moved swiftly. Imperial Oil was by far the dominant player in the Alberta petroleum sector, and Premier Manning was anxious to avoid direct confrontation. Manning wrote to C.D. Howe again to urge that the minister use his influence to persuade the Imperial Oil Company to accept the Weymouth plan.[207] The Alberta premier warned that his government would not relent, and appropriate legislation would be introduced if necessary at the next session of the Legislative Assembly. On the same day, the Conservation Board released Order No. 300. The new order allowed the Royalite wells to continue producing at the higher and excessive rate so that the minimum requirements set by the "federal authority" for the war effort could be maintained. To help cover for the anticipated criticism by the independents, the order also stated that this latest reallocation should "not be interpreted as in any sense condoning or forgiving the overproduction."[208] Of greater satisfaction to Turner Valley operators was the order-in-council that followed on August 9. It listed the amount of gas produced by the Royalite Oil Company in violation of Board Order No. 139 and also the estimated amount of excess gas that the company would produce before 31 October 1943, when both Order Nos. 139 and 300 would expire. Since this over-production was "at the expense of all the wells in the Turner Valley field," the Alberta government held that it was "only just and equi-table that the owners of the entire field should be compensated by sharing rateably in the proceeds of the sale of the gas produced or to be produced in excess of the allowables fixed by Board Order No. 139."[209] Royalite was instructed by order-in-council to provide the Board with a complete return showing the money earned from the sale of the excess gas as well as the "reasonable" costs incurred before sale. After collection from Royalite, the Board would allo-cate and distribute the compensation among Turner Valley well owners. Manning followed the order-in-council with a press release explaining the government's action. He took dead aim at Royalite. Noting that the company had failed to comply with the request of the Conservation Board, he announced that his government was "determined that no such monopoly shall be permitted to continue to overproduce its own wells to meet wartime requirements while at the same time other producers in the field are forced to burn mil-lions of feet of cubic feet of gas daily which is being thus wasted simply because they have no access to these markets."[210] Manning told Albertans that, as an initial step, the Conservation Board had prepared a compensation scheme requiring Royalite to share its proceeds earned from the overproduction, and that this would be followed by legislative action to do away with the existing monop-oly and open natural gas markets to all producers. The new Alberta premier had taken the decisive, if belated, action that many inde-pendent Turner Valley producers, most Conservation Board staff, and various outside experts had been demanding for more than a decade.

Although he had played a crucial part in the struggle to bring conservation to Turner Valley, J.J. Frawley's participation in the fi-nal phase would not be as Board chairman. In September 1943, he resigned to return to the attorney general's department, which he had been supervising, since early August, as acting deputy attorney general. On the eve of his departure, he wrote to Premier Manning reflecting upon the industry's reaction to the Board's most recent measures. Frawley wondered if he might be

pardoned for one observation that in the Oil Controller's eyes and in the eyes of Imperial Oil I am most unreasonable insofar as that company is concerned. To Mr. Mayland [Gas and Oil Products Company] it appears that my every act is done to permit Imperial Oil to do whatever they want.[211]

It had been a tough fight, and the varied reaction of the Turner Valley producers only underlined the certainty that, given the nature of the resource and the diversity within the industry, no conservation scheme would ever achieve universal approval.

The legislative step that Manning promised Albertans was revealed in the 1944 spring session of the Legislature. Based upon an idea put forward by Model Oils and pursued by Frawley and Weymouth with Imperial Oil,[212] the *Natural Gas Utilities Act* contemplated the formation of a public utility company. This company would act as a common purchaser of all gas held to be economically available at a price set by a utilities board and would then transport the gas as a common carrier at a fixed charge to the Calgary Gas Company, if for immediate use, or to Royalite for storage in the gas cap, both at prices to be fixed by a utilities board. To achieve this end, the Act decreed that every pipeline, every scrubbing plant and every well, works or plant for the production of natural gas was a public utility.[213] The Natural Gas Utilities Board provided for in the Act was given wide powers in regulating construction and operation of pipelines and equipment necessary for conserving and handling natural gas, the production from wells, the return of gas to underground formations, and the price paid for gas at any point before resale to the ultimate consumer. This considerable responsibility was assigned to a Board comprising just two high-profile members, the chairman of the Board of Public Utility Commissioners and the chairman of the Petroleum and Natural Gas Conservation Board.

For its part, Royalite transferred all its wet gas gathering lines, compressor equipment and scrubbing plant to an existing wholly owned subsidiary, the Madison Natural Gas Company Limited. The British American Oil Company Limited similarly secured the incorporation of British American Gas Utilities Limited to which it passed its wet gas pipelines. Before the enactment of the statute, Gas and Oil Products Limited sold its refinery, absorption plant, and wet gas gathering lines to Gas and Oil Refineries Limited. In May 1944, after a preliminary hearing, the Natural Gas Utilities Board, consisting of G.M. Blackstock from the Public Utilities Board and Dr. E.H. Boomer from the Conservation Board, directed Madison and British American to proceed with the construction of designated gas gathering lines, residue gas lines, and the installation of the machinery required to bring the system into operation. When this work was nearly complete, the Board began on 15 January 1945 a lengthy series of hearings, popularly known as the Blackstock enquiry and lasting until 21 June 1946, to work out such complex matters as the price to be paid for gas at the wellhead and a fair rate of return for the carriers. This in turn called for the careful assessment of a range of problematic issues, including estimation of the gas reserves in the Turner Valley field, present and estimated future market demand, market sharing and pooling arrangements for conserved and repressured gas, rate bases for the two pipeline companies, a rate of return on the established rate bases, and a method of computing depreciation and the cost of delivering gas. Consequently, it was not until March 1947 that the Board was finally able to complete the burdensome task and determine that the wellhead price would be 3¢ per Mcf and the wholesale price 9¢ per Mcf.[214] Meanwhile, in March 1945, the Conservation Board purchased Carleton No. 1 and Pacalta No. 1 (see Map 4.3) to use as input wells in the repressuring of the south end of the Turner Valley field, thereby completing the critical first phase of what at last would bring the Weymouth plan fully into play.[215]

The essential purpose of the 1944 statute was to effect the conservation of Turner Valley gas by giving all producers a share in any market for natural gas. In this, it complemented and completed the purpose intended in the 1938 conservation acts. Although the combined impact of the two acts was 10 years too late to make a really profound difference in the life of the Turner Valley field, the legislation did lay a firm foundation for the postwar phase of the petroleum industry in Alberta.

CONSERVATION IN THE FIELD

What was happening on the conservation front was more than that revealed in the extended public debate on natural gas conservation. During these first years, the Board was gradually building the critical foundation of field practice. Less noticeable, this was no less important than conservation legislation and gaining consensus on production controls. The evolution of drilling and production regulations and the changes that occurred in field practice during the Board's first decade must be understood not just as the products of a specific political and economic environment. Development in these areas also has to be seen in the context of scientific discovery and technological change during the 1930s and 1940s. Throughout these decades, petroleum engineering and reservoir engineering emerged as established disciplines supported by both pure and applied scientific research in university and corporate laboratories. Through such research, the physics of fluid behaviour in reservoir rocks and in producing wells, processing vessels and pipelines became much better understood. Knowledge of the properties of the various kinds of reservoir rock also expanded greatly. In large part, this was due to more extensive coring and core analysis, as well as the development of new geophysical diagnostic tools such as reflection seismic, electric and radioactivity logs.

The differing sound transmitting quality of different types of rock and rock-bearing fluids is the principle upon which reflection seismic is based. Shock waves generated at the surface, such as by the detonation of explosives, are reflected back where sharp changes in the nature of rock occurs. Reflection seismic uses this behaviour to determine the contours of underground rock formations by measuring the time it takes such shock waves to travel from the surface to the rock formation and back. Initially, only major structures could be identified, but as the science developed it became useful for identifying minor structures, and still later the potential of the formations as oil and gas traps.

The bases underlying electric well logs are the varying resistance (resistivity) and varying natural potential (voltage) of different rock types and their contained fluids. Electric well logs measure the resistivity and the spontaneous voltage of the strata adjacent to the wellbore. By moving the logging device from the bottom of the wellbore to the surface, a record of these characteristics of the strata traversed is obtained. The first electric logging devices only indicated qualitative differences in formation resistivity and spontaneous voltage, but nonetheless permitted identification of formation tops and bases. Later, tools were developed to measure resistivity accurately at various distances into the strata beyond the near-wellbore region, which is affected by drilling fluids. Equations were also developed to enable the porosity of the strata and the nature of the contained fluids to be determined. Electric logs are effective only in uncased wells.

Radioactivity logs utilize the different natural radioactivity, and the different radioactivity when bombarded with gamma rays or neutrons, of different strata. The gamma ray log measures the natural radioactivity of rocks. The scattered gamma ray log or density log measures the rock-deflected scattered rays when the strata are irradiated from a source in the logging device. It is effective in measuring the density or porosity of the strata. In neutron logs, the strata are bombarded from a source in the logging device and the gamma radiation generated in the strata and returned to the detector in the logging device is measured. Since the amount of radiation returned to the detector is a reflection of the hydrogen richness encountered, neutron logging is effective in measuring porosity and, to some extent, the nature of the contained fluids. Radioactivity logs can be run in cased or uncased holes, whether or not they contain fluids, although the wellbore conditions need to be taken into account in interpreting the logs.

Sonic or elastic waves propagation logs utilize the relationship between the sound travel time of strata and their porosity. Sonic logs have received application in determining the adequacy of cement bonding behind the casing, in identifying the fluid content of strata, and especially in determining porosity. Normally, they are run in uncased holes, but can also be used in cased holes. As with most logs, however, the wellbore conditions need to be taken into account in interpreting sonic logs.

Each of the above logs has deficiencies, especially when run on exploratory wells. These deficiencies can be largely overcome by running several types of logs, especially on exploratory wells where data on strata are limited. Well logging serves three essential purposes. It is of great assistance to geological mapping; it is a vital aid in reservoir definition; and the taking of logs helps to avoid the bypassing of thin "pay" zones. On occasion, in their rush to drill down to a known target zone and to avoid additional costs required for detailed zone evaluation, operators will bypass zones that might have production.

Use of the reflection seismic technique to detect underground structures was introduced to Alberta in 1932,[216] but this type of geophysics was employed only intermittently, and it was not until the mid-1940s that seismic information was extensive enough to contribute significantly to the mapping of the province's subsurface geology. The technique of electric logging was initiated in France in 1927. The first electrically logged well in Alberta was the Paragon Spring Coulee No. 1 in 1939.[217] Noting that this type of log usually could identify sandstone, shale, coal and carbonates, many operators by the early 1940s were using this new technology for estimating pay thicknesses and as techniques were refined, porosities and water saturations.

The scientific and technological advances of the 1930s and 1940s offered the possibility of vastly improved information bases. From the beginning, the Board had seen the organized collection of geological, well and reservoir data as crucial to its ability to establish effective conservation measures. The Board responded quickly, amending its field procedures to incorporate the new data sources into its information gathering system. Among the changes was a renewed interest in core collection and analysis. The Board had the authority to have core sent in, but this was not being done on a systematic basis. In 1942, Board geologist Ian Cook complained to Board Chairman J.J. Frawley that under the present system often months elapsed before cores could be examined, by which time their analytic value was much reduced. He suggested that operators be required to send core from all wildcat wells, but not field core, "where the geology is known."[218] Acting on Cook's suggestion, the Board had architectural plans prepared for a core storage warehouse, and in 1943 introduced the provision that the operator pay the expense of transporting Board-required core to the Board's offices. Although wartime stringencies postponed the construction of the core storage facility, a much more active interest in core retention was pursued from this point.

Electric log technology also attracted the Board's attention during this period. Acting even before it had the regulations formally amended, the Board began to insist on the running of electric logs in certain instances. Once again, it was the Board's geologist Ian Cook who sought a more comprehensive approach. Reminding the Board chairman that "it is now the practice of the Board, although not covered by the regulations, to require an electrolog to be taken on wildcats," including "outpost wells in proven fields or wells far removed from production," he noted that one of the beneficial consequences was that in many instances electrologs had found producing zones that otherwise would have been passed by.[219] The problem, Cook explained, was that not all operators were prepared to comply with Board requests, and it was difficult to get copies of the logs from some companies. In February 1943, the Drilling and Production Regulations were amended to say that operators would take an electrolog when directed to do so by the Board and that copies of all logs, whether or not taken at the Board's direction, had to be submitted to the Board.[220]

Just as industry adopted the scientific knowledge gained from university research and from their own laboratories, along with information gathered from new technologies and techniques in the field, as the foundation for increasingly sophisticated analysis, the Board and its senior technical staff also realized that the first critical requirement of analysis is access to the best available information. In the early 1940s, the Board moved to ensure such access, and in this regard, relative to its sister agencies on the continent, was clearly ahead of its time. The conservation authority in the largest

TABLE 4.5   Turner Valley Gas-cap Raw Gas Reserves Estimates
(billions of cubic feet)

Date	Initial in Place	Produced	Remaining in Place	Initial Recoverable	Remaining Recoverable	Estimator	Estimation Method
Aug/34		610		920	310 (to 400 psig aband. pres.)	F.K. Beach et al.(Dept. staff)	Material Balance
Aug/35	1,250	710	537	1,010	300 (to 400 psig aband. pres.)	F.K. Beach et al.(Dept. staff)	Material Balance
June/36	1,240	800	443	1,013	213 (to 400 psig aband. pres.)	F.K. Beach et al. (Dept. staff)	Material Balance
Dec/38	1,300	960	341			D.L. Katz & G. Granger Brown, (Univ. of Mich.)	Material Balance
Dec/45	1,420	1,115	300	1,362	247 (to 100 psig aband. pres.)	G.E.G. Liesemer et al. (Bd. staff)	Material Balance
Dec/52				1,349	213 (to 200 psig aband. pres.)	Podmaroff, Warke et al. (Bd. staff)	Material Balance, Production Decline
July/54	1,500	1,149	311	1,400	265 (well economic limit)	Podmaroff & Warke (Bd. staff)	Material Balance, Production Decline
Dec/88	1,493	1,315	178	1,344	29	Industry & Bd. staff	Decline Projection

SOURCE: ERCB, G.A. Warne

oil-producing state, the Texas Railroad Commission, to this day does not have authority to acquire an operator's core permanently or to require the protection and integrity of core, and only since February 1986 has logging been compulsory in Texas.[221]

The Conservation Board's determination from the early 1940s to expand its information-gathering capacity both quantitatively and qualitatively was substantially reinforced by a series of important advances in reservoir engineering theory that occurred between 1930 and 1950. In 1936, R.J. Schilthuis of Humble Oil and Refining Company introduced material balance equations that offered a

TABLE 4.6 Turner Valley Crude Oil Reserves—Rundle Pool
(millions of stock tank barrels)

Date	Initial in Place	Produced	Remaining in Place	Area Acres	Initial Recoverable	Remaining Recoverable	Estimator	Estimation Method
1939				4,950	99		Dr. G.S. Hume, Dr. G.A. Brown (Geol. Survey)	Geological & Correlation
1939	230	11	219	4,950	57	46	Dr. G.G. Brown (Univ. of Mich.)	Material Balance
1939	210			4,950			Dr. B.B. Boatright (Houston consultant)	Rough Volumetric
Dec/54	964	97	885	About 6,800	113	26	E. Stoian (Bd. staff)	Material Balance Production Decline

SOURCE: ERCB, G.A. Warne

means of estimating reservoir size by correlating measured reservoir production and the accompanying change in reservoir pressure. Also in 1936, University of Michigan Professor D.L. Katz offered a tabulation means of determining oil in place. His equations accommodated many different situations, including those where natural water drives and gas caps were present. Equations to show the effect of increasing water saturation in reservoir pore space on oil flow capacity when an oil zone is being invaded by water were developed in 1941 by S.E. Buckley and M.C. Leverett, both of the Humble Oil and Refining Company. Based on industry research, E.C. Babson of the Union Oil Company and J. Tarner of the Phillips Petroleums Limited published methods for predicting solution gas-drive performance of oil reservoirs in 1944.[222]

Possession of the best available and constantly updated database was particularly important in the Board's efforts, either for use by its own professional staff or with the assistance of outside experts, in developing estimates of the amounts of gas and oil recoverable from particular reservoirs. Once this was determined, the Board

had its own reference against which it could judge industry estimates and proposals and which it could use as a base for determining reservoir withdrawal rates that were consistent with conservation principles. In the 1940s, the Board's energies in this regard
were focused upon Turner Valley. The tabulations for reserves data
for the Turner Valley Rundle reservoir, shown in Tables 4.5 and
4.6, indicate the nature of the work done on the pool since 1934.

The Turner Valley Rundle pool is a particularly complex and difficult reservoir (see Map 4.2). Its interpretation and management
presented a great challenge to all who were concerned with improving the efficiency of its production. It served through the 1930s and
1940s as a good training ground, where technical experience was
enriched and procedures and policies were developed that could be
applied to advantage when new reservoirs were discovered in the
late 1940s.

It was during the chairmanship of the fourth Board chairman,
Dr. E.H. Boomer, that these recent scientific and technical advances
began to have a noticeable impact on Board activities. While
Frawley, Manning and Tanner worked to tidy up the loose ends of
conservation legislation, Boomer was left to concentrate on the
Board's day-to-day responsibilities. Boomer was a good choice to
follow Frawley. Whereas the former chairman was at home in the
legislative and legal environment in which the Board was completing its baptism of fire, Boomer was a master of the technical environment in which the Board also had to find its place. A Vancou-
verite, Boomer graduated from the University of British Columbia
in 1920 with a degree in chemical engineering. Obtaining a Ph.D. in
physical chemistry at McGill in 1923, he went on to study at Cambridge under the direction of Lord Rutherford. In 1925, he joined
the Department of Chemistry at the University of Alberta, and by
the mid-1930s he had established himself as a noted researcher in
the fields of high-pressure reactions, natural gas hydrogenation,
and liquid-vapour equilibrium in hydrocarbon systems. With the
onset of war, Boomer was called upon by the Canadian government
to advise the Allied War Supplies Corporation, and he was largely
responsible for designing the ammonia and nitric acid plant operated by Alberta Nitrogen Products in Calgary. He was also called to
Ottawa at various times to advise on certain aspects of the government's atomic energy research program. The fourth chairman came
to the Board as a Fellow of the Royal Society of Canada and with
an international reputation. He was also familiar with the situation
in Turner Valley.[223]

Boomer directed much of his attention to field practice. His engineers monitored well production closely, and through repeated visits maintained a record of each well's progress. Wells were shut in for overproduction in October and November 1943. In January, Model Oils was called to account for exceeding the allowable of two of its wells.[224] This action was followed a few months later by a program of production tests on all wells at which production figures had become erratic. To help tighten administrative procedure in the field, Boomer advocated the creation of an "Official Well Name Register," and following lengthy discussion with the Alberta Petroleum Association advisory committee, he had this measure incorporated into a revised edition of the Board-administered "Drilling and Production Regulations."[225] Approved by the Alberta cabinet in May 1945, the new regulations included another significant addition. Henceforth, each application for a "Licence to Drill" a well had to be accompanied by a deposit of $2,500 to be held in trust until the well was completed or abandoned. If the well and wellsite were left in a manner that satisfied the Board's requirements, the $2,500 was returned, if not, the regulations gave the Board authority to use whatever portion of the deposit was necessary to defray the costs of any work required to bring the completion or abandonment up to standard.[226] Still in pursuit of responsible field practice, Boomer secured, a few months later, another amendment to the regulations requiring anyone operating drilling equipment to have a permit.[227] From this point, any drilling contractor or driller who did not comply with the "Drilling and Production Regulations" could have his permit cancelled. The Board's presence in the field was strengthened further in June 1945 when it took on responsibility for the regulations governing the operation of production tank farms.[228]

These initiatives were motivated largely by the need to respond more effectively to the increasing pace of exploration activity and to counter the increasingly notable effects of poor production practice, particularly in areas more remote from Turner Valley and the Board's surveillance. The seriousness of the latter situation had been graphically drawn to attention by the Board's Chief Inspection Engineer D.P. Goodall. After a tour of the Vermilion oilfield, Goodall reported, "In general, surface conditions are deplorable. Most of the active wells have been producing into earthen pits and in many cases the pits are poorly constructed of loose top soil which has been built up as the oil accumulation increased." In many instances, a slough or depression in the vicinity of the wells had simply been used as a reservoir, and often spring water runoff had carried the oil farther afield. Goodall described one wellsite as "a sticky mess and it is necessary to walk over planks in order to reach the wellhead without bogging down in oil." The inspector's near disbelief at the mess he confronted at the Vermilion Consolidated Oils No. 8 location is apparent in his comment, "This well is pumping steady to shallow pits, slough or wherever the oil has an inclination to run since there is very little evidence of any pretence at directing the flow." Added to the sticky unpleasantness were the assorted "carcasses of wild ducks," which Goodall found "in nearly every oil pit."[229]

Unacceptable production practice in the Vermilion field, as with most conservation problems, was connected to market conditions. Development of the Vermilion and Lloydminster fields had been stimulated by the market opportunity that emerged as Canadian National Railways began to switch its locomotives from coal to low-priced oil. The flashpoint of Vermilion oil was high enough to permit using the oil as locomotive fuel directly after sand and water had been removed. The trouble was that the heavy-oil producers were unable to adjust easily to the railway's variable monthly demand. Heavy crude produced from shallow Vermilion wells usually included a lot of fine sand. When wells were shut down in response to an ebbing market demand, they "sanded up," and before production could be recommenced all pumps and some production equipment had to be dismantled and cleaned thoroughly. The meagre dollar per barrel that operators generally received for their product through the early 1940s made it difficult to pay for well maintenance; hence, operators preferred to keep their wells going, even if this meant producing into pits. The marginal economics of the production operation were such that the Board had been inclined to overlook the unfavourable conditions that had developed over the years, even though it had maintained an office in Vermilion since 1941. Goodall's report, however, exposed the conditions that had built up. As a consequence, steps were initiated that forbade production into earthen pits and resulted in the eventual clean-up of the area.[230]

After taking this step and having its regulations amended in ways that would strengthen its hand, the Board launched a concerted effort to bring Alberta's other oil and gas fields under its close supervision. After Vermilion, on the Board's priority list, were the Redcliff and Medicine Hat gas fields. Many of the wells in this area dated from the turn of the century. The deplorable and dangerous condition in which they were often operated and abandoned had been a source of criticism for decades. Leaking gas wells, some within the city of Medicine Hat, grew to be of particular concern in 1944, and the Board began to apply greater pressure upon indifferent owners. When the owner of one such well failed to act on requests to take appropriate remedial action, he was given a notice saying that if repair was not under way on the well within 48 hours, the Board would initiate the necessary work and take legal action to recover the costs.[231] Following an assessment by a team of its engineers, the Board decided in April 1945 that immediate steps should be taken "to inaugurate a program of conservation in the Medicine Hat-Redcliff field."[232] The first of these steps was to send T.M. Geffen to open a Board office in Medicine Hat. Geffen was given an assistant and the temporary support of the Board's chief inspection engineer to supervise preliminary tests and to help work out a program of operations. The extension of the Board's presence into the southern part of the province not only began the process of bringing field practice in the Redcliff-Medicine Hat area up to a minimum acceptable standard but also enabled the Board to direct closer attention to the several small newly discovered fields in the vicinity. Operators in the nearby Conrad field were among the first to feel the closer presence of regulatory authority. Found to have "oil waste strewn around the lease" of its Mid-Continent East Crest No. 1 well, and to have been preparing the drilling site for another well for which it did not yet have a licence, Mid-Continent Oils had the offending well sealed and permission to drill any further wells cancelled until all its well leases in the area had been cleaned up to the Board's satisfaction.[233]

Another part of Boomer's effort to establish a stronger presence outside Turner Valley was the Conservation Board's continuing endeavour to enforce 40-acre well spacing. Although the restriction of one well to a 40-acre tract had been included in the first regulations issued in January 1939, operators in the Vermilion field had been allowed to continue with a 10-acre spacing pattern. Earlier, when this had been drawn to Robert Allen's attention, he had immediately informed oilmen in the Vermilion area that the Board would not issue drilling licences on a spacing pattern of less than 40 acres.[234] This was followed by a circular to operators elsewhere in the province that drew attention to the existing policy and reminded them that the surface location of the well had to be as nearly in the centre of the 40-acre parcel as surface conditions permitted.[235] Allen was convinced that, with the odd exception, 40-acre spacing was the minimum required to promote uniform withdrawal and to ensure efficient oilfield drainage, to prevent the economic

waste of overdrilling, and to protect the rights of competing owners. James Frawley followed Allen's lead. When it was discovered in
August 1941 that the Conestoga No. 1 well had been drilled and
brought into production on a 10-acre plot without a drilling permit, Frawley delayed issuing a retroactive licence to legitimize the
well until 7 May 1942, when wartime demand dictated that any
well capable of production be allowed to produce.[236] Regulations
were not inflexible, the Board could make exceptions where it
seemed to serve the interests of good conservation practice best.
When California Standard and Pacific Petroleums requested that
Boomer and his colleagues allow 20-acre spacing in a portion of the
small Princess oilfield, the Board agreed.[237] Similarly, in heavy crude
oil areas, such as Vermilion and Lloydminster where low oil viscosity and fluid transmissibility made closer well spacing appropriate,
the Board authorized spacing as close as 10 acres per well; however, in most cases the Board remained adamant, and in one instance it was compelled to endure a lengthy court battle with a
Turner Valley operator who insisted that he should have the right
to drill on a 20-acre lease.[238]

In its approach to the matter of well spacing, the Board was establishing an important precedent through this period. Although it
has little effect by itself on ultimate recovery in many pools, it could
substantially affect the rate of production. Well-spacing policy is
part of the critical core on which effective and equitable conservation policy must be built. Unlike in Texas, where the number of exceptions granted, particularly to protect the "vested rights" of small
landowners, was so extensive as to undermine the purpose of wider
drilling patterns,[239] deviations allowed by the Alberta Board from its
established well-spacing program were rare. This would remain
true in the postwar period.

Of the various measures carried forward and initiated during
Boomer's administration, perhaps the most important was the promotion and approval of an agreement to operate the province's
newest and most promising gas field in a unitized fashion. Discovered in 1944 by the Shell Oil Company of Canada, a relative new-
comer to the Alberta exploration scene, the Jumping Pound gas
field was located on the eastern edge of the foothills, about 20 miles
west of Calgary and approximately 30 miles north of Turner Valley.[240] By the spring of 1945, it was apparent that Jumping Pound
might be the largest gas field yet discovered in Alberta, thus bringing Shell and the Conservation Board together to discuss eventual
production arrangements. Both parties were attracted to the idea of
operating the entire prospective field area identified by Shell's seis-
mograph studies as a simple unit. To this end, the Board made

known its support of Shell's efforts to consolidate and merge the petroleum and natural gas rights of all of those who had an interest in the field.[241] In accordance with the "Drilling and Production Regulations," the Board later approved the boundary of the area to be covered by the unit agreement, set spacing at not more than one well for every 80 acres, and finally gave its blessing to the contract that Shell proposed to take to the various rights holders.[242] Unlike lands on the prairies to the east, but like Turner Valley, the Jumping Pound area had been settled by ranchers and some homesteaders in the early 1880s when property titles still conveyed both surface and mineral title. Thus, while Shell was able to acquire the petroleum and natural gas rights to a portion of the field from the provincial government, it also had to persuade a number of small freehold owners of petroleum and natural gas rights that unitized operation of the field and a 12½% royalty was to their advantage. Until this was accomplished, there existed the possibility that the roughly 6,000-acre field might be developed by competing interests. Nearly four months later, when Shell returned to the Board to obtain a permit to proceed with development, it still lacked agreement from the owners of the rights to 760 acres; nevertheless, the Board was happy to exclude the 760 acres from the designated "Jumping Pound Unit Area" and issue "Pool Permit No. 1" to the Shell Oil Company as operator of the "Jumping Pound Unit Agreement."[243]

This first unit agreement in Alberta set the precedent of joint participation and co-operative development, an ideal that promised an escape from the endless controversy and wasteful practice that had attended development in Turner Valley. It offered an example and a formula that were fitting tributes to Shell Oil Company and particularly to Board Chairman Dr. Edward Boomer, who died of a heart attack just four days after the agreement was announced.

Boomer's passing was mourned by the academic community, who recognized the loss of one of Canada's outstanding research scientists, and by the petroleum industry which had come to appreciate the extent of his technical knowledge. Calgary oilmen also paid high tribute to Boomer's personal style. In the words of Alberta Petroleum Association President F.M. Graham, the late chairman would always be remembered for "his fair-mindedness and extremely courteous manner."[244] With Boomer's death, the Alberta government had lost a chairman who had begun to make the Conservation Board an effective presence province wide, not just in Turner Valley.

The government did not appoint a successor immediately. The Conservation Board was left under the direction of A.G. Bailey, who had been deputy chairman since July of the previous year. Bai-

Having outgrown its first Calgary head-quarters, the Board moved to this building on 11th Avenue in southwest Calgary in 1945. ERCB.

ley was from Montreal and was with the F.W. Woolworth Company before he joined the War Supply Board shortly after World War II began. It was with this agency that Bailey met and became friends with Dr. Boomer, who subsequently persuaded him to come to Calgary to work for the Petroleum and Natural Gas Conservation Board. Boomer had been impressed by Bailey's organizational and administrative skills, and given his own busy schedule as university lecturer, researcher and government consultant, he was anxious to be relieved of the growing day-to-day administrative burden that the Board chairmanship entailed. Therefore, he had prevailed upon the Alberta cabinet to appoint Bailey as his deputy chairman.[245]

When Bailey assumed full responsibility for the management of the Conservation Board in November 1945, he had had but 15 months apprenticeship as Boomer's assistant and no prior experience in the oil and gas industry. Throughout this initial period, and after his appointment as chairman in October 1946 until his resignation in June 1947, Bailey was compelled to rely heavily upon the advice of D.P. Goodall, the remaining Board member. In contrast to Bailey's background, Goodall had spent almost his entire career in various oilfields since graduating from the University of Alberta as a geologist in 1926. He had been with the Conservation Board from its formation in 1938 and had been appointed to the Board with Bailey in the autumn of 1944. Goodall could be relied upon

for sound practical advice, particularly in matters of field practice. In addition to Goodall, the Board possessed several experienced and competent field engineers, but there was no one who could command the same technical or professional authority as the late Dr. Boomer. Bailey's administration is characterized therefore by no major new policy departures. Possessing the instincts of a good administrator, Bailey relied upon close consultation with the industry and upon seeking the advice of experts when difficult technical questions arose.

In January 1946, Bailey called Dr. D.L. Katz at the University of Michigan to come to Calgary. He wanted Katz to review the Board's application of the Brown plan and to address a complex debate centred about what should be the appropriate allocation for the Home-Millarville No. 18 well.[246] The Home Oil Company Ltd. and the Anglo-Canadian Oil Company Ltd., two of the largest independent Turner Valley operators and the owners of the well, argued that because of the well's unique geology and production

During the postwar years, the Board began to make its presence felt throughout the province. Among the Board's field staff were (l-r) Alex Essery, Dick King, Lloyd Hicklin, Pat Webster, and George Horne, in front of the "new" Black Diamond Office in 1946. ERCB, Bohme Collection.

FIGURE 4.2   Petroleum and Natural Gas Conservation Board
Organization Chart, 1946

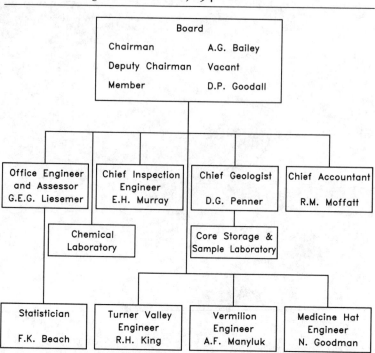

characteristics a larger allowable production was warranted. Convinced of the validity of their claim, the companies produced the well in excess of the allotment assigned by the Board. Lacking the in-house expertise to be confident in its decision, the Board relied on the advice of both Dr. Katz and the Alberta Petroleum Association advisory committee before eventually settling upon a revised allowable.[247]

In other areas, Bailey followed the course set. He continued Boomer's crackdown on production violations in the Vermilion, Wainwright and Lloydminster fields. Board field staff were instructed that any well found to be storing production in pits was to be shut in and not reopened until the lease was cleaned up to the Board's satisfaction.[248] To reinforce the Board's efforts to bring acceptable engineering practice to the area, Bailey initiated discussion with the Saskatchewan supervisor of mines to see if common regulations might be devised for both the Saskatchewan and Alberta parts of the Lloydminster oilfield.[249]

The main legacy of Bailey's stewardship was not in the field but in the Calgary office (see Figure 4.2). Building upon the changes

that he instituted as deputy chairman, he pressed on with the refinement of various Board procedures and policies. Existing well-spacing practice, for example, was codified in a precise policy statement that distinguished Board requirements as they related to farm gas wells, municipal gas wells, and all other gas wells.[250] In the area of personnel administration, Bailey also clarified management expectations. An office attendance register was introduced, compelling employees to sign in each morning and after returning from lunch.[251] The Board's irregular approach to salaries and job classification was dismissed in favour of a formal salary schedule that established minimum and maximum salaries for a discrete set of described job positions.[252]

The changes reflect not only the personal influence of A.G. Bailey but also that the agency was maturing in light of its own experience and in response to the changing economic and social environment of the immediate postwar period. Instead of the return to economic stagnation that many predicted would follow at war's end, the pent-up demand suppressed for nearly two decades, along with external factors, pushed the Canadian economy into an expansionist phase. Nowhere in the country was this change more apparent than in Alberta, where the petroleum industry was in the midst of a new and larger surge of exploration activity. This climate of rising expectations and rising wages was not without an impact

As the Board settled into its second decade, Imperial Oil had emerged as the leading oil player in Turner Valley. Its Royalite Gas Plant is shown here in 1945. ERCB.

Old-style riveted separators used in Turner Valley and other areas in the 1940s and 1950s. ERCB.

upon the Board. When Turner Valley operators granted their employees a 10% wage increase as of 1 January 1947, the Board's field staff were quick to take note. It made the $10-a-month increase that they had been offered seem inadequate, and they threatened to strike unless the Board's offer was raised to $25 a month. A hastily called meeting at the Turner Valley field office on February 12 brought forth a $15 increase for most of the 12 incipient strikers.[253] This new experience in labour-management relations was immediately overshadowed by an event just one day later that dramatically changed the world that the Board, the Turner Valley operators and Albertans had known. On February 13, near Leduc, Imperial Oil discovered the riches hidden in the rocks of the Devonian age. Alberta and the province's petroleum industry burst into a new phase.

By 1946, most of the producers in Turner Valley had come to see the Board's role in a positive light, and even the most reluctant independents had become resigned to its regulatory authority. In the

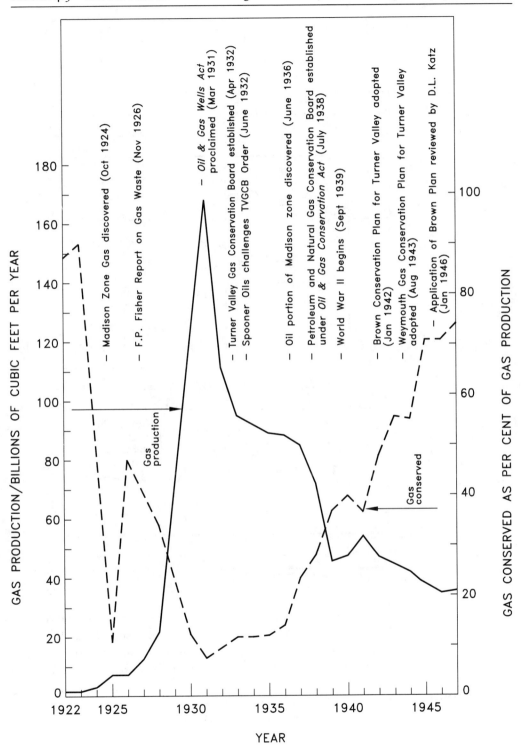

MAP 4.4 Turner Valley, 1938–1948

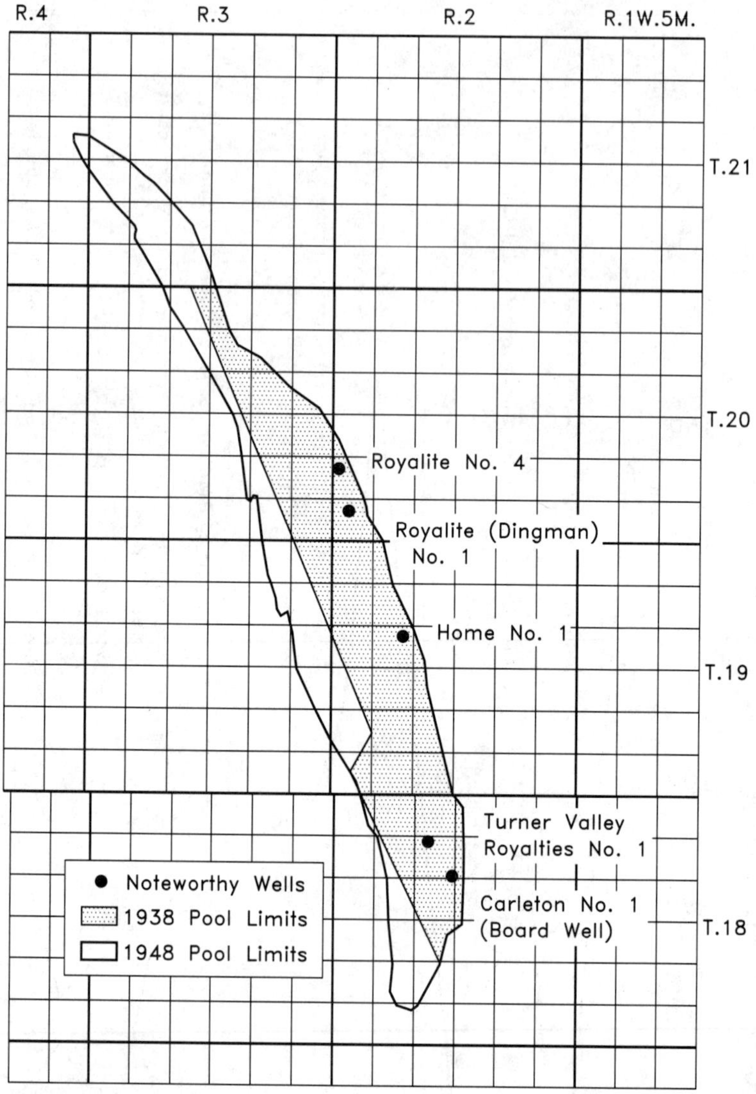

outlying fields, the most glaring abuses were giving way before the Board's growing and more assertive field presence. It had been a long struggle against formidable obstacles that included small independents desperate for maximum cash flow that could only be achieved through unrestrained production, a single dominant producer unwilling to give up monopoly privilege, politicians reluctant to act for fear that they might drive off potential investment and de-

MAP 4.5   Producing Oil and Gas Fields of Alberta, 1946

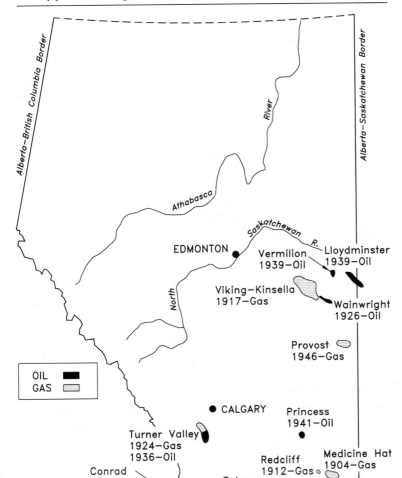

velopment, excessive wartime demands and a complex geology that yielded only gradually the secrets of the producing formations. To the end of 1946, 75% of the natural gas produced in Turner Valley was wasted. Moreover, this extravagance reduced the amount of oil that could be recovered from the reservoir by millions of barrels,[254] and must stand by itself as a harsh indictment of the interminable waste gas debate. Yet, judgement of what was going on in Turner Valley throughout this period is best set in a larger perspective.

Conservation was slow in coming, but the battle being fought in Alberta was not happening in isolation, it was but a lesser engagement in a conflict under way in hundreds of oil and gas fields throughout North America. When considered within this context, it is apparent that the conservation forces in Alberta had made impressive gains.

In Texas, the fight to prohibit the flaring of natural gas began in 1899. In the face of constant resistance from operators, some progress had been made by the late 1930s, particularly as it related to the stripping of wet gas to produce natural gasoline. Instead of simply allowing the burning of large volumes of gas after retrieving a small amount of condensate gasoline, the Texas Railroad Commission began to insist that operators recycle the dry gas back into the reservoir, and by the mid-1940s recycling schemes had become common. But the flaring of huge volumes of the casing-head gas produced in conjunction with crude oil production continued. Throughout the early 1940s, the Commission attempted to control the flaring of casing-head gas by statewide orders forbidding production from wells with gas-oil ratios of more than 2,000 cubic feet per barrel. These efforts had little impact as the Commission lacked sufficient staff to enforce its orders. Operators in Texas were lax about keeping records and the Commission did not have enough people to check; most of the flared gas did not even show up on commission reports. As anxiety over wartime petroleum supply grew, concern mounted in Washington about gas waste in Texas, and the Federal Power Commission was known to be considering imposing its regulatory authority to address the problem. To forestall federal interference, the Texas Railroad Commission, after special hearings to discuss the gas flaring problem, announced in late 1944 that less than 1% of the 400 billion cubic feet of casing-head gas produced in Texas during 1943 had been flared.[255] This assertion caused a furore, and in November 1945 the embarrassed commissioners were compelled to accept an authoritative report prepared by an independent committee of engineers that showed that Texas was wasting nearly 1.5 billion cubic feet of casing-head gas daily, or nearly 57% of the state's total production. The big producers immediately warned that they would be ruined if they were faced with an order to stop flaring. In Texas, the waste gas issue was still some years away from resolution.[256]

Even though a review of the situation in Texas does offer some insight into the common character of the problems facing conservation authorities and the nature and timing of remedial measures, the vast difference of scale renders questionable all but the most general conclusions. A brief assessment of the evolution of oil and gas conservation in Montana during the period offers the advantage of comparison with a jurisdiction situated geographically next door and where the scale of the industry's activity was much closer to that in Alberta.

The discovery and development of Montana's hydrocarbon resources began, as in Alberta, on the eve of World War I. In 1921, authority to make regulations to prevent oil and gas waste was given to the Montana Board of Railroad Commissioners. Found to be largely ineffectual, the authority of the three elected railroad commissioners was supplemented in 1933 by the creation of an Oil Conservation Board. Not intended in any way to compromise the existing powers and duties of railroad commissioners, the new Board was the direct product of the widely held fear that the lack of an acceptable state conservation law would expose local operators to stiffer federal regulation.[257] The Board consisted of five members appointed by the governor for two-year terms. All were required to be bona fide residents of Montana, and four of the five members had to be crude oil producers. Given their concurrent functions, both the Commission and the Board issued rules, regulations and orders. Regulatory authority was diffused further by the existence of a third agency, the State Board of Land Commissioners, which issued regulations, including those intended to prevent the waste of oil and gas on state oil and gas leaseholds. Although industry expressed a preference to have regulatory authority assigned to one body removed from "political influence," such as the Board,[258] this was balanced by a popularly held counter view that has always had a great impact upon U.S. administrative practice. This was the conviction that divided responsibility was the best way to protect the public interest.

Two attempts were made to achieve comprehensive reform of Montana's conservation statutes. In 1937, the Drumheller Bill proposed the creation of a Petroleum Commission of the State of Montana. Fearful of the powers to be accorded the proposed commissioners, industry strenuously objected and was successful in having the bill withdrawn. A decade later, the Montana House of Representatives produced a bill promising to create a State Oil and Gas Conservation Board, consisting of five gubernatorially appointed members with exclusive jurisdiction over regulations to initiate and enforce conservation measures. With the defeat of the bill, Montana backed away from modern concepts of conservation and efficient oil and gas production.

The situation in Montana in 1947 stands in sharp contrast to that in Alberta. In Montana, effective conservation regulation was compromised by the overlapping jurisdiction of several agencies, each of which lacked adequate funding to discharge its responsibilities.[259] Montana statutes and regulations in 1947 offered neither a definition of waste nor any specific provision for the repressuring, recycling and controlling of water and gas production, the preservation of reservoir energy, the disposal of salt water, open flow restrictions, or limitations based upon market demand, as was typical of conservation legislation in more progressive jurisdictions.[260]

The relatively greater success of the petroleum and natural gas conservation movement in Alberta to 1947 was largely a product of the leadership offered by several key individuals and the different political and economic environment that distinguished Alberta from comparative U.S. states. In all the oil- and gas-producing regions, government regulatory intervention came only in the wake of public outcry. The form of this intervention in Alberta was shaped first by the U.S. experience. It was here that Alberta engineers, civil servants and politicians looked for both technical guidance and the specific regulatory mechanisms designed to curb wasteful production practices. The structure of the agency created to give effect to these specific measures, however, was much more the product of the British-Canadian legal and political environment in which the drafters of the Alberta legislation had been nurtured. Alberta's appointed Petroleum and Natural Gas Conservation Board was given wide-ranging powers and was less open to legal challenge than its U.S. counterparts. How successful this agency would be in meeting its responsibilities rested upon the dedication and competence of those appointed to the Board and the political will of the province's political leadership.

Once the Board was established, its leadership was crucial, and the initiative for change shifted from the public to the new agency. No industry group or political party came forth to promote the conservation cause. The Board continued the efforts of a few clear-sighted civil servants to pull both industry and cautious politicians to act responsibly in the public interest. In this task, six individuals stand out. The contributions of Charles C. Ross, Charles Dingman, William Knode, Robert Allen, James Frawley and Edward Boomer are of special significance in the history of petroleum and natural gas conservation in Alberta. Unlike the others, Ross never headed the Conservation Board, but he did lay the foundation for oil and gas conservation in Alberta. He opened the Calgary office of the Department of the Interior's Mining Lands and Yukon Branch in

1916, and until control over natural resources was surrendered to the province in 1930, he directed the struggle to conserve Alberta's oil and gas resources. After helping to draft the new provincial administration's oil and gas regulations, Ross watched most of the men he had trained and supervised join the Petroleum and Natural Gas Division of Alberta's new Department of Lands and Mines. Later, as Aberhart's minister of lands and mines, he tried to achieve from the vantage point of the provincial cabinet what had eluded him as a professional engineer in public service—an effective conservation program for Turner Valley. In this, he was once again to be disappointed. Dingman, Ross's long-time colleague, also spent most of his professional career trying to promote conservation. Never quite able to gain the full confidence of his political masters, Dingman won nonetheless the grudging respect of even the most recalcitrant operators. Dingman was never Board chairman, but it was he who soldiered on through the 1930s, leading the conservation forces in the province during the darkest period. For their combined efforts, spanning 25 years, Ross and Dingman deserve to be known as the fathers of oil and gas conservation in Alberta. Knode, as first Board chairman, had to guide the Board in its first tentative steps and to withstand the wrath of those opposed to government regulation. He possessed the commitment and the personality to give "as good as he got" in the vigorous debate about regulation. Allen helped the conservation cause and the Board gain credibility and technical authority at a time when they were badly needed. Frawley's contribution extends well beyond his role as Board chairman. He led conservation's defence in the courts. His enduring legacy is Alberta's oil and gas conservation legislation. Boomer lent the Board an authority that came with his reputation as a distinguished scientist. He gave meaning to the tradition of an active field presence supported by technical excellence in Calgary and, particularly, at the Board level. The achievement of these Board chairmen is all the more remarkable given the short period of their respective tenures, and stands in the face of what at the same time was the Conservation Board's greatest handicap. In the eight-year interval that followed the agency's creation, the government let it drift for three years without a chairman, and during the five years that the Board was privileged to enjoy a chairman, five different men held the office. In the effort to establish its presence and meet its legislated responsibilities, the Board was never able to draw upon the advantage of leadership continuity and administrative stability. If not absent, leadership was usually in a state of flux.

That Board chairmen were able to make significant headway, especially relative to their U.S. counterparts, had also to do with factors other than their individual abilities and the structure of the agency that they supervised. There was the important difference of scale. The Alberta Board's energies were focused essentially on one field and a few hundred wells, rather than many fields and tens of thousands of wells, as was the case with the Texas Railroad Commission or the several oil and gas fields that occupied the attention of the Commission and Board in Montana. Another important aspect of the more simplified environment in western Canada was that Alberta's regulators did not have to contend with a variety of powerful major oil companies, not to mention strong independent companies supported by thousands of ranchers and farmers who, unlike most of their Canadian neighbours, usually possessed mineral rights along with their surface title.

Alberta's conservation officials enjoyed the advantage of being able to draw from U.S. technical and regulatory experience, and then concentrate their attention on one field as it was developed over the decade (see Map 4.4). A single but geologically complex reservoir, Turner Valley became the training school for a growing number of Alberta engineers at the Board and in the industry. (See Figure 4.2.) In their efforts to understand better the dynamics of the Turner Valley reservoir, Board engineers began the systematic collection and assessment of well data and the potential applications of new technology. This was done not only in the belief that the Board needed to stay on the leading edge of knowledge if it was to supervise development and production but also in the belief that if this information was made available to the industry, especially the smaller operators, the interests of conservation would be served by more intelligent development and production practice. This started a tradition that in time would make the quality of available technical information on the Alberta portion of the great sedimentary basin superior to that available in any other jurisdiction. In itself, this would become an ever more important factor, fostering the objective of efficient and maximum recovery.[261]

All things considered, the single most important consequence of the Turner Valley years was not the eventual accumulation of knowledge about the reservoir, or the volume of oil eventually coaxed from the formation, or even the appalling volume of gas that was wasted. The significance of Turner Valley is to be found in the regulatory framework that emerged in the course of the field's development. This was the critical legacy carried forward into the next phase of the petroleum industry's development in Alberta. The guiding principles shaping Alberta's postwar oil boom were fashioned in Turner Valley.

To state that a regulatory framework supported by an agency with experienced personnel was in place is not to suggest, however, that the conservation battle was over. Although the Board had fought for and won industry acceptance of certain broad conservation principles, the rigour with which these would be defended and the manner in which they would be refined to meet changing conditions would depend, as it always had, on the quality of the Board's leadership and the strength of political will to be found in Edmonton.

The discovery of oil at
Imperial-Leduc No. 1
in February 1947
heralded a post-war
petroleum boom and
the coming of age for
the Board. Imperial
Oil Limited.

FIVE

The Leduc Discovery and
the New Regulatory Environment

CARS BEGAN ARRIVING in the late morning at Mike Turta's farm near Leduc, about 15 miles southwest of Edmonton. In the February chill, the growing crowd drank coffee, shuffled their feet in the snow and waited, their eyes fixed on Vern "Dryhole" Hunter and his drilling crew at work on the Imperial rig. They had been drawn by word that Imperial-Leduc No. 1 was going to be "brought in." The crowd continued its vigil through the middle afternoon, and was reassured by the arrival of Minister of Lands and Mines N.E. Tanner, Imperial Oil's Calgary General Manager Walker Taylor, Assistant Manager Vern Taylor, and others of what clearly was an official party. As the afternoon light was about to fade, there was suddenly a sign of life. The well blew in with a roar, snorting and puffing great bursts of gas and watery oil. The flare line was lit, and, in Hunter's words, "the most beautiful smoke ring you ever saw went skyward."[1]

Imperial's next well, Imperial-Leduc No. 2, completed three months later just southwest of the discovery well, found an even bigger producing horizon 300 feet lower in the Devonian limestone[2] (see Maps 5.3 and 5.8). The Imperial discoveries were dramatic, and by the end of 1947 the Leduc area boasted 28 producing wells, with 17 more in the drilling stage. As they converged on central Alberta, after the Imperial discovery on February 13, the exploration and drilling crews of the Turner Valley companies—now able to attract investment capital—were joined by those representing a host of new companies from Oklahoma, Texas and California. The real oil boom that a generation of Albertans had dreamed of was under way. Leduc roused the interest of international capital in Alberta's potential, and long-awaited development funds poured into the

The Honourable N.E.
Tanner, Minister of
the Department of
Mines and Minerals
from 1936 to 1952,
strongly supported the
Board's role as the
province's petroleum
regulatory agency.
Provincial Archives of
Alberta, Public Affairs
Bureau Collection,
PA166/2.

province. Although the large inflow of development capital brought
new discoveries that quickly made the province a major producer, it
also took Alberta back to the position of a decade before when production capacity greatly exceeded market demand. By early 1949,
the production potential of Alberta oilfields had shot past market
demand in the prairie region.[3] Prorationing to bring production and
the available market into balance followed, and the provincial government once more took up the crusade for access to more distant
markets, picking up where it had left off in the late 1930s. Surplus
capacity in both oil and gas is the dominant theme marking the two

TABLE 5.1 Number of Wells Drilled Annually in Alberta, 1945–1952

Year	No. of Wells	Year	No. of Wells
1945	130	1949	798
1946	128	1950	1044
1947	224	1951	1268
1948	378	1952	1662

SOURCE: Alberta, Petroleum and Natural Gas Conservation Board, *Alberta Oil and Gas Industry, 1945–1952.*

TABLE 5.2 Alberta Crude Oil Production, 1943–1951
(millions of barrels)

Year	Turner Valley	Leduc-Woodbend	Redwater	Other Alberta Fields	Total Alberta
1943	9.0			.2	9.2
1944	7.9			.4	8.3
1945	7.0			.6	7.6
1946	5.9			.8	6.7
1947	5.0	0.4	—	1.0	6.4
1948	4.4	4.7	—	1.4	10.5
1949	3.8	9.7	4.8	1.5	19.8
1950	3.3	10.6	10.7	2.5	27.1
1951	3.0	13.7	23.0	6.0	45.7

SOURCE: Alberta, Petroleum and Natural Gas Conservation Board, *Alberta Oil and Gas Industry, 1951.*

decades that followed the Leduc discovery. Everything was conditioned by this.

To deal with this situation, the Petroleum and Natural Gas Conservation Board started with the advantage of its Turner Valley experience. Now, however, it had to be addressed in a postwar environment that was much more complex. The rapidly expanding petroleum industry comprised a much more diverse array of multinational, medium-sized and small independent companies. No longer was there just one pool in which to establish and supervise equitable production rates. After 1948, there were not only many pools for which individual well allowables had to be determined but also the difficult task of determining equitable market share between pools. Given the rapid sequence of discoveries and the dramatically mounting productive capacity, combined with a market that was growing at a much slower rate, the hallmark of this period is competition for market share. Companies jockeyed to gain any advantage that would increase their well allowables, an advantage that always had to come at the expense of someone else. The Board's essential role would be that of refereeing to ensure equity among resource owners, co-ordinated development, and good production practice in the public interest.

LEDUC

Following the end of the European war in May 1945, the domestic oil supply situation in western Canada continued to deteriorate. Turner Valley production remained in decline, whereas local demand maintained its upward momentum. Despite aggressive exploration in the foothills and further searching on the plains, no major fields had been found. Imperial's Regina refinery was already relying on Wyoming crude oil, and it was anticipated that imported oil would soon be required to supplement Turner Valley crude oil in Calgary refineries. With natural gas, it was completely the reverse. Established reserves were more than adequate to meet gradually increasing consumer demand, yet the search for oil had brought forth numerous gas discoveries to add to the existing surplus. Faced with this bleak picture, the more active companies began a serious reappraisal of their prospects in Alberta during the winter of 1945–46. The critical and most eagerly awaited decision was that of the predominant company, Imperial Oil. Its eventual conclusion was twofold: it would look more closely at the utilization of Alberta's growing natural gas supply and it would make one more effort to find a significant oil pool.[4]

Imperial Oil's interest in natural gas at this point was stimulated by the emerging awareness of how German scientists had been successful in developing the technology to produce synthetic gasoline during the war and specifically by research under way in the laboratories and refineries of the parent company, Standard Oil of New Jersey.[5] Projecting that a gas-synthesis plant would require a reserve of about 450 billion cubic feet to meet a throughput of 50 million cubic feet per day required to assure production of 5,000 to 6,000 barrels of synthetic gasoline daily, Imperial began to evaluate the

TABLE 5.3 Production and Distribution of Alberta Petroleum, 1945–1951
(barrels)

| Year | Total Production | Crude Refined in Alberta | Crude Delivered to | | Total* Exported |
			SASKATCHEWAN	MANITOBA	
1945	8,062,512	6,420,884	—	—	1,586,544
1946	7,140,256	6,236,628	—	—	955,787
1947	6,809,284	5,667,237	—	—	1,106,156
1948	10,973,583	7,718,722	2,365,298	726,248	3,091,546
1949	20,246,392	11,616,485	6,821,584	1,196,671	8,018,255
1950**	27,149,232	13,857,106	9,405,333	1,586,970	10,992,303

1951

Opening inventory	2,955,952	
Total production	45,915,384	
Available for sale		48,871,336
Sold: Alberta refineries	15,220,312	
Alberta miscellaneous	27,699	
Saskatchewan	11,083,893	
Manitoba	4,584,940	
Ontario	13,666,328	
USA	463,000	
Total	45,046,172	
Losses and adjustments	−15,291	
Closing inventory	3,840,455	
Total		48,871,336

SOURCE: Adapted from Alberta Petroleum and Natural Gas Conservation Board, *Alberta Oil and Gas Industry, 1945–1951.*
* The difference between Total Production and Total Deliveries in the examples shown above is due to losses or other adjustments in storage and/or in the pipeline.
** Before 1950 Total Production includes natural gasoline.

natural gas supply in the Viking-Kinsella area 80 miles southeast of Edmonton.[6] After completing an exploratory well in November 1945, the company began a drilling program to prove-up the reserves. Directed by the same thinking, the McColl-Frontenac Oil Company and Union Oil of California were trying to define a large gas reserve in the Pakowki Lake region south of Medicine Hat.

As drilling results continued to expand the boundaries of the Viking-Kinsella natural gas field, senior technical and management people from Imperial and Standard met in Toronto on 9 April 1946 for a discussion on whether or not to continue oil exploration in Alberta. After a review of comparative prospects in the western Canadian sedimentary basin, the central Alberta plains region, particularly the Edmonton area, was judged to be the most promising. Standard's assembled executives and technical officers decided that the company's "last" oil-directed exploratory effort should be here.[7] Imperial then acquired the petroleum and natural gas rights to nearly 200,000 acres of crown reserve lands spread throughout 24 townships around Edmonton (see Map 5.1).[8] Seismic crews traversed the area to provide data from which favourable subsurface structures or anomalies might be identified. By late fall, Imperial's exploration staff had identified such an anomaly near Leduc that looked promising. The request to begin a drill test initiated a protracted debate at Imperial. Reflecting on that debate, Director, later President, W.O. Twaits recalled

> there was a major argument among the Imperial Board, ending in a reluctant agreement to proceed with the test, rather than immediately developing final design on a gas synthesis plant for Viking-Kinsella. The word was that "after $23 million this is final."[9]

Imperial-Leduc No. 1 was spudded in on 30 November 1946.[10] Although the prime target horizon was the Cretaceous, the intent was a deep test, to drill through to the Silurian to an estimated depth of about 7,000 feet (see Figure INT.1). Gas blows with water and traces of oil were picked up in the Lower Cretaceous sandstones between the 4,200–4,300-foot levels. In early February, just beyond the 5,000-foot depth, the drill cuttings revealed porosity in a dolomitic limestone, with yellowish fresh oil in the rock pores. The decision was made to run casing, and on February 13 the well was brought into production. When the well was abandoned in 1974, it had produced more than 318,000 barrels of oil. Whereas the discovery well had encountered Devonian oil in what was first called the D-2 and later the Nisku formation, it was the second Imperial well that found the Leduc reef, which proved to be the principal producing horizon. Imperial-Leduc No. 2's story is one with greater drama. Begun in January 1947, when Leduc No. 1 was encountering some encouraging shows in the Lower Cretaceous sands, the second well became the subject of even more intense in-

MAP 5.1   Townships Containing Crown Acreage Acquired by
Imperial Oil, 1946

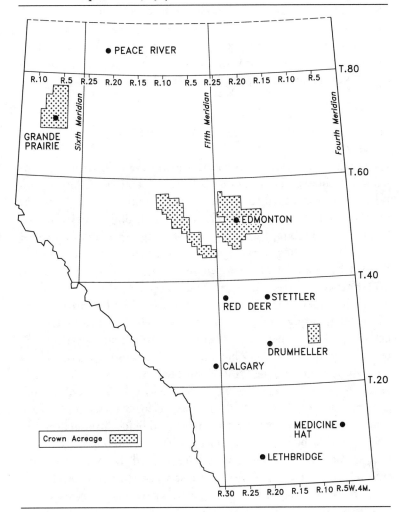

terest after the D-2 discovery. The well's progress, however, was
disappointing. First, noncommercial gas and oil shows were encountered in the Cretaceous sands, and when the D-2 target horizon was penetrated, it was barren. The decision to stop drilling was
almost automatic. Some wanted to pull the rig off location, but other Imperial staff wanted to continue drilling to complete the deep
stratigraphic test that had been the original intention with the No. 1
well. This was a pivotal moment. It was resolved to keep drilling.
The verdict on this decision came quicker than anyone could have
imagined. After drilling a mere 150 feet farther, porous dolomitic

limestone was encountered. Imperial Leduc No. 2 had found the Leduc reef, and when placed on production, on 27 May 1947, it soon became the province's largest producer. By year-end, the well had produced 35,699 barrels.[11] With the discovery of the prolific D-3 horizon, the race for petroleum leasehold was on, and almost at once Alberta was lifted from its dependent status on the periphery of North America's agricultural hinterland.

SHAPING THE REGULATORY ENVIRONMENT

Faced with a clamorous demand for exploration rights and petroleum leaseholds, the Alberta government and its Conservation Board were compelled to respond quickly with a series of regulatory adjustments. Anticipating a rush of drilling activity in the Leduc area, the Board issued a well-spacing order within days of the discovery, setting one well per 40 acres as the limit for oil wells.[12] Board Chairman A.G. Bailey recognized that, if effective conservation practices were going to govern development of the Leduc field, it was necessary "that immediate steps should be taken to control the drilling, testing, and production of all wells."[13] He initiated discussion with the Alberta Petroleum Association's advisory committee, and by the end of May he had gained consensus regarding casing requirements and well testing. Surface casing was to be set to a minimum depth of 250 feet and cemented to the surface, production casing was to be set with sufficient cement to seal the Viking sand, and all wells were to be electrologged before production casing was run or before abandonment.[14] The initiative was as important as the measures themselves. It reinforced the Board's leadership role as Alberta's petroleum industry entered a new phase. For his part, Bailey resigned to cultivate emerging opportunities in the private sector.[15]

The response of the Alberta government followed closely upon that of the Conservation Board. In formulating their regulatory approach, Premier Manning and his cabinet relied not only upon the advice of their own professionals at the Conservation Board and J.J. Frawley in Ottawa, but also upon that of N.E. Tanner, their well-informed colleague. Tanner had just returned from a month-long tour of Colorado, Oklahoma, Texas, New Mexico, California and other oil-producing states.[16] His purpose had been to familiarize himself with U.S. oil and gas legislation, to establish government and industry contacts, and to encourage U.S. investment. Tanner felt comfortable in the United States: he had been born in Utah, and his religious world remained centred in Salt Lake City. His trip in 1947 was but the first of many that bolstered his deep free enterprise convictions and soon made him a well-known figure in Tulsa, Denver, Dallas and Houston.[17]

The *Right of Entry Arbitration Act* was the first of a series of bills that delineated over the subsequent 24 months, the essential framework that would govern the shape of the province's contemporary petroleum industry. The *Right of Entry Arbitration Act* was intended to deal with the common situation where land surface title and mineral title to a particular property were held by different parties, one of the several key variables that distinguishes early development in Alberta from that in the U.S. oil states. First, the Act reaffirmed the tradition that no operator had the right of entry, or to use the land surface for the removal of minerals, or for drilling, or for laying pipelines, or for the erection of tanks or any other purpose, without the surface owner's consent. Second, it provided for an appointed three-member board to resolve instances where an agreement on the conditions of entry could not be reached.[18] An operator who was unable to negotiate access to a drill site on a farmer's or other surface owner's property could make an application to the Board of Arbitration, which would hold a hearing and decide upon the appropriate amount of land required to meet the operator's needs and the amount of compensation due to the farmer, as well as any other conditions it might deem necessary.[19] The Board's decision regarding the terms under which entry was to be allowed was final; the Act did not allow appeal. In the sequence of government legislation, this was a logical first step. Provincial authorities knew that hundreds of farmers would soon have oil company representatives at their doorsteps seeking access for their drilling rigs. Right of access and the compensation arrangements presented by the Board of Arbitration provided ground rules, but did not prevent embittered disputes and occasionally angry confrontations between landowners and would-be drillers, as they attempted to move their equipment onto a lease.[20] That the government decided to assign this responsibility to a new agency, rather than simply hand it to the body that actually prescribed drilling locations, was fortunate, at least from the Conservation Board's perspective. It meant that the Board was largely insulated from such controversy.

Next on the government's priority agenda was reform of the existing regulations governing the reservation and disposal of petroleum and natural gas rights. Recognizing that the Leduc discovery had turned Alberta into one of the most attractive exploration areas on the continent, and sensitive to the emerging popular realization that Imperial Oil held the petroleum and natural gas rights to the thousands of acres surrounding the discovery well,[21] the Manning government made a key and innovative change in the regulations governing the acquisition of petroleum leasehold. When the Leduc field was discovered, crown-owned petroleum and natural gas rights could be acquired in two ways: through the direct purchase of a petroleum and natural gas lease or through a two-stage process that involved obtaining a petroleum and natural gas reservation (PNGR) first and later converting all, or a portion of, the reservation to a petroleum and natural gas lease. The two-stage process was intended to encourage wide-ranging exploration, by allowing the holder to concentrate resources on the investigation of a large tract and to be assured of subsequent access to portions of the explored area that seemed promising. PNGRs could range up to 10,000 acres in size, and there was no restriction on the number of reservations. The term was for 60 days and subject to the filing of satisfactory progress reports; it was renewable in 90-day intervals for up to one year. A required cash deposit was returnable on proof that a proper geological or geophysical program had been undertaken, part of which included the submission of a full geological report to the Alberta Department of Lands and Mines. At any point before the term expired, it was the reserve holder's exclusive right to apply for the lease of petroleum and natural gas rights to all, or a portion of, the reservation. If oil was discovered in commercial quantity, however, the reserve holder was compelled to apply for the desired portion of the reservation within 30 days.[22]

The first changes to the existing reservation system came in July 1947. Henceforth, the maximum size of a PNGR was set at 100,000 acres, but the maximum number of reservations that a company or individual might hold was limited to two.[23] Although the change was presented as a device to ensure the wider distribution of crown reservations, the *Oil Bulletin* rightly noted, "While Reservations are not assignable, it will still be possible for operators willing or able to work more than 200,000 acres to carry out the obligations on reservations issued to subsidiaries or individuals."[24] To ensure that the reservation system was not abused, the new regulations stipulated that no geological reservations would be granted in areas where the government believed adequate geological data was already available. Also, with the intent of encouraging legitimate exploration, the government lengthened the reservation period to four months, with possible retention for up to three years, through extensions, if obligations had been fulfilled.

It was government's innovative revision of the regulations governing oil and gas leases that had the most profound impact upon Alberta's post-Leduc oil boom. Presented in mid-August 1947, the new "Petroleum and Natural Gas Regulations" initiated a creative ongoing expansion of the province's "crown reserves."[25] The new regulations provided that wherever a PNGR holder was encouraged by his exploration results[26] and wanted to convert his reservation to a lease, not more than 50% of the reservation of crown rights in any township was available. The remainder reverted to the government and became crown reserve. Further, it was specified that leases could not be larger than 10,240 acres or 16 sections, and that the maximum length of any lease could not be greater than four miles. Although more than one maximum-sized lease could be acquired from within a reservation larger than one township, the boundaries of these, or smaller leases, could not touch other than at the corners. The intended result was a checkerboard-like distribution of crown-held petroleum and natural gas rights. This assured that publicly held oil and gas rights would be retained in the vicinity of almost every discovery. With the value of such rights dramatically increased by virtue of their location in "proven" territory, they could be offered for sale to advantage under the government's previously established bonus bidding system. Divided into acreages of various sizes, these parcels were offered for sale by public tender several times a year. Prospective buyers filed sealed tenders with a certified cheque for the amount of cash bonus they were prepared to pay for the lease on a given parcel.[27] The province reserved the right to reject bids that were in its judgement too low and did not reflect the value of the leasehold offered. Alberta's unique crown reserves system proved to be one of the most important influences shaping the development of the province's postwar petroleum industry. It assisted small companies, or combinations of small companies, to gain access to prime oil and gas leasehold, and thereby contributed to the diversity of ownership characteristic of most of Alberta's oil and gas pools. Although it was probably not realized by those who drafted the legislation, the new crown reserve system, by promoting the entry of more producers, accentuated the need to have a tighter regulatory structure to prevent "excess" production.

But for one other change, the remaining provisions of the new regulations were in keeping with earlier practice. The lease term remained at 21 years and was renewable as long as the location was capable of commercial production, and although the rental stayed at $1 per acre per year, the credit for approved drilling and exploratory expenditures that could be deducted from the rental was reduced from 100 to 50%. Similarly, the provision allowing for a royalty rebate of up to 50% for discovery wells was dropped. Drilling still had to be begun within one year of acquiring a lease, but the obligation to begin drilling a second well, after the first was completed or abandoned, was shortened to 90 days from six months. Also retained was the provision that

> the Minister may at any time assume absolute possession and control of any location acquired under the provisions of these Regulations if...such action is considered necessary or advisable, together with all building, works, machinery, and plant upon the location....[28]

New to the regulations, but nonetheless the product of an old concern raised in the 1920s and a harbinger of the vigorous debate to come, was the proviso that should natural gas be discovered it had to be "used or processed within the Province." Leaseholders were informed that only in "special circumstances," which the regulations left undefined, would the Alberta cabinet consider authorizing the export of natural gas.[29]

The Social Credit government's new regulations drew varied criticism. On the one side, in a feature editorial, the *Calgary Herald* warned, "Don't Strangle Alberta's Oil Industry," and cautioned against the restrictive measures advocated by "anti-business groups." The editorialist reasoned that, whereas the government collected a royalty, the equivalent of one barrel in eight, "the Crown takes no risk, expends nothing in exploration or drilling, and loses nothing if the well is a duster."[30] The Manning government also faced the charge that it was not administering the province's natural resources in the best interests of the people. During the debate on the Speech from the Throne that opened the 1948 spring session of the Alberta Legislature, Calgary CCF member A.J.E. Liesemer attacked Social Credit petroleum policy. He scoffed at the government's vaunted "checkerboard" system, alleging that it barely affected Imperial Oil. Warming to his attack, Liesemer asserted that

> when God placed beneath our feet these great pools of wealth for the enjoyment of all of us, he did not put up a sign "Reserved for the Imperial Oil." But our premier did![31]

Of as much concern to the government as the CCF championship of the "public interest" was the widely held feeling among independent operators that the revised regulations still allowed the large

MAP 5.2   Petroleum and Natural Gas Reservations, Leduc Area,
18 February 1947

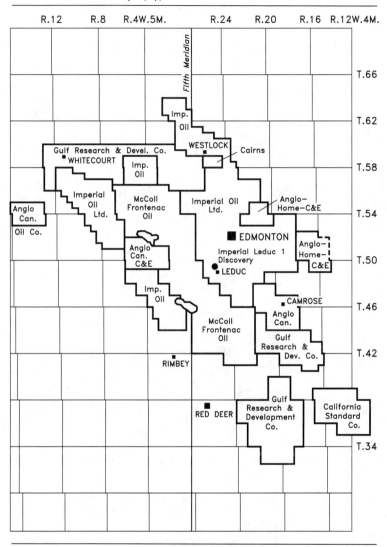

SOURCE: *Daily Oil Bulletin*, 21 February 1947.
NOTE: Companies listed have bulk of rights within areas outlined, but substantional
amounts of unassigned patented rights lie within many blocks.

companies to protect too great an acreage in the vicinity of a discovery. In the Woodbend discovery township, Imperial Oil had
been able to retain 16 sections surrounding the discovery well.[32] On
29 March 1948, less than a month after the debate in the Legislature, the government amended the regulations. The maximum size

MAP 5.3 Distribution of Petroleum and Natural Gas Rights, Leduc Discovery Township

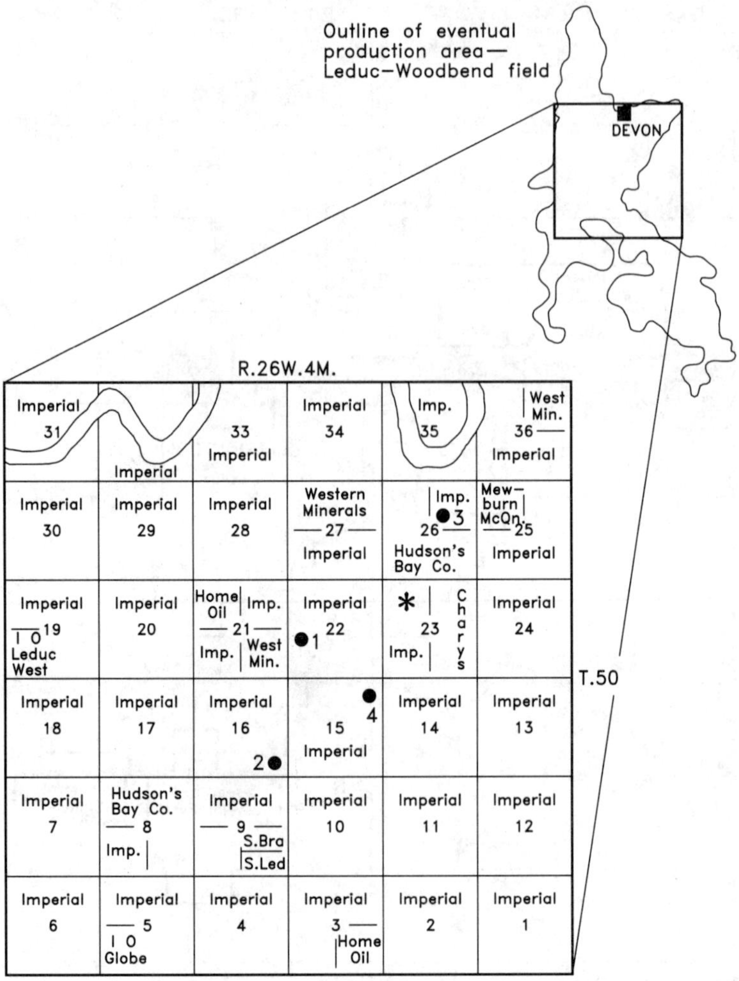

SOURCE: *Daily Oil Bulletin*, 28 March 1947.
● Imperial Leduc No. 1 to No. 4
✻ Shown as part of the Imperial Oil acquisition, the NW1/4–23–50–26 belonged to the Rebus family and the drilling rights were obtained by Atlantic Oil Company.

of any area within a reservation that could be leased was sharply reduced from 16 to nine sections.[33] The government also terminated the long-standing provision for satisfying annual lease rentals from drilling credits.

The evolution and impact of the government's regulatory response, especially the provision requiring the conversion of at least

50% of petroleum and natural gas reservations to crown reserve status, is illustrated in the pattern of development in the discovery townships of the three Imperial strikes. Imperial's activity at Leduc through the winter of 1947 generated a land rush that covered almost six million acres and extended nearly 125 miles along a northwest-to-southeast axis running through Leduc (see Map 5.2). Under the existing regulations, companies could convert as much of their reservation to petroleum and natural gas leasehold as they chose. Although the companies shown held the bulk of the rights under reservation within the areas outlined, in some of the reservation areas there were also patented rights held by individuals and corporations—especially the Canadian Pacific Railway and the Hudson's Bay Company—that remained unassigned. Imperial converted the petroleum and natural gas rights that it held under reservation in the Leduc township to petroleum and natural gas leasehold (see Map 5.3).

This pattern of petroleum and natural gas rights distribution was fundamentally altered by the amended petroleum and natural gas regulations. Coming into effect in August 1947, the impact of the crown reserve provision can be seen in the distribution of petroleum and natural gas rights that occurred in Woodbend, the second discovery township. Woodbend township, north of the North Saskatchewan River and immediately north of the Leduc discovery township, was also contained within a PNGR held by Imperial Oil. Perhaps because of the sandy and hilly nature of the land, the Canadian Pacific Railway had not selected sections, and the settlement that had occurred came late, which meant that Imperial's reserved rights here covered the entire township. When Imperial-Woodbend No. 1 struck oil near the middle of the township in January 1948, the subsequent transfer of land from reservation to lease status called for the first application of the 50% reversion formula (see Map 5.4). Imperial acquired under lease a four-mile-square block of 16 sections centred on its discovery well plus two additional sections for a total of 11,520 acres. The 18 sections returned to comprise the first increment to Alberta's crown reserves were situated about the periphery of the Imperial block.

Within six weeks of the Woodbend township split, the petroleum and natural gas regulations were again revised. Although the amended regulations of March 1948 did not represent a change of approach, the reduction from 16 to nine sections (10,240 to 5,760 acres) of the maximum single-block area within a reservation that could be protected had a significant impact upon subsequent development.[34] This is demonstrated in the distribution of petroleum and

MAP 5.4 Distribution of Petroleum and Natural Gas Rights,
Woodbend Discovery Township

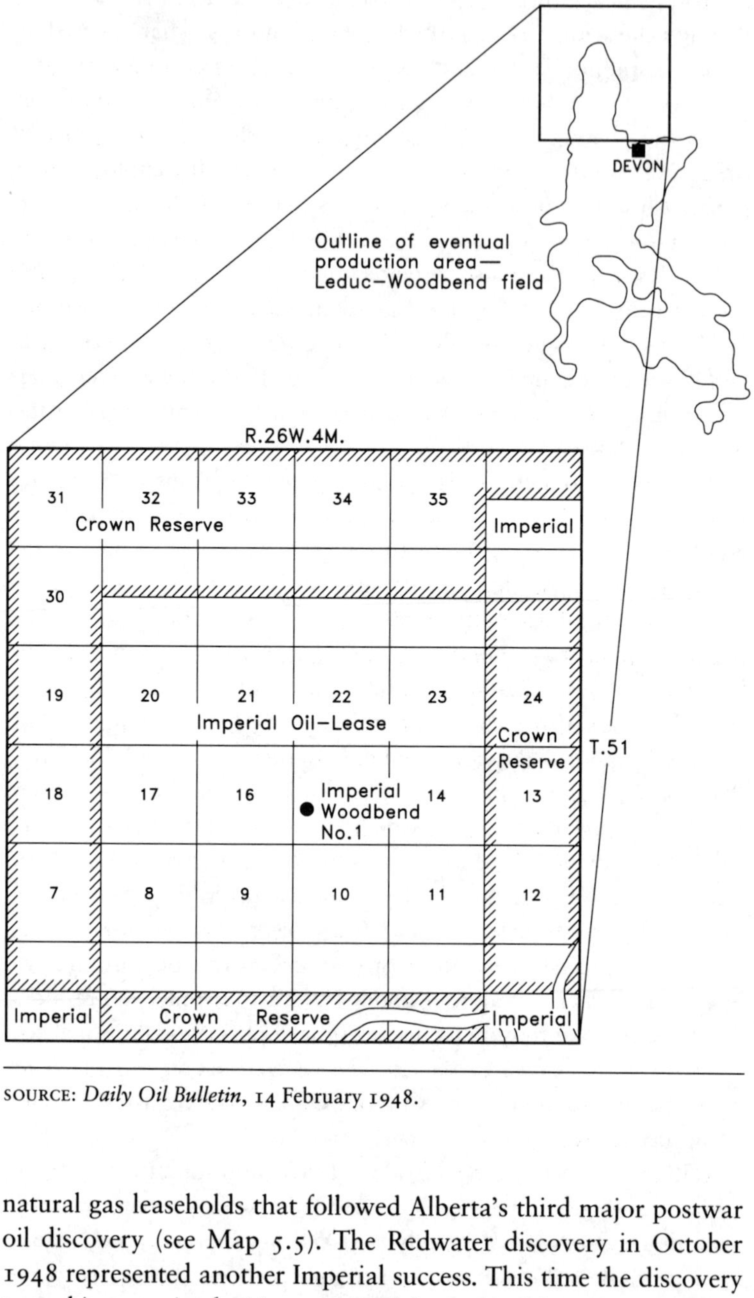

SOURCE: *Daily Oil Bulletin*, 14 February 1948.

natural gas leaseholds that followed Alberta's third major postwar
oil discovery (see Map 5.5). The Redwater discovery in October
1948 represented another Imperial success. This time the discovery
township contained 6½ sections of freehold rights (Canadian Pacific Railway and Western Leaseholds) and 29½ sections of petro-

MAP 5.5 Distribution of Petroleum and Natural Gas Rights,
Redwater Discovery Township

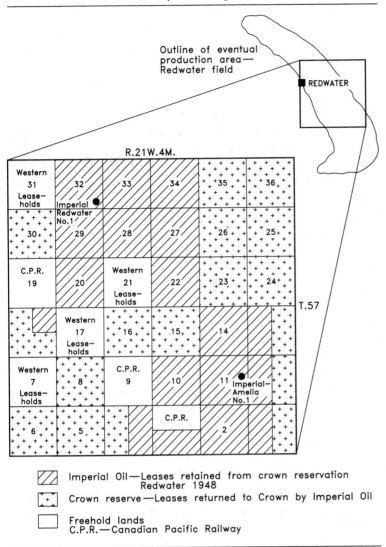

SOURCE: *Daily Oil Bulletin*, 31 December 1948.

leum and natural gas rights held under reservation by Imperial Oil.
Imperial-Redwater No. 1, the discovery well, was in section 32, and
by the time of the lease-crown reserve split in December 1948, there
was, in addition to a dry hole drilled in section 29, a second successful well in section 11. The information provided by these three

wells, combined with the geophysical data and the revised regulations, produced a much more dispersed pattern among the 14¾ sections that Imperial was permitted to select.

The 50% reversion principle introduced in August 1947 and refined in March 1948 greatly expanded the opportunity of access for other companies, and thereby checked the threatened dominance of the handful of companies that by midwinter 1947 had established a reservation stranglehold on the most promising exploration lands in central Alberta. Examination of lease sale records for the Leduc, Woodbend and Redwater discovery townships shows clearly that the overwhelming majority of purchasers were independent Canadian companies. Their purchases of crown reserve leasehold in the three townships returned cash bonuses of $446,176; $3,402,645 and $10,486,066 to the Alberta treasury.[35]

While the Alberta government was concentrating upon the urgent need to revise existing petroleum and natural gas regulations, the Conservation Board was adjusting to the escalating pace of exploration, discovery and development. By the end of 1947, steadily rising average daily crude oil production in the Leduc field reached 3,222 barrels, for a cumulative total production of 372,427 barrels in the 10½ months following the discovery.[36] No sooner had the Board begun to shift the centre of its attention from Turner Valley and to start thinking about an acceptable development plan for the Leduc field when the success of Imperial-Woodbend No. 1 in January 1948 revealed the Leduc pool to be considerably larger than had been realized. With the discovery of the prolific Redwater field about 25 miles northeast of Edmonton in September 1948, the whole of central Alberta was thrust into a frenzied oil exploration rush.

Acting Board Chairman D.P. (Red) Goodall concentrated his energy, as well as the time of his technical staff, on the task of keeping up with the inspection of drilling sites and "educating" operators, drilling contractors and field managers who were often new to the Alberta scene. This was Goodall's forte, because he identified with and understood the practical concerns of the operator. With the exception of Dingman and Knode, his 20-year familiarity with the industry at the field level exceeded that of all his predecessors. After graduating in geology from the University of Alberta in 1926, Goodall spent almost his entire career in the oilfields. After he joined the Conservation Board in 1938, he spent most of his time as an inspection engineer outside Turner Valley, where he developed an instinct for determining what was achievable given the circum-

The Board's first Field
Office at Leduc, with
engineer-in-charge
Nate Goodman (l) and
Calgary head office
engineer, Ted Baugh
(r). The Board later es-
tablished a permanent
office in Devon. ERCB,
Essery Collection.

stances of the moment.[37] This, combined with a somewhat retiring
and cautious temperament, meant that Goodall was not one to offer leadership initiatives outside matters related to field operations.
He began to feel increasingly uncomfortable with the growing
weight of his administrative responsibilities.[38]

At the same time, the Manning government's need for strong and
respected Board leadership was becoming critical. Not only did the
government require authoritative technical advice to help fashion
petroleum policy in a number of controversial areas but also the old
and difficult problem of market allocation had emerged once again.
The Board's first attempt, on 12 November 1947, to regulate crude
oil production in Leduc met stiff resistance.[39] Several of the independents announced that they would ignore their assigned production
quotas, forcing the Board, as the *Herald* reported, to "backdown
[sic]."[40] After rescinding the offending order, Goodall called a meeting of the Alberta Petroleum Association's advisory committee to
discuss the "factors to be used to obtain maximum efficient rate of
production commensurate with good conservation practice."[41] At
the meeting, Board engineer G.E.G. Liesemer conceded that, in establishing its limit of 150 barrels per day for each well producing
from the D-3 zone, no formula had been used. Rather, this was a
"safety figure," intended to apply until more information was
obtained. After considerable discussion, that included the usual
mention of the need to take "economics as well as conservation effi-

ciency" into account and the need to allow wells capable of efficiently producing more oil to expand their production more quickly, it was agreed that the "allowables set by the Board be adhered to, but that the matter be reviewed within two months."[42] During this period, the Board would gather all the necessary information from each well, and from this "a proper formula" was to be determined and discussed. It seemed like Turner Valley all over again.

Faced with the immediate need to find a production formula for Leduc and already alert to the impending debate on the export of natural gas, Premier Manning announced the appointment of two new Board members in February 1948.[43] In Ian McKinnon and Dr. George Govier, the premier believed that he had found the right combination of expertise to command the respect of the increasing array of powerful independent and major oil companies. As much as protecting the Board's hard-won status, Manning was anxious also that the new Board be seen by Albertans as an effective guardian of the public interest. Social Credit had to go to the electorate soon for a renewed mandate, and the session of the Alberta Legislature then under way made it clear that petroleum policy would be one of the central issues.

Named chairman, Ian McKinnon was a senior civil servant who carried with him to his new position his status as assistant deputy minister of lands and mines. Even more important, McKinnon came with the close personal trust of the premier. It was a confidence shared by only a small group that included one of McKinnon's predecessors, J.J. Frawley, who remained Manning's adviser on petroleum legislation. The son of a solicitor, McKinnon was born in Edinburgh in 1906. He immigrated to Canada in 1923, his passage being paid by the Canadian Bank of Commerce on the understanding that he would work for the bank for five years. After gaining experience as a teller in small Alberta towns, he was promoted to accountant at the bank's main Edmonton branch. In 1930, he joined the newly created provincial Department of Lands and Mines, where his accounting skills were put to use building the administrative structure required after the province assumed responsibility for lands and resources. Despite the security and preferment of what had become a middle-level position in the civil service, McKinnon served in the Royal Canadian Air Force and soon became one of the principal officers in charge of the procurement and supply of new aircraft. He later rose to the rank of wing commander, was made a Member of the Order of the British Empire, and in 1946 was offered a promotion as an inducement to remain in the peacetime service.[44] McKinnon elected to resume his service with the Alberta government, however, and after discharge in 1946 he returned to the Department of Lands and Mines. His promotion to assistant deputy minister followed almost immediately and to deputy minister shortly thereafter. When he assumed the Conservation Board chairmanship in February 1948, the 42-year-old McKinnon soon showed the marks of character that were to distinguish his career. He put a high value on personal integrity, commitment to principle, public service and devotion to duty. All these attributes, combined with his professional bias as an accountant, led to an elevated sense of stewardship, particularly as it related to the public purse. With McKinnon, everything had to be accounted for precisely, including, for example, a 10-by-12-foot canvas tarp worth $8.75 that had been "worn out" in Turner Valley and the eight-inch crescent wrench valued at $1.05 that unfortunately had been "lost," also in Turner Valley in 1950.[45] The new chairman was a modest man, not inclined to the perquisites of rank or office. These dispositions he shared in common with his political master, Ernest Manning, and his friend and predecessor J.J. Frawley. He was also quiet, somewhat shy and cautious, and inclined to worry over important decisions.[46]

George Govier, McKinnon's new colleague on the Board, was a man of a different personality and background. His immediate contribution was to enhance the Board's technical credibility. Govier was a professor and recently appointed head of the University of Alberta's chemical and petroleum engineering department. Of keen mind and possessed of great energy, Govier had a scientific background that was directly relevant to the activities of the Conservation Board. Probably no other individual in the Canadian academic community possessed the background and training so closely suited to the needs of Alberta's conservation authority.

Born in Nanton, Alberta, in 1917, Govier spent all his school years in Vancouver. Graduating in chemical engineering from the University of British Columbia in 1939, Govier was confronted with the Depression's still-lingering impact on the job market. Although he was fortunate to find employment with the Standard Oil Company of British Columbia, the position offered little opportunity to apply newly acquired engineering skills. Having decided that graduate study might offer more interest and challenge, he was alerted to a University of Alberta advertisement for an instructor in chemical engineering that offered the attractive prospect of being able to combine employment with graduate classes. The successful applicant, Govier began work on a master's degree in physical chemistry under Dr. Edward Boomer, the university's senior scholar in the field. It was a turning point in Govier's dawning professional status, and but the first of a series of remarkable circumstances that seem to have shaped the course of his career.

Although Boomer's contact with Govier spanned only four years, cut short by his death, the influence upon his graduate student was great. As a professed Albertan, as a scientist, and later as a regulator, Boomer was preoccupied with the problem of Alberta's growing natural gas surplus. Significantly, he was part of the continuing applied science tradition that founding President Henry Marshall Tory had established at the University of Alberta. Govier's introduction to the natural gas issue came therefore almost as soon as his arrival in Edmonton, and it was not long before he was attracted to his mentor's vision of a petrochemical industry based upon the province's abundant natural gas resources.[47] In pursuit of this objective, the project selected for Govier's master's degree research was an investigation of the feasibility of using a low-pressure technique to produce formaldehyde from natural gas. Beyond focusing the further development of his scientific and engineering background in the natural gas sector, Govier's close association with Boomer exposed him, after 1943, to the regulatory side of oil and natural gas production. Sometimes a participant in discussion with Boomer and A.G. Bailey, when the latter came to Edmonton to consult with Boomer on Board matters, Govier became interested in the application of physical chemistry and chemical engineering principles to oil and gas recovery. Shortly before Boomer's death, it was arranged that, following the completion of the university's spring term in 1946, Govier would take on a research assignment with the Conservation Board.[48] Throughout the summer of 1946, Govier worked with Dick King, the Board's senior field engineer at Turner Valley, and Vern Horte, a chemical engineering student also employed for the summer, to try to find practical ways to reduce the flaring of solution gas.[49] Work in Turner Valley provided a firsthand opportunity to observe the practical application of the reservoir engineering principles being advanced by two of the leading men in the field—George Granger Brown and Donald Katz. The summer experience stimulated Govier's interest in research centred upon the recovery of oil and gas from underground reservoirs enough that he arranged leave to embark upon doctoral study under Brown and Katz at the University of Michigan.[50]

While Govier was in the United States, the Leduc oil discovery dramatically changed the oil and gas situation in Alberta. Through his continued contact with Bailey, Govier kept informed of the Board's efforts to manage the emerging oil boom, but by the time he returned from his studies to resume his position at the University of Alberta, Bailey had been gone from the Board for some months and the need for a fully manned and technically informed Board was urgent. When the Manning government finally acted on Bailey's departing recommendation[51] to appoint George Govier, the Board regained the technical expertise that it had lacked since Boomer's chairmanship.

Eleven years McKinnon's junior and youngest of the three Board members, Govier was a remarkably energetic, disciplined and confident personality. Committed to the principle of hard work and able to function simultaneously and effectively as a university administrator, lecturer, publishing scientist and Board member, Govier expected the high standards that he set for himself to be met by those he supervised or with whom he worked.

The new Board, composed of three different yet complementary personalities, was unusually well balanced. McKinnon was a proven administrator, who by nature and training placed high value on efficient organization. His sense of public accountability and his accountant's background led him to emphasize the value of keeping good records that could stand the test of any kind of exam-

In 1948, the Board comprised three very different yet complementary personalities: (l-r) Deputy Chairman D.P. Goodall, Chairman I.N. McKinnon, and Board Member G.W. Govier. Harry Pollard Collection, Provincial Archives of Alberta, P1545.

ination. This was an attitude that bolstered the Board's tradition of systematic data collection. Govier brought a scientific background that was on the leading edge of current reservoir engineering research. He was inclined to focus on the agency's technical expertise, and as the Board's responsibilities grew in response to the industry's rapid development after 1948, he concentrated on hiring the best young engineers for new professional positions. Beyond having a sophisticated ability to evaluate the Board's technical requirements and the technical abilities of those seeking positions with it, Govier's own active research was of particular value during the early post-Leduc hearings that centred upon the formulation of long-term policy. Deputy Board Chairman Goodall brought a detailed working knowledge of the industry in Alberta as it functioned in the field. More than his Board colleagues, Goodall had long experience from which to interpret the practical implications of policy or regulatory change and to anticipate industry response.

The first test of the revitalized Conservation Board came quickly. It was literally a baptism of fire.

ATLANTIC NO. 3

The most pressing task facing the new chairman was that an acceptable production schedule had to be established for the Leduc field. As was typically the case, the various producers were known to have different ideas regarding how allowable well production should be determined. McKinnon announced at the first Board meeting under his chairmanship that one week hence, on 9 February 1948 at the Calgary court-house, the Conservation Board would hold a public hearing to allow "all owners and other interested persons an opportunity to put forth their views with respect to the regulation of the production of oil and gas" from the Leduc field.[52] Speaking at the hearing for Imperial Oil, J.D. Gustafson noted that there was already some decline in pressures as a result of production, and affirmed his company's view that the existing Board quota of 100 barrels a day for wells producing from the D-2 zone and 150 barrels per day from the D-3 was as high as Leduc production should go.[53] He recommended, however, that these rates be set as a monthly average, rather than a daily rate, so that daily rates might be stepped up by as much as 25 barrels to allow producers to make up for shut-in production when wells were not operating. In turn, Dr. J.O.G. Sanderson presented the classic concern of

the small independent operators. He reminded the Board that shareholders had invested their money in the ventures of independents and wanted returns. The prospect of improved recovery in the long term through conservation practice did not ease Sanderson's worries. He wondered who could predict the price of oil five years hence, and speculated that the synthetic fuel industry might cut the price of oil to 75¢ a barrel. Instead of a blanket maximum quota for all wells, geologist Sanderson argued that, within the limits of a maximum efficient flow rate that might be established for each well, the "Law of Capture" should prevail.[54] After considering the hearing presentations, and after a subsequent meeting with the advisory committee of the Western Canada Petroleum Association, McKinnon and his colleagues amended the Board's existing Leduc crude oil production order to conform to the recommendation of Imperial Oil at the February 9 hearing.[55] During the 69 days the allowable was to be in effect, Goodall and Govier would examine oil

Atlantic No. 3 out of control. Oil and gas can be seen escaping from below the production casing, although the rig is still standing. The 1948 blowout tested the Board's authority in a major crisis. Alberta Government photograph; ERCB No. 87.022/040.

and gas production in the Leduc field and prepare a report to be used as the basis for a new and more refined production order.[56]

Before Goodall and Govier could report, another matter intervened to preoccupy the Board and the industry's attention totally. In the pre-dawn darkness of March 8, the Atlantic Oil Company's third well in the Leduc field roared out of control. The wellbore had barely penetrated the D-3 reservoir when the ground shook and a terrific surge of pressure shot a 150-foot gusher of oil, gas and drilling mud up through the drill pipe. The immediate and traditional response of trying to stem the flow by pumping tons of drilling mud into the hole proved futile. After a few days pause, the tremendous—but now released—reservoir pressure began to force oil and gas through the shallower formations and to the surface. Miniature geysers of mud, oil and gas began to erupt out of hundreds of craters in a wide radius around the drill hole. The fear of fire now became extreme, and there was the problem of runaway crude oil production. Although earth and snow dikes were bulldozed to contain the acres of accumulating oil, cold winds continued to blow oil spray far from the rig, thereby adding to the mess and making working conditions more difficult for relief crews.

The blowout and its publicity only dramatized the already controversial presence of Frank McMahon's Atlantic Oil Co. in the Leduc oil play. McMahon was the quintessential oil company promoter. From the early 1930s, he had charmed Vancouver investors into supporting a series of mainly unsuccessful wildcat drilling ventures. His first success in Turner Valley in 1938 only whetted a restless entrepreneurial spirit that gained full rein with the Leduc discovery. In 1947, he learned somehow that one of the leaseholds (NW ¼ 23 50 26 W4M) in the heart of the discovery area had been obtained from a family where there were competing claims of title (see Map 5.3). McMahon won the race to the owner who had the superior claim, and with a cash offer of $175,000 and 25,000 company shares he snatched from Imperial one of the most promising quarter sections in the discovery township. The manner of McMahon's arrival on the Leduc oil scene might have coloured the minds of some of the key players who watched Atlantic try to deal with the disaster.

As March passed, with the rogue well still blowing thousands of barrels of oil to the surface each day, enough to maintain a crude oil lake, the immensity and seriousness of the situation became apparent. The Board could no longer avoid becoming directly involved. All efforts made by the well owners and the drilling contractor to stop the flow had failed, including a last-ditch effort to pump a

10,000–bag mixture of cement slurry down the well. The cement had disappeared just like everything else that had been pumped into the well. As one of the participants summed it up, "You might as well have pumped it into the North Saskatchewan River for all the good it did."[57] The press was beginning to suggest the potential for heavy surface loss claims, apprehension was being raised about oil-contaminated spring runoff reaching the nearby North Saskatchewan River and endangering Edmonton's water supply, and there was the fear of permanent damage to the reservoir. Faced with the Atlantic Oil Company's failure to bring the well under control, the Board took the initiative to hire the well-known American wild-well fighter Myron Kinley on April 12.[58] His efforts proved fruitless as well. After failing to clear the drill pipe, as a prelude to lowering and setting off a charge of explosives to block the hole, Kinley tried to seal the wellbore with heavy mud weighted with barite. Not only was this effort unsuccessful but also oil and gas began to erupt from a new band of craters north of the rig and oil production began to increase. In early May, Kinley decided to try to stem the flow by pumping down a mix of "roughage." His exotic recipe of 16 tons of redwood fibre, 43 tons of cottonseed hulls, 21 tons of sawdust, ¼ ton of feathers, 490 sacks of mud and 10 sacks of lime[59] simply disappeared down the hole without effect like all the previous offerings. Although Kinley was convinced that if "not interfered with" he could bring the well under control,[60] what credibility he had on arrival was now largely dissipated.

There was now great pressure on the Board to come up with a response that promised some certainty of success. Untamed for 10 weeks, the well had quickly gained national and then international attention. Albertans had not experienced anything like this before. At first, the blowout was incorporated into the popular excitement that still lingered from the Leduc discovery and the hectic development that had fallen in its wake. By late April, however, the excitement had begun to change to unease, especially in the farm community around Leduc. The "experts" were obviously having no success in bringing the well under control. The *Calgary Herald* agreed that, although the situation was serious, it did "not warrant the hysteria engendered by highly coloured news reports," which suggested that communities in the Leduc vicinity were likely to be "blown sky high."[61] The Board felt compelled to reassure Albertans that "excellent progress was being made."[62] The public's emerging restiveness was complemented by growing unease in another quarter. As owner of 80% of the field's production, Imperial Oil was growing increasingly worried that if an effective means to bring the well under control was not found soon the field would be ruined. From the outset, Imperial had monitored control efforts closely at Atlantic No. 3 and had consistently lent whatever material assistance was requested. As the weeks passed, discussion regarding the Atlantic problem took on a more organized form and moved to a higher level within the Imperial organization, and conversations with the Conservation Board regarding what might be done at Atlantic No. 3 also became more focused.[63] The thought that Alberta's newest and long-awaited oilfield might be in jeopardy was just as disquieting to the Alberta government. So too was the rising public restiveness in the Leduc-Edmonton area, especially since Premier Manning and his colleagues were contemplating an election call that would have to be announced within a few weeks. Technical, economic and political circumstances thus combined to demand bold action on the Board's part. At this juncture, the Board's credibility also became directly bound to the outcome at Atlantic No. 3.

Following discussions with Imperial Oil on May 11, the Conservation Board directed wells in the Leduc field to cease production at 8 a.m. May 13. When, for obvious reasons, the Atlantic Company proved unable to stop production from its No. 3 well, the Board then announced that, by authority granted under section 46 of the *Oil and Gas Conservation Act,* it was assuming control of Atlantic No. 3.[64] This procedure was more than just a necessary legal strategy to prepare the way for takeover,[65] the field shut-in order recognized that the runaway well was producing an estimated 50 million cubic feet of gas and 10,000 barrels of crude oil daily. It anticipated the Board's subsequent notice of May 18, stating that all production and accumulated oil from Atlantic No. 3 would be transported by Imperial Oil's pipeline to Nisku, loaded on tank cars, and sold on the Board's behalf.[66] Only by shutting in the field could Imperial Oil's pipeline system be made available for clearing the growing lake of oil surrounding the Atlantic well. With notice of its takeover of the wild well, the Board also announced that it had retained the services of V.J. (Tip) Moroney, Imperial Oil's western Canada operations manager, to supervise operations at the well until it was brought under control.

Tip Moroney was a Standard Oil of New Jersey career man, and had been with the company since 1926 after graduating from Georgetown University in Washington, D.C. His apprenticeship had been served in Oklahoma's Seminole field and in the Brea-Parinas fields in Peru. In late 1947, when Imperial's parent company realized that the Leduc discovery represented a significant oil strike, Moroney was sent from Venezuela to bolster the still limited

FIGURE 5.1   Atlantic Oil Company Petroleum and Natural Gas
Lease, Summer 1948

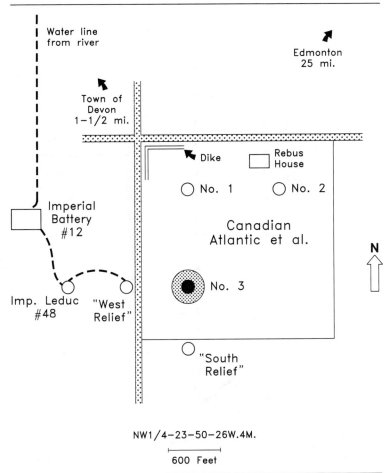

NW1/4−23−50−26W.4M.

600 Feet

SOURCE: Aubrey Kerr, *Atlantic No. 3* (Calgary, 1986).

professional staff that Imperial Oil had in Calgary. A man of broad
field experience and some familiarity with wild wells, Moroney was
the logical member of Imperial's senior resident engineering staff to
be nominated to try to tame Atlantic No. 3. Under no illusion about
the difficulty of the situation, Moroney accepted the challenge only
on the condition that he be given absolute control.[67] This agreed, he
returned six days later to seek the Board's approval for his plan.
There would be no attempt to continue Kinley's efforts to remove
the drill pipe. Moroney proposed a more sophisticated approach.[68]
He advised the drilling of two relief wells to intersect the D-3

FIGURE 5.2   Atlantic No. 3 Relief Wells, Summer 1948

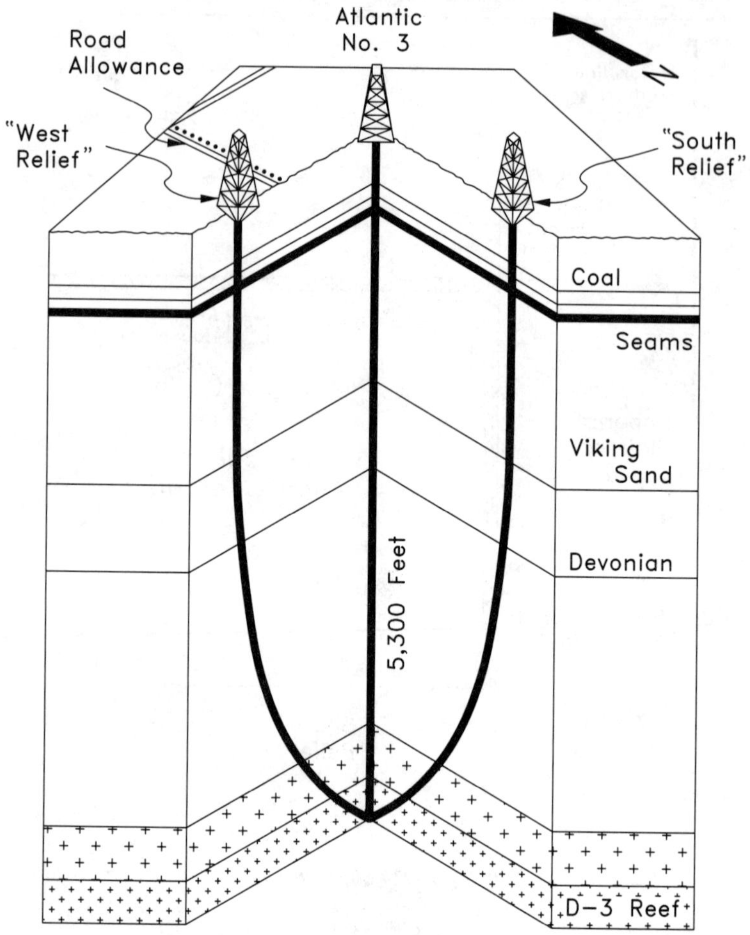

SOURCE: Aubrey Kerr, *Atlantic No. 3* (Calgary, 1986).

porous zone as near as possible to the point where the Atlantic No. 3 wellbore entered the same zone (see Figures 5.1 and 5.2). While this was under way, the nearby Imperial No. 48 well would be prepared as an injection well to take water from the North Saskatchewan River to begin an attempt to flood the wild well.[69] Atlantic's Nos. 1 and 2 wells were also to be prepared as injection wells to accept the excess oil produced by Atlantic No. 3 that could not be handled by pipeline facilities. In addition, Moroney wanted more stringent steps taken to reduce fire hazard and the maintenance of a first-aid and ambulance station near the danger zone. Accepted by

Atlantic No. 3 caught fire on 6 September 1948,
just days before flow was shut off, and six months
after blowing wild. Provincial Archives of Alberta,
Public Affairs Bureau Collection, PA585/1.

the Board, the plan marked the efficient judgement that characterized Moroney's leadership and eventually brought success months
later.

With Moroney directing efforts at Atlantic No. 3, Imperial was
seen to be in charge, and public confidence began to lift.[70] Within
the industry, the situation also became less tense, particularly after
June 3 when it became possible to reopen the field for production,
thereby enabling the independent companies to take advantage of a
strong market and to re-establish badly needed cash flow. The now
muted concern about Atlantic No. 3 came just in time for the Alberta government. With a provincial election called for August 18,
Premier Manning and N.E. Tanner were busy enough defending
their government's petroleum policy without having to address
awkward questions about the six-month-old Atlantic No. 3 blowout.

Drilling of the west and south relief wells proceeded slowly
through the summer. After departing from the vertical at about the
2000-foot level, directional surveys had to be taken constantly to
ensure that the sloping wellbores remained precisely on target.[71] It
was not until early September that the west relief well neared its target. On September 5, Board Chairman Ian McKinnon arrived for
the critical and long-awaited moment. Tension at the site was
heightened by serious new cratering that had begun to occur in the
immediate vicinity of the rig. Braving pieces of rock and loose bits
of iron being hurled into the air by the force of the oil and gas escaping from the nearby craters, Imperial's "danger pay" gang tried
unsuccessfully to prevent the derrick from toppling. The drilling
crew at the west relief well was also experiencing difficulty, they
had missed their target by 500 feet, and only after a second "acid
job" on the morning of September 6 were they able to increase the
formation permeability sufficiently to permit effective water access
to the No. 3 wellbore. Moroney's trial, however, was still not over.
His men had just started pumping water down the relief well when
Atlantic No. 3 caught fire. McKinnon, who had stayed up all night
with Moroney at the wellsite, was on hand to witness the huge fireball that engulfed everything at once. From acres of burning oil and
gas, flames shot 700 feet into the air. The blaze could be seen from
20 miles away, and soon a pall of black smoke stretched for nearly
100 miles across central Alberta.[72] What caused the fire remains
conjecture, but such a disaster had always been Moroney's great
fear. Still, he was lucky, no one had been killed and, at this juncture, he possessed effective means to deal with the crisis. By the second day, the presence of grey smoke with the flames revealed that

the massive injections of water down the west relief well were mixing with the oil and would eventually stem the flow of oil and gas
and smother the fire. On September 10, the fire was quenched, but
not before pictures of the inferno had made the front pages of newspapers throughout Europe and North America. Commenting on
the dramatic scene near Leduc, Alberta, the *Financial Post* reported
that

> Death came to Atlantic No. 3 in a blazing 59 hour funeral pyre
> that marked a spectacular end to a six month rampage that fo-
> cused world-wide attention on Alberta's 200 million barrel
> Leduc Woodbend field.[73]

The struggle at Atlantic No. 3, however, was not yet over. The
resumption of oil and gas flow was prevented only by the hydrostatic head or weight of the water column produced by the continuous injection of enormous amounts of water. The cavity at the bottom of the No. 3 wellbore had to be plugged, and it was for this
reason that Moroney had ordered the drilling of two relief wells.
While the injected water flow was maintained at the west relief
well, the crew at the south relief well began cementing and plugging
the No. 3 wellbore. This essential but almost forgotten second part
of the Atlantic No. 3 struggle took remarkable ingenuity and another two months to achieve.[74]

It is apparent that the Atlantic No. 3 crisis caught both the government and the Board unprepared. Both Manning and McKinnon
seemed uncertain at first about the kind of leadership they should
provide. For the first two months of the wild well's rampage, they
mainly stood by in hopeful anticipation that the industry would
solve the problem. In good part, this is understandable, both were
confronted with a new situation and some time was required to assess the capability of the well owners to deal with the situation. The
previous incident from which direct experience could have been
drawn was the Royalite No. 4 blowout, but that occurrence was
separated not only by different geological and technical circumstances but also by the 24 years that had passed.[75]

Given the government's confidence in the industry and its Board,
Manning's inclination to stand aside is easily understood. The
Board's reluctance to intervene, however, was more complex. It
was not just because of inexperience and faith in industry expertise,
the Board's situation was more complicated because it had been
party to the flawed decision that precipitated the blowout. Atlantic
No. 3 had provided good evidence of its troublesome nature long

E.C. Manning, Pre-
mier of Alberta,
1943–1968. Provin-
cial Archives of Al-
berta, Public Affairs
Bureau Collection,
PA1655/13.

before the celebrated blowout of March 8. Loss of drilling mud circulation occurred first on February 17. All subsequent efforts to re-
establish satisfactory circulation by injecting assorted mixtures of
mud, sawdust, oats and gel-flake, as well as running cement plugs,
proved unsuccessful. Moreover, from February 22, the well was
"alive" with escaping natural gas. Yet, despite all these warning signals, the Atlantic Oil Company, after consultation with the Conservation Board's Leduc area field engineer, Nate Goodman, and approval of Board member Red Goodall, elected on March 6 to make
"a run for it" and drill "dry" to the producing zone estimated to be
about 30 feet beyond. The plan to drill ahead with water circulation only until the pay depth was reached, and then cement the drill

pipe in place as production casing, was a calculated cost-saving risk. That the decision was flawed is apparent not only given the advantage of hindsight but also because there was a wide expression of concern and even direct opposition to the proposal to continue drilling.[76] That the most experienced member had given his consent to continue made it extremely difficult for the Conservation Board to question Atlantic's procedure, and offered the Board little alternative other than to stand aside in the hope that the company would be able to assemble the expertise necessary to bring the well under control soon.

The criticism levelled by the *Edmonton Bulletin* just a few weeks after the well began its rampage was close to the mark.

> The fact that the damage may be restricted is no reason to excuse the fact that the well was permitted to get out of control. The highly competitive nature of oilfield development makes well drilling a race against time. Under these circumstances it is essential that the government exercise strict supervision of the work.

> Standards should be set and rigidly maintained with regard to equipment used and precautions taken to prevent a catastrophe in case of trouble such as that which plagued Atlantic No. 3 at Leduc. A proper trained and efficient staff can prevent a repetition of the calamity.[77]

The reason that Premier Manning and his minister of lands and mines did not become directly involved sooner than they did was not just because of their confidence in the government agency that dealt with such oilfield problems. An earlier initiative on their part was also precluded by the political climate in the spring of 1948. In addition to their interest in the impressive production statistics and geological reports that followed the Leduc discovery, industry attention was focused on the petroleum legislation that began to emerge from the Alberta Legislature. Given the fiscal and economic "heresy" that had attended the birth of Social Credit in Alberta, and the subsequent struggle with Ottawa over matters of money and banking, potential investors watched anxiously for signs of a lingering interventionist philosophy. By early 1948, the industry, particularly the interested U.S. component, began to interpret the signs in a negative manner and industry journals started to express concern about the Manning government's approach.

It was Edmonton's new petroleum lease policy that brought industry apprehension into the open. At issue was the question of

"appropriate" royalties. For some time, the industry had been lobbying to persuade the government to remove from the petroleum
and natural gas regulations an enabling clause that provided that
the Alberta cabinet could at any time set whatever royalty it might
see fit.[78] The oil companies argued that they required a more predictable future, and therefore they pressed for a binding commitment
from the government never to raise the royalty rate beyond a specified maximum. In January 1948, when the government altered its
new competitive bidding system for crown reserve leasehold to allow bidders to offer a royalty bonus in addition to the formerly required cash bonus, it became apparent that Premier Manning and
his colleagues had not been moved by the lobby's arguments. The
industry's worst fears seemed confirmed by the results of the first
sale of petroleum rights under the new formula. For the first of two
40-acre Leduc parcels, the government accepted the tender of New
York investor Renzo Falco, who offered a cash sum of $10,000
plus a royalty bonus of 58½% over the royalty ordinarily payable
on crown leases. The second parcel went to Saskatchewan Feder-
ated Co-operatives of Saskatoon for $10,000 and a 50% royalty
bonus.[79] Warning that such "unrealistic" royalties would deter most
substantial outside investors, oilmen began a much more aggressive
campaign to win the government to its point of view. In March, after discussions with the industry, the government assented to the request to set a royalty ceiling that it was bound not to exceed.
Henceforth, the royalty maximum payable on petroleum produced
from crown leasehold during the first 21-year term of the lease
would be one-sixth or 16⅔% of the gross value of production.[80]
This important concession, however, did not preclude the possibility of having to pay a higher royalty as long as the royalty bonus
bidding system remained a part of the government's method of releasing crown reserves. In addition, the new royalty limit was partly
offset by another amendment to the petroleum and natural gas regulations that sharply reduced the maximum acreage that could be
contained within any petroleum and natural gas leasehold.[81] Therefore, oilmen were not inclined to relax from the pursuit of what had
now become their major objective—the termination of the royalty
bonus system. They resolved to send the Manning government an
emphatic message at the next offering of selected petroleum and
natural gas leases from the crown reserve.

On May 4, the Department of Lands and Mines held its second
sale of crown leasehold under the bonus royalty system. Offered
were four 160-acre tracts in the highly regarded Woodbend township. Only four tenders were submitted, all for one parcel. The successful bidder was Fred A. Schultz, a Turner Valley operator who

offered $25,000 cash plus a 38% bonus beyond the normal crown royalty.[82] The industry's boycott was intended to warn Alberta's Social Credit government that continued development should not be taken for granted and that an "excessive" royalty structure would drive away foreign capital. Significant support for the industry's position came the day following the sale when G.M. Blackstock, chairman of the Alberta Board of Public Utilities Commissioners, which also served as the province's Securities Commission, announced that, perhaps with the exception of Turner Valley, no Alberta oilfield had demonstrated its capacity to ensure that wells subjected to a royalty of more than 15% could return a net profit. Therefore, as a general principle, he explained, his board would not approve offerings of oil company shares or other securities to the Alberta public if the rights involved were bound to a royalty obligation of more than 15%.[83] Helping to sustain the pressure, Carl Nickle, the editor of the *Oil Bulletin,* devoted the front page of his May 7 issue to warn the Manning government that "the flow of U.S. money to Alberta will be sharply curtailed, however, regardless of favourable findings from exploration, if the government's policy becomes too greedy and the prospects of ultimate profit from exploration are eliminated."[84] Reporting on a recently completed tour of U.S. oil cities, Nickle explained that U.S. oilmen were greatly disturbed by changes in the Alberta government's oil policy, particularly the changed reservation and lease regulations and the royalty bonus system used to dispose of crown reserves. Alberta, Nickle interpreted, did not want to risk gaining a reputation such as that of certain Latin American republics.

> U.S. oil companies have had some bitter experiences, notably in Mexico where the government took over the oil industry after British and U.S. capital had developed it. The changes in Alberta's policy, started last summer and continued this spring, have roused some fears as to the future.[85]

Impressed by the boycott and fearful that U.S. investment might be jeopardized, the government quickly changed course. It was announced that on May 25 the Woodbend leases would be reoffered on a straight cash bonus basis. The decision to back away from the royalty bonus system was probably assisted by an awareness of the Conservation Board's concern that "excessive" royalty bids might lead to premature abandonment of less productive wells.[86]

To be certain that this policy change was properly noted, the minister of lands and mines took the message to U.S. oilmen gathered at the International Petroleum Exposition in Tulsa, Oklahoma. Speaking on May 21, four days before the rescheduled crown reserves sale, N.E. Tanner explained that the Alberta government's "first and greatest responsibility" was to encourage development so that Canada would not remain dependent upon other countries "for such an essential product."[87] Alberta's petroleum policy was governed by the philosophy that "only through individual enterprise where there is a wholesome competitive system in operation will the greatest development take place and the greatest good be accomplished."[88] While asserting that the public must be assured of a fair share of the returns from the development of the province's natural resources, he emphasized that the Alberta government was emphatically opposed to government-financed exploration or development. Tanner informed his audience that, under Alberta's newly adopted lease form, the maximum royalty that could be collected during the 21-year life of a lease was one-sixth. Further, he assured that any changes made in legislation or regulations dealing with oil reservations or leases would never be made retroactive, as was possible in the past. On the subject of crown reserve disposals, the minister informed the oilmen that Alberta had experimented with offering some of these rights on a fixed cash bonus plus competitive royalty bonus but that parcels being offered later in the month would be on a straight cash bonus basis. Tanner also discussed the organization and operations of the Conservation Board, stressing that the Board worked closely with an advisory committee that represented the oil industry. With some pride, he asserted, "I think it is generally admitted that our conservation program is one of the best on the American continent."[89] The *Oil Bulletin* concluded, "Mr. Tanner today did a good job of 'selling' Alberta as a good land to explore in."[90]

The first weeks of May 1948 represent, if not a turning point, at least an important shift in Social Credit's approach to the petroleum industry. First, the government backed away from the royalty bonus idea; then, a few days later, its Conservation Board moved decisively to deal with the Atlantic No. 3 crisis.[91] These quite disparate acts are related because, under pressure in both cases, the government and the Board moved to accommodate industry concern and because the second act necessarily had to follow the first. In the difficult environment that existed up to May 4, with the industry railing against Edmonton's "greedy" and interventionist approach, Manning and his ministers were doubly sensitive about how government seizure of an oil well and the shutting in of an oilfield might be interpreted. After about May 7, by which time the government had made its decision to abandon the bonus royalty system and Imperial Oil had begun to express some anxiety and exasperation with the ineffectual efforts under way at Atlantic No. 3, the perceived environment seemed much less threatening, and the Alberta government could contemplate more readily its Conservation Board acting to shut in an oilfield and seize private property, something no conservation agency had done before in North America.[92]

The first weeks of May also confirmed the theme of the coming provincial election. Social Credit proposed to rally Albertans in the fight against socialism. Although the Co-operative Commonwealth Federation (CCF) had been represented by just two members in the previous Legislature, the party's impressive breakthrough to 24.9% of the popular vote in the 1944 election gave the Manning government pause for concern. In the campaign, the CCF continued its offensive against the Social Credit government's petroleum policy. CCF leader Elmer Roper warned against allowing large U.S. and Canadian oil companies to control such large amounts of crown land. He promised Albertans that, if elected, his party would "break the ironclad monopoly position" of Imperial Oil by retaining 50% of oil production for sale by the province.[93] In addition, Roper promised to raise oil royalties to 25% on privately produced oil and to involve the government in petroleum exploration. On the hustings, Alberta CCF candidates, supported by national CCF leader Major James (M.J.) Coldwell and Saskatchewan Premier T.C. Douglas, also campaigned for public ownership of the province's electrical utilities and for a government-managed program that would provide Albertans with free hospital and medical care. Sensing that there was an especially strong interest in the public power idea in the Social Credit stronghold of rural Alberta, Manning had astutely put the question of private versus public power to the electorate in the form of a separate election day plebiscite. This permitted the Alberta premier much more political freedom during the campaign. When announcing that before determining policy his government was anxious to hear "the people's voice in the matter,"[94] Manning informed his election audiences that the disadvantages of government ownership seemed to outweigh the advantages; yet, he was careful to stress his commitment to rural electrification. The challenge to its petroleum policy, on the other hand, Social Credit confronted directly, although the emphasis was less on policy than on the more general "evil" of socialism. Manning frequently defended his government's oil policy by saying that $17 million in revenue had been collected from the industry without the expenditure or risk of 1¢ of the taxpayers' money.[95] Greater effort was devoted to warning that the CCF would put Albertans on "the slippery road of socialism down to communism."[96] In keeping with the extreme cold war rhetoric that dominated the North American press, Manning informed a Calgary audience that "the only difference" between Nazi and communist philosophies and the socialism of the CCF was "the degree of centralized control of industry by the government." He pointed out that Canadian communists supported the CCF because they understood that socialism effectively conditioned people to be more receptive to communism and ominously concluded, "You only need to look at geography to know why communists want socialist states in Western Canada."[97]

Joining the attack on the CCF's proposed petroleum policy, a group of independent oil operators purchased space in principal Alberta newspapers to give wider audience to a feature editorial published earlier by Carl Nickle in his *Daily Oil Bulletin*. The advertisement assured that "the quickest way of turning Alberta's current oil 'boom' into a 'bust' would be to elect a socialist government." Those who doubted were asked to reflect on the state of affairs in the CCF-administered province next door and to "contrast the huge oil program in Alberta with the small scale work in Saskatchewan, a neighbouring province which possesses millions of acres of potential oil land."[98]

Albertans proved receptive to the message. On 17 August 1948, they voted overwhelmingly for Premier Manning and his brand of Social Credit. Gaining 55.6% of the popular vote, the highest percentage achieved by any party in Alberta since 1909, Social Credit won 51 of the 57 seats in the Legislature. The CCF party retained the two seats it had held prior to the election and attracted only 19.1% of the province-wide vote. Two Liberals, one Independent and one Independent Social Creditor rounded out the opposition forces that Albertans sent to Edmonton.[99] Although five of the six newly elected opposition members hailed from Edmonton or Calgary ridings, small-town and rural Alberta remained almost entirely in the Social Credit fold. The plebiscite on public versus private power, however, produced a more ambiguous verdict. By a close margin, Albertans voted in favour of private power. Voters in Social Credit's rural heartland leaned in favour of a publicly operated system, whereas city dwellers, especially those in Calgary, registered greater support for private ownership.[100]

The 1948 election in Alberta represents a lesser, but still important, turning-point in Alberta's political history. The election confirmed acceptance of Manning's recast Social Credit party. Several "Douglasites" were returned, but they were relegated to the back benches in the Legislature where they never again comprised a voice that carried weight in the party caucus. The election also confirmed that the widening base of popular support for the CCF party in western Canada, following the party's 1944 victory in Saskatchewan, had been stemmed, at least in Alberta. With their tiny representation in the Legislature, the opposition parties would continue to be fatally handicapped in their search for public credibility. With its own house at last in order, faced by an opposition too small to be truly effective, and blessed with burgeoning oil revenues, Social Credit emerged from the 1948 election with its power consolidated beyond anything that the party had achieved since it had been first elected in 1935. The massive support conferred upon the Manning government, combined with rapidly expanding revenues generated from the oil boom, assured that economic and social policy emanating from Edmonton would be characterized by caution and limited debate, both in the Legislature and among the public at large. Guided by a stable, development-oriented government, Alberta was about to enter a period of unparalleled growth and prosperity.

With the election over, the government and the Conservation Board moved quickly to tidy up the Atlantic No. 3 mess. As soon as the fire at Atlantic No. 3 was extinguished, the Board issued an order on September 10 permitting all but the closed Atlantic wells to return to the production of their normal quotas, thereby ending the 80%-of-quota restriction that had been in effect for nearly three months. A new uncertainty, however, emerged from the background to preoccupy the Leduc producers, the Conservation Board and the government. The critical issue was that of liability. In Toronto, the *Financial Post* was quick to point out that "'almost endless' claims for damages are expected to be filed," and uncertainty about who were likely to be the winners and losers began to be reflected in share values.[101] It was anticipated that the unknown, but presumed, substantial cost of extinguishing the well would be charged against the Atlantic Oil Company, and that the company would suffer some kind of penalty for overproduction. It was also presumed that there would be royalty claims from the freehold owners of the wellsite, that there would be claims by farmers for surface damage, that operators, forced to restrict production, would seek compensation, and that other operators would bring forward the complex question of damage caused to a common reservoir.[102]

Anxious to deal with matters outside the courtroom, in early October 1948 the Atlantic Oil Company expeditiously submitted to the Conservation Board a memorandum that the company hoped might provide the basis of a negotiated settlement among the involved parties.[103] The government had a similar anxiety; it did not want a series of complex interlocking court actions that might slow further development. As long as Moroney and his men were still trying to place a permanent seal at the bottom of the Atlantic No. 3 wellbore, the final cost of taming the rogue well could not be determined, and this gave the Conservation Board time to move potential litigants towards an out-of-court settlement.

In the negotiations that continued through the autumn of 1948, Atlantic tried to put the best face on the situation. At the outset, the company affirmed that it was prepared to bear "the entire cost of this whole unfortunate occurrence."[104] Although the actual cost of fighting the wild well could be known precisely and the much smaller amounts to be paid for surface damage were not expected to pose great difficulty, the company presumed that the Conservation Board would take the responsibility for determining and mediating the question of subsurface damage. Atlantic presented an opening proposal, supported by an assertion of "the absence of any negligence or misconduct" on its part.[105] The company argued that if its No. 3 well remained shut for the time that it would take the foregone daily production quota to equal the volume of oil produced during the blowout, there was no good reason why the company should be penalized further by restricted production at its nearby No. 1 and No. 2 wells. Atlantic even went so far as to suggest that there was at least one positive side to the disaster, by pointing out that the experience and information gained would reduce the risk of a future accident.

Farmers' claims, including that of $25 from Mary Yaremko for the damage suffered to the washing hung on her clothesline, were settled with little dispute; however, the Leduc operators were not satisfied so easily. Their losses were of a different magnitude, and they were not as readily quantified. By December, after several informal meetings and the continuing exchange of proposals and counter proposals had failed to bring agreement, the government and the Conservation Board began to lose patience. The government's impatience was accentuated by a second oil well blowout. Although the Mercury-Leduc No. 1 blowout of November 15 was brought under control in just four and a half days, it served to remind that decisions made regarding the settlement at Atlantic No. 3 would be recognized as precedents.[106] To bring negotiations to a conclusion, the Conservation Board called representatives of the 11 producing companies in the Leduc field to meet in its Calgary office on January 26. There, with McKinnon; Goodall; the Board's senior engineer, G. Liesemer; the Board's legal adviser, C.E. Smith; and

Alberta's deputy attorney general, the oil company delegates pounded out an agreement.[107]

At the meeting, with the prospect of an arbitrary legislated settlement hanging over their heads, the oilmen recognized that compromise would have to be the order of the day. At issue was how the money in the Board-administered trust fund that had accumulated from the sale of oil produced by Atlantic No. 3 should be distributed and the penalty that should be exacted from the Atlantic Oil Company. After some debate, it was concluded that $100,000 would be set aside for each of the two nearest wells, Imperial-Leduc No. 48 and Leduc Consolidated No. 2, as compensation for the damage caused. The Conservation Board was directed to retain in the trust fund a sufficient amount to ensure that all the costs that had accrued in bringing the well under control could be paid, plus a further amount to be set aside for two years to meet whatever additional expenses that the Board might feel appropriate to charge, including the cost of any investigations or conservation measures that the Board might deem necessary.[108] To calculate the production penalty to charge against the Atlantic Oil Company, it was first determined that Atlantic No. 3 in its uncontrolled phase had overproduced to the extent of 565,195 barrels, or approximately 72% of the actual production. From this, it was determined that further production from the Atlantic Nos. 1 and 2 wells would be restricted to two-thirds of the production that they would normally be allowed until the foregone production and the previous overproduction were balanced. In addition, Atlantic was restricted from drilling further wells on the leasehold where its Nos. 1, 2 and 3 wells were situated.[109] The agreement reached at the January 26 meeting was then confirmed and reinforced by a special act of the Alberta Legislature. Assented to on 29 March 1949, the *Atlantic Claims Act* assured the Conservation Board of whatever additional powers it might need to carry the terms and the intent of the agreement.[110] More important, the Act shut off the possibility of legal challenge by forbidding anyone who remained dissatisfied from taking legal action against Atlantic or the Conservation Board "unless such a person [had] first obtained the consent in writing of the Attorney General."[111]

The *Atlantic Claims Act* brought to a close what had been a long ordeal for the Alberta government and for Board Chairman Ian McKinnon. Both could now direct their full attention to the two most substantial matters remaining on their respective agendas. The primary issue of concern to McKinnon and his Board colleagues was the growing problem of crude oil allocation in a situa-

TABLE 5.4 The Petroleum and Natural Gas Conservation Board Atlantic No. 3 Well Trust Statement of Revenue and Expenditure to 27 February 1953

Revenue

Sale of crude oil (949,229 bbls.)	2,818,584.01		
Bank interest	13,276.48		
			2,831,860.49

Expenditure

Costs incurred to control the well		946,645.95	
Royalties: gross royalty Rebus	352,323.00		
Family Imperial Oil Ltd. (re: royalty agreement with Atlantic)	254,351.41		
		606,674.41	
Damage claims: surface damage	27,447.45		
Leduc Consolidated Oil Ltd. (re: Leduc Cons. No. 2 well)	100,000.00		
Imperial Oil Ltd. (re: Imperial-Leduc No. 48 well)	100,000.00		
		227,447.45	
			1,780,767.81
Excess of revenue over expenditure paid to Atlantic Oil Company Ltd.			*1,051,092.68

SOURCE: ERCB Library—Atlantic No. 3 file.
* The first payment of $500,000 to the Atlantic Oil Co., made on 11 May 1949, was followed by four lesser payments, the last of which was on 27 February 1953.

tion where production capacity was growing dramatically faster than market demand. For N.E. Tanner and his cabinet colleagues, the most pressing issue to be dealt with was the question of natural gas exports. Probably no one was more pleased to see the Atlantic No. 3 blowout behind them than Frank and George McMahon. As Aubrey Kerr notes in his study of the disaster, the penalties levied against the Atlantic Oil Company might have been more severe.[112] With the question of blowout liabilities settled and cash in their pockets, the McMahons could concentrate on turning what had been a struggling lesser company in the Alberta oil scene into a much more significant player.

THE SEARCH FOR A PRORATION PLAN 1947–1951

If just one producer is responsible for determining the overall production rate from a reservoir, it can be expected that preferred conservation practice will be regulated by economic incentives. The need to assign production quotas to ensure good engineering practice or conservation is more acute when a field has a number of operators. Hence, the success of Alberta's checkerboard leasing system, in opening the door to more small producers, was a factor that made it necessary to have strong government monitoring of the development of each oil pool.[113]

The evolution of Alberta's postwar crude oil prorationing program between 1947 and 1951 can be seen to have happened in two stages. Stage one, from 1947 to 1949, was shaped by a situation in which the productive capacity of the province's wells was less than, or only modestly exceeded, market demand. In this environment, the Conservation Board went about its assigned tasks of promoting

Discovered in October 1948, the Redwater oil field brought immediate changes to the town's rural vista. Note drilling rigs in background. ERCB.

TABLE 5.5 Major Alberta Crude Oil Discoveries and Field Production, 1947–1951 (barrels)

Field	Discovery Date	1947	1948	1949	1950	1951
Leduc-Woodbend	Feb 1947	372,427	4,657,371	9,688,784	10,589,466	13,743,118
Redwater	Oct 1948	—	36,875	4,793,491	10,745,538	23,177,607
Joseph Lake	Mar 1949	—	—	35,858	168,855	727,936
Excelsior	Nov 1949	—	—	1,616	272,186	723,005
Golden Spike	Apr 1949	—	—	85,081	292,873	640,972
Stettler	May 1949	—	—	15,725	246,198	606,068
Big Valley	Sept 1950	—	—	—	10,215	155,580
Wizard Lake	May 1951	—	—	—	—	190,595

SOURCE: Alberta Petroleum and Natural Gas Conservation Board, *Alberta Oil and Gas Industry*, 1947–1952.

good engineering practice to ensure maximum ultimate recovery from each reservoir and ensuring that, in the typical situation of multiple resource ownership, all were able to produce their equitable share. Although the market condition was not a factor in the Board's calculation of production quotas, it did shape the industry's response. Small Alberta operators and independent companies were inclined to resist production quotas set in the name of good engineering practice. With the market beckoning, such companies sought to produce their wells at a faster rate and argued for higher allowables. In contrast, large operators, particularly the integrated companies, were more inclined to focus upon the longer term, and thus looked kindlier upon production rates that promised to bring maximum ultimate recovery. They could afford to wait to achieve maximum return. This classic divergence was exemplified before the Board on 9 February 1948. J.D. Gustafson, chief petroleum engineer for Imperial Oil, argued that the present Leduc production allowables might be too high and warned of the danger of too rapid a withdrawal from the reservoir. Speaking in the interests of the Globe Oil Company, Dr. J.O.G. Sanderson pointed out that shareholders had invested their money in the ventures of independents and wanted returns sooner rather than later, for no one could be sure how much oil might be worth in five years.[114]

Separating stage one from stage two was a short transition period that lasted from the spring of 1949 to December 1950. In February 1949, Alberta's light crude oil production potential surpassed

MAP 5.6   Major Oilfield Discoveries in Alberta, 1947–1951

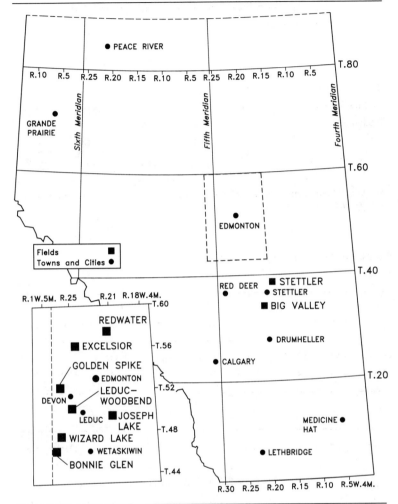

the requirements of the three prairie provinces for the first time (see Table 5.5). This capacity was largely a function of the development of one field—Leduc. But in late 1948, a prolific new field was discovered at Redwater, and this was followed in 1949 by the discovery of four additional reservoirs: Joseph Lake, Golden Spike, Stettler and Excelsior (see Map 5.6). Early in 1949, the region's crude oil refiners, by far the largest of which were Imperial Oil and the British American Oil Company, began to set crude oil purchase quotas based upon their projected requirements. Soon, some producers in the Leduc and Redwater pools were not able to sell all the

TABLE 5.6 Producing Oil Wells in the Turner Valley,
Leduc-Woodbend, Redwater and Other Fields,
1947–1951

Field	1947	1948	1949	1950	1951
Turner Valley	282	271	327	329	330
Leduc-Woodbend	28	167	351	519	800
Redwater	0	2	278	733	898
Other	138	155	286	414	710
Total (all Alberta)	448	595	1,242	1,995	2,738

SOURCE: Alberta Petroleum and Natural Gas Conservation Board, *Alberta Oil and Gas Industry,* 1947–1952.

TABLE 5.7 Crude Oil Prices, 1947–1951
($ per barrel)

	JAN	1947 APR	DEC	1948 JAN-NOV	DEC	1949 JAN	SEPT	1950 JAN-OCT	OCT-DEC	1951 APR
Turner Valley 33°	1.97	2.82	3.52	3.52	3.16	3.09	3.44	3.44	3.29	2.94
Gravity 64°	2.59	3.44	4.14	4.14	3.78	3.71	4.06	4.06	3.91	

	JUL	DEC	JAN-NOV	DEC	JAN-SEPT	DEC	JAN-OCT	OCT-DEC	APR
Leduc-Woodbend 39°-40° Gravity	2.67	3.45	3.47	3.00	3.00	3.20	3.20	3.05	2.61

	JAN-SEPT	DEC	JAN-OCT	OCT-DEC	APR
Redwater 35° Gravity	2.73	2.88	2.88	2.73	2.44

	SEPT-DEC	JAN-OCT	OCT-DEC	APR
Golden Spike 35° Gravity	3.18	3.18	3.03	2.59

SOURCE: Alberta Petroleum and Natural Gas Conservation Board, *Alberta Oil and Gas Industry,* 1947–1952.

allowable production assigned to their wells by the Conservation Board. More important perhaps, some producers were able to sell less of their allowable than others, and later entrants into the marketing sweepstakes, especially those who had just completed wells in new fields, often found it difficult to gain market access. In this environment, the issue of correlative rights, or fair market share, became a topic of debate.

The second stage began in late 1950 and lasted until November 1951 when the Conservation Board issued the first postwar proration order that took into account market demand as well as engineering factors. During these months, the productive capacity of Alberta's oilfields began to escalate beyond what the regional market could absorb. In this environment, the attitude of the producing companies was the reverse of that in stage one. With access to their own refineries and market outlets, the large integrated companies tended to be much less interested in the idea of sharing the available market. Independent companies, on the other hand, were inclined to be enthusiastic supporters of proration to market demand, without which they were not assured market entry and thus faced the prospect that their share of the resource would be drained by competitors who had better market access. Confronted with this situation, the Conservation Board was left with two possible courses of action. One alternative was to stand aside, in which case those companies with better marketing outlets would produce their quantities of oil up to any technical limits that the Board might set and the companies with inferior market connections would produce less, perhaps nothing at all. This approach was not considered acceptable, either in the interest of physical conservation, which is assisted by even rates of reservoir withdrawal, or in the interest of correlative rights. That the Board opted for the other alternative, proration to market demand, is not surprising. This course of action was consistent with its mandate and with the precedent set under similar circumstances in Turner Valley in 1938. What was different was that the situation in postwar Alberta was technically more complex and on a vastly different scale. Particularly, it was the Board's technical competence that would be put to the test.

The priority task that George Govier inherited on his appointment to the Conservation Board in February 1948 was to help formulate a fair plan of proration based upon sound engineering practices. His situation was, in some measure, eased by the Turner Valley experience. Prorationing was not just a theoretical concept— regulated crude oil production was accepted practice in Turner Valley. There, production was governed by Brown's plan of reservoir fluid withdrawal and allowables were set for individual wells to achieve a uniform rate of reservoir drainage. Although there were always operators prepared to argue, on technical or other grounds, why their particular well or wells should be assigned a higher allowable, the principle and the necessity of prorated production was almost universally conceded. Although Govier might not have had to fight, as did his predecessors, to gain acceptance of this essential principle, the task that he faced was daunting nonetheless. His challenge was more technically than politically complex. With the assistance of office engineer, G.E.G. Liesemer, he had to devise an acceptable overall plan for regulating production that was capable of taking into account differences in the reservoir characteristics of the several new fields but was still uniform in principle.

Proceeding from the initial efforts that Goodall had made in the fall of 1947, Govier, Goodall and McKinnon began meeting with Alberta oilmen in February 1948. The Conservation Board's ultimate objective was to determine for each oil pool the maximum efficient rate of production (MER) that would seem to offer the best assurance of achieving the greatest recovery from the reservoir in the long term. Both the industry and the Board recognized, however, the near impossibility of formulating an MER that had technical validity for oil pools that were in the earliest stage of their producing lives and for which reservoir data was still limited. There needed to be a temporary formula that offered a reasonably sophisticated means of estimating the maximum rate at which wells or pools should be produced, until such time as production and pressure history would allow the calculation of a valid pool MER. Govier's solution was to fashion a provisional maximum permissible rate of production (MPR) for each well based upon the following principal variables: acreage assigned to the well, average net thickness of the producing zone, pool average porosity, connate water (water present with oil and gas in rock pore spaces), shrinkage factor, expected recovery, expected producing life, and degree of development of the pool.[115]

In a subsequent address to the Canadian Institute of Mining and Metallurgy in Vancouver in November 1950, Govier outlined the technical rationale and the approach used to introduce the new MPR system.[116] The first essential task, he explained, was to identify which kind of drive mechanism was present in the reservoir. He hastened to remind his audience that, with the limited data provided from just a few wells in the discovery stage, this was not always an easy task. In support of this point, he might have added that nearly two years after the Leduc discovery there was still some controversy about the nature of the drive that existed in the D-3 pool.[117] Govier then proceeded to explain how the MPR approach was applied in each of the three main reservoir drive situations. Redwater was presented as a "a textbook example, of a water-drive field."[118] At a depth of about 3,100 feet, and under a pressure of 960 pounds per square inch, this field contained an undersaturated crude oil underlain by a huge volume of water. When a wellbore penetrates this type of reservoir, the body of compressed water expands and moves towards the region of pressure release, pushing the oil in front of it. Study had demonstrated clearly that the water-drive mechanism is efficient only when the rate of advance of the water front is even and slow.[119] To determine the rate at which the pool should be produced to avoid waste, Govier pointed out that it was necessary in the early life of the field to rely heavily upon the experience gained from similar types of reservoirs. Since the highest recoveries from water-drive reservoirs had been achieved when the total annual withdrawal was restricted to an amount not exceeding 2½ to 4% of the estimated volume of oil in the reservoir, the Board used this "experience" range as the base upon which to apply local variables and thereby determine the MPR for wells in the Redwater field.

Govier presented the Leduc D-2 pool as a typical example of a field with a solution-gas drive. Here, he explained, there was "no indication of underlying water or of an overlying gas cap."[120] This pool contained saturated crude oil at a depth of 5,300 feet under an original pressure of 1,760 pounds per square inch. In this circumstance, the driving mechanism is activated when a wellbore penetrates the reservoir and allows the reservoir pressure to drop to the point where gas starts to break out of solution from the oil. The expansion and flow of gas evolved from solution in the oil forces the displacement of the oil in the reservoir. Solution-gas drives provide less efficient displacement, and ultimate recoveries of only 20 to 40% over the eight-to-15-year life of a field were found to be normal. This suggested an annual maximum rate of withdrawal of 2½ to 3% of the estimated oil in place. Even though ultimate recovery from pools with solution-gas drives was known not to be as sensitive to withdrawal rate, the Board restricted production and based its MPR "on the 2½ to 3% figure adjusted to take care of local factors."[121]

The third type of displacement process that the Conservation Board engineers had to take into account in determining allowable production rates was the gas-cap drive. Rated somewhere between the solution gas and the water drive in efficiency, the gas-cap displacement mechanism is the downward, piston-like expansion of the overlying gas cap that occurs as the reservoir pressure declines. Like the water drive, the gas-cap mechanism is a rate-sensitive process and high efficiencies can be achieved only when the rate of expansion of the gas cap and the downward movement of the gas-oil interface is kept low. The other essential, one that had only been accepted after nearly 20 years of debate in Alberta, was the preservation of the gas cap that provided the driving energy. As with his other examples, Govier explained that, in the beginning stage of development of such a pool, one had to "be guided by the rates which elsewhere and in similar reservoirs, have yielded good results."[122] Experience seemed to demonstrate that an annual withdrawal of about 4% of the estimated oil in place would yield recoveries of 40 to 60% over a 10-to-15-year life. The Board used this general guide, qualified by local variables that included underlying water, to establish an MPR for the Leduc D-3 field. Govier was proud to point out the initial results. No production had been permitted from the gas cap, and wells with high gas-oil ratios had been assigned production penalties, so that, after 19 million barrels of production over three years, pressure in the D-3 had only declined from 1,894 to 1,795 pounds per square inch.[123]

Largely through Govier's efforts, by the late spring of 1948 the Conservation Board had in place an MPR formula to determine the maximum allowable production for wells in Alberta's new oilfields. It was a rational system, consistent with the best engineering practice of the day. Beyond this, it was a technically innovative response that departed somewhat from practice elsewhere.[124] It hinted at a level of technical expertise and suggested a quality of leadership that boded well for the future. Once set in early 1948, the Board held MPR rates for D-2 and D-3 wells in the Leduc field constant until 30 September 1950. Wells producing from the D-2 zone were restricted from producing more than 100 barrels of oil per day and those from the D-3 to 150 per day.[125] The latter limitation was soon refined to state not more than 150 barrels of combined oil and water production, thus penalizing wells with higher water-oil ratios that were less efficient in producing oil.[126] Apart from some further temporary adjustments necessitated by the Atlantic No. 3 crisis and the subsequent setting of an MPR of 125 barrels per day maximum for production from the Lower Cretaceous formation,[127] the Board's Leduc proration program functioned with typical but mild controversy. Imperial wondered if the allowable rate might not be too high, and some smaller producers were convinced that the rate was too low.[128] Real pressure upon the production regulation system did

MAP 5.7   Location and Cross-section of Leduc-Woodbend Field,
Nisku (D-2) Pool

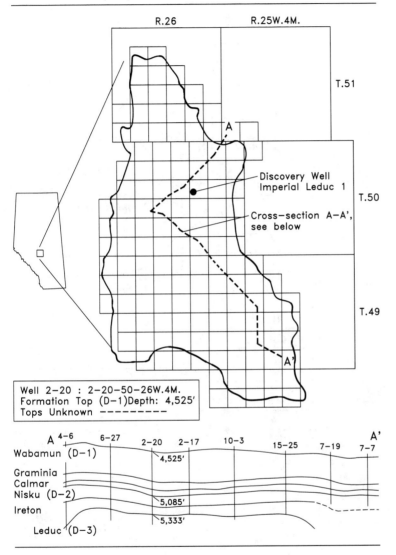

Well 2–20 : 2–20–50–26W.4M.
Formation Top (D–1)Depth: 4,525'
Tops Unknown ---------

not come until the spring of 1949, and it was the consequence of a
growing market strain. At the centre of the problem was the prolific
new Redwater field northeast of Edmonton. Discovered by Imperial Oil in October 1948, this field revealed a production potential
reminiscent of some of the big prewar discoveries in Oklahoma and
Texas. In February 1949, the light crude oil potential of Alberta oilfields passed beyond the requirements of the three prairie prov-

MAP 5.8   Location and Cross-section of Leduc-Woodbend Field, Leduc (D-3) Pool

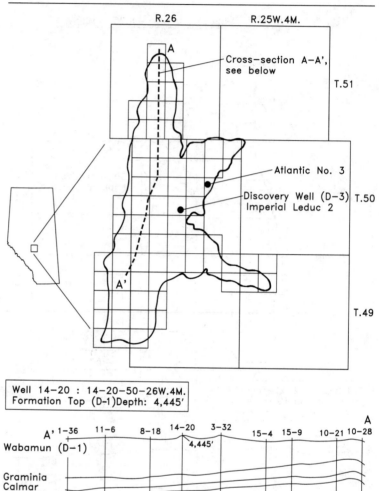

inces. Redwater's contribution to this situation is underlined in the field production figures for the week ending 13 March 1949. During this period, the average daily production of Redwater's 27 wells totalled 31,385 barrels, making Redwater the primary producing pool in Alberta.[129] This was nearly three times the daily production rate of 10,142 barrels from the 328 operating wells in Turner Valley, the province's oldest and now third-largest producing oilfield.

MAP 5.9   Location and Cross-section of Redwater Field, Leduc
(D-3) Pool

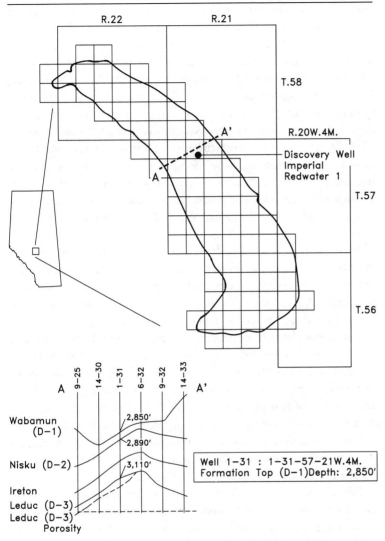

By coincidence, but nonetheless setting an appropriate backdrop
of urgency, Redwater's emergence as Alberta's most prolific oilfield
came during the week that the Conservation Board had earlier
scheduled a public hearing to seek producers' advice on the formulation of well allowables for that very field. According to the *Western Oil Examiner,* the March 9 hearing brought forth "spirited
argument," as various operators presented their views about what

allowable formula would be in the best interest of the Redwater field from the perspective of their own companies' situations.[130] Imperial Oil and Royalite argued in favour of a withdrawal quota based upon five barrels per day per foot of productive thickness up to a maximum of 400 barrels.[131] Anglo-Canadian and Home Oil objected on the grounds that such an approach did not take into account the varying porosity at different locations. Others, including British American and the Hudson's Bay Company, urged that a flat quota be set for all wells until field characteristics were better understood. Govier's compromise suggestion that a flat rate be set for all wells, with a bonus production in proportion to pay-zone thickness, found no support. In one area, there was consensus, and the hearing concluded with the expressions of concern about the danger of water intrusion. The general feeling was that production penalties should be fixed against any wells that started producing water.

Faced with such a wide difference of opinion, the Board reserved decision. Eventually, Redwater producers agreed to institute a voluntary program of maximum production rates as a temporary measure until May 1, by which time a Board-ordered conservation plan was expected to be ready.[132] Operators pledged to cut back production and not draw more than an average of 650 barrels per day from any well while Board and industry engineers continued their study. The verdict came on 23 April 1949. With its first Redwater production order, the Board set the maximum average daily allowable for each well at 550 barrels.[133] More than being just a simple reduction from the earlier voluntary quota, the new Board figure referred to combined oil and water production, thereby reflecting industry and Board concerns about water encroachment and support for the principle of uniform total fluid withdrawal.

For all the Board's efforts, its more sophisticated allotment, based upon preferred engineering practice, was largely academic. As spring passed, market limitations meant ever fewer wells were able to produce the maximum allowable that the Board had assigned. Companies with contracts to supply Imperial Oil's refinery found their wells limited to 300 barrels daily.[134] Although the British American Oil Company was taking a little larger daily production from the wells of its suppliers, all realized that with more wells coming on stream and no prospect of a significant increase in local market demand there would be continued pressure for reduced refinery allotments. Pipeline "market proration" rather than "conservation proration" was emerging as the dominant factor determining daily well production.

In this more complex environment, pressure began to mount to have the Conservation Board play an expanded regulatory role. Responsibility for prorating crude oil production to market demand was not a burden that the Alberta government and its Conservation Board were quick to embrace, at least without careful preparation. Anticipating, however, that it would eventually have to move in this direction, and with growing apprehension that Ottawa's recently passed *Pipe Line Act* might compromise the authority of the province's Conservation Board to regulate production, the Manning government moved on two fronts.[135] It engaged the continent's leading authority on oil and gas conservation law, Robert E. Hardwicke of Fort Worth, Texas, to come to Edmonton to review Alberta's oil and gas legislation, particularly the Conservation Act.[136] The call to Hardwicke came on Tanner's instruction. He had first met the Texan in early 1947 when visiting the Interstate Oil Compact Commission in Tulsa, Oklahoma, and was greatly impressed. In his diary, he described Hardwicke as "the best-informed lawyer in the industry I have ever met."[137] While Manning and Tanner waited to consult with Hardwicke, the Conservation Board initiated discussions with the Western Canada Petroleum Association (WCPA) to canvass industry opinion regarding possible amendments to the existing conservation legislation.

Hardwicke's visit to Edmonton in May 1949 was an important part of an ongoing fundamental reorganization of the administrative and regulatory environment in the resource sector. The Manning government had just dismantled the Department of Lands and Mines, creating in its place two new ministries: the Department of Lands and Forests and the Department of Mines and Minerals.[138] The latter could focus its administrative energy on the rapidly expanding petroleum industry, and to this department went the greater strength and continuity of political and bureaucratic resources, including McKinnon, the former deputy minister of lands and mines, and most of his key personnel. The split in civil administration did not extend to the political level, and the premier asked his trusted colleague N.E. Tanner to act as minister of both departments and thus continue his supervision of the entire resource field. Hardwicke's task was to help the Alberta government reconstruct the legislative foundation upon which the administration of the province's petroleum industry would rest. His attention was directed to: the *Public Utilities Act*, the *Pipe Line Act*, the *Repeal of the Natural Gas Utilities Act*, the *Pipe Line Regulation Act*, the *Gas Resources Preservation Act*; and *An Act To Amend the Oil and Gas Resources Conservation Act*.[139]

Of the draft bills under scrutiny, it seems to have been the proposed amendment to the *Oil and Gas Resources Conservation Act* that was most troublesome. Edmonton's main purpose in amending the existing Conservation Act was to prepare the way for the proration of oil production to market demand. Section 16 of the old Act simply gave the Conservation Board, with cabinet approval, the unqualified power "to control and regulate the production of petroleum." A more sophisticated administrative and legal instrument was now required. The difficulty inherent in the necessary first step of defining just what it was that was to be prorated is suggestive of the complexities faced by Govier and his colleagues. The problem was that, in certain situations, gas, oil and condensate can be produced in variable proportions from the same well.[140] In the first draft form, "oil" was defined as "crude petroleum oil and all other hydrocarbons regardless of gravity which are or can be produced from a pool in liquid form."[141] This definition drew Hardwicke's stern caution. He warned that if an act defined as oil "hydrocarbons which could be produced in liquid form" then

an operator of a well drilled to a pool which contained condensate gas could produce at a slow rate and his production would doubtless be gas or predominantly gas. He could produce at high rates, reducing pressure materially, and a good part of his production would be liquid (or oil) and the rest would be gas. The operator would have the power, to a great degree, to change at will the character of the hydrocarbons produced, and in that manner change the application of the laws and regulations to suit his own interests. I doubt whether that would be a desirable situation.[142]

Hardwicke's preferred definition was "Oil means crude petroleum oil and all other hydrocarbons which were, at the time of the discovery of the pool, in the liquid phase in the reservoir." He noted, in addition, that one of his colleagues, also a specialist in conservation law, believed that this definition should be strengthened by adding the final clause: "and which are produced at the well in liquid form by ordinary production methods." The definition ultimately agreed upon, and that appeared later in the new statute, was "oil means crude petroleum oil and all other hydrocarbons regardless of gravity which are or can be recovered from a pool in liquid form by ordinary production methods."[143]

This was but preliminary to the main areas upon which Hardwicke had been directed to concentrate his attention. Undertaking to prorate to market demand, to declare purchasers to be common purchasers and pipelines to be common carriers, was a potential legal minefield, and the Alberta government wanted the enabling legislation to be as tight as North American legislative and judicial experience would allow. Alberta's apprehension and purpose was largely shaped by the *Pipe Line Act* that had been passed by Parliament in April 1949. This Act assigned responsibility for the regulation of pipelines crossing national and provincial boundaries to the federal Board of Transport Commissioners and immediately aroused Alberta's acute sensitivity about matters of resource control and management. The province's misgiving was well represented by J.H. Blackmore, the Social Credit MP for Lethbridge, who stated during debate on the bill in the House of Commons:

> I am desirous that the bill should be entirely in accord with the wishes of Alberta. I hope that it will not be assumed that because Alberta has gone ahead and developed this oil somebody else should go in and run everything.[144]

A.L. Smith, the Conservative MP for Calgary West, addressed the two specific areas of concern. He urged that the St. Laurent government amend the *Pipe Line Act* so that all companies licensed under the proposed legislation could be declared not only common carriers, as was set out, but also common purchasers to ensure that all producers could sell their fair share. Smith also sought assurance that the bill not permit the subversion of the conservation efforts of Alberta's Petroleum and Natural Gas Conservation Board.[145] For the benefit of those House of Commons members outside western Canada who had little familiarity with the petroleum industry, federal Social Credit leader Solon Low sought to clarify the concerns raised by Smith with a hypothetical example.

> Let us say a pipe line company is organized and a pipe line built to convey gas out of the province of Alberta to the province of Manitoba. It begins to operate under this act as it now stands, and let us say it goes into the Kinsella field in Alberta and begins to purchase gas for transportation out of the province. The petroleum and natural gases [sic] conservation board find that the situation demands the return of that gas to the structure for conservation purposes, and they so order, but the gas is being purchased by the pipe line company under this act, and the dominion government through the Board of Transport Commissioners for Canada, is the only one that can regulate the movement of

that gas out of Alberta and into Manitoba. What we want to know is will the order of the conservation board apply? Will it not be interfered with? Will it be possible under the act as it stands for them to see that the gas will go back into repressure so as to make possible the optimum recovery of liquid petroleum or will it be possible for the company to flout the order of the board and say, "We are going to take this out because the act says we can"? That is the position we are in, and we want to know definitely where we stand.[146]

Although Prime Minister St. Laurent was prepared to affirm his view that provincial resources were controlled by the legislation of that province, he would not countenance amendment to the bill.

Hardwicke's task was to help construct Alberta's legislative counterattack. In rendering his advice, Hardwicke drew not only upon his own expertise but also upon the experience of various informed Texans, including William Knode.[147] When the draft legislation was finalized, the Manning government, in keeping with its cautious but prudent approach, sent it to the industry for comment.

On 23 and 24 June 1949, 51 individuals, representing almost all the larger oil companies, including Royalite, Imperial, Standard Oil of New Jersey, British American, California Standard, Socony Vacuum, McColl-Frontenac and Shell, plus certain banking and investment houses, met under the auspices of the Western Canada Petroleum Association to consider the government's impending legislation. After listening to the review by the association's legal committee, one member was quick to note that the proposed legislation seemed to be patterned closely on that in Texas, and he expressed the opinion that, given the extremely different situation in Alberta, such legislation might not be appropriate. Others took the view of "the less government intervention the better," and proclaimed their direct opposition to prorationing.[148] In varying degrees, most of the oilmen were sympathetic to prorationing, and even the less enthusiastic accepted the conclusion of D.P. McDonald and J.C. Mahaffy, both of whose experience included Turner Valley litigation and who were now deans of the legal section of Calgary's petroleum fraternity, that "the legislation would be enacted whether the Association approved it or not," and that it would be better to work with the government and suggest amendments.[149] Taking this cue, the oilmen passed motions recording the Association's general approval of the draft bill's provision for the "proration of oil to the reasonable market demand" and the right of the Board to declare any person who purchases oil or gas from any pool in the province to be a common purchaser. The oilmen also asked for one key amendment, an amendment they had sought since the original 1938 Conservation Act. They directed the government to make provision in the revised Act for a much broader right of appeal to Board decisions.[150]

That the draft amendments to the Conservation Act so easily gained the Association's general approval is not surprising, for the exercise was really only a final formal review. Key people in the industry had been consulted along the way, and there were no surprises being sprung. Although they might have been hopeful, they were probably not surprised when the *Act To Amend the Oil and Gas Resources Conservation Act* assented to on July 7 did not contain the cherished expanded right of appeal.[151]

Changes in the revised Act were concentrated in two areas. The stated objective of the 1938 Act "to effect the conservation of oil resources and gas resources" was expanded. Added were two key supporting objectives: "to prevent the waste" of the province's oil and gas resources and "to give each owner the opportunity of obtaining his just and equitable share of the production of any pool."[152] To effect the latter task, section 16 of the old Act was completely overhauled. Thirteen subsections were added, giving the Board the power to fix allowable provincial oil production based upon the "reasonable market demand as determined by the Board" and to allocate the provincial allowable "in a reasonable manner among the pools in the province by fixing the amount of oil which may be produced from each pool without waste."[153] Further, the Board was authorized to prorate the production of oil allocated to each pool among the producers "for the purpose of giving each producer the opportunity of producing or receiving his just and equitable share of the oil in the pool." In support of the proration objectives, other subsections added to section 16 gave the Board the power to declare pipelines to be "common carriers," and thus prevent a carrier from discriminating in favour of its own or anyone else's oil or gas. Similarly, the Board was empowered to name any purchaser of oil or gas produced in the province a "common purchaser." So-named, a purchaser could not discriminate in favour of his own production, or one producer over another in the same pool, or between pools in the province designated by the Board.[154] Also added under section 16 was a new subsection that specifically identified the only acceptable uses of natural gas as being: gas lift in reservoirs, repressuring, recycling, pressure maintenance, light or fuel. Any other use required a permit from the Board, and this could only be obtained by demonstrating that the proposed use was in the "public interest." To consolidate the Board's new presence in the pipelines area, the government also amended the *Pipe Line Act* to transfer from the Department of Public Works to the Conservation Board the responsibility for approving pipeline construction proposals.[155] In addition, the Board was authorized to make such orders and regulations as might be necessary to control the operation of any pipeline in the province.

Rounding out the amended Conservation Act was a more wide-ranging yet precise set of provisions that enabled the Board to fulfil its original priority mandate: the conservation and prevention of waste of natural gas resources. These provisions were also necessary to permit the Board to establish and supervise "measures to be taken insuring the orderly purchase, sale and transportation of gas," a responsibility acquired with the repeal of the *Natural Gas Utilities Act*.[156] More important, the Board required authority to manage an important new responsibility for approving any proposal for removing natural gas for sale outside the province, a duty assigned by the *Gas Resources Preservation Act*.[157] Finally, as a modest response to one of the oilmen's most long-standing criticisms, and perhaps in anticipation that its prorationing efforts might bring forth vigorous legal challenge, the government, while not broadening the grounds of appeal against Board decisions, did define much more precisely the process of appeal.[158]

With the amended Conservation Act in place, Premier Manning and Tanner stepped back to give industry opinion on crude oil prorationing still more time to coalesce while the Board continued to set maximum permissible rates of production for Leduc and Redwater wells that producers seldom reached for want of a sufficient market. As the industry deferred and continued through the autumn of 1949 with voluntary market-driven allowables superimposed upon the Board-inspired MPR system, Edmonton moved again to bolster the province's conservation legislation. This time, the purpose was to consolidate regulatory authority firmly within the embrace of the Conservation Board. To this end, the frequently amended *Oil and Gas Resources Conservation Act* of 1938 was completely rewritten. This involved more of a rearrangement than a deletion of the original clauses, all the important 1949 amendments were retained, and a number of new elements were added. In January 1950, Board Chairman McKinnon sent a copy of the draft legislation to the Western Canada Petroleum Association for the industry's consideration.[159] Through February and March, industry delegations conferred with Board members and Tanner. The more contentious matters having been debated in 1949, points of concern at this stage were relatively modest and agreement was reached without much difficulty.[160] With just a few minor revisions, the new conservation Act was tabled before the April 1950 sitting of the Alberta Legislature.

The *Oil and Gas Resources Conservation Act, 1950*, differed from the 1938 Act and the revised Act of the previous year in two important ways. First, all the important provisions of the Oil and Gas Wells Act were incorporated into the new statute, and the former Act, which had largely directed Board activity in the field, was repealed. This change represented more than just a transfer from one Act to another. Although the Board was still charged with supervising the same regulations in the field as before, in the changeover it gained the right to formulate regulations governing the drilling, completion and abandonment of wells as it might see fit. Under the former legislative framework, regulations governing oil and gas wells emanated directly from the Alberta cabinet, and the Board merely shared an advisory role with the Department of Lands and Mines.[161] Second, in addition to having this specific regulatory authority and initiative, the Board was given a significant general regulatory authority that it had not possessed before. Formerly, the Board was authorized to issue orders and regulations necessary "to control and regulate the production of petroleum." This section of the old Act was rewritten to say that the Board, with cabinet approval, might "make such just and reasonable orders and regulations as the Board deems requisite to effect the intent, purpose and object of this Act."[162]

Refashioned from the more limited and blunter instrument of 1938, the new Conservation Act gave comprehensive authority to the three men charged with administration of Alberta's oil and gas industry. Assented to on April 5, the *Oil and Gas Resources Conservation Act* was scheduled to come into force on 1 June 1950.

With its reconstructed legislative foundation in place, the Alberta government and the Conservation Board waited for industry to take the critical first step along the now shortened path to market prorationing. In preparation for the anticipated chain of events, McKinnon and Govier began discussion with the WCPA regarding the adoption of an MPR formula that could be used to determine well production in all the new postwar pools.[163] After some weeks of deliberation, the association's technical committee concluded that "the adoption of such a formula was not necessary at this time."[164] The committee chairman, Imperial Oil's Field Superintendent V.J. Moroney, explained that his committee felt that "the Board's suggested formula was impractical" and not suitable for "universal application." While association members contemplated appropriate production formulas, the situation in the field grew bleaker. Carl Nickle, oil editor of the *Calgary Herald,* wondered about the purpose of calling hearings to hear representations "from well owners and other interested persons with respect to the fixing of maximum efficient rates of production" when "the conservation quotas so established by the board following the hearings have no relationship to the market allowables set by the purchasing companies."[165] The Conservation Board maximum production rate set for the Leduc-Woodbend field was 115 barrels per day for wells producing from the D-3 zone and 75 barrels daily for D-2 wells. Imperial Oil's market quotas were 60 and 45 barrels daily from D-3 and D-2 wells respectively. The maximum rate set by the Board for Redwater wells was an average of 225 barrels daily over the calendar month, but wells delivering to Imperial were governed by a market quota of 60 barrels daily.[166]

Finally, in late August, one of the independent producers, the Continental Oil Company, broke ranks and made the question of "equitable market proration" a public issue. Managed by Alberta pioneer oilman Fred A. (Sugar) Schultz, Continental was one of the more aggressive and successful of the Turner Valley companies to move into central Alberta after the Leduc discovery. Schultz had first made his presence known in the Leduc area by ignoring the industry-wide boycott of the Alberta government's second cash and royalty bonus auction of crown petroleum rights. To his competitors' chagrin, Schultz bid for and obtained what was believed to be one of the most attractive properties. From this success, he had moved quickly. With the message, "You supply the land and I will drill wells at my own expenses, and you and I will share production equally," Schultz approached farmers with freehold petroleum rights and acquired additional Leduc acreage.[167] For small independent companies such as Continental, desperate for drilling capital, the ability to market oil from producing wells was always crucial. And by the summer of 1950, this ability seemed precarious. In a letter to the Conservation Board, the text of which was released to the press, Continental asked the Board to name Imperial Oil a common purchaser of oil from the Leduc-Woodbend field.[168] The company remarked that, over the preceding 11 months, other producers had been able to sell more oil per well. Continental was under contract to sell the bulk of production from its 15 Leduc wells to the British American Oil Company, but British American would not purchase as much per well as Imperial did from its contract companies. According to Schultz, had his company been able to sell to Imperial, its deliveries over the period would have been increased by 46,167 barrels and its revenue by approximately $140,000. But it was more than the loss of revenue that was the focus of the company's concern. If this situation was allowed to continue, Continental pointed out to the Board, its oil reserves would be drained by offsetting wells that were able to produce and sell a greater quantity of oil. Continental called upon the Conservation Board to take appropriate action, consistent with the purpose and object stated in section III of the Act, namely, "to give each owner the opportunity of obtaining his just and equitable share of the production of any pool."[169] The intent of Continental's application to have Imperial declared a "common purchaser" for the Leduc-Woodbend field was that all wells producing from a given producing formation in the field would be permitted exactly the same market quota based on the overall demand of all purchasers. British American and other purchasers would obtain their share of output from the common purchaser.

With the desired formal application in hand,[170] the Board sent a notice to all producers and purchasers of oil from the Leduc-Woodbend field that on 19 September 1950 it would hold a hearing to consider Continental's request. On the evening before the hearing, 51 oilmen, representing all the bigger companies active in the province, but not including the Continental Oil Company, met again to discuss and decide upon the official position that the Western Canada Petroleum Association would present at the hearing.[171] J.G. Spratt, WCPA president and former Department of Lands and Mines engineer, announced to the members that the Board of Directors had studied the report of the Association's technical committee and recommended that it be submitted to the Conservation Board subject to the approval of the general meeting. Perhaps it was appropriate—since his company's lower quota had precipitated Continental's application—that the spokesman for the British American Oil Company rose to speak first. He prefaced his remarks with the comment that his "company [had] always been in favour of proration to prevent waste," and that British American "would certainly not oppose any application [a company] made which might result in their obtaining a fair share of the available market for their particular type of crude."[172] Nevertheless, the company had two concerns that it thought WCPA members should hear. British American expressed the concern that prorationing might constrain refiners from getting the kind of crude that they could use most economically and that the purchaser's freedom of choice offered the best means of enlarging the market for Alberta crude oil. Second, the British American representative wondered whether or not it made sense "to recommend to the Board the adoption of a proration plan which according to our [WCPA] official legal opinion, is beyond the competence of the Board."[173] These observations did not overly disturb the great majority who were in favour of proration. The main debate centred about a suitable proration formula. On this question there was sharp difference. In the end, the meeting voted first in favour of the principle of province-wide prorationing and second, with some abstentions and opposing votes, in favour of accepting and submitting the technical committee's proration report to the Conservation Board. The next day, the Board listened to F. Schultz, general manager of Continental, give evidence in support of his company's application and J.H. Hamlin, divisional solicitor for Imperial Oil, speak in opposition.[174] It was apparent to McKinnon and his colleagues that there was still great diversity of opinion, and they decided to hold a second hearing to begin on October 23. Interested parties were asked to present their submissions to the Board in advance.

In the interval, the Conservation Board took another important step in the evolution of its MPR approach to determine allowable production. On 1 October 1950, the Board issued new "conservation allowables" for wells in central Alberta oilfields. Reflecting that sufficient time had elapsed to permit a more sophisticated understanding of reservoir behaviour in these fields, the new maximum production rates were expressed as a daily average rate of "clean oil" production. Clean oil was defined as that produced without significant amounts of natural gas or formation water, and penalty factor charts (gas-oil ratios and water-oil ratios penalty tables) were issued showing the reduction in clean oil production rates for increasing gas-oil and water-oil ratios.[175] Even though the "clean oil" initiative represented an approach that existed more on paper than in practice while market demand quotas remained substantially below MPR allowables, this served as a clear signal to the industry that the Board intended to relate production rates to total withdrawals of all fluids.

Also before the October hearing, international events conspired to lend additional urgency to settle upon a plan that would ensure the orderly marketing and further development of Alberta's oilfields. Just as the Board concluded its initial hearing on the Continental application, United Nations troops landed in South Korea. As the Cold War entered a new phase in the North Pacific, uncertainty about continental petroleum supply, especially in the Pacific Northwest, began to materialize.

The public inquiry regarding the advisability of restricting the production of oil by proration based upon market demand was held on October 23 at the Calgary court-house. Here, McKinnon, Govier and Goodall listened once more to the industry's ideas on how the available market for oil produced in Alberta could be shared among the several producing pools, and how the share of market allocated to each pool could be rateably distributed among the producing wells in each pool. The WCPA took the opportunity both to request that the Board introduce market demand prorationing and to recommend a formula that it believed could accomplish this objective. Under its proposal, the market demand for the period in question was determined, and from this demand was subtracted the production of all marginal pools and the allowable MPR production from new pools. The Association proposed that the remaining market be allocated among the remaining wells on the basis of three factors. First, a "floor" production of 25 barrels per day should be allocated to each prorated well to cover operating expenses. Since all the new fields had uniform 40-acre spacing, it was suggested that the 25 barrels could also be considered an acreage factor. Second, 70% of the remaining market should be awarded to wells on the basis of "so many barrels per foot of hole." This also was an economic factor intended to provide for the return of development cost and the encouragement of further exploration. The third factor governing the assignment of the final 30% of the remaining market was each well's MPR, which reflected both reserves and the ability of the well to produce.[176]

The WCPA plan was challenged by 16 independent operators with wells in the Redwater field. Headed by Pacific Petroleums Ltd. and Atlantic Oil Company Ltd., the independents argued in a joint submission that the current situation did not allow them their just market share, and to remedy their plight they proposed an "Alberta formula" as an improvement on the WCPA proposal. This formula rested upon the same three factors, but with different weightings given to each. Rather than a basic allowable for each well, they wanted a basic allowable that varied in accordance with the total market allowable, thus reducing the "excessive fluctuations in the barrelage [sic] dedicated to both the depth factor and the MER factor...."[177] These operators also felt that the depth factor proposed in the WCPA plan penalized producers from shallow fields. They argued that the depth factor was not worth twice the MER factor and submitted that the two factors should be given the same weight.

In its presentation, the Royalite Oil Company asserted that a proration plan was desirable, and in this regard it was consistent with the position taken by all the companies appearing at the hearing. Beyond recognizing the necessity of the Board's role in assuring that each producer got his just market share and seeing that this was done without creating injury to a common reservoir, Royalite expressed grave doubt about any plan that called upon the Board to give substantial weight to economic factors. Royalite held that "maximum efficient rate of production is the pertinent factor to be considered in all proration where allocations are granted to meet market demand."[178]

Imperial Oil also focused on this essential point in its submission. Conscious of its commanding presence in the Alberta oil industry, Imperial's statement was diplomatic yet revealing. It endorsed the WCPA plan, which it recognized was "a compromise of many conflicting views on the subject" and was being presented as "an initial approach which the Association considers worthy of trial."[179] That Imperial would offer its general sanction was never in doubt. Even if Imperial had been inclined to resist, it was apparent to the company that the forces pushing in favour of prorationing were too strong to be denied; moreover, given the position of its field superintendent V.J. Moroney as chairman of WCPA's technical committee, Imperial had presumably played an important indirect role in shaping the Association's proration document. What might not have been anticipated was that Imperial qualified its general approval with a tactful but fundamental criticism of both the WCPA and "Alberta" proration plans. Imperial argued that both formulas gave excessive weight to economic factors, and in doing so violated both an essential principle of petroleum conservation and an essential principle of the free enterprise ethic. Presenting its corporate philosophy to the Board, Imperial explained that

> Fundamentally we believe that the business of exploring for and developing crude production is the employment of risk capital. In undertaking a drilling operation an experienced operator knows that he may either find no oil at all, or he may find marginal production, or he may find prolific production. These results represent the success and hazards of the risk incurred. If the operator is successful in his gamble and finds prolific production, he should be afforded the opportunity to produce the prolific wells at much higher rates than other less productive areas. In short, the operator should be allowed to produce commensurate with his ability to produce. *Provided, however,* that he does not produce at wasteful rates (exceeds MPR) and provided correlative rights are recognized by rateable taking in each Pool.[180]

At the conclusion of the hearing, McKinnon announced that the Board would call a "nomination meeting" in November to enable the oil purchasers to inform the Board of their future crude oil requirements. The province's new crude oil proration plan would be announced shortly afterward.[181]

The long-debated and much-anticipated proration plan was announced on November 29 and scheduled to come into effect on December 1. It was propitious timing. Canada's first major crude oil pipeline was on the verge of completion and scheduled to begin delivery of Alberta oil by year-end. Initiated by Imperial Oil, the Interprovincial Pipe Line from Edmonton to Superior, Wisconsin, was capable of handling up to 90,000 barrels a day. From Superior, the oil was shipped by tanker across the Great Lakes to Imperial's Sarnia refinery. This meant a significant expansion of the now-to-be-shared market, although this benefit was tempered by the anticipation that access to this more competitive market was going to mean lower per barrel prices.[182]

The Petroleum and Natural Gas Conservation Board presented its approach as "a simplification of the procedures inherent in both the Western Canada Petroleum Association and the 'Alberta' plans."[183] The Board plan, which gave greater weight to engineering factors, represented a strong move in the direction of the philosophy articulated by Royalite and Imperial Oil. Govier and his colleagues combined the idea of a "depth factor," which had been proposed as a measure to compensate for the pay-out of drilling costs, and the concept of a "floor," which had been presented as compensation for operating expenses, into a single factor. Termed an "economic allowance," this factor provided an allowance for individual wells within the various producing pools of 30 to 50 barrels per day. The precise allowance was indicated graphically (see Figure 5.3), as a function of the average well depth in each pool. Production beyond the "economic allowance" would be assigned on the basis of maximum permissible rates (MPRs). The remaining available market would be allocated among pools according to their average well and pool production characteristics.

The operation of this two-tiered allocation system required first that the Board obtain "purchasers' nominations" each month for the crude oil produced in the province and then that it hold a public hearing to consider the nominations from which the "provincial demand" for light, medium and heavy crude oil could be determined. Once the market demand was established, this could be expressed as the "provincial allowable." The provincial allowable production was then divided among the appropriate pools according to the

FIGURE 5.3 Economic Allowance Chart, Schedule No. 1, 1950
Proration Plan

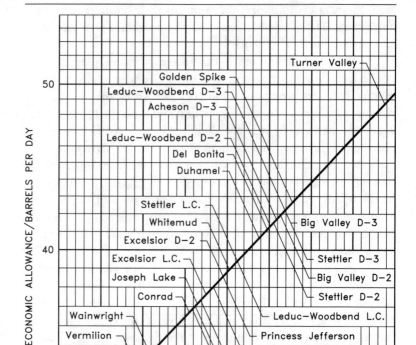

type of crude oil produced, the economic allowances of wells in each pool, and the pools' MPRs. Next, each pool's allowable, adjusted for the incapability of some wells, was distributed equally among the wells in the pool. The resulting maximum allowable rate per well for each pool was then published in a Market Demand Order (MD) issued by the Board following cabinet approval, a copy of which was sent to all operators in the province.[184] Given limited demand, however, the allowable rate assigned by the Board for wells in many pools was substantially below the MPR rate. Except for an adjustment in 1957, this plan, employing the MPR as the proration

FIGURE 5.4 Proration of Crude Oil to Market Demand

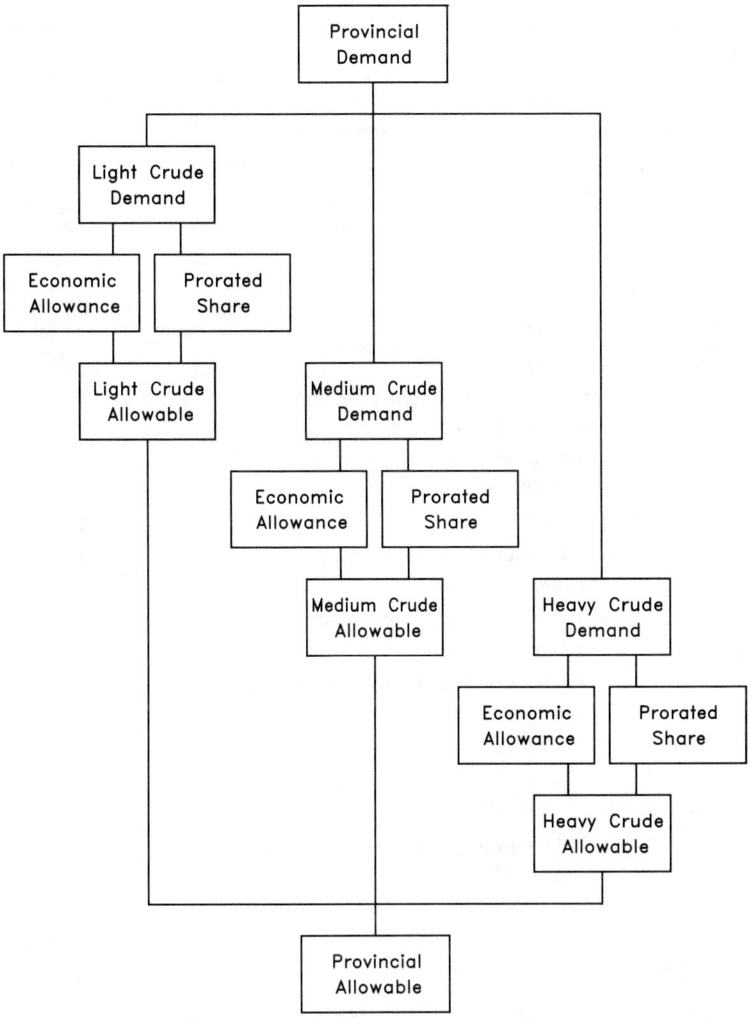

base and well economic allowance as an overriding factor, continued with modest modifications until 1 May 1965.

Set at 81,855 barrels per day, the provincial allowable was divided among 21 fields and pools, and on 1 December 1950 Alberta oil wells for the first time began production on a "rateable take" basis (see Table 5.8). For the independent producers, this represented a critical turning point in the history of the postwar petroleum industry in Alberta. They did not gain the economic emphasis they would have preferred in the Board's allocation formula, but

TABLE 5.8 First Market Demand Order Issued by the Petroleum and Natural Gas Conservation Board for December 1950

Field or Pool	Number of Barrels Per Day	"Clean Oil" Daily Average	Allowable Per Well Maximum in any 24 Hrs.	Penalty Subject to Gas-Oil Ratio	Factor Water-Oil Ratio
Leduc–Woodbend Field					
Lower Cretaceous Pool	620	45	60	yes	yes
D-2 Pool	9,610	49	65	yes	yes
D-3 Pool	14,970	53	70	yes	yes
Lloydminster Field	3,785	45	65	yes	—
Redwater Field	35,970	53	70	—	yes
Turner Valley Field	8,600	(individual well production governed by separate Board Order allowables calculated on the basis of 25 reservoir barrels per day)		yes	yes
Campbell Field	470	44	50	yes	yes
Conrad Field	330	(no well limit subject to good production practice)		yes	yes
Del Bonita Field	20	(no well limit subject to good production practice)		—	—
Dina Field	45	(no well limit subject to good production practice)		—	—
Excelsior Field	960	45	60	—	yes
Golden Spike Field	590	122	140	—	yes
Joseph Lake Field	1,085	43	60	—	yes
Princess Field	400	(no well limit subject to good production practice)		—	—
Stettler Field					
Lower Cretaceous Pool	125	43	60	—	yes
D-2 Pool	670	49	65	—	yes
D-3 Pool	420	48	65	—	yes
Taber Field	600	(no well limit subject to good production practice)		—	—
Vermilion Field	200	(no well limit subject to good production practice)		—	—
Wainwright Field	50	(no well limit subject to good production practice)		—	—
Whitemud Field	175	45	60	yes	yes
Discovery wells and others in fields or pools not designated	2,160				
Total Provincial Allowable	81,855				

SOURCE: ERCB Minutes, 22 November 1950, Order No. MD 1.

they did gain assured market access. In the situation where productive capacity exceeds market demand, and this would continue to be the case for the next 20 years, market demand prorationing meant that correlative rights were protected and that the small independent producer would continue to play an important role in development. By assuring that the lesser independent companies would be able to market their equitable share, prorationing gave them access to development capital they would not have had otherwise. A proven reserve plus a market quota, or even a promising property, combined with the assurance of market access if drilling proved successful, was sufficient collateral to attract bankers and investors. The Canadian independent companies, Atlantic, Western Decalta, Hudson's Bay Oil and Gas, Home, Dome, Calgary and Edmonton Corporation, Canadian Oil Companies, and others competing with the major companies entering Alberta after 1947, grew upon the proration foundation. At the same time, by permitting the Board to maintain uniform reservoir withdrawals more effectively, the proration formula promoted physical conservation in the public interest. In its various phases, the evolution of prorationing had extended over four years. It had been a difficult process, worked out with imperfect scientific knowledge at the intersection of private and public interest.

The Natural Gas Export Debate

W HILE IAN MCKINNON and his Board colleagues were concentrating upon the management of the increasing Leduc
and Redwater crude oil production, Tanner and his cabinet colleagues were increasingly preoccupied with the natural gas
question. The provincial government was anxious to see the expansion of its natural gas industry, but there was little incentive for
further development unless an export market could be secured.
However, the Alberta public was hostile to the idea of export commitments, unless it could be established that there were sufficient
reserves to meet their long-term local requirements. The question of
gas exports was a delicate matter, complicated not only by competing industry interests but also by a charged political history that
went back to the 1920s.

In a few areas, drilling had been undertaken to develop gas fields.
This was the case with the Medicine Hat, Bow Island and Foremost
fields, which had been developed to supply natural gas to Calgary
and other southern Alberta communities. Similarly, the Viking-
Kinsella field in the east-central part of the province was developed
to serve Edmonton and central Alberta towns as far south as Red
Deer. By far the greatest amount of drilling was done in search for
oil, but this also had led to several important natural gas discoveries, particularly at Turner Valley and Jumping Pound in the Alberta
foothills and at Princess on the southeastern prairie. Elsewhere on
the plains, there had been a number of smaller discoveries. Following World War II, an interest in the possible production of synthetic
gasoline from natural gas led to a widened and successful search for
additional natural gas reserves. Drilling in the Pakowki Lake region
of southeastern Alberta, McColl-Frontenac Oil Co. Ltd. and the
Union Oil Company of California found four dry gas fields: Pen-

dant d'Oreille, Manyberries, Black Butte and Smith Coulee. Also active in the southeast, California Standard Co. undertook an extensive drilling program in the Princess region and established the natural gas potential of the Foremost, Dunmore and Brooks-Bantry areas. Further north, exploration work by Imperial Oil extended the Kinsella portion of the Viking-Kinsella field and established the potential of the nearby Provost field. Thus, on the eve of the Leduc oil boom, the considerable extent of Alberta's natural gas resource was also beginning to be apparent.

Although the number of companies involved in natural gas exploration dwindled after Leduc, a new interest in the longer term potential of Alberta's seemingly abundant natural gas supply began to materialize. This interest was given a strong push with the release in April 1948 of a "Special Report" on the "Natural Gas Reserves of the Prairie Provinces" prepared for the federal Department of Mines and Resources by Dr. G.S. Hume, of the Geological Survey of Canada, and A. Ignatieff, of the Bureau of Mines. In their report, Hume and Ignatieff divided their reserves estimate into three parts, with the explanation that different types of gas reserves estimates varied in precision. In the "proven" category, where assessment could be based upon sufficient drilling and production tests, they estimated reserves of 1.433 trillion cubic feet. In less developed areas, but where there had been enough drilling to use the "porosity-area" method of evaluation, they arrived at an estimate of 2.185 trillion cubic feet of "probable" natural gas reserves (see Table 6.1). Although a numeric value could not be assigned to the "possible or potential" reserves category, on the basis of general geological appraisal, Hume and Ignatieff were optimistic. "There can be no doubt," they assured, "the potential gas reserves in Alberta are exceedingly large."[1] In their view, given the favourable geological conditions, all that was required was a large market and a reasonably attractive price to stimulate a vigorous search that would certainly lead to the discovery of vast additional reserves. Hume's assessment carried great weight; his western Canadian experience went back to early Turner Valley days, and he was highly regarded in industry and academic circles. The release of a supplementary report in December 1948 lent further credence to this optimistic evaluation. In light of extensive new drilling data gathered since their first publication, Hume and Ignatieff increased their estimate of proven natural gas reserves in Alberta to 1.46 trillion cubic feet and of probable reserves to 2.8 trillion cubic feet.[2]

Hume and Ignatieff's positive report coincided with a growing Ontario interest. The natural gas reserves in southwestern Ontario were near exhaustion, and the main distributor, the Union Gas

TABLE 6.1 Proven and Probable Natural Gas Reserves in Alberta, April 1948

	Proven Reserves		Probable Reserves	
AREA	RESERVE ESTIMATE (BILLIONS OF CUBIC FEET)	AREA		RESERVE ESTIMATE (BILLIONS OF CUBIC FEET)
Viking-Kinsella-Fabyan	994	Foremost		9
Medicine Hat-Redcliff	78	Pendant d'Oreille		260
Bow Island	25	Manyberries		31
Foremost	26	Black Butte and Pinhorn		39
Brooks	11	Smith Coulee		2
Vermilion	9	Princess		405
Turner Valley	290	Dunmore		8
		Hanna		12
		Provost		52
		Elk Point		8
		Jumping Pound		920
		Leduc*		389
		Athabasca		16
		Battleview		18
		Lloydminster		15
Total	1,433	Total		2,185**

Alberta Proven and Probable

Proven	1,433
Probable	2,185
Total	3,618

SOURCE: G.S. Hume and A. Ignatieff, "Natural Gas Reserves of the Prairie Provinces," Ottawa: Department of Mines and Resources, April 1948, p. 10.
* Does not include gas dissolved in the oil in D-2 and D-3 zones.
** Rounded volume.

Company, was in an almost desperate search for a new supply. In 1944, Union had contracted with Panhandle Eastern Pipeline Co. in the United States for the sale of natural gas by way of a pipeline connection to be constructed under the Detroit River. The U.S. Federal Power Commission (FPC) approved the contract in 1946, but placed such severe restrictions on Panhandle's deliveries that only negligible amounts of natural gas were exported to the Canadian company during two weekends in 1947.[3] Having cut off service to 13,000 domestic customers since 1940 and having spent over $1.5 million constructing facilities to receive the gas,[4] the Union Gas

Company, after repeated unsuccessful attempts to persuade the FPC to ease its restrictions, called upon the Canadian government to intervene on its behalf. The prospect of a growing natural gas shortage in the nation's industrial heartland did arouse Ottawa's concern, and the federal government was moved to make a strong effort in support of Union's position. Secretary of State for External Affairs Louis St. Laurent suggested to H.H. Wrong, the Canadian ambassador in Washington, that he should "indicate to the Federal Power Commission that the critical shortage of natural gas in southwestern Ontario has considerably handicapped the industrial output in that district with resulting effects on the whole national economy."[5] On 17 May 1948, a Canadian Embassy spokesman appeared before the FPC to impress this point upon the U.S. commissioners. He explained that

> the Canadian Government is concerned with the possibility that when a commodity of prime importance to the economies of both countries becomes in short supply in one country, the result of such a shortage will be to reduce very drastically, or even to cut off entirely, the export of that commodity to the other country.[6]

The FPC was reminded that such a decision could be devastating. The Canadian representative predicted that, if the Union Gas Company could not obtain an adequate supply of natural gas, the important industrial area served by the company would undergo an "economic dislocation." The prospect of such a serious disruption, the embassy argued, was a matter of "mutual" not just Canadian interest. This presentation was followed by a remarkable petition from the Canadian ambassador arguing that, given the seriousness of the situation, political boundaries should be ignored and that Canada's need should be treated by the Commission exactly as it would treat a similar dislocation within the United States. "Such equal treatment," Ambassador Wrong argued, "would be consonant with the principles established by the Hyde Park Agreement of April 20, 1941."[7] The serious economic dislocation argument and the claim of mutual interest failed to move the FPC. In July, the chairman proclaimed that, "In light of the acute shortages now existing in the United States and the demands upon the Panhandle System, we believe it impossible to give Union the relief requested at this time."[8]

Given the deteriorating natural gas supply picture, by the summer of 1948 Alberta's apparently vast natural gas potential began to excite entrepreneurial minds within the province and beyond. This interest was soon translated into three specific proposals to the Alberta government by several groups anxious to find out the terms under which natural gas might be exported from the province. Western Pipe Lines, a Winnipeg group headed by Osler Hammond and Nanton Ltd., proposed to build a line from Alberta to Winnipeg, with branch lines connecting Saskatoon and other Saskatchewan communities. A U.S. group, in association with the British Columbia Electric Co. of Vancouver, known as the Northwest Natural Gas Company, wanted to build a pipeline from southern Alberta through the Crowsnest Pass to Spokane, Washington, then west to Portland and Seattle, with a northern branch extending to Vancouver. The third proposal, known as the McMahon Project, combined the interest of Calgary oilman Frank McMahon and a number of mainly West Coast U.S. companies who encouraged an "all Canadian" route from Alberta, through the British Columbia interior to Vancouver, and then south to Seattle. Existing regulations stipulated that any natural gas found in Alberta had to be used in the province, but in "special circumstances" the Alberta cabinet would consider export outside the province.

Sensitive to the active concern that Alberta communities had long shown about natural gas price and supply, the Manning government decided to appoint a commission to investigate the natural gas situation in Alberta. Called under the provisions of the *Public Inquiries Act* in November 1948, the Natural Gas Commission was directed to investigate the existing and proven reserves of natural gas in Alberta, to inquire into and estimate the potential reserves of natural gas in Alberta, to estimate Alberta's future requirements of natural gas for domestic, commercial and industrial purposes, and to make whatever further inquiries regarding the use of natural gas as might seem appropriate to the best interests of Albertans.[9] Appointed commissioners, and charged with the task of holding hearings to obtain the desired information, were Robert J. Dinning, a prominent Calgary businessman; Andrew Stewart, an economist who was head of the Department of Political Economy at the University of Alberta; and Roy C. Marler, a well-known Edmonton area farmer who was president of the Alberta Federation of Agriculture.[10] D.P. McDonald, long-time advocate of the independent producers, was named counsel to the Commission. Known as the "Dinning Commission," the commissioners set about advertising for written submissions and between December 1948 and February 1949 listened to the presentations of interested parties at public

hearings held in Medicine Hat, Calgary and Edmonton. The commissioners also called several expert witnesses, including Dr. G.S. Hume, H. Zinder, a natural gas utility consultant from Washington, D.C., and J.R. Donald, a chemical engineer from Montreal whose expertise centred upon the petrochemical industry.

The first matter to be addressed was the reserves question. As Conservation Board engineer G.E.G. Liesemer and various company representatives presented estimates of natural gas reserves in the fields about which they had special knowledge, it became apparent that there was broad agreement with Dr. Hume's revised province-wide total of 4.26 trillion cubic feet of "proven and probable" reserves.[11] Of this, it was estimated that 3.49 trillion cubic feet of dry clean gas were ultimately recoverable. With regard to the estimation of future prospects, no one was prepared to come up with a number on what was an inherently speculative exercise, but all were once again in agreement with Dr. Hume's assertion that "... the potential reserves, will exceed those now proven by many times."[12] The Commission also found the tabulation of Alberta's current domestic, commercial and industrial consumption of natu-

R.J. Dinning, prominent Calgary businessman and president of Burns and Company Ltd. among other business interests, was appointed Chairman of the Natural Gas Commission to enquire into Reserves and Consumption of Natural Gas in the Province of Alberta. The commission commenced public hearings in December 1948. Courtesy of J. Dinning.

TABLE 6.2 Alberta Gas Consumption in 1948 and Estimated
Requirements in 1960

	1948 Mcf	1960 Mcf
Canadian Western		
Domestic	6 699 000	8 400 000
Commercial	4 066 000	4 500 000
Industrial	3 027 000	3 600 000
Imperial Oil Refinery, Alberta Nitrogen and Taber Sugar Factory	5 817 000	6 180 000
Subtotal	19 609 000	22 680 000
Additional possibilities		2 420 000
Total—Canadian Western	19 609 000	25 100 000
Northwestern		
Domestic	5 894 000	11 400 000
Commercial	4 602 000	7 600 000
Industrial	1 538 000	2 200 000
Imperial Oil Refinery and Edmonton City Power Plant	156 000	3 780 000
Subtotal	12 190 000	24 980 000
Additional possibilities		4 780 000
Total—Northwestern	12 190 000	30 740 000*
Remainder of province (esp. Medicine Hat)	5 466 000	6 000 000
Province generally		8 200 000
Total—Alberta	37 265 000	70 040 000

SOURCE: Alberta, *Report of the Commission,* Edmonton: King's Printer, 1949, pp. 82–84.
* The original estimated requirement of 29,760,000 Mcf was subsequently raised to the figure shown.

ral gas to be a straightforward task. At the request of the Commission, Northwestern Utilities Ltd., Canadian Western Natural Gas Co. Ltd. (formerly Canadian Western Natural Gas, Light, Heat and Power) and Medicine Hat submitted data showing Alberta's 1948 consumption to be 37.26 billion cubic feet, along with the projection that this would increase to 70 billion cubic feet yearly by 1960 (see Table 6.2).[13]

The real burden of the Commission's work came as it tried to address the question of Alberta's future requirements. After Alberta's natural gas utility companies had given their estimates of the province's future needs, two of the three companies seeking to export natural gas filed submissions indicating the volume of gas that they expected to purchase. The Northwest Natural Gas Company informed the Commission that, within two years of the completion of its transmission line to the Pacific Northwest, it anticipated a demand of 56 billion cubic feet of gas per year at an average of 153 million cubic feet per day.[14] Western Pipe Lines estimated that it would need an average 72 million cubic feet per day or 26 billion cubic feet per year to supply customers in Saskatchewan and Manitoba.[15] In support of their proposals, the would-be exporters presented a wide-ranging list of benefits that would result from natural gas exports, including the stimulation of exploration that would lead to the discovery of additional reserves; a significant annual operating expenditure on supplies, wages and salaries, and royalty payments to the province;[16] and the opportunity for the owners of natural gas rights to begin to recover exploration and development costs. This theme was reinforced by spokesmen representing many of the independent oil operations. Speaking for one group of smaller companies, Clifton Cross informed the commissioners that a lack of a market for natural gas meant that the precious capital invested in drilling could not be recovered. He argued that a market for gas would draw additional investment capital to the search for oil.[17]

Almost all the others appearing at the hearings were actually hostile, or at best cautious, in their acceptance of the idea of exporting Alberta's natural gas. Representatives speaking for towns and cities already supplied with natural gas wanted to be assured of continued long-term supply. Speaking for Edmonton, City Commissioner D.B. Menzies advised that Northwestern Utilities' estimates of future population increase and gas consumption in the capital city were too conservative.[18] Medicine Hat claimed that others had misrepresented its future requirements, and Stanley Davies argued the same for Calgary. Davies's views carried particular weight. His experience in Alberta's oil and gas industry went back to the discovery period in Turner Valley. As a young University of Alberta engineering student, he worked in Turner Valley under S.E. Slipper of the Canadian Geological Survey. Graduating in "the technology of oil" from the Royal School of Mines in London after World War I, Davies worked in Mexico, Trinidad and Romania before returning to Alberta in 1924 to take employment as a petroleum engineer in the Calgary office of the Department of the Interior's Petroleum and Natural Gas Division.[19] A private consultant since 1926, Davies became one of the most active voices in the fight to bring natural gas conservation to Turner Valley. In the ensuing years, he was a significant presence at every inquiry relating to the oil and gas industry and he acquired a reputation as the foremost defender of Calgary's natural gas consumers. In his presentation of Calgary's submission to the Natural Gas Commission, Davies described the Canadian Western Natural Gas Company's estimate of Calgary's future needs as "extremely conservative."[20] Not only did Calgarians think that their city's requirements had been underestimated but also Davies articulated their view that it was necessary to predict and ensure that an adequate natural gas supply existed much further into the future than current gas company projections. Calgary consumers thought that the appropriate span of assurance should be 40 to 50 years. Davies informed the commissioners that just over 20,000 newly constructed dwellings had been added to the Calgary and Edmonton systems during the previous four years and that the average cost of natural gas installation for these homes was $1,500, representing a total capital outlay for Alberta families of $30,000,000. Since this equipment could not be used to burn any other fuel and since most of these houses had been built with 30- to 40-year mortgage loans, the assurance of a 40-year natural gas supply for these new users, as well as for more established consumers, was presented as a reasonable request. Davies took the commissioners through the available data on each of Alberta's gas fields to establish the conclusion that there was "not sufficient proven and probable recoverable natural gas reserves to adequately supply" the people of Alberta and an export market.[21]

Towns not served by natural gas were even more vociferous in their opposition to natural gas export. Airdrie, Crossfield, Carstairs, Didsbury, Olds, Innisfail, Bowden and Penhold were quick to insist in a collective brief that their present and future needs had to be met before the province could legitimately allow natural gas to be exported from the province. By 1950, there was an emerging sense that Alberta was on the verge of dramatic growth, and underlying the concern of those communities without natural gas was the fear that without access to this cheaper, cleaner and more efficient "modern" fuel their towns would be left behind.[22] Even Leduc added its voice to the chorus of opposition. Situated virtually on top of one of the province's natural gas reservoirs, the town of Leduc remained unable to persuade any company to supply its homes and businesses with this preferred fuel and therefore refused to endorse any natural gas export proposal.[23]

Leduc's submission raised another more commanding issue. The Leduc town council pronounced their belief that the apparent abundance of natural gas offered Alberta an unparalleled opportunity to rebuild and reshape its economy. For the commissioners' instruction, Mayor F. Johns contended:

> It is practically indisputable that industrial expansion bolstering our basic agriculture and to some extent cushioning its vicissitudes would be of immense benefit to the economy of the province. Many parts of Canada—the Prairie Provinces in particular—have suffered much in past years because of the fact that their prosperity was based almost exclusively and directly on agriculture, while it is probable that agriculture will, for many years at least, supply the mainstay of our population, it now appears that we have within our grasp a stabilizing factor in the nature of secondary industry.[24]

Other towns and cities, individually or collectively through the Union of Alberta Municipalities, along with the Alberta Branch of the Canadian Manufacturers' Association, also informed the Commission that they believed that Alberta could use its cheap fuel to promote industrial development.[25] With the traumatic experience of the depression decade still fresh in Albertans' minds, this was a powerful argument. It was, moreover, one with roots that went back to the first debate about natural gas exports in the 1920s, and it would remain part of the core discussion surrounding the formulation of Alberta energy policy to the 1980s.

To address the relationship between natural gas supply and industry location, the Commission called three special witnesses. The question had been debated earlier in the United States. H. Zinder, a natural gas utility consultant from Washington, D.C., and the first expert called to give evidence, drew upon research that he had prepared for the U.S. natural gas industry's submission to the Federal Power Commission.[26] He pointed out that a study to determine the ratio of cost of fuel to the value of product produced by the 20 industry groups reported in the United States Census of Manufacturers found that only in two industry groups—iron and steel, and stone, clay and glass products—did fuel costs contribute more than 4% to the total value of the product.[27] This finding was held to ex-

plain why the availability of a cheap and abundant natural gas supply had not motivated the migration of industries from the northeastern to the southwestern states. Zinder also drew attention to a
case study of fields in the midcontinent region that demonstrated
that it was more economical to transport natural gas to industrial
markets than to move industrial plants to sources of natural gas,
and thus presented a cautionary message to regional interests that
might be prone to "place undue emphasis on holding back the production of natural resources for possible but uncertain future local
demand."[28]

The other witnesses were more optimistic regarding the local potential of Alberta's natural gas. J.R. Donald, one of Canada's most
distinguished chemical engineers, was well-acquainted with Alberta's natural gas potential. As wartime director general of explo-
sive and chemical production for the Department of Munitions and
Supply, he had been responsible for the building of the federal government's ammonia plant in Calgary. Donald was an unabashed
champion of Alberta's prospects. "Alberta," he assured the commissioners, "is probably the richest, the most richly endowed Province in Canada in the matter of natural resources."[29] After reviewing
the chemical industries that required natural gas as a raw material
or as a fuel, he concluded that Alberta's unique situation in the Canadian economy made certain that some would come to the province. Addressing the commissioners, Donald assured:

> I would like to say to you that this question of the United States
> chemical industry considering establishments in Alberta is not
> just a matter of fancy. A number of the larger chemical compa-
> nies have been studying this Alberta situation. My own organi-
> zation has had a request from one of the larger users of petro-
> chemicals to advise them in general on this situation and they
> talk in very substantial quantities of natural gas and propane
> and butane.

Donald went on to note that the development of a chemical industry would require large capital investment, and with this observation he added the powerful warning that such investment could
"only be made where there is an assured source of raw material
supply over a considerable period and consequently establishment
of this type of chemical industry requires very adequate reserves of
natural gas."[30] W.A. Lang, of the Alberta Research Council, drew
similar attention to the importance of natural gas as a source for a
wide variety of chemical products. He added that, because the

province was also blessed with other abundant raw materials required by industry, the great potential of natural gas as the foundation element for the industrial diversification of the provincial economy went well beyond the production of industrial chemicals.[31] The testimony submitted by others to the Commission, showing that nine companies, including brick, pottery and glass works, had located near Redcliff and Medicine Hat because of a virtually unlimited supply of cheap natural gas, gave credence to Lang's confidence.[32]

Although the Medicine Hat-Redcliff experience and the evidence of the two Canadian experts more than countered Zinder's reservations about the likelihood of industry moving to Alberta, the submission of other groups helped to build what became an overwhelming case against natural gas export, at least in the short term. Alberta coal producers pointed out that natural gas transmitted to Saskatchewan, Manitoba and the Pacific Northwest would displace Alberta coal and thereby cripple an industry that employed several thousand Albertans.[33] The Alberta Federation of Agriculture, representing well over half the province's population, pronounced that it was not opposed in principle to export, as long as a 50-year reserve supply was set aside for Albertans and the province received an adequate royalty for the exported gas.[34] Even the appeal of Saskatchewan and Manitoba that, since the prosperity of all three prairie provinces was linked, their needs should be given priority over any other export market lent implicit support to the notion that an assured supply of natural gas offered a means of breaking the bonds of a dependent agricultural economy.[35]

On 8 March 1949, the commissioners released their report. The main finding was much anticipated but of little surprise to those who had followed the hearings. Alberta's "existing and proven" natural gas reserves as at February 1949 were stated as 4.26 trillion cubic feet, the generally accepted estimate of Dr. G.S. Hume. Rather than offer a speculative figure on the province's potential reserve, the commissioners would only restate conventional wisdom that "substantial additions" would undoubtedly be made to existing reserves. The central question, and the aspect of the report that Albertans awaited with greatest interest, was the matter of Alberta's future requirements. After reviewing the assembled evidence, the commissioners concluded that not only was the Alberta gas utilities' projected annual provincial consumption of 70 million Mcf by 1960 too conservative but also it was necessary to think of the province's needs well beyond 1960. Noting that "references, too numerous to cite, were made before the Commission to a period of 50 years,"[36] the commissioners adopted this as the time span over which Albertans might reasonably expect to be assured that there was sufficient natural gas for their needs. Conscious that they had not been directed to make a recommendation on the question of natural gas exports, the commissioners nonetheless presented their evidence in a manner that provided an implicit but obvious answer to this question. They presented a series of hypothetical situations, showing the relationship between different rates of consumption, withdrawal and addition to reserves over a 50-year period, to demonstrate that the province's current level of reserves was inadequate to sustain Alberta's long-term needs if combined with any of the projected export commitments.[37]

The commissioners also laid out in their report a number of recommended principles and guidelines. First was the principle of local priority in the matter of supply. Also given the Commission's blessing was the principle of priority to Canadian users. The commissioners had noted that the U.S. Federal Power Commission had followed without deviation a policy of assuring priority to U.S. users and drew attention to Ontario's recent experience.[38] The commissioners took a similar position on price. Responding to the "consistent and emphatic representation of consumer groups," they advised that natural gas exports should not be allowed to disturb the gas price structure within the province and that the price to Alberta users should be protected.[39] The Commission advised that a further prerequisite to the sale of gas outside Alberta be the assurance that the province derive "maximum revenues consistent with continued and sound development of the fuel resources of the province."[40] To reinforce the general theme of maximum benefit, the Commission felt constrained to remind the Manning government that, given the province's substantial ownership of mineral rights and its jurisdiction over production, it was in a strong bargaining position. "It is clearly in the interests of the people of the Province," the commissioners reported, "to use such bargaining strength as the Government may possess" in concert with sound development principles.[41] The report advised that particular attention be given to the impact of natural gas development upon the coal industry, so that both these energy resources would be used wisely. Also singled out for special attention was the fuel need of Alberta's rural population. It was acknowledged that the direct benefits derived from natural gas production through the availability of a cheap and convenient fuel largely accrued to urban users. Evidence brought forward during the course of the hearings pointed to the increasing use of propane in rural areas in the United States, and the commissioners recommended that, as more propane became available through increased natural gas production, special effort be devoted to finding ways of extending its use on Alberta farms.[42] Consistent with the great value that Albertans attached to assured natural gas supply, the commissioners commended the achievements of industry and the Petroleum and Natural Gas Conservation Board in reducing gas waste. In the interest of furthering this objective, they proposed that it was not unreasonable to expect oil producers to absorb the costs of natural gas conservation.[43]

In addition to offering support for the Conservation Board's continued role in the conservation area, the Natural Gas Commission also recommended that the Board be given new administrative responsibility in another area. It recognized that the availability of detailed statistical data and sophisticated technical expertise were essential to the wise administration of Alberta's petroleum resources. So that there would be "technical personnel of the highest competence" plus "full and up-to-date information, not only on the Alberta situation, but also on developments in all other oil- and gas-producing areas in North America" available, it recommended that the Petroleum and Natural Gas Conservation Board be given additional personnel and adequate finances to manage this vital responsibility.[44] The Dinning Commission recommended, as the McGillivray Commission had in 1940, that the Conservation Board be in a position to provide at any time, and to publish annually, comprehensive statistical information on the petroleum industry.[45]

The Natural Gas Commission made one further recommendation that proved to be of far-reaching importance. Dinning, Stewart and Marler were attracted to an idea presented by one of Edmonton's most prominent lawyers and businessmen, H.R. Milner.[46] In his joint submission for the Canadian Western Natural Gas Company Ltd. and Northwestern Utilities Ltd.,[47] he pointed out that the natural gas export proposals under consideration rested upon the assumption that the pipeline companies had free rein to enter into any kind of long-term supply contract with producing companies. To highlight the implications of this situation, Milner pointed to the three contracts that the Northwest Natural Gas Company (NWNGC) had signed with Shell, Imperial and California Standard. Under the contract with Shell, the entire production of the Jumping Pound field was dedicated to NWNGC. California Standard pledged production to the extent of its interest in the Princess, Dunmore and Foremost fields, and Imperial promised to make its deliveries mainly from the Viking-Kinsella field—the one that Edmonton had always considered the source of its long-term supply.[48] To get be-

H.R. Milner, President
of Northwestern Utili-
ties and Canadian
Western Gas, Light,
Heat and Power Com-
pany Limited and ap-
parently the first to
propose a trunk
pipeline system to
gather provincial gas
production for
markets outside Al-
berta. Northwestern
Utilities Ltd.

yond the equity and other conflicts that such contracts promised to
assure, Milner suggested that the interests of all would be best
served if the province's natural gas reserves were "pooled" and exporters secured their supplies out of the general pool. This would
require the construction of a pipeline grid to link producing fields
and the existing pipeline systems. Admitting that such a grid or
trunk pipeline system would only be feasible if the export of gas
outside was contemplated, the commissioners nonetheless alerted
the Alberta government to the many advantages of such an approach. Foremost of these advantages was a means of producing
the resource in a way that promised to maximize conservation ob-

jectives. The report explained that a grid system of pipelines tying together all the province's producing fields would

> make it possible to ensure priority to the utilization of residue gas; dry gas fields could, if necessary, be held in reserve; and such fields might, if circumstances require, be used as storage reservoirs. In other words, gas required for any market, at any time, could be drawn from a source determined by conservation considerations.[49]

Other benefits included the elimination of unnecessary pipeline duplication, thereby conserving capital and reducing operating costs. Also, more communities in the province could be served, and those communities already served with natural gas would gain additional assurance of uninterrupted supply. To supervise the operation of such a system and to manage the withdrawal of natural gas from particular fields, the report suggested the "Natural Gas Conservation Board or some other public regulatory body," and pointed to the common purchase arrangements already in effect in Turner Valley as an appropriate working model.[50]

Milner was unable to persuade the Alberta government that his two Alberta natural gas utility companies offered the ideal vehicle to build and manage such a trunk line system; nevertheless, he gave the Manning government the key that could unlock traditional objections to natural gas export once additional reserves had been confirmed. It was, moreover, an insightful contribution that marked the beginning of Milner's role as perhaps the most influential of Premier Manning's private sector advisers on natural gas matters.[51]

The Natural Gas Commission hearings and submissions gave the government a strong reminder of where the majority of Albertans stood on the question of natural gas export. Medicine Hat residents had been using natural gas to heat their homes and to cook their meals since 1903, Lethbridge and Calgary residents since 1912, and Edmontonians from 1923—long enough that Albertans were convinced that they had been blessed with access to this superior fuel. It was clear that politicians who might dare to take up the cause of natural gas export would have to do so cautiously. The guidelines that would anchor the province's natural gas policy for the following 50 years were found in the Natural Gas Commission's report.

Evidence of both the influence of the Commission's report and the premier's growing concern about Ottawa's emerging presence in the regulation of Alberta's oil and natural gas resources is apparent in the legislation put before the "Special Session" of the Alberta Legislature that began on 4 July 1949. The province's apprehension was reinforced by Parliament's subsequent consideration of a related matter: the incorporation of a number of pipeline companies intending to remove natural gas from Alberta. On April 28, the day after the federal *Pipe Lines Bill* passed second reading, six private member pipeline company incorporation bills were presented for Commons approval.[52] Five of the bills were passed before the House adjourned two days later. Despite only perfunctory debate, Social Credit House Leader Solon Low was able to articulate Alberta's fear that the province's legal right to deal with its resources as it saw fit was being eroded by federal intrusion. Low explained that, in passing these bills, Parliament was telling Alberta in effect:

> Yes, you have control over your resources but we, the dominion parliament, are going to tell you how you may be allowed to export your resources. We have not sufficient confidence in your judgment to make the decision.[53]

At issue was that the terms of incorporation for companies proposing to export Alberta's oil and natural gas were being decided by the federal government. Moreover, once incorporated, these companies were compelled by the recent federal *Pipe Lines Act* to apply to a federal regulatory authority, the Board of Transport Commissioners, which had the power to make approval conditional upon whatever terms and conditions it chose.[54] The pipeline and incorporation issues touched upon a sensitive nerve in a province where the memory of the nearly 30-year struggle to wrest control of lands and resources from federal administration was still fresh. Low's comment in Parliament was but a hint of the vigorous opposition and filibuster that would occur when the next group of pipeline company incorporation bills were brought before the House of Commons in October and November 1949.[55]

Alberta responded with a battery of legislative measures especially prepared under the guidance of Texan Robert E. Hardwicke. The centrepiece of the legislation presented to Alberta MLAs in July was the *Gas Resources Preservation Act,* which delegated an important new responsibility to the Conservation Board. To ensure "the preservation and conservation of the oil and gas resources of the Province" for "the present and future needs" of Albertans, the Act provided that anyone wishing to remove natural gas from the province had to apply to the Petroleum and Natural Gas Conservation Board for permission. In turn, the Board was obliged to consider the application at a public hearing after which, with the approval of the Alberta Cabinet, it would grant or refuse the application.[56] Further, the Act specified that the Board would only permit the removal of natural gas that in its judgement was surplus to the province's present and future needs. In the event that the Board did see fit to recommend an export licence, the Act authorized the Board to set whatever time period, terms and conditions it might see as appropriate. Among the terms specifically mentioned were the right to designate the pool or pools from which the natural gas might be produced and purchased, to prescribe that the natural gas be taken rateably from the wells in designated pools, to set the maximum monthly and annual quantities of natural gas that might be produced, and to insist that the exporting company supply natural gas from its pipeline to any nearby Alberta community or consumer at a reasonable price. Finally, as an ultimate reassurance to Albertans, the Act contained an override clause. It provided that, regardless of the terms or conditions that might be set, if in the opinion of the Board any emergency should arise where additional natural gas was required to meet the needs of Alberta consumers, the shortfall could be made up by a Board order diverting whatever amount might be necessary from the export stream.[57]

The counter measures put before the Alberta Legislature also included a further measure to strengthen provincial control over natural gas exports and which promised to add to the Conservation Board's growing list of responsibilities. Manning sought amendment of the province's existing *Pipe Line Act,* and proposed the transfer of responsibility for the evaluation of all provincial pipeline construction proposals from the Department of Mines and Minerals to the Conservation Board.[58] With the mandate to review pipeline applications, the Board was given authority to recommend cabinet rejection or approval, subject to whatever changes it thought appropriate, including naming the company a "common purchaser."[59] The last of the bills put before the Special Session of the Alberta Legislature was an amendment to the *Public Utilities Act.* With the proposed amendment, the government intended to strengthen the authority of the Public Utilities Board to fix "the just and reasonable price or prices to be paid for gas in its natural state," regardless of "the terms of any contract" a company might have signed for the sale or purchase of natural gas.[60] In addition to giving its Utilities Board the potential power to control the field price of natural gas, the government also wanted the Utilities Board to be able, after hearing the interested parties, to fix "just and reasonable" pipeline charges for gathering and transporting oil or gas for any company defined as a common carrier by the Conservation Board.[61]

After surveying the bills presented for first reading in the Legislature on July 5, the *Calgary Herald* offered its readers a succinct interpretation of the government's strategy.

> Well head control of the resources is what the province is aiming at because pipeline legislation and export matters are federal jurisdiction. This way the province is making certain of its right to control production and purchase in any area of Alberta.[62]

The *Edmonton Bulletin* also saw the proposed measures as a "formidable line of defence against possible encroachment of the province's natural resources,"[63] but Opposition politicians were not as impressed, and their comments underlined Albertans' apprehension about the very idea of natural gas exports. Debate in the Assembly centred upon the *Gas Resources Preservation Act* and focused upon two main issues: the general question of natural gas exports and concern about assigning such an important responsibility to the Petroleum and Natural Gas Conservation Board. Liberal leader J. Harper Prowse led the attack. He argued that the matter of natural gas supply was so important to the everyday lives of Albertans and the province's future development that decisions about natural gas export could not be handed over to an appointed board. That Board decisions would require cabinet approval was dismissed by Prowse as undemocratic and unacceptable. He insisted that individual Albertans should be able to have their voices heard and called upon the government to hold a plebiscite on the general question of exporting Alberta's natural gas and to make a commitment to Albertans to hold plebiscites on every individual export permit.[64] Calgary Liberal Hugh John MacDonald picked up on his leader's assertion that the Manning government was giving itself powers that properly belonged to the Legislature. If the Assembly passed the government Bill, MacDonald warned

> We will have given up any right to approve or disapprove of the handling of our natural resources. We will be at the complete mercy of the cabinet and the big utility companies.[65]

CCF leader Elmer E. Roper expressed his party's objection to legislation that allowed the government to run the province by order-in-council and urged that the final decision on gas export applications be left for the Legislature. Roper also left no doubt that the CCF party would oppose export outside Canada. Permitting export of gas to the United States, Roper declared, would be

> adding a pipeline to the economic shackles with which this country is bound by American economic imperialism. One of the world's rich oilfields in Alberta already [has] been given into the hands of foreign imperialists.[66]

With party positions outlined on the first day of debate, the Legislature began the second day of what one newspaper described as a "tooth-and-nail fight" that lasted through the evening and into the early morning of July 8.[67] Tempers flared frequently, Social Credit back benchers sometimes sided with the Opposition, and in one instance the Independent Social Credit MLA for Banff-Cochrane, Arthur Wray, was ejected from the Legislature for refusing to withdraw an accusation hurled at the attorney general. The handful of Opposition members had come well primed and encouraged by the *Calgary Herald* and the *Edmonton Journal*. In strong editorials, the newspapers supported the prime objection raised by Prowse, Roper and their colleagues. They too challenged the government's plan to remove natural gas export decisions from the Legislature.[68] In the *Herald's* view, the government was saying, "Give us a free hand and we will look after the details. You can trust us to do what is best for the province."[69] The *Herald* was not sure that either the Conservation Board or the Manning cabinet would look after the interests of Calgary natural gas consumers. Manning, Tanner and Attorney General Lucien Maynard carried the debate for the government side. Manning reminded the Legislature that natural gas export companies were about to be incorporated in Ottawa, at which time they would be free to begin constructing gas pipelines from Alberta fields. The premier ruled out plebiscites and calling the Legislature each time a gas export licence was recommended by the Conservation Board as impractical.[70] In addition, he tried to assure his critics that the proposed legislation would assure the protection of Alberta natural gas consumers. This did not convince Calgary MLA Hugh John MacDonald, perhaps the most effective of the Opposition critics. He discussed the much-touted last line of defence that the government had built into its legislative arsenal, the clause empowering the Conservation Board to divert gas from export if necessary, as but the illusion of protection. MacDonald argued that this was entirely unrealistic, because it was "unthinkable" that out-of-province cities could have their supplies cut off, especially in midwinter.[71] The hardships that such an action promised to inflict would make it next to impossible in practice. As the one real protection of Calgary's future needs, he urged that the Jumping Pound field be set aside for the city's requirements.

The debate was as vigorous as the Legislature had seen for sometime, but it lasted only four days. Given the government's overwhelming majority, the outcome was never in doubt, and all the bills were passed.[72] The Manning government was able to consolidate important new legislative powers within the cabinet, but it had been served another warning that it would have to continue to tread warily on the natural gas export question.

Now aware of the regulatory framework that the government intended to use to administer the production, transportation and export of natural gas, industry and the public began to press Premier Manning to state his government's policy on natural gas exports. Noting that it had not responded to Opposition demands that the government prohibit the export of natural gas, some Albertans were anxious to have the government state its position. On July 18, the Alberta cabinet met with representatives of the North West Transmission Company, incorporated to build a gas pipeline to the Pacific Northwest, and H.R. Milner, who represented Alberta's natural gas utility companies. The meeting was followed by a press release from the premier that told Albertans that in recognition that his government's "foremost responsibility" was to protect the interests and welfare of the people of the province, no natural gas export application would be given favourable attention until the government was "satisfied beyond question that under sound conservation and proration practises there [were] sufficient gas reserves to meet the present and future domestic and industrial requirements of" Albertans. For emphasis and reassurance, he added, "This condition definitely does not exist at the present time and is not likely to exist for some time to come."[73]

The premier's hope of putting the natural gas export issue to rest was immediately dashed. His press statement was interpreted in some newspaper reports as ending hopes of export; further explanation was offered therefore the following day. Although the government was not satisfied that present gas reserves were adequate to permit export, Manning assured that "it did not mean that gas export projects have been thrown out the window."[74] The government's commitment to the principle of gas exports and to giving Canadian requirements first priority once reserves were sufficient was reaffirmed. It was also noted that the Conservation Board would maintain a thorough surveillance of reservoir development.

The first application to export natural gas under the provisions of the *Gas Resources Preservation Act* was received by the Conservation Board on 19 August 1949. It was the application from the New York-owned Northwest Natural Gas Company and its local subsidiary, Alberta Natural Gas Grid Ltd., to transport natural gas from Alberta to the Pacific Northwest states and to Vancouver.[75] The request was for a permit to remove 220 million cubic feet of gas per day for a period of 30 years. It was proposed that Alberta Natural Gas Grid Ltd., an Alberta-incorporated company, would gather the natural gas from the main sources of supply in the province by constructing a grid pipeline system extending north and east from the Pincher Creek field in southwestern Alberta. From Pincher Creek, the gas would exit Alberta in a 24-inch pipeline through the Crowsnest Pass, cross the Canadian-American border at Kingsgate, B.C., and continue to Spokane, Washington, from which a branch line would be constructed to Trail, B.C. Continuing westward, the main line would proceed to the Seattle area and then split, with one branch running south to Portland, Oregon, and the other north to Vancouver. The Board was informed with the application that, in compliance with the federal *Pipe Line Act*, a subsidiary company— to be named the Alberta Natural Gas Company—would be incorporated to own and operate the Canadian section of the export line. A hearing of the application was set by the Board for 28 November 1949. In the interval, Northwest Natural Gas awaited passage of the *Alberta Natural Gas Company Bill* through Parliament.

Presented to the House of Commons in late October, the Alberta Natural Gas Company incorporation Bill immediately ran into trouble. The Bill was sponsored by Ralph Maybank, the MP for Winnipeg South Centre and parliamentary assistant to the minister of mines and resources. Sensitive that his Bill was the only one of the pipeline company incorporations that had failed to pass in the rushed closing days of the previous parliamentary session, Maybank tried to counter the concerns that he expected would be raised in a carefully argued introduction. He stressed that, like the other bills, there was nothing in this one that specifically identified the pipeline route. Maybank informed members that the company had surveyed five possible routes to the Pacific coast and that the Board of Transport Commissioners would have the final say about which route was appropriate. In addition to the control of the Board of Transport, Maybank called attention to Alberta's recent legislation. "The mere fact that a group of applicants are granted incorporation by this house does not mean that any gas can be taken out of Alberta without the consent of the government of that province."[76] Beyond this, he reminded that no gas could be taken out of the country except under an annual permit granted by the Department of Trade and Commerce under the *Electricity and Fluid Exportation Act*. The Bill, Maybank insisted, was "a mere incorporation." He concluded:

> Those who are concerned with respect to the route this pipeline may take need have no more fear concerning this particular company than any of the other companies this parliament has already set up, and in addition there are the several controls I have mentioned.[77]

Maybank could scarcely have guessed the extent of the opposition. Led by Calgary Conservative MPs Arthur L. Smith and Douglas Harkness, the debate on the incorporation of the Alberta Natural Gas Company turned quickly into a filibuster that dragged on through the fall of 1949 and into the 1950 winter and spring session of Parliament.

The initial debate on the Bill's second reading set the stage, the key arguments and debaters changed little from those introduced at this point. Douglas Harkness, the member for Calgary East, launched the Opposition challenge. He was not prepared to accept Maybank's assertion that five routes to the coast were being considered. Harkness and his Calgary colleague, Arthur Smith, knew that the application before Alberta's Conservation Board identified only one route—a route that left Alberta through the Crowsnest Pass and crossed into the United States at Kingsgate. This route was the critical flaw, which Harkness believed was "prejudicial to the interests of Alberta, British Columbia, the city of Calgary and Canada as a whole."[78] To make his case, he directed the attention of his parliamentary colleagues to three areas of concern. First was the extremely negative impact this pipeline proposal would have upon the long-term supply and price of natural gas to Calgary and southern Alberta consumers. And even if in time it was determined that a gas surplus did exist in southern Alberta, Harkness believed it should go to Regina, Winnipeg and points east.[79] Second, of all the possible routes to the Pacific coast, Harkness pointed out that the one favoured by the Alberta Natural Gas Company would serve the fewest people in Alberta and British Columbia. Moreover, in the small region that the line would serve, the Crowsnest Pass and East Kootenay area, the line was more of a curse than a blessing because it threatened the region's main industry—coal mining. Not only

was this pipeline proposal presented as a detriment to the people of
southeastern British Columbia but also, at the other end of the
pipeline, Vancouverites were given notice to be concerned. Harkness warned that Maybank's suggestion that regulations established by the Board of Transport Commissioners and by the Alberta
government would protect Vancouver consumers was misguided.
"Once that gas has passed out of this country and it is in the United
States," he explained, "if there is a matter of allocation of that gas
as between United States cities which need it, and where there are
people who are likely to be cold if they do not get it, and Canadian
cities, I think one would need to be naive indeed to accept the suggestion that Canadian cities would get it."[80] As a general observation, he added, that since, under this proposal, Vancouverites were
at the end of a transportation system, they could expect to pay
more for their gas than U.S. users. This was a point to which most
westerners could relate, and Harkness's concluding railroad analogy came easily from the wisdom of his own western experience.

> Where you are situated on a railroad, a road, or gas line and oil
> line or anything else, it is much better for you to be near the start
> of it or along the route of it rather than to be at the end of it. If
> you are at the end of it, you are always likely to get the worst of
> the deal.[81]

Harkness's third area of concern was the negative influence that the
granting of an Alberta Natural Gas Company charter would have
upon the development of northern Alberta and northern British Columbia. It was apparent that only one pipeline was needed to serve
the Pacific Northwest market, and Harkness argued that if this line
was built from southern Alberta through the United States it would
undermine incentive to continue exploration in the northern part of
either province for years to come.

CCF members also took the floor to speak in favour of an all-
Canadian route and to announce their preference for pipelines to be
operated as crown-owned utilities.[82] Alberta Social Credit members
in the House of Commons, in contrast to their active role in the debate surrounding the *Pipe Line Act*, were a negligible presence in
the incorporation debate. Their only extended contribution was
that of one of the members from central Alberta, Ray Thomas, who
asked leave to read into the record of debate a telegram that Premier Manning had recently sent to Senator James Turgeon, one of
the Alberta Natural Gas Company sponsors. The telegram stated
that the Alberta government was not opposed to the incorporation

of the Alberta Natural Gas Company or any other company; that it felt "very strongly that no *bona fide* company should be refused incorporation"; that the Alberta government had not decided to prohibit the export of gas from any particular area of the province but rather to prohibit export until reserves were sufficient to assure Alberta's future needs; and that export approval would come only when there was a sufficient surplus to justify export under sound conservation and proration practices and after priority consideration of Canadian requirements.[83] Thomas concluded with a rambling statement on the importance of natural gas conservation and his party's confidence in the ability of the Alberta Conservation Board to determine how much gas could be exported while allowing for Alberta and Canadian needs.[84]

The debate dragged on through October, November and into December. Supporters of the Bill argued that, since a number of almost identical pipeline company incorporations had passed without opposition at the end of the previous session, it was only fair that the Alberta Natural Gas Company be treated in the same manner. Picking up the cue from Social Credit, they also insisted that acceptance by Parliament was largely a formality, and that it should be left for Alberta to decide which company it wanted to deal with. The Opposition remained insistent that, as with a railway company incorporation, no pipeline company incorporation should be accepted without a precise statement of the route to be traversed. On 10 December 1949, Parliament prorogued and the Bill lapsed, having failed to come to a vote.

When Parliament resumed sitting on 16 February 1950, Maybank had to start all over. He introduced a new Bill to incorporate the Alberta Natural Gas Company on February 23. Opponents to the Bill continued their stubborn resistance, bolstered by widespread popular sympathy, especially in British Columbia.[85] As mid-May approached and the Bill remained stalled, the minister of trade and commerce gave vent to his frustration. One of the most powerful wartime ministers, and now a senior member of the St. Laurent cabinet, C.D. Howe was not used to obstruction. Even though this was not a government Bill, it was by virtue of its connection with the recently passed *Pipe Line Act* at least of indirect concern to his Department of Trade and Commerce. In any case, Howe had had enough. Asserting that never in the history of Canada's Parliament had members been compelled to endure a filibuster running through two sessions, he charged that this was all a devious ruse.[86] Under the pretence of holding out for an all-Canadian route, the opponents of the Bill, according to Howe, were in reality fronting for a competing U.S. pipeline company with an interest in the West-coast Transmission Company. "The matter of routes has been brought in as a red herring to cover up the effort to shut off competition for this United States pipeline company."[87] From this point, the debate took on a rancorous and personal tone. After several hours of vigorous slanging, the Alberta Natural Gas Company Bill finally came to a vote on the evening of May 15. Supported by Social Credit, the Liberal majority pushed the Bill through.[88]

The 1949–50 pipeline debates brought several critical energy resource issues to national attention for the first time. Questions about the ownership and control of petroleum resources, access priority and security of supply, pricing and regulatory policy appropriate for a new transportation technology all figured directly or indirectly in the concerns raised by MPs. These sessions defined the issues and set the stage for the famed pipeline debate of 1956 and initiated discussion that led to the calling of the Borden Royal Commission on Energy and the establishment of the National Energy Board in 1959.

Although it would take more than half a decade to clear the regulatory hurdles that emerged in the wake of Alberta's caution, it was little more than a matter of time. Ironically, the usual paramount obstacle to be overcome in the development of hinterland resources, namely, the cost of delivering a competitively priced product to a distant market when faced with the expense of building or paying for the use of a vast transportation infrastructure, was of minor significance in the case of natural gas in the late 1940s and early 1950s. A ready market existed, as did the financial resources to build pipelines spanning the continent.

Southern Ontario utilities found it almost impossible to replace diminishing local natural gas supply because potential U.S. sources could not keep up with the burgeoning demand on the U.S. East Coast. Prices rose as Appalachian gas fields neared depletion, and by 1947 Texas natural gas could be delivered in New York for less than one-third the cost of West Virginia coal.[89] Beyond this, the national coal strike of November 1946 underlined the obvious but important fact that unlike West Virginia and Pennsylvania coalfields, Texas and Oklahoma oil and gas fields were much less likely to have to contend with the disruptions of a militant labour force.

Given the existence of the strong market that materialized after the war, how was the remaining problem of financing expensive long-distance pipelines to be overcome? It so happened that the pipeline interests and the major life insurance companies discovered at this juncture that they had needs and interests that could be harmonized. New York life insurance companies (which controlled more than 80% of all life insurance assets in the United States) were in a bind after 1945. The insurance companies had vast pools of capital, but they lived on actuarial time, and by 1947 the average yield on their portfolios had dropped to only 2.88%. Nearly one-third of their investments were earning less than the contractual interest on outstanding policies.[90] In good part, the problem was a consequence of New York state insurance laws, which in practice set the standard of fiduciary responsibility for the industry. Equity investment was strongly discouraged in favour of high-quality debt investment. The difficulty was that quality corporate bonds with attractive rates of return were in limited supply and the bonded indebtedness of life insurance companies was in sharp decline. Just as the disparity between what the life insurance companies were earning on their investment portfolios and the interest they had contracted to pay on outstanding policies began to raise serious concern, the prospect of pipeline bonds appeared. Pipeline bonds presented a next-to-ideal debt investment for life insurance portfolio managers. Organized on the basis of long-term natural gas supply contracts at one end of the line and long-term delivery contracts at the other, pipeline companies offered precisely defined pay-out schedules and near absolute security, combined with a greatly improved rate of return. For their part, pipeline companies gained immediate access to the vast capital resources that they required without the difficulty of having to market their bonds through public offerings. The difficulty was that pipeline bonds, limited by the Federal Power Commission to a 6% rate of return would have had to compete with common stocks that promised more attractive rates of return. A measure of the mutual attraction is that by 1949 over 90% of new pipeline bonds issued were being absorbed by life insurance companies, mainly New York's "Big Five": Metropolitan Life, New York Life, Prudential Life, Equitable Life and Mutual Life.[91] While Alberta's Conservation Board pondered the natural gas reserves questions, U.S. pipeline promoters, supported by the vast capital resources of the North American life insurance industry, waited impatiently.

The protracted debate over the Alberta Natural Gas Company incorporation Bill compelled the adjournment of the first hearing which, under the provisions of the *Gas Resources Preservation Act*, the Conservation Board had called for 28 November 1949. It was not until late May 1950, after the Bill cleared Parliament, that proceedings finally began on the Alberta Natural Gas Company's application. In the interval, the Board began with the second natural gas export application, that of the Westcoast Transmission Company Ltd., one of the six pipeline companies that had cleared the federal incorporation hurdle in the dying hours of the 1949 spring session of Parliament. Westcoast was the creation of Calgary oilman Frank McMahon, and in addition to McMahon the company was sponsored by Pacific Petroleums, the Sunray Oil Corporation of Tulsa, Oklahoma, and Eastman, Dillon and Company, a New York investment house.[92] Like the Northwest Natural Gas Company and its subsidiary, the Alberta Natural Gas Company, McMahon's company was organized to supply markets in the Pacific Northwest. For this purpose, Westcoast sought an export permit for up to 200 million cubic feet per day for 30 years. The initial proposal called for the construction of a pipeline gathering grid starting at the Pincher Creek field and running northward to the Edmonton area, where it would connect with the main export line going west into the British Columbia interior through the Yellowhead Pass. Once in British Columbia, the planned pipeline ran west to the Thompson River, where it intersected with a feeder supply line from the north bringing natural gas from the Peace River area. From Kamloops, the proposed export line followed a southerly route through Merritt, Princeton and Hope, then west to Abbotsford, where the main line turned south to the international boundary to connect with a U.S. pipeline supplying the Seattle and Portland areas. The Vancouver region was to be served by a branch line from nearby Abbotsford.

With the first hearing under Alberta's new natural gas export legislation set for 12 December 1949, the Alberta government could no longer delay giving precise meaning to its promise to protect the province's "future requirements." The government's first statement on the necessary minimum life of Alberta's proven natural gas reserves was made at the annual meeting of the American Petroleum Institute in Chicago. Alberta's Minister of Mines and Minerals N.E. Tanner informed the assembled members that his province's industrial and domestic needs would have to be guaranteed by available reserves for 50 years.[93] The Westcoast hearing began therefore with the Manning government's affirmation of the 50-year reserve supply recommended earlier in the Dinning Commission report.

Representing Westcoast Transmission at the hearing was one of the most experienced members of the petroleum industry's legal fraternity in Calgary and ex-counsel for the Dinning Commission, D.P. McDonald. After presenting an outline of his company's proposal, McDonald asked for an adjournment to allow more time for intended submissions to be completed. Finally, on 30 January 1950, the first of the long-awaited hearings began. Added to the array of lawyers representing the various interests, assorted expert witnesses, city aldermen, newspaper reporters and interested members of the public whose presence helped to dramatize the moment were seven members of the Alberta Legislature, including Tanner.[94] The noticeable extent of public, business and political interest focused upon this hearing reinforced the Conservation Board's inclination towards caution. From the outset, the Board set a high standard, both for the quality of the technical information it expected and for the scrutiny with which such information was reviewed. Indeed, the Board proceeded with such deliberation that it was not long before the industry, and the government more discreetly, began to express concern about the time being taken. Nonetheless, the procedures established by the Board during the spring and fall hearings in 1950 set the pattern and the standard for natural gas export hearings that followed.

The routine, which was already well-established practice and which the Board intended to follow through the course of the natural gas export hearings, was reviewed by Chairman McKinnon at the initial Westcoast session.[95] He reminded those assembled at the hearing that the company had been required to publish notice of its application, along with date, time and place of the initial hearing, in each of the province's 114 weekly and six daily newspapers, and also that the applicant would soon provide for the Board and all others who were interested copies of the submission that Westcoast intended to present, so that everyone might have time to prepare for the cross-examination of the applicant's witnesses or prepare submissions of their own. Once the applicant's submissions were in hand, McKinnon explained, any other party intending to present a submission had to distribute copies to all the interested parties at least seven days before it was given as evidence.[96] McKinnon also announced that, despite this requirement, the Board would retain the discretionary option to allow any person unable to prepare a written submission to make a verbal representation to the Board.

The chairman announced that the Board's practice was to hear all the applicant's evidence first. This was usually done by presenting expert witnesses who, under the guidance of the applicant's counsel, attempted to lend authority to the important components of the application. Witnesses were subject to cross-examination by any participant at the hearing. Following the review of the applicant's presentation, the hearing was open to the submissions of interveners. The presentation of additional witnesses and extensive cross-examination were also characteristic at this stage. Having heard the applicant and all the interested parties, the Board might at this point recall certain witnesses for further questioning. Usually, the intent was to seek clarification of complex technical matters and to review some of the more contentious issues. With all parties heard and Board questioning complete, the applicant was given the opportunity to present closing argument. The hearing was then adjourned and Board members began the task of reviewing the evidence to prepare a written decision which, in the case of natural gas export hearings, was sent as a recommendation to the Alberta cabinet.

In keeping with this format, McDonald began Westcoast's presentation by calling three expert witnesses: Dr. Arthur Nauss, a Canadian consulting geologist with experience estimating reserves in South America and Alberta, who was currently chief geologist for Pacific Petroleums; Dr. J.F. Dodge, a former professor of petroleum engineering at the University of Southern California who had worked as a reserves consultant for both the private sector and the government of California; and Dr. Charles Hetherington, a chemical engineer with Ford, Bacon and Davis, an internationally known firm of consulting engineers from Monroe, Louisiana, that specialized in pipeline construction.[97] Nauss gave evidence on the central issue, the present level of Alberta's natural gas reserves; Dodge addressed the reserve situation in southern Alberta's two largest gas fields, Pincher Creek and Jumping Pound;[98] and Hetherington's testimony was focused upon the matters of natural gas deliverability and market requirements. The cross-examination of the witnesses was as vigorous as it was predictable. L.H. Fenerty, counsel for the city of Calgary, was unrelenting in his challenge of the Westcoast witnesses. He was not convinced that great trust could be put in reserve estimates and feared that natural gas exports would mean increased prices to Alberta consumers.[99] Early on, Fenerty got Dodge to admit that the development of the Jumping Pound and Pincher Creek fields had not been sufficient to eliminate the possibility of "barren areas," that the possibility of "barren areas" had not been taken into his calculations, and that his reserve estimate was at best a "respectable guess."[100]

He also led Dodge to remark that since Alberta had enormous coal reserves, "long after the gas has gone, the people in Alberta won't be cold...."[101] This was hardly the reassurance Calgarians were seeking. Dodge's unfortunate comment was reported by the

Calgary Herald, which continued to provide its readers with close daily coverage of the hearings, giving particular attention to Fenerty's questioning.[102]

From the outset, the hearings took on a direct political overtone, not only had a number of politicians come to attend the opening day of the hearings but also several stayed to participate in the cross-examination of witnesses. Liberal leader J. Harper Prowse asked certain witnesses to clarify the manner in which reserve estimates were determined, to comment upon the impact that exports would likely have on local pricing, and to explain what benefits Alberta firms and manufacturers might expect from such a large pipeline construction project. A.H. Wray, Independent Social Credit MLA for Banff-Cochrane, saw the Conservation Board hearings in a different light. He wanted to look at the Westcoast application, but not within the confines of the *Gas Resources Preservation Act.* Wray was convinced that there was a plot to get Westcoast's application approved before any of the other applications were heard and that he was doing his public duty in bringing this to light.[103] He had questions about the apparent discrepancies between the Westcoast Bill presented earlier to the Alberta Legislature, the promotional literature being circulated by Westcoast, and the company's submission before the Conservation Board. He wanted to know about the participation of Sunray Oil Corporation, if prominent businessman Max Bell was associated with Westcoast, what role Senator Harris had played in getting the Westcoast incorporation Bill through Parliament, and what prices the company intended to pay Alberta gas producers.[104] McDonald protested at Wray's line of questioning and Board Chairman McKinnon agreed that most of the questions were not pertinent to the consideration of the application.

So that politicians interested in the hearing could attend to their legislative duties, the hearings were recessed for the duration of the spring session of the Alberta Legislature. This simply allowed the members participating in the Conservation Board hearings to transfer the pursuit of their concerns about natural gas exports and Westcoast's application to a larger, and what they probably thought would be a less restrictive, forum.

The session was barely under way when Wray had the Legislature in an uproar over his announcement that he had been informed by "a man in Calgary" that "$200,000 in bribes [had] been paid out by the Westcoast Transmission Company to pave the way for its franchise." Wray also charged that he had been treated unfairly at the Board hearing, that his questions had been ruled out of order by the Board counsel C.E. Smith "whose brother [A.L. Smith, MP for Calgary West] was in on the filibuster" in Parliament that blocked the incorporation of Westcoast's rival, the Alberta Natural Gas Company.[105] In the ensuing clamour, Attorney General Lucien Maynard rose to shut off further debate, saying that the Westcoast application was still before the Board and that the matters raised were *sub judice* and not to be discussed for fear of prejudicing the outcome of the current hearing. Opposition members immediately took up the defence of Wray's right to speak. Prowse argued that if newspapers were allowed to make editorial comment on information given at the Board hearing, and if Westcoast could distribute brochures in support of their project, it hardly seemed correct that MLAS should be muzzled. His Liberal colleague, Hugh John Mac-Donald, insisted that the Board was set up by the Legislature and was not a court of a judicial body. In moving to sit down, he forgot that he had moved his chair and sat down heavily on the floor, contributing much laughter to the general commotion.

Able to speak only on natural gas exports, but still with reference to his observations at the Westcoast hearing, Prowse expressed his worry that the authorities being brought to give evidence at this and subsequent hearings were only interested in the case for the companies, not for the province, and he raised the question of whether or not the Board was adequately supplied with technical personnel to meet this important additional responsibility.[106] To assure that Albertans' interests were protected, Prowse also recommended that a "public counsel" be appointed to attend natural gas export hearings. He envisaged someone who would play a much stronger role of public advocate than the private counsel currently hired by the Board, who seemed more occupied with the management of the hearing process. Uncertain if the Manning government would act on these recommendations, Prowse tried to establish in provincial law an ironclad policy on the export of natural gas. Referring once again to the hearing under way before the Conservation Board, he informed his colleagues in the Legislature that those applying to export natural gas had based their estimates on 20 and not 50 years, and he speculated that gas export would likely mean a price increase for southern Albertans. Prowse sought an amendment to the *Gas Resources Preservation Act* that was intended to state government policy and give specific instructions to the Board. It read:

> The Board shall not grant a permit for the removal of any gas from the province unless such gas is surplus to the present and future needs of the province and the Board is satisfied that such

removal would not adversely affect, with regard to both supply and price, present and future consumers for a period of at least 50 years from the date of the Board decision.[107]

Both Manning and Tanner confirmed that this amendment did represent government policy, but they argued that it was imprudent to bind the Board so strictly, and that the granting of export permits should be left in the hands of the cabinet. The Legislature did not pass the Liberal amendment.

On April 11, the debating forum on natural gas exports shifted back to the Calgary court-house. The second round of the Westcoast hearing lasted just three days and closed with the question of Calgary's future gas supply still in the forefront, at least in the public mind. A brief submitted by the Canadian Western Natural Gas Company warning that it would be unlikely to have enough gas to meet peak Calgary demand next winter made front page headlines in the *Calgary Herald*.[108] Suddenly, the "future" was but a year away. The hearing closed almost as it had begun, with Calgary's counsel Fenerty taking another round out of one of the expert American witnesses.[109]

Fenerty's parting shots were aimed not only at deflating what he considered to be overly optimistic reserve estimates but also at the hearing procedure. As the hearing was about to be adjourned, Fenerty requested that in future the Board allow members of its own professional staff to take the stand and be available for cross-examination. This would allow others access not just to what knowledge the Board possessed but also an opportunity to evaluate and challenge Board evidence and interpretation. It was "desirable," Fenerty reasoned, "that all of the facts that are known and which might influence a decision should be known to all of the parties."[110] Others, it seems, had come to a similar conclusion, and McKinnon replied that the question of putting Board engineers on the stand had been considered and would likely be done at a joint hearing after the other applications had been heard. It was a step that would open hearings to the possibility of two-way exchange.

The first stage of the Westcoast hearing and the intervening session of the Legislature confirmed that the natural gas export hearings were going to have a difficult political dimension; however, the Manning government had been under no illusion about this, it was not unanticipated. What was also revealed, and perhaps not so much anticipated, by the government at least, was the technical aspect of the hearings and the demand that this was going to impose upon the Conservation Board. The government became convinced of the need to make Ian McKinnon a full-time Board chairman and receptive to the idea of the Board having its own lawyer and the need to bolster the strength of the Board's technical staff so that it could respond effectively to the parade of U.S. consultants who seemed likely to continue appearing at Board hearings.

With the completion of the initial stage of the first hearings under the *Gas Resources Preservation Act,* the Board turned its attention to the four applications outstanding.[111] Of these, the only applicant ready to proceed was Gordon M. Plotke. His application was distinguished from the rest in that it involved a much smaller volume of natural gas and was intended to bring gas export that was already under way into conformity with the new legislation. Plotke sought a permit to transport 600,000 cubic feet of natural gas per month into Montana from the Red Coulee field just across the border in Alberta. The gas from Plotke's oil well in Alberta was needed as a source of power to operate pumping engines at two of Plotke's Montana oil wells. It was pointed out at the hearing on April 17 that a permit to do so had been given annually by the Department of Trade and Commerce in Ottawa since 1934. The Conservation Board recommended approval, and on 29 May 1950 the Alberta cabinet concurred and Gordon Plotke earned the distinction of gaining the first permit to export natural gas issued under the provisions of the *Gas Resources Preservation Act.*[112]

In May, the Board began to hear the Northwest Natural Gas Company's application.[113] After taking this application through the introductory stage, the Board began to hear a new application from the Westcoast Transmission Company in June. Although it was not quite in the Plotke category, the new Westcoast application was more straightforward than the other applications under way or waiting to be heard. The proposal involved construction of a 17-mile pipeline from the Pouce Coupé field in Alberta to Dawson Creek, B.C., just on the other side of the border. Westcoast sought a permit to remove 207,165 thousand cubic feet annually for 30 years. Even though the hearing revealed that the Pouce Coupé field had ample reserves to supply Dawson Creek for the period requested, Grande Prairie and other communities in the Peace River area were opposed. Although further removed from the gas field than Dawson Creek, they insisted that no gas should be exported from the region until they were provided with gas.[114] That the opposition was led by the region's Social Credit MLA, Ira McLaughlin, did not deter Manning and his cabinet colleagues from accepting the Conservation Board's recommendation of approval. Westcoast gained its permit on 13 July 1950.[115]

While the Board was occupied through May and June, more applications had materialized, and by early July there was a backlog of two partly heard and five unheard applications. Growing industry and political frustration began to be apparent. In a front-page editorial on June 22, Calgary's *Daily Oil Bulletin* warned that the question of exporting Alberta natural gas to the Pacific Northwest had "been booted around Alberta and Canada now for over two years" and the need for a decision was now "urgent."[116] The *Bulletin* explained that the Fish Engineering Corporation of Houston had begun discussions with Spokane, Portland and communities in the Puget Sound area on a proposal to supply natural gas from Texas as early as the late fall of 1952. Editorialist Carl Nickle observed that it was "increasingly obvious that Alberta [had] gas aplenty to spare" and that, because the Conservation Board was concerning itself "with many matters not directly the concern of the Board or the Alberta Government," the hearings were proceeding far too slowly. There was a real danger, he warned, that Alberta would be shut out of the Pacific Northwest market and Pacific coast British Columbians would be left "sitting high and dry without natural gas."[117] Nickle urged the Manning government to speed up the decision. This was but the start of an escalating pressure upon the Board to proceed more quickly and to find that Alberta reserves were sufficient to allow natural gas export.

On 25 September 1950, the Conservation Board met to begin the scheduled hearing of the Western Pipe Lines application to export natural gas from Alberta for use in Saskatchewan, Manitoba, North Dakota and Minnesota.[118] After opening the hearing, Board Chairman McKinnon interrupted the usual procedure to bring to public attention three letters received by the Board. The first was from the solicitors representing Western Pipe Lines. They explained that the company "was somewhat concerned over the amount of time which [had] been devoted to the presentation and cross-examination of evidence dealing with pipeline routes, design and cost."[119] It noted that, even though the Board might require this evidence, it was the federal Board of Transport Commissioners that had to authorize the route and construction of any interprovincial pipeline in the end. This being the case, they advised that the proper place for a detailed discussion of such matters was before the Board of Transport Commissioners. The letter recommended that the Board consider conducting a joint hearing of all applicants focused upon the essential question "of the existence or otherwise of an exportable surplus of natural gas."[120] The second letter, dated September 20, informed the Board of the unanimously passed resolution of the Western Canadian Petroleum Association:

> That the Petroleum and Natural Gas Conservation Board be requested to come to a decision at the earliest possible date as to whether the gas requirements of Alberta and the reserves of the Province are such as to warrant the export of natural gas; and that in order to expedite such decision, they confine their enquiries for the present to these questions only.[121]

The third letter read by McKinnon was the most important. It was from N.E. Tanner, and it contained a memorandum from Minister of Trade and Commerce C.D. Howe. The memorandum explained that Ottawa had been informed by the U.S. Munitions Board that it had become concerned about a growing fuel shortage in the Pacific Northwest states, where industrial expansion induced by the Korean conflict seemed likely to accentuate the problem. "It is suggested," Howe wrote, "that the availability of large supplies of natural gas in the Province of Alberta is one source from which the scarcity of fuel and power in the Pacific Northwest could be alleviated. I am asked the question whether these supplies can be made available."[122] Howe went on to explain that the letter stated that, if supplies were not available, immediate steps would be taken to obtain natural gas from Texas. The minister advised that there was a need for Alberta to decide quickly, one way or the other, if the province had natural gas available for export. Perhaps to indicate what the decision should be, Howe concluded, "I see little prospect of a line being built from Alberta to the Canadian Northwest unless that line can be extended from Vancouver southward to serve the Pacific Coast cities."[123] In light of the urgency expressed in the minister's letter, Tanner informed McKinnon that he had been instructed by his cabinet colleagues to ask the Board to do all that it could to speed the hearings and determine "the amount of proven reserves of deliverable natural gas within the province and after assessing the province's foreseeable needs advise whether or not and to what extent there was an exportable surplus."[124] After discussion of the letters, McKinnon announced that the Board would hold a joint hearing of all applicants beginning on October 30 to answer these questions.

The questions that the Board was asked to answer were technical, but the letters read by McKinnon hint at the political environment that helped to shape the questions and sought to influence the responses. Although the letters suggested something of the political backdrop, they revealed little of the activity behind the scene, at least to most contemporary observers. The letters presented to the Board and made known to the Alberta public were only a select part of a correspondence web on the gas export question that connected Ottawa and Washington, the two Canadian levels of government, and the pipeline company applicants.

The first letter to urge the Alberta government to adopt joint hearings to speed matters along was not that of Winnipeg-based Western Pipe Lines. The Northwest Natural Gas Company had already argued in favour of this approach in a lengthy memorandum to Premier Manning.[125] Conscious of the hostility of some prominent Calgarians towards the Northwest Natural Gas Company's export proposal, Manning decided that this was a letter better kept in his file. While the Northwest Natural Gas promoters were advising Edmonton as to the proper course, a senior official of another U.S.-based contender, the Pacific Northwest Pipeline Corporation of Houston, was in Ottawa hoping to persuade C.D. Howe. Perhaps by coincidence, Howe's letter to Tanner on September 16 alerting Alberta to the growing shortage of fuel supplies in the Pacific Northwest, was sent just following a meeting with the vice-president of Pacific Northwest where the merits of a continental approach to the sharing and distribution of natural gas resources were discussed. When Vice-president T.H. Jenkins returned to Houston, he wrote to thank Howe for the opportunity to discuss his company's proposal "for the furnishing of gas to the various utilities and industries throughout your fine country." He also commented, "Since returning from Canada, I have had the pleasure of reading a copy of the letter which you sent to the Government of Alberta," which he noted had been read at the Conservation Board hearing in Alberta.[126] To remind Howe of the continental exchange that his company had in mind, Jenkins explained that they needed 100,000,000 cubic feet per day from Alberta for the Pacific Northwest. When more gas was available, they would contract for a volume in excess of 500,000,000 cubic feet, and when the exportable volume from Alberta passed the 250,000,000 cubic feet a day mark, Jenkins told Howe, "we will bring gas to your good people in Eastern Canada."[127] Jenkins's argument was that the U.S. Federal Power Commission would never allow his group to extend their pipeline to supply eastern Canada with gas unless a nearly equivalent volume of natural gas was being made available to the Pacific Northwest from western Canada.[128]

Tanner's reply to Howe's letter of September 16 is revealing. On the same day that he sent the trade minister's letter to the Conservation Board, Tanner informed Howe that the Alberta government was prepared to co-operate in every way to make gas available. He added, "Though the Conservation Board has not completed its hearings, I might say that the Government is satisfied that the present proven reserves *are sufficient* to meet Alberta's requirements."[129] Howe's response contained the reminder that the federal Board of Transport Commissioners could not deal with the route of any pipeline until the applicant could produce a permit to export gas from the Alberta government, but he assured Tanner that, once the Conservation Board had done its part, he would see that the federal Board lost "no time in disposing of such an application." To maintain the pressure, Howe concluded, as he had done in his first letter, "It seems to me," he wrote, "that time is of the essence in obtaining suitable outlets for such gas from your Province as may be available for export."[130]

Both governments it seems sat back in anticipation that the Conservation Board would deal quickly with the reserves questions and come to a "favourable" decision. When nearly two months passed without a ruling, Howe became impatient. On November 22, Jenkins wrote to Howe to say that an Alberta gas export permit was the critical factor in his company's plan to bring natural gas to eastern Canada. He thanked Howe for any assistance in bringing to the attention of Alberta officials the "great urgency" in quickly reaching a decision.[131] Howe replied that he had reason to believe that Alberta would "authorize export from that Province before the end of this year."[132] A timely dispatch from the Canadian ambassador in Washington gave Howe the opportunity he needed to impress upon the Alberta government once more the need to come to a quick decision. The telegram from Washington that Howe passed on to Premier Manning disclosed that the U.S. Munitions Board was urging Canadian federal authorities to take action "to obtain an early and favourable decision by the Gas Conservation Board of Alberta permitting the export of natural gas from Alberta to Montana."[133] The ambassador noted that the Montana Power Company was particularly concerned about the hearings under way in Calgary, because it desperately needed more gas to supply expanding copper, lead, zinc and manganese refining operations in the area. In addition, the Canadian government was advised that the Department of the Interior's Petroleum Advisory Board for Defense was anxious to encourage the export of gas from Alberta to Mon-

tana and the Pacific Northwest, and to this end had included a provision for a pipeline from Alberta in their petroleum steel program.
In his reply Manning informed Howe that he had discussed the ambassador's dispatch with Tanner and McKinnon and that he expected to have the Conservation Board report on the province's
natural gas reserves by January 20. Ever cautious, but also anxious
to see gas moving out of Alberta, Manning offered Howe a suggestion. "It is possible that the report might be modified if we are officially advised by the Government of Canada that natural gas is urgently needed in the interests of national defence in either Montana
or the Pacific Northwest or both."[134] Undoubtedly, Manning knew
that, despite all the pressures that had been put upon it to find otherwise, Alberta's Conservation Board was going to report that the
province's natural gas reserves were as yet insufficient to permit export. Alberta's premier was aware that he would need to have a
strong and specific emergency case to be able to sidestep Albertans'
traditional hostility to natural gas export, a hostility soon to be reinforced by the Conservation Board's decision. Since Howe's reply
did not come before the release of the Board report, the precise nature of the "modification" that Manning had in mind will never be
known. The sequence of events that followed through the spring of
1951, however, reveal how the "emergency case" scenario evolved.

The interim report on Alberta's natural gas reserves and requirements presented by the Conservation Board on 20 January 1951
was based upon an analysis of the evidence presented by the five
companies seeking a natural gas export permit and seven other interested parties.[135] Presentations before the Board were much like
those before the earlier Dinning Commission. Many of the same experts reappeared, but with revised evidence based upon more extensive drilling. There was, however, a critical difference. In this instance, the presentations were before a professional and technically
competent panel. This meant that witnesses faced, and had to deal
with, questions of more penetrating technical depth. In addition,
the Board undertook its own study of Alberta's natural gas reserves
and, in keeping with Fenerty's request, made an important adjustment to its hearing format. After the applicants had presented their
experts for cross-examination, the Board put its chief engineer,
G.E.G. Liesemer, on the stand to present his estimate of Alberta's
natural gas reserves and to answer questions from his peers. The
differing estimates of Alberta's current "disposable" natural gas reserve are shown in Table 6.3. It must be noted that, since two of the
applicants presented estimates for only part of the province, and as
the number of fields included in the other estimates varied, the esti-

TABLE 6.3 Estimates of Alberta's Disposable Natural Gas Reserve
Submitted to the PNGCB Joint Hearing Oct.-Nov. 1950

Estimate by	Volume of Disposable Gas (MMMcf)
Westcoast Transmission	7,023[1]
Northwest Natural Gas Company	5,006[2]
Prairie Pipe Lines	4,684[3]
Western Pipe Lines	5,614[4]
McColl-Frontenac and Union Oil Co.	6,284[5]
G.E.G. Liesemer	3,635[6]

SOURCE: PNGCB, Interim Report, 20 January 1951.
1. Estimate for entire province, includes 90 MMMcf beyond economic reach.
2. Estimate for 11 fields, excludes Medicine Hat and other local reserves.
3. Estimate for nine fields with over 50 MMMcf each.
4. Estimate for entire province, includes reserves beyond economic reach.
5. Estimate for entire province, includes 660 MMMcf beyond economic reach.
6. Estimate for entire province, but excludes reserves less than 20 MMMcf, and includes 191 MMMcf beyond economic reach.

mates are not directly comparable. Although the table gives only an indication of the range of estimates, it does underline the difficulty faced by the Board in assessing the mass of submitted technical data.

Although the companies differed in their estimates of established reserves of disposable gas, they were unanimous in their opinion that already proven reserves were sufficient to assure Alberta's long-term needs and that, given the nature of the vast sedimentary deposits underlying the majority of the province, future oil and gas discoveries were assured. They warned, however, that the pace of discovery would not be accelerated or maintained without the incentive of an export market. In its assessment of the evidence, the Board acknowledged that, although "all the engineers and geologists who submitted reserve estimates were sincere and endeavoured to provide the Board with all available information at hand," the paucity of solid factual data meant that in many instances the chief factors in their calculations were "judgment" rather than scientifically justified figures.[136] The Board weighed the company evidence accordingly, reviewed the data presented by its own senior engineer, and came to an independent conclusion regarding the province's gas reserves. It found the established reserve of disposable natural gas within economic reach to be 4,439 billion cubic feet.[137] In the Board's judgement, company estimates of reserves

TABLE 6.4 Estimate of Alberta's Natural Gas Requirements
1 January 1951–31 December 1980
(billions of cubic feet)

Year	Domestic	Commercial	Industrial	Total
1951	20.1	14.5	19.0	53.6
1952	21.4	16.1	21.1	58.6
1953	22.2	17.2	24.0	63.4
1954	22.9	18.6	26.8	68.3
1955	23.6	19.6	30.3	73.5
1956	24.1	20.4	33.2	77.7
1957	24.9	21.2	36.9	83.0
1958	25.7	22.2	40.5	88.4
1959	26.5	22.9	44.2	93.6
1960	27.3	23.8	47.7	98.8
1961–62*	27.6	24.1	48.3	100.0
1963–64	28.6	25.0	49.9	103.5
1965–66	29.4	25.7	51.4	106.5
1967–68	30.2	26.4	52.9	109.5
1969–70	31.2	27.3	54.5	113.0
1971–72	32.0	28.0	56.0	116.0
1973–74	33.1	29.0	57.9	120.0
1975–76	34.2	29.9	59.9	124.0
1977–78	35.0	30.6	61.4	127.0
1979–80	36.2	31.6	63.2	131.0
Total	873.1	751.7	1,434.5	3,059.9

SOURCE: PNGCB, Interim Report, 20 January 1951.
* Please note that the data for each year after 1960 are not shown, whereas totals include data for all the years 1951–1980.

were unrealistically high, and it found the presented estimates of Alberta's future needs, including those of the province's gas utility companies, to be too low. Table 6.4 shows the Board's yearly and cumulative estimate of Alberta's requirements to 1980. (Also see Figure 6.1.) To ensure that the requirement of 3,059.9 billion could be delivered, the Board estimated that established reserves should be "in the order of four and one-half trillion cubic feet."[138] Although the difference between estimated reserves and requirements was small, it was because the Board had projected the province's needs only 30 years into the future rather than 50 years as was the government's stated policy. This, along with the recognition that the reserve estimates for some of the fields rested upon insufficient

FIGURE 6.1   Estimated and Actual Alberta Natural Gas
                      Consumption, 1951–1980

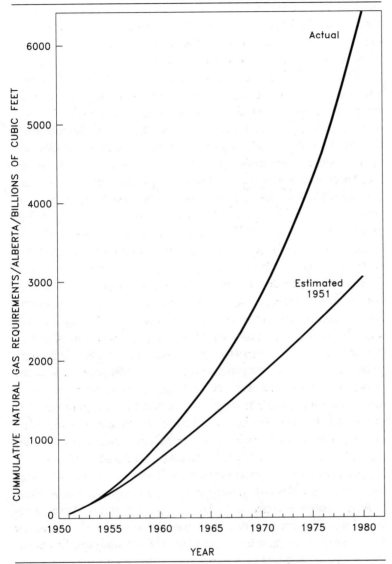

data, led the Board to advise the Alberta cabinet that the province's natural gas reserves were not yet large enough to issue export permits to any of the applicants. They recommended that further gas export hearings be adjourned until 4 September 1951.[139]

Bound by the government's policy to come out against the recommendation of gas exports, the Board was concerned nonetheless about the potential impact of its decision and offered its opinion that development of the province's gas reserves had been hindered by the government's long-standing restrictions on export. McKinnon and his colleagues were conscious that, over the previous 20 years, there had been a number of promising gas strikes, but because of limited markets, there had been little follow-up drilling. They advised the government to reduce the 50-year reserve requirement to a rolling 30-year supply, insisting that this would protect Alberta consumers adequately and have a less stifling impact upon the development of natural gas and its related industries.[140] This was welcome advice, for Manning and Tanner had been reluctant from the outset to tie the government to a natural gas export policy that rested upon a reservation time span of 50 years. They had acquiesced only under the intense pressure of popular opinion. As Tanner had told Howe the previous fall, the Alberta cabinet was convinced even then that Alberta's proven gas reserves were sufficient to meet anticipated needs. The 30-year reservation suggested by the Board was greater than the 20 years preferred by industry, but it still reduced the reserve threshold enough to encourage continued development. At the same time, given the sanction of an independent authority such as the Board, it would be easier for Manning and Tanner to present the 30-year time frame to Albertans as "reasonable" assurance. Outside the featured recommendations, tucked into the body of the report, was the observation that there were scattered reserves on the provincial periphery that were beyond the economic reach of the province's existing utility systems and could be considered available to supply markets outside the province. The Pakowki Lake area on the Alberta-Montana border was identified as containing several small fields from which export might be considered, were it not for the government's enunciated policy of giving preference to Canadian consumers. It was a passing statement, the significance of which would not be apparent for several months.

At a press conference on 25 January 1950, Premier Manning released the Conservation Board's "Interim Report" and announced his government's complete agreement with its findings. Declaring that a lack of adequate proven reserves at present meant that no export of natural gas could be allowed, he emphasized his belief that once sufficient reserves existed to assure Alberta's needs, natural gas export would be in the province's best interest. To reassure the oil and gas industry, Manning indicated there would be abandonment of the 50-year reserve requirement in favour of the 30 years suggested by the Board, and he noted that this new reserve level was now close to being achieved.[141] The premier also called attention to the report's promise that export hearings could be reopened immediately if new evidence warranted, and to stimulate the search for new "evidence" he announced that the existing oil and gas regulations would be amended to make them more suitable for natural gas exploration.[142] Finally, Manning informed Albertans that there was one additional factor that could have a bearing on the gas export situation. He pointed out that, just as the Conservation Board had had to adjust its preferred conservation program during World War II to increase oil production for the war effort, one could not discount the possibility that developments in the international field might again compel a change of policy.[143]

Reaction was immediate and varied. Oil and gas company shares slumped. The *Calgary Herald* denounced the premier's decision as "the worst that could have been made" and predicted that Texas companies would capture the Pacific Northwest natural gas market.[144] Carl Nickle, editor-owner of the *Daily Oil Bulletin*, tried to put a better face on the situation. He hailed the government's decision as a "limited victory." Nickle saw "victory" lying "in the clear-cut statement by Alberta's Conservation Board and Premier E.C. Manning that they favour[ed] export" and in the new 30-year reserve period, which now left a relatively small gap of additional reserves to be made up before export hearings could recommence.[145] Alberta's Opposition politicians were roused once more to denounce what they perceived as the potential sell-out of Alberta's long-term interest. In Ottawa, attention focused on a different part of Premier Manning's press release. Although Howe hardly needed coaching on the matter, an astute observer directed his attention to "the important part of the Premier's statement." The reference was to Manning's concluding comment.

> If due to the uncertain international situation it should develop that Alberta gas is needed urgently for defence purposes the Government will give immediate consideration to any official request from the proper authorities knowing that the people of Alberta are ready and prepared to play their full part in any program necessary to the security and defense of this continent.[146]

On February 21, Howe wrote to Premier Manning, enclosing a specific request from Charles E. Wilson, director of the Office of Defense Mobilization in Washington.[147] In his memorandum, Wilson explained that he had resisted pleas from interested U.S. groups that he use his influence to persuade the Canadian and Alberta governments to allow the construction of a natural gas pipeline to the U.S. Pacific Northwest. Now, he was faced with an urgent and specific need. The limited natural gas supply available to the Montana Power Company would be insufficient to meet the needs of the Anaconda Copper Mining Company after 15 September 1951. Wilson noted that in anticipation of this deficiency the Montana Power Company had purchased, subject to the acquisition of necessary export permits, virtually all the gas reserves in the Pakowki Lake area. The pipe required to connect the Pakowki field with the Montana system was already stockpiled, and Wilson urged that an export permit be allowed since Anaconda's copper production was "badly needed."[148] Howe reminded Manning of Alberta's commitment to give serious consideration to any urgent need of natural gas for defence purposes. "It seems to me," Howe asserted, "that the situation of Anaconda Copper Mining Company outlined in Mr. Wilson's letter constitutes an urgent need for defence purposes."[149] Howe asked Manning for a prompt reply, since any curtailment of Anaconda's production "would be a crippling blow to the United States war effort," and if gas could not be obtained from Alberta it was necessary that Wilson have as much time as possible to find an alternative source.

For Manning, the request came at a good time. The spring sitting of the Alberta Legislature had just begun, and therefore the appeal from Washington via Ottawa would not involve calling a special session. Although the "emergency" request was hardly unexpected, Manning knew that any debate on the matter of natural gas exports would generate a good deal of heat. Cautiously, he set about further preparation of the ground. In a telegram to Howe, he advised that special legislation would have to be prepared, and he directed that Wilson be informed that a condition of any export permit would be that the gas be used exclusively to meet Anaconda's requirements for defence purposes over a five-year period. He asked also that Wilson prepare a document stating the precise daily volume required.[150] Once the desired documentation was in place, and after the draft legislation had been submitted for Howe's consideration, on 22 March 1951 Premier Manning presented to the Alberta Legislative Assembly his *Act to Permit the Temporary Export of Gas to Montana for Essential Defence Production.*[151]

The legislative environment into which Manning introduced his Bill could hardly have been more conducive to the support of a measure intended to assist the United States counter communist aggression. While the actions of allied forces in Korea occupied the front pages of the daily newspapers and fed a growing popular anxiety, Alberta's own variant of the anti-communist hysteria that gripped the North American continent was manifest in the Alberta

Legislature. At the beginning of the 1951 session, discussion in the Legislature had been sidetracked over an observation by one of Manning's cabinet colleagues that 500 to 600 communist spies were active in Alberta.[152] Later, a Social Credit back-bencher's assertion that an Alberta teacher was "exalting to the skies conditions in Russia and was flouting Canadian institutions" led to the demand that teachers in the province be compelled to take an oath of allegiance.[153] This subcurrent of paranoia within the Legislative Assembly surfaced to engulf one of the sitting members. When Arthur Wray, the Independent Social Credit MLA for Banff-Cochrane and an active member of the Canadian Peace Council, spoke to question Canadian support of U.S. intervention in Korea, he aroused the vitriolic censure of virtually the entire Assembly and was denounced as "a Communist of the deepest dye."[154] Yet, it is remarkable, given this atmosphere and predisposition to look favourably upon any measure judged important for continental security and defence, that the Manning government's attempt to link the export of natural gas to a specific wartime emergency was still insufficient to overcome the traditional caution of Albertans.

Although there had been no advance warning in the Throne Speech of the government's intention to introduce a special gas export Bill, there had been several rounds on the gas export question that helped to prime the Assembly for the ensuing debate. At the beginning of the session, Liberal MLAs had warned that the government appeared to be abandoning its commitment to a 50-year protection policy. They charged that, despite the findings of the Dinning Commission and the recent Conservation Board report, Manning and his cabinet had already decided in favour of natural gas export and were unsuitable trustees of Alberta's resources.[155] The CCF members expressed their displeasure by calling for a complete and indefinite ban on gas exports to the United States. As A.J.E. Liesemer told the Legislature, "If Americans want our natural gas let them come here and build their factories and we will get the payroll benefits."[156] Later, in a combined motion, the Liberals and CCF tried to persuade the government to engage an economist to undertake a comprehensive study of the Alberta economy and to report upon the economic effects of gas export before the Conservation Board reopened hearings. The minister of mines and minerals countered to say that the Conservation Board had taken economic factors into consideration in the preparation of its report. In defending the idea, Liberal leader J. Harper Prowse responded that "the Board's considerations dealt with physical qualities and requirements, not the economic effects of export upon all phases of

Alberta life," and he noted that "most of those at the hearings were interested in selling gas. No substantial evidence was presented on the economic effects on Alberta consumers."[157] A few days after this debate, Opposition members learned that not only would there be no special study on natural gas and the Alberta economy but also that the government intended to bypass its newly established hearing process for at least one of the current gas export applications.[158] The proposal was to remove the application of McColl-Frontenac Oil Co. Ltd. and Union Oil Company of California to deliver southern Alberta gas to the Montana Power Company from assessment before the Conservation Board as required under the *Gas Resources Preservation Act* and to give it speedy approval in the Legislature with the aid of special legislation.[159]

The "emergency case" foundation upon which justification for this quick approval strategy rested did not prove as compelling as Manning might have hoped. He began by explaining that his Bill was motivated by the request that his government had received from "authorities in Washington and Ottawa," and he read from the Wilson and Howe correspondence to establish that the proposed export of natural gas in this instance was a "defence emergency measure." Manning relied upon Howe's more forceful words to lend credence to the "crisis" claim, and he quoted the minister of trade and commerce to inform the legislature that "it would be a crippling blow to the United States war effort were production of Anaconda Copper seriously curtailed."[160] After making the case for a vital need that required extraordinary treatment, Manning assured that this should not be seen as a precedent. Rather, it was a unique case, which was reflected in that the Bill restricted use of the exported gas to Anaconda. The premier reminded the Legislature that, when the Conservation Board's interim report had been released in January, he had clearly stated then that his government would consider defence requests apart from the normal manner through the Conservation Board.[161] By making the point that Albertans had been given prior notice of how the government intended to deal with such an application, Manning hoped to deflect criticism that this was not merely an expedient approach dictated by a desire to avoid delay. The remaining part of the premier's defence centred upon the question of Alberta's natural gas supply. He noted that, in addition to being beyond the economic reach of Alberta's utility systems, the fields from which the gas was to be drawn could not be classed among Alberta's major fields. Moreover, even though the natural gas in the Pakowki Lake area was by location and volume peripheral to Alberta's needs, Manning pointed out that his Bill placed an export ceiling of 10 billion cubic feet annually for a five-year term.

Opposition speakers immediately expressed their doubts about whether Alberta's interests were being protected. The Liberal leader pointed out that daily allowable export to Anaconda of 40 million cubic feet was not insignificant; it equalled nearly one half the daily volume required by that area of the province north of Red Deer.[162] Prowse also complained that MLAs lacked sufficient information to be able to judge whether or not the requested gas was an essential defence requirement. All they had was a request from the U.S. government on behalf of Anaconda. Prowse suggested that the people concerned with the needs of Anaconda and Montana Power be brought before the Legislature to prove their claims. "Nobody is going to worry about Alberta," he insisted, "we have to do it ourselves."[163] Perhaps the most revealing question came from CCF leader Elmer Roper. "It is strange," he noted, "that if we are engaged in a joint defence effort and copper is so vital, Canadian copper should have such high tariffs against it. It would be well to know if it is possible to have Canadian copper placed on the free list going into the U.S., if we are to export our gas."[164] Following this line of thought, Liberal Hugh John MacDonald asked about Consolidated Mining and Smelting at Trail, B.C. This company was engaged in wartime production; he wondered if it would get Alberta natural gas. As Alberta's leading daily newspapers noted, no one on the government side chose to respond to these points.[165] Put off by Opposition sniping, Premier Manning reminded them that the United States had not restricted the movement of gasoline to Canada during the war. He drew debate to a close with the rejoinder:

> Some Albertans who are adamant in their ideas of what the government should do in a case like this one should recall the treatment the United States afforded Albertans and Canadians during the Second Great War.[166]

The outcome of the debate was never in doubt. In the end, the government's massive majority prevailed, and on 6 April 1951 the first significant export of Alberta's natural gas was approved after a truncated discussion in the Legislature rather than after a hearing before the Conservation Board.

A month after the approval, the president of Northwestern Utilities and Canadian Western Natural Gas wrote to Premier Manning. Milner stated his distress over the government's decision to authorize the export of gas to Montana. "It seemed to me," Milner wrote, "that Anaconda was being used as a screen to enable McColl-Frontenac, or rather the Texas Company and Union Oil of California, to make a ten million dollar profit."[167] He then called Manning's attention to the Montana Power Company's recently released "Annual Report," which revealed that the company had brought six new natural gas wells into production and also that new reserves in southern Montana were also being used. In addition, the report stated that five new communities, a large cement operation and another part of the Anaconda company's reduction works had been connected to Montana Power's distribution system. Milner concluded:

> It is rather peculiar that the Montana Power Company was continuing to make commitments for the delivery of gas up to the end of last year if its gas position was so shaky.[168]

Milner was not misled by the Anaconda "emergency," but he was perhaps less aware that the export of Alberta gas was but part of a larger agenda. During the period that the special gas export Bill was before the Alberta Legislature, Manning, Tanner and Howe were also thinking about a continental exchange of crude oil. Early in February, Tanner wrote to Howe to remind him that the potential market for Alberta oil in the United States was hindered by an import duty, and he suggested that some arrangement be worked out whereby Alberta oil could go there on an exchange basis. Howe replied that such discussions were currently under way. He explained that Canada was meeting with the Americans "to work out a continental supply system adequate to war needs." Howe assured:

> The U.S. will expect Canada to play a large part in expanded production, transportation, and refining projects designed to bring the continental situation into balance. In that connection, an oil pipeline from Alberta to the Pacific Coast is taking a leading part in our discussions.[169]

Given the discussions under way, Alberta was anxious to accommodate. Manning and Tanner did not want to be seen as obstructing U.S. access to Canadian natural gas, especially if such access might help the Americans look more favourably upon the entry of the province's crude oil and given their presumption that approval of natural gas export was just months away in any case.

This was not an opportunity to be missed, even if it meant promoting an "emergency" case to circumvent existing natural gas export legislation.

As it turned out, the pipe assembled to build the 90-mile link to bring "urgently" needed natural gas to Montana remained stacked on the ground for months before the U.S. Federal Power Commission saw fit to approve the necessary import licence.[170] Natural gas did not begin flowing across the Montana border until 7 February 1952, almost 11 months from the vote of approval in the Alberta Legislature.

Primed by the Anaconda "emergency" debate in the Legislature, proponents and opponents of natural gas export waited for the next round of hearings scheduled to begin before the Conservation Board on 10 September 1951. In the interval, companies seeking a coveted export licence devoted as much of their energy to the intensive lobbying of politicians and courting the Alberta public as they did preparing their technical briefs for McKinnon, Govier and Goodall. The old contenders were Northwest Natural Gas Company, headed by New York engineer-promoter Abner Faison Dixon, who proposed to sell Alberta gas in the Pacific Northwest;[171] Frank McMahon and his Westcoast Transmission Company, which had also been incorporated to market natural gas in the Vancouver, Seattle and Portland areas; Western Pipe Lines, a company organized by several of the most prominent members of Winnipeg's financial elite, Lionel Baxter, Edward Nanton and Gordon Osler, to supply the Canadian prairie region plus a U.S. market centred in North Dakota and Minnesota; and Prairie Pipe Lines Ltd., headed by Ray C. Fish of Houston, Texas, who hoped to tie Alberta natural gas into the vast continental grid that he was building under the corporate umbrellas of the Tennessee Gas Transmission, Trunkline Gas, and Panhandle Eastern Pipeline companies.[172] In addition, there was Canadian Delhi Oil Ltd., and its subsidiary Trans-Canada Pipe Lines Limited, an aggressive new applicant headed by Clint Murchison of Dallas, a Texan of immense energy and wealth. Anticipating that it would be almost as much a fight to get any scheme to import Canadian natural gas past a variety of hostile U.S. interests as it would be to persuade Canadians to allow export before their own needs were fully determined and served, Murchison offered what he believed would be a more politically acceptable and achievable alternative. The plan he presented did not involve dependence upon a U.S. market; he proposed to take Alberta gas over an all-Canadian route to serve southern Ontario consumers. Each group set about to try to change the climate of popular opinion in Alberta, to try to persuade Manning, Tanner and Howe of the merits of their particular schemes, and to discredit the claims of their rivals.

Given the continental scope of their ambition, the Fish organization directed most of its attention towards Ottawa. They reminded Howe not only of the benefits of continental energy exchange but also that they were the only organization with the capital resources and the pipeline infrastructure in place that could make such a plan a reality.[173] Murchison also concentrated on the minister of trade and commerce. In January 1951, he took Howe on a winter holiday, found the minister receptive to his ideas and a firm friendship ensued.[174] Perhaps because they were more sensitive to regional political realities, the Calgary- and Winnipeg-based applicants directed more of their lobbying efforts towards Premier Manning and Tanner. Ray Milner, a director of Western Pipe Lines and the most prominent figure in Alberta's natural gas utility industry, for example, warned Manning of those who professed to be "gas exporting friends." To support his point, Milner passed on to Manning part of a letter allegedly written by the senior lawyer working for the Northwest Natural Gas Company. The letter discussed how opinion might be mobilized in Washington state to thwart the McMahon proposal.

> We have come to the conclusion that it would be advisable that the responsible groups in the State of Washington indicate that they will not take any gas at the tail end of a Canada-for-Canadians deal with the proposed differential in price when, as everyone knows, it will be impossible to build any line unless at least 75% of the output is purchased in the States. In other words, the largest purchasers, who alone can make any project feasible, should start to say that they are not going to take the crumbs after the Canadians play their game of Canada-for-Canadians.[175]

In contrast to the latecomer Americans, who wanted to cash in on Alberta's resources, the Winnipeg group presented themselves as deserving pioneers who had invested and persevered through all the industry's ups and downs since 1914. The president of Western Pipe Lines reminded Premier Manning that this should not be overlooked when the time came for a final decision on the merits of the various gas export applications.[176]

The almost desperate scramble for preference was a reflection that the stakes were high. By 1951, Alberta's exciting prospects had become the subject of world attention. In July, the *Daily Mail* presented Alberta to British readers as "The Empire's Oil Klondike." Caught up and carried away by the boom atmosphere, the *Mail's* travelling correspondent declared, "It is a pity the Persians are not with me here in Alberta. Perhaps if they were they would realize that one day they may have to peddle their oil for pennies, because this is where most of the Commonwealth's and Western World's oil is going to come from."[177] In a more informed but no less enthusiastic manner, *Time* magazine offered its readers a different perspective. In a feature presentation, titled "Texas of the North," that included a map and several pages of colour photographs, Americans learned that on this new northern frontier, "The brisk, winy aroma of prosperity is in the air."[178] *Time* acknowledged that with discoveries of iron ore, nickel, copper and uranium all of Canada had expanded dramatically since the war, but declared that the biggest boom of all was oil in Alberta. The Alberta discovery was heralded as "the most significant new find on the continent since Texas' Spindletop roared in, 50 years ago."[179] In addition, *Time* reported that it was expected that Alberta would soon agree to release natural gas for export and set off "a rush of pipeline building that will rival the railroad era."[180]

This heady atmosphere contributed to the sense of excitement and urgency that began to build in anticipation of the next round of Conservation Board hearings. Like those of the previous autumn, these were joint hearings, held essentially to determine whether or not the province's expanding natural gas reserve was sufficient to permit favourable consideration of one or other of the pending export applications. But it was not just a matter of going over the same ground once again. Of the 50 fields now to be assessed, many were newly discovered, and in almost all the others, recent development work meant that there had to be a reworking of the data. From September 10 to December 14, 1951, McKinnon, Govier and Goodall listened to what must have seemed like an endless parade of company engineers and geologists present detailed, often complex and sometimes controversial analyses of the reservoir characteristics and production potential of each field. In turn, Alberta utility companies and others offered their projections of Alberta's needs over the next 30 years. As the Board members listened, their growing technical staff considered the data submitted at the hearings along with that generated by the Board itself to come to their own conclusions about reserves and requirements.

As the weeks passed, the anxiety level began to rise, again mutterings were heard among the applicants and within the industry that the Board was being overly cautious and taking too long. At the same time, public unease became more vocal, especially in Calgary, to the extent that the minister of mines and minerals felt constrained to counter the "biased" and "false propaganda" regarding the question of gas export.[181] Over the radio in late October, Tanner attempted to reassure Albertans that they could rely upon the government and particularly its Conservation Board to make the proper decision. He reminded his listening audience that, "when one realizes how costly it is to establish natural gas reserves, it is not difficult to understand that markets must be available for the gas if exploration and development is to continue." But as he and Manning had done so often in the past, Tanner was also quick to insist that, although the government was in favour of export, only when there was found to be a surplus beyond Alberta's own long-term requirements would gas export be permitted. He added an important new qualification that "any export permit will provide that all constituents of the gas, with the exception of methane, must be processed within the province if the government should so decide," and he closed with, "Surely the action of the government and the thoroughness with which the Board has gone into this whole matter . . . can be taken as assurance that the interests of the people of Alberta are being fully protected."[182]

If the radio address helped to convince some doubting Albertans that they could rely on the Conservation Board, those who were following the gas export hearings in Calgary might have had second thoughts following the presentation of a startling and unexpected submission just as hearings were about to conclude. Ray Milner, president of Alberta Inter-Field Gas Lines Ltd., stepped forward to express his fear that Alberta was in danger of losing control over gas exports. He argued that, if a company exporting Alberta gas or its subsidiary constructed a pipeline system to collect gas from various fields right at the wellhead and then tied this system directly to its export pipeline, such a system would come under the jurisdiction of the Government of Canada since the senior level of government had exclusive authority to regulate interprovincial trade and commerce. Milner suggested, as he had before the Dinning Commission, that this situation could be avoided and the transportation of gas within Alberta retained under provincial authority only if exporters were not allowed to build their own integrated gathering and export systems. Instead, a provincially incorporated company, such as Inter-Field, should operate all the gathering facilities within the province and deliver gas to exporters at appropriate points on the Alberta border. Operating only within the province, such a company would function as a public utility fully under provincial control. Finally, Milner warned that such an approach was essential given that the primary legislation under which the province presumed to regulate natural gas exports, the *Gas Resources Preservation Act,* was fatally flawed and could not be relied upon to defend the interest of Alberta consumers. Milner's point, supported by a written opinion from Alex Smith, a recognized authority in constitutional law at the University of Alberta, was that the province, through the *Gas Resources Preservation Act,* or any other Act for that matter, could not regulate interprovincial trade or interprovincial pipelines. He directed attention to section 9 of the Act, which provided that in any emergency the Board, with cabinet approval, could "divert gas from the holder of an export permit to consumers within Alberta."[183] Milner pointed out that in all likelihood this would be interpreted as an infringement upon the federal government's exclusive authority to regulate trade and commerce. In the submitted brief, Smith reviewed the relevant court decisions that had upheld Ottawa's undivided authority in the trade and commerce area and concluded that, just as section 9 was invalid, so was section 4, requiring every person before exporting any gas to apply to the Conservation Board for a permit to do so, and section 13, which prohibited export except under permit.[184]

By the time the joint hearings ended on December 14, McKinnon and his fellow Board members had on hand the vast body of technical evidence that they would need to prepare an informed recommendation to the Alberta cabinet. Milner's last minute intervention, however, doomed what little prospect there might have been to keep the impending debate on natural gas exports focused upon the technical evidence expected in the Board's report. The question of the *Gas Resources Preservation Act*'s validity raised to public notice a sensitive issue that undermined Premier Manning's contention in the Legislature and elsewhere that, in the unlikely event that the province's reserves or needs had been misjudged, section 9 of the Act provided the ultimate protective backstop.[185]

While the Conservation Board went about preparing its report and recommendation, a final flurry of lobbyists' memoranda were despatched to Edmonton and Ottawa. Among these was a reminder for Howe from the Office of Defense Mobilization in Washington. The director, Charles Wilson, again wrote to say that "the United States [had] an interest in assuring an adequate supply of power for the Pacific Northwest in which are located an increasing number of industrial installations important to the national defense," and that he was sure that Canada was "keeping in mind the needs of this

whole Western area."[186] Of a somewhat different class than the
other memorialists seeking to influence Manning and Tanner was
the Alberta and North West Chamber of Mines and Resources.
This body of prominent businessmen and professionals outside the
petroleum sector, mainly from the Edmonton region, presented to
Tanner a carefully prepared commentary on the gas export question that they hoped might serve to supplement the mass of evidence that the Board had compiled for the government's consideration. Included with the submission was notice of the resolution
they had passed, stating that the society was "opposed to the export
of natural gas at this time."[187] With supporting documentation from
technical bulletins and commission studies, the Chamber sought to
make two essential points: that examination of the estimates of natural resources computed by the best authorities at the particular
time of development showed that overestimation was the rule and
that regulatory authorities in the gas-producing regions of the
United States were alert to the danger of not protecting a future
supply of this vital resource. On the first point, the Chamber drew
attention to the decline in the estimates of Alberta's coal reserves
from 1913 to 1946, on the overestimation of natural gas reserves in
southern Ontario at the turn of the century, and the unanticipated
decline of Alberta's first important gas field at Bow Island. In support of its second point, the Chamber called upon evidence from a
recent investigation by the Federal Power Commission of U.S. natural gas resources.[188] It quoted the Commission's affirmation regarding submitted evidence.

> In weighing the evidence as to the extent and probable life of the
> country's natural gas reserves, as contained in the investigation
> record, it must be borne in mind that the oil and gas industry
> witnesses were engaged in special pleading. They were openly
> determined to prove that the country had no need for concern
> for the future adequacy of this resource. . . . [189]

Also presented for Tanner's attention were selected comments of
regulatory authorities in the more important gas-producing states.
Speaking of the consequence of unrestricted gas export from Louisiana, a senior state official told the FPC

> We have learned our error. We have watched out-of-state capital
> buy our raw products, turn them into higher value goods, and
> get rich. We have observed our State's producers of raw prod-
> ucts cultivate crops, chop trees, peddle petroleum, sell gas for
> export—and stay poor.[190]

Variations of this theme of the importance of natural gas in building a diversified economy were included from the Alberta Conservation Board's sister agencies in Kansas, Oklahoma and Texas. Tanner acknowledged the Chamber's submission and assured that it would be given "careful consideration."

The Conservation Board's long-awaited report was tabled in the Alberta Legislature by Premier Manning on 2 April 1952. It stated that Alberta's natural gas reserves established as of 31 December 1951 stood at 6.8 trillion cubic feet, up from the 4.7 trillion cubic feet estimated in the interim report for the previous year. The Board's estimate, like that of the previous report, was much more conservative than those submitted by the applicants.[191] Consistent with its cautious approach in estimating Alberta's gas reserves, the Board's evaluation of the province's 30-year natural gas requirement was more liberal than the calculations of the applicants. The Board revised its estimate of the province's needs to 31 December 1981 upwards to 4.2 trillion cubic feet from the 3.1 trillion stated in the interim report.[192] Taking into account the variable character of seasonal and peak daily demand, the Board set Alberta's requirement at 6.5 trillion cubic feet. From its assessment of how and where this projected demand was distributed, and from its analysis of the capability of the fields currently supplying Alberta's needs, the Board noted that by about 1956 substantial further reserves would be required for the area tributary to the Calgary-centred Canadian Western Natural Gas system. Similarly, by about 1963, the Edmonton-based Northwestern Utilities system was expected to require additional large reserves. The Board concluded that *all* the established reserves in central and southern Alberta had to be retained for provincial use. The only gas surplus to the province's needs, some 300 billion cubic feet, existed in the northern Peace River area. This was the estimated surplus after some 200 billion cubic feet of the established reserves of the three Peace River fields—Pouce Coupé, Tangent and Whitelaw—were set aside for local use. The Conservation Board recommended that all applications, except that of the Westcoast Transmission Company, be denied.[193] In the case of Westcoast's application, the Board's recommendation was sharply qualified.

In the interval since it had first been put forward in October 1949, Westcoast's application had undergone several revisions. The proposal considered by the Board in the fall of 1951 still sought permission to take Peace River gas for use on the Pacific coast as before, but now Westcoast also wanted to take gas from the Pincher Creek field in southern Alberta for distribution in southeastern British Columbia, particularly to serve the smelter facility in Trail and several nearby U.S. cities, including Kalispell, Montana, and Spokane, Washington. The volume of gas requested for the southern scheme was 25 billion cubic feet annually for 25 years, and for the coastal area, 70 billion cubic feet for 30 years. In its report to the Alberta cabinet, the Board dismissed Westcoast's Pincher Creek export proposal and recommended in favour of allowing export from the Peace River area, but for a period and rate that was greatly reduced from the applicant's request. The Board advised an annual maximum withdrawal of 42 billion cubic feet for 22 years. In putting forward its positive recommendation, the Board noted that, although the Peace River surplus gas supply was remote from the rest of Alberta's population centres, allowing gas export from this region would make it economically feasible to supply natural gas to most Peace River communities.[194]

Having dealt with the questions of Alberta's natural gas reserve requirements and exportable surplus, the Conservation Board acknowledged, but chose not to respond to, the delicate question of whether or not the *Gas Resources Preservation Act* might be unconstitutional. The Board observed that the matter was "entirely beyond its jurisdiction,"[195] and it was certainly not a matter that the Alberta government intended to explore. If the question of the Act's constitutional status was going to be put to the test, the challenge would have to come from either Ottawa or the private sector.

Heated debate on the natural gas export question had become a standard feature of the spring sitting of the Alberta Legislature. The 1952 session built upon that tradition. By the time Manning tabled the Board report, the Legislature had already been through one round of stormy debate in anticipation of what the report might say. Early in the session, the Liberal Opposition denounced the Conservation Board's gas export hearings, charging that they were "stacked in favour of companies favouring export." J. Harper Prowse insisted that decisions of the Board were influenced by the evidence presented. The problem, in his view, was that the export interests were represented by "27 to 30 high-priced lawyers" at each hearing, whereas no one was "present to give the case of the Alberta people."[196] It seemed to CCF member A.J.E. Liesemer that "not enough attention was being given to the possible uses of natural gas right here in Alberta," and he spoke in favour of the Liberal motion that a special study be undertaken to determine "provincial uses" for the gas from Gulf Oil's vast Pincher Creek field. "It is up to the Legislature," Calgary's Hugh John MacDonald concluded, "to protect the people of Alberta. Gulf Oil Company can protect its own interests. We have heard all the arguments in favour of gas export. Now let's hear the arguments in favour of the people of Alberta."[197] Described by Tanner as "unnecessary and foolish," the motion was soundly defeated, although one Social Creditor broke party ranks to vote with the Opposition. R. Earl Ansley, in whose constituency Alberta's postwar dream of prosperity was born, announced that since his constituents in the Leduc area did not favour gas export, he had to vote in favour of the special study.

This preliminary debate merely reinforced Premier Manning's natural caution. Before presenting the Board's report to the Legislature, he sought confirmation on a related matter of special importance. Tanner had gone to Ottawa in January 1951 to inform the minister of trade and commerce of progress on natural gas export applications and to relate Alberta's concern about jurisdiction over branch pipelines of companies seeking to export gas from the province. Now, Manning instructed Attorney General Lucien Maynard to get Howe's assurance that, for administrative purposes, gas export pipelines would be said to commence three miles within the Alberta border, at which point federal authority would also begin, thereby confirming that the federal government would not challenge Alberta's jurisdiction over branch or gathering lines.[198] Manning, it seems, was not prepared to proceed without such an understanding, even though he was fully aware of Howe's keen interest in getting Alberta gas into the Pacific Northwest and therefore reasonably certain of his concurrence. Following the receipt of Howe's positive reply, Manning tabled the report in the Legislature. Still the epitome of caution, however, he announced that since government policy had not yet been decided there would be no accompanying statement. Manning offered the report merely as an item of information. Few were surprised by the contents, for a week the stock market had been fuelled by rumours that the Board was going to recommend in favour of gas export from the Peace River area. As to the government's response to the Board's recommendation, political observers were of the opinion that given the likelihood of a summer election, gas export would not be permitted just yet.[199]

Although Opposition members of the Legislature might have been relieved that applications to export gas from southern Alberta, particularly from the Pincher Creek field, had been turned down, the president of Gulf Oil, S.A. Swensrud, was appalled. He telegraphed Premier Manning to challenge the Conservation Board's recommendations, and complained that the Board "attached undue importance to presently discovered versus future potential reserves." He warned that it was wrong to assume "that without additional markets to provide incentives discoveries would go on at the rate estimated by the Board." To hold Pincher Creek gas in the ground to 1968, Swensrud insisted, "would be practically tantamount to destroying all present worth of this great reserve," and he urged Manning not to approve any of the Board's recommendations "until we have had an opportunity to discuss this matter with you personally and at greater length."[200] On learning that the premier and his minister of mines and minerals would be happy to meet, but also that there was no prospect of having the Board's recommendations rejected or amended, Swensrud replied that there was no purpose in meeting and that there appeared "no alternative for us now but to curtail our whole Pincher Creek program."[201] Unable to move the premier, Gulf put its case to the Alberta public. By way of a press release, Gulf informed Albertans that if the Conservation Board's recommendation was accepted it would "retard or even stop a program involving an expenditure of $365,000,000 now under way in the Pincher Creek area," of which "at least $215,000,000" would come to the Alberta Treasury in the form of royalties and income taxes.[202] The notice pointed out that Pincher Creek, "Alberta's largest gas field was unlike most of the other fields in the province in that it was a condensate or wet gas field and that production in this instance offered a real opportunity to begin the development of a petro-chemical industry." But Gulf pointed out that, before the "gasoline, propane, sulfur and other chemicals" could be extracted, there had to be a market for residual dry gas. Having found such a market, Gulf was now being denied the opportunity to proceed. The consequence, according to the press release, would be to "clamp a tight lid on the search for additional gas fields in the south." Gulf warned that markets now available could be lost and advised that it would be wise for Alberta "to seize the market now offered and available rather than hopefully to look at some theoretical future."[203] This was also the editorial opinion of the *Calgary Herald*. Noting that the Board had presented evidence to show that, at the present pace of discovery, Alberta would soon have more than an ample gas supply for future needs, the paper blasted the Board's "interminable investigations" and "excessive caution." It condemned the Board for apparently thinking that "we should leave the gas in the ground and sit on it, like broody hens on a clutch of sterile eggs," thereby missing an opportunity that might not always be available.[204] This concern was no doubt stimulated by reaction to the Board's report in the United States. H. Henry Gellert, president of the Seattle Gas Company, announced his disappointment, and said that five times the amount of gas recommended for export would be needed to justify a pipeline from Alberta to Seattle. In Houston, Ray Fish, chairman of Pacific Northwest Pipeline Corporation, was quick to say that his firm was drawing up plans to build a $130-million pipeline to carry Texas gas to the Pacific Northwest.[205] The *Herald* pointed out that it was widely predicted that atomic power would soon be an important source of industrial energy and warned that "the one great fallacy in Alberta's approach to the gas export question is, it seems to us, the assumption that we can sell our gas as easily in 20 or 30 years' time as we can sell it now."[206] The *Daily Oil Bulletin,* which had been campaigning for gas export from the Pincher Creek field for months, was equally harsh.[207] The Pincher Creek and Cardston chambers of commerce expressed their conviction that the Conservation Board's decision was "discriminatory and not based on sound judgement" and if accepted would deal a "death blow" to their communities.[208] Given that their MLA, N.E. Tanner, was the minister of mines and minerals, Cardston residents felt particularly aggrieved.

The premier scheduled an address on the Conservation Board's report for April 8. Conscious that he would be presenting one of the most far-reaching policy decisions that his government had made since the Leduc discovery, Manning decided that his presentation would be more than just a commentary on the position his cabinet intended to take. Rather than simply leave the Board recommendation for cabinet decision, as prescribed by the *Gas Resources Preservation Act,* Manning thought it better to put the matter to a vote in the Legislature. Alert to what was likely to be the main issue in a soon expected provincial election, the premier was beginning to prepare the ground. His speech to introduce the motion calling for the Assembly's approval of the Board recommendation and the granting of a permit to the Westcoast Transmission Company was planned with more than the usual care.[209]

Manning began what turned into a nearly one-and-a-half-hour oration by emphasizing that the overriding principle directing his government's policy was the protection of Alberta's interests. It was to ensure that this interest was never overlooked or underestimated, he reminded the Legislature, that the *Gas Resources Preservation Act* had been passed in 1949. In keeping with the requirement of this Act, hearings had been held, and on the basis of these hearings, which the premier described as "exhaustive," the Conservation Board had determined in favour of "limited" export. The "facts" Manning argued, "left only one sensible conclusion and that was to support the Board's recommendation." The premier insisted that rejection would have a "disastrous" effect upon oil and gas exploration province wide, whereas approval of the Westcoast application would benefit the "entire" province. He stated that approval would mean extensive development in the Peace River area and allow many northern communities the same benefit of domestic natural gas service now enjoyed by most southern towns and cities. Also not to be ignored was the $600,000 annual royalty that approval would mean to the provincial treasury. Beyond these benefits, Manning drew attention to what he believed was a commitment owed to those who had risked their capital. The government had promised investors that when Alberta's future requirements were secured, surplus gas would be available for export. A surplus was now at hand, and Manning insisted that the province must not "break faith" with those it had invited to invest. This was the only way to assure continued exploration which, he pointedly noted, would soon "result in a situation where the Board could tell the government Pincher Creek could be used for export."[210]

Manning's motion launched a seven-hour wrangle that kept the Legislature sitting until midnight. The handful of Liberal and CCF members, supported by Leduc Social Creditor Earl Ansley, spoke against the motion. Liberal leader Prowse tried to argue the merits of having the Supreme Court of Canada rule on the validity of the *Gas Resources Preservation Act,* so that Albertans might be certain they were protected before they bound themselves to a long-term export contract. Unable to make any impression, Prowse abandoned the debate, saying that in years to come it would be known as "the great betrayal." Premier Manning responded that he was "quite content to leave to history and the good judgement of the public as to whether or not our decision was a sound one."[211] Both leaders knew that Albertans would soon have a chance to cast their verdict and both began to prepare for what would be the final battle in a dispute that had begun nearly 30 years before. As for the skirmish at hand, there was never any doubt about how it would end—the Legislature voted its approval of Alberta's first significant natural gas export proposal.

While Alberta's politicians began to organize for the yet uncalled but anticipated summer election, supporters and opponents of natural gas export prepared a last round of appeals. The United Farmers of Alberta sent a letter to the premier and every MLA stating farmers' "wholehearted opposition to any policy of gas export at this or any other time."[212] President Henry Young charged that the government was pursuing a "short-sighted policy of immediate gain," and he warned that "once a pipeline is built and market outlets established, then we have a moral and probably a legal obligation to supply those markets no matter how that may affect our industrial development here."[213] On the other side, the *Daily Oil Bulletin* lamented the government's caution and warned of possible lost market opportunities. Faison Dixon went to Calgary so that he could speak directly with Ian McKinnon about the probability that gas exploration would cease in southern Alberta if export was not allowed.[214]

As each side promoted its case, the government remained aware that Alberta householders and farmers were more numerous than Calgary oilmen and that foreign investors were not eligible to vote. Sensitive to the possibility that the general public might not be getting its message on the natural gas question, the government advertised that the minister of mines and minerals would speak to Albertans in a province-wide radio broadcast. N.E. Tanner informed his audience that it was necessary for him to speak, first because the government felt the people should be kept fully informed about major decisions relating to natural resources, and second because "there had been excessive loose talk and misleading statements bandied about" by "representatives of special interests" and by "those who oppose anything... for the sake of opposition."[215] The essential thought that Tanner sought to leave with Albertans was that they had every reason to trust the government and its Conservation Board to make the right decisions. Tanner wondered how anyone could say, "without a measure of dishonesty that this government [was] not interested in the conservation of the resources of Alberta." According to Tanner, the government's record showed otherwise. He reminded his audience that, "One of the actions taken by your present administration was to put a stop to the flagrant waste of gas in the Turner Valley field" by setting up the Conservation Board in 1938. More recently, he added, the government had passed the *Gas Resources Preservation Act,* giving the Board authority to control gas export and to ensure that only gas surplus to Alberta's required 30-year supply was available for export. Tanner also pointed out that this was not a fixed 30-year period; as each year went by, the 30-year time frame also moved ahead. Above all, Tanner emphasized that the Conservation Board could be relied upon to execute this policy. "The competency of the Conservation Board has never been questioned," he assured, "the sincerity of the three members of the Board is recognized by everyone."[216]

Tanner also went to some lengths to explain why, in the government's view, it was necessary to export Alberta's surplus gas. "What has happened elsewhere when export of gas has taken place?" he asked. He used the Texas experience to answer his question. Texas was anxious to export gas, Tanner explained, because it found that export promoted exploration, which in turn led to the discovery of new and larger gas reserves, and because "it brought to that State the greatest industrial expansion in the petro-chemical field ever to be witnessed anywhere." The idea that gas should be hoarded to attract industry to Alberta, Tanner said, represented a view that could only be termed "ridiculous."[217] The minister pointed out that valuable raw products such as propane, which was used in Alberta farms and rural homes that were distant from natural gas pipelines, would be extracted from exported gas and would be more readily available as export volumes increased. Finally, Tanner laid out the terms and conditions that would apply to the Westcoast Transmission Company's permit. He stressed that no gas would leave the province until the company had built distribution lines and made gas available to Peace River communities. Tanner concluded, saying that the government had carefully studied the Conservation Board's report and agreed that its recommendations were sound and that he believed Albertans would accept the government's judgement, since it was not

> given to rash decisions, but has tried conscientiously for the past sixteen and one-half years to administer the affairs of this Province with one thing only in mind, and that is the welfare of the people who live here.[218]

The immediate effect of Tanner's radio broadcast was to generate a response from the Liberal party. Just a week later, J. Harper Prowse explained on the radio that it was necessary to speak to Albertans directly since the Legislature was filled with "Manning's yes-men," who would not think for themselves on the gas export question.[219] Prowse expected, on the other hand, that Albertans would listen to the evidence and conclude, as the Liberal party had, that it was in the province's best interest to keep "our gas here for our own use." The Liberal leader went on to challenge Tanner's advice that Albertans could put their trust in the Conservation Board. Prowse commented that at the Board's inquiry he had observed teams of American lawyers, geologists, engineers and financial experts, with years of experience in arguing similar cases before the U.S. Federal Power Commission, who had been hired by "promoters who wanted to export our gas...."[220] "Let me ask you," he continued,

> do you believe that even the most impartial judge could give a fair break in a case after he had listened only to the evidence for the prosecution, and heard nothing for the defense.[221]

Prowse expressed his concern that estimating reserves involved only "educated guessing." The proper course, he argued, was to keep Alberta gas at home, which would mean certainty of supply and the offer to industry to come to Alberta to access the gas supply. He cited the recent decision of Sherritt Gordon Mining and Smelting Company to bring ore from their mines in northern Manitoba to the Edmonton area for processing. Also, he drew the example of Medicine Hat to his listeners' attention, noting that proximity to a cheap and plentiful natural gas supply accounted for the city's glass, brick and tile factories, which now supplied the whole prairie region. Prowse ventured, "If our gas became available in Winnipeg, Medicine Hat would not be able to compete for markets east of Regina." In addition, he alleged that gas export would disrupt Alberta's traditional coal market and put "500 Alberta miners out of work." Albertans, he said, would "have to migrate to the end of the pipeline in order to find work." Finally, Prowse charged that Social Credit's attempt to draw comparison with Texas was to construct "a dishonest case." He said that when Texas embarked upon large-scale gas exports, the state had 15 times as much gas as Alberta and that it was not gas export sales nearly so much as the frantic search for oil that accounted for the great increase in Texas gas reserves. Prowse concluded by asking his audience to examine the evidence presented by either side. As for the Liberals, he insisted, "We are satisfied that the case is so obvious that to export our gas is to throw away our future."[222]

The radio addresses were the second salvo in the pre-election skirmish for positional advantage. In the false calm that followed, Manning moved to tidy up a number of loose ends, the most important of which was the Westcoast Transmission application. On May 10, he wrote to Westcoast President Frank McMahon to say that the cabinet would issue the permit as recommended by the Conservation Board just as soon as Westcoast could demonstrate that it had concluded arrangements that assured the financing of the entire project, and that Westcoast had until August 1 to produce such evidence. McMahon was also informed that the permit, once given, would be cancelled if Westcoast failed to obtain the necessary permits and licences from the Canadian Board of Transport Commissioners and the U.S. Federal Power Commission by 1 November 1952.[223] Financing arrangements were completed in early June, and Alberta issued its first significant natural gas export permit on 16 June 1952. The permit also specified that Westcoast could take gas only from a designated area in the Peace River region along the British Columbia-Alberta border, that no gas could be removed unless the Alberta incorporated member of the Westcoast organization had constructed pipelines and made gas available to the communities mentioned in the Conservation Board report by 31 December 1953, and that removal of gas had to begin before 31 December 1954.[224]

The other matter Manning had to attend to was a delicate matter involving the minister of mines and minerals. Tanner was Manning's most able and trusted colleague. The two had worked closely together as cabinet colleagues in the Aberhart government and had become close personal friends. He was, in Manning's judgement, "one of the most capable ministers in the Government. . . . He was absolutely dependable."[225] For 14 years, Tanner had managed one of the government's most demanding and important cabinet portfolios. Dedicated to the principle of hard work, he had supervised the development of the petroleum industry from its Turner Valley infancy, through the Leduc boom, to emergent international status. Having proved himself a devoted churchman, an able politician and a remarkably capable administrator, Tanner's stature in Alberta's political firmament was but second to that of the premier. But now, Tanner wanted out; he was tired, and he wanted to try his lot in the private sector—in the petroleum industry that he had come to know so well. When approached by Tanner, Manning recognized both the tremendous loss that his colleague's departure would mean to the cabinet and also, with an election almost under way, the awkward timing. But Tanner's mind was made up; he was determined to leave. It was agreed, however, that he would assist in the campaign and would continue his supervision of the Department of Mines and Minerals until after the election when Manning himself would take over the position.

August 5 was the date set for Alberta's twelfth general election. The Liberals, to no one's surprise, centred their campaign on opposition to Social Credit's gas export policy and Ernest Manning's "one-man" rule. Elmer Roper's CCF followers also denounced the government's gas export policy, but used most of their energy advocating government funding for an accelerated program of rural electrification, low-cost rental housing in the cities, plus hospitalization and automobile insurance schemes modelled upon those implemented by Saskatchewan's CCF government. New to the Alberta political scene after a 17-year hibernation was the Conservative party, whose handful of candidates spoke in favour of "immediate" natural gas export but mainly of the need for a stronger opposition in the Legislature.

Social Creditors campaigned mainly on their government's record in office, and promised to begin "the most extensive program of highway and road improvements in Alberta's history," to assume a larger share of education costs, to improve hospital care programs, and, like the CCF, to offer greater funding for rural electrification.[226] Rather than say a lot about the natural gas export issue, Social Creditors preferred to talk more generally of their program for the "orderly development of Alberta's natural resources, through individual and competitive enterprise." Nonetheless, when pressed, the party was prepared to defend the idea of "limited" gas exports, as the premier did in Calgary on election eve. Before one of the largest crowds of the campaign, Manning admitted that his government's administration of the province's natural resources was one of the major issues of concern to the electorate. He said that Social Credit favoured rapid orderly development without waste and with a fair share of the revenue for the people of Alberta. He reviewed his government's gas export policy, and described it as "the only sensible" approach, lamenting that "we've got a lot of cheap, pipsqueak politicians running around saying 'whatever you do, don't let a cubic foot of gas go to the United States.'"[227] Then he reminded his audience of how the United States had continued to export gasoline to Canada during the war, even though it was being rationed there.

That the Liberals and the natural gas export issue were the most worrisome of the premier's campaign concerns is revealed further in Manning's "Personal Message to the Citizens of Alberta," the party's last campaign advertisement published the day before the election. Manning's message warned:

> Those, who would put this Province back under Liberal domination or sacrifice its welfare to the idol of Socialism, have done everything in their power to undermine your confidence in the Government responsible for Alberta's enviable record of progress and in those of us who have tried to serve you faithfully and well these past 17 years. The tactics used have shown a flagrant disregard for the truth and a readiness to resort to any means to gain political ends. This is a situation that can and should be dealt with by you whose sense of decency it offends.[228]

Albertans proved receptive to the call and Social Credit rolled forward to a massive victory, capturing 52 of the Assembly's 61 seats. For their efforts, the Liberals saw their percentage of the popular vote climb from 17.9 to 22.4%, but this translated into only four seats, a gain of but one. Electors were even less receptive to the CCF appeal, and the party's meagre presence in the Legislature remained as before—only Roper and one of his colleagues were suc-

cessful. Rounding out the Opposition numbers were one Conservative and one Independent Social Creditor, Earl Ansley from Leduc.[229]
The *Calgary Herald* offered its verdict.

> The liberals got precisely what they asked for. The ridiculous op-
> position to the export of natural gas which we are convinced was
> only bait for the rural vote, got them nowhere at all. Most Al-
> bertans now know that this province will never reap the full ben-
> efit of this vast resource until gas is exported.[230]

The election result was received with equal enthusiasm by the industry. In the *Daily Oil Bulletin*, Carl Nickle predicted that the re-
sounding defeat of the "liberals and socialists" would mean that the
Manning government would move more aggressively towards
broader gas export approval and that a likely first step would be to
order the "Conservation Board to reappraise rapidly enlarging gas
reserves."[231] Nickle also noted that the Liberal government in Ottawa could now give full support to natural gas exports without
fear of embarrassing Alberta's anti-export Liberals and anticipated
that Ottawa would now give stronger support to the Westcoast
project at hearings in Washington.

The 1952 election drew Alberta's multichaptered natural gas export debate to a close. It seems that Albertans now largely accepted
the government's declared policy of exporting gas surplus to the
province's 30-year requirement and were prepared to trust the Conservation Board's ability to manage effectively its important new
mandate—to monitor the province's natural gas reserve situation,
to forecast future requirements, and to adjudicate natural gas export proposals. The election revealed that most Albertans were
comfortable with Premier Manning's approach, not only to the natural gas export question but also to the whole matter of administering the province's oil and gas resources. Such confidence is apparent in the words attributed to a farmer's wife in the Medicine Hat
area, who proclaimed, "God knew that Mr. Manning would use
the oil wisely, so He let it be discovered."[232]

Once the election was over, Tanner was free to embark upon his
planned business career. In September 1952, he emerged as president of Merrill Petroleums, a position held formerly by his son-in-
law, Clifford R. Walker. Formed to participate in Alberta's burgeoning petroleum industry by New York broker Charles Merrill of
Merrill Lynch and Company, the firm had been created with Tanner's administrative talents and his industry and government connections specifically in mind.[233]

The transition at the Department of Mines and Minerals was smooth. It had been common knowledge that the premier himself was going to take over the department's supervision and this was of great reassurance to the industry. But Manning's decision to place Mines and Minerals under his personal direction was not merely a calculated political response designed to signal continuity and to reassure the industry that "theirs" was a first ranked ministry with a direct line to the premier's office. The decision had as much to do with Manning's genuine personal interest in the industry, and that his related technical, legal and administrative knowledge was far greater than that of any of his cabinet colleagues or Social Credit back-bench supporters.

As a student and employee of William Aberhart's Prophetic Bible Institute between 1927 and 1935, Manning had been a personal witness to the gas conservation debate that had raged in Calgary. Later, as minister of trade and industry in the Aberhart cabinet, Manning had been involved with the drafting and passing of the 1938 Conservation Acts in the Alberta Legislature. He remained part of the group that had overseen subsequent amendments to the Act, and after he became premier in 1943 he played a direct and usually leading role in the formulation of all legislation relating to the petroleum industry. Working closely with N.E. Tanner, Manning always played the primary part in discussions with Ottawa regarding oil and gas matters. The Department of Mines and Minerals gained the leadership of one whose knowledge of the industry was at least as strong as that of the respected but departing minister. If there was a discernible difference in the remarkable continuity of personal ability, style and guiding philosophy that would characterize Manning's administration, it was only that the premier was somewhat less of an avowed continentalist in his thoughts regarding energy resource relationships with the United States.[234] As with his free enterprise instincts generally, Manning was a little more cautious, and insofar as this was reflected in his supervision of the Department of Mines and Minerals, it was reinforced by the overriding responsibility that Manning shared for government initiatives and policies in all departments.

For the Conservation Board, Manning's assumption of Tanner's portfolio represented even less of a change than it did for industry and civil servants in the Department of Mines and Minerals. Although Ian McKinnon acted both as the deputy minister of mines and minerals and Conservation Board chairman until June 1952, and in each of these capacities worked closely with Tanner, from the time he had been appointed Board chairman in 1947 McKinnon had also reported directly to the premier. Now the relationship between the premier and his Board chairman became even closer.

Although the election had confirmed broad public support for Social Credit natural gas export policy, the government nonetheless remained deeply involved with the natural gas question. The critical matter of pipeline jurisdiction had yet to be resolved, and the obstacles that remained to be overcome before natural gas could begin to flow from Alberta, as it turned out, would take much longer to surmount than either the industry or the Alberta government had anticipated. For the Conservation Board, the natural gas question also remained a central focus. Its new mandate to monitor the development of the province's natural gas reserves and to keep current estimates of future needs, so that it could evaluate all export applications properly, represented a commitment that absorbed a substantial part of the agency's technical resources.

While Manning had been managing Social Credit forces in Alberta's summer election campaign, Frank McMahon had launched Westcoast Transmission's assault against the three remaining regulatory barriers. By the end of August, it was apparent to McMahon that he now faced a far more formidable array of adversaries and victory was going to be much more difficult to achieve than he had anticipated. The Canadian part of the campaign to get approval from the federal Board of Transport Commissioners to construct a pipeline that crossed provincial boundaries and a permit to export natural gas beyond Canada's borders went well enough, but it was a different story in the United States. Westcoast discovered that the opponents it had to face before the Federal Power Commission were much more powerful.

Frank McMahon explained to Premier Manning that, on the basis of preliminary questions raised by opponents of the Westcoast project at the initial FPC hearing, it was apparent that his company could expect a "Motion to Dismiss" their application on the grounds that the Alberta permit did not assure a supply beyond the first five years.[235] The problem was the permit provisions, which said that the Conservation Board would hold a hearing at the end of the first five years to review the situation and—by Board order—set the amount of natural gas to be removed annually for the balance of the 22-year contract. This, it was contended, did not constitute security of supply. There was similar objection to section 9 of the *Gas Resources Preservation Act,* which provided that, regardless of the provision of any permit, should an emergency occur, the Board with cabinet approval could order the diversion of all or any portion of gas under export contract for the needs of Alberta consumers.[236] McMahon judged correctly that this was going to be a difficult problem to surmount, and he suggested that Alberta's regulations be changed. "The overwhelming support given the Government at the recent election," he wrote, "must be interpreted as approval by the electorate, as a whole, of the export of gas from the Peace River area." Given this support, McMahon argued that the government would be "justified in amending the Permit" and taking such measures that would strengthen Westcoast's application to the FPC "for purpose of securing the United States markets for Alberta gas."[237] He concluded his request with the warning that opponents of Canadian gas, supported by the president of the Seattle Gas Company, were gaining strength.

Manning was alert to the danger inherent in McMahon's request. Cabinet amendment of the Westcoast permit would not only mean overruling the recommendation of the Conservation Board and thereby undermining the Board's credibility, but would also leave the government open to the charge that it was subverting its own recently established and much-touted public hearing process. Manning informed McMahon that any change was out of the question.[238] But this was not the last of the matter. A few days later, C.D. Howe contacted Manning to express his concern that the FPC might turn down the Westcoast application unless section 9 was amended. He expressed his view that the FPC would probably be prepared to accept Ottawa's *Electricity and Fluid Exportation Act,* which required annual renewal of export permits, were it not for the additional limitations imposed by the Alberta government. "I am writing you," Howe explained, "to suggest that possibly the rights reserved by the Federal Government may be sufficient to protect both British Columbia and Alberta, without the need for the additional restrictions contained in your permit."[239] Alberta's premier remained unmoved. If there was going to be any amendment of the regulatory structure, it would have to be in Ottawa's sphere. Howe and his cabinet colleagues still did not fully appreciate this, but Manning understood that his massive election victory did not mean that the deep sensitivity of Albertans towards natural gas exports could be discounted. He understood that Albertans had signalled their support for his government's cautious approach that put the emphasis on protecting the security of Alberta's long-term natural gas needs. His decision to stand firmly in support of his Board's decision regarding the Westcoast export permit and the established regulatory process in face of intense pressure from both industry and the federal government established a vital precedent in the crucial formative phase of the province's natural gas export policy. In April 1953, the federal government amended the *Electricity and Fluid Exportation Act* so that export permits no longer had to be reviewed on a yearly basis.[240]

McMahon's and Howe's hope that Ottawa's co-operation might ease the opposition to the Westcoast application was quickly dashed. By the time the FPC hearings resumed in February 1953, only two major contenders were left in the field: Westcoast Transmission and Pacific Northwest Pipeline. Pacific Northwest was headed by Ray Fish, president of Fish Engineering Corporation of Houston, Texas, and a 25-year veteran of the pipeline business. Wealthy, aggressive and used to thinking in continental dimensions, Fish had just finished construction of the Transcontinental Pipeline System to bring Texas gas to New York, and he was now determined to pipe gas to the Pacific Northwest, the only significant population area in the United States yet to be supplied with natural gas. Supporting the Fish proposal to supply the region with gas from the San Juan basin in New Mexico was Henry Gellert, head of the Seattle Gas Company. Westcoast's ally in the struggle was Charles Gueffroy of Portland Gas and Coke Company, the region's largest utility. Through the second round of hearings, Fish and Gellert kept the debate centred upon the "uncertainty" of Canadian supply. The campaign to marshal support was carried on as intensively outside as within the FPC chambers. In radio broadcasts and in full-page newspaper advertisements, Gellert warned:

> The American market—our Pacific Northwest—would dangle at the end of the Canadian pipeline after all Canadian needs were satisfied.... We, an American utility would actually become an economic vassal of a foreign power.[241]

Countering for Westcoast, Gueffroy asked:

> Are we economic vassals of Canada because we're dependent upon that thriving young nation for 90% of our nickel, 75% of our wood pulp, and 80% of our newsprint...?[242]

Various western states' governors entered the fray to state publicly and before the FPC that they favoured the Fish proposal. Governor Thornton of Colorado informed the Commission that he favoured an "all-American" pipeline that would be "regulated by U.S. agencies, not by Canada."[243] Going still further, Senator Lester Hunt of Wyoming tried to introduce in the Senate an amendment to the U.S.

Natural Gas Act that would declare the international movement of gas inconsistent with the public interest if the FPC found it would "result in economic dislocation, unemployment, or injury to competing fuel industries of the U.S."[244] In support of Westcoast's application, Portland Gas presented the FPC with the results of a commissioned study from Bechtel Corporation of San Francisco. After studying both proposals, Bechtel concluded that the Westcoast alternative offered two important advantages. It had access to a cheaper source, and it required a pipeline of only 1,000 miles, compared to nearly 1,500 miles for Pacific Northwest, and could deliver gas at 20% less cost. Moreover, delivery through the larger diameter Westcoast line could be expanded from the initial requirement of 200 million cubic feet per day to 400 million, compared to a maximum of 300 million for Pacific Northwest.

The debate raged on for months. Finally, on 1 June 1954, the FPC got down to final oral arguments. The Fish group held to their well-worked theme. Gellert persisted.

> We are fearful that our company's financial integrity and the industrial development of our city is to be subject to the control of a foreign power which will not only act, but has in the gas and electric issues (on the Skagit River) already acted in the Canadian public interest to the detriment of U.S. interests.[245]

Westcoast supporters tried to direct attention to economic factors. The commissioners were known to be sceptical about whether or not Pacific Northwest had sufficient proven gas reserves to supply its intended market. At one point, Commissioner Draper had asked the New Mexico Oil Conservation Commission whether they had "geologists on their Commission or just cowboys," and whether, "if they ever came to Washington, they would arrive by train or on horseback."[246] The McMahon group made a final attempt to discredit Northwest's San Juan reserve estimates and to emphasize the cost advantage of Canadian gas. On June 18, the FPC delivered its verdict. The Commission authorized Pacific Northwest Pipeline Corporation to deliver natural gas to U.S. Pacific Northwest markets. Among the reasons given for the decision was that a larger U.S. population would be served by the Pacific pipeline route and that the chosen pipeline would traverse "at least three large undeveloped sedimentary basins in Colorado, Utah and Wyoming," thereby spurring exploration and development in these areas.[247] Such reasons, however, were but the background in which the main reason was situated. The FPC decision report emphasized that secure supplies were essential for the protection of U.S. consumers. "Such protection," the commissioners explained,

> would not be afforded to any segment of the American people if the sole source of essential natural gas were through importation from a foreign country without some intergovernment agreement assuring the continued adequacy of the supply. Otherwise, all control over the production, allocation, and transportation to our border of such natural gas would be in the hands of agencies of foreign governments, whose primary interest would of necessity always be in the needs and advantages of their own people, and whose judgments and actions would be essentially dependent upon public opinion within that country, rather than upon the interests of American consumers.[248]

The pursuit of an "intergovernmental agreement" that would eliminate the ability of Canadian agencies, particularly the Alberta Petroleum and Natural Gas Conservation Board, to discriminate in favour of Canadian consumers remained a central objective of U.S. energy policy until it was achieved under the Free Trade Agreement of 1988.

The *Calgary Herald* announced that the fault for Westcoast's failure lay with the Alberta government. Four years previous, at a time when Alberta natural gas was eagerly sought in the Pacific Northwest, "the Alberta government was muttering something about inadequate reserves," and saying that no export could be permitted until provincial reserves were "big enough." "The truth," the paper concluded," is that a golden opportunity has been thrown away by the government's inability to make up its mind."[249] Calgary MP and petroleum industry pundit Carl Nickle was also quick to identify "narrow nationalism" as the reason for the failure of the Westcoast bid, even though he was only able to see evidence of narrow nationalism on the Canadian side of the border. It was, he held, Canada's oil and gas policies that were "primarily responsible" for the rejection of Canadian gas for use in the U.S. Pacific Northwest.[250]

It was soon apparent that the FPC decision had produced a stalemate. Fish had a U.S. permit, combined with an inadequate and costly natural gas supply and little prospect of gaining direct access to Canadian gas. McMahon had the necessary Canadian permits, combined with a large and relatively inexpensive natural gas supply but little prospect of gaining direct access to the U.S. Pacific North-

west market. To break the deadlock, a deal was hammered out between the two parties. Westcoast would build a 30-inch line from
Peace River to Vancouver and nearby Sumas on the American-
Canadian border. Pacific Northwest would buy Westcoast gas at
the border for distribution through the region, and San Juan gas
would be freed for the expanding California market.[251] The second
run through the regulatory obstacles necessitated by the Westcoast-
Pacific Northwest agreement culminated with a positive decision
handed down by the FPC on 25 November 1955. On this occasion,
the FPC chose not to make an issue of the dependence of U.S. consumers upon foreign energy supply. This was probably because a
few months previous the Alberta government had made a small but
important change to the "offensive" emergency clause in its *Gas
Resources Preservation Act*. Whereas section 9 in its original form
provided that in an emergency the Conservation Board could command the diversion of *"all or any portion of the gas* to which a permittee is entitled," the amended wording provided for "the diversion of any *gas intended for industrial use* outside the province."
This change was meant to undercut the argument of U.S. opponents
to Canadian natural gas imports that U.S. residential consumers
would always be at risk of being left in the cold.[252] Henceforth, only
gas destined for industrial use, which through the 1950s was usually a little more than 50% of the exported volume, remained subject to possible diversion.

Finally, on 6 October 1957, British Columbia Premier W.A.C.
Bennett pulled a lever to start the first flow of Westcoast gas across
the border (see Map 6.1). While the Vancouver ceremony marked
the beginning of a new phase in the continental exchange of energy
resources, it also signalled the willingness of Canadian governments
to amend their regulatory frameworks to forms more acceptable to
the United States. At the time of the FPC hearings, one observer had
noted that for Canadians "one of the most interesting developments" to arise was the question of the U.S. Federal Power Commission's jurisdiction over foreign sources of natural gas supply. He
suggested that the Westcoast decision would be precedent setting.[253]

While the McMahon and Fish groups had been sparring in
Washington before the FPC a more far-reaching, if less dramatic,
matter was being decided in Edmonton. Throughout this period,
Premier Manning was working out the details of the last critical
component of Alberta's natural gas export policy—a provincially
controlled natural gas pipeline gathering system. The practical and
constitutional rationale for this had been around since 1949 when
H.R. Milner presented the idea to the Dinning Commission. At the

MAP 6.1   Oil and Gas Pipelines Leaving Alberta, 1960

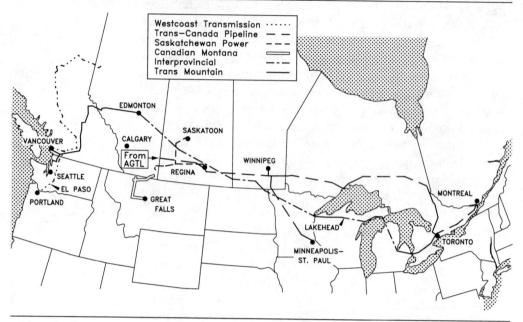

SOURCE: Canadian Oil & Gas Industries, May 1960.
NOTE: South of the U.S. border, Westcoast Gas was transplanted by El Paso.

first joint hearings on natural gas export applications before the Conservation Board, Milner was there to argue the case for not allowing the export applicants to own and operate natural gas gathering lines within the province. He offered his own, Alberta Inter-Field Gas Lines Limited, as the appropriate vehicle to protect the province's interests. During the third round of joint hearings, Milner took the stand to speak in support of a new brief submitted by Alberta Inter-Field. Again he stressed the necessity of having the gas gathering facilities within the province owned by an Alberta company such as Inter-Field, which would function as a common carrier and operate as a public utility under provincial control. This, he insisted, was particularly urgent, since section 9 of the *Gas Resources Preservation Act,* in his opinion, was open to constitutional challenge and could not be relied upon as an ultimate defence for Alberta consumers. This issue was of critical concern to Premier Manning. The question in his mind seems to have been the kind of organizational structure appropriate for such a company. His deep personal commitment to the free enterprise ethic made the idea of a government-owned utility abhorrent, but the Alberta Inter-Field alternative also left him uneasy.

Alerted to the issue but not yet decided on how to respond, Manning took the precaution of obtaining C.D. Howe's written assurance that Ottawa would not challenge Alberta's jurisdiction over Westcoast's natural gas gathering lines in Alberta before agreeing to accept the Conservation Board's recommendation to approve Westcoast's export application.[254] With Howe's assurance, Manning went ahead, but it was understood from the outset that this was only a temporary expedient. This pipeline was local, of limited extent, and located in the distant Peace River country. When it came time to permit natural gas export from the southern part of the province, a more enduring and formal mechanism would have to be in place to enforce Alberta's undivided control over removal.

By the summer of 1953, it was apparent that Manning would soon have to come to a decision. The Conservation Board was about to begin another round of gas export hearings, and it was widely anticipated that Alberta's proven natural gas reserve had grown large enough to allow export to eastern markets. Two of the three contenders from the previous round of hearings returned to promote their respective schemes. The Fish group (Prairie Pipe Lines) dropped from the Alberta contest, but only to launch a counter scheme designed to outflank their rivals. Fish proposed to attach the southern Ontario market to his Tennessee Gas Transmission network at Niagara. This left Trans-Canada Pipe Lines (the Murchison group) and Western Pipe Lines (the Milner-Williamson group) as contenders for Alberta natural gas.[255] Organized as Trans-Canada Pipe Lines Limited., Trans-Canada Grid of Alberta Ltd. and Canadian Delhi Oil Ltd., the Murchison group proposed to serve Saskatchewan, Manitoba, Ontario and Quebec markets by way of a 30-inch 2,240-mile trunk line following the Canadian Pacific Railway main line from southern Alberta to North Bay, Ontario, before turning south towards Toronto and Montreal. To meet their estimated requirement, they sought permission from Alberta to remove 365 million cubic feet of natural gas per day for 25 years. Western Pipe Lines, the creation of Edmonton, Winnipeg and Toronto interests, presented an amended two-phase proposal. Phase one involved the construction of a 30-inch trunk line from southern Alberta to Winnipeg, with a lateral extension going south to the U.S. border at Emerson, Manitoba. Here, the Canadian gas would be purchased by the Northern Natural Gas Company for distribution in North Dakota and Minnesota. To supply this prairie

and U.S. market, Western Pipe Lines requested permission to remove 275 million cubic feet per day for 30 years. The Western application gave notice that, once the first phase was operating and
the market established, in any event not more than three years after
the beginning of gas transmission, the company intended to launch
the second phase of its project. This would involve extending the
pipeline eastward from Winnipeg to Toronto and Montreal and an
application to remove an additional 275 million cubic feet of natural gas per day. In addition to a trunk line running east from Alberta, the Trans-Canada and Western proposals both included their
own pipeline gathering network within Alberta.

In early May 1953, McKinnon, Govier and Goodall once again
began to hear the parade of expert witnesses summoned by Trans-
Canada and Western Pipe Lines. Aware that it would be several
years from the time an export permit was given before Alberta gas
would begin moving eastward, Premier Manning was anxious to
see things move along. He was convinced that, given the magnitude
of the project, its cost, the nature of the regulatory obstacles to be
overcome and the uncertainties of market pricing, the companies
should combine their resources and submit a single export application. Unknown to all but those involved, and perhaps Ian McKinnon, early in January 1953, months before the Conservation
Board's summer hearings, Manning persuaded Ray Milner and
Alan Williamson of Western Pipe Lines to meet with Clint Murchison and his associates in Vancouver. The specific purpose of the
meeting was to explore the possibility of a joint proposal.

Williamson reported back to Manning on January 29, explaining
that Trans-Canada was insistent upon a commitment that they
would press ahead jointly with the eastern Canada scheme regardless of the findings of independent engineering reports.[256] Western
Pipe Lines' eastern associates, Wood Gundy of Toronto and Nes-
bitt Thomson of Montreal, knowing or perhaps anticipating that
the engineering reports were going to raise grave doubts about the
economic viability of any project that did not tap into a U.S. market, were not prepared to endorse Murchison's terms of union.
Manning declared his great disappointment, and he bluntly informed Williamson of the course to be taken.

> ...we should not regard the breakdown in negotiations be-
> tween your group and Trans-Canada as final, but that further ef-
> forts should be made to arrive at a mutually acceptable arrange-
> ment that will enable the interested parties to appear before the
> Regulatory Boards concerned on behalf of one comprehensive

joint project emphasizing the use of Canadian resources for Canadian people with export to the United States [a] secondary rather than the major consideration on which the project rests.[257]

He warned that two groups appearing before the Conservation Board and the Board of Transport Commissioners, each arguing for their proposal and against each other, would only result in long drawn-out hearings. For concluding emphasis, Manning laid out the stark reality of the political landscape. He quoted to Williamson from a recent letter from C.D. Howe.

> ... gas from Alberta must be available to potential customers in the Province of Ontario and Quebec and in the intervening Provinces before further export of gas could be authorized. This attitude is consistent with the position that the Government has taken with regard to the export of hydro-electric power for many years. You will appreciate that the supply of gas through a pipeline is analogous to the supply of electric power in that both demand continuity of supply.[258]

Premier Manning's warning was confirmed publicly a short time later in the House of Commons when Howe announced that an all-Canadian pipeline was government policy.[259] Although, in this instance, Howe's intent was as much to deter the potentially serious competing scheme being developed by Ontario's Consumers' Gas Company and Fish's Tennessee Gas Transmission to bring U.S. gas across the border at Niagara.

Trans-Canada and Western continued their independent ways, although Western Pipe Lines did amend its application proposal, adding emphasis to the second phase that would take Alberta gas into central Canada at a later date and thereby bring its scheme more in line with what Premier Manning had suggested. Although the outcome was reasonably certain, the hearings provided the Board with an opportunity to assemble and assess the most recent data on the current state of Alberta's natural gas reserves. This set of hearings was also important in another way. Buried within the high-profile public debate between the contending groups was a persistent and more restrained second level of discussion—the question of gathering natural gas within the province.

The key person in the gas gathering debate was Calgary lawyer J.C. Mahaffy. As counsel for Alberta Inter-Field Gas Lines, he attended the hearings with a specific objective. He wanted to undermine the Trans-Canada and Western proposals for provincial natu-

Calgary lawyer and businessman, J.C. Mahaffy, became Secretary Treasurer of Alberta Gas Trunk Line Ltd. in 1955, later General Manager and President. Glenbow Archives, NA-2345-45.

ral gas gathering systems controlled by affiliated companies incorporated in Alberta. At the same time, he worked to build the case for a provincial gathering grid owned and operated by a completely independent Alberta company such as Inter-Field. For example, when Trans-Canada's senior pipeline consultant, F.E. Warterfield from Dallas, Texas, appeared to be questioned on the company's proposed gathering system, Mahaffy noted that the plan presented to the hearing showed a line to pick up flare gas from the Leduc-Woodbend oilfield. He asked Warterfield if any provision had been made "for picking up substantial quantities of [flare] gas from the

TABLE 6.5 Evolution of Conservation Board Estimates of Alberta's Thirty-year Natural Gas Requirements

	Interim Report, 20 Jan 1951 (Period, 1951–1980) billion cubic feet	Report, 29 Mar 1952 (Period, 1952–1981) billion cubic feet	Report, 24 Nov 1953 (Period, 1953–1982) billion cubic feet
Domestic	873.1	1,114.7	1,216.7
Commercial	751.7	763.5	807.6
Industrial	1,434.5	2,248.7	2,421.0
Total Requirement	3,059.3	4,126.9	4,445.3

SOURCE: PNGCB, Interim Report, 20 January 1951; Report, 29 March 1952; Report, 24 November 1953.

Rimbey-Homeglen area?"[260] In an instant, it became apparent that Warterfield did not know where Rimbey-Homeglen was. He was informed by Mahaffy that the field was only about 30 to 40 miles southwest of Leduc, and from that moment Warterfield's and his employer's credibility were suspect. In his cross-examinations, particularly that of D.G. Hawthorn, a petroleum consultant from Houston who appeared for Western Pipe Lines, Mahaffy was able to draw forth expert testimony regarding the advantages to be derived from a province-wide gathering system operated by an independently owned common carrier.[261] Trans-Canada spokespersons remained hostile to the idea of an independent gathering system, claiming that such an approach would increase the cost of gathering the gas they required, make the financing of the entire project more difficult, and result in continual disputes over cost allocation. Western Pipe Lines, on the other hand, was more amenable to working with an independent provincial gatherer. Perhaps this had something to do with H.R. Milner being a key figure in both Western Pipe Lines and Alberta Inter-Field.

The Conservation Board concluded its hearings in late September and presented its report to the Alberta cabinet on 24 November 1953. Of critical first concern was the question of the province's proven natural gas reserves. According to the Board's estimate, this essential reserve, as of 30 June 1953, equalled 11.5 trillion cubic feet, up from the previous estimate of 6.8 trillion cubic feet in December 1951. The report also revealed that the Board had revised its estimate of Alberta's natural gas needs for the next 30 years upwards, from its earlier figure of 4.2 trillion cubic feet to 4.45 trillion cubic feet (see Table 6.5).[262] It was acknowledged that a large enough supply of surplus gas now existed to meet the request of one of the applicants. Despite this, the Board declared, and it came as no surprise to Manning, that it found both proposals sufficiently flawed that it was unable "to state whether or not the granting of an export permit to the applicants would be in the best interests of the Province of Alberta."[263] With reference to the Trans-Canada application, the report explained that the Board was not convinced by the evidence presented that the applicants could sell enough gas in Ontario and Quebec at prices that would make the project economically viable, or that the gas storage fields necessary to the project in southern Ontario were adequate or even available for their use. In the case of the Western Pipe Lines application, the Board stated that it wanted to see more clear-cut evidence that Western could sell sufficient gas to the Northern Natural Gas Company at a price that would make the project economically sound. The Board also insisted that it would have to receive much stronger assurance that "Western Pipe Lines could and would perform its commitment to proceed with Phase II" of its project.[264] Having identified their failings, the Board assured both companies that they would be given the opportunity to reveal how these shortcomings had been addressed at hearings to be reconvened in time for the Board to render a decision before the beginning of the 1954 pipeline construction season.

Although it generated less immediate public comment, the most significant part of the 1953 report was the Board's stated opinion that "efficient development, gathering and utilization of the gas resources of the Province would be promoted by the institution of a 'Trunk-Line' system operating within the Province as a common carrier under full Provincial jurisdiction and control."[265] McKinnon, Govier and Goodall explained that a common carrier trunk line system would allow the interconnection of diverse fields with a pipeline network serving *both* the Alberta and export markets. The trunk line would connect the main producing areas with the main market areas of the province and with a "gate" on the provincial border to markets east of Alberta. Access to Alberta gas for eastern markets would be permitted only through the trunk line's border gate. As a common carrier, it would accept for delivery to Alberta's markets or to export markets gas delivered to it by other lines owned and operated by producers, the local utility companies, export companies or collected by the trunk line itself. Such an approach, the report informed, promised numerous advantages. It would provide the best means of addressing one of the Board's most difficult problems, the conservation of unavoidably produced

oilfield gas. A common carrier grid offered a much greater likelihood of market access for such "waste" gas. Such a system would
permit the widest flexibility by allowing ready interchange of supply sources and the ability to make any necessary diversion of gas
from provincial to export use or the reverse. Joint use of facilities
(field, plant and pipeline) would save costly duplication for the mutual benefit of consumers, producers and distributors. Similarly, a
common carrier network would assure consumers in the province
access to that gas that could be delivered most economically. The
report also pointed out that a common carrier pipeline network
was prerequisite to any effective plan of market sharing or a market
proration scheme. Whereas the board members began their list of
benefits with their own priority concern, they ended with the concern uppermost in the government's mind. This approach, they
said, would "strengthen the control of gas within the Province by
provincial authorities."[266] It had taken years to germinate, but Milner's idea now had a priority place on the government's policy
agenda.

Within a day of receiving the Conservation Board's report, Manning was on his way to Ottawa to discuss its contents with Prime
Minister Louis St. Laurent and Minister of Trade and Commerce
C.D. Howe. Both were anticipating Manning's visit, for the Alberta
premier had been there just a month previous to prepare them for
the decision that he knew was coming.[267] This time, Manning
brought his technical experts with him, Conservation Board Chairman Ian McKinnon and Board member Dr. George Govier, to lend
support to the case Alberta intended to make. The problem to be
addressed was Ottawa's stated policy which, as Howe had reminded Calgarians earlier in July, was that "Canada's natural gas
[would] not cross the U.S. border as long as there is a market for it
in the Dominion," and that, in the case of the Toronto and Montreal markets, the required gas had to be brought from Alberta by a
pipeline situated entirely within Canada.[268] This was not incompat-
ible with Alberta's stated position that, after the province's long-
term natural gas needs were assured, Canadian interests had to be
given priority over export to the United States.[269] But during the
course of the hearings, the marginal economics of the Howe-
Murchison all-Canadian route became more apparent. It became
clear that, to deliver gas to the Toronto area at a price even closely
competitive with alternate fuels or U.S. gas that the Consumers'
Gas Company of Toronto proposed to purchase from Tennessee
Transmission at Niagara, the Trans-Canada Pipe Lines proposal
would require a considerable subsidy. This would have to be some

combination of federal assistance to compensate for pipeline construction costs over the longer northern Ontario route, higher consumer pricing in the eastern market, and a lower price paid to Alberta producers. Also not far from the premier's mind was that lower prices to Alberta's natural gas producers meant lower royalty payments to the Alberta government. Manning drew upon compelling evidence gathered by the Conservation Board to demonstrate what had already emerged as the consensus in financial and industry circles—that the Western Pipe Lines proposal made greater economic sense.[270] The Albertans also came with an economic and political solution that was attractive both to the government that controlled the route and to the government that controlled the gas. It was a solution much like that urged upon Trans-Canada and Western Pipe Lines by Manning the previous January. After gaining the support of St. Laurent and Howe, Manning presented it to the public along with the Board's report on December 3, after he returned to Edmonton.

In a press release, Manning summarized the Conservation Board's findings. He then explained that, although the Board found that the province did have surplus gas available for eastern export, it was not convinced that either Trans-Canada or Western Pipe Lines had "established that it had assured markets which will provide the volumes and prices necessary to enable the payment of fair and equitable prices to Alberta producers and ensure that its proposed line can be successfully financed."[271] Manning expressed his government's concurrence with the Conservation Board's recommendation that neither company be given an export permit. It was not a surprise for Albertans to learn that the province now enjoyed a sizable natural gas surplus, but it did come as a surprise to many that neither company would be allowed to export Alberta gas. The decision could hardly have come as a surprise to Murchison, Milner and their associates. They had been told by Manning in January what the requirements were, but they had chosen to ignore the premier's "advice." Now, they were being told publicly that, in the interests of the producers, consumers and Canada as a nation, it would be best if they combined their proposals into one project. This project was stated to be "the construction of an all-Canadian pipeline from southern Alberta through Regina and Winnipeg to the markets of Ontario and Quebec with a connecting line running south from Winnipeg to the Minneapolis market area of the United States."[272] This time, the message was not only public but also reinforced by a joint statement from C.D. Howe. He proclaimed that the Alberta proposal "was a wise move and that the federal govern-

ment [would] try to expedite the amalgamation as soon as possible."[273] For his audience, Howe added that the all-Canadian pipeline would have to be under construction in Ontario before any gas
could be exported to the United States by the connecting line.

Still, the embattled companies proved reluctant, and it was soon
clear that Howe would have to intervene directly. Near the end of
his patience, he called each group to meet with him in Ottawa. Each
delegation was ushered separately into the minister's office to receive the same blunt message. The time for lobbying was over; there
was only going to be one all-Canadian transcontinental pipeline.
Whether or not the pipeline was going to be built depended upon
the contenders agreeing to contribute their experience and resources to a single plan. Howe directed them in the national interest
to settle their differences and to come back "together" to see him
the following morning.[274] An all-night session brought the realization that neither party could be the dominant partner and that
an equal division of shares would have to be the basis upon which
the new single company could be created. Informed of the agreement, Howe telephoned Manning to announce "mission's accomplished!"[275] On 8 January 1954, he told Canadians of the "big
merger," which meant, he explained enthusiastically, that the
$350-million all-Canadian pipeline project at last could go ahead.[276]

The forced merger proved to be just the beginning of the Trans-
Canada pipeline saga. This was but the first of several key government interventions required to keep the project from foundering.
This second intervention came quickly after the first. From their
agreement in principle, the Trans-Canada and Western Pipe Lines
groups moved speedily to draft a formal memorandum of agreement. Having agreed on the share split, they now had to reach an
accord on the composition of a board of directors. It was easy
enough to accept that each group would have four seats and to
agree on the addition of several prominent and representative lawyers, but they could not agree upon who should be president of the
new company. This impasse resulted in another telephone call to
Manning. Howe explained that he had been asked to find a suitable
outsider without links to either party. Asked for his advice, Manning suggested N.E. Tanner. Howe immediately recognized that
Tanner possessed a unique combination of qualities that made him
an almost ideal candidate. "Will you call him?" Howe asked, "and
see if he would be interested, or would consider it? I'll call him in an
hour's time; you have a chat with him first."[277] They agreed that the
best way to "sell" Tanner on the idea was to make the case that the
project was "vital to Canada's national interests" and that his lead-

ership was necessary to break the impasse and move the project forward.[278] The persuasion worked. Tanner accepted the challenge, but
not without some prodding and second thought.

The next step was taken before Alberta's Conservation Board.
After a short hearing in March, the Board concluded that the
merged application of the new Trans-Canada Pipe Lines Limited
had "a good chance of being economically feasible" and recommended that an export permit be granted.[279] At the federal level, the
Board of Transport Commissioners followed to give their blessing
to Trans-Canada's pipeline plans. The smooth passage through the
two levels of Canadian regulatory authority belied the extent of the
difficulties that still lay ahead.

Trans-Canada's natural gas and pipeline permits required the
company to have the necessary financing for the project in place no
later than the end of 1954. This was in keeping with Trans-
Canada's intention to begin pipeline construction in early 1955. To
secure financing, most of which would be obtained from insurance
companies in the form of first mortgage bonds, the company had to
have in hand long-term sales contracts from western Canadian producers and long-term purchase contracts from eastern natural gas
utility companies. The conundrum faced by Trans-Canada was
soon apparent. Most southern Alberta producers, led by Gulf Oil,
which controlled the largest single source of supply, would not accept the prices Trans-Canada was prepared to offer for their gas.
They had always favoured the economics of a pipeline route that
took Canadian gas through the United States and dedicated a much
greater percentage of its volume to the closer and larger American
market, thus promising a much better price to the producer. Eastern utilities, particularly Consumers' Gas of Toronto, were not
pleased to bear the added transmission costs of the longer and more
costly all-Canadian route. Even under strong pressure both from
Ottawa and the Ontario government, the company refused to sign a
purchase contract. Without contracts, there was little hope of selling Trans-Canada bonds in a New York market that already had
little enthusiasm for the project. Earlier, the Metropolitan Life Insurance Company had warned Trans-Canada's New York brokers,
Lehman Brothers, that the Alberta Conservation Board's authority
under section 9 of the *Gas Resources Preservation Act* to divert
contracted gas in an emergency was a "substantial impairment" to
the attractiveness of the company's bonds.[280] Preceded by newspaper
reports of its impending collapse, Trans-Canada was soon back to
seek Howe's assistance. First, there was talk of a government guar-
antee for Trans-Canada bonds. Howe's cabinet colleagues were not

enthused about this idea. Then, as new extensions were granted to Trans-Canada's financing deadline, attention focused upon Howe's idea of creating a crown corporation to build the northern Ontario portion of the pipeline. Once Trans-Canada was on its feet, it would take over the corporation and reimburse the federal government for all costs. Eventually, on 15 March 1956, Howe asked Parliament's approval to establish a crown company, the Northern Ontario Pipeline Corporation, to build, at an estimated cost of $118 million, the 675-mile portion of the pipeline from the Manitoba-Ontario border to Kapuskasing, Ontario. Ottawa also proposed to lend $80 million to help Trans-Canada finance the balance of its pipeline system.[281] A raucous 10-week wrangle, known popularly as "the Great Pipeline Debate," followed. In the end, Trans-Canada got the assistance it sought, construction began in June 1956, and natural gas delivery to Ontario began in the fall of 1958. This story is well known,[282] but the parallel and perhaps more important Alberta chapter of the pipeline story has received much less attention.

The provincially controlled natural gas gathering system recommended by the Conservation Board called for the construction of a large-diameter trunk line running south from the Leduc-Woodbend field near Edmonton to Calgary, where it would veer in a southeasterly direction to Arrowwood. Here, in south-central Alberta, the projected trunk line changed in diameter to 36 inches and headed east to the Alberta-Saskatchewan border (see Map 6.2). From the north to south running trunk line, branch or feeder pipelines would be extended, first to the large Pincher Creek and Cessford fields. Conditioned to the idea by the preceding series of proposals from Alberta Inter-Field Gas Lines, the Alberta cabinet accepted the Board's affirmation that an Alberta-controlled trunk line could play a vital role in the conservation and efficient use of the province's natural gas resources. Work began on drafting the necessary legislation as soon as Manning returned from Ottawa. Under the premier's supervision, senior officials in the Attorney General's department and at the Conservation Board in Calgary laboured to fashion the last critical measure necessary to complete the core of statutes upon which Alberta's contemporary oil and gas policy remains anchored. Although Milner contributed the central idea, his company, Alberta Inter-Field, was not destined to be the chosen beneficiary. Manning wanted Alberta control exercised through the private sector, but he was uneasy about relying upon a company dominated by the president of Alberta's two major natural gas utility companies.[283] The solution to his dilemma, it seems, was sug-

MAP 6.2   Alberta Gas Trunk Line Proposals, 1954–1955

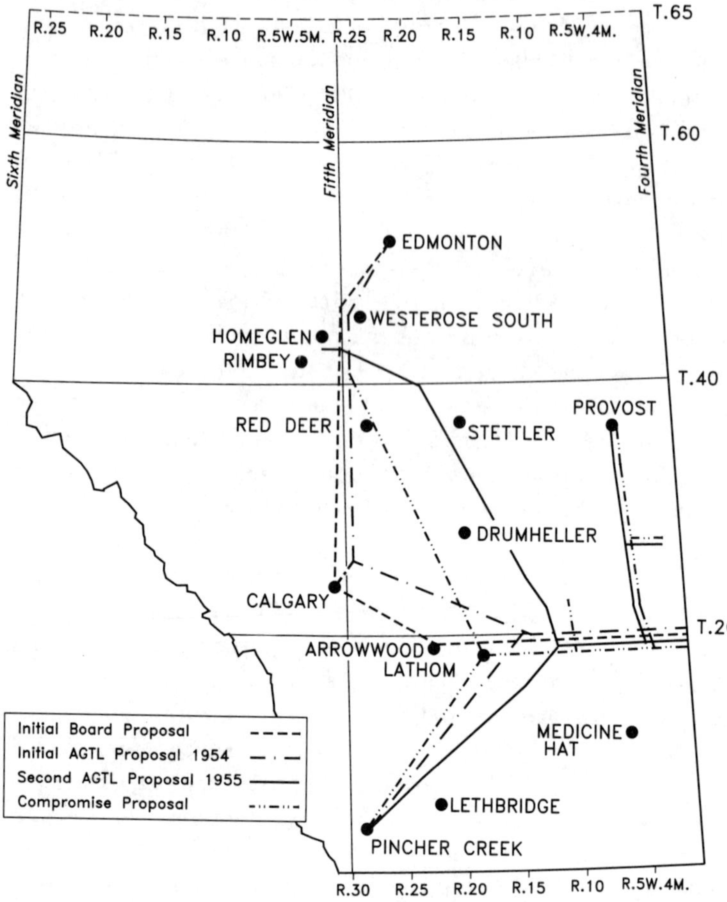

SOURCE: Exhibits 2 & 3 AGTL Hearing, November 1955.

gested by Carl Nickle. In a lengthy letter to the premier, reviewing the Conservation Board's November Report, Nickle declared his strong support for a gathering system under provincial control, but warned that neither the local utility nor the export transmission companies should be allowed to control the trunk line system. He advised the ownership and control be fixed in a "broad cross-section of gas producing interests in Alberta."[284] Manning took the cross-section idea a step further.

The draft *Alberta Gas Trunk Line Company Act* specified that the proposed company's purpose was to act as a common carrier and purchaser of the province's natural gas. To achieve this objective, the intended legislation empowered the company to construct transmission lines, to install compressor stations and scrubbing plants, and to lease or purchase storage fields and whatever other facilities that might be necessary in Alberta. The company's affairs were to be managed by a board of seven directors. Two of the directors were to be appointed by the Alberta cabinet and the remaining five were to be elected: one by the utility companies, one by the export interests, and three by the gas producers or processors. Voting, or class B, shares were distributed and fixed among the four groups so that no individual could ever acquire control.[285] Directors were required to be Canadian citizens, and resident in Alberta. The proposed company's capital stock was set at eight million class A shares with a par value of $5 and no voting rights. Initially, these shares were to be made available to Alberta residents only.[286]

The proposed Bill, once drafted, was circulated for industry comment and suggestions in keeping with past practice. A select group, that included Gulf and Imperial Oil, were summoned to the premier's office for a personal briefing.[287] The trunk line idea was endorsed, and no significant changes were suggested. Producers were attracted to the scheme, not only did it promise greater assurance of market access but also provincial jurisdiction promised, as Carl Nickle observed, "better assurance of fair well head gas prices over the long term, than would apply if complete control between wellhead and consumer were vested in a Federal board where there might be a tendency to weight decisions on Rate Hearings in favour of the mass of consumers outside of Alberta."[288] Comment from the industry was supplemented by the Conservation Board's detailed examination of desirable pipeline size and location, construction costs, and the appropriate transmission charges that formed the basis of a final recommendation to the premier on the technical guidelines that should govern the building of an Alberta trunk line network.[289] On 8 March 1954, Board member Govier gave the draft legislation a final review. A few days later, Premier Manning introduced the *Act to Incorporate a Gas Trunk Pipe Line Company* in the Assembly.

On cue, the Legislature launched into what had almost become a spring ritual—the annual natural gas export debate. The actors remained the same, and they stayed closely to their well-known scripts. Liberal leader Harper Prowse denounced the proposal, claiming that, even if one accepted the Bill's questionable constitutional validity, the two government-appointed directors could do little to protect the interests of Albertans because the proposed legislation "would put five directorships in the hands of New York finance."[290] CCF stalwart Elmer Roper dwelt upon the same point and argued that the government's plan would merely establish a "private monopoly" with little concern for the consumer. His preference for a crown-owned trunk line was dismissed by the premier, who reminded the members that "public ownership is bad in principle and worse in practice."[291] The debate ended as it had in previous springs with the large Social Credit majority imposing its will.

Manning knew that he had little to worry about in the Alberta Legislature. Ironically, it was the Ontario Legislature that was the focus of his concern. The Alberta government was somewhat apprehensive that its clever idea might be stolen, that Ontario might decide to follow the Alberta example and set up a trunk line system built and operated by a provincial authority that would accept Alberta gas at the Manitoba-Ontario border for distribution throughout the province. It was believed that if this were to happen it would greatly reduce the incentive of private interests to build a pipeline just from Alberta to Ontario.[292] As with almost every other facet of Alberta's economic development, at some point Ontario always seemed to figure in the picture.

Once approval was gained in the Legislature, Alberta Gas Trunk Line (AGTL) was duly incorporated, and on 5 June 1954 the company's provisional board of directors gathered to attend what would prove to be the most important inaugural corporate meeting in Alberta's history. The government's appointees were the president of Trans-Canada Pipe Lines, N.E. Tanner; Ralph Will, a Calgary drilling and well service contractor; Vernon Taylor, a senior Imperial Oil Co. executive; George Church, a prominent farmer-businessman from Balzac, near Calgary; Ronald Martland, an Edmonton lawyer; and R.H.C. Harrison, QC, president of the Canadian Petroleum Association.[293] A few months later, the provisional board was expanded to include Robert J. Dinning, a Calgary businessman and former chairman of the important 1948 Alberta Natural Gas Commission. The presence of Conservation Board Chairman McKinnon and Premier Manning at the first meeting marked the importance that the government attached to the task. For the next several years, Manning watched closely over his creation, and even though there were two government-appointed board members, he often preferred the more direct approach of telephoning AGTL's president or general manager with "suggestions."[294]

The essential first step facing the directors was to find a general manager. To fill this key position, they turned to William F. Knode, who was well known in industry and government from his earlier experience as the founding chairman of Alberta's Petroleum and

Natural Gas Conservation Board. Knode began the task of assembling administrative and technical staff and initiating engineering studies of alternative pipeline routes. James C. Mahaffy, who had so capably managed the promotion of Milner's trunk line idea at Board hearings, was named AGTL's secretary-treasurer. In November 1954, with Mahaffy's assistance, Knode took AGTL's first pipeline application before the Conservation Board.[295]

There was no analogous precedent in the natural gas industry for building so large a gathering system as a common carrier, and it was the Board's engineering philosophy that set the ground rules. AGTL's proposal was worked out in consultation with the Board, so that in its design and routing the pipeline represented an attempt to harmonize three functions.[296] The primary function was to provide the most efficient transportation for the gas. Second, if the system was going to fill its conservation function, it had to be sufficiently flexible to handle such quantities of residue gas from *oilfields* as the Conservation Board might designate to permit optimum operation of oilfield facilities.[297] This condition implied that there would be variations in the rates of deliverability from the various *gas fields* and that these rates would be determined by the Conservation Board. Third, the system had to be sufficiently flexible to balance gas between Alberta utilities during individual peak load conditions. Only by taking this requirement into account would a trunk line system designed primarily for export enhance rather than hinder the existing distribution of gas within the province. The proposed trunk line routing between Calgary and Edmonton (see Map 6.2) also served two other functions. The region to the west of the line possessed some of the best geological prospects in the province for future natural gas discoveries. This route offered the advantage of close proximity to a region of anticipated future natural gas reserves and the opportunity to serve consumers in the numerous communities along Alberta's most important population corridor. Extending 315 miles, with an estimated cost of $37,250,000, AGTL's proposal largely duplicated that submitted to the government earlier by the Board as part of its recommendation in favour of a trunk line common carrier system.[298] Not surprisingly, the Board passed on its approval of AGTL's plan to the government.

A year later, AGTL was back before the Conservation Board. It now sought to abandon the approved pipeline route in favour of a new proposal. In the interval between hearings, the administrative, economic and political environment had shifted, and the Board was pushed hard to defend the threatened conservation component of the trunk line plan. The nature of the challenge that the Board was about to face was foreshadowed by a delegation of the Canadian Petroleum Association, which had appeared before AGTL in November 1954. On the basic question of whether the trunk line's operating policies would "be controlled and dictated by sound economics or other factors," the association gained AGTL's assurance that "to the extent the management of the trunk line is permitted to do so, sound private business economics will be the desired policy to follow."[299] It would not be the first time that "sound" economics pushed conservation to the margin.

Early in 1955, AGTL's provisional directors were replaced by a permanent board. Robert J. Dinning and George Church remained as the two government-appointed members. Austin Brownie, the new president of Canadian Western Natural Gas Company, Northwestern Utilities and Canadian Utilities, was elected to represent the utilities group. The elected representative of the export sector was the president of Trans-Canada Pipe Lines (TCPL), N.E. Tanner, who had also served on the provisional board. Two of the three chosen to speak for the producing interests, Vernon Taylor and Ralph Will, were also former members of the provisional board, and Will retained his position as president. The third, Carl Jones, was an engineer with the Hudson's Bay Oil and Gas Company.

The priority item on the new board's agenda was to make sure that the first phase of the trunk line system required to deliver gas from southern Alberta fields would be ready in time to meet Trans-Canada's delivery schedule. The problem was that TCPL had run into serious financial difficulties and had not yet begun pipeline construction. To get its bond financing in place, the company needed supply and purchase contracts, but it was being squeezed on price at both ends of its proposed pipeline and few contracts were falling into place. With little room to manoeuvre, TCPL desperately sought government assistance and cost-saving transmission efficiencies. Faced with this situation, Tanner seems to have prevailed upon his AGTL colleagues, none of whom had any substantial pipeline experience, to review their pipeline project to find a route that would ensure the lowest possible transmission costs.

After a new engineering study, AGTL applied to the Conservation Board on 31 October 1955 to have its existing pipeline permit cancelled in favour of a new proposal. The new application did not come as a surprise. Several weeks previous, the Conservation Board had been called to meet with a delegation from the utility companies, TCPL, AGTL and the Canadian Petroleum Association, where it learned that AGTL intended to apply for a change of route and that there was widespread support within the industry for such a change.[300] Consideration of AGTL's new pipeline application began on November 22. AGTL counsel Mahaffy began by reminding the Board that Trans-Canada "was the only customer of any size in sight for Alberta Trunk."[301] Hence, the company had not proceeded with pipeline construction because it could only finance and build its line after TCPL had begun its pipeline, and it was clear that the need for the trunk line was imminent. Continuing with the preparation of his case, Mahaffy also called attention to several important things that had occurred since the first hearing. There had been an increase in the province's established natural gas reserves, and a portion of this increased reserve was within economic pipeline distance from Calgary. Also, the estimated cost of the project had risen to approximately $58,600,000. From this starting point, Mahaffy tried during the hearing, with the assistance of expert witnesses and a new engineering report, to make and sustain the argument that, given the reality of the current situation, "Trans-Canada and its producer suppliers [had to be] in a position to move gas out of Alberta with the least possible transportation cost."[302] The corollary was that AGTL should be permitted to adopt a new trunk line route that cost $11 million less and would mean reduced transmission charges.

The engineering report submitted at the second hearing was a different document from that presented the year previous. In keeping with the spirit of the Conservation Board's November 1953 recommendation to the Alberta government in favour of a common carrier system, the first engineering report put forward a trunk line design intended to meet the three objectives. The second report, in keeping with the explicit instructions of the new AGTL board of directors, offered a design based upon a single objective—the least costly route and capacity to meet TCPL's immediate requirements.[303] Instead of a central spine, running south from Edmonton to Calgary before turning east to the Saskatchewan border, the proposed :w route took a more direct route to the export gate. It ran from ie Homeglen-Rimbey area south of Edmonton southeast to the big ?essford gas field before turning directly east (see Map 6.2).

The new report, along with its supporting witnesses, was received with dismay. McKinnon, Govier and Goodall were under no illusion as to what was at stake. The Board's preferred trunk line system was being challenged by a proposal that threatened to turn the project into simply a supply gathering system for TCPL. In their questioning, McKinnon and Govier did their best to lead witnesses to admit the deficiencies of a common carrier system based almost exclusively upon the lowest cost requirement of a particular user.[304]

Govier was successful in getting AGTL General Manager William Knode to acknowledge that the full board of directors had not met to give consideration to any factors other than which route would provide the most economical service to Trans-Canada.[305] This established, the Board persuaded AGTL to call a special directors' meeting at which McKinnon and Govier urged consideration of some of the advantages of an admittedly more costly western route. Knowing that they could not hope to have the directors go back to the original location, they tried to interest AGTL in a compromise alternative route that, for an estimated additional cost of 1$ to $2 million, would move the line a little farther west and thus partly salvage some elements of the original proposal.[306]

The meeting proved to no avail. This might have been predicted, for TCPL had consistently opposed the trunk line proposal at the gas export hearings on the ground that such a scheme would mean added costs. TCPL had imposed its agenda on AGTL.

The Board's report to Minister of Highways Gordon Taylor mirrored its unhappy situation. A carefully crafted document, it gave the minister numerous reasons not to accept the approval that it recommended. McKinnon informed Taylor that "the proposed route had been designed to give the most economical transportation for Trans-Canada's present requirements" and that the design engineer for Alberta Trunk "had given no consideration to future deliveries of gas from other fields which might be required to meet any future requirements of Trans-Canada or to provide interchange with gas utilities in the Province."[307] Moreover, the report pointed out that the proposed route was locked in place. The design of the line was such that it prevented the shifting of the main junction point farther west and the rerouting of the Pincher Creek and Homeglen-Rimbey lateral lines if it was found during the first year of construction that TCPL's needs were greater and required the connecting up of other fields. The Board's report also pointed out that the proposed routing would not, and indeed was not intended to, serve the needs of Alberta communities. AGTL had not studied the feasibility of supplying towns in the vicinity of the Pincher Creek lateral but nonetheless took the view that it would be uneconomic to do so. "The Board," the report declared, "does not agree with this contention."[308] In addition, the minister was advised that, whereas the original trunk line route "was designed to transport gas to the Calgary area, to receive surplus summer gas from the Calgary area and to supply gas to the towns between Red Deer and Calgary," the new proposal left these towns to be served by more expensive gas sometime in the future.[309] Perhaps to attach a political

British American's Pincher Creek Gas Plant under construction in 1952. Gulf Canada Resources Ltd.

implication, McKinnon added that these communities were represented at the hearings and had expressed their feeling that the new "route to the Saskatchewan border would be subsidized by the extra cost that they had to bear."[310] Although Calgary did not take a stand in direct opposition, the chairman chose to mention the city's observation that "the proposed pipe-line would be so closely integrated with the Trans-Canada line that there would be no basis for maintaining it within Provincial rather than Federal jurisdiction."[311] After using the greater part of the report to reveal what it saw to be the deficiencies of the new route, the Board recommended that approval be given and briefly stated the reasons. These were that the gas supply situation in the Calgary area had improved; the Canadian Western Natural Gas Company preferred to build its gas lines to serve future requirements rather than to pay its proportionate share of the costs of transportation through the original Alberta trunk line, and without a Calgary connection it was not economically feasible to move the proposed line to serve a few small communities; the new route had the unqualified support of the producers; and the AGTL board of directors did not favour the Board's alternative route.[312]

The minister did not rise to the bait, even though the *Pipe Line Act* specified that the Conservation Board's recommendation should be only one of the factors taken into account and that "any public interest," along with "the needs and general good of the residents of the Province as a whole," was to be considered.[313] Conscious of the political reality, Board members probably realized that, even though they had provided the minister with a rationale to reject the application, the likelihood of his doing so was remote. On 29 December 1955, the minister of highways issued the requested pipeline permit to AGTL, and in doing so gave tacit recognition that the trunk line ideal had been largely sacrificed, in the short term at least, to the immediate interest of TCPL.

Although it is clear that the industry majority supported the new route, a few recognized the decision for what it was. Oldtime oilman Walter Campbell lamented the change in the trunk line's route to his Ottawa friend Dr. George Hume, the director of general scientific services at the Department of Mines and Technical Surveys. Having been a notable presence at the Canadian Geological Survey for a generation, Hume was perhaps Canada's most distinguished geologist. He had been a long and sophisticated observer of the petroleum industry in western Canada, and his reply to Campbell's letter is worth quoting at some length.

I have your letter of January 6th and am very disappointed that the Alberta Petroleum and Natural Gas Conservation Board agreed to the new Alberta Gas Trunk Line proposals. The Trunk Line is in no sense now a Grid System as it is only being built for the present to take gas from the areas where it is now available and only for a few of these. In my opinion, although it may be cheaper to do it now in the way it is being done, it will be a lot more expensive in the years to come. The main Trunk Line should have paralleled the Foothills, where the gas prospects are so good that the major supply of gas will eventually be produced. I cannot see why a pipe line should be built from Pincher Creek direct to Princess through no fields whatever, just for the benefit of the Canadian Gulf Oil Company. Had the pipe line been built north from Pincher Creek to the Calgary area, it would have had a position such that new fields developed along the Foothills, for example, Savanna Creek, could have taken advantage of it. The Savanna Creek gas field may be just as big as the Pincher Creek for all we know. It is left now entirely without any hope of an outlet.

Numerous other prospects are in the same category and, consequently, there is no encouragement for development at all of the
gas resources of Alberta in the present Trunk Line system. It
would have seemed to me that one of the objections of the Government people would have been to see that the developments
were encouraged. The present Trunk Line system takes gas in
large quantities from a few areas and leaves the others without
hope of a market.[314]

Could the Board have done otherwise? Two factors made its situation difficult: one was the immediate political environment and
the other was the legislation that defined the Board's responsibilities. By the time McKinnon was writing his report in November
1954, the question of natural gas export to eastern Canada had
been under active debate for nearly five years. Industry and political
frustration had nearly reached their limits, as each began to charge
the other with responsibility for the continued delay. The intensity
and sensitivity of such feelings are revealed in an occurrence coinci-
dent with the writing of the report. It developed from what on the
surface seemed to be a perfectly innocuous and informative public
lecture delivered by Dr. George Hume on 1 December 1954 at the
National Museum in Ottawa. Towards the end of his lecture, "Natural Gas for Canada," Hume commented on the formation of the
"government-sponsored" grid system in Alberta and the pipeline
hearings just completed in Calgary. He concluded, "There is no
doubt that the delays caused by these arrangements have complicated the completion of contracts for the Trans-Canada pipeline at
a time when Alberta badly needs markets for gas."[315] Further, he observed that while the delay continued the burning of "waste" gas
was accelerating. The Calgary Herald's headline the next day informed Albertans, "Gas Pipe Line Work Slowed; Alberta Grid Plan
Is Blamed: Federal Expert Doubts if Target Can Be Met."[316] Hume's
remarks and the extensive coverage that they received in Alberta
drew forth an angry letter from Alberta's premier. Writing to Prime
Minister Louis St. Laurent, Manning charged that it was "absurd to
suggest that the Alberta Gas Trunk Line company [was] responsible
for delaying the progress of the Trans-Canada project," and he appealed for the prime minister's intervention to ensure that "in future Dr. Hume refrain from speeches embodying incorrect and un-
warranted statements relating to this important project in which
provincial as well as national interests and policies are involved."[317]
The response was not what Manning might have hoped. Replying

British American's Pincher Creek Gas Processing Plant came on stream in 1956, two years after the Board decided that Alberta's natural gas reserves were sufficient to permit major marketing outside Alberta. Gulf Canada Resources Ltd.

on behalf of the prime minister, Minister of Mines and Technical Surveys George Prudham, an Edmonton MP, pointed out that both N.E. Tanner and TCPL's counsel at the recent pipeline hearing had commented upon the difficulties they faced waiting for the feeder line situation to be cleared up in Alberta. "It seems to me," Prudham replied, "that Dr. Hume's statement about these arrangements complicating the completion of contracts for the Trans-Canada pipeline, is correct."[318] This sharp exchange, under way as McKinnon was completing his report, underlines the political reality of the moment. Those close to the premier, such as McKinnon, understood that the Manning government was bent upon moving the project ahead and not in the mood to allow Conservation Board reservations about AGTL's proposed pipeline route to stand in the way.

The government's inclination to be co-operative on the trunk line route was reinforced by a simultaneous skirmish with TCPL over another matter that the company complained was compromising its ability to succeed. Again the point of contention involved a permit.

TCPL's complaint centred upon two sections in Alberta's natural gas export permit: section five, which identified the fields from which the exported gas was to be drawn, and section six, which stated that, notwithstanding the provisions of section five or of any contract for the purchase of any gas, the Conservation Board might direct TCPL to purchase "other gas which reasonably and economically, in the opinion of the Board, may be utilized for the requirements of TCPL."[319] TCPL protested that the clause was "a stumbling block in financing and in the negotiation of contracts."[320] With the assistance of producers and AGTL, TCPL pressed the government to "clarify," and by implication narrow, the interpretation of section six. In late December, the Conservation Board wrote to Tanner to say that, given the expanded market for gas that might reasonably be expected, it did not expect that the diversion and required marketing of "waste" oilfield gas would have a detrimental effect on present negotiated contracts. Having reaffirmed this essential principle regarding the marketing of residue gas, the Board went on at length to reassure, saying that it anticipated that arrangements for the interchange of gas in most cases would be worked out "between TCPL and other directly interested parties through normal business negotiation." "In no case," the Board insisted, "would any diversion or reduction be directed by the Board," unless equitable arrangements had been made to ensure that "prudent investment" would not be materially affected.[321] Finally, the Board confirmed that new gas discoveries in pools not covered by the permit would share in *new* markets and the growth of the TCPL market, but not in the estimated TCPL export market covered by the permit. Premier Manning forwarded a covering letter, stating that he fully agreed with the Board's clarification of section six.[322] Nonetheless, the "stumbling block" section was eventually dropped. In 1959, a new natural gas export permit form was introduced, wherein section six was eliminated in favour of a new section requiring the permittee to be prepared to supply gas from the AGTL pipeline "at a reasonable price to any community or consumer within the province that is willing to take delivery of gas at a point on the pipe line" in any instance where the Conservation Board judged appropriate.[323] Although the amendment added a positive reassurance of natural gas service to Alberta communities in the vicinity of the AGTL pipe line, it also removed the Board's stated authority to substitute "other" or conserved gas for contracted gas, where it thought necessary, to facilitate the proper management of the province's oil and gas resources. The change pushed the idea of AGTL as a vital link in the province's natural gas conservation program into the background and was further confirmation of AGTL's primary role as a marketing device to hold Ottawa's regulatory sway at Alberta's borders.

Assuming that the Conservation Board had been prepared to discount the immediate political environment and to stand firmly in defence of the pipeline route that maximized conservation potential, the Alberta *Pipe Line Act* provided an insecure foundation upon which to stand. This was in marked contrast to the *Oil and Gas Resources Conservation Act* and the *Gas Resources Preservation Act,* which explicitly defined the Board's responsibilities and powers and thereby provided a much more administratively secure refuge for unpopular decisions. The *Pipe Line Act* assigned to the Board only the general obligation to notify the minister whether it approved or disapproved of a pipeline application.[324] There were no guidelines regarding what factors were to be considered. It seems to have been expected that the Board's review would centre upon the narrower matter of technical design. The public interest factor, insofar as it was to be the concern of the Board, had to do with safety, whereas responsibility for the general public interest was assigned by the Act to the minister. The statutes that defined its role made it a lot easier for the Board to deny an export permit, by saying that there was insufficient gas, than it did to deny a pipeline permit by saying that it did not like the route, especially when the fate of a "national project" hung in the balance.

The pipeline permit in hand, AGTL eventually commenced construction of phase one, the most easterly portion of the trunk line, in December 1956, and the first natural gas was delivered to TCPL at the export gate in July 1957. This delivery signalled the functional beginning of a common carrier system unique to the North American experience, and was an experiment that showed great promise. With its recommendation accepted, the Conservation Board now had in place the beginning of a grid network. Its central spine was not in the Board's preferred location, but it still provided Alberta with a promising vehicle with which to promote the efficient development of the province's oil and natural gas resources. The degree to which the AGTL system would live up to this potential remained to be seen, however, particularly as it related to natural gas conservation.[325] The network core was in place; the question was what priorities would govern its use and further development. One thing was clear from the experience surrounding AGTL's creation, the Board's full commitment to conservation objectives would be required to ensure that this aspect of the public's interest received a fair hearing. After all, the province's interest in a provincially incorporated common carrier derived from its desire to keep Ottawa at a distance rather than a deep attachment to natural gas conservation.

The necessity for the Board's vigilance had much to do with the nature of the corporate structure that the Alberta government had devised for the management of its common carrier system. The problem centred upon the composition of AGTL's board of directors, an issue raised by J. Harper Prowse and Elmer Roper when the AGTL Bill was being debated in the Alberta Legislature. It was a matter given explicit emphasis later by AGTL's first general manager in light of his personal experience. In November 1956, William Knode wrote to Premier Manning to convey his concern that AGTL was being directed in the interest of too small a group. He explained that, given the pro rata distribution of shares, seven companies elected four of the directors and had voting control of AGTL. "If it was the intention of the Act to give control of Trunk Line to a very limited number of qualified companies in the Province," he facetiously conjectured, "then I can see no objection to the Act."[326] He urged Manning to consider changing the Act to ensure a broader base of control. Nothing happened, and a few months later Knode resigned.

It was not just the corporate structure of AGTL that would pose a challenge for the Conservation Board. Its determination would also be tested fully by the economic environment that formed the general background to its relationship with AGTL. This environment was conditioned by one overriding factor—a surplus supply of natural gas. This, combined with the great distance that separated the producers from their major market and the higher transmission costs that this entailed, kept the price paid to producers low. The result was that any amount of residue or oilfield gas that the Board might want to introduce into the AGTL system always threatened to displace dry gas produced elsewhere and to trim the already thin profit margins of Alberta natural gas producers.[327]

In the late 1950s, Alberta Gas Trunk Line's place in Alberta's natural gas picture remained to be fully determined. What its role in natural gas conservation would be would depend almost entirely upon the initiative of the Petroleum and Natural Gas Conservation Board. Still, it was only this detail that remained to be filled in; the broad outline of that picture could at least be sketched (see Map 6.3). No one, however, foresaw that one of the most significant outcomes of the government's 1954 decision to establish a provincially incorporated common carrier would be the eventual creation of a substantial Alberta-based capital pool. This is of particular significance, given the nonresident domination of the province's natu-

MAP 6.3   Alberta Gas Trunk Line Company Limited Pipeline
System, 1958

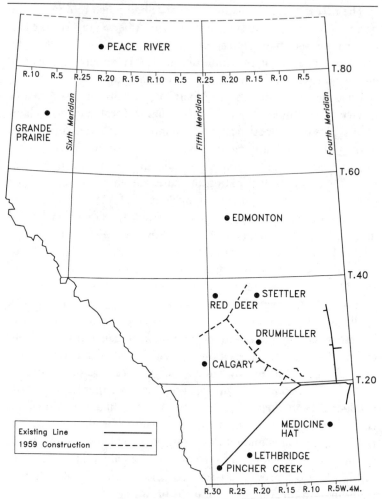

SOURCE: NOVA Corporation of Alberta.

ral gas utility sector and its oil and gas industry. The development
of Alberta's oil and gas resources, though contributing handsomely
to the public sector, was mainly contributing to the growth of capital pools outside the province and region. Alberta's largest provincially incorporated and predominantly Alberta-owned company,
NOVA Corporation, emerged in 1980 from the AGTL capital base.

The gas flowing eastward through TCPL's pipeline brought to a
close the formative and most important period in the evolution of

Alberta's natural gas export policy. Anchored by two statutes, the *Gas Resources Preservation Act* and the *Alberta Gas Trunk Line Company Act*, the essential objectives of this policy were to forestall regulatory encroachment by the federal government and to ensure that Alberta's present and future natural gas requirements would take priority. Yet, while Alberta was bolstering its defences against Ottawa, it was gradually yielding ground to another more distant authority. Initiated in Washington by the Federal Power Commission and in New York by the insurance and banking community, the slow erosion of Alberta's regulatory authority had already begun. This can be seen in Alberta's response to the U.S. assault on section nine of the *Gas Resources Preservation Act* and to some degree in the reassurance offered TCPL with regard to the interpretation of section six in its export permit. Both represented an effort to limit Alberta's legislated flexibility to restrict the flow of natural gas in favour of assured uninterruptible long-term supply to distant markets. This also was a manifestation of a much older, larger and more important phenomenon—the inexorable process of Canadian adjustment of their regulatory structures to bring them into conformity with U.S. practice. Although the history of Alberta's relations with Ottawa conditioned the province's Social Credit politicians to be ever wary of any perceived threat to Alberta's undivided control over the administration and development of the province's lands and resources from the federal capital, historical experience had not led them to be as alert to the less direct, more piecemeal and distant erosion of provincial sovereignty. There was a price to be paid for access to foreign capital and a foreign market, but if noticed it perhaps seemed a small price.

SEVEN

Conservation and
the Struggle to Expand the Market
for Alberta Crude Oil

WITH THE LOCATION OF the Alberta Gas Trunk Line (AGTL) pipeline decided at last, the Conservation Board was able to return to another issue of mounting concern— Alberta's still growing surplus oil production capacity. The proration of Alberta crude oil production to market demand had begun on 1 December 1950, and with a few minor adjustments it functioned well. But in early 1957, as market demand began to shrink, producers began to voice concerns about certain features of the plan. These concerns, combined with some reservations at the Board about the way the plan was operating, prompted the Board to hold hearings for the first comprehensive review of Alberta's proration system.

The need for a review underlined how Canada's oil industry had changed since the proration plan was introduced in 1950. In 1947, before the discovery of Leduc, Canada's oil production supplied only 9% of Canadian refinery needs, whereas by 1957 it supplied 50% of the increased requirements. By 1956, Canada had sufficient productive capacity to supply all its requirements, and by late 1957 Alberta by itself had sufficient capacity to meet the nation's crude oil needs (see Figure 7.1).

Before 1951, Alberta oil was marketed only in the prairie region. With the completion of extensive pipeline facilities in 1951 and 1953, Alberta oil became available to markets in Ontario, British Columbia and the United States. The total export market for Alberta oil reached a peak during the first quarter of 1957 because of the disruption in normal supply patterns brought on by the Suez Crisis of the previous summer. When the Suez Crisis passed, de-

FIGURE 7.1 Canada Productive Capacity Versus Production and
Refinery Operations, 1950–1961

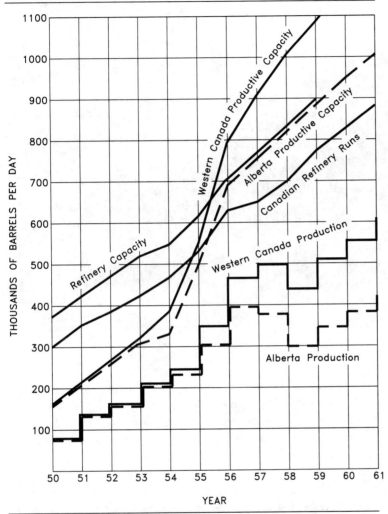

SOURCE: R.J. Cooper and W.D. Keller, *Outlook for Alberta Oil–OGCB* (1958),
Graph 3.

mand for crude oil from western Canada began to shrink in Ontario, as well as in the U.S. midcontinent and U.S. West Coast areas.
This resulted from the surplus productive capacity worldwide, excess tanker capacity, and more competitive prices partly because of
declining tanker tariffs. The decline in shipments from Alberta was
severer than for western Canada as a whole because of the balanc-

Laying the Trans
Mountain Pipe Line in
1953 to carry Alberta
oil westward. Imperial
Oil Limited.

ing role played by Alberta production in meeting market requirements. Since British Columbia, Saskatchewan and Manitoba producers were closer to export markets, their production of light and
medium crude oil had first access to available demand. Alberta producers, however, had to share the remaining balance.

Alberta's production was being depressed both by increasing
production from other provinces and by shrinking markets. In
1950, Alberta had some 97% of Canada's productive capacity of
oil and natural gas liquids, but by 1957 this had declined to 83%.
Its share of Canadian production had dropped from some 95% in
1950 to 76% in 1957, mainly because of increased production
from Saskatchewan, British Columbia and Manitoba. Alberta's
shut-in productive capacity, which was 40% in 1950, had increased
to 51% by 1957. This trend had an impact on the Alberta daily allowable production set by the Conservation Board (see Figure 7.2
and Table 7.1).

At 1957 year-end, Canada, with about 1½% of the world's
crude oil reserves, was producing 3% of world production. By com-

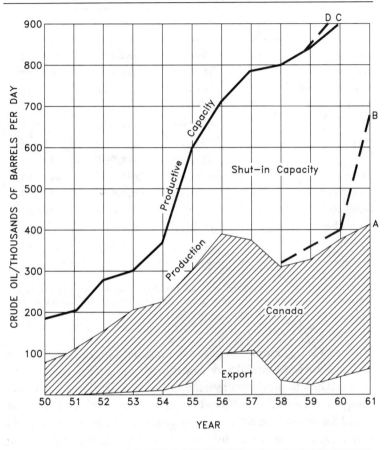

FIGURE 7.2 Alberta Productive Capacity Versus Production, 1950–1961

A — Conservative Forecast
B — Includes Montreal Market & Additional Exports
C — Expected Growth—Existing Markets
D — Expected Growth with Added—in Markets

SOURCE: R.J. Cooper and W.D. Keller, *Outlook for Alberta Oil—OGCB* (1958), Graph 4.

parison, Venezuela, with 6½%, produced 17%; the United States, with 15%, produced 48%; and the Middle East, with 70%, produced only 21%.[1]

Increasing competition for Canadian crude oil in U.S. markets, and the resultant decline in monthly allowable production, brought a dual response from Alberta producers. A group of independent

TABLE 7.1   Alberta Allowable Monthly Production, 1956–1958
(daily allowable in barrels for the month)

	1956	1957	1958
January	389,313	442,372	346,833
February	413,191	429,406	368,217
March	392,624	403,884	313,115
April	356,515	423,373	272,259
May	326,562	435,684	266,775
June	*395,301	418,703	275,841
July	*444,174	417,639	337,167
August	448,698	380,334	293,969
September	421,931	347,600	289,718
October	412,871	310,619	289,082
November	397,642	290,421	304,262
December	448,474	301,609	348,358

SOURCE: ERCB, Minutes.
* In June and July 1956, the market responded to growing tension in the Middle East, and this was reflected in increased allowable rates through the autumn months.

producers led by Home Oil began to press hard for the construction of a pipeline to Montreal and allocation of a greater share of Canadian markets to producers in western Canada. The large, integrated companies, on the other hand, began to press the Conservation Board for revision to the proration system in the hope of gaining a new formula that would increase their share of the still existing markets.

The battle for market share was being waged on a variety of fronts when the Board met to hear submissions in May 1957. A close look and brief analysis of the more important submissions reveals the diversity of interests within the industry's crude oil production sector, the great range of technical evidence that the Board had to evaluate, and just how complex a task it was to try to establish the right balance between the entangled, yet often disparate, demands of conservation and equity.

The 1950 proration plan provided for a two-tier system of allocating market demand. Allocations were first made to each pool sufficient to allow each well to produce up to the lesser of its capacity or the economic allowance (see Figure 5.3). The remaining market demand was allocated among pools on the basis of each pool's calculated MPR (maximum permissible rate). The total allowable of a pool comprised the sum of allocations made to it based on the economic allowance of its wells and the MPR of the pool.

Producers' ideas about how the 1950 plan should be altered reflected their situations in the industry and the nature of the pools in which their principal interests were located. Among the major companies, British American, California Standard, Canadian Fina, Union Oil Company of California and especially Imperial Oil and Texaco had large interests in high MPR Devonian pools and considerable interests in Pembina, a field with high reserves but relatively low productivity. Hudson's Bay, Mobil and Pan American had less interests in high MPR pools, but did have interests in a large number of pools, especially in the areally large Pembina Cardium pool, which had modest well MPRs. Small and independent companies generally wished assurance that they would be able to recover their invested capital, as would be assured for successful wells in pools of shallow and medium depth if the existing economic allowance were continued, whereas larger companies, with large amounts of production from pools with high MPRs, sought to have a bigger share of the market assigned to such pools. Many producers also wanted an adjustment of the economic allowance schedule that would increase its value at greater well depths. A considerable number also favoured some reduction at shallower depths, and there was much support for the existing level for wells in the 4,000- to 6,000-foot depth range.[2]

In accordance with the Board's request, producers focused their submissions on two issues: a review of the main principles of the existing proration plan and the role and method of computing the economic allowance. The latter generated most discussion.

VIEWS RESPECTING THE ECONOMIC ALLOWANCE

The views of the large, integrated companies with substantial amounts of production in high MPR pools were set forth in the submissions of Canadian Fina, California Standard, Imperial Oil, Union Oil Company of California and Texaco. Imperial and Texaco were major marketers of petroleum products in Canada as well as being major producers; California Standard marketed petroleum products in British Columbia and, through Irving Oil Company, in the Maritime provinces; and Canadian Fina marketed production through an affiliate in eastern Canada. All these companies were affiliated with major petroleum-producing and petroleum-marketing companies with operations in the United States and, in most instances, also overseas. For a number of reasons, these companies had limited need for an economic allowance and each submitted that the role of this factor in proration should be reduced by lowering its amount and by reducing its significance in the allocation of production. California Standard argued that the existing high level of the economic allowance gave a disproportionate share of production to those wells having low reserves, and that too large an economic allowance confined the wells in reservoirs with greater reserves by reducing the provincial allocation factor that allocated production on the basis of MPRs. Texaco submitted similar criticisms, noting that a Pembina Cardium zone well, drilled on 160-acre spacing, received an allowable of 81 barrels per day, corresponding to 74% of its MPR and giving the typical well a remaining uniform rate life of nine years. In contrast, a Bonnie Glen well received an allowable of 204 barrels per day, corresponding to a remaining uniform rate life of 28 years. It added that a more reasonable approach would be to lower the existing schedule for all wells at depths less than 10,000 feet, which would still be sufficient to cover operating costs, to pay out development costs, and to provide a 10% rate of return. Imperial contended that the function of the economic allowance should be to prevent premature abandonment of existing wells, and that it should not be designed to pay out capital costs. In its view, too generous an economic allowance tended to divert effort from the search for new reserves and other useful projects, and in the long run would increase the cost of producing oil in Alberta and damage the competitive position of Alberta oil in world markets. Union Oil took a more moderate position in submitting that marginal fields should receive a high enough economic allowance to permit developers not only to get back the money they invested in a reasonable time but also to make a reasonable profit where it was feasible through the existence of sufficient reserves. It believed that this could be achieved by a lower economic allowance than prescribed at present for wells to 6,000 feet and would require a higher economic allowance for wells deeper than 6,000 feet.[3]

Among the major companies that supported the concept of the economic allowance were British American, Hudson's Bay Oil and Gas, Pan American (later Amoco) and Shell. Mobil Oil also supported the concept but favoured utilizing "investment" and "operating" economic allowances, with the investment economic allowance applicable until six years following drilling of 50% of the wells in any newly discovered pool. Pan American submitted that the existing economic allowance, which provided approximately a three-year pay-out, with sufficient profit to make the drilling of a well attractive, furnished the necessary incentive for continued exploration and development. It observed that, with the markets available under the existing proration plan, about half the production was allocated on the basis of the economic allowance and about half on the basis of MPRs that were related to reserves. It contended that overemphasizing reserves in the proration plan would result in further exploration for thick reef high-reserve pools, and that many low-reserve zones would be rendered uneconomic to explore and develop. Pan American believed that the existing plan accommodated all types of fields and provided for maximum recovery of Alberta's oil reserves; however, it agreed that some adjustment of the economic allowance schedule should be made, downward for extremely shallow wells and upward for extremely deep wells. The position of Hudson's Bay Oil was similar.[4]

The independent oil companies, including Bailey Selburn, Canadian Oil Companies, Dome, Great Plains, Home, Husky, Merrill, Okalta, Triad and Western Decalta, as well as the small oil companies, supported continuation of an economic allowance at about the present level. A considerable number agreed upon the need for a reduction for wells of shallow depth and an increase for deeper wells. Merrill Petroleum suggested that the economic allowance should be higher for new wells relative to those that had been paid out to facilitate obtaining development loans from chartered banks to finance new developments. It submitted data showing that to meet a three-year pay-out, required by the banks for purchase of a typical proven Pembina Cardium zone lease, an allowable of 97 barrels per day was required, compared with the seven-year pay-out provided by the current economic allowance of 42 barrels per day. It added that the independents played a major role in development, provided a truly national core to the industry, and had a future that was irrevocably linked to the future of Canada. Dome noted in its submission that nine large pools, Acheson D-3, Bonnie Glen D-3, Fenn-Big Valley D-3, Golden Spike D-3, Leduc D-3, Redwater D-3, Sturgeon Lake D-3, Westerose D-3 and Wizard Lake D-3, had only 32% of the province's wells but produced 54% of Alberta's oil in 1956. Moreover, these fields had 66% of the province's recoverable reserves, and thus would return their development costs many times over, whereas wells with low reserves would return only one to two times the original investment. Dome contended that in these circumstances it was unreasonable to lower the level of the economic allowance or to reduce its significance in the proration plan. Bailey Selburn added that most independents supported the existing proration plan and that the pay-out provided through economic allowance rates was important in purchasing leases with proven reserves. Home Oil submitted that if anything the economic allowance should be increased, and that if it were lowered it would seriously affect those wells least able to stand a reduction. Lowering the allowables of low reserve pools would not be in the interests of conservation, Home argued, because it would render some pools uneconomical to develop.[5]

Western Decalta, in a comprehensive submission, stated that it had invested $6,500,000 in acquiring leases at crown sales on the premise of certain specific rates of return, with the economic allowance as a minimum production level. Most banks required a three-year pay-out using the economic allowance as a minimum; hence, Western Decalta pointed out that reducing the economic allowance would reduce the revenue from crown lease sales, since smaller Canadian companies would be unable to raise bidding capital. It added that junior companies faced higher borrowing costs, alleging that banks required junior oil companies to pay 5 to 6% on a 10-to-15-year loan, whereas major companies got money at 3½ to 4¼% over 25 to 30 years. Western Decalta presented data indicating that, in the previous four years, independent companies had been responsible for 50% of the drilling, even though they received only 24% of provincial production. This compared with the 17% of exploratory wells drilled by the three largest majors, which produced 49% of provincial production. It added that reducing the economic allowance to 20 barrels per day, as proposed by Imperial, would have the effect of reducing the cash flow of independent companies by $7,000,000 per year, which would make a corresponding reduction in their ability to explore. It proposed that no change be made to the economic allowance for wells up to 5,000 feet deep and that the trend of the schedule be steepened beyond that depth to give an allowance of 105 barrels per day for a 12,000-foot well.[6]

The submission that aroused the Board's particular interest and scrutiny was Mobil Oil's. It suggested that the economic allowance be retained but divided into two parts. The proposal called for pools to be granted an "investment economic allowance," about equivalent to the existing economic allowance during the early years of a pool's life, followed by a lower "operating economic allowance." The investment economic allowance would be designed to permit a producer to retire his investment in a well and production facilities exclusive of basic acquisition costs in a six-year period. The operating economic allowance would apply following the period of the investment economic allowance and would be intended to cover operating costs.[7]

RESPECTING PRORATION PLAN PRINCIPLES

The views respecting the proration plan were also split. Major companies with large interests in high MPR pools favoured a procedure that would diminish the role of the economic allowance. British American, California Standard and Union Oil of California proposed adoption of a "residual proration plan" under which pools with MPRs less than the economic allowance would not share in the allocation of the residual demand—the demand remaining after accommodating the economic allowances of all wells. British American and California Standard objected to the existing proration plan in which pools with low MPRs achieved rates up to and exceeding the calculated MPR, whereas pools with high MPRs had producing rates representing only a small fraction of their MPRs. They said that providing a lower limit of production for low MPR pools, as would be provided by a residual proration system, would reduce these inequities. California Standard also suggested that an individual well allowable system be considered for distributing production in a pool. Such a system could distribute the pool allowable determined on an MPR basis to individual wells in the pool on a schedule of barrels per day per foot of pay or producing zone. It added that, although such a plan would be more difficult to administer, it would establish individual well equities in the pool and discourage drilling fringe wells that were not needed but were added to boost allowable production.

Imperial Oil, Texaco and Canadian Fina and associated companies all favoured a "floor plan." Under such a plan, allocations would first be made on the basis of pool MPRs. Where pools received allocations insufficient to allow each well in the pool to produce the lesser of its economic allowable or productive capacity, the allowable would be increased to the economic allowance rate. It was argued that this would provide results about equivalent to a residual proration plan for pools with MPRs significantly less than the economic allowance, but would distribute more of the remaining allowable to high MPR pools and less to pools with MPRs near the economic allowance than a residual proration plan. The criticisms these companies made of the existing proration procedure were similar to those of companies favouring a residual proration plan. Canadian Fina thought that a floor plan provided sufficient incentive for purchase and development of proven properties and that bigger returns should be offered those making bigger discoveries to compensate for the increased cost and risk in exploration. Imperial was concerned with the high allowables accorded low MPR pools and noted that pools with MPRs as high as 55 barrels per day at the current allocation factor of 0.25 received allowables equivalent to their MPRs, whereas high MPR pools received allowables of as little as one-third of their much higher MPRs. Texaco also expressed concern that under the existing proration plan too much of the available market was being allocated to low MPR pools. It submitted that a floor proration plan would satisfy the objective of providing each and every well the lesser of its economic allowance or its productive capacity and would distribute the available market more reasonably.

The floor proration plan idea was undermined by Mobil's submission. It presented data that showed that, with the two-level system of economic allowances it proposed, a floor system of proration would almost double the level of the allocation factor and distribute a much larger share of the available market to high MPR pools. It presented a table that gave data for pools with 95% of the oil wells existing in May 1957. It showed that a floor system, even with the existing economic allowance schedule, would drastically shift production from low MPR to high MPR pools and have an adverse effect on a large number of pools with MPRs near or modestly above the economic allowance. At one end of the scale, the low MPR Pembina Cardium pool area on 160-acre spacing would have its current allowable (May 1957) of 23,515 barrels per day reduced by 33% to 15,792 barrels daily, whereas at the opposite end, the high MPR Golden Spike D-3 pool would have its allowable of 9,275 barrels increased by 101% to 18,621 barrels daily.[8] Mobil's analysis demonstrated that the floor plan proposal would have a negative effect on 50% of existing wells and would impair the economic foundations of companies operating in many fields, especially companies that had acquired costly properties on the premise that the existing proration system would be continued. Mobil concluded that, although a floor plan appeared reasonable in principle, application at that time would have too drastic an effect on the distribution of market demand and that consideration of such a floor plan should be deferred for six years.

The major companies that supported continuation of the existing proration plan included Hudson's Bay Oil and Gas, Pan American and Shell. Hudson's Bay said that the present proration formula was equitable and encouraged development of all the reserves that were economical to develop. It added that a major change in the proration formula could alter the crude oil supply pattern within the province. Pan American noted that under the existing proration plan more than half of the allowable in Alberta was distributed to wells on the basis of their MPRs and MERs, which were roughly proportional to reserves. If the market were distributed only on the basis of reserves, the change from the existing method would be severely damaging to the oil industry and the economy of Alberta. Future exploration would be directed only towards finding high-reserve fields, such as those in thick reefs, and many low-reserve formations that might be found would never be developed because of the poor economic return.

The independent and small oil companies almost all supported continuation of the existing proration plan. Dome stated that it supported the existing plan, while recognizing that, with its large reserves in the Redwater D-3 pool, it would benefit from a system in which greater emphasis was given to allocations on the basis of MPRs. It stressed, however, that the prime consideration in the proration plan should be to allow all companies in the industry to obtain a reasonable pay-out on their development wells and to provide them with sufficient funds to do further exploration. Western Decalta observed that the independent oil industry had invested more than $114,000,000 in proven and semi-proven parcels of land, and that it alone had invested $6,500,000 in such purchases in the previous five years. These investments had all been made on the basis of expected specific rates of return on the total acquisition and development costs of leases as provided under the existing proration plan. Any reduction in allowables of leases would make financing by junior companies more difficult. Western Decalta argued that a system of allocation of allowables that favoured the larger, more concentrated oil reserves would be one that favoured the major oil company at the expense of the junior company. Merrill Petroleums was most concerned with the opportunities for independent companies to develop new production. It was concerned as to how developing low MPR pools, such as the Pembina Cardium pool, would be treated, and submitted that such new pools should be provided with preferred status under the proration system relative to long established and paid-out fields such as Redwater, Leduc and Stettler. It suggested including a factor in the calculation of the economic allowance of a pool that would take into consideration the investment that had been recovered to date.[9]

THE 1957 PRORATION PLAN

The complexity of the task facing the Board was daunting. A need to revise the existing proration plan was clear enough, for the economic allowance was too generous and the attractive pay-out encouraged the drilling of unnecessary wells (in a constant or declining market about the only way to gain market share was to drill a new well that, under the plan, had a guaranteed economic allowable production). Most of the small, independent companies were opposed to change, and those—mainly larger and major companies—that were not, were divided on the kind of revision that would be appropriate. Although they tended to fall into groups, there were almost as many ideas as there were companies, and each submission was framed from the perspective and evidence drawn from a particular base of producing properties. What was equity to one was theft to another. The Board was conscious that whatever approach or formula it adopted could mean the gain or loss of thousands of barrels of daily production to dozens of companies. In the wings, the companies waited; their interest intensified by a declining market. The Board's decision came in the form of a letter, "To All Operators," on 30 August 1957.[10]

The Board and its staff made a detailed analysis of the economic allowance and floor allowance proposals and the pay-out periods at different depths, considering drilling, completion and operating costs. In addition to preserving incentives for exploration, it was also anxious to provide an incentive to encourage introduction of enhanced recovery programs, which were then recognized as being needed in pools such as the Pembina Cardium pool that had inefficient natural drive systems. The Board concluded that an economic allowance should afford a prudent operator the opportunity to meet his operating costs and to recover both his drilling and completion costs in a reasonable period. There seemed little justification, however, for providing for the indefinite continuance of an economic allowance that would permit recovery of drilling costs several times over. The Board decided therefore on a two-stage economic allowance system similar to that proposed by Mobil Oil, with the two levels defined as the "initial" and "operating" economic allowances (see Table 7.2).

The initial economic allowance would reflect the operating costs, a five-year pay-out of drilling costs, and a two-and-a-half-year pay-out of completion costs. This would be satisfied by a new schedule, which was lower than provided in the previous economic allowance schedule at shallow depths and higher at deeper depths. It would

TABLE 7.2 Economic Allowances Schedules: 1957 Proration Plan

Initial Economic Allowances

Depth-Interval	bbls/day	Depth-Interval	bbls/day	Depth-Interval	bbls/day
0–3,600	30	7,701–7,900	60	12,051–12,250	121
3,600–3,800	31	7,901–8,100	62	12,251–12,500	125
3,801–4,000	32	8,101–8,300	64	12,501–12,700	129
4,001–4,200	33	8,301–8,500	66	12,701–12,850	133
4,201–4,350	34	8,501–8,700	68	12,851–13,050	137
4,351–4,550	35	8,701–8,850	70	13,051–13,200	141
4,551–4,700	36	8,851–9,050	72	13,201–13,400	145
4,701–4,850	37	9,051–9,200	74	13,401–13,600	149
4,851–5,050	38	9,201–9,350	76	13,601–13,750	154
5,051–5,200	39	9,351–9,550	78	13,751–14,000	159
5,201–5,350	40	9,551–9,700	80	14,001–14,150	164
5,351–5,500	41	9,701–9,800	82	14,151–14,350	169
5,501–5,650	42	9,801–10,000	84	14,351–14,550	174
5,651–5,800	43	10,001–10,200	87	14,551–14,750	179
5,801–5,950	44	10,201–10,400	90	14,751–14,900	184
5,951–6,050	45	10,401–10,600	93	14,901–15,000	190
6,051–6,200	46	10,601–10,800	96		
6,201–6,350	47	10,801–11,000	99		
6,351–6,550	48	11,001–11,200	102		
6,551–6,800	50	11,201–11,400	105		
6,801–7,050	52	11,401–11,550	108		
7,051–7,250	54	11,551–11,700	111		
7,251–7,500	56	11,701–11,900	114		
7,501–7,700	58	11,901–12,050	117		

Operating Economic Allowances

Depth-Interval	bbls/day	Depth-Interval	bbls/day	Depth-Interval	bbls/day
0–4,300	25	7,851–8,100	37	10,701–11,000	49
4,301–4,650	26	8,101–8,400	38	11,001–11,400	51
4,651–5,000	27	8,401–8,650	39	11,401–11,800	53
5,001–5,400	28	8,651–8,900	40	11,801–12,150	55
5,401–5,700	29	8,901–9,150	41	12,151–12,500	57
5,701–6,050	30	9,151–9,400	42	12,501–12,850	59
6,051–6,400	31	9,401–9,650	43	12,851–13,200	61
6,401–6,700	32	9,651–9,900	44	13,201–13,500	63
6,701–7,000	33	9,901–10,100	45	13,501–13,800	65
7,001–7,300	34	10,101–10,300	46	13,801–14,150	67
7,301–7,600	35	10,301–10,500	47	14,151–14,450	69
7,601–7,850	36	10,501–10,700	48	14,451–14,700	71
				14,701–15,000	73

SOURCE: ERCB, Adapted from Appendix A of letter to operators, dated August 30, 1957, Re: Proration Plan and the Economic Allowance.

apply for a period of seven years, following the designation by the Board of a pool or field, or after drilling a few productive wells into any new pool. For pools and fields that were already designated by Board order, the initial economic allowance would apply for a shorter period.

The operating economic allowance would come into effect after expiration of the initial economic allowance and was intended to cover normal operating costs and to continue the two-and-a-half-year pay-out of completion costs. Continuation of the pay-out of completion costs was intended to cover equipment replacement and other additional costs during the life of the well.

The Board's letter went on to address the critical matter of the weight it intended to assign to the economic allowances and the MPR factor in the new proration plan.[11] The Board's decision was in favour of the "residual MPR plan" that several companies had proposed, and which it saw as a fair compromise between the industry extremes. It seemed to the Board that, with the adoption of the two-stage economic allowances, the MPR of a well should have greater weight in determining its share of the residual demand. Introduction of the new economic allowance was scheduled for 1 January 1958, and the residual MPR plan was to come into effect on 1 January 1960. The different impacts of the 1950 and 1957 economic allowances are shown in Figure 7.3.

The new proration plan promised to apportion the available market more equitably, and by giving more weight to the MPR factor it also moved production distribution more in line with preferred conservation practice. The prospect of the new plan, however, did little to ease the industry's situation. While it waited for implementation, the market continued its descent. Each month, refinery nominations decreased, and in turn the Board set lower allowables (see Table 7.1). By October, the production from many wells was already at or near their economic allowable. This meant that the more prolific high MPR fields, such as Golden Spike, Wizard Lake and Bonnie Glen, had to bear most of the further production cuts.[12] In November, the Board was compelled to cut provincial daily allowable production by a further 20,000 barrels. *Oil in Canada*'s November 4 headline, "Alberta Hits the Skids," captured the prevailing mood.[13] Carl Nickle added his sombre assessment, saying that, unless something could be done to resolve the marketing problem, there was "certain to be a sharp decline in the flow of cash into the Alberta Treasury."[14]

The emerging consensus in the industry that there was little prospect of relief brought forth an urgent letter from Board Chairman

FIGURE 7.3   Comparison Between 1950 and 1957 Proration Plans
(Smoothed-out Data)

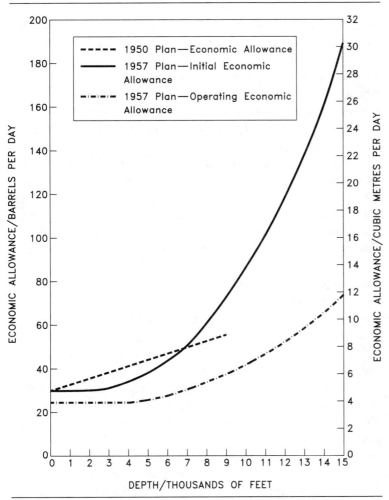

McKinnon to Premier Manning. McKinnon explained that current
allowables were so low that it was getting to the point where exploration and development would come to a near standstill, especially
among the independents. "I had a phone call recently," McKinnon
added, "from one of the leading independents asking for assistance
in estimating future allowables. The company is preparing its 1958
budget and apparently some large cuts were contemplated."[15] The
letter gave Manning a careful review of the crude oil-marketing situation faced by the province. The immediate problem was that the

increasing production of unprorated Saskatchewan crude oil was cutting further into Alberta's share of a declining Canadian market and the continued shrinkage of exports to the United States. But most alarming was McKinnon's report that the U.S. Pacific Northwest and California market, the part of the U.S. market that had been thought most secure, was also vulnerable. The Board chairman disclosed that he had met recently with Walter Levy, the former head of the oil division of the U.S. Marshall Plan and one of the petroleum industry's most highly regarded consultants, who was in Alberta preparing a market analysis for a group of independent Canadian companies. Levy had pointed out that the U.S. companies, who were the main importers and refiners on the U.S. West Coast, also controlled vast supplies of Middle East crude oil. Levy anticipated that they would begin shipping oil to the West Coast, and thus Canadian crude currently moving into the region could expect stiff competition. Moreover, according to Levy, Alberta producers could not realistically look to the Montreal area as a replacement market, since imported Venezuelan and Middle East crude oil had too much of a price advantage. Still worse, Levy's analysis indicated that, unless there was an emergency, like the Suez Crisis, the outlook for the next decade was dull. Alberta would not likely be able to market more than 50% of its potential production.

McKinnon suggested that Manning seek a meeting with the federal government to discuss Alberta's situation. If representatives from the two governments should decide that curtailment of drilling and development was not in the national interest, McKinnon offered several responses for consideration. The Canadian government could approach the U.S. government to work out a continental energy-sharing program that would have Canadian crude oil placed in the same category as U.S. domestic crude oil, or the Canadian government might place restrictions on the import of foreign crude oil into the eastern Canadian market. As a final alternative to protect the important West Coast market, McKinnon advised that if it were determined that Alberta crude oil was being displaced in the Pacific Northwest by Middle East or Venezuelan crude, Manning or the federal government could "call a meeting of Canadian companies who are importing into Montreal and who also through their own or parent companies are exporting crude into the Pacific Northwest supplanting Alberta crude oil, to see if some *quid pro quo* might be established."[16]

The nature and speed of Manning's response is a mark of the province's concern. A face-to-face meeting with the prime minister was what Manning wanted, a desire accentuated no doubt because the government in Ottawa was new.[17] Within a few weeks of receiving the Conservation Board assessment, Premier Manning and Chairman McKinnon were in the capital to discuss the "grave situation of national concern" with Prime Minister John Diefenbaker and Minister of Finance Donald Fleming.[18] Manning urged Diefenbaker to act upon one or the other of the suggestions raised earlier by McKinnon, and in consultation with Diefenbaker he took the initiative himself to arrange a meeting with senior officials of the major oil companies. Advised by their invitations that "the luncheon and discussions will be held in private and no publicity of any kind is intended," a select group of 13 presidents or senior vice-presidents of the major oil companies assembled in Toronto, on 19 December 1957, to hear what the Alberta premier had to say.[19] With Manning was Ian McKinnon, and representing the federal government, since both Prime Minister Diefenbaker and his minister of trade and commerce were out of the country, was acting Minister of Trade and Commerce J.M. MacDonnell and Deputy Minister Mitchell Sharp. Manning's purpose was to express at the highest level Alberta's deep concern about the rapid decline in crude oil production in the hope that he might elicit some kind of remedial response. The presidents, in their comments and subsequent letters to the Alberta premier, offered explanations that tried to place Alberta's situation in the international crude oil picture, but nothing in the way of concrete support. Imperial Oil's view was shared by all. W.O. Twaits, Imperial's executive vice-president, explained that the problem of Alberta's surplus producing potential was "part of a world-wide condition aggravated also by the development of the super tanker which has greatly increased the marketing limits for tidewater oil."[20] He went on to advise against looking to what he described as the "sub-marginal" Montreal market as the solution to Alberta's market shortfall. Movement of oil into the Montreal market area would require the construction of an approximately 200,000 barrels per day capacity pipeline from Edmonton. The problem, Twaits explained, was that the estimated $200-million cost could be financed only by long-term throughput guarantees, which Montreal refineries would not give unless they were "assured of complete duty protection against cheaper foreign crudes." Such an approach, he advised, would raise serious political and economic difficulties. The proper course was to leave Alberta crude oil to find the natural economic range and price, and in this regard Twaits tried to be reassuring. He observed that there had never been any question that North America could utilize all the oil that had been, or would be, found. "Only the question of time is in-

volved, and it seems to us that this will take care of itself, if Canadian oil is allowed access to 'economic' markets in the U.S."[21] Imperial recommended that Manning concentrate his efforts, first on
pressing for the exemption of Canadian oil from U.S. import duties,
and second on encouraging the growth of refinery capacity in Ontario that would assure continued growth in the demand for Alberta crude oil in this important market.[22]

How Manning evaluated Imperial's response is not revealed in
his papers. Twaits had done his best to give his observation and recommendation a mask of objectivity, with the preface that Imperial
itself would "benefit greatly" by a forced opening of the Montreal
market because the company's "high proportion of flush production," meant that a "new outlet including the Montreal market
is . . . more critical to us than to most of the rest of the industry."[23]
But Manning must have been aware that Imperial operated the
largest of the six Montreal area refineries and that, at 70,000 barrels a day (1957), it was by far the largest importer of duty-free
Venezuelan crude oil.[24] Moreover, the imported oil came from one
of Imperial's sister subsidiaries in the Standard Oil of New Jersey
family, Creole Petroleum. In contrast to the oil drawn from its Alberta fields, the Venezuelan oil brought to the Montreal refinery
had the great advantage of coming from an unprorated source. (Although Imperial was the dominant producer in Alberta, any boost
in Alberta production, since it was prorated, meant that the increase had to be divided among all those who shared an interest in
the various fields.) However the Alberta premier chose to credit Imperial's advice, and that offered by the other luncheon guests was
essentially the same, two things, at least, were clear. The major oil
companies would not willingly be part of any campaign to gain the
Montreal market for Alberta crude oil, although it would take a
royal commission rather than a Toronto meeting to make this clear
to the public and other interested parties. Manning also found out
that none of the invited companies was prepared to offer any adjustment in either their production or their purchases elsewhere on
the continent or in the world to help relieve Alberta's plight.

Premier Manning was more successful in his attempt to press the
federal government to act in the province's favour. Prime Minister
Diefenbaker had been a prominent figure in the pipeline debate the
previous year, and he had spoken in Parliament of the urgent need
for a comprehensive review of the principles and procedures that
should be applied in the formulation of national oil and gas policies. He was also aware that the necessity for such a review had
been stressed in the December 1956 "Preliminary Report of the
Royal Commission on Canada's Economic Prospects."

On 15 October 1957, just the second day of his government's first parliamentary session, Diefenbaker announced the formation of a royal commission to enquire into and to make recommendations on a broad range of energy matters, including the desirability of forming a national energy board to administer such aspects of energy policy as fell within federal jurisdiction.[25] Chaired by Toronto industrialist Henry Borden,[26] the Royal Commission on Energy was just getting organized as the Alberta crude oil market began to collapse and Manning came to seek Ottawa's assistance. Hence, although the federal government was not going to take any policy initiative until it had received and considered the Commission's report, it was prepared to press hard in Washington to try to persuade the U.S. government to exempt Canadian oil from its import substitution program.[27]

As Alberta waited, its situation grew more desperate by the week. By December, it was apparent that a major new oilfield had been discovered in the Swan Hills north of Edmonton and that there would soon be dozens of new producing wells to add to those with which the declining market had to be shared. Then, Albertans learned that the Eisenhower administration had dismissed the "strenuous arguments" put forward by Canada's ambassador and had extended the "voluntary" quota system for oil imports to District No. 5, the Pacific West Coast area. In effect in the rest of the United States since July, implementation of the quota system in District No. 5 signalled the impending further erosion of Alberta's best export market. Reflecting the mood in the province, the *Calgary Herald* denounced the decision saying that, since the U.S. government had decided to cut back Canadian oil imports, Canada was left "no option but to prohibit the entry of foreign oil into Canada."[28] It lashed out too at the West's traditional target. Noting that the consequence of banning Venezuelan oil from the Montreal market might be higher prices for petroleum products in eastern Canada, the paper lectured, "In this regard, it must be remembered Westerners have been paying heavy tariffs on freight rates for years for the benefit of Eastern industry. It is time now for the debt to be repaid."[29]

PROJECT OILSAND

In the midst of this gloom, it was not the hoped-for notice of a new market that miraculously appeared, rather it was notice that the miracle of modern science might dramatically increase the province's oil glut. After gaining tentative approval of the U.S. Congress

Atomic Energy Commission, representatives of the Los Angeles-based Richfield Oil Corporation appeared in Edmonton in June 1958 with a proposal to detonate a nuclear device in Alberta's northern oil sands. Code-named "Project Cauldron," the scheme was presented to Deputy Minister of Mines H.H. Somerville and George Govier by company officials, who explained that their proposal was an "experiment in the peaceful use of nuclear energy as an aid in producing oil from the McMurray oil sands buried too deeply to permit economic extraction of oil by mining methods."[30] Richfield planned to drill a well at a test site in "a remote undeveloped area" about 60 miles south of Fort McMurray and, at a depth of 1,250 feet, about 20 feet below the oil-sand formation, detonate a nine-kiloton nuclear device (see Map 7.1 and Figure 7.5). It was expected that the resulting explosion would create a 230-foot-wide cavity, into which would drain several million cubic feet of oil released by the explosion's tremendous heat. Also, it was claimed that there would be no radioactive fallout since the explosion would be contained underground. It was the brainchild of M.L. Natland, who explained that he had been struck by the idea one evening in the desert of southern Arabia where, as he "sat watching a spectacular sunset from a small hill overlooking a flat, endless sea of sand, the sun looked like a huge orange-red fireball sinking gradually into the earth."[31]

Even though the province was awash in oil, Natland's unusual sunset-inspired vision generated great interest in the Alberta capital. For over 30 years, government-sponsored research had been trying to develop a viable oil-sand separation technology that would permit commercial development of the McMurray oil sands. Known to be one of the world's two great bitumen deposits, the sands underlay about 30,000 square miles of northeastern Alberta and were estimated to contain upwards of 800 billion barrels of oil (see Figure 7.4). But this was hinterland science on a shoestring budget. Now came a proposal supported by the world's leading scientific establishment to apply an awesome new technology that might solve the problem that had thwarted the efforts of local entrepreneurs and scientists for so long.

The first extensive research on McMurray oil-sand reserves and extraction was carried out between 1913 and 1916 by Sidney C. Ells, a geologist with the Mines Branch of the federal Department of the Interior. Ells's interest in the oil sands was taken up in 1920 by what became the Research Council of Alberta. Supported by council funding, chemist Karl A. Clark began the search for an extraction technology and for uses for bitumen.[32] In 1924, Clark and

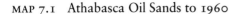

MAP 7.1   Athabasca Oil Sands to 1960

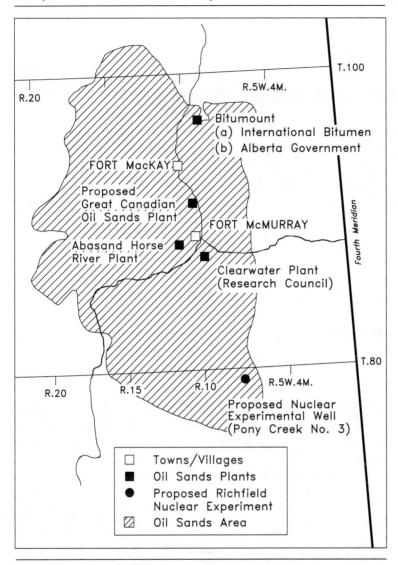

his associates constructed in the north Edmonton Dunvegan rail-
yards the first pilot separation plant to test their laboratory conclusions. A refined version of their hot water separation process was
tested at a second plant built in 1929 on the Clearwater River near
Fort McMurray. The first successful commercial activity in the oil
sands began in the late 1920s with R.C. Fitzsimmons and his International Bitumen Company at Bitumount, about 40 miles down-

FIGURE 7.4   Oil Sands Cross-section

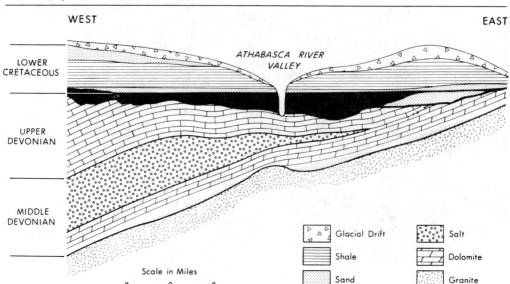

SOURCE: M.A. Carrigy and J.W. Kramers, *Guide to the Athabasca Oil Sands Area*, Edmonton Research (Alberta Research Council).

stream from Fort McMurray. Between 1930 and 1937, Fitzsimmons, with a series of plant variations using the hot water process, intermittently produced roofing tar and eventually some oil from the separated bitumen before his company collapsed. The second significant commercial extraction attempt, that of Abasand Oils Ltd., began about the time International Bitumen began to fade. Based upon the technical foundation established by petroleum engineer James McClave at a prototype hot water separation plant in Denver, and upon the engineering and financial skills of American Max W. Ball, Abasand completed construction of the first important oil-sands facility near the town of Fort McMurray in 1936. After a series of design adjustments and the building of a small refinery, operation on a regular basis began in May 1941. In November, the separation plant was destroyed by fire, but in the short interval 41,265 gallons of gasoline, 70,700 gallons of diesel oil and 137,550 gallons of fuel oil had been produced.[33] This hint of Abasand's potential, plus the emerging wartime shortage of petroleum products, combined to arouse federal interest, which led to the restructuring of Abasand Oils as a joint government and privately owned enterprise. Constructed largely with federal money, the rebuilt separation plant and refinery started intermittent trial runs in December 1944. Operations seemed promising, but on 16 June 1945 the plant was razed once again. Minister of Reconstruction C.D. Howe assessed the situation and concluded that a new plant should not be built at government expense since the wartime emergency was over.[34] In late 1946, what remained of Abasand's assets, mainly its oil-sand leasehold, were returned to its private shareholders.

This left initiative with the provincial government, and it was not an unwelcome prospect. From the outset, Manning and his cabinet colleagues had been uneasy about the federal government's new and direct presence in the oil sands.[35] The Alberta government had already moved, well before the second Abasand fire, to bolster the province's position in the oil sands with its own government-supported enterprise. In December 1944, it had entered into a joint partnership with Lloyd Champion of Oil Sands Ltd., the corporate successor of R.C. Fitzsimmons' International Bitumen. The agreement called for the Alberta government to contribute $250,000 towards the cost of constructing a new experimental scale separation plant at Bitumount. A board of trustees, composed of cabinet ministers W.A. Fallow and N.E. Tanner, and Lloyd Champion from Oil Sands Ltd., was appointed to supervise the construction and operation of the plant. Designed by Born Engineering of Tulsa, Oklahoma, construction of the plant was completed in September 1947, but not before several transfusions of additional government money. Still, it was not until the late summer of 1948, after a series of start-up problems that included a fire and the financial collapse of the government's private partner, that the plant began limited production. Full-fledged tests of what was now a government-owned operation were carried on through 1949. Unfortunately, the results did not yield a clear verdict on the feasibility of commercial production. Faced with the prospect of a significant cost to turn the plant into an all-season facility, the board of trustees engaged Sidney Blair, formerly one of Karl Clark's Alberta Research Council assistants, to conduct a thorough evaluation to resolve the critical question of the commercial feasibility of existing oil-sands technology.

Blair's positive report was completed in December 1950. It concluded that bitumen could be "processed by established methods."[36] Basing his calculation upon a 20,000-barrel-a-day operation, which he concluded was the minimum economically efficient size, Blair estimated that the per barrel cost of mining, separation, refining and delivery to the Great Lakes would total $3.10.[37] He cautioned, however, that further testing at the Bitumount plant was required before industry was likely to be convinced. In the interval, he recommended holding a public conference to review oil-sand science comprehensively and to alert the scientific community and the petroleum industry to the current state of oil-sands technology. Blair and Karl Clark were asked to organize a conference. Hosted at the University of Alberta in September 1951, senior oil company executives and related industry and government scientists gathered in Edmonton to hear and discuss papers on the geology of the oil-sands formation, possible mining methods, the technology of oil-sands separation, and the refining and transport of synthetic crude oil before being flown to Fort McMurray to see the pilot separation plant at Bitumount in operation.[38] To lend positive encouragement to oil-sands development, Minister of Mines and Minerals N.E. Tanner took the opportunity afforded by the conference to announce a new and much more attractive bituminous sands lease policy. Henceforth, exploration permits of up to 50,000 acres could be obtained upon payment of a fee of $250 and a cash deposit of $50,000 to guarantee the carrying-out of a satisfactory exploration program, upon completion of which a 21-year bituminous sands lease could be obtained. The leaseholder's obligation, beyond an annual rental of $1 per acre and an eventual royalty of up to 10% of the extracted products, was to have a plant operating within five years.[39]

Beyond achieving its immediate purpose of bringing key industry representatives up-to-date on current oil-sands technology and development regulations, the impact of the conference was limited. Partly, this had to do with the release of a second report on the oil sands, which stood as a negative backdrop to the conference proceedings. Prepared by Dr. D.A. Howes for Anglo-Iranian Oil (later British Petroleum), this report concluded that oil production from Alberta's bituminous oil sands would not be economic and was unlikely to be in the near future. Howes's estimates of mining, separation and refining costs were much higher than those calculated by Blair, and he concluded that oil drawn from the sands could not be delivered to the Great Lakes at less than $3.80 a barrel.[40] He argued that, given the particular quality of the oil produced from bitumen, its real value in the Great Lakes market area was in the vicinity of $3.06 a barrel, rather than the $3.50 Blair had claimed. Although the merits of either estimate could be debated, one thing was clear, oil produced from the sands was going to be expensive relative to conventional oil. This came to stand in even higher relief later when conventionally produced Alberta oil began to face the market challenge of even more cheaply produced offshore oil.

Although Howes's report helped to dampen industry interest in the oil sands after 1951, the attention of the Alberta government was also diverted by the dramatic development of the conventional oil sector in the early 1950s. Nonetheless, interest in the Fort McMurray oil sands did remain, if only on the periphery. After all, everyone understood that locked within the sands was one of the world's largest oil deposits, if only the problem of a cost-competitive technology could be solved. Alberta kept the Bitumount operation going on an experimental basis until 1955, and private interests continued modest test-drilling programs. In late 1954, Oil Sands Ltd. was re-formed as Great Canadian Oil Sands, and the new company acquired Oil Sands' primary asset, bituminous sands lease No. 14, an area of some 4,177 acres along the Athabasca River about 50 miles north of Fort McMurray. Centred about the Toronto investment firm of Fry and Company Ltd., which contributed two members to the directing board, Great Canadian also struck an agreement with Abasand Oils Ltd., Canadian Oil Companies Ltd. and Sun Oil Company of Philadelphia, thereby acquiring access to bituminous lease No. 4.

Once the corporate reorganization was complete, Great Canadian initiated discussions regarding oil-sands development with the Alberta government. The main item on the company's agenda was Alberta's system of prorating the available market among producing wells. Although there was usually no difficulty about increasing or decreasing the flow of oil from an oil well to meet the prorated schedule set by the Conservation Board, this was not the case for an oil-sands facility. Great Canadian pointed out that oil-sands production involved mining and an industrial process, and it could not operate a processing plant properly on an economically efficient basis with a fluctuating production schedule. The company advised that it could enter the *pro rata* scheme only if it could be guaranteed a minimum production of 20,000 barrels per day, which it estimated as being the smallest efficient operating volume.

In addition to consideration within his own Department of Mines and Minerals, Premier Manning took Great Canadian's proposal to the Conservation Board and to the oil and gas community for consideration. Discussion with the Board followed an earlier consultation regarding the question of the Board's jurisdiction in the matter of oil-sands recovery. Although the Board had not yet been involved with oil-sands development, it was concluded that the definition of "oil" in the Conservation Act would include oil contained in the oil sands, that recovery by well was covered in the existing regulations, and that, if the oil were produced by other means, the Board had clear authority to establish the regulations necessary to control operations.[41] It was determined that if the sands had to be mined and transported to a processing plant to recover the oil, market share probably would be determined by establishing the plant's capacity as the MPR and calculating the economic allowance as a percentage of capacity.[42]

The hostile attitude of the conventional industry to the Great Canadian scheme was conveyed by the Canadian Petroleum Association (CPA), which formed a special committee with instructions to "prepare a vigorous brief protesting the applicant's submission."[43] Despite the CPA's strong objection, Manning pressed ahead in the belief that, after nearly two decades of government-supported research and encouragement, several company groups were on the verge of launching major oil-sands projects.[44] His solution was the segregation of oil-sands production from oil well production, thus removing synthetic crude oil from the proration system as requested. But the separate or special status designation given oil-sands production came at a price. In deference to the conventional industry's unease, Manning decreed that, just as oil-sand producers would carry on outside the proration system, so too would they have to find and negotiate their own markets. They would have no guaranteed market share. Just how difficult this was going to be was not immediately realized.

The new approach gained formal sanction with the passage of the *Bituminous Sands Act* by the Alberta Legislature in April 1955. Most important was the new Act's stipulation that any operation for the recovery of hydrocarbons from the sands or any production or the disposal and transportation of such production would not be subject to any of the provisions of the *Oil and Gas Resources Conservation Act*.[45] Among other things, this assured that oil produced from the sands would be excluded from the province's oil market prorationing scheme. Although left to their own devices as far as finding a market, the Act did lend potential oil-sands operations one critical assurance. It provided that any pipeline operating with a permit under the *Pipe Line Act* could not discriminate between oil from bituminous sands and conventional oil. From this point, the government waited in anticipation, exploration work continued, but no concrete development proposal appeared.

This was the situation when the Richfield Oil Corporation appeared in June 1958. The old dream of oil-sands development was rekindled despite the current oil glut. In the 1950s, nuclear science held the world in awe, and it was generally believed that mankind was on the threshold of a new age—the atomic age. In this atmosphere, the eagerness of many Albertans to embrace a totally new and exotic technology that held out the promise of sharply reducing the still prohibitive cost of oil-sand separation is hardly surprising.

Richfield's proposal certainly roused Govier's scientific curiosity, and he was quick to take the initiative on the province's behalf. He informed the Richfield party that the Conservation Board "would be prepared to recommend the issuance of the necessary well licences and to support the project assuming that it received the necessary support from the appropriate Federal Government agencies and from any Provincial departments that might be involved."[46] After learning that Richfield had met with and gained preliminary approval from the Mines Branch, Atomic Energy of Canada and Defence Research Board officials in Ottawa, Govier advised Premier Manning that it was time for Alberta to define its position. To this end, he offered to call together senior civil servants from the relevant departments and agencies to examine the proposal and to prepare a document outlining the terms and conditions under which the project might be safely approved.[47] Given Manning's assent, the meeting was held on 9 October 1958. Govier reported to the premier the group's consensus

> that there was no reason not to approve the pilot test provided that the Conservation board, after receiving the advice of the joint technical committee, was completely satisfied that the fullest possible precautions would be taken to ensure that there would be no . . . radioactive contamination of the oil . . . waters [or surface].[48]

The joint technical committee to which Govier referred, and that was intended to review and oversee the project, was to be composed of key people from Richfield, the U.S. Atomic Energy Commission, Atomic Energy of Canada, the Department of Mines and Technical Surveys (Ottawa) and the Conservation Board. Expressing his own feelings, Govier concluded that, with the joint technical committee's presence to ensure that the proper precautions were taken, there was "nothing to be lost and much to be gained from the proposed test."[49]

Early in 1959, Richfield Oil's proposal was presented to Premier Manning and his cabinet. Rumours that consideration was being given to the detonation of an atomic device in the Athabasca oil sands had already begun to circulate, and public interest was stimulated further when it was learned early in 1959 that talks were under way in Ottawa between representatives of Richfield Oil and federal bureaucrats. The enthusiastic comment by Dr. John Convey, director of the Mines Branch of the Department of Mines and Technical Surveys, that the atomic blast, if successful, would "at a single stroke double the world's petroleum reserves" was interpreted by the *Calgary Herald* to mean that an "Oil-Sands Atom blast" was virtually certain.[50] If all went well with the experiment, the paper added, "it will give the Western world a measure of independence from the huge Middle East oil deposits."[51] Finally, to stem the rampant speculation, the Alberta government called a press conference for February 13. The press conference was carefully organized and hosted by Alberta's Minister of Economic Affairs A.R. Patrick. Assembled with the minister to inform Albertans of the plan were Deputy Minister of Mines and Minerals Hubert Somerville and Dr. George Govier. Present on the special invitation of the Conservation Board were Doctors Gerald W. Johnson and Garry Higgins, who were from the University of California Radiation Laboratory, the research home of a group of scientists contracted by the U.S. Atomic Energy Commission to investigate and plan for the peaceful use of atomic energy. Richfield Oil Corporation was represented by S. Stewart, manager of the Company's Canadian division and Dr. M.L. Natland, Richfield's director of production research.[52] It was a high-powered assembly, and each had a designated role to play in the presentation.

Patrick explained the purpose of the press conference, gave a brief general background on the Richfield proposal, and announced that each of his panel members would speak on different aspects of the plan. Somerville and Govier outlined the manner in which federal and provincial authorities would be involved. Govier explained that Edmonton and Ottawa would set up a joint technical committee to review the project and that the work of this committee would be screened by an Alberta technical committee, which in turn would submit a recommendation to the Conservation Board. The latter would study the recommendation, set out specific terms and conditions, and forward its recommendation to the minister of mines and minerals, who might set certain conditions before approval was given. Govier emphasized that only if both these committees, the Board and the Minister were satisfied that there was no danger to public health or safety or to the province's natural resources would the project be approved. Moreover, if the project were approved, he assured that "a senior representative of the Oil and Gas Conservation Board [would] live with the project through to the completion."[53] The rest of the panel were left mainly to tell more about how the experiment would be carried on and to add re-assurance on the safety question.

Sam Stewart began his statement with the announcement that the Cities Service Oil Company of Tulsa, Oklahoma, was joining Richfield as a partner in the project, and he said that he expected other U.S. and Canadian oil companies would also be joining. Stewart added that the discouraging results of his company's experimentation with existing separation technologies had led to the "conclusion that the tremendous amount of heat required to liquefy the oil in the Athabasca Oil-sands [could] only be obtained from a nuclear device."[54] Dr. Natland followed. With the aid of a film, he explained to the press exactly how it was hoped that the nine-kiloton explosion would "wrest the very stubborn oil from the Athabasca sands."[55] The greater part of the conference was left to the scientists Johnson and Higgins. Johnson described the overall character of the search for peaceful applications of atomic energy under Project Plowshare, and he placed the oil-sands experiment, along with the still embryonic idea of setting off a nuclear device in one of the U.S. oil shale formations, into this framework. He also described in some detail the setting off and scientific conclusions drawn from underground nuclear explosions at the Nevada test site. Johnson explained that they had found that underground firing had not triggered earthquakes, that the seismic effects were relatively "quite small."[56] Another major concern was groundwater contamination, and Johnson assured that their experience showed that this was not a problem.[57] Higgins, a nuclear chemist, emphasized that the health and safety of people who lived in close proximity to the Nevada test site had always been a priority concern, but suggested that this would not have to be as much a factor in Canada. "We only have really something in the order of 60 or 70 miles of exclusion area in each direction," Higgins noted, "so that comparatively speaking our range is even smaller than this great wilderness we saw on the last couple of days...."[58]

After each of the participants had made their statements, Patrick distributed the government's press release. It informed Albertans that, "In view of the vast reserves of oil in the oil sands," the government was interested in seeing the proposed test proceed, provided it received adequate assurance that the test could be conducted with complete safety.[59] In the question and answer session that followed, reporters directed most of their attention to minor details. They drew forth the added information that Richfield hoped that the proper permits could be obtained in time to allow the detonation to take place by the end of the following winter drilling season and that the price for an explosion in the nine- to 10-kiloton range was about $500,000.[60] The topic that sparked most discussion was how an oil-sands field might be developed if the experiment worked in the anticipated way. Such interest allowed Dr. Natland's enthusiasm and scientific speculation free rein. On the subject of how and when commercial development might occur, Natland stressed at the outset that at present they were just entering into the experimental stage, but went on to present his ultimate vision of being able to "create an oil field on demand" in the Athabasca oil sands.[61] He explained that you "would go into the centre of the township and seed an area with all these heat cells [created by nuclear explosions] and then perhaps let it sit for a year. That would keep the heat completely spread around, and do as much good as it possibly could in fluidizing the oil and then we could come back later and [remove the oil by conventional means]."[62] Natland likened the heat situation to that contained by an exceptionally efficient thermos bottle. According to Natland, until such time as the nuclear-created chamber (see Figure 7.5) was tapped to release the heat and the oil that "had come tumbling down into the cavity,"[63] the heat would stay right in the explosion area for perhaps as long as 10,000 years.[64] Natland also addressed the only question on radiation. He said that some Athabasca oil-sand had been sent to the Oak Ridge reactor in Tennessee and exposed to radiation. It was now back in his laboratory in California, where it was not giving off any more radiation than his wristwatch.[65]

The press conference produced detailed front-page matter-of-fact accounts in provincial newspapers.[66] A few days later, Govier wrote to Manning with a list of suggested names for the Alberta Technical Committee and an offer to chair the committee if the premier so wished.[67] Manning accepted the offer and the list, although he added H.H. Somerville, Minister of Economic Affairs A.R. Patrick and Grant MacEwan, MLA and Liberal party leader in Alberta to Govier's group of nominees, which included two of the Conservation Board's most senior engineers, D.R. Craig and A.F. Manyluk. In March, federal Minister of Mines and Technical Surveys Paul Comtois announced the composition of the Joint Technical Committee responsible for assessing Richfield Oil Corporation's proposed "Project Cauldron." From his own department, the minister named Dr. John Convey, a chemist and the director of the Mines Branch, to chair the committee, and A. Ignatieff, a mining

FIGURE 7.5   Richfield Oil Corporation Pony Creek No. 3 Well,
Stratigraphy of 9-kiloton Nuclear Blast Location

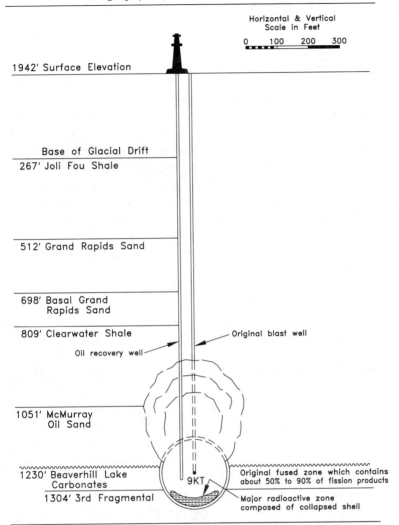

SOURCE: M.L. Natland, "Project Oilsand," 1963.

engineer from the Mines Branch fuels division. The other federal representatives included Dr. W.E. Grummitt, a chemist from Atomic Energy of Canada; Dr. J.M. Harrison, a geologist and the director of the Geological Survey of Canada; Dr. A.H. Booth, a physical chemist and radiation safety officer with the Department of National Health and Welfare; and Dr. R.J. Uffen, a geophysicist from the University of Western Ontario. Alberta's named representatives were Govier and Manyluk from the Conservation Board. Dr. M.L. Natland of Richfield completed Comtois's committee.

The Alberta committee got to work quickly. Its immediate and lasting focus was on the collection of data, both to evaluate the proposal and to gain assurance of access to all data gathered if the experiment was to proceed. Overriding all other matters was the general concern about the amount of radiation likely to be released and its containment, along with the particular fear of possible groundwater contamination. At the outset, and through the course of its subsequent meetings, the committee continued to seek reassurance in these areas.[68] New information on Project Cauldron was presented to the committee by a large delegation of Richfield engineers and scientists on April 22.[69] Dr. G.W. Johnson and radiation chemist Dr. R.E. Batzet from the University of California radiation laboratory came to Calgary a few days later to review the radiation question.[70] The next step, an important joint meeting of the Alberta and national committees, was held just before the scheduled visit to the test site in Yucca Flats, Nevada. After reviewing the most recent information obtained from Richfield and the University of California scientists, and agreeing on points where their knowledge was still deficient, committee members agreed that, to avoid raising unnecessary spectres in the public mind, "a less effervescent name" than "Project Cauldron" should be found.[71] "Project Oilsand" was the eventual choice. In Nevada, the joint committee met with U.S. Atomic Energy Commission (USAEC) scientists and were conducted through the labyrinth of tunnels and driven through the cavity areas created by underground nuclear explosions. Before returning to Canada, the committee attended the USAEC's Second Plowshare Symposium and met with scientists at the University of California's Lawrence Radiation Laboratory.

Although still concerned about the groundwater situation and radiation, the Alberta Technical Committee was ready to give its decision by mid-June. Even though it did not yet have its sister committee's verdict, Govier's committee voted unanimously to recommend in principle to the Conservation Board and the minister of mines and minerals that Richfield be permitted to proceed with its test.[72] The committee also agreed that, for the time being, their decision would be held in confidence, perhaps so that Govier could convey Alberta's position at the Joint Committee meeting in Ottawa the following week. As expected, the Ottawa meeting did produce agreement that the preapproval requirements had been satisfied and that the committee should proceed with the writing of a report recommending that Project Oilsand, as it was now called, be approved.[73] Before the federal report could be submitted, external factors emerged to push it into limbo. The Alberta committee proceeded on course, but not without some difficulty. Liberal leader Grant MacEwan, who had not been at the approval meeting, resigned and Dr. D. Dick from the Department of Public Health grew uncomfortable with the emerging tone of the committee's draft report. Dick's responsibility was the section of the report on public health and safety. After working through several drafts, Dick was persuaded to say that, "it would seem that the setting off of a nuclear device beneath the Oil Sands in Northern Alberta should pose little or no hazard to health or safety," but he also insisted on retaining the cautionary statement from his original draft that "what constitutes a safe level of radiation is not yet known."[74] Reluctant to concede that the blast could be conducted safely, Dick was anxious that the report simply recommend approval without encouragement.[75]

Agreement on the wording of the report was eventually reached, and in late August the Alberta Technical Committee submitted its positive recommendation to the minister of mines and minerals and to the Conservation Board. A public announcement was made a few weeks later. As letters opposing the test began to accumulate in the premier's office,[76] the Conservation Board continued to wait for the report of the Ottawa Joint Committee. It never came. The timing of such an initiative could hardly have come at a more inopportune moment. Just as the Joint Committee set about preparing its report, the prime minister appointed a new secretary of state for external affairs. Howard Green was a deeply committed advocate of nuclear disarmanent, and he believed that Canada should play a more aggressive part in the search for an international treaty to limit nuclear testing and the possession of nuclear weapons. As the would-be apostle for such a cause, it was clearly out of the question for Ottawa to give its blessing to Project Oilsands. As this became apparent, the Conservation Board saw little use in preparing a formal report for the Alberta cabinet on a controversial issue, the future of which had been already decided at the federal level. Richfield Oil Corporation continued its promotion of Project Oilsand without success until the summer of 1963. Premier Manning was disappointed, he later reminisced, "I was always sorry the [the experiment] didn't go through to fruition."[77]

The premier's disappointment might have been eased by the coincidence that, as the hope for Project Oilsand began to ebb, the promise of an earlier oil-sands venture returned. While the Richfield proposal preoccupied Alberta politicians and scientists through 1958 and 1959, Great Canadian Oil Sands Ltd. (GCOS) had continued research and exploration in the McMurray area. In late 1958, it was successful in assembling sufficient financial support to go ahead with preliminary engineering studies and to develop a project plan. The key player was Sun Oil, the 75% owner of the preferred Abasand lease at Mildred-Ruth lakes, about 20 miles north of Fort McMurray. Sun Oil agreed that it would take 75% of the synthetic crude oil produced by a GCOS plant and delivered to Edmonton and would pay a premium of 20¢ per barrel compared with conventional crude.[78]

The Philadelphia company would remain the critical element in the GCOS vision, and this was largely a function of the complementary personalities of the two primary players: Premier Manning and J. Howard Pew, Sun Oil's chairman. Founded, owned and managed by the Pew family, Sun Oil stood out as a fiercely independent and innovative company in an industry well known for its maverick elements. This had much to do with the dominating influence of J. Howard, who became president of the company in 1912. The refining techniques that enabled Sun to convert poor bottom-of-the-barrel heavy East Texas crude into high-quality lubricating oil and to gain market advantage were based largely on his research. He was also the father of the first commercially successful petroleum asphalt, and he was instrumental in Sun's completion of the first pipeline built specifically to move refined products in 1931.[79] It is not hard to see how his pioneering and entrepreneurial instincts were attracted to the Athabasca oil sands. As one of his close colleagues put it, "J. Howard Pew was enamoured of the enormity of the resource up there."[80] Beyond the oil sand's vast potential was the challenge itself. Speaking of the Athabasca oil sands, Pew at one point instructed his company's stockholders, "unless projects of this kind were periodically challenged and solved, our organization would become soft and eventually useless."[81] Pew was also a religious man and held high lay office in the Presbyterian church. His commitment to the Athabasca challenge was reinforced by the close relationship that he developed with Premier Manning.[82] The Alberta premier admired Pew's spartan style and the emphasis that he placed upon personal integrity. The two also shared a remarkably similar private enterprise philosophy. Their paternalistic utopia was one where the freedom of an unfettered market-place was balanced by the self-discipline and social responsibility of business and political leaders.[83] In the long and costly struggle to bring GCOS into production, Pew provided the capital and the entrepreneurial lead-

ership at the key decision-making level, while Manning held the opposing conventional oil lobby at bay.

Anticipating the GCOS application, and given that it had already
assigned to the Board the special responsibility for evaluating Richfield's proposed oil-sands project, the government at the 1960
spring session of the Legislature moved to regularize the assessment
of oil-sands proposals. A special oil-sands section was added to the
Oil and Gas Conservation Act, requiring all applications for experimental work in the oil sands to be heard by the Conservation
Board.[84] The last phase in the work that Karl Clark had begun 40
years before and the first round in J. Howard's last great project began before the Conservation Board in June 1960.

GCOS sought approval of a scheme to produce 11.5 million barrels (31,500 per day) annually of synthetic crude oil and other by-
products from oil sands in bituminous lease No. 4 (see Map 7.1).
The sands were to be mined by a bucket-wheel excavator and transported to a separation plant by a shiftable conveyer system. At
the plant, GCOS planned to use an improved variation of Karl
Clark's hot water separation process to extract the oil from the
sand. Following primary refining, the synthetic oil would be sent to
Edmonton by pipeline. In support of its application to undertake
the $110,000,000 project, GCOS also submitted documents to show
that it had a market. It showed that Sun Oil Company was committed to purchase three quarters, or 23,500 barrels, of the plant's projected daily production for its Sarnia refinery and that Canadian
Oil Companies had pledged to take the remaining 7,800 barrels per
day, also at its Sarnia refinery. At the hearing, a Sun Oil representative later elaborated on the important question of the displacement
of conventional by synthetic crude oil. He explained that the displacement would involve 72% Canadian and 28% foreign crude.[85]

Interveners at the hearing, including Cities Service, Richfield and
Imperial, were critical of almost every facet of the GCOS proposal.
Cities Service, for example, insisted that the oil-sands terrain could
not safely support the heavy mining equipment that GCOS expected
to use. It was argued that the proposal was based upon inadequate
research and grave doubts were expressed about the capabilities of
the proposed oil-sands processing systems. Most of the objections
were ably countered by oil-sands authority Karl Clark, who appeared on behalf of GCOS.[86]

The Board's report was submitted to the Alberta cabinet in November 1960; it offered a mixed assessment. On the conservation
and recovery aspects of the proposal, the Board expressed its satisfaction that the recovery of the crude hydrocarbon present in the oil

sands was at an appropriate level. It was unhappy, however, with the proposed disposition of gaseous waste, which would see some 300 tons of sulphur per day dissipated into the atmosphere from the top of a 300-foot-high incinerator stack. The Board recommended that, in the interest both of reducing air pollution and of conserving sulphur, any approval should be conditional upon the recovery of a "substantial portion" of the sulphur.[87] The Board decided that, although the overall process proposal was technically feasible, some of the individual steps were not yet fully proved and there should be greater recognition of the possibility of costly delays during the transition period leading up to continuous integrated operation. This observation supported the Board's overall doubt about the project's economic feasibility.

Although it was not differentiated as such, the most important aspect of the Board's assessment had to do with the projected impact of the proposal upon the market for conventionally produced crude oil. It was noted that the Sarnia market proposed by GCOS was one that otherwise would be supplied largely by conventionally produced crude oil from Alberta fields. The resultant decrease in demand for conventionally produced Alberta crude oil, according to the Board's calculations, would represent about 5% of the total market demand. In itself, this was not a large percentage, but the Board hastened to point out that this would translate into a 20 to 30% reduction in the proratable demand (total demand less economic allowance).[88] This meant that, since Sun Oil and Canadian Oil Companies, Ltd. had relatively small conventional production within Alberta, the burden of the market loss would be carried mainly by other producers, especially those with interests in the more prolific pools.

All things considered, the Board decided that it was not "in the best interest of Alberta" that the application be granted. But it did not recommend outright rejection; rather the Board advised that the application not be granted "at this time."[89] GCOS was told that, provided it was able to offer substantial new evidence of technical and economic feasibility along with a marketing plan that would have less impact on the conventional sector, the Board would be pleased to hear a new request anytime after 1 January 1962.

Deferral of the final disposition of the application was the only possible compromise between the province's long-standing desire to see commercial development of the oil sands and the still grim uncertain state of the conventional crude oil market. The cautious Manning government did not want an oil-sands decision until it was clear exactly how the federal government was going to respond to the recent recommendation of the Royal Commission on Energy for the formulation of a national oil policy.

More importantly, the Board's November report signalled a larger truth regarding the oil sands. The sands existed on the high-cost outer margin of world oil. Hence, even more than in the case of conventional Alberta crude oil, the ebb and flow of development interest would continue to exhibit the erratic pattern that marked the paramount influence of unpredictable outside forces.

THE BORDEN COMMISSION

By the time that the Conservation Board delivered its GCOS decision in the fall of 1960, the petroleum industry and the Alberta government had been waiting nearly three years for Ottawa's response to Alberta's crude oil crisis and the recommendations of the Borden Commission regarding Canadian oil policy. The delay had driven provincial frustration to new heights, and it seemed, as yet another reminder to many Albertans, not so much that the issues in question were complex but more so that, regardless of which party was in power in Ottawa, western interests were never a priority concern.

The Borden Commission was instructed to make recommendations on the policies that would best serve the national interest in relation to the export of energy from Canada; the policy that ought to be applied to the regulation of the transmission of oil and natural gas between the provinces or from Canada to another country, "in order to ensure the efficient and economical operation of pipelines in the national interest;" the extent of authority that might best be conferred on a national energy board, together with the character of administration and procedures that might be established for such a board; whether, in view of the special relationship that existed between the Northern Ontario Pipeline Crown Corporation and Trans-Canada Pipe Lines (TCPL), any special measures needed to be taken in relation to TCPL to safeguard the interest of Canadian producers and consumers; and on any other matters the commissioners considered appropriate.[90] Hearings began in Calgary on 3 February 1958.

The environment in which the commissioners started their deliberations could hardly have been more highly charged or politically complex. First, there were the Alberta crude oil producers whose declining fortunes had just been dealt another severe jolt. On 24 December 1957, the U.S. government announced a further "voluntary" 15% reduction in permissible crude oil imports. Even worse, this reduction was also to apply to the previously exempt District No. 5, the Pacific Northwest. In an instant, Alberta's real hope of an expanding market in the Puget Sound area was dashed, and Premier Manning wrote once again to the prime minister to plead that opening the Montreal refining area to Alberta oil had now "become a national necessity."[91] Then there was the natural gas question. Alberta was locked once again in another vigorous round of its seemingly endless natural gas export debate. The well-rehearsed ritual on this occasion was initiated by two new natural gas export applications given initial hearing by the Conservation Board just weeks before the scheduled meeting of the Borden Commission in Calgary. The first application was that of Westcoast Transmission, which sought a permit for the purchase of up to 55 billion cubic feet of natural gas annually for 25 years. Westcoast proposed to collect the gas from the Savanna Creek Field, the east Calgary field and other southern Alberta fields for a pipeline running through the Crowsnest Pass to the Canadian-American border at Kingsgate, B.C., where the gas would be picked up by the Pacific Northwest Pipeline Corporation. Southern Alberta and southeastern British Columbia communities were to be served along the route. The proposal drew widespread and hostile response. TCPL opposed on the grounds that it would require some of the gas requested by Westcoast to satisfy its projected future market growth. AGTL opposed Westcoast's plan to build a pipeline from the Savanna Creek field to the Crowsnest Pass in the belief that the building and operation of such a pipeline fell under its common carrier mandate. The most strenuous opposition came from the city of Calgary, which argued, on the basis of an independent study, that the export of gas from these southern Alberta fields would jeopardize its long-term supply and likely cause an increase in the cost of natural gas to Alberta consumers.[92] The second controversial application was from the Alberta and Southern Gas Company Ltd., a subsidiary of the San Francisco-based Pacific Gas and Electric Company, which sought a permit for the annual yearly export of up to 167.9 billion cubic feet of natural gas, most of which would go to California, and a much smaller amount to the Montana Power Company. This larger and more complex proposal drew even stronger opposition from TCPL and the city of Calgary.[93] In late January 1958, the Conservation Board decided to suspend further consideration of the natural gas export question until after the conclusion of the Borden Commission's Calgary hearings.

It was not just the especially vigorous oil and gas debate during the winter of 1957–58 that helped to establish the anxious environment in which the Commission met. The Calgary session, the longest and most important of the Commission's hearings, happened to coincide with the spring session of the Alberta Legislature. This provided a forum for partisan debate on petroleum policy that ran parallel to the Commission hearings.

As the commissioners met for their first session, the tension was palpable. Assembled before the commissioners were Premier Manning, flanked by several of his cabinet colleagues, and all the relevant senior civil servants, including Deputy Minister of Mines and Minerals Hubert Somerville and McKinnon, Goodall and Govier from the Conservation Board. Also, there was J.J. Frawley, the premier's trusted adviser on all matters involving the federal government. Called from Ottawa to represent the province at this and all subsequent commission hearings, Frawley's presence marked the great importance that Alberta attached to the Commission's examination. Added to the provincial delegation were rows of lawyers, representing other provincial governments; the city of Calgary; assorted associations, ranging from the Canadian Petroleum Association to the National Coal Association in Washington, D.C.; utility companies; and a host of producing companies intending to submit briefs. To accommodate all the interested parties, journalists and the public, the Commission was set up in Calgary's new 2,500-seat Jubilee Auditorium.[94]

In his welcoming address to the Commission, Premier Manning took the opportunity to outline the gravity of the situation faced by Alberta and stressed the importance that the province attached to the Commission's work. Hoping to set the Commission on the desired course, he reminded it that the substantial development that had taken place in the province over the previous 10 years was but a fraction of the potential. Established deliverable reserves of crude oil and reserves of natural gas already were far in excess of what could be absorbed by all present and future local markets. The lack of markets, Manning explained, had immobilized "millions of dollars of investment capital already spent in drilling wells which are now capped." He warned that, without improved market prospects, there was little incentive for continued development, which would have serious repercussions[95] upon both the provincial and national economies. Manning understood the importance of linking Alberta's plight to the national interest; the problems facing the province could not be solved without key policy decisions at the federal level. To this end, he asked for "a clearly defined national policy with respect to both domestic and export markets."[96] For the Commission's guidance, he promised a formal submission from his government and the full co-operation of the Department of Mines and Minerals and the Oil and Gas Conservation Board in supplying whatever information might be desired.

The Commission's first round of hearings in Calgary were devoted almost entirely to the gas question. Leading off with the first major submission was the Conservation Board.[97] Although the Board did not venture an opinion on the export question, it did put forward the statistical data that implicitly supported a policy of increased natural gas exports. According to the Board, Alberta's current established natural gas reserves stood at 21 trillion cubic feet and ultimate gas reserves of between 60 and 80 trillion cubic feet could be presumed. From this point, presentations and discussion at the inquiry assumed a predictable pattern and continued the debate that had been under way before the Conservation Board a few weeks before. Basing its brief on the work of the long-established petroleum engineer S.J. Davies, the city of Calgary took issue with reserve estimates of the Board, the utility companies, the Canadian Petroleum Association and Westcoast Transmission.[98] The strong argument, put forward by Canadian Western Natural Gas and Northwestern Utilities, Westcoast and others that export was required for the proper development of Alberta reserves and vital to the Alberta economy, was partly deflected by revelations in the cross-examination that followed the Westcoast submission. It was revealed that the contract between Westcoast and Pacific Northwest Pipelines Corporation in the United States called for a price of 22¢ per 1,000 cubic feet. This was not only less than the price currently being paid by Calgary consumers, and substantially less than the 32¢ per 1,000 cubic feet that Westcoast was charging for its gas in the Vancouver area, but was also a contract set for 20 years without an escalation clause.[99] After further probing, the Commission learned that promoters of Westcoast had received, for 5¢ each, 625,000 stock shares that were selling for $40 to $50 shortly thereafter.[100] The Westcoast revelations overshadowed the appeals of the more aggressive export enthusiasts, such as that of J.K. Horton, president of the Alberta and Southern Gas Company, who warned against delaying gas export on the mistaken assumption that U.S. markets would wait forever.[101] An ensuing side debate on whether Canadian customers were "subsidizing" the cost of providing gas to U.S. buyers served to reinforce the city of Calgary's worst fears.[102]

While the Opposition in the Alberta Legislature pressed with renewed enthusiasm for some plan of natural gas price protection for
Alberta consumers, the Commission moved on towards the conclusion of the first Calgary session with a marathon four-and-one-half-
day examination of TCPL. Although the company did not express direct opposition to the export plans of Westcoast or Alberta and
Southern, its brief to the Commission emphasized that TCPL had underestimated its potential market and would soon have to apply to
the Alberta Conservation Board for an increase in its natural gas
allotment—an increase nearly sufficient to absorb all Alberta's estimated surplus capacity. Close questioning followed on the company's supply-demand projections, on its financing arrangements,
and on what had turned into one of the hearing's central themes—
the question of "subsidized" gas export.[103] Judging that he had assembled most of the required information, Henry Borden called the
Calgary natural gas session to a close on February 29. As the Commission prepared to move on to Regina, he announced that an interim report on natural gas was being contemplated.

While the Commission continued its work in Regina and Victoria, the press, oilmen and politicians took stock—the considerations
of the latter being framed by the federal election currently under
way. The Calgary Herald advised that the hearings clearly demonstrated that Albertans should be wary of those who argued that
there was an abundant gas supply and that the industry should not
be burdened with "unnecessary delay." The paper noted that "unnecessary haste" already seemed to have proved costly. It condemned the sale of natural gas to Washington state at "bargain
basement" prices and declared that "the comparatively piddling
sums that are accruing to Alberta from the sale of gas from a southern field to an industry in Montana are certainly cause for thought
by Albertans." The paper concluded that the Borden Commission
had shown the need

> to establish an orderly plan under which Canada's gas will be de-
> veloped for Canada's interest, and under which terms favouring
> the United States without real advantage to Canada will not be
> allowed again.[104]

In Edmonton to support his party's campaign in the West, Liberal
leader Lester Pearson announced that he would not feel obliged to
wait for the Commission's report before establishing a national oil
and gas policy. He reasoned that it was clear that gas surplus to fu-

ture Canadian needs should be exported. But alert to the most controversial issue to emerge at the Calgary hearing, he asserted that his party was "opposed to price discrimination against Canadian consumers" and, if in power, would not hesitate to take legislative action should it be necessary.[105] The CCF party was also contemptuous of the Commission's efforts. At the beginning of the campaign, CCF leader M.J. Coldwell had called for Borden's removal as chairman of the Commission on the grounds that his position as head of a giant private utility in Brazil made him incapable of an "unprejudiced" decision.[106]

The election was over by the time the Commission returned to Calgary, but political tensions were only marginally less acute. Diefenbaker's Conservatives had been returned to power with a massive majority, part of which had been gained at the expense of the 13 Alberta Social Credit MPs who, along with all the Social Credit candidates in other provinces, were eliminated in the Tory onslaught. In addition to a renewed sense of isolation in the federal political arena, the Manning government had been hard pressed on the home front. It continued to face the Opposition parties' allegation in the Legislature that it had failed "to evolve a proper policy with regard to energy in order to protect the citizens of the province."[107] The Liberals noted that "the government had not yet bothered to present a formal submission at the Borden Commission's Calgary hearings and speculated that this was a major factor contributing to the delay of the Commission's urgently awaited interim report.[108]

When the seven-member Commission returned to Calgary, this time to gather information on the oil industry, Premier Manning was the first to take the stand. It was, however, the natural gas area that he wanted to address. His brief restated the detailed policy paper that he had read in the Alberta Legislature on April 10. In a long preamble, directed as much to Albertans as to the commissioners, Manning sought again to reassure that, even with substantially expanded export sales, the province's proven reserves were ample to meet Albertan and Canadian long-term needs. This was in keeping with what the Manning government had been saying for many years. What Manning really wanted to put before the public on this occasion was notice that Alberta's expanding common carrier natural gas pipeline grid made it possible for Alberta to integrate domestic and export market demand. He explained that the AGTL grid system of main gathering and transmission lines linking together all the major gas-producing areas of the province meant not only that wet and dry gas fields could be produced in a manner consistent

with preferred conservation practice but also that domestic users
would not have to compete with export interests on an individual
field basis and would not have to fear that their cost of natural gas
would be set by the export market (although many British Columbians might have thought this a preferable option). Manning insisted, moreover, that Alberta consumers, along with assured
supply, would gain a price advantage through market integration.
Alberta communities would simply tie into the grid network and,
according to Manning, it would only be

> reasonable for gas drawn from the grid system for export
> markets to bear the greater portion of total transmission costs
> within the grid system, thereby reducing costs to Alberta com-
> munities and thus providing Alberta consumers with an advan-
> tage equal to or greater than they could obtain by a specific local
> source of supply being dedicated to their present and future
> requirements.[109]

Conscious that price had become as important as the supply issue,
Manning gave notice that his government's objective of market integration would be pursued aggressively. He expressed his confidence that the government would have the full co-operation of
producers, the utility companies and the exporting interests in im-
plementing this policy. To make sure that the message was understood in industry circles, he presented the thinly veiled warning of a
worrisome alternative. Manning gave notice that his government
was also studying the possibility of establishing a provincial gas
marketing board. If such a scheme went ahead, all producers would
be required to sell their gas to this central agency, which in turn
would negotiate contracts for resale.[110]

Having completed his detailed review of the policies designed by
his government to protect the interests of Albertans, Manning went
on to address the larger question of Canadian interest. Given that
once natural gas has left the province in which it was produced it
has entered the field of interprovincial or international commerce,
and thereby become a commodity subject to Ottawa's jurisdiction,
Manning recommended the creation of a national energy board. In
doing so, he gave Alberta's important endorsement to the recommendation made earlier in the Gordon Commission reports of December 1956 and November 1957. Manning had some specific
ideas about the powers that should be assigned to such an agency
and the manner in which it should function. Unlike the Gordon
Commission's advice that a national energy board be given respon-

sibilities relating to the export of oil, natural gas and electric power,[111] Alberta's premier proposed that the mandate be confined to the transmission and marketing of natural gas. He advised that the new federal agency be given jurisdiction over the granting of permits for the construction of interprovincial pipelines, the earnings of such pipelines, and the conditions under which an interprovincial pipeline company might be declared a common carrier. The other major function that Manning thought should be assigned to a national energy board had to do with natural gas volume and price. He proposed that the federal board should collaborate with provincial boards, such as Alberta's Conservation Board, to identify which Canadian markets it was economically feasible and in the public interest to supply with Canadian gas, to determine what gas was surplus to the requirements of such Canadian markets, and then to approve the export of such gas to foreign markets. In addition, Manning wanted the board to be responsible for reviewing "all export sales contracts which establish gas prices at the international boundary to ensure that no Canadian gas, entering the United States, is sold...at a price which unjustly discriminates against Canadian consumers...."[112] Always sensitive to the possibility of federal power encroaching on the province's hard-won control of natural resources, Manning concluded his remarks before the Commission with the caution that the proposed national energy board should in no case infringe on the jurisdiction of the province or upon the duties and powers assigned to the Alberta Oil and Gas Conservation Board. Specifically, he warned against interference with the regulation of production, the regulation of wellhead or field prices, the regulation of oil or gas transportation within the province, and the regulation of ultimate consumer prices.[113]

Manning's submission to the inquiry was acknowledged as being significant, but far greater anticipation was reserved for that of Home Oil Company. In part, this was because it marked the turning of the commissioners' interest to the crude oil situation, the matter that was uppermost in the minds of most producers. Just three weeks earlier, Imperial Oil had reduced the wellhead price of Redwater crude by 7¢ to $2.56 per barrel.[114] Market contraction and expanding surplus capacity, it seemed, were moving price in a predictable direction. Industry and press anticipation was fed by the broad knowledge of the solution that Home Oil was going to propose.

Inspired by Charles Lee, president of Western Decalta Petroleums Ltd., and largely financed by Robert A. Brown, president of

Home Oil Company Ltd.,[115] a coalition of hard-pressed independent oil companies had commissioned a series of studies by three highly experienced and respected consulting firms.[116] W.J. Levy Inc. of New York prepared a detailed analysis of market outlets for Canadian crude oil; Dutton-Williams Brothers Ltd. of Calgary did an engineering study for a proposed crude oil pipeline system from Alberta to Montreal; and Dallas consultants Purvin and Gertz Inc. examined the possible use of Canadian crude oils in the Montreal area refineries. It was the detailed Levy report that formed the essential core of the Brown group's submission and the centre of subsequent discussion.

Beginning their report with an examination of the industry's current situation in Alberta, Levy and his associates reminded that the future of the Canadian oil-producing industry over the long run was closely tied to levels of production. They warned that "operations could not continue at around 50% of capacity without imposing seriously on the flow of investment funds necessary to sustain the vitality of exploration and development efforts."[117] Comparable production rates in the United States, the report noted, averaged 70%, and in Texas 60%. Levy's analysis showed that, with the exception of southwestern Ontario, existing Canadian markets for western crude oil offered little prospect of significant market growth over the coming decade. Moreover, the report pointed out that even the projected growth of the Ontario market had to be interpreted cautiously, since in this region Alberta crude oil ran into direct competition with both offshore oils refined in Montreal and mainstream domestic U.S. crude oils moving into the Detroit-Toledo region southwest of Sarnia, Ontario, the refinery destination of western Canadian crude oil.[118] The health of the oil-producing industry in western Canada was thus shown to depend upon the search for new or expanded markets on the periphery of the existing market orbit. Levy's report offered a close examination of the three such markets that seemed to offer some potential.

As for the U.S. West Coast market, Levy concluded that the California portion, which Alberta had tapped during the Suez Crisis, could not be counted on, for even though Canadian crude could be laid down in the San Francisco refining area at prices comparable with Venezuelan and Middle East crudes, it remained nonetheless at a competitive disadvantage. The principal difficulty was that the San Francisco refineries were all operated by integrated oil companies, all of whom had direct access to vast reserves in the Middle East and Venezuela. Given the choice, these companies preferred to purchase their own oil. Owned production earned greater overall profits than what could be earned on Canadian oil, which was pro-rated and involved the purchase of a substantial volume of independent oil along with every barrel of owned oil.[119] Farther north, in the Puget Sound area, Canadian oil had established a preferred position by virtue of a price advantage gained by a relatively short pipeline distance from the production source. But even here, Levy judged the Canadian hold on the market to be uncertain for a similar reason. Curbed by quotas on the U.S. East Coast, Venezuelan and Middle East producers were seeking replacement markets. Moreover Indonesian production was increasing. Competition was bound to intensify, and Levy reminded that "the opportunity that an integrated company has for the use of its own crude may be a more important commercial consideration than *pro forma* cost comparisons."[120]

In the Minnesota-Wisconsin market, Canadian and especially Saskatchewan crude also enjoyed a geographical advantage and had established a market presence. It was an increasingly competitive market, however, marked by intense price competition among domestic producers. In this difficult market, not only was the growth prospect limited by the assurance of vigorous competition and the handicap imposed by the 10.5¢ per barrel duty levied against foreign oil but also the development of new fields in the U.S. portion of the Williston Basin in eastern Montana meant the likelihood of a pipeline being constructed to the Minneapolis-St. Paul region. If this were to happen, the one advantage enjoyed by Canadian oil—its shorter transportation route—would be eliminated.[121]

In its examination of these markets, the Levy report added firm documentation to what was already well understood in industry and government circles in Alberta—that, at least before 1965, not only was there little prospect of market growth but also Alberta producers would be fortunate simply to retain their existing market. For this reason, the report offered an even more detailed examination of the Montreal market, the one market into which expanded Alberta production could be directed as a matter of public policy.

According to Levy's comparative analysis, the cost of crude oil delivered to Montreal from Venezuela was $3.32 per barrel and from the Persian Gulf it was $3.28 per barrel. On the strength of a detailed evaluation of wellhead prices and pipeline charges, he estimated that Redwater crude could be brought to Montreal for $3.52 a barrel.[122] To be competitive in the Montreal market, Canadian crude oil had to overcome a cost disadvantage of about 25¢ per barrel. Normally, this would have to be accomplished by lowering wellhead prices or reducing pipeline tariffs, or some combination of the two.[123] Assistance in closing the price gap, the report pointed out, also could be provided as a matter of government policy. The 10.5¢ per barrel duty levied by the U.S. government to reduce the price advantage of foreign crude oil was presented as an example.[124]

If Canada decided that the uncertainties of the U.S. market were likely to inhibit the balanced development of the Canadian petroleum industry and moved to support the entry of western Canadian crude oil into the Montreal market, Levy cautioned that this would require more than simply the right formula of wellhead pricing, pipeline charges and export duties to bring western crude oil prices into line with those offshore. There remained a more formidable hurdle, the same obstacle that faced Alberta crude oil in other markets where it was price competitive. He reminded that "access to the foreign production of international companies with which Montreal refineries are affiliated offers opportunities of profit that Canadian crude cannot match."[125] Levy offered no direct recommendation, but he did point out that

> the United States—whose practice with respect to a similar problem may or may not recommend itself to Canadians—has acted to protect eastern markets for its domestic crude against excessive foreign crude imports by setting individual "quotas" for importing companies and calling upon them voluntarily to comply. Thereby, the domestic producing industry is supported by voluntary action of the refining industry, even though compliance by individual refiners involves the foregoing of potential advantages in each instance.[126]

The U.S. "quota" measure was in addition to its import duty on foreign crude oil entering East Coast ports.

With reference to the formulation of a national oil policy, Levy advanced one other consideration that stood outside the specific problem faced by the petroleum industry in the West. He noted that, in 1956, Canada's merchandise trade balance showed a deficit of $734 million. Over the same year, Canada had imported 390,000 barrels of crude oil daily, at a total cost of some $271 million. His report concluded that "the replacement of foreign crude by Canadian crude at Montreal would equally tend to alleviate substantially the unfavourable trade balance."[127] He noted in passing that the Canadian Board of Transport Commissioners had cited the positive impact upon the balance of trade of moving crude oil to refineries in British Columbia as a factor for speeding the construction of the Trans Mountain pipeline in 1951.[128]

Robert A. Brown took the stand confident that he had a persuasive document backing the appeal he was about to make on behalf of the independent Canadian producers. Setting the stage with a statistical overview of the problem faced by the industry, he informed the Commission that in 1956 western Canada's crude oil production potential was estimated to be about 900,000 barrels per day, whereas production had actually averaged approximately 470,000 barrels daily, of which about 120,000 barrels were exported to the United States. During the same period, Canada imported some 390,000 barrels of petroleum and refined products per day to meet a total Canadian demand of a little over 700,000 barrels per day.[129] Arguing that the prospect of increased exports to the United States could not be realistically expected for years, Brown asked that in the national interest western Canadian producers be given a larger share of the Canadian demand and, in particular, access to the Montreal market. To this end, he presented on behalf of Home Oil, and its 12 associated independent companies, the studies that they had commissioned to demonstrate that this was both a desirable and economically rational request.[130]

The reaction of the major integrated companies to the Levy-Brown plan was harsh but predictable. W.M.V. Ash, president of the Shell Oil Co. of Canada, led off the attack. He told the commissioners that the construction of a pipeline from Alberta to Montreal would be a "violation of natural economic laws."[131] He argued that the proper market for Canadian oil was the Toronto-Hamilton area, a market that would grow naturally as the region's refining capacity increased. In his view, the "despondency and alarm" of the independents was unjustified since the U.S. West Coast and mid-continent markets were only "temporarily depressed." Under cross-examination, Ash put his company's position even more bluntly. Shell, he said, would not agree to accept Alberta crude at its Montreal refineries unless it had no alternative, and he warned that if Canada forced its oil into Montreal the foreign crude oil thus displaced would have to find a market elsewhere, "maybe on Canada's own west coast." The *Calgary Herald* reported Ash's thinly veiled threat that "even today Shell could put foreign crude into its Vancouver refinery cheaper than Trans-Mountain [pipeline] can get it there—but isn't doing it because it feels the Alberta oil is an economic proposition in the long run."[132]

Shell's position on any matter relating to the petroleum industry in Canada was important, but the critical response for which the industry, the Commission and Albertans waited for was that of Imperial Oil—the clearly dominant producer, refiner and marketer of petroleum products in Canada. Addressing the Commission, Imperial's President J.R. White took an equally strong though much less confrontational approach in his opposition to the Home Oil proposal. In recognition of the inclination by some to present the Canadian market question as a contest between independent Canadian companies and foreign-controlled multinational corporations, White chose to begin by establishing Imperial's credentials as a "Canadian" company.[133] He pointed out that, even though Imperial was 69.8% owned by Standard Oil of New Jersey, the ultimate responsibility for the company's business decisions resided with its own board of 10 directors, eight of whom were native-born Canadians. This established, White had two primary messages that he wanted to deliver: that the proposed national energy authority and the Montreal pipeline were both unnecessary and potentially harmful.

The brief submitted by White stated that the suggested creation of a national energy board made little sense, since Canada's energy resources were abundant and the existing authority of the federal Department of Trade and Commerce was already sufficient to look after Canadian interests. It also pointed out that the energy industries—coal, petroleum, natural gas, hydroelectric power, solar and atomic energy—had so little in common that regulation by a single authority would be virtually impossible. It would be unworkable in a practical sense, and Imperial speculated that it was also constitutionally unfeasible. Given the distribution of federal and provincial powers, the creation of a national energy authority with real authority could be predicted to arouse intergovernmental conflict, which if anything would only impede the industry's progress.[134]

The principal part of Imperial's submission focused upon the matter of markets for western Canadian crude oil. Identifying itself as the largest, single exporter of western crude with a high percentage of shut-in production, Imperial declared that it had "reluctantly come to the conclusion that the present marketing limits of Canadian crude could not be economically extended eastward by direct delivery." Such an artificial extension, the company warned, "would carry grave risk to the long-term well-being of the industry."[135] According to Imperial, a pipeline to Montreal could not be financed as a normal commercial venture without a combination of government guarantees and protective measures, which would mean higher costs for Canadian consumers and taxpayers. Further, Imperial predicted that the requirement of a complex permanent system of protection would demand an increasingly high degree of government control at the expense of the industry's future freedom of action, which in turn would diminish the flow of investment capital into the industry. Like Shell, Imperial argued that the long-term prospects were bright. Therefore, rather than consider a Montreal pipeline, which could only be justified as a long-term measure, it was much more appropriate to examine what measures might be taken to expand the use of Canadian crude in existing markets in the short term. Imperial expressed its optimism about the growth potential of both the "intermediate" Ontario and Puget Sound markets. Imperial advised that the most effective immediate steps that the government might appropriately take first would be to bring the Canadian industry's depletion allowances in line with those permitted in the United States. More generous tax allowances would make western oil more competitive. Second, the Canadian government was advised to continue its efforts to have the import quota lifted on Canadian oil by pressing on with the "highest level" representations to the U.S. government.[136]

Although it was not featured in White's introduction and does not seem to have been part of the cross-examination or press evaluation that followed, there existed within the submission an important section that reviewed the regulatory framework governing the exploration and development of petroleum resources in western Canada. It pointed out that the land tenure system, which by virtue of the 50% crown revision accentuated multiple-lease ownership; conservation practice, which was built upon the principles of equitable share allowables and market demand proration; and government land policy, which included such features as the drill-pay-or-quit mineral lease, all encouraged rapid exploration and development well beyond that justified by existing market conditions.[137] The implicit conclusion that the problem of the region's vast surplus production capacity lay in good part at Alberta's door was put on the record but left to lie fallow.

Imperial's much awaited presentation to the Commission received swift judgement. The conservative and staunchly free enterprise Calgary Herald remained unswayed by the company's arguments. Although sympathizing with Imperial's objection to the idea of a national energy board, the editor concluded, "we still contend that the only protection Canadians have, at this point, over their foreign-owned oil resources comes from Canadian governments."[138]

On the question of marketing western Canadian crude oil, the paper came down firmly on the side of Robert A. Brown and his associates. The Alberta government was similarly unconvinced. Alberta's counsel at the hearings, J.J. Frawley, went after Imperial President J.R. White regarding his company's preference for Venezuelan over Alberta crude before a standing-room-only crowd, while Premier Manning, with the assistance of the technical staff at the Oil and Gas Conservation Board, put the final touches on his second submission.[139]

Premier Manning appeared before the second Calgary session of the Royal Commission on Energy on 16 August 1958. This was by prior arrangement with Borden; it having been agreed that Manning would not appear until he had heard what everyone else had to say.[140] Appearing with the premier, and reflecting the important roles that they had played in establishing the provincial position, were Board members Govier and Goodall; Board Secretary Vernon Millard; the upper echelon of the Board's engineering staff, D.R. Craig and J.G. Stabback; and statistician R.J. Cooper. Manning came to speak about crude oil marketing and his concern regarding the stability of the industry in western Canada, especially the independent Canadian sector. He noted that all those who had appeared before the Commission agreed on the seriousness of the problem. The real difference of opinion seemed to be whether or not these difficulties were of a transitory or permanent nature. Although conceding that this point was open to debate, Manning drew attention to the different situations facing the Canadian independent and integrated companies. The integrated companies, operating in all phases of the industry and through affiliated companies that had reserves in various parts of the world, were not as seriously affected by regional marketing problems and could afford to take a more detached and longer term view. In contrast, for the independent company that depended primarily on production revenue, low production rates represented a serious loss of revenue that could be fatal, even if the present marketing difficulty continued for a relatively short time.[141] Therefore, market expansion had to come sooner rather than later. Manning noted that there was broad consensus that some expansion was possible in the Ontario market and in the two traditional U.S. market areas, but there was a wide divergence of opinion about how much growth could be expected in these markets. He asked the Commission to consider the independent analyses of these markets prepared by Alberta's Conservation Board. The Board's conclusion, Manning instructed the commissioners, was that expansion in the Ontario market and the re-establishment of the U.S. markets, even anticipating reasonable growth beyond their former levels, would fall far short of what was necessary for an effective solution to the problem.[142] Access to the Quebec market or a much more significant expansion of the U.S. market, or a combination of both, were thus presented as an urgent necessity.

With these factors in mind, Premier Manning requested that the Commission advise the prime minister without delay to convene a meeting with senior officers of the importing, refining, transporting and marketing companies to advise them of the urgent need to increase the use of Canadian crude oil. At the meeting, Manning advised that the companies be provided with specific volume objectives to be fulfilled in accordance with an established time schedule or be faced with imposed import quotas. The Alberta premier explained that he hoped industry would seize upon the opportunity to solve the problem voluntarily, but if industry proved unable to act speedily or if its solution fell short, he advised that "the government of Canada must be prepared to take a firm stand."[143] Such measures at home, Manning believed, should be supplemented by cabinet-level discussions with the U.S. government regarding the entry of Canadian oil into the U.S. market.

On the grounds that the Commission had heard much testimony concerning the costs and estimated tariffs for a new pipeline to Montreal, Manning also devoted a special section of his submission to the pipeline question. First, he informed the Commission that the Alberta government favoured the construction of such a pipeline. He then asked that particular attention be given to the preliminary studies prepared by Conservation Board technical staff on the pipeline proposal submitted by the Home Oil group. Manning pointed out that these studies indicated

> that it is economically feasible to deliver Western Canadian crude to the Montreal refining area at prices roughly the same as those now being paid in that area for imported crude oil provided that the volumes permitted under voluntary contracts or an enforced import quota system are sufficient.[144]

Conscious that all would rest or fall on opinion in central Canada, Manning was quick to emphasize that this meant that the use of Canadian crude oil in Montreal "should not necessitate any increase in price to the consumers in Eastern Canada."[145]

In his appearance before the Commission, Manning made public the policy he had been urging upon the Diefenbaker government for months. Now he moved to mobilize public support as the commissioners moved on to complete their schedule of hearings in Winnipeg, Toronto and Montreal. Backing was first sought within the Alberta business community. Led by the arguments of L.H. Fenerty, a prominent Calgary lawyer and past president of the Calgary Chamber of Commerce and Agriculture, delegates gathered for the annual convention of the 130 associations that comprised the Alberta Associated Chambers of Commerce and Agriculture and unanimously passed a resolution calling for Ottawa to take such steps as necessary to "provide for the entry of Alberta oil into the Montreal market."[146] The provincial business community solidly on side, Manning called a press conference to inform Albertans of the continuing crude oil market deterioration and to affirm his government's position. He informed the assembled reporters that oil production in the province had plunged to just 37% of that permitted under existing conservation requirements, and that incoming refinery nominations for Alberta crude oil suggested a continuing downward drift. Noting that the United States had taken action to restrict oil imports when domestic production in that country had fallen to 65% of the permissible amount, Manning challenged that surely protective measures in Canada were "long overdue."[147]

As if to confirm the Alberta government's bleak prediction and to provide authentic evidence in support of the solution proposed by the independent companies, Shell Oil, with remarkable insensitivity, announced just two days after the Manning press conference that it was stopping the importation of Alberta crude to its refinery in Anacortes, Washington. Asked to explain, Shell Canada's President W.M.V. Ash said that the decision was made by the parent company in the United States "as a matter of simple economics."[148] The company could purchase oil from producers in British Borneo at a better price than from Canadian producers. Shell's announcement, plus information that the Texas company planned to supply its new Anacortes refinery with Arabian crude oil, effectively undermined the notion put forward at the Borden Commission hearings by those who spoke positively about the prospect of market growth in the U.S. Pacific Northwest.

McKinnon immediately sent a revised Conservation Board market forecast to Premier Manning, and conveyed his colleagues' opinion that the province's production decline had now reached the point where it was "clearly dangerous to the stability of the oil industry."[149] (See Table 7.1.) Three days later, Manning was on his way to Ottawa for the third time to discuss the province's crude oil crisis with Prime Minister Diefenbaker. Manning hoped to persuade Diefenbaker that the time had come for an immediate federal response.[150]

The renewed sense of urgency regarding Alberta's crude oil markets added to the sense of anticipation surrounding the Borden Commission's Toronto hearings. Of the many submissions presented, a second brief from Imperial Oil was one of the more important. In its introduction, Imperial observed that "the Calgary hearings reflected the interest of a producing area where more than normal surplus crude oil producibility has developed after 10 years of extremely rapid growth in crude oil production." It went on, "the Commission is now sitting in Canada's largest consuming area."[151] Albertans could readily anticipate what was coming next. Imperial stated that it was appearing to provide additional information, since it was apparent from some of the commissioners questions and from the transcripts "that some of the basic mechanics of the industry [had] received insufficient attention or [had] not been placed in proper perspective to the marketing problem."[152] The point that the company wanted to direct the commissioners towards was the inability of the producing industry in western Canada to control the development of producing capacity in response to demand variations. This had been touched upon in the first submission, but in passing and without emphasis. Now, Imperial advised, "this is a matter that should be given close consideration by the Commission." Imperial's position was that much of the surplus production capacity was a function of particular aspects of Alberta's regulatory approach.

Imperial's evaluation of the Alberta variant of the North American production system acknowledged that this system allowed the adoption of sound conservation measures under a complex land tenure system. It was anxious, however, that this significant achievement not be allowed to obscure what it saw as important negative side effects. To set the latter in high relief, Imperial drew comparison with production practice under the concession system that operated in most areas outside North America. The standard oil and gas lease was presented as one of the main culprits. Typically, such leases contained a "drill, pay or quit" provision that required the leaseholder to drill or develop his leasehold according to a specified time schedule that did not take market factors into account and therefore could force drilling even in times of oversupply. Imperial pointed out that, under the concession system where one lease might cover an entire oilfield, drilling in times of oversupply was unusual.[153] According to Imperial, this bias towards overdrilling was accentuated by two other features of Alberta's land tenure and production system.

The discoverer of an oil pool on Crown land can lease only up to 50%, the remainder being disposed of to purchasers by sealed bid. This policy, coupled with prorationing, means that *any* purchaser, even one previously unconnected with the industry, can buy oil production and automatically participate in [a] market outlet provided by the investments already made by others.[154]

Imperial observed that under prorationing there was not equal pressure on all producers to develop new markets. Many simply rode on the backs of the few, whereas under the concession system producers had to find their own markets. The North American system was presented as one that tended to protect marginal producing properties at the expense of more prolific wells. Imperial alleged that in some instances the principle of an economic allowable, which guaranteed a floor production rate or minimum volume to any producer, encouraged the development of marginal properties. "The end result," Imperial asserted, was "a substantial volume of 'protected' high-cost production being produced and prolific lower cost properties are produced at severely restricted rates."[155]

Imperial argued that, given the regulatory framework under which oil was produced in Alberta, it made little sense to build a crude oil pipeline to Montreal, since the guarantors of a new pipeline would simply be providing "a new outlet for new producers and [thus stimulate] a further build up of capacity, which could well result in a repetition of this situation within a few years." "Consumer interests in eastern Canada," Imperial concluded, "can best be protected by freedom of competition."[156]

Imperial's second submission struck at the very heart of Alberta's regulatory policy. Provincial reaction was quick, and in some quarters angry. Drawing upon the expertise of Dr. Govier and his Conservation Board technical staff, Premier Manning prepared a detailed rebuttal and sent it directly to Commission Chairman Henry Borden. "I believe," Manning wrote, "your Commission would be interested in a comparison of operations in Alberta and in a Concession area in order to place the points brought out by Imperial in their proper perspective."[157] Venezuela, the preferred source of supply for Imperial's Montreal area refineries, was the chosen concession area. The main issue that Manning wanted to address was Imperial's assertion that the concession operator was not pressed to drill wells beyond what was required by market demand, as was the case in Alberta. Manning admitted that the offset provisions in the normal Alberta oil and gas lease did impose drilling commitments, but he pointed out that Alberta's Conservation Board had made a number of changes in the last two years that would reduce the impact of some of the forces that tended to push the pace of development and about which Imperial had complained. He noted that the economic allowance schedule had been altered, as had the proration formula, so that greater weight would be given to a well's ability to produce. As well, sympathetic consideration was being given to applications for wider than normal 40-acre well spacing. All these changes aside, Manning countered that it was in any case unrealistic to believe that foreign governments were any less inclined to exert pressure to assure production of substantial rates. Further, he noted that "in areas of uncertain political stability the exporting companies themselves have considerable reason for producing at the maximum rates."[158] In support of his rejection of the argument that the North American system tended to result in overdevelopment when compared to the concession system, Manning provided a table showing that, under the concession system in Venezuela between 1956 and 1957, the number of producing oil wells had increased by 1,855 or 13.3%. In Canada, the number of additional producing wells totalled 1,501 or 12.4%. In the United States, over the same period, the number of producing wells increased by less than 2%, whereas in the Middle East under the concession system the increase was 20%.[159] Finally, Manning alerted Borden to Conservation Board calculations that indicated that Venezuelan concession oil was being produced at a more rapid rate than that in western Canada and Alberta.

Agitation felt by the Alberta cabinet and the senior civil servants involved with the presentation of Alberta's case was stimulated further by what they perceived as the "Imperial inspired" analysis of central Canadian newspapers regarding the proposed Montreal pipeline.[160] Alberta's irritation is well illustrated by the government's reaction to a speech delivered in early August by Vernon Taylor, Imperial Oil's Western Production Manager. Taylor used the opportunity of an address before the Edmonton Kiwanis Club to advance the argument that the oil surplus problem was largely a problem of inappropriate regulation.[161] As the speech was prominently reported in the *Montreal Gazette,* Manning was furious. A press release was quickly drafted to denounce Taylor's criticism of Alberta's land and conservation policies. The statement expressed distress at being told by one of the major land-holding oil companies that further exploration should be discouraged. Particular offence was taken to the suggestion that access to the Montreal market would only compound the problem of "premature" development, and it drew forth a scarcely contained anger. In part, the proposed statement read:

It would be a disservice to the national interest for me to ignore such a specious and short-visioned argument. It is like telling a starving man that you cannot give him a loaf of bread today because that will only whet his appetite and he will be wanting three square meals tomorrow. If Western Canada's reserves of oil are doubled in the next five years (as every recognized geologist expects them to do under normal exploration policy) what harm has been done to Canada?[162]

After consultation, Manning decided in the end that it would be imprudent to be dragged into a press war and dropped the press release idea. Instead, Frawley informed McKinnon that, since he had a meeting already scheduled with Borden, "Mr. Manning felt that it might be more effective if you told Mr. Borden at that time what we thought of this continued Imperial propaganda...."[163]

As the weeks passed, frustration in Alberta continued to mount. Ottawa seemed indifferent, and the major oil companies remained unrelenting in their desire to push Alberta to modify its proration system substantially, or to step around it completely. Imperial Oil offered to open up a new market area by shipping crude oil from its prolific Golden Spike field to Vancouver, and then on to Halifax if the field were removed from Alberta's proration schedule.[164] Even when the Texas Company informed the Conservation Board in August that it did not intend to use Canadian crude oil in its Puget Sound refinery as long as it could purchase cheaper oil elsewhere, the company still suggested that it would be prepared to consider using Alberta oil, "if a special allowable, not subject to proration, was granted to it for two high productivity fields in which it [had] virtually the controlling interest."[165] Manning and the Conservation Board remained unmoved. They understood full well that one such concession would bring down the system they had built with such care and would place the independent companies in an even more precarious situation.

In October 1958, attention shifted from the depressed crude oil market to the just-released "First Report" of the Royal Commission on Energy. Its primary focus was on the question of natural gas export. Borden and his Commission colleagues recommended that the export of natural gas surplus to Canadian requirements be permitted, but under licence. The issuance of such a licence, the report advised, should take into account such factors as the quantity of natural gas available for export, the terms and conditions of the export contract, and the contemplated arrangements regarding the disposal of by-products and the avoidance of waste. It also recom-

mended that export contracts not be granted for periods longer
than 25 years. To manage this licensing system for natural gas and
all other energy exports, the commissioners recommended the creation of a permanent federal agency to be known as the National
Energy Board (NEB). In addition to enabling the Canadian government to exercise effective control over the export from and import
into Canada and the movement across provincial boundaries of all
forms or sources of energy, the "Report" advised that the proposed
Board should have a general mandate to advise the federal government on "policies designed to assure the people of Canada the best
use of the energy and sources of energy in Canada."[166] To discharge
this responsibility, it was suggested that the NEB be given authority
to compile and maintain an up-to-date inventory of Canada's energy resources. In the natural gas and oil sector, a more comprehensive mandate and authority was suggested. The Commission advised that the NEB should be engaged in the "continuing study and
appraisal of all matters relating to the exploration for, production,
processing, transportation and marketing of natural gas and oil and
by-products thereof in Canada and elsewhere."[167] It was perceived
that, upon this foundation of gathered information and analysis,
the NEB might play the same important advisory role at the federal
level as that played by the Conservation Board in Alberta. To this
end, the "Report" advised that the three-to-five-member Board be
empowered to employ the necessary staff of professional and
technical advisers, to compel attendance of witnesses and the
production of documents, and to enforce obedience to its orders,
regulations and licences. Finally, as a measure to support the
agency's independence, the Commission urged that the NEB be set
up as a separate agency not subject to the direction of any specific
ministry.[168]

The Commission's "First Report" also dealt with two additional
matters that had generated a good deal of public interest at the
hearings. To arrest the popular perception that the interests of Canadian consumers and producers had been discounted in favour of
private and corporate gain,[169] the Report commented specifically on
the activities of Westcoast Transmission and Trans-Canada Pipe
Lines (TCPL). With regard to Westcoast, it advised that should it or
any affiliated company apply for a licence to export additional
quantities of natural gas, before giving approval the Canadian government should

ensure that the aggregate of natural gas to be exported by
Westcoast Transmission Company Limited, under *all outstand-*

ing and proposed contracts for the sale of such gas, is being sold at prices which, when averaged, are fair and reasonable after taking into account the price at which natural gas is being sold to Pacific Northwest Pipeline Corporation under its contract with Westcoast Transmission Company, Limited, dated December 11, 1954.[170]

In the case of TCPL, the Report commented more extensively. Partly, this was because the Commission had been directed to determine if "any special measures" needed to be taken to safeguard the interests of Canadian producers or consumers.[171] Although the recommendation to establish a National Energy Board addressed the general question of safeguarding Canadian interests in the future, the Commission made one specific recommendation that applied to TCPL retroactively. Recalling TCPL's concern expressed at the hearings that the natural gas requirement for eastern Canadian consumers had proved to be much greater than anticipated, the commissioners advised that the Canadian market required all the gas reserves that the company possessed. They advised that the Canadian government cancel Trans-Canada's so-called "Emerson contract." This was not a formal contract, but rather a promise issued to TCPL in a letter from C.D. Howe in 1955 saying that once a U.S. permit had been obtained, the Canadian government would authorize the export at Emerson, Manitoba, of 200,000,000 cubic feet of gas per day for 25 years.[172] The "Report" also reviewed and commented critically upon the controversial stock options given to and the large capital profits subsequently made by certain TCPL officers, including N.E. Tanner.[173]

Reaction to the Commission's Report focused mainly upon the National Energy Board recommendation. It was predictable, and the *Calgary Herald* came out strongly in support of the Commission's findings. It acknowledged that "normally the less government interference in business the better it is for all concerned," but argued that the Commission hearings had demonstrated clearly enough that the interests of promoters were not always the interests of the Canadian public and that the situation had to be "considered in the light of the fact that Canada's natural gas industry is controlled by financial interests beyond our borders."[174] Robert A. Brown announced that he was particularly pleased that the Commission had decided in favour of a National Energy Board, which he termed "an absolute necessity."[175] Perhaps Brown thought that the Commission's advocacy of Canadian interests boded well for his group's proposed Montreal pipeline. Spokespersons for the major companies were less complimentary. E.D. Brockett, president of British American Oil, warned that the extensive regulations proposed by the Borden Commission should be approached "with great caution."[176] The most outspoken opposition to the National Energy Board idea was taken up by the industry journal *Oil in Canada*. In a feature editorial, entitled "And Borden Created Chaos," that followed immediately upon the release of the Report, the journal argued that the Commission's only "achievement" was to create a new and serious level of uncertainty within the industry.[177] With a week longer to consider the implications of Borden's recommendation, *Oil in Canada*'s alarm took on a more extreme dimension. It concluded:

> If M.J. Coldwell and Solon Low had sat down around a table with the president of Mexico to draft the most streamlined, efficient and speedy technique to nationalize the petroleum industry, their deliberations would probably have yielded only the palest carbon copy of the scheme which has been presented in the report of the Borden Royal Commission on Canada's energy resources.[178]

The editors grimly warned that "the composite result of all the most important recommendations is to show the nearest structure to a perfectly nationalized petroleum industry that has ever been created outside of an avowedly socialistic nation."[179] In a series of subsequent editorials, the journal went on to warn the provinces of the looming danger threatening provincial control of natural resources, but pointed out that they were in a strong constitutional position to defend their interests. The onus was put upon Alberta to stand up to Ottawa's encroachment upon provincial rights.[180] *Oil in Canada*'s unrelenting campaign carried on into December when it drew attention to the "highly critical comment" that the "Report" had drawn at the recent annual meeting of the American Petroleum Institute. U.S. oilmen were said to have "grave doubts on the state of the political climate in Canada."[181] A chief executive of one U.S. company "explained to us," the editors wrote, "that all the physical factors of Canadian oil and natural gas are adverse." This meant, he went on to say, that "Canada has only one argument in favour of its oil and gas—and that is its warm, even in the past almost tropical, political climate. If you lose that, what have you left?" It seemed clear to the editors that the government needed to move with haste to assure "the foreign and the domestic backers of our industry that the political climate has not cooled as much as they fear."[182] Following the editorial was an even more dramatic appeal for Albertans to make their "opinions known and felt in Ottawa." The appeal came within a reported speech given by J.M. Pierce, president of Ranger Oil (Canada) Ltd. to the Calgary Gyro club. Pierce had begun:

> Imagine, if you will, the sad state of affairs that would exist if the industry were forced to retrogress 10 years to the status of 1948. In Calgary and Edmonton numerous great office buildings would be empty. Broad streets would be travelled by a few tired pedestrians. Our four-lane highways would have but a few farm trucks and the odd car. The innumerable retail stores which have prospered on this great influx of workers and money would be empty, and would soon close their doors. Alberta would become to Canada what Nordegg is to Alberta. It would be a ghost province.[183]

Pierce informed his listeners that he was not suggesting that "the interim report of the Borden Commission threatens to obliterate the industry immediately," but it was necessary to emphasize just what the industry meant to western Canada.

At the national level, the campaign against "excessive regulation" was directed by the Canadian Petroleum Association (CPA). In a special submission to the prime minister, the CPA warned against the regulatory system implied in the Borden Commission Report. The Association advised that existing regulatory powers should remain centred in the Board of Transport Commissioners supplemented by the addition of an oil and gas division, and an "energy board could best serve the national interest if limited to fact finding and advisory responsibilities."[184]

In the world of Alberta politics, the Liberal party had little trouble finding an election issue in the Borden Commission Report. Indeed, they were delighted to find that they could still go one more round with the old and well-rehearsed natural gas export drama. They began to talk of the need for a natural gas marketing board to protect Alberta interests in the face of export proposals that had been given new sanction by the Commission.[185] Through the fall and winter, Albertans warmed to the ritual of their almost annual natural gas export debate.

Ignited by the "First Report's" general statement in favour of natural gas exports, the debate was fuelled by the Conservation Board's subsequent decision to reopen hearings on revised natural gas export applications from Westcoast Transmission and Alberta and Southern Gas Company Ltd. (A&S). At this juncture, Trans-Canada, a strong opponent of the Westcoast and A&S applications on the first round, also submitted an application for the Board's consideration. Both Westcoast and A&S had made adjustments in their applications that they hoped would counter the strong objections raised the previous year by the city of Calgary. Westcoast sought a permit to export up to 165 million cubic feet of natural gas per day and up to one trillion cubic feet over a 20-year period. The key feature of the application was a contract giving Calgary's utility, Canadian Western Natural Gas, first call on up to 50 million cubic feet per day of the export total. Moreover, this supply was offered without a price adjustment for variation in seasonal load.[186] Alberta and Southern's revised application was for a much larger 500 million cubic feet per day or 4.2 trillion cubic feet over a 25-year period. In addition to recognizing a larger role for AGTL, A&S also submitted an agreement that it had signed with Canadian Western, whereby the Calgary utility could purchase gas under certain conditions, but for which it would pay 1.3 times the average field price for gas taken at a load factor of less than 70%.[187] Trans-Canada's application came in two parts: "Part A" was for an amendment to its original permit that would allow an increase in the amount of gas permitted to be removed and the removal of the restriction on the amount of gas that it could take from any of the fields named in the original permit; and "Part B" was for a new permit authorizing gas removal from a new group of fields. There was no specific attempt to address Calgary's concerns.[188]

The January 1959 hearings quickly turned into the most turbulent in the Board's experience. Anxious to protect both long-term supply and price, the Calgary delegation, led by R.H. Barron, QC, took a particularly aggressive position from the outset. Taking the Westcoast proposal as a model, Barron confronted A&S President J.K. Horton as soon as he took the stand. "You know your agreement is not acceptable to the City of Calgary and the consumers," Barron charged. Horton dismissed the point, saying that he had come to that conclusion at the previous Board hearing. At this point, the counsel for Alberta's two major utility companies interjected, saying that Barron was there to represent City Council, not the consumers. Barron replied that he was there to represent natural gas consumers because the utility companies were not doing so. Barron pressed on to ask Horton if A&S was prepared to revise its agreement along the line of Westcoast's. He got a flat "no" in response.[189] This exchange set the tone for the duration of the hearings. Towards the end, petroleum engineer Stanley J. Davies presented Calgary's submission, featuring the city's objection to the A&S and Trans-Canada applications. Under cross-examination, Davies received a stern lecture from the A&S counsel, who demanded to know on what authority Davies, a nonelected person, could claim to say that this or that contract was "not acceptable to the City of Calgary."[190] Trans-Canada closed its presentation with an appeal to the Board to dismiss Calgary's unfounded objection.[191]

As the proceedings closed, debate swirled into the larger public domain. Calgary was not the only community concerned about the prospect of increased gas rates. This was a reality that Edmonton was already confronting. It was contesting, before the Board of Public Utility Commissioners, Northwestern Utilities' application for a revised rate schedule that would increase the company's gross revenue by 29%.[192] Meeting to reflect upon the outcome of the hearings, Calgary City Council decided upon a threefold response: to marshal the support of Calgarians, to challenge the objectivity of Alberta's Oil and Gas Conservation Board, and to send a delegation to Ottawa to persuade the government to withhold the necessary federal approval.[193] At the city level, staff began preparation of a statement to go out with the city tax notice in order that the public might "be informed of the serious situation with which they are faced" should foreign export be approved.[194] At the provincial level, City Council proposed to advance its opinion that, since the Conservation Board's salaries were 50% paid by the Alberta government and 50% by the producers, it was "partly in the pay of the producers" and therefore "not in a position to protect the best interests of Alberta gas consumers.[195] At the national level, the Calgary delegation prepared to meet with the prime minister and Calgary and area MPs to emphasize the importance of setting up "a National Energy Board to cope with the export problem."[196] In a brief submitted to Prime Minister Diefenbaker, the City complained about "its" local natural gas supply being "drained off to export lines," thus forcing the acceptance of more distant costly gas in the future.[197] The submission also outlined the City's objections to the A&S—Canadian Western agreement. According to the submission, local consumers were "being penalized because they live in a geographic area where the variable climate makes a minimum 70 per cent annual load factor incapable" of being achieved.[198]

The next phase of the contest took place on radio and television. On February 4, Calgary Mayor Don MacKay spoke to Calgarians about the serious problem. It was not that the city was opposed to gas export, the mayor informed; rather, the issue was the character of the proposed export contracts and the implications that these contracts held for the future price of natural gas in Alberta.[199] MacKay's public challenge caused great alarm and brought quick broadcast rebuttals from the vice-president of Canadian Western, from Carl Nickle, and from A.G. Bailey, president of Bailey, Selburn Oil and Gas Ltd.[200] The latter were presented to Calgarians as the views of "Independent Canadian Oil Operators," but were organized by the Canadian Petroleum Association. As the Association's general manager explained to Premier Manning, "It was decided that the Association should not sponsor these broadcasts officially as that might attract undue attention from press wire services, etc., and build this local controversy into a national one.[201]

Both the Canadian Petroleum Association and Premier Manning were nonetheless compelled to respond at the national level. The Association and Manning each wrote to the prime minister to take issue with certain aspects of Calgary's submission. Manning began his memorandum saying that "normally" he would not comment on such a presentation, but since some of the facts had been misrepresented he felt impelled to bring these to attention.[202] What Manning mainly wanted to impress upon Diefenbaker was that the Alberta government possessed the necessary regulatory powers to ensure the protection of provincial natural gas consumers in whatever way necessary. He explained that

> the Alberta *Public Utilities Act* and the Alberta *Oil and Gas Conservation Act* contain a number of provisions under which marketing of gas within the Province is regulated and which specifically provide that orders for this purpose override the terms of any contracts.[203]

Meanwhile the ever-expanding natural gas export debate had engulfed the spring session of the Alberta Legislative Assembly. Liberal leader Grant MacEwan promised that his party would stand with "the more than half the people of the province" who were natural gas consumers, and he urged that to receive a "fair hearing" the cities be allowed to bring their concerns directly before the Legislative Assembly.[204] The debate ran its course, and on 8 April 1959 Premier Manning announced in the Legislature that his cabinet had accepted the Conservation Board's recommendation that, since a somewhat scaled down version of each of the applications could be met from natural gas reserves surplus to Alberta's needs, export should be permitted.

Now the arena of decision shifted to Ottawa—to a government that seemed to be having difficulty making decisions in many areas.

Pressure mounted quickly. Pipeline companies and western natural gas producers hastened to complain that they had been held back for years, and they demanded that there be no further delays. Alberta consumer groups, on the other hand, demanded that their interests be reconsidered. At the same time, the crude oil sector was still waiting for the Borden Report on oil, and its frustration with Ottawa's procrastination was becoming more audible. This was especially the case after March 11, when President Eisenhower's announcement that the United States was abandoning its voluntary quota system in favour of a more restrictive mandatory system of import controls gave the industry another jolt. In Alberta, monthly production allowables continued to decline after a brief and modest winter resurgence. The bleak picture was reflected in that nearly half the drilling rigs in western Canada were idle.[205] Still Ottawa hesitated. The problem was that the Conservatives were badly divided on both the national energy board proposal and the Montreal market question.[206]

Pushed by Alberta's controversial decision to allow a significant expansion in the volume of its natural gas exports, the Diefenbaker government finally introduced its *National Energy Board Bill* in the House of Commons on April 23. In late May, as the Bill neared completion of its passage through the House, Ian McKinnon left Calgary ostensibly to discuss Alberta's views on the federal government's energy board with Minister of Trade and Commerce Gordon Churchill. The Alberta papers were quick to speculate, correctly as it turned out, that the real reason for McKinnon's trip was to decide whether or not to accept the federal government's extended call to become the new board's founding chairman.[207] With McKinnon in the capital and presumably after obtaining unofficial notice of his agreement to take on the responsibility of chairmanship, Diefenbaker took the opportunity to announce his government's plans for and commitment to the soon-to-be established National Energy Board (NEB). In a speech that made headline news in Calgary and Edmonton, he promised that there would be "no delay on gas export plans," that the processing of export applications would be the NEB's first chore. He promised also that the new board would carefully guard the public interest and have the means to deny "fly-by-night promoters" the opportunity to reap inordinate profits on pipeline projects.[208] The formal announcement of McKinnon's appointment did not come until 11 August 1959.

It was a hard decision for McKinnon. He did not want to leave Alberta, and it would appear that it was only after a good deal of "persuading" that he agreed to accept. McKinnon was not one who could shirk what he felt was a duty, and perhaps it was the suggestion that he could serve both the provincial and national interests that moved him to arrange a two-year leave of absence from Alberta's Oil and Gas Conservation Board to help get the important new federal agency organized and functioning.

The nature of Ottawa's interest in McKinnon is obvious. First, there was no one in Canada whose experience was more relevant and who was better qualified for the position. Second, the issues on the sidelines waiting to be addressed by the new board were complex and politically sensitive. Beyond his experience, McKinnon was known to have the confidence of the Alberta government, Calgary's Conservative MPs and the industry, the key constituencies upon whom all the initial decisions would bear. Indeed, no one could have more familiarity with the natural gas export applications that were to be the first items on the NEB's agenda. There was perhaps a less obvious reason why the Conservative government might have found it attractive to have an Albertan as head of the NEB. Diefenbaker probably knew that the much awaited final report of the Borden Commission was going to come out against an Alberta-Montreal crude oil pipeline. The appointment of a Manning confidant as head of the federal government's new energy authority could be expected to temper Alberta's response.

It was not just from Ottawa's perspective that McKinnon's appointment could be seen in a positive light. There is little doubt that Premier Manning was also pleased to see McKinnon accept the chairmanship of the NEB, and this was possibly a factor in McKinnon's decision. The Alberta government had taken a public position in favour of the creation of a national energy board, but both before the Borden Commission and after the release of the Commission's "First Report," Manning had warned that this new federal body must not infringe upon provincial jurisdiction and especially upon the duties and powers his government had assigned to its Oil and Gas Conservation Board. McKinnon's appointment therefore was reassuring. He was one of Manning's most trusted senior officials, and he would be sensitive to the boundaries where the two boards' jurisdictions touched. McKinnon would be predictable and, as the correspondence in the Manning papers reveals, he could be relied upon to consult and to stay in close touch with Edmonton.[209] With both Frawley and McKinnon in Ottawa, Alberta's interests and point of view would not be overlooked.

The National Energy Board that McKinnon was called to head did not vary greatly from that recommended by the Borden Commission, and in its structure, in its operating procedure, and in some of its responsibilities it resembled Alberta's Conservation Board. The *National Energy Board Act* provided for a board of five members with seven-year terms, although it specified that members of the first Board might serve for a shorter period.[210] Established as a "court of record," the Board was given the usual power to call witnesses and to compel the presentation of documents. The Board was accorded the power to issue mandatory orders, licences and permits in its areas of jurisdiction. The Board's mandate included a broad advisory function. It was to keep under review all aspects of the energy industry, so that it might advise and make recommendations it considered "in the public interest for the control, supervision, conservation, use, marketing, and development of energy and sources of energy."[211] Responsibility for the location, construction and operation of interprovincial pipelines and international power lines was transferred from the Board of Transport Commissioners and became one of the more important elements of the new Board's role.[212] In addition, the Board was given authority to make orders regarding all matters relating to tariffs and tolls. Although the Act authorized the Board in the public interest to direct a pipeline company to construct branch lines to service communities adjacent to its pipeline, it also specified that the Board had no power to compel a company to sell gas to additional customers if to do so would impair its ability to render adequate service to its existing customers. For the transmission of oil, the Board's authority over pipelines was expanded to say that it could declare pipelines to be common carriers. Finally, the Act provided that natural gas or power could not be exported or natural gas imported without a licence from the NEB. For the time being, licences for the import or export of oil were left as a declared option for the federal cabinet to insert into the Act by order-in-council if it chose.[213] The *National Energy Board Act* was passed by the House of Commons on June 3 and came into effect on 2 November 1959.

In the interval, Alberta's attention shifted back to the still urgent matter of depressed crude oil markets and the settling of political accounts. The rising intensity of the ancient gas export debate through the winter and spring was partly a consequence of the realization that a provincial election was imminent. It seemed to some, given the apparent concern of citizens in the larger urban centres about natural gas pricing, that at last the political moment for the gas export questions had come. Manning, like the other party leaders, was anxious to find a suitable election issue. But he was particularly cautious, since Social Credit had fared poorly in the 1955 election. Alberta electors had trimmed the party's overwhelming majority to a modest 37 seats in the 61-seat Legislature and unexpectedly rewarded the Liberals with 15 seats.[214] Having stretched the province's customary four-year mandate almost to the limit, Premier Manning called a provincial election for June 18. Perhaps Manning chose the date anticipating that by this time the *National Energy Board Bill* would be passed and the gas export question could be deflected into the federal realm. In any case, the natural gas issue did not emerge as a major issue, nor, for that matter, did any other single issue rise to gain dominant attention. The election result, however, was clear enough. Albertans returned to the Social Credit fold, giving the party 61 of the enlarged Legislature's 65 seats. Only one Liberal, one Progressive Conservative, one Coalition member and one Independent Social Creditor survived the Opposition parties' rout.[215] The unspoken verdict, it seems, was a profound province-wide confidence in Premier Manning's leadership.

With renewed authority, Manning could now return to the most important outstanding issue facing the province. J.J. Frawley wrote Manning a few days after the Alberta election. "I realize that the word 'mandate' is a much-abused word in political circles," he stated, "but if anyone ever had a mandate from Alberta and its people on any and every aspect of the oil and gas industry, you certainly have it now, and I assume that in due time you will again communicate to the Prime Minister your views respecting the urgent need for action to move our surplus oil and gas to market."[216] Frawley also included word that the second Borden Report was expected to be tabled in the House of Commons in a few weeks. An astute and well-connected observer of the Ottawa political scene, Frawley also sent to Manning's and McKinnon's attention a pamphlet, entitled "Oil and Canadian-United States Relations," prepared by John Davis for the Canadian-American Committee.[217] He noted that Davis's conclusion was "that on balance the building of a pipeline to Montreal would not be a good thing," and that the study was being given a lot of press publicity in eastern Canada.

Frawley was telling Manning what to expect. The Diefenbaker government was divided on the question of moving western oil to the Montreal market. The objection of the major oil companies and eastern business leaders was well known, but there was also strong opposition among some influential senior-level civil servants. In a selectively circulated confidential memorandum, one analyst concluded that "our oil industry is indeed at a crossroads and that a sharp expansion in the sales of crude is essential if the industry is to grow and if the independent companies are to continue to survive and grow with the industry."[218] The suggested advice, however, was to have the government go after increasing U.S. sales, "using the pipeline if necessary as a threat." If such efforts failed, then a Montreal pipeline should be considered. There was a caveat: "before granting the import restrictions which are so essential to the construction of the pipeline, there should be some provision for tying Alberta prices to world prices generally. In this way the consumers in the captive eastern market would have some protection against exploitation by the western producers."[219] These were words, had they known of them, that westerners could have related to from firsthand experience. But given the source, they would not have engendered much sympathy. The conclusion drawn was prophetic nonetheless.

The Borden Commission's "Second Report" was released on 28 August 1959. It acknowledged that the petroleum industry in western Canada was in difficulty but advised the government against taking action at this time to enable the construction of an oil pipeline to Montreal. Rather, it advised that the industry be given 12 to 18 months "to demonstrate that it can find markets elsewhere in Canada and the United States sufficient to sustain a healthy and vigorous Canadian oil industry with the incentive for further exploration and development."[220] The Report provided a production target of about 700,000 barrels per day (of which an estimated 482,000 barrels would be for the domestic market and 218,000 barrels for export), which it claimed could reasonably be expected by the end of 1960.[221] The commissioners agreed with the major oil companies that the Ontario market offered the strongest growth potential for western crude, and they predicted that it could be expanded from an average acceptance of 155,000 barrels of western crude oil daily in 1958 to 221,000 barrels by 1960 and 409,000 barrels in 1967.[222] In addition, the commissioners identified a specific place to start; they directed the industry "as soon as possible to displace with products refined from Canadian crude" the 50,000 barrels moving into the Ontario market from Montreal refineries.[223] The job of keeping the situation under review, as well as the option of supplying Canadian crude oil to the Montreal refinery area, was suggested as a task for the new National Energy Board.[224]

The Borden Commission's report on crude oil has been seen by some as a masterly compromise that broadly satisfied all the interest groups involved.[225] This is hardly the case. In the west, few were surprised by the report's conclusion and the reaction was predictably mixed. The majors saw the report as reasonable, whereas many of the independents remained unconvinced that the near-term growth potential envisaged for the Ontario and U.S. markets was realistic. Even if it was, the 700,000-barrel target still represented less than 70% of the industry's production potential. The disappointment of the Alberta government was real but tempered. It could only wait and see what was going to happen over the coming months, clinging to the knowledge that the Montreal pipeline idea had not been completely thrown out and that the situation was likely to be placed under the review of an agency headed by an Albertan who was strongly disposed to the placement of western crude oil in the Montreal market.[226] Reflecting upon the issue, Manning later explained:

> The federal government of course was scared stiff of it [the Montreal pipeline] because it had political connotations. They wouldn't antagonize the voters in Quebec and Ontario by jacking up the price to have them buy Alberta oil.[227]

In the end, the market problem faced by western Canadian crude oil producers in the late 1950s and early 1960s served only as another reminder to many Albertans of their region's inferior political and economic status.

The most important product of the lengthy period consumed by the Borden inquiry and the preparation of its written conclusions was the creation of the National Energy Board. It fundamentally altered the regulatory framework within which Canada's rapidly developing energy sector would evolve. In turn, the selection of Ian McKinnon as the first chairman was a critically important factor in shaping how the provisions of the Board's governing statute would be translated into practice. McKinnon's experience with Alberta's Oil and Gas Conservation Board naturally carried over and conditioned his approach in Ottawa. The connection, however, was more than just that of past experience. In the first years, at least, the NEB relied heavily upon Conservation Board data, and as it set about developing its own technical expertise, it drew directly from the Alberta Board's pool of professional talent. Gas engineer J.R. Jenkins, who later became one of the NEB members, left to go to Ottawa almost immediately following McKinnon's departure. Later, J. Stabback left his position as the Conservation Board's chief gas engineer to join; eventually, he became chairman of the NEB. Hence, in character, expertise and approach, the NEB was influenced greatly by the experience of Alberta's Conservation Board that went back to Turner Valley days. In a distant and unlikely way, the taking of this experience to Ottawa brought full circle a process that had begun more than a generation earlier when, in the public interest, engineers with the federal Department of the Interior arrived in Calgary to try to curb the waste of energy in Turner Valley.

The Petroleum and Natural Gas Conservation Board: Organization and Regulation of Field Development, 1948–1959

IN EXAMINING THE EVOLUTION of oil and gas conservation, particular attention has been focused upon the Conservation Board's statutory responsibilities and the influential part that it played in the formulation of provincial petroleum policy. The overriding conclusion is that, in the context of the period and relative to other jurisdictions and conservation authorities, Alberta and its Conservation Board almost from the beginning stood at the leading or progressive edge of North American and indeed world petroleum conservation policy. It is, however, one thing to have progressive legislation and a body of supporting regulations on the statute books and quite another to have effective implementation in the field. Ultimately, the critical test is the translation of policy into practice. It is necessary therefore to shift the focus of examination and to look at conservation policy and the Conservation Board from a field perspective. Moreover, it was in the field that the Board and the industry met on a daily basis.

The general background of industry development after 1948, the statutory responsibilities and powers assigned to the Board under the revised 1957 Conservation Act, and organization and procedure in the field are outlined. Examination of the Board's field performance is centred upon its efforts to reduce natural gas flaring and to promote solution-gas conservation. Since this was the issue that motivated the creation of the Board in the first instance and since it has remained the critical element in the Board's conservation program through the post-Leduc decade, it is an appropriate focus.

It must be emphasized, however, that the Board's responsibility and activity in the field was by no means limited to natural gas conservation. It kept close watch upon drilling operations, and thus directed its attention to such matters as blowout prevention, casing requirements, deviation surveys, spacing of facilities, and well-abandonment procedures. Through the creation of a policy web that included regulations governing well spacing, compulsory pooling, unitization, maximum rate limits and proration, concurrent depletion and multizone completions, the Board also tried to manage crude oil production. Production facilities were included as part of the Board's field surveillance system, with particular attention being paid to battery sites and the disposal of produced water, as well as lease clean-up and maintenance. An examination of the full measure of the Conservation Board's field presence, and related topics, is presented in Appendix X.

DEVELOPMENT OF THE OIL AND GAS INDUSTRY IN ALBERTA, 1948–1959

Looking at the industry's development during the decade after Leduc, the first significant measure of the post-1948 environment is the dramatic increase in the pace of drilling. (See Appendices V and VI.) In 1948, the number of wells drilled jumped to 378 from 224 the previous year, thus raising to 2,194 the total number of wells drilled in the province since the turn of the century, and boosting the number of producing wells to 1,623. By comparison, 1,667 new wells were drilled during 1958. The dimension of change during this decade is suggested in that several hundred more wells were completed in 1958 than were drilled in the entire Turner Valley period from 1914 to 1946. By the end of 1958, the total number of wells drilled had climbed to 16,173 and the number of producing wells stood at 8,386. The increased number of producing wells reflected the remarkable growth in the number of oil and gas fields. There were only three significant gas fields and two oilfields of any consequence in 1948. A decade later, there were more than 25 important producing gas fields and more than 75 oilfields (see Map 8.1 and Table 8.1). The obvious consequence is manifest in the province's oil and gas production figures. In the case of oil, annual production climbed during the decade more than tenfold, from 10.9 to 113.5 million barrels. For natural gas, annual production rose from 55.7 to 286.7 million cubic feet, approximately a fivefold expansion. During the course of the decade, the petroleum industry changed from a localized business centred upon Turner Valley and several gas fields to a vigorous province-wide industry that com-

MAP 8.1 Alberta Oil and Gas Fields, 1958

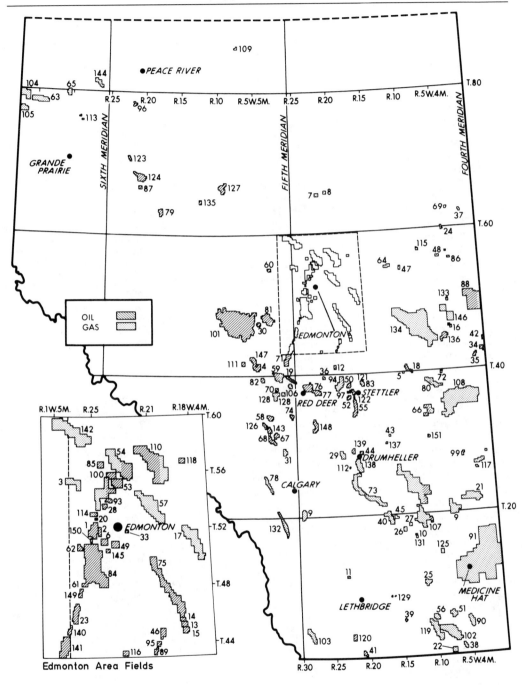

NOTE: For key to numbers, see Table 8.1

TABLE 8.1 Alberta Oil and Gas Fields, 1958

No. on Map	Field	No. on Map	Field	No. on Map	Field
1	Acheson	52	Ewing Lake	103	Pincher Creek
2	Acheson East	53	Excelsior	104	Pouce Coupe
3	Alexander	54	Fairydell-Bon Accord	105	Pouce Coupe South
4	Alhambra	55	Fenn-Big Valley	106	Prevo
5	Alliance	56	Foremost	107	Princess
6	Armisie	57	Fort Saskatchewan	108	Provost
7	Athabasca	58	Garrington	109	Red Earth
8	Athabasca East	59	Gilby	110	Redwater
9	Atlee-Buffalo	60	Glenevis	111	Rocky Mountain House
10	Bantry	61	Glen Park	112	Rosebud
11	Barons	62	Golden Spike	113	Rycroft
12	Bashaw	63	Gordondale	114	St. Albert
13	Battle	64	Hairy Hill	115	St. Paul
14	Battle North	65	Hamelin Creek	116	Samson
15	Battle South	66	Hamilton Lake	117	Sibbald
16	Baxter Lake	67	Harmattan East	118	Skaro
17	Beaverhill Lake	68	Harmattan-Elkton	119	Smith Coulee
18	Bellshill Lake	69	Harold Lake	120	Spring Coulee
19	Bentley	70	Hespero	121	Stettler
20	Big Lake	71	Homeglen-Rimbey	122	Stettler South
21	Bindloss	72	Hughenden	123	Sturgeon Lake
22	Black Butte	73	Hussar	124	Sturgeon Lake South
23	Bonnie Glen	74	Innisfail	125	Suffield
24	Bonnyville	75	Joarcam	126	Sundre
25	Bow Island	76	Joffre	127	Swan Hills
26	Brooks	77	Joffre South	128	Sylvan Lake
27	Brooks North East	78	Jumping Pound	129	Taber
28	Campbell	79	Kaybob	130	Thompson Lake
29	Carbon	80	Kessler	131	Tilley
30	Carnwood	81	Keystone	132	Turner Valley
31	Carstairs	82	Leafland	133	Vermilion
32	Cessford	83	Leahurst	134	Viking-Kinsella
33	Chamberlain	84	Leduc-Woodbend	135	Virginia Hills
34	Chauvin	85	Legal	136	Wainwright
35	Chauvin South	86	Lindbergh	137	Watts
36	Clive	87	Little Smoky	138	Wayne-Rosedale
37	Cold Lake	88	Lloydminster	139	West Drumheller
38	Comrey	89	Malmo	140	Westerose
39	Conrad	90	Manyberries	141	Westerose South
40	Countess	91	Medicine Hat	142	Westlock
41	Del Bonita	92	Morinville	143	Westward Ho
42	Dina	93	Namao	144	Whitelaw
43	Dowling Lake	94	Nevis	145	Whitemud
44	Drumheller	95	New Norway	146	Wildmere
45	Duchess	96	Normandville	147	Willesden Green
46	Duhamel	97	Oberlin	148	Wimborne
47	Duvernay	98	Okotoks	149	Wizard Lake
48	Elk Point	99	Oyen	150	Yekau Lake
49	Ellerslie	100	Peavey	151	Youngstown
50	Erskine	101	Pembina		
51	Etzikom	102	Pendant d'Oreille		

SOURCE: ERCB

prised nearly all the major companies active in the North American oil and gas scene as well as numerous smaller independent operators.

What the change in the petroleum industry between 1948 and 1958 meant for the province of Alberta can be adequately suggested in a single statistic—the amount of the industry's direct contribution to the provincial treasury relative to the government's total annual expenditure. In 1948, revenue collected from the oil and gas industry was $1.7 million and the government's actual operating budget for the year was $36.9 million, whereas for 1958 these totals were $122.5 and $173.3 million respectively (see Appendix VIII).

For the Conservation Board, the decade's change brought immense technical and administrative challenge. In no other resource sector industry was there such a number and range of producing and servicing companies. On a day-to-day basis at the Board's Calgary office, the industry's expansion meant a huge increase in the volume of information that had to be collected, stored and interpreted to ensure that dozens of new fields were managed properly in both the public and private interest. Work centred upon the increasingly complex matter of prorating oil production and upon the close monitoring of the ever-changing balance between Alberta's required natural gas reserves and exportable surplus. At the more routine level, expanding industry activity meant the evaluation of thousands of drilling applications. The impact upon the Board's responsibilities in the field was similar. Within just two or three years, the Board had to organize its field resources to manage the thousands of drilling rig inspections, battery inspections, site inspections, gas meter checks, production checks and abandonment inspections that effective regulation required.

THE *OIL AND GAS CONSERVATION ACT*, 1957

One of the most important early consequences of the rapid development of the industry was the redrafting of the *Oil and Gas Resources Conservation Act*. The Act was the statutory embodiment of the province's oil and gas conservation philosophy and the critical core that defined and gave authority to Board activities; it is therefore a proper starting place for an analysis of the Board's effectiveness as a conservation agency.

By the mid-1950s, both the Conservation Board and industry recognized the need for a comprehensive review of the *Oil and Gas*

Resources Conservation Act. Passed originally in 1938, the Act had been revised in 1950 to keep abreast of the rapid development that followed the Leduc discovery. The industry's dramatic growth through the ensuing years brought about numerous amendments to the Act, but by the middle of the decade it had become obvious that yearly ad hoc adjustments had to be abandoned in favour of a thorough redrafting. Although the Board and industry shared some thoughts on the direction the redrafting should take, their different perspectives also assured that there would be sharp differences in some areas.

The Board's attitude was conditioned by the redrafting that had taken place in 1949 and 1950. A great deal of time and care had been spent working through this revision, extensive review of U.S. legislation had been undertaken, and the special advisory assistance of noted Texas lawyer Robert Hardwicke had been obtained. The Board was thinking more of redrafting to update and integrate the numerous amendments of the preceding years rather than a re-examination of basic principles and approaches. For example, the Board wanted an expansion of the definition of "waste," and to help manage the growing number of hearings it sought the right to appoint "examiners" to act in the place of Board members at hearings. Reflecting the maturing character of the petroleum industry in the province, the Board wanted explicit authority to control the drilling of multizone wells, the location and equipment of production batteries, and the reconditioning of wells. Also on the Board's agenda was the wish for greater authority to promote unit operation and to facilitate pooling (the combining of ownership interests in a spacing unit to permit its operation as a unit).

The Alberta oil companies had more expansive ideas about revision of the province's *Oil and Gas Resources Conservation Act*. Insofar as it was represented by the Canadian Petroleum Association, the industry wanted to address some of the more fundamental components of the Act. In preparation for this round of discussions regarding the Act, it was the industry that undertook a careful review of U.S. conservation legislation, canvassed state conservation agencies and the Interstate Oil Compact Commission, and interviewed leading oil companies. This marked the industry's intention to make a special effort to have the legislation shaped more to its liking. It underlines also just how much the industry had changed over the previous half decade. Rather than the Western Canada Petroleum Association, an indigenous organization whose roots were embedded in the Turner Valley experience, it was now the Canadian Petroleum Association (CPA), an organization dominated by

U.S. major companies that spoke for the industry. Tied to companies whose interests spanned the continent, or even the globe, the CPA was a more formidable lobbyist and was able to mount the most serious challenge yet to the unique central provisions of Alberta's Conservation Act. In addition to promoting numerous but relatively minor changes in the detail of the Act, the CPA directed its primary energy to the achievement of two overriding objectives: to have conservation requirements in the field guided by the economic standard of "what a reasonably prudent operator would undertake under similar circumstances" elsewhere, and to widen the grounds for appeal of Conservation Board decisions.[1] If not in detail, the CPA sought to bring the Act's guiding principles into conformity with U.S. practice.

The delegation from the CPA's committee on economic conservation returned from their trip to Texas, Oklahoma, Louisiana and Arkansas in August 1956 and at once began to draft a detailed report on U.S. oil and gas conservation legislation and practice. The completed document, presented for discussion at a special meeting of the CPA, was prepared with a revealing section devoted to general observations of special interest. Here the authors pointed out that large-scale development of oil and gas resources in many states had taken place before detailed conservation legislation was considered. As a result, such legislation was often a compromise between what was recognized as desirable and what was practical in light of existing development. Such compromise, the report explained, often took the form of desirable legislation modified to allow "exceptions to the rule." Curiously, although the committee noted that it was generally conceded by both operators and the commissions that the provision for exceptions often limited the effectiveness of sound conservation legislation,[2] they proceeded to draw favourable attention to some of the features of U.S. practice that supported the exceptionalist doctrine. Given particular emphasis was the premium that U.S. practice placed upon economic criteria in the evaluation of any conservation scheme. "Time and again," the authors reported, "we were advised that no Conservation Commission would attempt to compel, either directly or indirectly, an operator to undertake a scheme which was not economically sound and which would not produce a fair return to such operator."[3] When waste existed and the operator was able to demonstrate that it was not economically feasible to eliminate such waste, the report informed that such an operator would not be "forced" to take remedial measures and would be permitted to continue his operations, even though they resulted in physical waste. This principle was supported by the wide right of appeal afforded to those affected by commission decisions. Even though the report admitted that lengthy and expensive litigation did not always encourage proper conservation, it maintained that it was "very important that all interested parties have aright [sic] of appeal from or review of Orders of a Commission where important rights or large sums of money were involved."[4] How wide a right of appeal should exist? The report suggested a middle ground between the extensive right typical in the United States and the Alberta Act that was seen to be "so restrictive that it virtually denies the right."[5] Adding further, if implicit, emphasis on the need to broaden the avenue of appeal, the committee concluded its observations with the affirmation of what had been commonly understood, but was now supported by a comparative analysis, that the powers delegated to the Alberta Board were in excess of those granted to state commissions.

A major factor stimulating Alberta operators' interest in stemming "their" Board's power, by promoting decision criteria based upon the economic judgement of the "prudent operator" and broadening the grounds of appeal, was existing apprehension about how the Alberta Board was going to deal with the growing volume of oil well or casing-head gas that was being flared. Under the Alberta Conservation Board's general authority to prevent such waste was the specific power to order operators to deliver produced gas to any specified plant. Such a power, it was pointed out, was not possessed by U.S. commissions. The report explained that, although most states had the authority to order the conservation of casing-head gas, as a matter of practice this was not done by actually ordering the gas to be conserved by processing, sale or reinjection.[6] Instead, conservation was addressed first by the issuance of oil allowables subject to a maximum or limiting gas-oil ratio (GOR), usually 2,000 to 1. For some states, such as Oklahoma, the effort to control the production of oil well gas ended at this stage and flaring was permitted when the allowable was produced subject to a GOR penalty.[7] In other states, where commissions considered that an excessive amount of gas was being flared, a second step was typical. Hearings were held at which operators were asked to show cause why a field should not be shut in if the gas was not put to beneficial use. If the operators failed to establish that conservation of the gas was "uneconomic by industry standards," the Commission could shut down the field. The report emphasized, however, that there was *no known case* of a field being shut in to compel the establishment of an uneconomic conservation program.[8] Perhaps to fix the issue in a continental perspective and to gain the sanction of accepted practice in the industry's North American heartland, the report observed that there were "still considerable amounts of oil-well gas being flared in fields in various States where it was not reasonably economic by industry's standards to gather and process [such gas]."[9]

Although the review of continental practice in the prevention of indiscriminate natural gas waste dominated the report, other areas, such as allowables, commingling of production, unitization, well abandonment, salt water disposal and well spacing, were also examined. Overall, the assessment made apparent that the Alberta legislation, if not already more stringent than U.S. regulation, had the potential to be so in practically every area.

Using the report as its basic reference, at its August meeting the CPA went through the 1950 *Oil and Gas Resources Conservation Act* clause by clause to prepare a draft bill that would be more in line with U.S. practice.[10] The second section of the 1950 Act included a multipart definition of "waste," and the CPA's proposed amendment to one of these parts is indicative of the direction taken by the numerous changes proposed. As defined in the Act, "waste" included

> the locating, spacing, drilling, equipping, operating or producing of any well or wells in a manner which results or could result in reducing the quantity of oil or gas ultimately recoverable from any pool.

To these words, the CPA wanted to add "under prudent and proper operations conducted in accordance with generally accepted oil field engineering practices."[11] Further on in the Act, where it stated " . . . any person who commits waste is guilty of an offence against this Act," the CPA sought to insert the word "knowingly" after "person."[12] In a similar vein, the Association sought the removal of such discretionary phrases as "in the opinion of the Board," or "as the Board deems necessary," wherever they were found. The oilmen had a phrase of their own to promote. They submitted that conservation requirements should be "subject to the test of the reasonable and prudent operator," and to this end added the preface "where it has been established that a reasonable and prudent owner or operator could be expected to undertake the same" to various sections of the Act, particularly those giving the Board the power to require repressuring, recycling or pressure maintenance of any pool.[13] The CPA's "reasonable and prudent owner" test also had special relevance to the part of the Act that the Association was most determined to change—the section dealing with appeals. In all those sections, such as the regulation of oil and gas production, the construction of facilities to support natural gas conservation, the purchase of residue gas, and the sale of gas to purchasers designated by the Board, where the "reasonable and prudent owner" clause had been inserted, the Association recognized that the clause left much room for interpretation. Therefore, as a backstop to the individual "prudent operator" clauses, it sought the right to appeal Board orders in these areas. Also, in keeping with U.S. practice was the associated request that the operation of any order of the Board under appeal to the Appellate Division of the Supreme Court of Alberta be suspended until the court had reached its decision.[14] As a further means of "opening" the Act, the oilmen sought an expansion of the hearing initiative. Their proposed amendment provided that any person affected by a Board order or regulation have the right to apply to the Board for a hearing.[15]

The CPA's submission, with respect to "Suggested Amendments to *The Oil and Gas Resources Conservation Act, 1950*," was presented to the Conservation Board on 31 August 1956, and from this point the industry lobby was unrelenting. In a series of meetings that continued until the eve of the spring 1957 session of the Alberta Legislature, the CPA pressed its case. Discussions on the proposed revisions began with the Board in October.[16] Gradually, the lesser amendments were worked out, in some areas general agreement came without much difficulty, in others concession and compromise by both parties eventually brought agreement. On 23 January 1956, the CPA's Board of Directors, other industry representatives and the Board met to try to resolve the last points at issue. These, of course, were the areas of most far-reaching significance. New section 46 dealt with the conservation of natural gas. In this area, as elsewhere, the oilmen had been compelled to abandon their preferred "where it has been established that a reasonable and prudent owner" phrase and were now holding firmly to the words, "where it is economically feasible." In opposition, the Board remained stubbornly attached to the wording "where in the opinion of the Board it is ..."[17] Pushed hard on this point, and after much discussion, the Board finally insisted that "there was no point discussing this matter further" and recommended that the matter be taken up with Premier Manning. No greater success was had with new section 119, the appeals section, and it too was left for further discussion. A second meeting also failed to find agreement on section 119, and again Board Chairman McKinnon was compelled to suggest that Premier Manning be consulted.[18]

Accompanied by his deputy attorney general, Premier Manning went to Calgary to resolve the impasse. Waiting for him at the Conservation Board were McKinnon, Goodall, Board solicitor N.A. Macleod and 13 representatives from the CPA. The meeting began with discussion on section 46. T.W.G. Thomson of Texaco, chairman of the CPA board of directors, reviewed the Association's reason for the requested changes. Manning replied, saying that the words "'economically feasible' were too broad since a court might say that an order must not interfere with the economics of anybody."[19] After considerable debate, it was concluded that the opening words of section 46 would read, "Where the Board finds it is reasonable and practicable and in the public interest." With regard to section 119, Manning accepted the need to expand the basis for appeal, but he was not persuaded that Board orders should be suspended while they were under appeal.[20] In less than an hour and a half, Premier Manning tidied up the outstanding details of the previous half year's debate and had in hand the draft legislation that he intended to bring before the spring session of the Legislature.

Passed with little discussion in the Legislative Assembly in April 1957, the new *Oil and Gas Conservation Act* gave the Board the changes it wanted. In addition to those discussed earlier, the revised Act gave the Board authority to close an area to all but authorized persons when hazardous conditions in a field or at a well warranted. The most important amendments from the Board's perspective were the new section on pooling and the greatly expanded section on unitization. The pooling provision applied to spacing units, the surface and subsurface area allocated to a well for the purpose of drilling for and producing oil or gas (normally 40 acres for oil and 640 acres for gas). It provided that, where a spacing unit comprised tracts held by several owners who were unable to come to an agreement on the operation of the tracts as a unit, any of the tract owners could apply to the Board for an order that all tracts within the spacing unit be operated as a unit to permit the drilling for and production of oil and gas from the spacing unit.[21] The consequent Board order would appoint an operator to be responsible for the drilling and operation or abandonment of the well and it would allocate to each tract owner his share of costs and production if the well was successful. This provision was intended essentially to protect correlative rights and as a practical measure to facilitate development in special instances. The related section on unitization was intended to promote the more efficient and more economical recovery of the oil and gas resources in a pool. Pending approval of the lieutenant-governor-in-council, the Board could order owners to consolidate their interests and operate a field or pool or portion thereof as a unit. Given the power to impose compulsory unitization, the Board could not, however, undertake the initiative on its own. The Board was empowered to proceed only upon receipt of an application from the owners of more than 50% of the working interests in a field or pool.[22] Nonetheless, the compulsory aspect of the unitization section was a matter of much controversy. When Manning presented the Bill in the Legislature, he explained that there had been "tremendous divergence of opinion on the rightness or wrongness of compulsion."[23] This explains why this section of the Act was set aside from all the other sections and was scheduled to come into force at a date to be determined by the government later. The real intent was to lend significant weight to the Board's ability to promote voluntary unitization. It was understood that if the Board determined that the unitized operation of a particular field or pool was necessary in the public interest and could not obtain satisfactory co-operation it had only to go to the cabinet. A quick and positive response seemed assured in Premier Manning's public declaration that the unitization section of the Act could be proclaimed in effect and the cabinet could pass the compulsion order "within 12 hours."[24]

Passage of the 1957 *Oil and Gas Conservation Act* completed the third round of debate regarding the appropriate legislative framework for oil and gas conservation in Alberta. It brought to a close serious challenge to the fundamental character of the Act. Those aspects of the original Act that distinguished the Alberta legislation in essence remained unchanged. Even though the appeal provision had been broadened somewhat, its practical impact proved negligible, as the small handful of appeals over the subsequent years attest. In part, this was because the expanded appeal provision was limited to several select sections of the Act. Of greater significance, however, was the CPA's failure to gain the other key part of its revision package—the suspension of an offending order while it was under appeal in court. This meant that there was little incentive to appeal any Board action but that for which there was the most compelling evidence. In the maintenance of the Board's undiminished authority, Premier Manning played the vital role at this critical juncture. This was because of the continuity of Social Credit administration that extended from the passage of the original Conservation Act nearly 20 years before and Manning's personal involvement from the beginning. Manning had been part of the government team that had "court-proofed" the Act in the first instance. He had retained a direct personal interest in the evolution of Alberta's conservation legislation over the subsequent years, and his presence in Calgary in February 1957 was more than simply that of the responsible minister facilitating resolution of the final difficult details.[25] Manning understood precisely what could be surrendered and what had to be maintained. The important consequence of continuity in government thinking about the Board's management of the province's oil and gas resources was that engineers rather than lawyers remained the final arbiters of conservation practice in Alberta. It meant that the Alberta Board's yearly review of activities would continue to be distinguished from that of the Texas Railroad Commission by the absence of a listing of scores of court cases that marked a typical year's progress.[26]

The evolution of Alberta legislation and practice through the 1950s is important in another context. Being the first, and by far the largest, producer, precedents and procedures set by Alberta established the standard and became the reference model for Canada's other oil- and gas-producing regions. Saskatchewan, the only other province to incorporate its conservation objectives and procedures into a formal Conservation Act in this period, assigned similar responsibilities over the prevention of waste, engineering procedures, drilling, production and proration to that of Alberta.[27] The stance taken by Alberta and its Board on fundamental issues such as gas disposal or the appeal question made it easier for Saskatchewan to maintain a similar position.

BOARD ORGANIZATION AND ADMINISTRATION

Throughout the decade that followed 1948, Alberta experienced its first genuine oil boom. Within a few years, the province changed from being a region of peripheral interest in the continental oil and gas scene to the foremost area of drilling and exploration activity. This brought proportionate change at the Conservation Board as it struggled to meet the regulatory responsibilities assigned to it under the *Oil and Gas Conservation Act*. A general measure of the Board's effort to keep pace with what quickly became the fastest growing sector of the Canadian resource economy is the growth of its staff and budget through the period (see Table 8.2).

More than simply an expansion of numbers driven by the dramatic expansion of the industry, the nature and the extent of employee recruitment was significantly influenced by the Board's Turner Valley experience and by the perceptions of the men who formed the new 1948 Board, particularly Dr. George Govier. The

Turner Valley experience established a tradition that placed a strong emphasis upon the close supervision of activity in the field combined with careful collection of drilling, production and geological information. This bias was clearly reflected in the Board's staff balance (see Table APP.I.I). Added to these existing habits was Govier's sense that the petroleum industry in the United States was undergoing rapid change as it absorbed new scientific and technical applications. He was convinced that one of the great challenges facing the Board would be to stay abreast of technological change and to ensure that Board activities were carried out with the advantage of an ever current understanding of advancing technology.[28] The need was not just to add more technicians to read more meters and to ensure that proper practice was followed in the field or to expand office staff sufficiently to handle the mounting volume of applications and tabulate the growing flow of statistical data. What was needed in addition were professional staff who could provide sophisticated interpretation of the data being collected, who could speak with knowledgeable authority to the industry's best technical people, and who could help establish production practice on a foundation of superior and independent analysis. This emphasis on technical capability is illustrated in the growth of engineers and geologists in relation to total staff shown in Table 8.2. The bias of experience and Govier's inclination to build the Board's engineering and technical strength were reinforced by two important new responsibilities that were thrust upon the Board in the immediate wake of the Leduc discovery—the need to devise and maintain an equitable method of crude oil market sharing, and to supervise the export of Alberta's natural gas in a manner consistent with assuring the province's long-term supply. Accordingly, the Board began in 1948 with an aggressive campaign to hire the best young engineers and geologists it could find.[29] Govier used his position at the University of Alberta to advantage, and until the mid-1950s he personally arranged for the Board to offer employment to select chemical and petroleum engineering candidates from each year's graduating class whom he believed had the qualities that the Board required. Unconsciously, Govier was helping to sustain the role that founding President Henry Marshall Tory had foreseen for "his" university. While University of Alberta graduates were by no means the only engineers and geologists hired by the Conservation Board, they soon emerged as the dominant managerial group, and they were a key element forming the distinctively Canadian core that distinguished the engineering and technical environment at the Board from that of most of the larger oil companies (see Table APP.IX.2).

TABLE 8.2 Board Staff and Budget, 1948–1958

Year	Staff Number	Engineers	Geologists	Budget (dollars)
1938–39	9	6	0	50,157
1948–49	31	8	1	138,087
1949–50	44	14	2	252,125
1950–51	55	14	3	190,942
1951–52	65	19	3	357,227
1952–53	79	20	5	504,295
1953–54	101	25	5	612,394
1954–55	116	32	5	762,371
1955–56	139	35	5	869,965
1956–57	187	41	7	1,091,944
1957–58	182	41	6	1,304,846
1958–59	209	52	8	1,384,751

SOURCE: ERCB

The second part of Govier's approach in building the Board's technical strength also stemmed from his university background. He placed high importance on continuing education as a means of keeping current and in raising general competence levels. To this end, he encouraged the Board to provide funding for professional staff to attend conferences and specialist workshops, and to take advanced courses, such as those offered at the Advanced Reservoir Engineering School in Austin, Texas. In addition, to upgrade the general level of technical competence of Board employees, an ongoing selection of in-house seminars and short courses given by senior staff and occasional outside specialists were arranged for junior engineers and field technicians. For the latter, these evolved into formalized courses with specifically prepared training manuals and examinations.[30]

Having invested much effort in the search for and the education of technical staff, it was necessary to ensure that good people stayed. To assist in this direction, Govier persuaded the Board of the necessity to adopt a salary policy that separated Board professional staff from civil service scales and attach Board salaries to industry averages. Although the Board witnessed a fairly high attrition of engineering and technical staff, generally comparable salaries and a progressive technically challenging working environment seem to have produced a high level of commitment and *esprit de corps* that held an important nucleus of first-rate career professionals at the Board (see Table APP.X.2). This was of special importance. In a steady growth environment, where experienced professionals were often in short supply, the Board managed to hold the key people essential to the effective management of the province's oil and gas resources.

Just as the accelerating pace of exploration, development and the assignment of new responsibilities brought a continued expansion of Board staff through the 1950s, so too did it bring great change to the Board's organizational structure. The administrative framework that McKinnon found in place when he was appointed chairman in February 1948 (see Figure 8.1) differed little from that that had existed before the Leduc discovery (see Figure 4.2). Existing departments had been enlarged and the field office in Medicine Hat had been closed in favour of a new office in Leduc. The first real hint of impending structural change came in 1949 with the appointment of J. Patrick as secretary to the Board. To relieve Board members from the growing burden of hiring and evaluating nonprofessional staff, Patrick was also assigned the duties of personnel officer. Gradually, a separate personnel division that included responsibility for office services began to take shape around the Board secretary. Substantial reorganization began the following year, and by 1952 an almost entirely new administrative structure—more appropriate to the scale and complexity of the post-Leduc phase of the Alberta petroleum industry—was in place.

The Conservation Board's engineering section was divided into three separate departments, each dedicated to one of the core areas of Board responsibility. A reservoir department, later known as the oil department, was established in 1950. Headed from 1951 by D.R. Craig, this department carried on in a more sophisticated form the kind of reservoir analysis that Board engineers had done from the beginning; however, with the discovery of new fields in quick succession after Leduc and the initiation of prorationing, res ervoir engineering required an even greater Board commitment. Also created in 1950, the gas department emerged as a direct response to the heavy obligation assigned to the Board by the *Gas Resources Preservation Act*. Headed first by R.H. King, and from 1953 by J.G. Stabback, the gas department kept a continual measure of Alberta's natural gas reserves, along with a current estimate of the province's future requirements, and provided the technical scrutiny required in the evaluation of natural gas export applications. The third major responsibility of the engineering sector

FIGURE 8.1   Petroleum and Natural Gas Conservation Board
Organization Chart, 1948

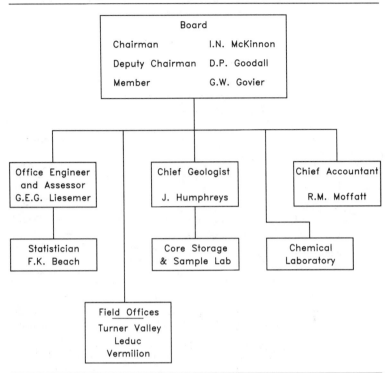

centred about operations in the field and the supervision of drilling and production activities. Generally speaking, approvals, including licences relating to these activities, were issued by the development department, which was formed in 1952. Directly associated with this department and responsible for ensuring that operations in the field were carried out satisfactorily were the field offices, which by this time were supervised by two district engineers: one stationed in Calgary, the other in Edmonton. In 1956, the three administration units were combined into an expanded development department under A.F. Manyluk, the chief development engineer. R.H. King retained the responsibilities of the old development department, and field duties were supervised by M.R. Blackadar, Southern District engineer, and V.E. Bohme, Northern District engineer (see Figure 8.2).

The reorganization of the engineering section also involved the statistics group. Charged initially with the maintenance of well records, the work of this group expanded rapidly after the discovery

FIGURE 8.2   Oil and Gas Conservation Board Organization Chart, 1958

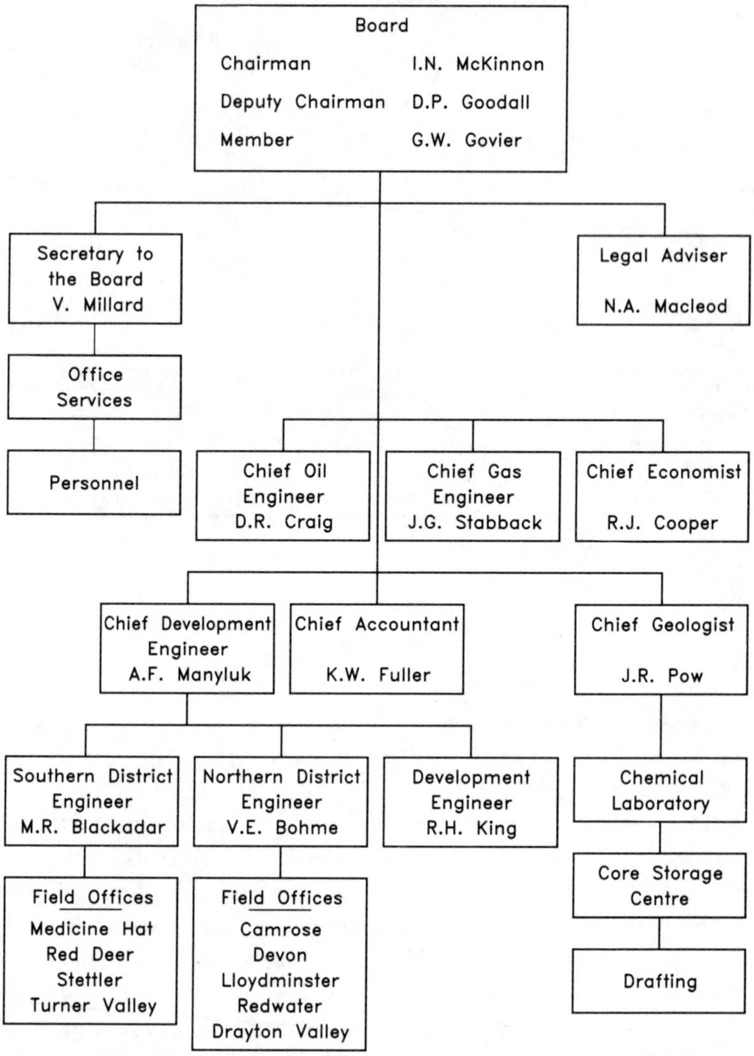

Board

Chairman   I.N. McKinnon

Deputy Chairman   D.P. Goodall

Member   G.W. Govier

- Secretary to the Board — V. Millard
  - Office Services
    - Personnel
- Legal Adviser — N.A. Macleod
- Chief Oil Engineer — D.R. Craig
- Chief Gas Engineer — J.G. Stabback
- Chief Economist — R.J. Cooper
- Chief Development Engineer — A.F. Manyluk
  - Southern District Engineer — M.R. Blackadar
    - Field Offices: Medicine Hat, Red Deer, Stettler, Turner Valley
  - Northern District Engineer — V.E. Bohme
    - Field Offices: Camrose, Devon, Lloydminster, Redwater, Drayton Valley
  - Development Engineer — R.H. King
- Chief Accountant — K.W. Fuller
- Chief Geologist — J.R. Pow
  - Chemical Laboratory
    - Core Storage Centre
      - Drafting

of Leduc, Redwater and other prolific fields. The growing demand imposed by this traditional duty combined with the new responsibility of preparing and maintaining the statistical foundation required to operate a proration to market demand scheme led to the designation of the statistics group as a separate department in 1951 under the direction of V. Millard. Although the collection of statistics on drilling was left with a well records section in the oil depart-

ment, all the Board's other important statistical and certain accounting functions, including the maintenance of production records, were assigned to Millard's group. In 1957, the Board decided to place all accounting functions in an expanded accounting department, thereby leaving the statistics department free to concentrate on prorationing and economic evaluations of various submissions presented at Board hearings. In recognition of the growing importance of the latter responsibility, the department's name was changed to economic studies in 1958.

Examining drill cuttings in the Visitors' Examination Room in the Board's Geology Department in Calgary in the 1950s. ERCB, Bohme Collection.

The geologists, the other core group among the Board's professional staff, also gained the distinction of being named a separate department in 1951. This was simply a reflection of the vital role that geologists played in Board reservoir analysis and the rapidly expanding dimension of this responsibility as new fields were discovered. Central to this role was the geology department's supervision of the Board's core and sample laboratory. The laboratory had operated as an important part of the Board's data gathering and analysis system almost from the beginning, and after J.R. Pow became head of the geology department in 1953 it gained an even higher priority so that the laboratory and core storage facility even-

FIGURE 8.3   Locations and Durations of Board Offices, 1938–1960

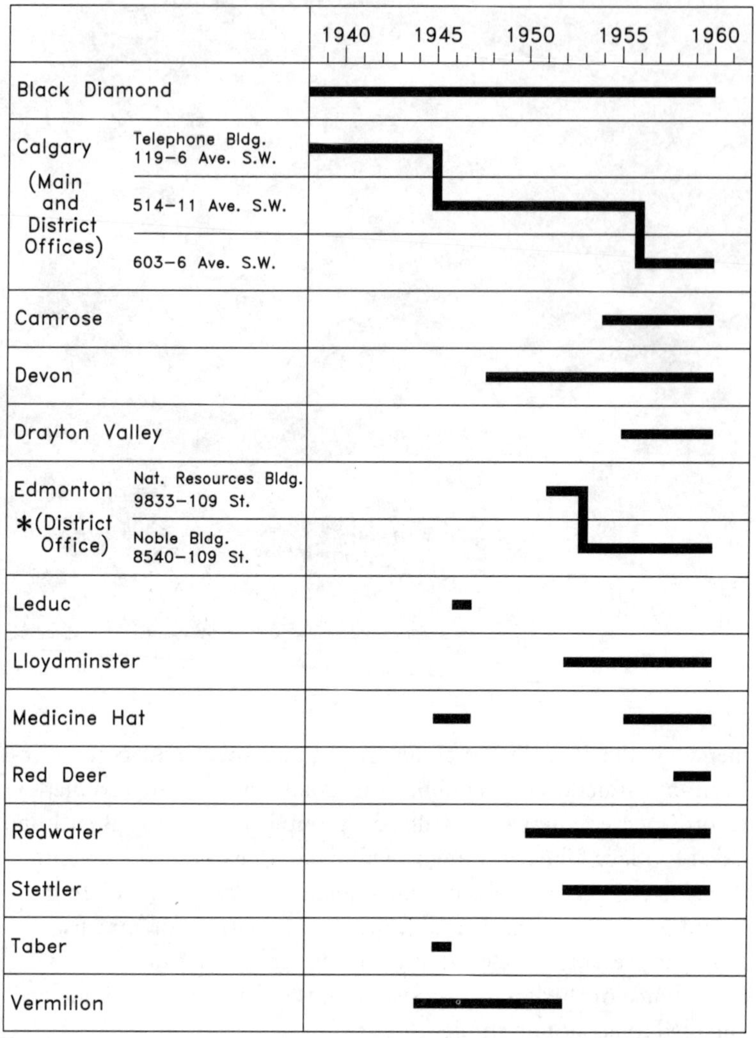

		1940	1945	1950	1955	1960
Black Diamond						
Calgary (Main and District Offices)	Telephone Bldg. 119–6 Ave. S.W.					
	514–11 Ave. S.W.					
	603–6 Ave. S.W.					
Camrose						
Devon						
Drayton Valley						
Edmonton *(District Office)	Nat. Resources Bldg. 9833–109 St.					
	Noble Bldg. 8540–109 St.					
Leduc						
Lloydminster						
Medicine Hat						
Red Deer						
Redwater						
Stettler						
Taber						
Vermilion						

* First District Office in Edmonton, not shown above, was set up in 1950, in a house near the Natural Resources Building.

tually became one of the Board's most important contributions to the public and private management of the province's oil and gas resources.

While the Board's central office organization underwent a fundamental restructuring in the wake of the Leduc discovery (see Figures 8.1 and 8.2), the Board's organization and approach in the field re-

MAP 8.2 Petroleum and Natural Gas Conservation Board Field
Offices and Districts, 1956

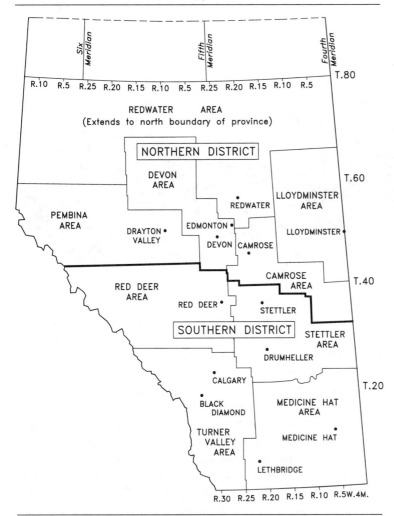

mained essentially as before. This reflected that the Board had an effective system of field supervision in place. Therefore, its response was to increase the number of field offices (see Figure 8.3 and Map 8.2) and to expand the number of field staff. Although it was not an explicit policy, the Board nonetheless maintained its traditional ratio of roughly one staff person in the field for every three employed in Calgary (see Table APP.IX.1) and the persistence of this balance stands as a measure of the Board's determination to retain an effective presence in the field.

Renamed the Oil and Gas Conservation Board in 1957, the Board had moved to this building on 6th Avenue in southwest Calgary the previous year. Provincial Archives of Alberta, PA839/2.

The reorganization and staff expansion at Alberta's Conservation Board came at a growing expense, and it marks the commitment that the provincial government was prepared to make towards the regulation of the petroleum industry in the province.[31] What this commitment, reorganization and expansion of Board staff represented, as an effective regulatory presence and especially as it related to the possibility of on-site attendance to monitor drilling and production practice in the field, is more apparent when examined in a comparative context. Table 8.3 shows the greater scale of the industry in Texas and the proportionately greater resources that the Alberta government was prepared to devote to the supervision of the industry in the public interest. In 1958, there were approximately 180,000 producing wells in nearly 10,000 Texas oil and gas fields. During the course of the year, an additional 20,000 wells were drilled. Responsibility for ensuring that drilling and production activities met the standards of good engineering practice, as laid out in state regulations, rested with the Oil and Gas Division of the Texas Railroad Commission (TRC). Although a record of the number of staff hired to meet this considerable obligation does not seem to exist,[32] it is known that the Oil and Gas Division operated in 1958 with a budget of $1.3 million, of which just over 1.1 million was devoted to salaries.[33] This is virtually the same budget and salary component expended by Alberta's Oil and Gas Conservation

TABLE 8.3 Comparison of Alberta Oil and Gas Conservation Board and Texas Railroad
Commission Annual Budgets, Number of Wells Drilled and Producing Wells,
1948–1958

Year	OGCB* Budget	No. of wells drilled in Alberta	Total No. of Producing Wells in Alberta	TRC* Budget	No. of Wells Drilled in Texas	Total No. of Producing Wells in Texas
1948	138,087	378	1,023	718,225	12,245	109,643
1949	252,125	798	1,221	770,511	13,558	115,483
1950	190,942	1,044	1,975	872,761	15,975	123,271
1951	357,227	1,268	2,696	918,235	17,671	130,309
1952	504,295	1,662	3,557	917,145	17,462	136,398
1953	612,394	1,410	4,272	914,294	18,383	142,159
1954	762,371	1,185	4,893	1,008,316	20,123	149,142
1955	869,965	1,625	5,856	1,141,927	23,540	158,598
1956	1,091,944	1,890	7,110	1,175,813	25,764	168,930
1957	1,304,846	1,430	7,558	1,155,160	24,134	176,705
1958	1,384,751	1,667	8,386	1,357,455	20,537	182,633

SOURCE: See Appendices III and V and *Annual Report of the Oil and Gas Division of the Texas Railroad Commission 1948–1958.*
* Budget figures are for the fiscal year ending March 31.
** Budget figures for TRC Oil and Gas Division are for the fiscal year ending August 31.

Board (OGCB) in 1958 as it watched over the drilling of 1,667 new
wells and the production of 8,386 wells in 138 fields.[34] Assuming
that the staffs of the two agencies were roughly equal,[35] it is clear
that the TRC, even if it had the desire, did not have the manpower resources either to make the once-a-week visit to every drilling site, as
was the practice in Alberta,[36] or to undertake the level of individual
reservoir analysis that was part of the Alberta Board's routine. Perhaps this accounts for the feeling of some U.S. drilling foremen and
oilfield superintendents operating in Alberta during the 1950s that
the OGCB was unreasonably zealous in its enforcement of regulations. Typically, they came from regions where the regulatory tradition was much more casual.

The governing philosophy and manpower resources bestowed
upon Alberta's OGCB allowed the evolution of a tighter and comparatively more effective administrative structure centred upon an
elaborate applications system. As a critical element in the translation of statutory obligation into administrative practice, the
Board's application system is worth a closer look. The number and

variety of applications received by the Board grew rapidly in the late 1940s and the 1950s. The applications for well licences and new well production allowables, for example, were largely routine and were processed in a few days. Applications for changes in well spacing, pooling orders, enhanced recovery schemes, gas processing schemes, subsurface disposal of water, approval to remove gas from the province, or to initiate oil-sands development projects had the potential to affect people other than the applicant and required special attention. The more complex applications, and those having the greatest potential to affect others, were considered at public hearings. Initially, such hearings were held before members of the Board. Later, as they became more numerous, many were heard by Board-appointed examiners.

The decisions of routine applications were normally indicated through issue of a licence approval or order. The more complex and controversial application decisions were communicated through reports. These reports summarized the evidence presented, the views of the applicant, interveners and the Board or examiners on the issues raised, and the Board's findings and decision. An example of the treatment of a more complex matter is indicated in the reports concerning the applications of Westcoast Transmission, Trans-Canada Pipe Lines and others to remove gas from Alberta.

Gradually, through the 1950s, the broad procedures for receiving and processing applications were formalized, and these are best illustrated in the procedure followed for the most common type of application—that for a well licence. Figure 8.4 presents this procedure, from application for a licence to drill to completion or abandonment of the well, in a flow chart format.

The application to drill a well, followed by the issuance of a licence, was the necessary first step in what was hoped would result in a producing well. An application to drill a well had to be submitted by the owner of the proposed well or his appointed agent. Upon receipt, the application was reviewed carefully by the drilling section of the development department. If the application was satisfactory, it was passed to a Board member for approval and signature, and then to the Department of Mines and Minerals which issued all drilling licences. If either the Board or the Department required more information, the application was sent back to the applicant and the process resumed. In many cases, a phone call would be sufficient to clear up problems. Licences were issued by the Department a day or two after receipt of the application by the Board, and the licence was immediately forwarded to or picked up by the owner's representative. Ordinarily, it was posted at the drilling rig

FIGURE 8.4   From Drilling Licence to Production Oil Well,
Regulatory Procedures, 1957

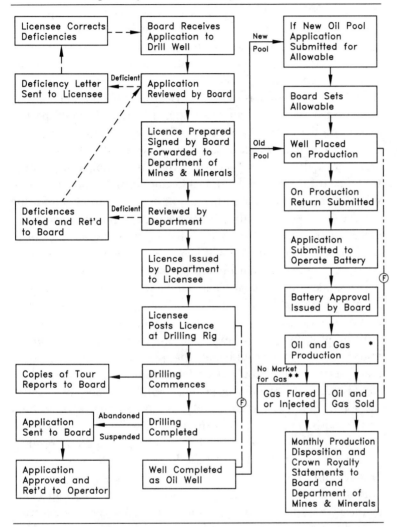

SOURCE: ERCB, M.R. Blackadar.
* Solution gas usually needed further processing before marketing. The step is not
shown herein.
** Solution gas was required to be conserved where conservation was economic.
Ⓕ Operations subject to periodic field staff surveillance.

soon after. Since time lost often meant money being spent on an idle
drilling rig, many licences were picked up almost before the ink had
dried on the minister's signature and couriered to the field for post-
ing at the drilling rig, since the well could not begin drilling until the
licence had been posted.[37]

Tour reports were prepared by the operator at the drilling rig for each day's operations, and the Board's copy was mailed to, or picked up by, a Board representative on a weekly basis, up to the time the well finished drilling or was abandoned. The well might turn out to be nonproductive, marginally productive and not immediately worth completion for production, or of possible future value as a service well. In the first case, an application to abandon would be made and approved if no clear need for the well existed. In the other two cases, an application to suspend would be forwarded to the Board for approval. The hope, of course, was that the well would strike oil or gas. If it were an oil well, the owner would then place the well on production and complete and submit to the Board the required "on-production form." A new field well was allowed a maximum of five days test production before being subject to a pool allowable. If the well was a discovery well, the owner had to apply to the Board's oil department for a new pool allowable. The application would be supported by reservoir data for consideration by the Board in setting the allowable. After the pool was designated by the Board, data on the wells would be made public and the allowable calculated on an average well basis. (Information on wildcat wells was kept confidential for up to one year.)

If the well owner had more than one well producing from the same field or pool, he could apply for approval to operate a battery under section 97 of the *Oil and Gas Conservation Act,* which would reduce the amount of facilities required. Upon approval of the application, production could be measured and reported on a battery basis. The production from individual wells producing to the battery would be determined by prorating total group production to the well on the basis of periodic monthly tests for each well. Production of both oil and gas each month was required to be reported to the Board with details of the disposition of each product, by the 15th of the following month. If the well was on crown land, a royalty statement had also to be submitted to the Department of Mines and Minerals. Official forms were provided for each of these purposes.

The Board's applications procedure stands as evidence of how much both the Board and the industry had changed over the decade that followed the Leduc discovery. Apart from the first few years after its creation in 1938, these were the years when the Board faced its most critical test. This was a period of dramatic growth and the challenge was whether or not the Board would be able to assemble and organize the administrative resources necessary to manage an aggressive international capital-rich resource industry where rate-sensitive production demanded a level of public supervision well beyond that of any of the other resource sector industries. That the Board was able to meet the challenge was a function of three crucial factors: the responsibilities defined by and the powers assigned to the Board by the Conservation Act, the legacy of the Turner Valley experience, and the remarkable combination of abilities possessed by the three men who guided the Board through this period.

The first two factors do not require further comment; however, it is appropriate to say something more of the leadership provided by Goodall, Govier and McKinnon, for it is the most important of the three variables. Each contributed a vital and complementary strength to Board leadership. Goodall was the embodiment of the Turner Valley experience on the Board. His long apprenticeship in the field made him the natural person to take charge of the Board's activities in this area. He was the persistent advocate of practical procedures. Govier was an immensely energetic scientist-builder who devoted himself to establishing and organizing the Board's technical strength. He was the forceful advocate of the importance of advanced engineering and scientific analysis. McKinnon was a cautious, able administrator and committed public servant. He was the unrelenting advocate of the necessity of being attentive to detail in the interest of careful management and the maintenance of public trust. Not only was there an almost ideal blend of abilities and interests among the three but also there was a harmony of personalities as well. The respect that each held for the other's expertise allowed them to work effectively as a team. Together, they established an administrative foundation that served to anchor the Board in the even more technically and politically complex environment of the following decade.

FIELD ORGANIZATION AND PROCEDURES

The post-Leduc Board inherited a legacy of hands-on supervision of field activities from the Turner Valley period. Accordingly, field offices were opened at principal centres of activity to keep a closer watch on operations. Thus, the Vermilion office and a temporary office at Taber were opened in 1941. In 1946, an engineer was assigned to the Medicine Hat area. If discoveries had continued to increase at the then expected rate, the Board would not have encountered any serious difficulties in maintaining proper surveillance of industry field operations; however, the discovery of major fields at Leduc in 1947 and Redwater in 1948, followed by many more in quick succession through the early 1950s, stretched to the limit the

TABLE 8.4 Southern and Northern Districts Field Offices, 1958

Southern District (Calgary)	Northern District (Edmonton)
Medicine Hat opened in 1945 by T. Geffen, closed in 1947, and reopened in 1954 by E.A. Moore	**Camrose** opened in 1953 by L.A. Bellows
Red Deer opened in 1957 by E.A. Moore	**Devon** opened in 1948 by N. Goodman
Stettler opened in 1951 by V.E. Bohme	**Drayton Valley** opened in 1954 by H. Coffin
Turner Valley located originally in Black Diamond and supervised by G. Connell, who took it over from Oil & Gas Division of the Department of Lands and Mines in 1938	**Lloydminster** opened in 1951 by E.M. Berlie, replacing a previous office located in Vermilion
	Redwater opened in 1949 by A.F. Manyluk

SOURCE: ERCB

capacity and commitment of the Board to maintain proper surveillance of operations in the new fields. In 1948, there were only eight engineers on staff, other than board members. By 1949, there were 14, located mainly in the field to accommodate the rapidly increasing workload. Subsequent field office openings and closures reflect the geographic growth pattern of the industry.

By 1958, the Board's field offices were sufficiently numerous to warrant grouping into two large administrative districts, a southern district under the supervision of a southern district engineer in Calgary and a northern district supervised from the Edmonton office (see Map 8.2). The field offices were maintained in the oil industry centres identified above and for the periods shown in Figure 8.3.

Each office was supervised by a senior field engineer, who in the 1950s was invariably a graduate engineer. The policy of having a graduate engineer in charge was formalized in 1948. Before then, graduate engineers had usually filled this position, although the distinction between a graduate engineer and other technical personnel was less clear. After Leduc, the Board recognized that oil company field offices were usually supervised by university-trained engineers, and the Board believed that its supervising field engineer should be of equivalent status and training.

In addition to regular reports, contact with the main office in Calgary was maintained both through discussions and correspondence with the relevant departments or through the district engineer. The district engineers made regular visits to each field office to communicate policy changes, to discuss problems with the staff, and to advise of administrative requirements, which required clarification from time to time. Regular meetings of field office managers at the district offices in Calgary and Edmonton were also effective in communicating field office concerns to head office and policy developments to the field staff. The heads of engineering and geological departments from Calgary provided perspective and contributed to the discussions. A meeting of all field office managers and both district engineers with Calgary office personnel was held annually.

During the 1950s, the field offices were in large part extensions of the development department in Calgary, and direct contact between that department and the field was frequent. Most of the field office reports were directed to the development department.

By 1960 the Conservation Board had developed a confidential field office manual that was provided to each member of the field office staff. This manual included guidance on oil and gas reservoirs, geology, functions of the Board and the Department of Mines and Minerals, drilling and production operations, and duties required of the field staff. The following list of duties is illustrative of those performed by the Board's field staff:

a. A well could not commence drilling until a licence had been issued and displayed at the well site, usually in the doghouse of the drilling rig (see Figure 8.5). On learning of a new well, the field staff would go to the location and, if there was a drilling rig there, check the doghouse for the licence to drill. Normally, the licence was on hand and posted. If there was no licence and the well was drilling, the well would be shut down until the licence was produced. The drilling report, maintained in triplicate for the contractor, well owner and the Board, noted all significant activities that had occurred in the period covered, including shut-downs.

b. If the licence was posted, the field man would check the daily drilling reports for completeness and forward copies for the elapsed days to Calgary.

c. Smoking on the rig floor and the use of hazardous heaters within a specified distance of the wellbore were not permitted, and the Board inspector would watch for any such infractions. Personnel and companies violating these requirements were liable to court action, and fines if found guilty.

d. The field man would also inspect the rig to ensure that there were no obvious shortages of equipment. With regard to blow-

FIGURE 8.5   Drilling Rig

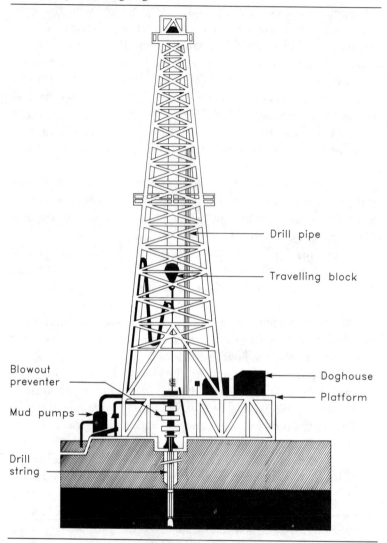

out prevention, in addition to a visual check, the driller could be required to demonstrate that the blowout preventer could be activated in an emergency (see Figure 8.6). Where the preventers were found defective, the driller was required to make the necessary repairs or adjustments before proceeding with drilling.

e. Deviation surveys were required to be taken each 500 feet of depth, or, if the well was being directionally drilled, directional

FIGURE 8.6   Blowout Prevention Equipment

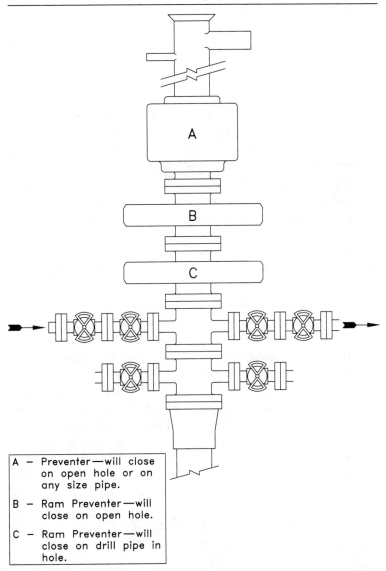

A – Preventer—will close on open hole or on any size pipe.

B – Ram Preventer—will close on open hole.

C – Ram Preventer—will close on drill pipe in hole.

surveys were required to determine the location at which the wellbore entered the top of the producing zone. To qualify for a full allowable, the well had to be completed within a specified target area.

f. After casing strings were cemented, the cement had to be given sufficient time to set, pursuant to the regulations, before drilling could resume. Periodically, field inspectors would arrange to be present to witness the operations to ensure that the regulation was met.

g. If a well was found to be "dry" (nonproductive), a proposed abandonment program was required to be submitted for review in Calgary or at the field level. These programs consisted of placing cement plugs so that fluids could not migrate from one zone to another and possibly damage a potential producing oil or potable water zone. Each plug required an eight-hour setting time, after which the plug had to be "felt for" by touching the top of the cement with the drill pipe to ensure it was satisfactory. Unannounced inspections were made from time to time to ensure that the rig remained over the wellbore for the period necessary to complete both the cementing and the feeling of the various plugs.

h. Periodically, Board inspectors witnessed casing cementing and abandonment operations. For wells in coal-mining areas, where it was imperative that coal seams be protected, all such operations were witnessed by field inspectors.

i. After drilling operations had been completed, the regulations required that the wellsite be restored as nearly as possible to its original condition, whether the well was productive or was abandoned. A Board field inspector examined the site to ensure restoration requirements had been met. Where approval was deferred, periodic visits to the site by the field inspector and letter queries from the Board's Calgary office persisted until final approval was given. Field staff often had to arbitrate disputes between operators and landowners, particularly in determining the adequacy of lease restoration. The responsibility for inspection of lease restoration was transferred to the Surface Rights Board about 1965.

j. After a well was completed and placed on production, daily production records were required to be kept either at the wellsite or at the company field office. Normally, the field inspector would check to see that this requirement was being met and that the oil, gas and water produced were accurately measured and reported. From time to time, certain wells were also subjected to production practices checks. An abbreviated check could include matching data in production records on site with the production figures submitted by the company to the Board to determine inconsistencies. Occasionally, the Board staff would undertake a meticulous examination of the company's producing program at the wells and battery site to ensure that everything was being done in accordance with the rules. In a few cases, flagrant violations were uncovered, but operations were usually found to be in accordance with the requirements of the Board. Therefore, unscheduled field inspector visits were helpful in maintaining a reasonable standard of production practices.

k. Gas measurement practices were checked on a periodic basis to ensure accuracy of meters. Usually, this was a full-time job for one or more of the field staff at each office.

l. The location of equipment used in the production operations, both at the wellsite and the battery site, were checked closely to ensure that the minimum distance requirements were met (see Figure APP.X.2).

m. Housekeeping practices at wellsites and batteries were monitored to ensure that a good standard of cleanliness and tidiness was maintained.

n. The Board staff included a number of wire-line bottom-hole pressure technicians and one acoustic well-sounder operator to obtain or ensure that up-to-date and accurate pressure readings were obtained for the Board and industry reservoir engineers.

o. Much of the field office staff's time was taken up with problems related to drilling and production in established areas, but wildcat drilling operations were not overlooked. Regular visits were made to outlying areas, including northern Alberta. Some of the inspections in remote areas had to be made by aircraft.

p. Gas, water or oil samples were often obtained to meet requests from the Calgary office or from the Board's laboratory in Edmonton.

q. The Board's gas department also arranged for the field staff to monitor flowing and shut-in pressure tests on gas wells to ensure that operators were following accepted procedures in determining the flow capabilities.

r. Where major problems such as a well blowout or extensive escape of hydrocarbons from a pipeline break occurred, Board field staff would be on site continually to keep the Board fully informed, to convey Board directions to the operator, and to ensure remedial operations proceeded as expeditiously as possible.

s. Occasionally, senior Board field staff were directed to complete the abandonment of old wells that had been abandoned unsatisfactorily by operators who were no longer in existence. Special funds were obtained from the Department of Mines and Minerals for this work.

t. Where a well was found to be operating with inadequate equipment or in hazardous circumstances, or without a licence, the field representative was obliged to shut down the well immediately if it was safe to do so. In most cases of contravention of the regulations, the circumstances were discussed with the district engineer before shutting down operations.

u. Serious breaches of regulations with respect to production operations were referred to the Calgary office, and an enquiry, including testimony by the Board field staff, was usually held to ascertain whether or not the well or wells should be shut down and whether or not charges should be laid in a criminal court. In serious cases, the shut-down preceded the enquiry.

Many of the above duties are described further in Appendix X. The staff implications of the above responsibilities are suggested in Table 8.2 and Table APP.IX.1.

SOLUTION-GAS CONSERVATION—LEDUC, REDWATER AND OTHER FIELDS

Gas produced might have evolved from solution in crude oil, as free natural gas or in combination, depending upon the reservoir in which it is produced. A reservoir containing free natural gas is only produced when a market for the gas exists. Dry gas, that is gas having a low content of heavier hydrocarbons, might meet market specifications with little or no processing. Wet gas, containing heavier hydrocarbons and perhaps sulphur compounds, must be processed before being marketed. Gas processing plants have been operated in Alberta for many years, beginning with the absorption plants in Turner Valley. Solution gas, or oilfield gas, on the other hand, is the wet gaseous component of petroleum, and whenever crude oil is produced, solution gas is unavoidably produced with it. Before the 1950s, solution gas was being gathered and conserved only in the Turner Valley field. The other oilfields, except those producing heavy oil with low gas-oil ratios, had flares burning solution gas, and of course the greater the gas-oil ratio, the larger the flare.

The Conservation Board was born in the controversy surrounding the wasteful burning of natural gas, and the control of the oilfield flaring remained central to the Board's moral and practical mission. Given the broadly accepted definition that the flaring of natural gas, which could reasonably be conserved, was a wasteful practice, the Board was locked almost constantly in debate with the

Minister of Mines and Minerals N.E. Tanner (left) turns the valve at the opening ceremonies for the Devon Conservation Plant in 1950 as Imperial Oil's M.L. Haider looks on. Imperial Oil Limited.

industry throughout the 1950s about what conservation measures were "reasonable" or "economic." The decision was seldom an easy one.

After the Leduc discovery, as the number of producing oilfields began to increase, the Conservation Board became more concerned about the escalating volume of solution gas being flared in central Alberta. At first, there seemed grounds for cautious optimism. Imperial Oil took the initiative, and in May 1950 placed on stream the first field gas conservation program outside Turner Valley. The scheme featured the construction of an absorption plant at Devon to process solution gas collected from wells in the still developing northern portion of the Leduc oilfield. In hope, the Board waited for other operators or groups of operators to follow suit and present applications for similar programs. Under the Conservation Act, all applications for repressuring, processing or disposal of produced gas had to be approved by the Board. Where the economics of gas conservation seemed borderline and the oil companies were reluctant, the Board was prepared to initiate discussions with the operators to assess whether the solution gas could be conserved economically. To assist in the promotion of schemes that appeared to be close to the economic margin, the Board could advise operators that a reduction in royalties on natural gas sales could be approved by the Department of Mines and Minerals if such a reduction was necessary to make a gas conservation scheme feasible.

Most of the operators of the wells in the southern part of the field, beyond the area served by the Imperial gathering system, believed that the quantities of gas produced and the cost of a gathering system rendered gas conservation uneconomic. Pushed along by the Board's encouragement, however, attempts were made by these operators in 1952 and 1953 to interest a company, Leduc Southern Absorption Ltd., in installing a processing plant in the area. The application made by that company was approved by the Board, but to its disappointment the project failed to proceed and the approval was cancelled. Finally, in the fall of 1953, the Conservation Board decided that the matter had been delayed long enough and that the time had come to take a more forceful stance. On 9 December 1953, the Board issued its first Gas Conservation Order (GCI). This marked a significant step beyond the policy of persuasion that had been followed up to this point, and it sent the industry a compelling reminder of the Board's authority. The order declared that since it was "in the public interest that waste of natural gas in the Leduc-Woodbend Field should be prevented," *all* gas produced in the field after 1 December 1954 had to be processed to remove natural gasoline, other hydrocarbons, or other substances.[38] The products of such processing, including the residue gas, were to be disposed of in a manner satisfactory to the Board or stored in an underground formation. Exemption from the order was possible, but required an application from the owner and satisfactory proof that it was not economic to gather, process or store the gas. A new round of discussions with the field operators followed. At a meeting held on 31 March 1954, an application by RKR Syndicate to construct a processing plant was discussed. Although the Board approved this application on 1 April 1954, this company also failed to proceed. Next, Imperial Oil appeared before the Board with a proposal to inject Leduc South gas into the D-3 Gas Cap; however, the approval issued on 12 November 1954 went only a small distance towards addressing the problem. It applied to only 22 Imperial-owned wells, leaving most wells in the southern part of the field outside any conservation scheme. In December 1954, the GCI deadline passed and the Board moved quickly to the next step in the administrative process. It held an enquiry and called operators in the Leduc field to explain why they had not complied with the order. After listening to the field operators, the Board decided to exempt 116 wells from the provisions of GCI, on the understanding that, should future conditions alter the economics of conserving gas, the Board might reconsider the exemptions. Not exempted from the order were some 112 wells, all but one of which were in Township 49. Hence, on 18 February 1955, the Board issued orders shutting those wells down, unless the operators came forward with a processing, storage or disposal scheme scheduled to go into operation by 1 December 1955.[39]

The magnitude of the shut-in order involved wells belonging to several dozen companies and seems to have come as something of a shock. Convinced of their case, the companies immediately asked for a meeting to allow their technical committee another opportunity to present evidence to demonstrate that a conservation scheme in the area proposed by the Board would be uneconomic. Within days of the February 28 meeting with the area operators, the Board was also in possession of a thorough new study prepared by its own gas engineers. Perhaps to the Board's surprise, the study "strongly supported" the operators' conclusion, and in keeping with its insistence on decisions consistent with the best technical evidence, on March 14 the Board rescinded its shut-in order.[40]

Although the Board was forced to back away from imposing a conservation program upon the operators in the southern part of the Leduc-Woodbend field, at the same time it was pressing ahead with the bigger and more significant challenge to establish solution-gas conservation in Alberta's most prolific oilfield. Virtually at the moment it was rescinding its well shut-in orders for Leduc-Woodbend, the Board's 1 March 1955 deadline, set by its GC3 order for operators in the Redwater field to come forward with a conservation plan, had just passed without any response. This order had been the result of discussions regarding gas conservation in the Redwater field begun in May 1953. As with Leduc-Woodbend, the Board in the end had endeavoured to move Redwater operators

Opening day at the Redwater Gas Conservation Plant in 1956. Imperial Oil Limited.

along with a Gas Conservation Order (GC3).[41] This order had advised operators, Imperial Oil being by far the largest, that they had until 1 March 1955 to devise to the Board's satisfaction a gathering, processing or storage scheme that would process or conserve "all gas" produced from their Redwater wells. The industry's strong resistance was made clear on 9 June 1954 when producers requested that the GC order be withdrawn. Although the request was denied, the producers' decision to ignore the deadline was hardly surprising. The Redwater solution-gas conservation question had become a contest of wills and not simply a competition of interpretations of technical data. All parties realized that what happened here would set the pattern. It was one of the most important tests that the Alberta Conservation Board had faced since Turner Valley days.

The Board's response was to call the Redwater producers to an enquiry set for 15 June 1955, at which they were instructed to "show-cause" why they had not taken action pursuant to the Board order. The answers given were judged unsatisfactory; hence, the Board followed the inquiry by issuing orders for the shut-down of hundreds of wells as of 1 November 1955 unless definitive action had been taken to provide conservation facilities. The letter accompanying the order sent to Imperial Oil, which was similar to the others, declared that the company had contravened the provisions of the Board's earlier conservation order and listed the 382 Imperial wells that were scheduled to be shut in.[42] It was a persuasive document. From the Board's perspective, the debate had been protracted long enough, it wanted action; in effect, it was giving notice that it was prepared to shut in one of the province's most productive oil-fields and to take on the country's dominant oil company. In late August, Imperial Oil submitted a conservation scheme for the Board's consideration. It was approved on 5 October 1955, and as a result the Redwater Conservation Plant went on stream in November 1956.[43]

During the 1950s, the Conservation Board issued Gas Conservation (GC) orders to initiate solution-gas conservation programs for six of the province's more important oilfields. In chronological order, they were Leduc-Woodbend, Glen Park (program approved November 1954), Redwater (program approved October 1955), Pembina (program approved August 1957), Innisfail (program approved March 1960) and Harmattan-Elkton (program approved November 1960).[44] In each case, satisfactory arrangements for the conservation of solution gas were achieved, but not without the Board's persistent prodding in some instances. The GCs stand as clear evidence of the importance of the Board's role. Where the industry dragged its feet, the Board issued an order for a conservation scheme to be prepared within a specified time limit. In this way, the Board continued the old struggle that dated back to 1932 and the Turner Valley Gas Conservation Board, Alberta's first regulatory authority. It was an important new chapter, for in the 1950s the petroleum industry was entering a new phase, and the course set by the Board was crucial. It made clear to the industry that, where in its judgement the conservation of solution gas could be achieved economically, the Board intended to have its way.

The pattern of production and disposition of Alberta's natural gas over the decade from 1949 is shown in Figure 8.7. The wasted portion declined from 22.2% of the total volume of gas produced in the province in 1949 to 20.3% in 1958. Given the time span, it might appear that the Board's achievement, at best, should be described as modest. In this instance, the appearance is deceiving, the Board's accomplishment was in fact substantial. First, it must be noted that the number of oil pools producing solution gas increased from the beginning to the end of the period. Not only were there many more pools but also they were much more geographically dispersed. In short, this meant that the solution-gas problem in the later 1950s was of a quite different magnitude than that in the early years of the decade. Considering these factors, it was evident that by 1958 the Board had made good progress. The appropriate reference year from which to measure this progress is 1956. Reflecting

FIGURE 8.7   Production and Disposition of Alberta Natural Gas, 1949–1958

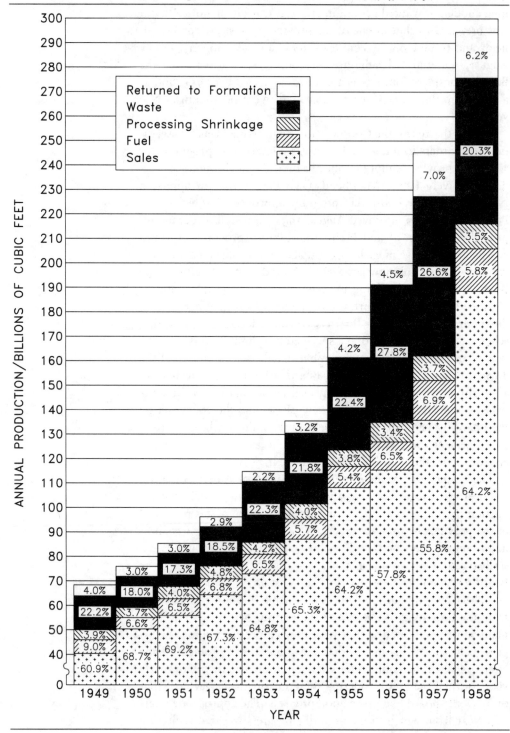

the large number of new fields that had come into production, the volume of waste gas as a percentage of total production peaked during this year at 27.6%. The trend from this point shows significant improvement and is evidence of the Board's efforts.

The decade that followed the Leduc discovery marked a crucial period in the Board's history. During this interval, it faced the formidable challenge of managing an industry that had transformed from a handful of minor local companies whose efforts were concentrated on one small field into an industry that comprised hundreds of companies and included many aggressive mid-sized independents and most of the major multinational oil companies all engaged in the rapid development of numerous oil and gas fields. The question was, could the "Turner Valley Board" manage in the Big League? In good part, the management test had to be passed in the field and the evidence suggests that the Board proved up to the challenge.

Following the Leduc discovery, the Board was quick to refine its organizational structure and to adapt its regulations to cope with its rapidly expanding responsibilities. It did this with reference to conservation legislation and regulations in place elsewhere, with ongoing industry consultation, and with special attention given to the building of its own technical expertise. But it was more than just putting in place a regulatory framework that can be seen as progressive when compared to conservation practice elsewhere. The aspect of the Board's effort of parallel or perhaps even greater importance was its ability to commit well-trained and effectively supervised manpower resources in the field. This is one of the more important features that seems to distinguish the approach of the Alberta Board from similar agencies in other jurisdictions.

The Board's presence in the field centred upon an evolving foundation of comprehensive regulations supported by an effective field inspection program. The system worked because it was comparatively thorough and because of the manner in which the regulations were created in the first instance. Building upon a practice initiated during the Turner Valley period, the Board's approach to improving field practice was to remain in constant dialogue with the industry and the government. In this manner, the Board constructed a regulatory framework that rested firmly on a basis of informed consensus. It must be noted that such consensus was not always easy to achieve, but in the end the differences between the Board and the industry that remained unresolved were usually matters of detail or timing rather than principle. Moreover, companies new to the province quickly learned that, after discussion and once decisions regarding good field practice took regulatory form, the Board was insistent, it was thorough, it followed through, and it was not averse to shutting in wells to win its point. The ultimate reality in this regard was that, when push came to shove, the Board would win. Unlike the situation in Texas or Oklahoma, where every regulation was open to court challenge and legal interpretation, Alberta's inspection system was supported by an uncontestable legal authority. The overall result was a field environment characterized by general co-operation. Compliance was the norm, and where Board field inspectors called attention to infractions, industry was customarily quick to address the problem.

PART III

Conclusion

The Petroleum and Natural Gas Conservation Board in Retrospect

T HE HISTORY OF OIL and gas conservation in Alberta divides naturally into relatively distinct periods that are shaped by market, technical and political factors. Free enterprise values, characteristic of frontier resource development throughout North America, defined the character of the discovery and pioneer phase that lasted until 1938. Largely dependent upon local capital resources and marked by an unstable economic setting, the generally chaotic environment of this period was symbolized by the "rule of capture" doctrine. Assuming that an individual possessed the mineral rights, whatever he could get out of the ground was his, and how he went about getting it was up to him. The market both dictated the focus of the search and shaped production practice. It offered a reward for oil. For the oil's associated product, natural gas, the market displayed profound indifference, and thereby decreed that it should be burned as waste. Accordingly, the struggle to advance the conservation idea begins in the pioneer period and is one of the most important elements in the story of oil and gas development in Alberta through the first and subsequent periods of development.

The second period marks a short but important interval of transition between the pioneer phase and the contemporary industry of the post-Leduc period. This transition phase begins with the creation of the Petroleum and Natural Gas Conservation Board in 1938, which points to the most important of the features that distinguish this period from the one that preceded it, namely, the emergence of the effective presence of a regulatory authority in the everyday operation of the Turner Valley oilfield. The principle of conservation having gained acceptance, it remained to work out the practical application of the principle in the field. It was the good fortune of those entrusted with the task that their initial efforts, guided by experience from elsewhere, could be concentrated upon a relatively small industry centred upon a single oilfield in a rare period when there was a ready market for virtually all the oil the field could produce. These few years, before the Leduc discovery rapidly and massively altered the scale of the industry's operations in the province, allowed time for industry and Conservation Board engineers to reach broad agreement on the basic elements of appropriate conservation practice. Time permitted the Board's administrative authority to gain credibility and acceptance and the core components of a regulatory structure to be set in place. Along with and directly related to the imposition of active regulation in the public interest came important advances in reservoir engineering that began to reshape ideas about what was acceptable production practice. Business organization in the Alberta petroleum sector was also in transition, and during this period it moved much further away from dependence upon local capital and entrepreneurship.

The Leduc discovery in 1947 pushed the Alberta petroleum industry dramatically into a third phase. The primary hallmarks of this period are political continuity and economic change. Within a year, the industry was transformed from an enterprise of modest regional importance to one of continental significance. Nearly all the big multinational and major U.S. companies hastened to the province. International, and especially U.S. capital, followed in their wake, along with a host of U.S. oil executives, technical professionals, experienced oilfield managers, drillers and other skilled workers. From its base in a single important oilfield and several gas fields, the industry rapidly expanded from its southern Alberta heartland to encompass many fields and to become a truly province-wide enterprise, although Calgary retained a tenacious hold upon its traditional position as the industry's administrative capital.

The dimension and pace of change in the industry and its impact upon the province are easily conveyed with just a few statistical reference points. In 1938, only about 60 wells were being drilled each year, compared with nearly 400 in 1948 and almost 1,700 by 1959. More than 60% of Alberta's current (1992) conventional crude oil reserves and more than 40% of natural gas reserves were discovered before 1960, mainly in the 1948–59 period. In 1938–39, the energy resource industry contributed $1.3 million towards the total provincial budget expenditure of $11.4 million. By 1948–49, these figures had grown to $14.6 and $46.5 million respectively, and by

1959–60, revenue from the energy sector's direct contribution to the provincial treasury equalled more than 50% of the provincial budget total, $143.3 of $228.2 million.

Such changes meant that the Conservation Board's mandate to ensure development of the province's oil and gas resources in the public's best interest had to be pursued in a much more complex environment. This was not just a function of the change in scale, which now required monitoring production in numerous fields as well as the activities of a vast array of companies new to the scene that, in many cases, were not used to firm regulatory supervision. There was another important though less tangible factor that can best be described as the spirit of the time. There is no question that the Alberta government was sincerely committed to the development of the province's oil and gas resources in the public interest and for the long-term advantage of Albertans, but this was, of course, from the perspective of its own experience. Like that of Albertans generally, this was from the perspective of hard times. For all but a few prosperous years in the late 1920s, the years since World War I had been differentiated only by the degree of difficulty and hardship. With the boom that began to build following the Leduc discovery, an old but long repressed feeling of optimism began to re-emerge. There seemed good reason to believe that at last the province might escape from dependence and find the illusive prosperity that had been the beacon of promise in the homestead era a generation before. It was a heady, but still cautious, enthusiasm that stood somewhat in awe of the province's newfound resource wealth. Their hopes for the future conditioned by their experiences in the past, Albertans were desperately eager that nothing hinder the development of an industry that brought so many jobs, contributed so handsomely to the provincial treasury, and underpinned so much hope for the future. This background mentality shaped all debate in Alberta regarding petroleum policy, and initiatives taken or not taken by the Conservation Board must also be understood within the context of this prevailing sentiment.

The other overriding factor that remained a shaping influence upon government policy and the Board's approach was the market, which within 18 months of the Leduc discovery became one of chronic oversupply. Development within the province and the search for expanded markets beyond it led ultimately to the creation of the National Energy Board in Ottawa and the subsequent formulation of Canada's first national oil policy in 1961. The central feature of the latter was the allocation of markets west of the Ottawa Valley to western Canadian oil. It was but one of the signals that the region's oil and gas industry was again entering a new phase.

The most important theme that runs through all these periods is conservation or, to state it another way, the search for maximum efficient but equitable production. Most often this was manifest in the clash between short- and long-term objectives, both public and private. It was in the arbitration of this ongoing contest, characteristic of oil-producing regions everywhere, that the Petroleum and Natural Gas Conservation Board played its essential role. In the competition of interests, it determined where the balance would rest. The significance of this contribution is properly measured with reference to the Board's legislated mandate. According to the 1938 and 1950 *Oil and Gas Resources Conservation* acts, the agency's principal responsibility was to ensure "the conservation of oil and gas resources" in the province. Given the unique physical properties of these resources, this responsibility necessarily meant the application of various measures to control the manner and the amount of production. Therefore, the legislation charged the Board with the linked responsibility of assuring "each owner the opportunity of obtaining his just and equitable share of the production of any pool."

The mandate, assigned by the Conservation acts, was the direct legacy of several far-seeing and dedicated individuals, in particular Charles Dingman. He witnessed firsthand the shifting combination of unchecked physical, market and personal forces that were destroying the Turner Valley oilfield. It was Dingman, with fellow public service engineers Charles Ross and William Calder, who largely bore the burden of the decade-and-a-half-long struggle to educate the small operators, the public and politicians on the basic principles of reservoir behaviour, and to convince them that the indiscriminate burning of "waste" gas in the production of naphtha was diminishing the field's ability to produce in the long run. It was mainly their efforts that led cautious Alberta politicians to accept the idea that it was necessary for the state to supervise oilfield development and production in the public interest. As the duration of the contest attests, it was no small achievement.

The Board's supervisory efforts were first concentrated upon the maintenance of reservoir energy and thus upon the elimination of the flaring of "waste" gas in the Turner Valley field. If production practice in Turner Valley never quite reached an ideal state, the Board managed nonetheless to solve in short order the problem that had plagued the valley for nearly two decades by imposing an acceptable balance between the difficult requirements of conservation and equitable division of market share. Oilfield gas production remained a matter of constant concern to the Board throughout the period. As the record shows, the Board vigilantly monitored gas production. It consistently prodded and ultimately, when such prodding failed, compelled operators to introduce oilfield conservation programs. To this end, the Board was constantly engaged in debate with the industry on the question of "What was economically reasonable?" On the crude oil side, the Board built and devoted its technical expertise to develop, with industry consultation, production practice to ensure maximum reservoir recovery. The formulas that it developed to balance equitable production sharing with uniform reservoir withdrawal and overall market demand were at least as good as, and usually superior to, those existing in other progressive oil-producing jurisdictions.

Far from being the price-fixing device of a de facto oil cartel, as is sometimes alleged, the Conservation Board's proration system was an indispensable tool of reservoir management required to achieve maximum long-term economic recovery. Finding hydrocarbon resources is not something that can be scheduled at so many barrels per day to balance real market demand in the future. Alberta, like jurisdictions in oil- and gas-producing regions elsewhere, determined rationally that it was in its own economic interest, and in the national interest, to maintain a healthy industry able to continue the search for these resources so that when they were required they could be produced. What was desired inevitably happened, at least in the period under discussion. In a region where there are adequate economic incentives to continue exploration, reserves and productive capacity will grow, possibly exceeding current market demand or alternatively creating new market demand beyond the existing marketing radius. Where productive capacity exceeds the available economic market, two courses of action are possible. One is to leave the industry unrestrained, in which case those companies with stronger marketing facilities will produce their quantities of oil up to any technical limit that might be imposed, assuming there is a regulatory authority of some kind to set such limits; however, companies with inferior marketing connections will be unable to market. The second course is to divide equitably the existing market and, in turn, to help maintain a broader-based exploration sector. Evidence that prorationing was not seen by the Board simply as a device to maintain price indirectly by holding production near to the level of market demand can be seen in the consistency of the Board's response in a situation where demand was in excess of production capacity. During World War II, the Conservation Board tried at first to operate the Turner Valley field at a rate below market demand and resisted pressure to allow the wells to produce at maximum capacity. Although it submitted to the federal government's declaration of an "emergency need" in the end, the Board persisted with its argument that accelerating production beyond each well's technically determined maximum efficient rate of production would compromise the field's oil recovery in the long term.

The Board's managed reservoir withdrawal also had an impact that went beyond determining the reservoir's ultimate production and life expectancy. Its proration system protected the small independent producers, that sector of the industry in which almost all Canadian companies were concentrated. The proration system assured these small producers that their legitimate share of the pool would not be "captured" by those producers who also owned refineries and pipeline systems.

The Board's supervision also expanded upon another important tradition begun by Dingman and his colleagues. From the beginning, they insisted that the industry keep accurate records. The Board understood that access to a thorough and reliable information base was essential to efficient nonwasteful development, and thus was an integral part of its mandate. Under the initial guidance of Floyd Beach, whose public service career also went back to the earlier days of Turner Valley, the Board began not just to collect but also to analyse and publish a broad range of statistical information. The assembly and provision of this information was of great value to industry, particularly to smaller companies that often had neither the in-house capability to collect nor the range of professional expertise to interpret such data. More than just a tool of scientific management, the collected data was also an essential component in the Board's ability to provide an independent technically sophisticated assessment of industry proposals at hearings. It functioned similarly as the information base upon which the Board drew to advise on government policy and to educate the public. The emphasis placed upon data gathering and processing at the Board could be one of the most important legacies of this period. Collected geological and reservoir production data relevant to the Alberta portion of the vast North American sedimentary basin is superior to that of any other part of the continental interior. Such information, gathered from thousands of wells and made readily available to the public, is a crucial factor in determining the initiation of secondary and tertiary recovery schemes, and thus maximum economic production.

To gain a sharp perspective upon the Board's overall contribution to oil and gas conservation in Alberta, it is necessary only to contemplate the continued development of the Turner Valley field, the heavy oilfields in east-central Alberta, or the Medicine Hat and Redcliff gas fields, without an effective regulatory presence. The ultimate result of the reservoir and surface degradation already apparent in these fields before the Board commenced supervision can be seen in the examples of numerous U.S. oil and gas fields where the deleterious influence of the rule of capture remained unrestrained. Although this offers the advantage of a stark and flattering comparison, it is not altogether appropriate, for no informed observer within or outside the industry since the late 1930s would argue that oil and gas field development should be left simply to the regulation of market forces. A better perspective is offered by measuring the Alberta Board and its approach against those of similar agencies elsewhere. By this measure, the Board also compares favourably. Starting a little later than similar U.S. bodies, the Board operated within a legislative framework that, with the guidance of Robert Hardwicke and other U.S. authorities, incorporated elements of the most progressive U.S. legislation. Nonetheless, by the late 1950s in a few areas, the Conservation Board had passed beyond relying on past U.S. experience. From an established tradition of a strong field presence, and an emphasis upon data collection and its own developing technical expertise, the Board moved on to conservation initiatives of its own. This was of great importance, because the foundation that was being built in Turner Valley and during the first decade of the post-Leduc oil boom would largely determine the manner in which the petroleum industry in Alberta would function as it moved into its mature phase.

The impact of the Board's presence is even more far-reaching than that already suggested. This is because its influence extended well beyond the conservation mandate assigned by the conservation acts. Almost from the moment it was created, the Board was called upon by the government for information and advice on petroleum policy. By the early 1950s, the Board had virtually assumed most of the advisory function that had formerly rested with the Petroleum and Natural Gas Division of the Department of Mines and Minerals. Under McKinnon's and Govier's capable guidance, the Board became, outside the cabinet itself, the most important influence upon the formulation of Alberta petroleum policy. This was a consequence of circumstance, not design. It had to do first with the important factor of McKinnon's dual responsibility; he was both the chairman of the Conservation Board and the deputy minister of the Department of Mines and Minerals. More important is that the Board had simply emerged as the government body with by far the greatest relevant expertise. It was an expertise that came both from the technical strength that Govier had built at the Board and from the Board's direct day-to-day working contact with industry in the field, in the Calgary office and at hearings.

Another of the Board's entirely circumstantial and important, though admittedly less tangible, contributions came from the distinctly Canadian character of its professional staff. In a period when Alberta's petroleum industry was rapidly falling under foreign, largely U.S., dominance, large foreign companies and their "experts" were compelled to argue their cases before an agency dominated by competent Canadian engineers and geologists, most of whom were educated at the University of Alberta. In an increasingly U.S. business environment, the Conservation Board was one of the few important bodies actively involved in the industry where a distinctly Canadian perspective remained. The Board's unique perspective also reached out into the industry in another important but unmeasurable way. In a critical period, it provided a unique public sector training environment for dozens of young Canadian engineers who later moved on to careers in the industry.

The broad-ranging character of the direct influence exerted by the Board upon conservation practice and upon the development of petroleum policy in Alberta, combined with the influence that it had upon the industry by virtue of its close ongoing working relationship, meant that the Conservation Board played a critical role in shaping the manner in which the industry developed in Alberta. In this sense, the Board must also be seen as one of the most important forces marking the course of the province's economic history in the postwar period.

It is not just the impact of the Board presence within Alberta that commands attention. The influence of conservation practice in Alberta extended also to Canada's other western oil- and gas-producing provinces, especially Saskatchewan and British Columbia. As the first, and by far the largest, producer of oil and gas, regulations and procedures set by Alberta established the standard and became the reference model for other Canadian jurisdictions. Although their regulations were presented under different administrative frameworks, each of the western provinces drew heavily upon Alberta's experience. For example, the extremely important definition of "waste" included in the British Columbia regulations, which came into force in 1958, was identical with that in the 1957 Alberta *Oil and Gas Conservation Act*.

The primary factors that account for the Conservation Board's progress and influence have structural and human dimensions that fall broadly into four areas: the legal-structural environment, the Turner Valley experience, the provincial political environment, and the quality of leadership exercised at the Board. Of particular importance in defining the legal-structural environment was the Board's founding statute. The structure, role and powers assigned to the Board gave it several advantages. Members of the Alberta Board were government appointed and given a wide-ranging mandate supported by near absolute authority. Unlike the legal tradition that governed the actions of its U.S. counterparts, the Alberta statute contained no provision for the appeal of Board decisions, except on questions of law. The deliberate decision of the Social Credit government to "court-proof" its Board produced a different history of oil and gas conservation, as even the quickest glance at the situation in Texas and Oklahoma will reveal. In the United States, almost every conservation initiative brought forth challenge and delay in the courts. Consequently, conservation practice in the United States has proceeded from a foundation of court decisions. Court challenge and delay are virtually unknown in Alberta, where conservation practice has had much more to do with the technical expertise at the Board than with legal expertise in court. Endowed with immense power, the Board was nonetheless cautious. Initially sensitive to industry and political objections to its largely unassailable authority, the Board was successful in cultivating the industry's reassurance through a formal commitment to continuous consultation and by its quick adoption of a hearings procedure that promoted broad industry and public involvement.

Also of immense benefit to the Board in the initiation of conservation measures, and in distinct contrast to the situation existing in the United States, was that mineral rights throughout the province were mainly in possession of the Crown. Hence, there were not thousands of farmers and other holders of subsurface rights clamouring for production allowances, and this minimized the prospect of broad public challenge to the Board's fiat. A further advantage was that by far the most important part of the oil- and gas-producing area in Canada was concentrated within the compass of the Alberta Board's jurisdiction, unlike the situation in the United States, where the primary producing area stretched across a number of state boundaries. Alberta did not face therefore the inherent problem of different and sometimes conflicting sets of regulations emanating from adjacent jurisdictions. Initiatives could be taken without fear that they might somehow tip the balance in favour of development in a neighbouring province. Together with all these advantages, it was the Board's good fortune to be spared administrative responsibility for a number of related areas where the potential for confrontation was high, particularly surface rights arbitration, natural gas utility pricing and the collection of royalties.

The Board's progress and success in the 1950s also owed much to the valuable experience that it gained in Turner Valley before the Leduc discovery. Turner Valley provided an almost ideal schooling for the novice Board. It contributed a complex geological and engineering environment and a small corps of hostile producers, and in the background stood a government preoccupied with other issues. All the usual problems associated with oil and gas conservation were there, but on a small scale compared to Oklahoma and Texas or Alberta after Leduc. It was the Board's good fortune to have nearly a decade in Turner Valley in which to gain basic technical experience and to develop administrative procedures.

The Board gained additional strength from the high level of political support upon which it could call and from the continuity of the political environment in which it functioned. The Social Credit party, supported by a large and usually massive majority in the Legislature, governed Alberta uninterruptedly during the period of the Board's history discussed in this book. The continuity that characterized petroleum and conservation policy in the province simplified the Board's task. A further aspect of continuity, also representative of the concentration of power within the Social Credit party that worked to the Board's advantage, was that from 1938 there were only two political figures of consequence in the Board's world. E.C. Manning and N.E. Tanner were at the pinnacle of power in the Social Credit government. Tanner was minister of lands and mines when the Board was formed. When he resigned from office in 1952, Premier Manning took over that portfolio. Both men had been involved in the Board's creation and both remained well informed on conservation legislation and issues. They also had high regard for the technical expertise of the Board they had created. Perhaps this had something to do with the technocratic bent of the original social credit theory they had learned from William Aberhart and C.H. Douglas. Whatever the reason, it meant that the Board could almost always rely upon the government's acceptance of the recommendations it put forward.

Although the Board's political world was framed by two politicians, its administrative world was shaped primarily by two civil servants, James Frawley and Ian McKinnon. Frawley, from the attorney general's department, was the public official most important to Board activities during its first decade, and was primarily responsible for drafting the Board's founding statute. He served as Board chairman for a time, and from behind the scene he watched over the evolution of conservation legislation and regulations. McKinnon, a senior member of Alberta's lean civil service establishment, was appointed Board chairman in February 1948 and guided the Board through its second decade. It is not just the small number of individuals in the Board's political and administrative firmament that accounts for the efficient and even character of its progress; this was also a consequence of the striking unity of values that distinguished the group. This handful of politicians and public servants viewed the world through similar eyes. They were cautious and austere men, both in their personal habits and as administrators. In their relationships, they were inclined to be formal, at the same time they were never ones to play upon rank or the superior status of their office. Integrity and hard work were attributes that they held in high regard, and all possessed a deep sense of public stewardship and responsibility. These were men who would not take a government pencil home from the office for personal use. Their sense of public stewardship, however, did not translate into a preference for public ownership. The proper role of government in their view was to guide, or provide a framework for, the development of the province's resources in the public interest. In addition to their shared convictions and responsibilities in the administration of the province's oil and gas resources, Manning, Tanner, Frawley and McKinnon were also good friends. More than being just another factor that explains continuity, this remarkable combination of men and shared views also meant that there was little threat that the Board's authority might be undermined for want of political support. Industry soon realized that little was to be gained by trying to outflank the Conservation Board by direct appeal to the premier or minister of mines and minerals. The other side of the coin, however, is that, given the common set of views, one does not see the Board pressing ahead with initiatives that were too much out of step with government thinking. It must be understood, however, that this was a function of common view more than any reluctance on the part of the Board to press for any measure or initiative that it deemed important.

No less important to the Board's overall success than the remarkable combination of individuals was the nature of the leadership and the body of professional staff developed at the Board. The contribution of the post-Leduc Board was crucial. Called together in February 1948, the new Board, perhaps as much by good fortune as design, proved to have the almost ideal balance of expertise to carry the Board into a new and much more demanding period in Alberta's petroleum history. The new chairman, who happened also to be the deputy minister of mines and minerals, brought with him the advantage just described. In addition, he contributed the methodical background of his accounting training, supplemented by a natural caution that was part of his Scottish upbringing. McKinnon was quiet, authoritative and inclined to run a "tight ship." Primarily, he built the Board's credibility as an agency that could be relied upon to consult and to provide a full fair hearing. Second on the Board in rank, but first in experience, was "Red" Goodall. Having been with the Board as a field engineer almost at its founding, Goodall was the first of the career professionals who would come to characterize the Board's technical staff and come to be one of the important factors accounting for its engineering competence. Goodall's laconic style masked a profound common sense, and his vital contribution was an ability to translate what was theoretically desirable or possible from an engineering standpoint into what was practically feasible in a field setting. Goodall built the Board's reputation in the field.

Joining the Board with McKinnon in February 1948 was Dr. George Govier, head of the Department of Chemical and Petroleum Engineering at the University of Alberta. Just like the new chairman, Govier carried with him the demanding responsibility of an existing full-time position, but Govier's university connection, like McKinnon's primary responsibility outside the Board, provided a link that worked to the Board's advantage. Having completed his doctoral study at the University of Michigan under two of the best men in the field, Govier brought to the Board a scientific background that was on the leading edge of current reservoir engineering research. Devoted to technical excellence and the principle of hard work, Govier set high standards for himself and expected the same from those with whom he worked or supervised. The impact of his presence at Board hearings was immediate and far-reaching. Industry, and particularly its representatives from farther afield, soon learned that Alberta, insofar as it was represented by the Conservation Board, was not a naïve backwater jurisdiction. Shoddy data, inferior analysis and technical obscurantism did not pass

Board scrutiny, and purveyors ran the risk of withering dismissal. Govier was responsible for building the Board's technical credibility. Along with his responsibilities as a Board member, Govier continued his role not only as a university administrator but also as an active publishing scientist, thereby lending further authority to his position at the Board. Articulation of the oil and gas conservation message to both industry and the public fell upon his shoulders.

Govier's work to establish the Board's technical reputation was of crucial importance, and it went well beyond the prestige lent by his individual stature as an active research scientist and his role at Board hearings. As the Board expanded to keep pace with the industry's rapid development after Leduc, Govier kept a watchful eye on the recruiting and development of the Board's professional staff until the mid-1950s. As a result, the Board was successful in hiring each year many of the best University of Alberta graduates. Most were engineers, and many were his former students. Govier sought well-prepared students who were attracted to a career in the technical administrative side of the industry and could identify with a public service responsibility. Anxious to earn the industry's technical respect, Govier made a special effort to ensure that his recruits could take in-service and outside courses to keep abreast of developing technology. To hold good people, he pushed hard and successfully for competitive professional salaries in excess of the normal levels set for those in the provincial civil service. In addition to retaining experienced professionals, this policy helped to reinforce an idea that the Board had tried to instil for many years—its staff was a unique body, neither civil servants nor private sector employees. Such efforts, combined with the dynamic growth of the industry and the Board's rush to stay apace, produced a stimulating, technically challenging working situation that had an important and lasting impact. This was an environment that helped to build high morale and the feeling that one was part of a team, making an important contribution to the province's development. Such positive identification with the Board and its role was a key factor that led a vital core of its recruits to resist alternative careers in the private sector and to become career public service professionals at the Board. Ultimately the Board's success depended heavily upon people like Frank Manyluk, Doug Craig and Vern Millard who built their professional careers at the Board through the 1950s.

Such career commitment was important to the building of the Board's technical expertise, and it also contributed to the emergence of a distinctive regulatory tone. Unlike other regulatory bodies, such as the Canadian Board of Railway Commissioners or the

Texas Railroad Commission, the Alberta Conservation Board was dominated by engineers rather than lawyers and politicians. The Conservation Board approached the world with an engineering cast of mind. Its primary focus was on "good engineering practice" in production and field management. Industry's focus was somewhat different; it was inclined to argue for field practice based upon "reasonable cost" and production practice that would enhance the individual company's market share. The emphasis on one side was engineering principles, on the other, cost factors. Typically, decisions reflected a compromise somewhere between these two positions. The quality of the compromise had a lot to do with the ability and commitment of Govier's professional staff.

The Board's overall approach to the industry in the post-Leduc period was a composite of the individual styles of the three Board members. It was cautious and pragmatic, but balanced by a clear sense of mandate. The Board knew where it was going and what it wanted to accomplish. Its style was to lead rather than to push or be pushed. McKinnon and his colleagues tried to educate and persuade from a foundation of technical evidence. They consulted and tried to build consensus. If the process was slow, it also tended to be unrelenting. Behind the Board's preference for persuasion lay an awesome and recognized legal authority that it was prepared to use as a last resort. For example, when the Board's patience was pushed to the limit after several years of discussion regarding the implementation of a residue gas conservation program for the prolific Redwater oilfield, it gave operators notice of the impending shut-in of hundreds of wells. In short order, the ultimatum produced an acceptable program for the processing of the field's "waste" gas.

The leadership style that evolved at the Conservation Board through the post-Leduc period is important in that it helped establish a distinctive and enduring agency culture. During the 1950s, the Board's public profile expanded beyond Calgary to gain province-wide recognition. This new public image derived from the regional and national significance of many of the Board's decisions, rather than from the personalities of its members. For industry, the Board's maturing image rested upon a growing perception of the Board as a guarantor of stability and as an open forum for argument and decisions based upon technical merit. Although it was gaining stature and respect, the Board triumvirate remained personally aloof from the public, from industry and, in some measure, from its own staff.

The distinctiveness of the Board's emerging character is more apparent when set against that of the Texas Railroad Commission.

Conservation regulation in Texas throughout the 1940s and 1950s was personified by the Commission's dominant voice, Colonel Ernest Thompson. A decorated World War I veteran, lawyer and former mayor of Amarillo, he was a charming self-promoter, an almost charismatic figure who was well attuned to the Texas social and political environment. Working upon a foundation of social contact with everyone of consequence in the industry, Thompson built regulatory consensus in the most difficult of settings. One scholar concludes that

> Ernest Thompson's private influence kept the Commission tranquil and supreme amid the constant public guerrilla warfare of the domestic petroleum industry. It is a fine point, therefore, whether the industry coopted Thompson or Thompson coopted the industry.[1]

The product of their own unique settings, the Alberta Conservation Board and the Texas Railroad Commission reveal two extremely different styles of consensus building and regulatory cultures.

By the end of the McKinnon period, Alberta had, for its day, a progressive program of oil and gas conservation. Against the difficult backdrop of a vast and growing surplus of productive capacity, the Board "cut its teeth" on prorationing and the careful monitoring of the province's natural gas requirements and reserves. Through a period of market stress, in the face of a powerful and dynamic industry and in tension with the still vibrant development ethos of a frontier region, the Board more than held its own. The record, of course, is not entirely unblemished; the Board's emphasis upon persuasion was slow to achieve results in some areas. In the case of the decision regarding the location of the pipeline for Alberta Gas Trunk Line (AGTL), the Board's technical rationale did not win the day. In this instance, the immediate overriding interest of development, which the Board was usually able to hold in check, broke through. Although in the AGTL example the political and legal circumstances that confronted the Board varied considerably from the norm, it underlines that the Board did not function in an ideal world. In varying degrees, it was confined by the limits of developing technology, its own limited experience until the late 1950s, and the values and politics of the time. Nonetheless, the Board remained upon an unswerving course and set an enviable standard that is apparent when measured against sister agencies in other oil-producing regions.

It is well known that after the Leduc discovery U.S. oil companies rapidly gained dominance in the Alberta petroleum sector. It is sometimes suggested that, since these companies were by far the most important source of investment capital, they largely wrote their own ticket as far as the development of the province's oil and gas resources were concerned. Although such an interpretation is undoubtedly simplistic and the ultimate merit of this argument must be discounted until such time as it is supported by a thorough study of the postwar petroleum industry in Alberta, it is clear that in one important area U.S. capital did not dictate. In the matter of conservation and production practice, U.S. companies found, and some were initially reluctant to accept, that the lax legislation and weak regulatory environment characteristic of many U.S. oil-producing states was not the case in Alberta. The Alberta Conservation Board was careful to cultivate and maintain good relations with the industry, but it is quite wrong to conclude from this, as the evidence clearly shows, that the Board was simply an agent of company interest.

By the late 1950s, the standard of oil and gas production in Alberta had come a long way from early Turner Valley days. By the time McKinnon's chairmanship drew to a close, the Board had emerged as a mature organization, led by a body of engineer-technocrats with a concentrated vision and clear sense of purpose. Moreover, there was an impressive legislative and regulatory framework in place, supported by a tradition of field supervision. Just how well this administrative structure would work, the Conservation Board's commitment as well as its technical competence would be tested severely in the 1960s, a decade of chronic and dramatically increasing surplus production capacity. It could be expected that the Board's technical competence would prove equal to the challenge; its technical expertise, after all, was the bedrock of its strength. Yet, this core strength contained a potential Achilles heel. The Board was the product of its time—the 1940s and 1950s. The ultimate question was how effectively would it manage to change with the times, not just changing technology? Time would tell, but there was a danger that the Board's vision might remain too focused, that by remaining professionally preoccupied with staying on the leading edge of reservoir engineering and petroleum technology there lay the danger of not being alert to the changing social and political environment. An engineering mind-set, like that centred upon any discipline, has its blind spots. Nonetheless, as McKinnon left for Ottawa to take up his responsibilities as the founding chairman of Canada's National Energy Board and as the mantle of leadership passed to Dr. George Govier in Calgary, Albertans had good reason to feel confident that their oil and gas fields were being capably managed in the public interest and that maximum economic recovery would be achieved.

Albertans had every reason to feel confident, because the essential components of sound conservation practice were in place. The story of how they had come to be in place has little to do with far-sighted politicians, it has little to do with industry leadership, and it does not have a lot to do with the public's clamour for reform. It is much more the story of a handful of committed public servants who worked hard for a long period to coax industry and government to recognize their obligation to ensure the responsible development of a vital, but easily wasted, resource.

Charged with managing Alberta's wondrous new resource, the Conservation Board was one of the more important agents of, and indeed the product of, change in Alberta. There are two great transition periods in the province's contemporary history. The first is the decade that begins at the turn of the century. Spurred by federal initiative and the pull of the wheat boom, these were the years of the homesteader and massive settlement. Provincehood came as a consequence, and the population's political energy was absorbed in creating the broad range of institutions and administrative structures required of modern government. The second is the decade that begins with the Leduc oil discovery. Dramatic development and population growth also mark this decade, and again the economy was propelled by distant forces and outside capital. The more obvious similarities, however, are less important than the differences that distinguish these historic turning points. The first transition period was directed largely by federal initiatives and institutions; but the second was governed mainly by provincial leadership and institutions, the most important of which was the Petroleum and Natural Gas Conservation Board. From this distance, it is apparent that McKinnon and his Board colleagues were helping to build the foundation for a new Alberta—a prosperous and confident Alberta that shared less and less in common with the prairie region and the rest of Canada. The wheat economy was general to the prairie region and, from the outset, an agent of East-West integration, whereas the petroleum economy was of consequence only in Alberta, but its outlook was continental. Just as the Conservation Board was not stimulated by the approach of Canadian sister agencies, the oil-centred business and technical class rising in Alberta had relatively weak links—professional and otherwise—with the rest of Canada. In short, the geography and unique characteristics of the petroleum industry continued, along with the Social Credit party, to remove Alberta further from the Canadian economic and political mainstream and to reinforce the broad sense of alienation that had developed through the preceding agrarian decades.

Board Chairmen,
Deputy Chairmen and Board Members,
1938–1962

TABLE APP.I.1 Board Chairmen, 1938–1962

Name	Appointed	Left Position	Comments
W.F. Knode	2 July 1938	15 July 1939	Consultant to Board July 1939–Nov 1940
R.E. Allen	18 Sept 1940	7 June 1941*	Returned to Washington and became Assistant Deputy Petroleum Administrator for Oil.
J.J. Frawley	26 May 1942	7 Sept 1943	Returned to Attorney General's Department, Edmonton
Dr. E.H. Boomer	7 Sept 1943	27 Oct 1945	Deceased
A.G. Bailey	15 Oct 1946	4 June 1947	To private industry
D.P. Goodall	12 June 1947	1 Feb 1948	Acting Chairman. To Deputy Chairman
I.N. McKinnon	1 Feb 1948	10 Aug 1962	On leave of absence Sept 1959 to Aug 1962 to chair National Energy Board, Ottawa. Continued NEB chairmanship to 1968.

SOURCE: *Alberta Gazette*, Orders-in-council 815/38, 1299/40, 761/42, 1417/43, 1865/46, 603/47, 112/48, 1336/62.
* Date last shown in the position in Board Minutes.

TABLE APP.I.2 Deputy Chairmen, 1938–1962

Name	Appointed	Left Position	Comments
C.W. Dingman	13 Dec 1938	4 Sept 1940*	Resigned
F.G. Cottle	25 Nov 1940	16 June 1941	To Ottawa, Oil Controller's Office
J.J. Frawley	22 July 1941	26 May 1942	To Chairman
G.W. Northfield	7 Sept 1943	28 Mar 1944	To Income Tax Department
A.G. Bailey	7 July 1944	15 Oct 1946	To Chairman
D.P. Goodall	6 Feb 1948	22 Sept 1959	To Board Senior Adviser
Dr. G.W. Govier	22 Sept 1959	15 Aug 1962	To Chairman
A.F. Manyluk	22 Sept 1962	30 June 1971	To Board Senior Adviser

SOURCE: *Alberta Gazette*, Orders-in-council, 1547/38, 1590/40, 1022/41, 1417/43, 1043/44, 137/48, 1464/59, 1337/62.
* Date last shown in the position in Board Minutes.

TABLE APP.I.3   Board Members, 1938–1962*

Name	Appointed	Left Position	Comments
C.W. Dingman	2 July 1938	13 Dec 1938	To Deputy Chairman
F.G. Cottle	2 July 1938	25 Nov 1940	To Deputy Chairman
J.J. Frawley	18 Sept 1940	22 July 1941	To Deputy Chairman
M.D. Kemp	16 June 1941		Acting member
	1 Oct 1942	30 June 1944	Resigned
J.W. Kraft	16 June 1941	30 Sept 1941	Resigned
D.P. Goodall	30 June 1944	12 June 1947	To Acting Chairman
Dr. G.W. Govier	2 Feb 1948	20 Sept 1959	To Deputy Chairman
A.F. Manyluk	22 Sept 1959	1 Sept 1962	To Deputy Chairman
V. Millard	9 Sept 1959		Acting member
	1 Sept 1962	1 June 1971	To Vice-chairman

SOURCE: *Alberta Gazette,* Orders-in-council, 815/38 1249/40, 817/41, 1564/42, 818/41, 992/44, 137/48, 1464/59, 1337/62.
* There were also a number of members appointed pro tem who served for varying periods, some for a few months and others for a meeting or two. The above list does not reflect these arrangements.

Turner Valley Oil and
Gas Production and Disposition,
1922–1960

TABLE APP.II.I Turner Valley Oil and Gas Production and Disposition, 1922–1960, Quantities-Gas-Mscf Oil-bbl

	Raw Gas Production				GAS BOW ISLAND STORAGE
YEAR	SHALLOW WELLS	LIMESTONE WELLS	TOTAL	GAS SALES	
1922	902,699		902,699	791,686	
1923	1,217,361		1,217,361	1,101,476	
1924	1,541,830	1,275,000	2,816,830	1,324,376	
1925	758,035	6,205,000	6,963,035	797,173	
1926	378,782	6,788,000	7,166,782	3,299,614	
1927	432,952	12,719,500	13,152,452	4,437,430	
1928	727,124	20,804,829	21,521,953	5,394,403	
1929	1,143,081	67,490,028	68,633,109	7,109,451	
1930	150,531	125,248,430	125,398,961	7,550,814	715,684
1931	142,413	168,527,549	168,669,962	6,897,801	1,768,358
1932	241,185	111,373,043	111,614,228	7,047,919	1,626,460
1933	290,744	94,958,366	95,249,110	6,910,152	1,727,987
1934	203,493	91,738,660	91,942,153	6,515,378	1,714,925
1935	381,731	88,581,007	88,962,738	7,252,842	1,578,045
1936	441,440	88,032,888	88,474,328	7,802,049	1,519,222
1937	213,690	84,707,143	84,920,833	7,932,493	1,514,955
1938	186,247	71,924,487	72,110,734	7,608,019	1,482,669
1939	41,728	45,072,857	45,114,585	7,572,884	164,629
1940		47,888,718	47,888,718	8,121,645	
1941		53,847,932	53,847,932	9,185,717	
1942		47,260,390	47,260,390	12,004,838	
1943	45,789	44,194,402	44,240,191	13,797,503	
1944	42,840	41,335,827	41,378,667	15,283,650	
1945	26,000	37,643,096	37,669,096	16,010,296	751,392
1946	14,500	35,617,205	35,631,705	15,202,095	938,603
1947	1,200	36,071,282	36,072,482	16,648,620	851,904
1948		37,747,233	37,747,233	18,386,520	351,930
1949		37,082,519	37,082,519	19,841,344	658,329
1950	447,385	38,022,910	38,470,295	22,894,719	995,771
1951	794,340	34,926,321	35,720,661	21,567,464	1,104,606
1952	641,168	31,427,917	32,069,085	18,610,595	1,347,836
1953	342,478	29,908,279	30,250,757	18,527,147	1,503,756
1954	384,962	29,436,884	29,821,846	26,291,716	
1955	324,240	31,643,360	31,967,600	26,485,370	
1956	406,585	29,011,919	30,569,100	25,159,108	
1957	345,134	26,804,638	37,149,772	23,857,401	
1958	295,078	24,078,856	24,373,934		
1959	254,099	25,721,331	25,975,430		
1960	20,251	10,954,283	10,974,534		
Total	13,781,115	1,816,072,089	1,829,853,204	425,231,708	

SOURCE: ERCB

GAS TURNER VALLEY STORAGE	GAS FIELD & PLANT USE	GAS WASTE ETC.	Oil Production		
			LIMESTONE	SHALLOW TOTAL	LIMESTONE SHALLOW TOTAL
		111,013			
		115,885			
		1,492,454	1,689	2,932	4,621
		6,165,862	169,008	2,926	171,934
		3,867,168	203,725	2,609	206,634
	909,892	7,805,130	284,595	38,808	323,403
	1,999,287	14,138,263	410,448	70,910	481,358
	9,157,707	52,365,951	908,411	73,181	981,592
	8,502,000	108,630,463	1,316,102	50,896	1,366,999
	3,811,280	156,192,523	1,345,310	26,936	1,372,246
	2,240,694	100,699,155	854,517	21,757	876,274
	2,572,532	84,028,439	766,755	23,915	790,617
	2,797,940	80,913,910	796,140	22,307	818,447
	2,406,791	77,725,060	711,451	18,903	730,354
	3,284,561	75,868,496	671,948	13,011	684,959
	10,659,499	64,913,886	2,098,970	10,589	2,109,559
	11,196,876	51,823,170	6,150,512	9,192	6,159,704
	9,161,220	28,215,852	7,251,063	8,431	7,259,494
	11,094,196	28,672,877	8,173,016	7,309	8,180,325
	10,813,564	33,848,651	9,531,207	6,014	9,537,221
	11,268,118	23,987,434	9,695,913	5,806	9,701,719
	9,510,652	20,932,036	8,986,663	4,865	8,991,528
	7,835,387	18,259,630	7,874,919	3,209	7,878,128
3,053,089	6,975,510	10,878,809	7,005,589	3,932	7,009,521
4,457,588	4,786,758	10,246,661	5,928,444	8,918	5,937,362
3,730,444	5,782,993	9,058,521	5,017,292	5,058	5,022,350
3,135,821	6,143,947	9,729,015	4,428,688	3,396	4,432,084
2,049,568	4,763,059	9,770,219	3,825,345	1,198	3,826,543
1,286,826	3,669,422	9,623,557	3,340,894	3,113	3,344,007
1,417,179	3,639,890	7,991,522	2,948,633	3,674	2,952,307
1,427,196	3,162,985	7,520,473	2,652,299	2,708	2,655,007
731,244	2,563,893	6,924,717	2,402,396	2,571	2,404,967
	2,690,204	2,690,204	2,135,799	2,108	2,135,799
	3,219,338	1,070,376	2,054,840	1,599	2,054,840
	3,096,700	780,027	1,775,523	870	1,775,523
	3,537,726	245,355	1,594,220	862	1,594,220
			1,445,621	846	1,446,467
			1,337,613	791	1,338,404
			1,199,199	645	1,199,844
	173,254,621	1,127,302,764	117,294,757	466,795	117,761,552

Petroleum and Natural Gas Conservation Board (PNGCB) Summary of Expenditure and Revenue, 1938–1958

TABLE APP.III.1 Petroleum and Natural Gas Conservation Board (PNGCB) Summary of Expenditure and Revenue, 1938–1958

	1938/1939	1939/1940	1940/1941	1941/1942	1942/1943
Expenditure					
Special Organization Costs	6,503	—	—	—	—
Salaries	23,783	21,781	25,379	22,831	27,779
General Operating	10,128	9,796	13,055	18,914	26,925
Capital	9,743	2,243	5,521	2,642	2,520
Total	50,157	33,820	43,955	44,387	57,224
Revenue					
Industry	50,157	33,820	43,955	44,387	57,224
Government	—	—	—	—	—
Sundry	—	—	—	—	—
Total	50,157	33,820	43,955	44,387	57,224

	1949/1950	9 mo. to March 31 1950/1951	1951/1952	1952/1953	1953/1954
Expenditure					
Special Organization Cost	—	—	—	—	—
Salaries	144,796	120,513	205,146	289,600	389,652
General Operating	80,003	63,617	103,207	155,712	178,252
Capital	27,326	6,812	48,874	583	44,490
Total	252,125	190,942	357,227	504,295	612,394
Revenue					
Industry	184,316	104,773	196,537	277,174	331,469
Government	52,234**	70,329	131,432	185,220	221,191
Sundry	15,575	15,840	29,258	41,901	59,734
Total	252,125	190,942	357,227	504,295	612,394

SOURCE: ERCB

* Extraordinary capital purchases in this year were:
  (a) A residential property in Calgary.   Value   6,100
  (b) A well in Turner Valley-Carleton Royalties No. 1.   Value   7,500
                                                 13,600

** Government contribution to revenue:
  (a) From 1939 to 1949, the government made no direct contribution.
  (b) From 1950 to 1970, the government made a 40% contribution.

1943/1944	1944/1945	1945/1946	1946/1947	1947/1948 June 30	1948/1949
—	—	—	—	—	—
30,058	34,615	38,957	37,580	44,738	71,512
19,246	21,196	22,620	21,607	27,430	44,575
2,338	717	15,124*	789	10,223	22,000
51,642	56,528	76,701	59,976	82,391	138,087
51,642	54,992	75,212	58,217	78,196	126,833
—	—	—	—	—	—
—	1,536	1,489	1,759	4,195	11,254
51,642	56,528	76,701	59,976	82,391	138,087

1954/1955	1955/1956	1956/1957	March 31 1957/1958	1958/1959
—	—	—	—	—
482,490	589,329	727,373	874,540	995,284
216,611	239,446	281,488	308,046	331,334
63,270	41,190	83,083	122,260	58,133
762,371	869,965	1,091,944	1,304,846	1,384,751
419,819	476,042	600,939	721,599	768,968
280,017	317,728	401,621	480,998	512,420
62,535	76,195	89,384	102,249	103,363
762,371	869,965	1,091,944	1,304,846	1,384,751

The Petroleum and
Natural Gas Conservation Board:
Order No. 63

THE OIL AND GAS RESOURCES CONSERVATION ACT
THE PETROLEUM AND NATURAL GAS CONSERVATION BOARD
ORDER NO. 63
THE CONSERVATION OF PETROLEUM RESOURCES

WHEREAS, Conservation as applied to petroleum and natural gas resources means, "efficient production without waste for beneficial use", and

WHEREAS, effective petroleum and natural gas conservation requires:

1. The protection of the correlative rights of the common owners of each single source of supply so as to insure ratable withdrawals and realizations therefrom.
2. The prevention of both underground and surface waste of petroleum, natural gas, reservoir energy, capital, and labor so as to insure the maximum economic recovery of natural resources of each reservoir.
3. The prevention of such acts and practices as might endanger life or property or be inimical and contrary to the public interest; and

WHEREAS, the successful achievement of equitable conservation depends largely upon the general application of a policy involving:

1. The adoption and use of the optimum rate of production as the basis for determining the proper production quota of each pool.
2. The use of reservoir pressure in conjunction with acreage, productivity and flow efficiency as determinants of individual well quotas that will assure ratable withdrawals.
3. The consistent maintenance of reservoir energy and the restoration of energy to the reservoir when necessary.
4. The development of additional reserves by drilling at such a rate that they become available at a rate substantially equal to the rate of depletion of older reserves.
5. The use of market demand quotas and proration thereto in normal times only when the optimum rate of production exceeds the market demand.
6. The application of a system of checks and balances to prevent economic disturbances within the industry and to stabilize the flow of capital to the industry, and

WHEREAS, the public interest in the present and future petroleum industry of Alberta requires the economic stabilization of that industry on a long-term basis, and

WHEREAS, the Petroleum and Natural Gas Conservation Board has made an exhaustive study of the aforementioned principles of conservation and has given due consideration to the effect that the application of such principles may have on the economic recovery of the petroleum and natural gas of Alberta.

NOW, THEREFORE, by virtue of the authority and powers conferred upon The Petroleum and Natural Gas Conservation Board by the Oil and Gas Resources Conservation Act of November 22, 1938, the said Board doth hereby order as follows:

The development and production of the petroleum and natural gas resources of Alberta shall henceforth be conducted in such a manner that:

1. The drilling of unnecessary wells and the expenditure of unnecessary capital may be avoided.
2. The maximum economic recovery of oil and gas may be obtained.
3. The reservoir energy of the field may be most efficiently used and effectively maintained.
4. The oil and gas resources of each pool and the revenue therefrom may be individually or collectively recovered ratably by each property.
5. The production of oil and gas from each pool may be maintained as nearly as possible at that constant rate or percentage of its indicated ultimate production as the reservoir characteristics of the pool may require for maximum economic recovery within the adopted limits of tolerance.
6. The production of oil from each pool at a uniform monthly rate may be maintained so as to avoid the effect of seasonal fluctuation in demand by running oil to storage during lulls in demand and withdrawing it during the season of peak demand.
7. The known petroleum reserves of Alberta may not be depleted at a rate exceeding the rate at which new reserves are discovered or made available unless production of known reserves at the optimum rate is required by the economic condition of the Province.

MADE at the City of Calgary, in the Province of Alberta, this seventh day of June, A.D., 1941.

THE PETROLEUM AND NATURAL GAS CONSERVATION BOARD,

R. E. Allen,
Chairman.

Province of Alberta, Development and Exploratory Wells Drilled or Completed, 1900–1959

TABLE APP.V.1   Province of Alberta, Development and Exploratory Wells Drilled or Completed, 1900–1959

YEAR	Total		Development		Exploratory		Exploratory %	
	WELLS	FEET	WELLS	FEET	WELLS	FEET	WELLS	FEET
1900	0	0	0	0	0	0	—	—
1901	0	0	0	0	0	0	—	—
1902	3	3,424	0	0	3	3,424	100	100
1903	0	0	0	0	0	0	—	—
1904	6	4,430	2	1,959	4	2,472	67	56
1905	0	0	0	0	0	0	—	—
1906	1	2,219	0	0	1	2,219	100	100
1907	6	7,630	1	1,199	5	6,431	83	84
1908	3	4,760	1	988	2	3,772	67	79
1909	7	10,029	3	2,740	4	7,289	57	73
1910	6	7,540	2	1,332	4	6,207	67	82
1911	11	19,704	9	15,532	2	4,172	18	21
1912	17	29,508	13	23,370	4	6,138	24	21
1913	26	39,347	18	22,687	8	16,660	31	42
1914	26	40,878	11	15,835	15	25,043	58	61
1915	21	42,794	4	9,879	17	32,915	81	77
1916	12	26,803	4	8,215	8	18,588	67	69
1917	18	36,457	17	35,659	1	798	6	2
1918	16	30,921	9	19,230	7	11,691	44	38
1919	4	6,367	0	0	4	6,367	100	100
1920	14	20,288	7	10,292	7	9,996	50	49
1921	11	19,927	2	1,826	9	18,101	82	91
1922	9	14,321	1	714	8	13,517	89	95
1923	23	48,999	9	18,916	14	30,083	61	61
1924	12	19,904	10	16,347	2	3,557	17	18
1925	12	25,615	6	10,141	6	15,474	50	60
1926	20	54,663	12	33,648	8	21,015	40	38
1927	32	98,727	19	56,130	13	42,597	41	43
1928	34	107,124	20	68,013	14	39,111	41	37
1929	65	179,969	48	146,065	17	33,904	26	19

1930	107	345,129	70	292,637	37	102,492	35	26
1931	45	178,648	37	146,226	8	32,422	18	18
1932	19	64,377	9	36,289	10	28,088	53	44
1933	15	57,815	8	37,159	7	20,655	47	36
1934	18	67,409	16	62,650	2	4,758	11	7
1935	11	47,880	10	45,756	1	2,124	9	4
1936	19	82,576	10	56,594	9	25,982	47	31
1937	45	265,926	30	199,850	15	66,076	33	25
1938	60	349,102	45	302,694	15	46,407	25	13
1939	59	359,253	40	289,105	19	70,148	32	20
1940	68	401,982	45	306,096	23	95,886	34	24
1941	87	473,318	70	414,389	17	58,929	20	12
1942	99	488,437	53	301,995	46	186,442	46	38
1943	122	488,372	90	378,596	32	109,776	26	22
1944	145	645,942	87	459,522	58	186,420	40	29
1945	130	519,751	77	297,945	53	221,806	41	43
1946	128	487,398	83	307,970	45	179,428	35	37
1947	224	766,608	155	532,421	69	234,187	31	31
1948	378	1,622,676	255	1,107,875	123	514,799	33	32
1949	798	3,249,674	590	2,289,851	208	959,823	26	30
1950	1,044	4,167,125	856	3,261,683	188	905,442	18	22
1951	1,268	5,395,263	850	3,532,297	418	1,862,966	33	35
1952	1,662	6,996,151	1,107	4,604,970	555	2,391,181	33	34
1953	1,410	6,407,458	947	4,326,262	463	2,081,196	33	32
1954	1,185	5,673,773	779	3,654,228	406	2,019,545	34	36
1955	1,625	8,445,885	1,220	6,184,221	405	2,261,664	25	27
1956	1,890	10,044,714	1,423	7,460,409	467	2,584,305	25	26
1957	1,430	7,521,871	885	4,650,999	545	2,870,872	38	38
1958	1,667	9,046,731	1,170	6,333,123	497	2,713,608	30	30
1959	1,601	8,836,375	1,060	5,895,185	541	2,941,990	34	33
Total	17,774	84,154,716	12,305	58,289,753	5,469	116,160,197		

SOURCE: ERCB

APPENDIX VI

Summary Tables

TABLE APP.VI.I   Annual Production

Year	Crude Oil* & Equivalent 10^3 bbl.	Natural Gas MMcf	Propane 10^3 bbl.	Butanes 10^3 bbl.	Sulphur 10^3 short tons
1938	6,743	76,165	—	—	—
1943	10,295	57,213	—	—	—
1948	10,974	55,744	8	—	—
1953	77,304	118,506	433	198	18
1958	113,544	286,716	1,058	663	122

* Includes field condensate and pentanes plus.

TABLE APP.VI.2   Annual Disposition, 1938–1958

	1938	*1943*	*1948*	*1953*	*1958*
			CRUDE OIL AND EQUIVALENT 10^3 bbl.		
Alberta	5,770	9,140	7,719	21,454	23,221
Other Canada	—	—	3,092	46,031	75,615
U.S.	—	—	—	2,192	13,479
Offshore	—	—	—	—	—
Total	5,770	9,140	10,811	69,677	112,315
			NATURAL GAS MMcf		
Alberta	8,034	36,084	63,759	66,567	124,009
Other Canada	—	—	—	457	32,425
U.S.	—	—	—	10,059	43,488
Offshore	—	—	—	—	—
Total	8,034	36,084	63,759	77,083	199,922
			PROPANE 10^3 bbl.		
Alberta	—	—	—	n.a.	728
Other Canada	—	—	—	n.a.	224
U.S.	—	—	—	n.a.	23
Offshore	—	—	—	n.a.	—
Total				n.a.	975
			BUTANES 10^3 bbl.		
Alberta	—	—	—	n.a.	482
Other Canada	—	—	—	n.a.	—
U.S.	—	—	—	n.a.	135
Offshore	—	—	—	n.a.	—
Total				n.a.	617
			SULPHUR 10^3 SHORT TONS		
Alberta	—	—	—	16	92
Other Canada	—	—	—	—	—
U.S.	—	—	—	—	—
Offshore	—	—	—	—	—
Total				16	92

TABLE APP.VI.3   Oil and Gas Wells Completions in Year Shown

Year	Oil	Gas	Capped Gas
1938	37	1	—
1943	59	16	—
1948	217	22	—
1953	885	53	142
1958	959	12	215

TABLE APP.VI.4   Cumulative Summary of Total Wells Drilled,
Producing or Capped, 1938–1958

Year	Total No. Wells Drilled	Producing*		Capped**
		OIL	GAS	GAS
1938	754	195	99	—
1943	1,189	281	—	—
1948	2,194	1,424	199	—
1953	8,376	4,000	272	393
1958	16,173	7,811	575	871

* In actual operation; capable somewhat higher.
** No record of capped gas wells for 1948 and previous years.

TABLE APP.VI.5   Cumulative Summary of Gas Plants Operated in
the Province, 1938–1958

Year	No.
1938	3
1943	4
1948	3
1953	5
1958	17

Alberta Milestone
Oil and Gas Discoveries,
1897–1959

TABLE APP.VII.1   Alberta Milestone Oil and Gas Discoveries, 1897–1959

Year Discovered	Name and Location of Field/Pool Discovered	Productive Zone	Hydrocarbon
1897	Pelican 79–24W4	Viking	gas
1904	Medicine Hat 13–3W4	Medicine Hat	gas
1909	Bow Island 11–11W4	Bow Island	gas
1913	Turner Valley 20–3W5	Blairmore	gas
1914	Turner Valley	Blairmore	light oil
	Viking-Kinsella 47–10W4	Viking	gas
1924	Wainwright 45–6W4	Sparky	heavy oil
1927	Skiff 51–14W4	Sawtooth	heavy oil
1936	Turner Valley 20–3W5	Rundle	light oil & gas
1940	Lloydminster* 50–1W4	Sparky	heavy oil
	Princess 20–12W4	Arcs ("Jefferson")	medium oil
1942	Pouce Coupé 80–12W6	Cadotte	gas
	Taber** 9–17W4	Taber SS	heavy oil
1947	Leduc-Woodbend 50–26W4	Nisku & Leduc	light oil
1949	Joarcam 48–21W4	Viking	light oil
1951	Okotoks 21–28W4	Wabamun	gas
1952	Minnehik-Buck Lake 46–6W5	Pekisko	gas
	Sturgeon Lake 71–23W5	Leduc	light oil
1953	Pembina 48–7W5	Cardium	light oil
1954	Pembina	Belly River	light oil
1955	Elmworth*** 70–11W6	Falher	gas
	Sundre 34–5W5	Elkton	light oil
1956	Bellshill Lake 41–12W4	L. Blairmore	light oil
	Medicine River 39–3W5	Nordegg	medium oil
1957	Swan Hills 68–10W5	Beaverhill Lake	light oil
1958	Red Earth 88–8W5	Slave Point & Granite Wash	light oil
1959	Bigstone 61–22W5	Dunvegan	gas
	Waterton 4–1W5	Rundle & Wabamun	gas

SOURCE: ERCB, J.R. Pow
* Preceded by Wainwright and Vermilion in the same general area.
** Preceded by the similar but much smaller Skiff field, discovered in 1927 in 5–14W4.
*** Significance not generally recognized until the 1970s.

Alberta Revenue from
Energy Resources and
Total Provincial Expenditure,
1938–1961

TABLE APP.VIII.1   Alberta Revenue from Energy Resources and Total Provincial
Expenditure, 1938–1961 Fiscal Years Ending 31 March

Year	Mineral Taxation	Coal Royalty	Other Coal Revenue	Petroleum & Natural Gas Rentals, etc.
1938	47,524	158,227	114,985	284,844
1939	47,928	143,071	111,488	372,450
1940	47,729	168,803	117,550	340,688
1941	47,969	200,046	90,649	229,845
1942	75,121	216,778	85,028	189,437
1943	94,365	235,613	84,298	267,612
1944	102,075	239,591	93,413	239,859
1945	92,783	251,137	118,189	598,752
1946	160,126	260,145	155,677	550,340
1947	356,837	286,487	114,417	272,917
1948	473,835	261,555	123,585	707,501
1949	566,069	600,424	181,451	2,150,560
1950	751,008	600,494	185,412	5,741,112
1951	833,603	577,668	174,035	9,786,607*
1952	769,800	394,441	211,583	15,384,311*
1953	886,161	354,124	170,287	19,688,233*
1954	1,002,261	268,694	142,279	24,604,468*
1955	1,311,232	216,147	49,934	20,656,827*
1956	1,297,213	201,607	50,133	21,372,058*
1957	1,344,366	151,791	66,416	25,821,687*
1958	1,515,909	102,602	75,002	31,045,880*
1959	1,422,130	95,187	58,408	29,836,190*
1960	1,247,798	79,088	57,258	32,684,111*
1961	1,261,064	55,686	48,637	30,901,572*
Total	15,754,906	6,119,406	2,680,114	273,727,861

SOURCE: Province of Alberta, Public Accounts.
* Includes royalties, leases, rentals, etc., on school lands.

Petroleum & Natural Gas Royalties	Sale of Crown Leases	Bituminous and Oil-sands Fees & Rentals	Net Revenue	Provincial Expenditure
249,267*			854,847	21,866,370
654,521*			1,329,458	16,758,653
664,599*			1,339,369	18,075,821
639,335*			1,207,844	17,063,349
799,789*			1,366,153	17,783,749
761,497*			1,443,385	18,751,025
718,351*			1,393,289	20,119,494
935,145*			1,996,006	22,547,439
741,730*			1,868,018	25,079,005
746,654*			1,777,312	31,939,342
1,059,696*			2,626,172	36,989,598
2,418,004*	8,720,507		14,637,015	45,495,102
3,662,503*	23,180,999		34,121,528	53,009,208
5,238,860*	29,080,632		45,691,405	61,838,557
11,081,059*	13,211,289		41,052,483	73,220,037
13,536,321*	23,527,444		58,162,570	77,234,730
18,583,001*	53,236,117		97,836,820	92,029,485
20,229,339*	40,013,320		82,476,799	109,223,931
28,810,106*	76,074,733		127,805,850	132,028,521
37,305,858*	69,050,860	89,879	133,830,857	141,593,597
32,818,756*	58,641,836	479,048	124,679,033	173,325,452
25,450,994*	53,609,952	363,416	110,836,277	205,934,737
27,338,359*	81,322,977	544,076	143,273,667	228,156,832
27,758,171*	44,127,513	663,393	104,816,036	253,966,311
262,201,915	573,798,179	2,139,812	1,136,422,193	1,894,030,345

Board Staff Distribution and Professional Recruitment

TABLE APP.IX.I   Conservation Board Office and Field Staff, 1948 and 1958

	Sept 1948 No.	Sept 1958 No.
Calgary Office Staff		
Engineers	3	24
Engineering technicians	—	10
Other assistants	—	9
Geologists	1	7
Geological technicians	1	2
Sampling laboratory	1	6
Other assistants	—	1
Chemist*	1	1
Assistants	1	4
Lawyers	—	1
Stenographer	—	1
Accountant	1	1
Clerks and others	1	7
Assessment group	—	7
Statistician	1	1
Economists	—	3
Clerks and others	—	17
Well records supervisor	—	1
Clerks	1	5
Administration	—	1
Secretary to the Board	1	—
Other	8	37
Total	21	146
Field staff		
Engineers	4	26
Field technicians	5	24
Clerk-typists	—	10
Total	9	60
Total Staff		
Calgary office	22	141
Edmonton	—	5
Field	9	60
Total	31	206

SOURCE: ERCB
* Moved to the University of Alberta in Edmonton in 1953

TABLE APP.IX.2   Engineers and Geologists Hired by the Conservation Board, 1948–1955

Engineer or Geologist	Type	University	Name	Date Hired	Position-1955	Career with Board
Engineer	Chem.	Alberta	Bailey, H.	1954	Calgary Office Engineer	Yes, Ret'd
Engineer	Pet.	Alberta	Bellows, L.A.	1952	Engineer-in-charge, Black Diamond	Yes, Ret'd
Engineer	Mech.	Sask	Berlie, E.M.	1951	Assistant Gas Engineer, Calgary	No
Engineer	Chem.	Alberta	Blackadar, M.R.	1949	Southern District Engineer	Yes, Ret'd
Engineer	Pet.	Alberta	Bohme, V.E.	1950	Assistant Northern District Engineer	Yes, Ret'd
Engineer	Mech.	Manitoba	Brown, R.H.	1952	Engineer-in-charge, Stettler	No
Engineer	Chem.	Alberta	Brushett, E.R.	1954	Field Engineer, Lloydminster	Yes, Ret'd
Engineer	Chem.	Alberta	Coffin, H.A.	1952	Engineer-in-charge, Drayton Valley	No
Engineer	Chem.	Alberta	Craig, D.R.	1949	Chief Reservoir Engineer	Yes, Ret'd
Geologist		Alberta	Crockford, M.B.	1949	Geologist, left Board by 1955	No
Engineer		Alberta	Crowe, J.A.	1953	Calgary Office Engineer	No
Engineer	Pet.	Alberta	DeSorcy, G.J.	1955	Field Engineer, Drayton Valley	Yes
Engineer	Chem.	Alberta	Edgecombe, R.W.	1951	Calgary Office Engineer	Yes, Ret'd
Geologist		Alberta	Ellison, A.H.	1949	Geologist, left Board by 1955	No
Engineer	Pet.	Alberta	Foo, E.M.	1951	Gas Engineer	No
Geologist		Alberta	Fuglem, M.	1952	Geologist	No
Engineer	Chem.	Alberta	Greenwood, P.M.	1951	Engineer-in-charge, Devon	No
Engineer	Chem.	Alberta	Horte, V.L.	1950	Engineer, left Board by 1955	No
Engineer	Chem.	Alberta	Jenkins, J.R.	1954	Gas Engineer	No, to NEB
Engineer	Chem.	Alberta	Kidd, R.K.	1954	Calgary Office Engineer	No
Engineer	Chem.	Alberta	Lashuk, N.J.	1953	Field Engineer, Black Diamond	No
Engineer	Pet.	Alberta	Larbalestier, P.D.	1952	Engineer-in-charge, Redwater	Yes, Dec'd
Engineer	Chem.	Alberta	Matthews, M.W.T.	1952	Assistant Reservoir Engineer	No
Engineer	Chem.	Alberta	Mazurek, L.A.	1953	Engineer-in-charge, Lloydminster	Yes
Engineer	Pet.	Alberta	McEachern, F.D.	1952	Calgary Office Engineer	No
Engineer	Chem.	Alberta	Meisner, Geo.	1952	Engineer-in-charge, Camrose	No
Engineer	Mining	B.C.	Moore, E.A.	1953	Engineer-in-charge, Medicine Hat	No, to DIAND
Engineer	Chem.	Alberta	Morrison, A.G.	1953	Calgary Office Engineer	Yes, Dec'd
Engineer	Chem.	Alberta	Morin, E.J.	1955	Field Engineer, Stettler	Yes, Ret'd
Engineer	Chem.	Alberta	Nickoloff, G.D.	1949	Engineer, left Board by 1955	No
Engineer	Chem.	Alberta	Pawelek, J.	1954	Field Engineer, Drayton Valley	No
Geologist	Eng.	B.C.	Phillips, F.	1952	Geologist	Yes, Ret'd
Engineer	Pet.	Oklahoma	Podmaroff, P.	1950	Engineer, left Board by 1955	No
Geologist	Eng.	Sask	Pow, J.R.	1951	Chief Geologist	Yes, Ret'd
Engineer	Pet.	Alberta	Peterson, R.M.	1955	Reservoir Engineer	No
Engineer	Chem.	B.C.	Pletcher, J.H.	1955	Field Engineer, Medicine Hat	No
Engineer	Pet.	Alberta	Richman, C.D.	1952	Field Engineer, Devon	No
Engineer	Chem.	Alberta	Schmaltz, S.A.	1955	Field Engineer, Stettler	No
Engineer	Chem.	Alberta	Stabback, J.G.	1949	Chief Gas Engineer	No, to NEB
Geologist		Alberta	Stafford, G.H.	1954	Geologist	Yes, Ret'd
Engineer	Chem.	Germany	Stoian, E.	1952	Reservoir Engineer	No
Engineer	Chem.	B.C.	Sullivan, J.T.	1955	Field Engineer, Devon	No
Engineer	Elect.	Alberta	Warne, G.A.	1954	Reservoir Engineer	Yes, Ret'd
Engineer	Chem.	Alberta	Webber, H.J.	1953	Field Engineer, Redwater	Yes, Ret'd
Geologist	Eng.	B.C.	Williams, R.R.	1952	Geologist	No

SOURCE: ERCB

In addition to the above, there were a further 11 engineers and geologists hired who stayed with the Board less than two years. Of these, nine were University of Alberta graduates.

NEB: National Energy Board   Ret'd: Retired
DIAND: Department of Indian Affairs and Northern Development   Dec'd: Died while in Board service

The Conservation Board and Field Management, 1948–1958

Floyd K. Beach, statis-
tician and engineer,
who developed many
of the Board's early
geological and engi-
neering record-keeping
systems. He retired
from the Board in
1950 after 23 years in
federal and provincial
government service.
Harry Pollard Collec-
tion, Provincial
Archives of Alberta,
P5359.

THE CONSERVATION BOARD AND
FIELD MANAGEMENT, 1948–1958

DISCUSSION OF THE Conservation Board's regulatory presence
in the field is organized under three general headings: (1) Information Collection, Organization, Interpretation and Distribution; (2)
Resource Evaluation and Evolution of Resource Conservation Policies in Handling Oil Production; and (3) Resource Evaluation, Evolution of Resource Conservation, Safety and Environmental Management in Handling Natural Gas Production.

INFORMATION COLLECTION, ORGANIZATION,
INTERPRETATION AND DISTRIBUTION

Collection of Data

It was understood from the beginning that effective conservation
practice rested upon thorough collection of appropriate basic data.
Up to the fall of 1930, collection of data relating to oil or gas wells
in Alberta was the responsibility of the federal Department of the
Interior. Responsibility was transferred to the Alberta Department
of Lands and Mines in 1931 and to the Petroleum and Natural Gas
Conservation Board in 1938.

Much of the credit for the development of procedures for the collection of detailed statistical data relating to drilling and completion appears to be owed to Floyd K. Beach. He initiated the official
compiling and recording in the Schedule of Wells of complete records for each well drilled in Alberta. Before Beach's time such data
were not consistently summarized and published, and it was only
with great dedication and labour that he was able to compile his
schedule from numerous published and unpublished documents.
Such data on early wells continued to be accumulated as early reports of such wells came to the Board's attention in the 1950s and
1960s. Albert Johnson of the Board's well records group took a
special interest in accumulating such previously missing data following Beach's retirement in 1950. Published annually, the *Schedule of Wells Drilled for Oil and Gas in Alberta* has continued in
much the same format and is used by geologists and engineers
throughout the industry in exploration and pool development programs.

Although the *Oil and Gas Conservation Act*, 1957, and the Regulations under it contained a few additional provisions, they gener-

ally described more specifically the samples, tests, reports and records to be made available to the Board, and thus continued the
policies that had been in effect throughout most of the decade and
earlier. The Regulations gave the Board authority to access, examine and report on a variety of tests and samples and to obtain reports of analyses. Production tests, core samples and drilling and
production records, for example, were to be kept at or near the
wellsite, where they would be made available to the Board's representative when requested. Records were to be retained by the operator for one year. Under the 1957 Regulations, the results of
production tests taken before a well was officially placed on production had to be sent to the Calgary office on special Board report
forms. Daily drilling reports were to be forwarded to the appropriate Board field office on a weekly basis, and production reports for
each producing well were required by the 15th of the following
month. In 1951, a change in field procedure was approved that permitted group production of wells at central batteries. Disposition
and production reports required from the well or battery operator
permitted comparison and verification with quantities set forth in
receipt and disposition reports from pipeline, other transport and
gas plant operators. Three copies of electrologs were also required
to be sent to the Board, at the operator's expense, to assist with engineering and geological studies of individual wells and pools. In
1957, additional logs, such as radioactivity logs that provided information similar to that provided by electrologs, were deemed acceptable. Where requested, cores had to be shipped to the Board's
Calgary office at the operator's expense.

The Board required reports from every well, without distinction
as to whether the well was in a developing pool or an exploratory
well, and it recognized early the need to make such data available to
the industry and public. Generally, data for new field wells in designated pools were released immediately, and the period in which
data for exploratory wells could be kept confidential was restricted
to one year from the date drilling was completed. A memorandum
circulated to the staff in 1944 clarified the distinction between immediately available data and confidential data. It stated that geological information, analytical data, deviations, test results, and drillers' reports or well logs were considered confidential. Except for
data appearing in the monthly summary, all absorption plant and
refinery data were confidential and no internal records were available for inspection without authorization by a board member. The
memorandum also explained that, although confidential data became public one year from collection date, in special circumstances

information could be retained as confidential longer than one year and would be released only upon approval of a board member. Confidential material could be released earlier than one year, but only when special permission in writing was received from a representative of the owner who had provided the information in the first instance.[1]

As far as can be ascertained, these rules were continued, with some modest changes, until they were included in the 1957 Regulations. The rules were not without controversy, however, as illustrated in 1952 when concerns raised by industry led the Board to issue a Letter to All Operators that outlined specific steps to be taken by the Board and its staff before release of confidential data.[2] For this reason, special attention was paid in the 1957 Regulations to defining which classes of information were or were not confidential.[3]

Cores and Drill Cuttings

Regulatory policies relating to the taking and handling of cores and drill cuttings were included in policy documents issued by the federal Department of the Interior. When Alberta gained administrative control of resources in 1931, new regulations required an operator, unless otherwise directed, to take and preserve drill cuttings from each 10-foot interval of hole drilled, or shorter interval if requested. These cuttings were to be washed, dried, packaged and labelled and forwarded to the Department of Lands and Mines. When cores were taken, they were to be placed in core boxes, housed in a suitable locked building, and forwarded to the Department only upon its written request.

The Conservation Board, after taking over the responsibilities of the Petroleum and Natural Gas Division of the Department of Lands and Mines, introduced Regulations in 1939 that authorized the Board to order that cores be taken at a well. Such requirements by field or area appear to have been issued for the first time in the mid-1950s when coring programs for the Pembina field and part of the Medicine Hat field were required. Two important policy changes in 1950 respecting well cores required core analysis reports to be furnished to the Board and prohibited destruction of core without the Board's consent. Company-operated sheds, at which cores were stored for later reference, became numerous in the province. By that time, larger companies had developed their own facilities for the storage of core in each of the major fields. A central Board-operated core warehouse that would be conveniently accessible to all geologists was planned in 1942–43, but the war caused its deferral. The 1943 Regulations specified the size of core boxes and how cores were to be forwarded to the Board; however, only a small storage area was maintained in the Board's Calgary office and most of the cores remained in industry storage sheds.

Following the Board's move in 1945 to larger premises, more space became available for basement storage. By 1949, however, this area had reached its capacity, and in 1950 the Board built its first core shed in the Manchester district of Calgary. By this time, the Board was receiving most of the cores taken, and those from confidential wells were accommodated in the head office storage area. The first Manchester shed was soon filled to capacity, and additional storage was added through the 1950s. Space was provided in each storage location for the examination of cores by interested parties, usually industry and Board geologists. The series of storage facilities was replaced in 1962 by a 41,000-square foot storage centre located just north of the University of Calgary campus.[4] The new facility represented more than simply the provision of centralized storage, it manifested an important departure with far-reaching implications.

Before the building of the Board's Core Research Centre, about 70% of existing core was held by industry at various provincial sites. This posed a serious problem, as there was no effective means of releasing privately stored core to other operators for evaluation, even though the core had been assigned nonconfidential status by the Board. Sometimes the core was in the hands of an operator who placed limited value upon ongoing scientific core analysis and went to little effort to store the core in a manner that would protect its identity and prevent deterioration. Such was the case with the company that stored 300 boxes of core in a "lean-to" attached to a pig barn in a gas field east of Calgary.[5] More often the core was held by a major company. Imperial Oil, for example, stored its core in three large, heated Quonset buildings in Edmonton. Although most of the larger companies made an effort to maintain a proper facility, there was often limited enthusiasm for providing open access. This reflected the prevailing attitude brought from the United States that an operator had exclusive rights to his own core. These companies preferred making their core available to competitors for analysis on a selective reciprocal trading basis.

The Saskatchewan government was the first to challenge this attitude by constructing a central core storage facility in 1958 and insisting upon the submission of all core samples. Perhaps this was the motivation that led the Alberta Society of Petroleum Geologists (ASPG) to urge the Alberta Conservation Board to move in the same direction. Prompted by J.R. Pow, their chief geologist, the Board initiated discussion with the industry. Almost all independent companies favoured exclusive Board storage of core. The majors were divided, but most favoured "combined" storage, which would allow a company to retain its more valuable core.[6] Consensus was eventually reached, and in February 1961 the Board announced its intention to construct a central core storage research centre.

The new facility permitted geologists to examine cores and drill cuttings from wells throughout Alberta. It was equipped with convenient examination tables, upon which well cores could be laid out for examination. Rooms were provided for confidential examination of especially critical wells. The same storage facility included drill cuttings samples obtained from all wells in Alberta where such cuttings were required and tour reports detailing the operations at wells when they were being drilled.

The Board's core policy and its supporting storage research centre were unique and had a profound long-term significance. There is no comparable program or centre in the United States. The Texas Railroad Commission, for example, has always lacked the authority to prevent the destruction of core or to require its submission.[7] In Alberta, under the Board's direction, an invaluable body of raw data was being assembled that would become in time the largest centrally located inventory of core and drill cuttings in the world. Almost from the outset, this meant that scientific study of the sedimentary basin underlying Alberta could be conducted with the advantage of a database superior to any elsewhere. On a day-to-day basis, work done at the core research centre contributed directly to reducing the likelihood of drilling a dry well. Over time, elaborate core studies, made possible by the Board's collection, played an important role in optimizing the exploitation of ancient reservoirs and in determining the feasibility of secondary recovery schemes. In this manner, the centre assisted both the public and private interests, and that it has served as the prototype for similar facilities in some 27 countries stands as clear testimony to the regard in which it is held.[8]

Chemical Laboratory
The Chemical Laboratory was organized in 1937 and was part of the Petroleum and Natural Gas Division of the Department of Lands and Mines. In May 1941, the Chemical Laboratory was transferred to the Petroleum and Natural Gas Conservation Board and K.C. Gilbart, the chief chemist, became a member of the Board staff. The main function of the Conservation Board Chemical Lab-

The Board's Chemical Laboratory–first organized in 1937–shown here as it looked ca. 1950 in Calgary. ERCB, Bohme Collection.

oratory was to obtain and analyse representative samples of gas, oil and water from the underground formations of Alberta. By identifying and analysing such samples, it was anticipated that the important characteristics of Alberta's underground resources of oil, gas and water would become better known and geological and engineering correlations of underground strata and fluids would be assisted. It was expected that such information would aid in the evaluation of special engineering proposals related to conservation, equity and utilization of resources. With these objectives in mind, the Board's chemical laboratory did its best to correlate and generalize the results of analytical studies and to disseminate such information to the industry and to the public.

In accordance with the 1950 Drilling and Production Regulations, a letter sent to all operators reminded that they should be prepared to supply to the Board, in vessels furnished by the Board, samples of oil, gas and water obtained at any point in the drilling of a well. The need to obtain unpolluted samples of both oil and water was emphasized.[9] Additional samples were not required from established pools for which the composition of oil, gas and water was already determined.

The laboratory was equipped to perform most routine analyses of gas, oil and petroleum products as well as water and rock analyses, and its services were available to the industry. It was also called upon to analyse samples of fluids from surface locations where complaints had been received from landowners respecting the effects of surface pollution on farmlands and livestock. Although not always conclusive, the analyses assisted, for example, in determining if illness or death of farm animals was caused by products from drilling or other oilfield operations.

The functions of the Chemical Laboratory were taken over temporarily by the Department of Chemical and Petroleum Engineering at the University of Alberta in 1952 pending development of more permanent arrangements. By late 1953, the Laboratory was re-established as part of the Board's functions in the Engineering Building on the University of Alberta campus.

Some of the larger companies had their own facilities for sample analyses, and by the mid-1950s the services offered by a number of independent laboratories met an increasing portion of the industry's needs. Accordingly, in 1957, the revised Drilling and Production Regulations accepted that, whenever the Board required a sample to be taken and analysed, the analysis could be performed by outside laboratories. The Board did not wish to duplicate analyses being performed to required standards at industry facilities; however, the Regulations required the analysis and the accompanying data noted on the Sample Information Sheet to be furnished to the Board "forthwith."

Distribution of Information

A permanent record of each well that had begun drilling was maintained in the Calgary office. Important information from each daily drilling report, and from memos and supplemental reports, was summarized in an individual well record. This information reached the interested public by means of Weekly Drilling Summaries, available each week from the Board, and through the annual Schedule of Wells, which included all significant, nonconfidential information for each well drilled. Production and injection statistics were maintained on an individual well basis and could be accessed by the public any time.

The geology department determined from drilling samples, electric and other wire line logs the depth to the tops of principal formations encountered in each well drilled, and completed an official report filed in each individual well record. A copy was forwarded to the licensee of the well. The analysis of water samples made in the Board's chemical laboratory was also filed with the Board and a copy sent to whoever had submitted the sample.

From 1940, the Board issued *Alberta Petroleum Industry* (*Alberta Oil and Gas Industry* after 1950), an annual publication summarizing data. Reports published before the discovery of Leduc, included annual and cumulative production for each well in the province. Another table listed each well drilled during the year, including the location, the commencement and finished drilling date, the depth and the year-end status. With the rapid growth in activity after the discovery of Leduc, these tables were soon dropped from the publication; however, tables showing production and disposition and prices received for crude oil were retained. In the much larger editions from the mid-1950s, statistics provided were included under four main headings: comparative statistics, historical data, current year data, graphs and charts.

The April 1960 notice of publication for the *Alberta Oil and Gas Industry* report advertised that its contents included

> a map of Alberta showing oil and gas fields, pipelines, refinery and gas plant locations; a geological classification of oil and gas production; a history of production and disposition of oil, gas and natural gas liquids for the province and also their production and disposition during 1959; a comparison of producing wells; drilling activity past and present and tables for 1959 showing casing used, Alberta crude prices and pipeline tariffs.[10]

The raw data gathered provided the statistical base for much of the internal work of the Board. The technical departments— development, reservoir engineering, economics and geology—used the information in many ways, including updating surface casing policy, pool reservoir studies, geological correlations, future performance, and demand and supply forecasts. The drafting department provided current maps showing existing oil and gas fields, field and pool outlines, well locations, and a wide variety of other maps and illustrations of assistance to the industry. The value of these reports, maps and publications, such as pool performance charts to the industry, particularly smaller companies, was evident in the wide and regular list of company subscriptions.

RESOURCE EVALUATION AND EVOLUTION OF
RESOURCE CONSERVATION POLICIES IN HANDLING
OIL PRODUCTION

Evaluation of Reservoirs, Assessment of Reservoir Performance and Estimation of Crude Oil Reserves

The spectacular growth in crude oil development following the 1947 oil discovery at Leduc required strengthening of the Board's reservoir evaluation capabilities and special emphasis on collection of quality data. These requirements were further accentuated in December 1950 when the plan for proration to market demand with emphasis on maximum permissible rates of production came into effect.

The Board emphasized improving the quality of its own geological, reservoir fluid and pool performance data. The appointment of J.E. (Ted) Baugh as the Board's first reservoir engineer in 1950 marked the beginning of the Board's reservoir department. Initially, he assisted G.E.G. Liesemer, who had been responsible for drilling and production, reservoir engineering and assessment matters as the office engineer in Calgary. Baugh's initial reservoir engineering assignment had been to estimate the size of the Leduc D-3 pool in the process of evaluating in 1948–49 the impact on the pool of the Atlantic No. 3 blowout.

The decision to assign high priority to first-rate reservoir analysis on an ongoing basis was consistent with superior conservation practice, but to maintain this commitment in an environment where the number of reservoirs was increasing monthly required a level of staffing and commitment not sustained in most jurisdictions. This was a demanding priority, both in the number and the quality of professional staff required. It was not just quality in the sense of professional training, the situation demanded a high level of initiative and motivation. The record of the Board's staff in this regard is impressive.

The development of an effective assembly line mechanical system for washing drill cuttings under D.R. Craig's direction in the late 1940s reduced staff needs for that task. Additional geologists were hired to keep up with the review of well samples, cores, well and drill stem test data, and the development of reserve maps. Additional engineers and technicians were added through the 1950s to keep up with the interpretation of pressure charts taken with industry and Board bottom-hole pressure gauges and to assess and incorporate the extensive geological, engineering and economic data being received and to include them in the Board's analyses.

In the production accounting area, the Board purchased tabulating machines and added staff to handle the rapidly growing number of oil, gas and water production reports and to assist in administration of the system of crude oil allowables. The tabulating machines were replaced by a computer in January 1959.[11]

In the field, the addition of inspectors ensured production data gathered was of proper quality. Strong emphasis was placed on quality and training in recruiting staff and their development. In-house training programs and industry courses, on such topics as reservoir engineering, fluid measurement and well logging, were stressed.

Detailed procedures were developed for preparing pool evaluations. The engineers and geologists determined average pool porosity and water saturation data from core analyses and well logs. Correlations were developed to improve consistency between core analyses and well-log derived values. Fluid shrinkage factor determinations, indicating the change in fluid volumes as fluids were brought to the surface, were refined. These utilized additional pressure-volume-temperature (PVT) analyses of reservoir crude oil samples taken using pressurized sampling devices. Generally, the analyses were done in commercial laboratories, and copies of the reports were required to be provided to the Board. In some instances, recombined samples were analysed. Such samples were obtained by mixing crude oil and the natural gas that evolved from it in correct proportions and agitating the mixture at reservoir pressure and temperature until a condition was achieved approximating that in the reservoir. Such recombined sample analyses were less costly, but usually they were not as reliable as analyses of samples carefully taken at reservoir pressure and temperature. Generally, however, they provided better results than could be obtained by pool-to-pool correlations.

Recovery factors indicating the portion of the crude oil that is recoverable were estimated considering the quality of the reservoirs and the recovery mechanisms that were demonstrated to be in effect. Correlations of recovery factors determined for pools with considerable production history assisted in estimating factors for similar pools with little production experience.

The development of Cardium sand reservoirs, beginning in 1953, required a review of core analysis procedures utilized in commercial laboratories to assure representative data. The difficulties with Cardium cores related to the low permeability, difficulties in cleaning, and the content of shale in some sand sections. Heat applied in the cleaning process, if not properly controlled, drove off a portion of the moisture content in the shale, thereby inducing increased permeability and, to some extent, porosity. Where erroneous values resulted they could lead to errors in estimates of reserves of oil and gas. The Cardium sands were quite variable, with some having conglomerate sections that often exhibited much higher permeability than adjacent sand sections, making it difficult to displace crude oil from adjoining lower permeability sand sections by injecting gas, water or solvents. Good assessments of permeability and its variability in wells were thus critical in forecasting crude oil recovery efficiency.

In the mid-1950s, N.A. Strom did considerable work interpreting and classifying reservoir rock types with a view to improving the Board's assessment of reservoir factors. This marked the beginning of detailed petrophysical studies at the Board, which were to receive further emphasis in the early 1960s.

The introduction of enhanced recovery programs involving gas, solvent or water injection required assessment of the crude oil displacing efficiency of the fluid to be injected in different types of reservoir rock. Such programs were being applied regularly in the United States where appropriate, and many had substantially increased the crude oil recoverable. The first such field project in Alberta was the pilot water flood program undertaken by Royalite at Turner Valley in 1948. A pilot water and gas injection program was undertaken for the Leduc-Woodbend D-3A pool in 1953. This program was continued and expanded as a pressure maintenance program to supplement the natural water drive. By the mid-1950s, a number of reservoirs, especially those in the Viking and Cardium zones, where inefficient solution-gas drives soon led to high gas-oil ratio penalties, were strong candidates for enhanced recovery programs. Water injection programs, in which wells were converted to water injection, in a uniform pattern, or in line drives in which a series of wells in lines across a pool were converted to injection, were soon common and effective in reducing gas-oil ratio penalties and increasing productivity and oil recovery. Such programs increased the amount of oil recoverable from one and a half to twice that recoverable by solution-gas drive. In some reservoirs with good flow characteristics, only a small portion of the wells were needed for injectors, whereas half of the wells were completed as injectors in other poorer quality reservoirs. In better quality Viking reservoirs, for example, perhaps 10% of the wells would be converted to water injection, whereas in lower capacity Cardium reservoirs usually every other well had to serve as a water injection well to control reservoir pressure and producing gas-oil ratios. Even in high-quality Devonian reef-type reservoirs with natural water drives, some fluid injection was often required to supplement the natural drive; however, for the high-quality D-3 reservoir at Leduc, single water injection and gas injection wells provided sufficient capacity to ensure a highly efficient crude oil recovery system for 260 producing oil wells.

The proponents of such enhanced recovery schemes often performed laboratory studies of the displacing efficiency of the injected fluid to support planning and development of their projects. These studies utilized cores from the particular pools selected as being representative of the reservoir rock. The Board maintained and published lists of these special studies. Its staff and industry specialists did pool-to-pool correlations, using data from such tests and the early performance of existing projects, to estimate how other parts of the same pool and other pools would perform with similar programs.

Enhanced recovery programs were usually initiated by the well operators on their own; however, approval of each program was required. The Board also initiated hearings when it believed enhanced recovery programs that were needed were not being implemented. These "show-cause" hearings were to consider the well operator's views on whether or not an order requiring commencement of an enhanced recovery program should be issued. Such hearings were normally required only where the economic feasibilities of such programs were just marginally profitable.

Although extensive data on Alberta's oil pools was obtained, substantial differences of opinion often existed between companies and between industry and the Board respecting the performance of the pools and the ultimate crude oil recovery from them. These differences were publicly assessed at the annual, and in some cases semi-annual, hearings respecting pool MPRs and at hearings of proposals for changes in well spacing, introduction of enhanced recovery schemes, and conservation of solution gas. In this regard, the Board played an especially important role. By functioning as a forum where debate probed the leading edges of reservoir engineering, technology and production practice, the Board helped to educate the industry, especially the sector that had little or no research capability of its own.

The 1950s generated the greatest contribution to the province's crude oil reserves of any decade. A review in 1979 indicated that almost seven billion barrels of the initial crude oil reserves assigned in

Alberta to that time were discovered between 1 January 1950 and 31 December 1959. This comprises almost half the conventional crude oil reserves (1992) identified as recoverable at present.

The growth in initial and remaining established crude oil reserves, as estimated by the Conservation Board and the Canadian Petroleum Association, is compared in Figure APP.X.1. There is good agreement in the estimates of initial and remaining established reserves of the different estimators; however, greater differences exist in the estimates of the ultimate potential indicated in the upper portion of the figure. These estimates require assessment of the increases in reserves that might result from the discovery of new pools, as well as the growth in reserves expected from already discovered pools.

Development Programs

A conservation agency's general mandate is to ensure, given the unique physical properties of oil and gas, that the resource is developed under certain guidelines or controls intended to protect the public interest and correlative rights. The common technical implements or strategies of such control have been outlined in the Introduction. They are discussed here with reference to the application and practice of Alberta's Oil and Gas Conservation Board.

WELL SPACING

The discovery of additional heavy crude oil reservoirs in the Lloydminster-Wainwright area in the early 1940s and the market for crude oil as locomotive fuel created a stimulus for development; however, the production rates of the heavy viscous crude at these wells declined rapidly once production commenced. Also, since wells drilled a short distance away exhibited the same initial productivity, there was an incentive to drill more wells and at closer spacing than was utilized for light crude oil wells at Turner Valley. The Board was reluctant to approve closer spacing for heavy oil pools, but it was soon evident that light oil would not likely be encountered in the Lloydminster-Wainwright area, and that closer spacing was desirable for the viscous heavy crudes to achieve maximum recovery. Spacing of less than the usual 40 acres was prescribed outside the Lloydminster-Wainwright area, at Bonnyville in September 1951, when 20-acre spacing was approved by Board order.

The discovery of light crude oil in the Edmonton area at Leduc in 1947 led the Board to prescribe, through Regulation No. 14, one legal subdivision (40-acre) spacing for oil development in that area.

FIGURE APP.X.I Trends in Estimates of Alberta Crude Oil Initial
Established Reserves and Remaining Established
Reserves, 1948–1962

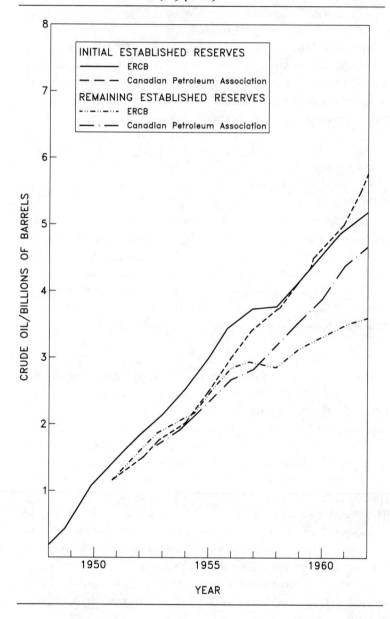

The same spacing was prescribed for Redwater and other pools encountered in the Edmonton region. Quarter section or 160-acre spacing was approved for oil production for a small area at Golden Spike, through spacing order SU-1, dated 6 February 1952. Imperial Oil appeared to control the full extent of the pool, and thus few wells were required to produce the extremely thick and highly permeable D-3 reef efficiently.

Province-wide spacing provisions were added to the Drilling and Production Regulations early in 1952.[12] They prescribed normal well spacing of one legal subdivision for oil wells and one section (640 acres) for gas wells. They also set forth specific penalties for wells completed outside the target areas.

With restricted well allowables because of low market demand and recognition of the good drainage characteristics of light and medium crude oil pools, there was recognition of the merits of wider spacing in the early 1950s. Following application by oil companies in 1953, two legal subdivision spacing was prescribed for the Drumheller D-2 pool in March. This was followed by requests for and approval of other two-legal-subdivision spacing orders for part of the West Drumheller field and the Westerose, Pembina, Sturgeon Lake and Stettler areas. The recognition of the merits of wider spacing for crude oil production and the frequency of special orders led to an amendment of the Drilling and Production Regulations on 2 July 1957 to prescribe two legal subdivisions (80 acres) as the normal well spacing in the part of the province West of the 5th Meridian where heavy oil was seldom encountered. By that time, still wider quarter-section spacing, which had already been approved at Golden Spike, was being sought for light oil production in additional areas.

COMPULSORY POOLING

Occasionally, ownership of a spacing unit might be divided among two or more owners. The incidence of divided ownership increased somewhat with the trend to wider well spacing. Even with one-legal-subdivision spacing, however, divided ownership of spacing units often occurred. In towns and villages, early subdivisions established before November 1887 conveyed title to both surface and subsurface rights for building lots to the new owners. Also, the government normally withheld mineral and surface rights under the portion of streams and lakes up to the high water mark. Hence, the likelihood of divided ownership increased wherever a lake or stream occupied part of a spacing unit.

Control of a full spacing unit by the applicant for a well licence is required before a well can be drilled, and where ownership is divided the parties might not reach agreement on sharing of costs and revenues in development of the mineral rights. In other instances, the owner of the rights to part of the spacing unit might be untraceable or deceased, without designation of an approving authority respecting disposition of property. Provisions were necessary therefore to allow the owner of the part of the spacing unit who wished to drill to do so even though he did not control the full spacing unit. This was achieved by including compulsory pooling provisions in the revised 1957 *Oil and Gas Conservation Act*.[13] The first pooling order was issued pursuant to an application by Canadian Prospect Ltd. and pooled tracts in and adjoining the community of Big Valley.[14] Although this was the first order designated as a pooling order, pooling of some tracts in other areas might have been approved earlier by the Board through another form of approval.

UNITIZATION

The difficulties in achieving gas conservation at Turner Valley in the 1930s and 1940s were complicated by the diversity of the interests involved and of opinion on the need for such a program considering its costs. A considerable number of people who obtained mineral interests prior to 1936 were deceased or untraceable. There was recognition of the need to combine the interests into units by allocating a share of production to each tract on a consistent basis to permit the majority interests to proceed with conservation programs. Such unitization of properties also would permit greater coordination of operations and a consequent reduction in total operating costs.

At this time, unitization was recognized as a device increasingly used in the United States to permit conservation programs and to reduce production costs. Those who opposed unitization wished to retain independence in determining the optimum operation of their individual properties. Although the merits of unitization were not embraced by all interests, there was considerable support for such programs. Provisions on unitization were added therefore to the new 1957 *Oil and Gas Conservation Act*. Sections 71 and 72 gave the Board direction to encourage units. Board approval of a unit was required before it could be put into effect, and the Board was authorized to vary or suspend any provision of the Act or Regulations or both regarding the development of oil or gas resources in the unitized area during the operation of the unit. The new sections also permitted delegation of individual owner responsibilities for

unitized lands to the unit operator, but provided that nonperformance by the operator would still be taken as nonperformance of the
individual tract owner. They did not, however, go so far as to provide for compulsory unitization as was provided for in the corresponding Saskatchewan Act.

Although specific legislation enabling unitization in the Turner
Valley field existed earlier, it was not until late in 1958 that the first
agreements were in place. The *Turner Valley Unit Operations Act*
discussed earlier was assented to and came into effect on 14 April
1950 to facilitate oil and gas conservation in the Turner Valley
field. It provided that the Board could require unit operation of
tracts in the pool in the Turner Valley Rundle group, or part of it, if
the owners could not agree or were unable to consolidate, merge or
otherwise combine their interests for the more efficient and more
economical operation of the pool or part of the pool. An application for such an order had to be made by or on behalf of tracts em-
bracing more than 50% of the lands containing the part of the pool
for which application was being made. With the coming into effect
of this special Act, areas were unitized along the length of the field,
the units approved and operations co-ordinated in the order shown
in Table APP.X.1. Each application for a unit was required to be considered at a public hearing and required approval by order-in-
council. Each order of the Board designating the unitized area was
required to set forth detailed provisions to facilitate allocation to
each tract owner of his appropriate share of costs and revenues.
Each amendment of a unit order was also required to be considered
at a public hearing and to be approved by order-in-council.

MAXIMUM RATE LIMITATION AND PRORATION
The evolution of the Alberta system of well allowables was discussed in chapter 5; however, a few additional comments on the application of the system and the impact on Board administrative
activities are in order.

The constraints of proration were confined to the light and medium crude oil categories, which as pipelines were extended became
more widely accepted by refineries throughout the markets served.
Heavy crude oil markets were more restricted. Such crude was processed mainly in local refineries for asphalt production, which was
in increasing demand for the rapidly expanding highway construction programs in western Canada in the 1950s. In producing asphalt, the lighter fractions were converted to gasoline and diesel
fuel. Some sour medium gravity crude oil, such as from the Stettler
field, was also not widely marketable; however, arrangements were

TABLE APP.X.1   Turner Valley Unit Operations Act, 1958–1962

Date	Unit	Operator
1/10/58	Turner Valley Unit No. 1	Home Oil Company Limited
1/12/58	Turner Valley Unit No. 2	National Petroleum Corporation
1/01/59	Turner Valley Unit No. 3	Western Decalta Petroleum Ltd.
1/07/60	Turner Valley Unit No. 4	Western Decalta Petroleum Ltd.
1/07/60	Turner Valley Unit No. 5	Royalite Oil Company Ltd.
1/07/60	Turner Valley Unit No. 6	Home Oil Company Limited
1/08/62	Turner Valley Unit No. 7 (Gas cap)	Royalite Oil Company Limited

SOURCE: ERCB

soon made to blend such crude oil with other light and medium crude oils to obtain a more marketable product.

Heavy crude oil production, while not prorated, was subject to maximum rate limitations. Such rates were determined as for light and medium crude oil using the MPR formula. Under that formula, the recoverable reserves were estimated and divided by a "uniform rate life" (determined considering the reservoir and fluid characteristics of the pool) to obtain a maximum daily rate of production. For heavy crude oil pools, this typically was less than the minimum MPR rate of 50 barrels per day, although higher rates did result for some wells.

There was special concern about gas conservation at the many light and medium crude oil pools, since gas-oil ratios were significant and led to considerable flaring. Progress in conserving gas continued, however, beginning with the Leduc-Woodbend gas conservation scheme that was completed in 1950. By comparison, the gas-oil ratios at heavy oil pools were low, and to minimize operating expense the Board exempted such pools from requirements for continuous gas measurement in most cases. In some areas, however, small gas caps existed in conjunction with the heavy crude oil, leading to significant gas production and the need for normal gas measurement facilities. The prescribed maximum rates of production were thus important in controlling gas production and providing for application of GOR and WOR penalties, especially in heavy oil areas.

Water production soon became a significant problem in a number of pools where water coning occurred. Credit was given for water returned to the producing zone in determining water-oil ratio penalties. This provided a strong incentive to install proper disposal facilities. The Board required continuous measurement of the amount of water produced where it became appreciable.[15]

CONCURRENT DEPLETION

In some pools with thin oil columns and a gas cap where oil could not be produced without substantial production of gas, provision was made to produce the gas cap and oil concurrently. Approvals of this manner of production were issued as a miscellaneous order and subject to specific conditions. Typically, a key condition was that all gas produced be conserved.

GOOD PRODUCTION PRACTICE

A considerable number of operators making submissions at the 1957 proration hearing supported establishment of a minimum rate of production at each well, at which gas-oil and water-oil ratio penalties would no longer be applied. After further discussion, it was concluded that granting such exemptions from penalties on a well basis in a pool could lead to inequities between operators. As an alternative, the concept of granting "Good Production Practice" status on a pool basis was developed. At oil pools in an advanced stage of depletion, where the operators generally supported relaxing of production regulations, the Board agreed to waive gas-oil ratio and water-oil ratio penalties, as well as individual well allowables. This manner of operation was referred to as Good Production Practice (GPP) status. Such relaxation of regulations was made to allow continued economic operations and increased recovery from marginally economic properties. The operating requirements prescribed for pools subject to GPP varied depending on the local circumstances.

MULTIZONE COMPLETIONS

In numerous oil and gas fields in the province, two or more productive zones can be penetrated by the same wellbore. In such circumstances, it is possible to produce more than one pool through the same string of casing. Such circumstances exist, for example, in the Leduc-Woodbend field, where the D-2 pool overlies the D-3 pool over much of the field area. The Drilling and Production Regulations issued in 1950 included a new regulation, section 45, which specifically addressed multizone completions. It stated that "No well shall be allowed to produce oil or gas from different pools or zones at the same time from the same string or column of casing, unless the permission in writing of the Board has been obtained in any particular cases."

There might be economic advantages to be gained by the producer in producing two (or more) zones from the same well; however, the larger interests of reservoir management deny such practice without segregation of production of each zone until it can be accurately measured. The main reason for segregating production from different producing zones is to prevent fluid migrating from one zone to another as naturally would occur given the varied pressures that may exist between different zones. Prevention of fluid migration is desirable for a number of reasons. With mixed or commingled production, oil, gas and water production derived from each zone could only be estimated, thereby compromising future performance studies of the separate reservoirs. Gas and water from one zone commingling with production from another could unduly penalize production from the better zone and possibly lead to earlier abandonment. Undesirable contamination of the receiving reservoir might occur. Another concern is that loss of reservoir energy from one zone to the other could lead to reduced recovery. Finally, commingled production can easily lead to inequities between interest owners.

Awareness of these problems led the Board in the 1950s to grant applications for dual-zone completions only where the completion program provided separate flow channels for each zone. These channels could comprise one through the casing-tubing annulus and the other through the tubing. Alternatively, two tubing strings could be used. Segregation tests were mandatory, both before final approval was granted to commence regular production and regularly, usually annually, thereafter. The Board often found proposed dual completions unsatisfactory. By the end of 1957, there were only 69 dually completed wells producing in the province, and a few of these were abandoned after presenting operating problems or difficulty in maintaining segregation between the zones.

In 1955, detailed requirements for applications for multizone completions, as well as operating instructions, were also provided to the industry. The role of the Board staff was also discussed in some detail. The main responsibility for assessing the acceptability of applications was placed with the reservoir engineering department, which made recommendations for approval or denial to the development department. The field staff were involved extensively in arranging for testing of the zones, in sampling of fluids, and in assessing the adequacy of the equipment installed and the various techniques used to assure segregation.

In February 1957, the Board issued a letter to operators of dually completed wells in which it emphasized the experimental nature of multizone completions and the need for comprehensive information on their performance. It also pointed out that, while the merits of such completions were being assessed, the Board would issue approvals only when development on a single-zone basis was not economical or when valuable information was expected to be gained from experimentation with important types of dual-zone installations. The Board had doubts as to the adequacy of the available segregation equipment, but it was prepared to approve some completions to develop experience with the equipment and procedures.

The Drilling and Production Regulations were amended in July 1957 by replacing section 45 of the former Regulations with several new sections under the general heading of Multi-Zone Wells. In addition to describing in detail the requirements that had to be met before approving an application to complete a multizone well, these regulations included a substantial list of conditions to be met where dual completion operations were to be continued. In assessing the merits of such a completion, the reservoir characteristics, including production potential, anticipated recoverable reserves of each pool, and the type and quality of equipment to be installed for production and the maintenance of segregation, would be considered. The applicant was required to substantiate the benefits of such completion in the particular circumstances encountered in the pools involved.

The revised regulations also took into account the recommendations of a special working subcommittee of the Interprovincial Petroleum and Natural Gas Committee, which was established in the mid-1950s with representation from government and industry to consider the regulation of multizone completions. It recommended, and the Board subsequently provided in its revision of the 1957 regulations, an addition to the provisions to allow the naming of parts of fields where a number of multizone completions could be made without further approval.[16] The first hearing with regard to a direction approving pools for multizone completion was held on 3 July 1958 and related to two zones in the Big Lake field.[17] In approving the application, the Board required that production from each zone be taken through a separate tubing string.

In 1959, the Regulations were amended to allow wells to be produced without segregation under certain conditions upon application and approval. Evidence was required from the applicant in each instance that ultimate recovery and the rights of owners would not be adversely affected under such operating conditions. The intent was to permit the continued economic operation of marginal pools through production of more than one zone on a commingled basis.

MEASUREMENT OF OIL AND WATER

From a conservation point of view, accurate production records of fluids produced from an oil or gas reservoir are essential to assess the adequacy of the depletion method in place and, where appropriate, to determine when improved recovery methods should be applied for oil or gas recovery. In most pools, gas-oil or water-oil ratio penalties are prescribed and require accurate measurement to be effective. Incorrect reporting for individual wells also could mask performance of the recovery mechanism and delay a proper understanding of the reservoir's performance. Accurate records are also important in assessing the profitability of alternative programs. For those receiving monthly royalty payments and minority working interest owners, it is also essential to have assurance that reported volumes are indeed correct. For these reasons, the Board has, since its formation, emphasized accuracy in data reported and has made regulations for this purpose. The 1957 Regulations provide a detailed summary of Board requirements at that time.

In accordance with these regulations, all measurements of oil, water or other liquids required by the *Oil and Gas Conservation Act, 1957,* were to be reported in either barrels or gallons. A barrel was equivalent to 35 imperial gallons, as that unit was defined by the Canada *Weights and Measures Act*. In practice, both crude oil and produced water were reported in barrels. Ordinarily, in the 1940s and 1950s, the oil was measured by manually gauging a tank, but approval could be obtained where adequate equipment was installed for measurement by other methods, such as metering. In addition, corrections were required to be applied to adjust the volumes produced to standard volumes at a temperature of 60° Fahrenheit.

Although it was a requirement that the total oil produced from a group of wells had to be measured by an approved method, most wells produced to a production battery, and their production was commingled before measurement. In these cases, oil production from each well had to be measured during periodic tests, and the total measured volume received at the battery prorated to the individual wells on the basis of the test rates to obtain individual well volumes.

At smaller batteries, oil production was measured by manual gauging of the oil level in the storage tanks. With the gradual increase in automation of battery facilities, together with improvements in automatic measurement equipment, the Board in most cases accepted use of positive displacement meters. Approval continued to be given on an individual battery basis. In issuing an approval to use a meter, the Board stipulated that the meter factor determined once a month by comparison with manually gauged volumes be used to correct the metered volume. This meter factor was the ratio of the amount of oil manually gauged to the amount determined for the same quantity by metering. The precision required for the measurement of water was less than for gas and oil. Only in wells in which water constituted 20% or more of the total liquid production was actual measurement of the water required. This was accomplished by separating the water from the oil and measuring it either by hand gauging or metering. Where water production was between 20% and 2% it was sufficient to estimate the production by centrifuging samples taken twice each week to determine water content, and to determine the amount on a ratio basis to the total oil production. For wells producing 2% water or less, the amount was determined using twice per month tests. When production was being calculated on a battery or group basis, the same limitation as described above for individual well reporting was applicable to the test production from wells producing more than 20% water. For the intermediate category, four samples taken at regular intervals were required to be centrifuged during each test period. Where water production was 2% or less, only one sample per test period was necessary. The total amount of water produced at each battery was required to be measured accurately, whenever it exceeded either 300 barrels per month or 2% of the monthly oil production at the battery.

RESERVOIR PRESSURES

Production allowables for Turner Valley wells under the Brown Plan were determined through the 1950s using a formula involving a number of reservoir factors. One of these factors was the reservoir pressure of individual wells. The Board's Black Diamond staff included a bottom-hole technician whose responsibility was to run bottom-hole pressure gauges down the tubing of each well to measure the pressure at the producing horizon. The pressure was in-

Board engineers run a bottom-hole pressure bomb in Turner Valley in the 1940s. ERCB.

scribed on a chart within the gauge by a stylus, and was later read at the Black Diamond office. This equipment and procedure was also used in other pools as they were discovered.

In 1948, the Turner Valley bottom-hole unit was used in the Leduc field to assess the effects on reservoir pressure at wells in the vicinity of the Atlantic No. 3 well, which was blowing out of control at the time. With the discovery of many oil and gas pools in the late 1940s and early 1950s, other bottom-hole units were soon added to complement surveys being run by oil company and service company personnel.

The bottom-hole pressure surveys were difficult at pumping wells where pumping equipment interfered with the running of the gauges. In such instances, pressures were frequently determined using echometers, which determined the fluid level in wells using sonic signals. Such measurements were not as accurate as those obtained with bottom-hole pressure gauges, but they provided a less expensive but acceptable substitute in most instances. Bottom-hole pressures were determined annually on a pool-wide basis for reservoir studies by both the Board and industry. Such data was espe-

Board field inspector, Bill Erkamp, installs an acoustic well sounder (echometer) used for bottom-hole pressure determinations in the late 1950s. ERCB, Erkamp Collection.

cially important in the early production history of pools where the recovery efficiency was being assessed.[18]

There were three types of bottom-hole pressure determinations, comprising static, flowing and build-up, of which the static determination was most common. This involved running the test after the well had been shut in for a defined period. The shut-in period required depended on the quality of the reservoir. Shut-in periods of a few hours were adequate for high-quality Devonian reef wells, whereas in low-quality reservoirs pressures still would not be fully built-up after wells had been shut in for several weeks. Pressure measurements by operators or service companies were required to be reported to the Board. The results of all surveys on wells included in designated fields were made available to anyone interested for use in reservoir studies.

Industry field operating committees, working in co-operation with the Board staff, had assumed much of the responsibility for scheduling pool surveys by the late 1950s. The Board staff devoted more of their time to maintaining standards for calibrating gauges and in co-ordinating survey scheduling among pools. The Board maintained calibration equipment and a provincial pressure standard. Other calibration equipment was required to be calibrated annually against this provincial standard. Board bottom-hole survey units were still used to compare gauge measurements, to complement industry surveys, and to survey isolated wells.

ENHANCED (IMPROVED) RECOVERY

Since promotion of maximum economic recovery from the province's oil and gas reservoirs was the fundamental reason for the Conservation Board's existence, measures that promised to stimulate or enhance recovery were a priority.

Enhanced recovery in Canada has been defined as measures taken to increase the amount of crude oil, natural gas, or natural gas liquids recovered beyond the amount recoverable by natural drive mechanisms. In the United States, the term "enhanced recovery" is sometimes used to refer to measures exclusive of natural depletion and gas or waterflooding. There, the term "improved recovery" is used to describe all processes beyond natural recovery.

Supported by the Board's active encouragement and guidance, enhanced recovery procedures were being applied in Alberta by the end of the 1950s. The first such program was the pilot project begun by Royalite Oil Company at Turner Valley in 1948 to improve oil recovery by injection of water to part of the Turner Valley Rundle pool. The pilot program was gradually expanded into a full-fledged project, and other similar projects were undertaken in other parts of the pool.

The 1950 version of the *Oil and Gas Resources Conservation Act* gave considerable attention to enhanced recovery, particularly to ensure that the Board had adequate authority to promote and monitor such projects.[19] This was reinforced in the 1957 Act, which provided that to prevent waste the Board could "require the repressuring, recycling, or pressure maintenance of any pool" and for such purpose "require the introduction or injection into any pool or portion thereof of gas, air, water, or other substance."[20] Board management in this important area was confirmed by the complementary provision that no scheme for repressuring, recycling, or pressure maintenance in any field or pool could proceed without Board approval.

The Board played an important role in working with industry to ensure that appropriate projects for increasing recovery were undertaken wherever practical. Determining the feasibility of enhanced recovery necessarily depends upon the availability of reliable reservoir data. In this regard, the Board's task of promotion was greatly eased, since from the beginning it had insisted upon the gathering of comprehensive geological, well completion, production and related reservoir information.

Along with the availability of quality data, direct economic incentives generally led operators to examine prospective projects promptly; but, where delays occurred, the regulatory structure provided other means of inducement. The system of gas-oil ratio and water-oil ratio penalties, which curtailed production of inefficient wells amenable to improved recovery programs, and declining productivity stimulated implementation of appropriate programs. The proration plan, which gave an immediate benefit to improved recovery projects through recognition of the increased recovery efficiency as soon as projects were in full operation, was also an important incentive. The increased efficiency was recognized through higher production allowables whenever the market for crude oil was satisfactory. Pressure maintenance orders issued for pools where pressure maintenance was shown to be feasible were another strong inducement for appropriate programs. These required operators to install an appropriate program or to demonstrate that such programs were inappropriate for their properties in the pools. Failure to do so could lead to the operators' wells being shut in by Board order.[21]

The requirement that operators submit their programs for approval before implementation ensured that where a program was controversial it was examined by the Board publicly and modified where appropriate before it proceeded. These public discussions also contributed to the exchange of experience and views among operators and accelerated the application of new technology.

In central Alberta in 1953, Imperial Oil initiated a pilot pressure maintenance scheme that featured both gas and water injection for the Leduc-Woodbend D-2A pool. Imperial Oil stated in its application that the program would increase recovery from 27 to 35 or perhaps 45%.[22] The program involved gas injection into the gas caps in the northern and northeastern parts of the pool and water injection in the northern and western parts of the pool. In retrospect, the anticipated performance of the Leduc-Woodbend D-2A pool proved conservative. The program was expanded and continued throughout most of the pool's producing life. It eventually resulted in recovery of 44% of the estimated initial oil in place in the reservoir, which is substantially better than the 30% recovery usually achieved from Alberta oil pools. The D-2A pool is of substantially lower quality than the deeper Leduc-Woodbend D-3A pool, which has recovered 65% of its estimated initial oil in place through a program of natural water drive assisted by injection of water and gas.[23]

The Pembina Cardium A pool in west-central Alberta was initially produced by the natural solution gas-drive recovery mechanism. In this lower quality reservoir, gas-oil ratios usually increased sharply after relatively small amounts of oil had been recovered. The concern about recovery efficiency and the increasing gas-oil ratio penalties pursuant to schedules set by the Board soon led the well operators to consider enhanced recovery methods. Pilot water injection schemes were started by Pan American (Amoco) in April 1956 and by Mobil in December 1956. Seaboard (Texaco) developed a full-scale pressure maintenance scheme by water injection in part of the pool in November 1956. The schemes were so effective in controlling gas-oil ratios and maintaining crude oil productivity that such programs were soon undertaken in most other permeable parts of the Cardium pool. In light of this experience, the Board held a hearing in 1958 to obtain the operators' views on the need for an order that would automatically require early pressure maintenance programs to be developed in the pool in an orderly manner wherever such a program was economically feasible.[24] The proposal was supported and an order was issued. Water flooding in the Pembina Cardium A pool increased recovery by natural depletion from about 11% to 24%.[25] The introduction of such programs in the Pembina Cardium A pool marked the real beginning of large-scale enhanced recovery programs in Alberta oilfields.

From the mid-1950s (see Table APP.X.2), the number of pressure maintenance schemes promoted and approved by the Board began to grow. The results were significant, and here the Board played one of its more important resource management roles. Although a few early programs involving gas injection proved to be of little benefit in Cardium sand pools, as did small slug solvent flood programs, the pressure maintenance programs by water flood and projects involving alternate injection of solvent and water in these

TABLE APP.X.2   Early Alberta Enhanced Recovery Projects, 1948–1960

Initiated	Field and Pool	Type	Remarks
1948	Turner Valley Rundle	Water flood	Pilot project
1953	Leduc-Woodbend D-2A	Pressure maintenance by water and gas	
1954	Leduc-Woodbend D-3	Pressure maintenance by water and gas	
1956	Pembina Cardium	Water flood	Two pilot projects, one commercial project
1956	Acheson D-3	Water injection	
1957	Pembina Cardium	Water flood	Eight projects
1957	Joffre Viking	Water flood	Two line-drive projects
1958	Pembina Cardium	Water flood	One pilot and 20 other projects
		Gas	One project
		LPG, gas and water	Two projects
1958	Hamilton Lake (Provost) Viking	Water flood	
1958	Turner Valley Rundle	Water flood	Two projects
1959	Pembina Cardium	Water flood	One pilot and 18 other projects
		Gas	Two projects
		LPG and gas	One project
1959	Gilby Viking	Water flood	Line drive
1959	Wainwright Sparky	Water flood	Pilot project
1959	Sundre Rundle	Gas	Partial pressure maintenance
1960	Pembina Cardium	Water flood	Nineteen projects
		Gas	One project
1960	Turner Valley Rundle	Water flood	Partial pressure maintenance
1960	Chauvin	Water flood	Pilot project

low-quality reservoirs usually increased the crude oil recovery as a portion of oil in place from 9 to 15% by natural means to 14 to 30% by water, or solvent and water, injection.[26]

Pipelines

Through the McKinnon period, a considerable amount of the Board's time and energy was devoted to assessing pipeline applications.

In 1949, the *Pipe Line Act* was amended to give a number of specific powers to the Petroleum and Natural Gas Conservation Board, although the final approval for a permit remained with the minister of public works. Undoubtedly, the reason for including the Conservation Board in the approval process was to enable the minister to draw on its extensive knowledge and expertise respecting the industry. In any case, the Conservation Board was required to examine the application for each pipeline and to advise the minister whether it approved or disapproved of it. It could also suggest changes and alterations to the plan. The minister would then take the Conservation Board's recommendations into consideration in coming to a decision. Further, the Act provided that the Conservation Board's approval was necessary before a permit was issued. Once the pipeline was completed, an order permitting its operation could be obtained from the Board of Public Utility Commissioners. This permission was subject to the condition that the operator could not own or have an interest in any petroleum or natural gas properties in the province until the pipeline had been declared a common carrier by the Conservation Board.

In the exercise of these new powers, the Board reviewed all applications coming before it. Most related to extensions to existing pipelines, such as laterals to field gathering systems. From time to time, however, major new pipelines were applied for, and in most of these cases the Board conducted hearings under the *Pipe Line Act*. Some of these were lengthy. Following the hearings, its recommendations were forwarded to the minister of public works.

In 1952, a number of changes were made to the *Pipe Line Act*. Henceforth, the minister of highways rather than the minister of public works was responsible for administration of the Act. In the granting of a permit, the minister was required to consider the recommendations of the Conservation Board, but was not required to have the approval of the Board before granting a permit. Expropriation of the land necessary for the placing of the pipeline was made subject to a review by the Board of Public Utility Commissioners. This review was followed by a hearing and an order outlining the terms and conditions of the expropriation. Thus, while the expropriation of land was still facilitated, the machinery put in place was intended to ensure adequate compensation and fair terms and conditions to the landowners.

In 1958, administration of the *Pipe Line Act* became the responsibility of the Department of Mines and Minerals. With respect to the transfer of authority, H.H. Somerville, then deputy minister of the Department of Mines and Minerals, recalls that the change was instigated by a meeting the industry had with Premier Manning.[27] According to Somerville, the industry expressed concern over the time taken to process applications and apparently convinced Manning that the complaint had merit. A reduced, although still important, advisory responsibility was left with the Board, and it continued to receive copies of pipeline applications. For a gas line, the Board was required to notify the Department whether it approved or disapproved, and it could recommend changes. For an oil line, or a secondary gas line, the Board had simply to notify the Department of any objections it might have.

The 1952 Act regulated flow lines for the first time. These are small-diameter pipelines that take oil from the well to field processing or storage facilities. Permits were not needed for their construction, as the owner was deemed to be the permit holder; however, a licence was required, and this necessitated the provision of certain details, such as a sketch showing the route, the size of the line, and the substance transmitted. Similar information could be requested for flow lines that had been constructed before the Act was enacted. Not all operators responded wholeheartedly to these requests, and the lack of full compliance resulted in little improvement to the incomplete record of flow lines in the province. Throughout this period, many flow lines were installed on the surface, because of a lack of regulatory control, and in the absence of an approval system (see COMMON FLOW LINES, this appendix).

At no time during this period did the Conservation Board have personnel in the field offices overlooking any aspect of pipeline construction or operations. Before 1958, provisions for inspection during construction and operations, and other essential checks, were included in regulations administered by the Board of Public Utility Commissioners. After that date, the superintendent of pipelines maintained a staff of pipeline inspectors for field surveillance.

Drilling Operations and Completions

BLOWOUT PREVENTION

Whenever or wherever oil and gas wells are drilled, blowouts are a possibility. In the days of cable tool drilling, blowouts were associated to a considerable extent with the discovery of gas. The first blowout in Alberta occurred in 1883 at a well near Suffield when, instead of water, the drillers found gas. A well drilled in 1897 in the Athabasca River valley found gas at 897 feet and blew out of control. The well flowed for 25 years before it was capped. Of 15 wells drilled along the Peace River from 1916 to 1920, eight that encountered gas, in some cases with salt water, were left flowing uncontrolled. One of these, Peace River No. 2, was finally plugged some 35 years later by Board staff under the direction of A.F. Manyluk.

The introduction of rotary drilling units, which began in the early 1930s, greatly reduced the frequency of blowouts. With rotary drilling, where mud is circulated through the drill bit and back to the surface, the hydrostatic pressure of the full column of mud is maintained against the formation pressure. The more rapid rates of drilling with rotary equipment also meant shorter drilling periods in which blowouts could occur; nonetheless, numerous blowouts have occurred since the 1930s. Atlantic No. 3 in the Leduc field was the most serious of those that occurred before 1960, and was caused by failure to control a lost circulation problem, aggravated by having an inadequate depth of surface casing, which in turn allowed well fluids to escape around the casing to the surface. Board policy on blowout prevention was made more stringent following this event.

By the early 1950s, major oil companies had developed drilling engineering departments with specialists, many of whom had worldwide experience. Many of these men were willing to share their knowledge and experience with Board engineers. From such sources, and from the experience gained by Board field men throughout the province, the Board put together its initial requirements for blowout prevention equipment, and in July 1957 they were included in the Drilling and Production Regulations for the first time. These Regulations were general, and the field staff were still required to exercise a great deal of judgement when determining whether the blowout prevention (BOP) equipment on any particular well was adequate. These regulations, however, did specify that the BOPs must be able to close on a pipe and open hole, have bleed off and kill lines installed, have controls outside the substructure, have been pressure-tested before drilling out of casing, be mechanically tested each day, and be heated during winter operations.

Each time a blowout occurred and a subsequent investigation identified a weakness in the blowout prevention system employed, requirements were reviewed and upgraded. For example, a principal source of ignition during the escape of gas or oil from a well was found to be the rig engines. Even though their normal fuel supply was shut off, the engines would gain fuel from the escaping gas entering the air intake to the engines. The rich fuel mixture would cause the engines to "run away," leading to red hot exhaust manifolds that would ignite the flow of gas from the well. To reduce the risk, the Board amended the Regulations to require all diesel motors within 75 feet of any well to be provided with an approved type of air intake shut-off valve. A readily accessible remote control or a suitable duct was also specified so that the air for the motor was obtained at least 75 feet from the well.

Generally, Board and industry had similar aims respecting well control and blowout prevention. The Board, however, found it necessary to spell out its requirements in regulation form to ensure that *all* operators and contractors met reasonable standards of safety. In October 1959, the Regulations were amended again, and this time requirements became much more detailed and rigorous.[28]

The type of blowout prevention equipment in use in the 1950s was considerably improved over that used earlier. The gate- or ram-type devices were usually dual units, equipped so that one set could close off open hole (blind rams) and the other could close on drill pipe or perhaps on tubing (pipe rams). Originally, these were mechanically operated, but later they were converted to hydraulic operation, at least for wells over 4,000 feet. The hydraulic preventer, then as now, consisted of a large donut-like rubber element that fitted into a steel cylinder with a piston at the bottom that could be moved up hydraulically to compress the rubber element against the drill pipe. Although not without shortcomings, it could close on any annular shape, and on itself if need be (see Figure 8.6).

When a Board representative inspected a drilling rig, part of his responsibility was to check the blowout prevention system and satisfy himself that the minimum requirements had been met and the equipment was in workable condition. Among other things, the daily drilling report was perused to ascertain if the BOPs were mechanically checked daily, and that a pressure test of the equipment had been run before the surface casing shoe had been drilled out. In addition, if operations at the rig permitted, the inspector could require a mechanical check of the equipment while he was present. If the equipment was inadequate in any way, the operator was required to remedy the problem, and for severe deficiencies the well could be shut down by the Board inspector until the condition had been corrected.

The knowledge that the Board representative could appear at any time on an inspection visit assisted the companies and contractors in enforcing acceptable standards among rig crews as well as staff involved in other drilling and production operations.

CASING PROGRAM
A typical casing program for most areas in the province included the setting of two strings of casing: the surface casing and the pro-

duction casing. For some areas, an intermediate string, that is casing set at a depth between the base of the surface casing and the top of the productive zone, was an acceptable alternative. Where production was assured, some operators preferred to set the production casing through the productive zone and then perforate it. In cases where special hazards existed, such as where high-pressure gradients might be encountered, the Board required, or operators opted for, intermediate casing at the top of the zone before drilling in, and to complete the well either "open hole" or by setting a liner (a short section of casing through the productive zone). In applying for a licence to drill a well, an operator was required to include details of his proposed casing program. This was reviewed by the Board staff to ensure that it met minimum standards for the area involved.

In the 1940s, before the Leduc field was discovered, casing programs were not significantly different than in the subsequent period. In Turner Valley in 1942, for instance, the length of surface casing was set at around 500 feet and production casing was cemented through to the producing zone. In the Princess area, surface casing was also set at around 500 feet in 1941; at Taber, a little later, surface casing was set at about 150 feet. The Board had no requirements on the diameter of casing. Usually, 10-inch surface casing and 7- or 5½-inch production casing was used.

With the discovery of the Leduc field in 1947, the Board assigned specific requirements for wells in that area. The minimum surface casing to be used was 250 feet, cemented to the surface. Production casing was to be cemented while having 25% in excess of the amount theoretically required to cover from the bottom of the hole to above the Viking Sand. The Viking Gas Sand, and prospective producing horizons in the Lower Cretaceous, were thus protected. The surface casing at Atlantic No. 3, was set at 296 feet, but the well lost circulation and blew out of control before the production string could be set. This catastrophe led to a quick review of the Board's casing policies and the introduction of revised casing requirements, which were applied for the decade that followed.

Surface Casing One essential role of surface casing is to protect near surface freshwater aquifers from contamination by drilling fluids and hydrocarbons. In addition, while the well is being drilled, surface casing is the primary anchor of a well's blowout preventers, and it is critically important that the surface casing be set in a competent (well-consolidated) formation and cemented full length with quality cement that bonds well with the pipe and the formations penetrated. Since the strength of formations and formation pressures increase with depth, it is necessary to increase the depth of surface casing as the depth of formations to be evaluated increases.

In the days of cable tool drilling, casing was set and cemented, usually with a dump bailer, whenever the penetration rate became unacceptably slow because of swelling clays and water influxes. Hence, by the time the well reached the expected producing formation, several strings of casing had often been cemented in the hole. Consequently, situations where hydrocarbons broke out around the outside of the last casing string and escaped to the surface in cable-drilled holes were rare. Rotary drilling became the dominant method of drilling by the time the Petroleum and Natural Gas Conservation Board was formed in 1938. The ability of a rotary rig to drill to substantial depths without the need for intermediate casing to protect the hole added to the importance of establishing minimum setting depths for surface casing.

Imperial Oil planned to test the Devonian formation in its Leduc discovery well and set 500 feet of surface casing. After Atlantic No. 3 in the Leduc field blew out of control and well fluids flowed to the surface outside the surface casing, the Board responded quickly and increased the minimum surface casing required from 300 to 600 feet for wells licensed to tap the Leduc formation. Operators who preferred not to set 600 feet of surface casing were required to set an intermediate casing string before penetrating the Leduc zone. In such cases, if production was found, the well could be completed with a liner. Surface casing setting depths have been increased as emphasis on well control and the prevention of the escape of gas containing deadly hydrogen sulphide (H_2S) heightened.

Surface casing was never required to be set in Alberta to below 1,500 feet. If it appeared that setting surface casing to 1,500 feet would not provide adequate protection from the pressures anticipated in the target zone, an intermediate string of casing was required to be set before the high-pressure zone was penetrated.

During the last half of the 1950s, the Board issued maps of the province that indicated the minimum surface casing requirements. The letter accompanying the map indicated that modifications to the standard requirements in a field or area would be considered if requested and if data supported such a change.

Cementing the casing in place is an important aspect of the casing procedure. The Board required that surface casing be cemented over its entire length. Since the cement slurry was circulated through the annular space between the casing and the wellbore from the base of the casing, returns of cement at the surface provided confirmation that this requirement had been met. For other casing, including production casing, cement had to extend high enough to seal off the shallowest sand that might be productive of oil or gas. Usually, this was ensured by using an excess of some 20% over the theoretical cement requirement. In coal areas, as an additional precaution, a temperature survey was required to show that the desired level of cement fill-up had been achieved.

Surface Casing Vents The Drilling and Production Regulations in 1951 included for the first time the requirement that a surface casing vent be included for wells drilled within two miles of coal mines, whether or not they were being worked or were abandoned.[29] The vent was included to avoid pressure build-up in the surface casing-production casing annulus and to reduce the risk of hydrocarbons escaping from the well. If pressure build-up did occur in the annulus, hydrocarbons could migrate through breaks in the surface casing or from below the casing shoe along the outside of the casing into a coal seam that was being mined or might possibly be mined in the future (see *COAL AREA WELLS*). The surface casing vent requirement was just one of several arrived at following meetings early in 1951 between Board Chairman McKinnon and J. Crawford, the director of mines in the Alberta Department of Mines & Minerals.[30] Oil and gas companies were seeking approval to drill for oil or gas within two miles of coal mines where drilling had previously been prohibited.

Although originally required only for wells drilled in coal areas, by 1957 the surface casing vent was recognized as being of special value as a means of minimizing the risk of hydrocarbons escaping undetected from a wellbore. Consequently, it was made a requirement for all new wells to protect near-surface aquifers and to avoid the escape of undetected hydrocarbons into other zones.

DEVIATION SURVEYS

The Drilling and Production Regulations required that a well be completed within its "target area" to qualify for a full production allowable. An oil well drilled for oil in a 40-acre drilling spacing unit (DSU), if it was to qualify for a normal allowable, had to intercept the producing horizon within an approximately 660-foot-square target area centred in the middle of the DSU, with its south dimension parallel to that of the DSU. Similar provisions were applicable to DSUs of other sizes.

A deviation survey was required at least every 500 feet, from the top to the bottom of the well, to ensure that the well was completed within its target area. As it implies, a deviation survey shows the deviation or divergence from the vertical of the well at the depth the survey is run. The surveys had to be done during the process of drilling and before drilling had proceeded more than 1,000 feet beyond the depth at which the previous survey had been run. It was one of the duties of the Board's field representative to ensure that these surveys were reported on the daily drilling logbook.

Whenever a well began drilling from a surface location outside the surface target area, a directional survey was always required. Other special circumstances could make a directional survey advisable. Regardless of the circumstance, the operator was required to report the results of the survey to the Board before the well was placed on production. The report had to be accompanied by copies of the directional survey. Failure to complete the well within the target area could result in a reduced well allowable. By running a directional survey, the operator could plot the path taken by the drill and determine accurately the location of the wellbore where it entered the producing zone.

COAL AREA WELLS

Special precautions for the protection of coal deposits in areas of workable seams have been included in Board-administered regulations since its inception in late 1938. The coal industry was extensive in the province, and coal seams were being worked in many areas where drilling for oil and gas could be expected. The Regulations under the *Oil and Gas Wells Act, 1931*, contained a section relating to wells licensed to drill through workable beds, or seams of coal that were being worked. Under this provision, both the Board and the chief inspector of mines had to approve the licence, including such conditions and precautions as they might find necessary for the protection of the coal seams and the persons engaged in mining them. The same authorities were required to approve any abandonment program for wells drilled in coal areas before the abandonment could be undertaken.

By 1951, the Drilling and Production Regulations included detailed requirements for the licensing of wells in coal areas, and even more stringent requirements were added by amendments in 1957. Under these regulations, no person was permitted to drill within two miles of any subsurface mine workings, either active or abandoned, unless certain provisions of the Regulations were complied with, or permission in writing had been obtained from the director of mines to proceed. Among the requirements were the filing of an up-to-date copy of the plan of survey of the mine workings or, if the mine was abandoned, the latest plan of the survey available; location of the well had to be at least 50 feet from any part of any active or abandoned mine workings; and a pillar of coal of a radius of 50 feet had to be left surrounding the wellbore. In certain cases, however, the Board could permit drilling through abandoned coal workings without the necessity for the pillar of coal surrounding the wellbore. There were also specific directions for the setting of surface and other strings of casing.

All requirements for the actual drilling, completing, producing or abandonment operations were the responsibility of the Board. Much of this responsibility was placed in the hands of the Board's field representatives, and particular attention had to be paid to special conditions that were a part of the licence to drill. Since a Board inspector had to be present at the lease for the cementing of all casing and also abandonments, it was necessary to maintain fairly close contact with the licensee's representative at the well. All programs had to be reviewed and approved by the Board, and this meant that the field inspector had to become familiar with the geological, engineering and other related aspects of the well being drilled.

On the other hand, the director of mines was responsible for approving the location of the well, since he had direct knowledge of the presence of coal mines. He also had to advise the Board of any significant information respecting the coal mine workings, including the depth of the workable coal seams. As well as administrator of the *Coal Mines Regulation Act,* he was responsible for ensuring that a 50-foot barrier of coal around the wellbore was maintained. Finally, he had the responsibility of approving beforehand any abandonment programs issued by the Board.

Overseeing coal area wells was one of the more demanding duties of the Board's field staff, because the schedules that had to be followed were beyond their control. The field inspector had to be available, on little notice at all times, night or day, regardless of weather or road conditions.

Despite the best efforts of all concerned, there is at least one record of a well being drilled inadvertently through an abandoned mine working. This well was drilled in the Big Valley area. The deficiency in the assessment came to light when the company tried to cement the surface casing in place only to find that inordinate quantities of cement disappeared.[3]

SPACING OF FACILITIES

Under the 1957 Drilling and Production Regulations, including amendments to 1959, there were a number of specific requirements regarding the spacing of equipment and facilities. These were intended to reduce the drilling hazards to the men involved in the drilling operation, to the well itself, and to the surrounding property.

There was a requirement that the remote control for closing the blowout preventer had to be located at least 100 feet from the well for wells of more than 4,000 feet. This would ensure that a person could operate the equipment even if it were not possible to get close to the rig. The other restrictions were intended to reduce the possibility of fire at the wellsite. No open-element electric heater or flame-type stove was permitted within 75 feet of any well. This regulation affected the type of heater used in the doghouse and on the drilling rig floor. Sometimes it was difficult to enforce in an Alberta winter. No boiler, steam generation plant, flare pit or open end of a flare line was allowed within 150 feet of the well. All motors within 20 feet of a well had to be spark-proof, as well as equipped so that air was obtained 75 feet from the well if a suitable air intake shut-off valve had not been installed. This would prevent the engines from running out of control in case of a blowout. The exhaust pipes from internal combustion engines within 75 feet of a well had to be coated or insulated to prevent ignition of flammable material.

Finally, smoking was forbidden within 75 feet of the well. The number of Board-initiated inquiries into such infractions attests to the difficulty of enforcing this prohibition.

During the process of drilling, dangers other than those addressed by the requirements noted above existed. For example, if the sump used to accumulate waste material and drilling fluids was improperly constructed or located, the possibility of environmental damage was quite real. To enable the Board to act quickly to reduce such potential dangers, the Drilling and Production Regulations also contained an important section that gave the Board and its field staff open-ended authority. It declared that

> where it appears to the Board or its authorized representative that a method or practice being employed in any drilling or abandonment operations is in any way inadequate, improper or hazardous, the Board or its representative may require that the operation be discontinued until approved methods are adopted.[4]

WELL ABANDONMENT

The abandonment of a well was also identified as an operation that required prior Board approval. An application for approval to abandon had to include a description of the condition of the well and the proposed program of operations. The Board approved the program as proposed or as altered to the extent that it considered necessary. In general, there were two categories of abandonment. The most common related to a well that had just completed drilling and had failed to find oil or gas, or had found amounts that could not be developed commercially. As a matter of practical necessity, the Regulations provided that the Board could approve a program verbally and confirm its approval after considering a subsequent written application. The second type of abandonment related to a well that the operator wished to abandon that had produced in the past but had subsequently become nonproductive or uneconomic to produce for whatever reason. In this case, an application submitted to the Board was processed in a more cautious manner before approval was given and operations began.

The issuance of programs for the abandonment of recently drilled wells usually fell to senior members of the development department in Calgary or to experienced personnel at the field office responsible for the area in which the well was located. In practice, the majority were issued by the field office. Before approval was given to proceed with an abandonment, the person approving the program (the issuer) had to consider a number of factors, including whether or not the well was on crown or freehold land, if there was any possibility that the well might be productive, and if there was a possibility that the well could be utilized for other purposes.

If the well was on crown land, the approval of the director of mineral rights in Edmonton had to be obtained before the abandonment could proceed. The director required all pertinent information relating to the well, but would normally be guided by the Board's opinion in making his decision. For a well on either crown or freehold land, the Board engineer had to establish to his own satisfaction that the well was nonproductive, since he was in part acting on behalf of the owners of the petroleum or natural gas rights. This was rarely a problem. A responsible operator would not knowingly abandon a potentially productive well; however, it was considered necessary to review the significance of drill stem tests and geological information provided to ensure that important information had not been overlooked.

Some wells being considered for abandonment were inside producing field areas, and the wells might be useful for other purposes, such as water disposal wells, gas injection wells, or one of a number of other service uses. If there seemed to be a real possibility of any use for the well, the abandonment could be deferred until that possibility had been considered by the Board's Calgary office.

With these possibilities resolved, the issuer of the program had finally to consider the regional geology of the area in which the well was located. He had to ensure that possible productive zones were sealed off in such a manner that they could not be invaded by fluids from other zones and thereby affect production from nearby producing wells. This could require either a cement plug extending through the formations that were productive in the area or, alternatively, plugs set to isolate them from other porous zones. To prevent fluid from reaching the surface, a cement plug was run through the base of the surface casing. To complete the process, the surface casing was cut off three feet below ground level and sealed either by a welded cap or by a five-sack cement plug.

A copy of the abandonment program issued or approved was sent to the development engineer in Calgary. If the operation had been approved by the director of mineral rights a statement was added to that effect.

Surface clean-up of the lease had to begin within three months of the date the subsurface abandonment had been completed (see Lease Inspections).

Production Operations

BATTERY FACILITY REGULATIONS

Before 1945, the administration of regulations for the operation of production tank farms was the responsibility of the provincial Department of Public Works. Under that arrangement, the regulations governing a producing well were the responsibility of the Conservation Board, whereas those for the tank farm, which in many cases was adjacent to the oil well, were the responsibility of the Department. The possibility existed for conflicting instructions to operators, and it was soon recognized that this introduced an unnecessary administrative complication. In discussions between the industry, represented by the Alberta Petroleum Association, the Board and representatives from both the Department of Lands and Mines and the Department of Public Works it was agreed that the responsibility for the administration of production tank farms should be transferred to the Board.

The Drilling and Production Regulations, as amended in 1945, included for the first time instructions relating to tank farms. These instructions stated simply that all oil tanks or batteries of tanks had to be surrounded by a dike or ditch of a capacity *equal* to that of the tank or battery of tanks, and the dike or ditch had to be maintained in good condition, free from high grass, weeds or combustible material. The 1951 Regulations changed the required capacity of the dike or ditch to *greater* than that of the tank or battery of tanks. Additional sections specified the minimum distances that tanks, flare lines and other equipment normally found within the confines of a tank farm had to be from rights-of-way, from occupied areas, and from each other. By 1957, this section had been further refined and made even more detailed. The 1958 distance requirements are shown in Figure APP.X.2.

The Regulations also required that flare lines be a minimum of 330 feet from railways, pipelines, road allowances and occupied areas. Depending on their total capacity, storage tanks were to be a minimum of 200 to 350 feet from other tanks. No minimum distances were specified between wells and separators, but separators could not be located within the diked area of the tank enclosure. Specific provisions governed the location of flame-type equipment.

An important part of the periodic checks carried out by the Board's field staff was the inspection of facilities at each battery to make sure that distance requirements had been met.

COMMON FLOW LINES

The central issue regarding common flow lines was the Board's need for consistently accurate measurement of the production from individual wells. The 1951 Regulations permitted, upon application and approval, the keeping of production records and the filing of production reports for a battery or group of wells on a group basis on the understanding that any proration of production to individual wells should represent as nearly as possible the production of each well. This required that there be test equipment available to test the rate of production of each well for which production was not measured continually. In the 1950s, a number of measurement and reporting infractions made the Board increasingly concerned over the adequacy of both the procedures and the equipment being utilized under the common flow line policy.[33]

By 1959, permission to operate a proration battery was withheld until the battery site, together with its components, had been approved by the Board at the time of installation, alteration, or whenever additional wells were tied into the battery. The policy required that the production from each well be transported to the battery installation through a separate flow line unless the Board had granted specific exemption from that requirement. In its Letter to All Operators dated 28 October 1959, the Board spelled out the conditions

FIGURE APP.X.2   Production Battery Showing Required Equipment Spacing, 1958

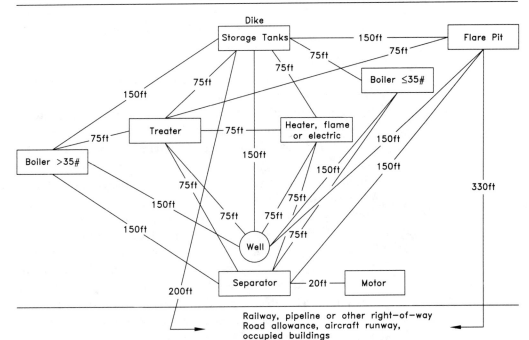

under which applications for a common flow line would be considered. These were the following:

(a) *Satellite batteries.* The Board was prepared to approve auxiliary batteries and transportation of total production to each battery through a common flow line where equipment was available for the separate testing of each well producing to it. Amounts of oil, gas and water had to be determined separately.

(b) *Parallel test lines.* Production of more than one well could be transported to the battery through the same line, providing that each well was served by a test line serving not more than four wells. Before a production test was run, the test line had to be flushed with at least one normal day's production from the well.

(c) *High-productivity wells.* Production from not more than four wells could be transported to a battery by a common flow line if the productivity of the wells was such that each was capable of successively producing its maximum daily allowable through the line and the operator agreed to produce the wells successively or, in other words, one at a time.

(d) *Incapable wells.* Production from two or more wells, each of which was incapable of producing its penalized allowable, could be transported by a common flow line, providing that each well was tested satisfactorily.

One of the problems associated with common flow lines was that each well had to be capable of producing at its normal rate against the pressure condition of the flow line. If a well was affected by this pressure, its production would either cease or be greatly reduced and test results would be meaningless. To obviate this problem, the Board required that each well be produced through a suitable choke to maintain its flowing wellhead pressure at at least twice the maximum flow line pressure. A pumping well, of course, would not require this control.

An important responsibility assigned to Board field staff as part of their battery-site inspection duties involved confirming the adequacy of the equipment installed for the particular circumstances. Initially, this required the witnessing of production tests and comparing the results with those of previous tests of the same well. If the two tests were inconsistent, the reasons were considered. Often the reasons were acceptable, and usually the differences were the result of inadequate procedures, rather than of equipment failure. One of the main sources of difference was the incomplete flushing of the test line. In later years, emphasis in inspections was made on production practice checks, which included a complete study of production practices at a battery.

DISPOSAL OF PRODUCED WATER

Oil is usually found in association with salt water. Often, salt water under pressure is the principal force displacing oil to producing wells. Even in reservoirs where the main driving force is gas expansion, free water underlying the oil may expand and migrate towards wellbores. Hence, as oil and gas are produced from the reservoir, the water-oil interface may advance and salt water enter producing wellbores where it is brought to the surface along with the oil. At this point, the salt water must be separated and disposed of.

Produced water from Alberta oil and gas pools exhibits a wide variation in salinities. Water produced from the shallow Milk River sands in southeastern Alberta contains relatively small amounts of dissolved salts and is commonly used by cattle and other domestic animals. Because of the long dry summers in this part of the province, evaporation is high and large quantities of low salinity water could be produced to surface pits without damage to the terrain. In the deeper producing formations to the west and north, salinities are markedly higher, and extensive production of salt water to surface pits was early recognized to have a deleterious effect on both the soil and the shallow sands that were a source of potable water.

The Regulations issued under the *Oil and Gas Wells Act, 1931,* which came into effect in January 1939, stated that "no salt water and no drilling fluid shall be permitted to flow over the surface of any land."[34] By implication this meant that salt water had to be contained in pits or other receptacles. The Board had no authority regarding production tank farms until late in 1945; hence, its control over produced water in that period was effected through regulation of its flow from wellsites. Even before 1945, however, it no doubt consulted with, and drew problems encountered at batteries to the attention of, the responsible body, the Department of Public Works.

Before the discovery of Leduc, the small Board field staff could not exercise the close and regular scrutiny applied in fields such as Turner Valley over other remoter areas of the province. Occasionally, however, the early Board minutes refer to poor operating conditions in those outside areas, requiring particular attention from the Board, such as producing operations in the Conrad field in the south and in the Lloydminster area. D.P. Goodall, after inspecting the latter area in 1943, reported on the appalling conditions there, relating mainly to poor practices in the pit storage of both oil and water.[35]

In 1945, with the Board assuming responsibility for tank farms, the regulation of salt water production and disposition was made applicable to water separated at the tank farm sites, replacing the separate regulation administered previously by the Department of Public Works.

Generally, until the end of 1950, producing oil pools were in their early stages of depletion, and the volume of water produced was small. Although occasional unsatisfactory practices did occur, such as those that demanded Board attention at Lloydminster in 1943, disposing of water produced in conjunction with oil did not create serious problems. Interestingly, serious water disposal problems first materialized at the Husky refinery in Lloydminster. The water produced in association with oil at Lloydminster did not lend itself to field separation. The oil had a low API (American Petroleum Institute scale) gravity, and the use of gathering lines to convey the production to a central battery was impractical. Gas production at each well was low and insufficient to run a treater to separate water from oil. Although some water separated out, and could be drained from field storage tanks, the bulk of the produced water was transported by truck to the refinery in the form of an oil-water emulsion. The refinery accumulated a large volume of water that could not be conveniently disposed of to a surface pit. Husky Oil Ltd. addressed the problem by completing the first subsurface water disposal facility in Alberta in July 1951. The water was injected into the Devonian formation, which is devoid of hydrocarbons in that area.

The second disposal well approved was in the Redwater field. Development drilling had resulted in some marginal wells being completed on the steep east flank of the Redwater D-3 zone close to the oil-water interface, and these soon produced large volumes of water. The Board had not prescribed surface disposal limits for water production up to that time.

The sandy soil was not suitable for farming in the area where the large amounts of water were produced; hence, it drained readily from the operators' pits to the adjoining sand. The Board, however, had established penalties for excessive water production to avoid excessive reservoir withdrawals and waste of reservoir energy. The penalty factor, $\dfrac{1}{1 + WOR/2}$, which was applied against the oil allowable of each well, provided an incentive for the owner, Dome Petroleum, to convert its wettest well, Redwater 16-31-56-20, to a disposal well. The penalty factor reduced the amount of oil that a well was permitted to produce by approximately the amount of water produced and thus penalized the well for inefficient use of reservoir energy. The produced water was returned to the zone of origin well below the oil-water interface and application of credits for such disposal exempted the oil production from a water-oil ratio penalty.

The first two subsurface disposal wells thus provided a satisfactory means to dispose of the water when surface storage became a problem and, in the case of the Redwater well, eliminated the waste of reservoir energy by returning the water produced to its zone of origin. Water disposal problems in Redwater and Lloydminster were solved relatively simply. The water injected from the first two disposal wells entered the porous and permeable limestone formations on a gravity basis without the need for pumps. When plugging did occur, an occasional acid treatment cleaned up the formation in the vicinity of the wellbore and solved most of the immediate problems associated with injection.

In the Redwater D-3 pool, however, with its active water drive, water production continued to increase. Recognizing the deteriorating situation, the pool operators established the Redwater Water Disposal Company, with the responsibility for providing a field-wide gathering and disposal system for the produced water. The first well designed as a disposal well was drilled and placed on injection in March 1952. As water production increased, more injection wells were added and the gathering system expanded. Fortunately, as a result of this carefully planned approach, water disposal at Redwater created few subsequent problems.

Water production associated with oil production had begun to become a problem in other pools by 1952. In the Joarcam Viking pool, the problem of disposal was more difficult to solve. Initially, surface injection pressures between 1,500 and 2,000 psi were required to inject water into the Viking zone, even at low rates. Such high initial injection pressures are not unusual in Alberta sand reservoirs, and they often decline as the injection continues.

Although some operators saw the advantages of subsurface disposal of produced water early in the life of a pool, many operators continued to dispose of produced water to surface pits. The Board field staff were required to check these pits at least annually, and wherever a pool's water production reached a point where disposal to surface pits was creating a problem (where the disposal rate to the pit exceeded the rate of evaporation and seepage), they would recommend the surface disposal volumes be limited. Unless an operator was able to find a subsurface disposal facility to handle his excess water production, he was forced to limit production of his water-producing wells once his pit limit was reached. In 1954, after field staff had completed an examination of salt water disposal volume and practice throughout the province, the Board began a more aggressive effort to shift the industry from surface to subsurface disposal. Through the summer and fall of 1954, a series of "W" orders were issued for various fields. The second of such Board orders is illustrative of their usual form. It informed operators in the Joarcam field that a recent survey conducted by the Board "indicated that the present condition and location of certain of the earthen pits used for the disposal of salt water are an immediate hazard to fresh water supplies in the area."[36] Operators were directed therefore that, effective immediately, not more than one earthen pit shall be used for disposal of salt water produced at any one production battery, and effective one month hence, not more than 300 barrels of salt water per month shall be disposed into an earthen pit. The order went on to identify specific production batteries where the disposal of salt water into pits would no longer be permitted. Other operators were told that they could no longer dispose of their salt water until their pits had been "repaired to the satisfaction of the Board's Camrose Field Engineer." Finally, all operators in the Joarcam field were given notice that after 1 January 1955 disposal or storage of salt water in open pits would no longer be allowed any-where in the field, and that they should prepare a scheme of under-ground disposal for the Board's approval. The numerous examples of wells being shut in for exceeding salt water disposal limits or poor containment bear witness to the Board's rigorous enforcement of its water orders.[37]

These measures, in conjunction with the WOR penalties applied to the oil production allowables, encouraged operators to establish a satisfactory water disposal scheme for each pool and proved ex-tremely effective. From 1952 to 1962, water disposal in the prov-ince changed from almost 100% surface to in excess of 99% sub-surface disposal. As a result, there are no areas in Alberta rendered sterile by the indiscriminate disposal of oilfield water. This stands in sharp contrast to the situation in many U.S. states, where salt-encrusted stream banks and sterile watercourses stood out as highly visible and disturbing reminders that produced brine was still being pumped directly into streams. In 1957, approximately one-third of the salt water produced in Louisiana oilfields was being disposed of in streams.[38] Only after the U.S. federal Department of Health, Edu-cation, and Welfare put strong pressure on Arkansas's Board of Conservation to stop pollution of interstate waters did that agency make a significant effort to end salt water disposal in streams. In Texas and Oklahoma, until the mid-1960s, salt water was still dis-posed of mainly in open surface pits.[39]

AUTOMATIC CUSTODY TRANSFER

A production tank battery receives the production from one or more wells, measures the incoming oil, gas and water, and stores the oil in tanks until it can be transported to a point of sale. The gas is either sold, injected or, if it is not large and there is no nearby pipeline or market, might be flared, and any water produced is re-turned to the reservoir or disposed of in another manner satisfac-tory to the Board. The act of transferring the oil from the battery to the transporting pipeline is referred to as "custody transfer."

Under Alberta regulations, production from each well had to be reported. Each well was tested at least monthly, and the total pro-duction at the battery, which was continually measured and ordi-narily transported to the battery through common flow lines, was allocated to the wells on the basis of the well tests. This manner of operation required extensive equipment and labour. Although pro-cedures had evolved to include some degree of automation, the op-eration was essentially manual and required the presence of skilled personnel to oversee every aspect of the process, including opening and shutting the valves, sampling the fluid stream, and manually measuring the amount of oil in the tanks. When the time came to transfer custody to the pipeline, the pipeline company also had personnel on hand to open and shut valves, to start the pumps, and to oversee the transfer of the oil.

In the less harsh environment of Texas and other southern states, great progress in automation had occurred in the late 1940s. Equipment that worked well in the south, however, did not necessarily function as well under Alberta's weather conditions. The first recorded experiment in Alberta with "automatic custody transfer" was conducted in the Stettler field in 1952.[40] In 1954, Imperial Oil installed newly developed equipment in one of its conventional tank batteries in the Leduc field to test and evaluate the component under winter conditions. The results were so promising that it was decided in 1955 to proceed with a further experimental project, this time at a Redwater field location. For this experiment, the company was joined by Imperial Pipe Line, the oil carrier, to test the viability of automatic custody transfer. Discussions regarding the proposal were held with the Conservation Board in late 1955, following which the Board granted an approval to proceed with a six-month trial to June 1956. The trial was subject to certain conditions, and the most important were that individual well test rates were to be confirmed by hand-gauging throughout the trial period, and the company was required to provide the Board with a statement of intent regarding the servicing, checking, replacing and calibration of the oil-metering devices.[41] The results confirmed that both automation of production batteries and automatic custody transfer were workable and desirable alternatives to the practices in common use.

The Board had received sufficient data by September 1956 to enable it to grant approval for the automatic production and automatic custody transfer at the experimental Redwater battery. Other similar installations were quickly proposed and approved in a number of pools throughout the province; however, the Board continued to require that each company wishing to install automatic custody transfer equipment apply for and receive approval to do so on an individual battery basis.

The changeover to automation of production batteries and custody transfer meant that all well production and battery operations could be supervised and controlled from a remote central location, from which telemetered instructions could, as necessary, be passed to the battery. On the whole, it was a significant improvement over the manual methods for larger batteries, and resulted in an operation that could meet the standards of accuracy and safety demanded by the Petroleum and Natural Gas Conservation Board. It also enabled the continuation of battery operations in all kinds of weather and road conditions and a significant saving of manpower requirements in the field.

Lease Clean-up and Maintenance

In his annual field report for 1957, Frank Manyluk, the Board's chief development engineer, discussed the number of lease, abandonment and battery inspections made during the year. He reported that Board field staff had inspected 4,727, or 49%, of the province's 9,637 leases.[42] He added that this high percentage would be reduced in subsequent years as more and more wellsites met Board requirements. Manyluk also noted that, at the beginning of 1957, there were 731 abandoned wellsites requiring inspection and that during the year, and counting reinspections, this had meant 1,044 visits by field inspectors. Finally, with regard to inspection of batteries, he observed that, in 1957, there were 1,694 operating batteries and that during the year 722, or 43%, were inspected. He mentioned again the probability of the percentage being reduced in subsequent years as batteries were brought up to Board requirements.

Despite this optimistic report, the Board had found it necessary, in November 1957, to issue a letter to the industry that discussed the policy of lease clean-ups and expressed concern about the industry's performance. First, the letter pointed out that the number of leases found in an unsatisfactory condition each year had been increasing steadily. Consequently, Board personnel were spending a great deal of time and public funds in supervising reconditioning of the leases. In addition, surface owners were denied the use of their land for several years, without receiving compensation beyond the first year. "Incidents such as these," the Board lectured, "give the Oil Industry a bad name and detract from the Public Relations efforts the industry puts forth in other phases of its work."[43] The letter instructed that, "effective immediately," all wellsites would have to be satisfactorily reconditioned within one year of the well's finished drilling date. To ensure that this was done, the letter announced the Board's intention to inspect each wellsite approximately three good weather months after the well had finished drilling. Operators of unsatisfactory leases would be given 30 days to complete the clean-up (weather permitting). A reinspection would follow at the end of the 30-day period and, if the wellsite was still unsatisfactory, an order requiring the lease to be cleaned up within 30 days would be issued. If the lease was still unsatisfactory, the Board would under-

Board engineer
Murray Blackadar
inspects an unusual
wellhead on a lease in
questionable condition
in the Princess Field in
1955. ERCB, Bohme
Collection.

take to recondition the lease and to prosecute the negligent company. In practice, the Board found that, where drilling resulted in a producing well, the most effective means of getting compliance from a negligent operator was to order the well shut in.[16]

From the foregoing, it is clear that the Board took its responsibilities in regard to both lease clean-up and maintenance seriously. Section 15 of the Drilling and Production Regulations, 1957, provided the authority for clean-up of wellsites after the wells were placed on production, suspended or abandoned. It provided for clearing the area of refuse material, debris, concrete bases and machinery, and restoring the surface to as near its original state as could reasonably be done.

During this period, an increasingly common field practice was to remove the topsoil from the leased area and pile it on a part of the lease where it would not be directly affected by the drilling operation. Shortly after drilling was completed and the rig removed, the water, mud and shale would be drained from the pit or sump and spread over the lease to dry. At the same time, the refuse material and other debris would be removed from the lease. After a time, which depended to some extent on the weather, the lease would become sufficiently dry to permit filling of the pits, contouring of the lease, and spreading the topsoil over the disturbed area. Once this operation had been completed, the company would contact the Board and a representative would be sent to inspect the site. If possible, the representative would contact the landowner at this time to obtain his opinion of the restoration. Usually, the condition of the lease spoke for itself. It was either satisfactory or more work needed to be done. In cases where the Board and the landowner could not agree, the Board representative was obliged to rely on his own judgement. Reinspections of unsatisfactory leases were sometimes required, usually for relatively minor problems.

The Board's responsibility for abandoned wellsites ended when approval of the restoration had been given. For producing wells and other leases, inspections were carried out on a regularly scheduled basis as insurance against sloppy or dangerous producing practices. It was against Board policy, for example, both from an environmental and safety viewpoint, to allow oil or water spills to remain unattended. Weeds had to be controlled both on the lease and the lease road, and well-name signs were required to be posted near the wellhead. In the latter part of the 1950s, the inspection procedure included checking for the presence of a surface casing vent open to the atmosphere.

BATTERIES

Lease inspections were also carried out on battery sites on a regular basis. The annual field report for 1959 stated that it was the Board's objective to inspect batteries once every three or four years. This was to ensure that distances between the various types of equipment used at a battery were maintained in accordance with the regulations (see Figure APP.X.2), to make certain that equipment used on the site was adequate for the particular functions being carried out, and to ensure that the battery site was maintained in a clean, safe condition and free of weeds. Although the housekeeping at some batteries was substandard, most were well-kept, and some operators took particular pride in battery maintenance. Usually, notice from the Board was sufficient to bring compliance, but if it did not the Board could and did shut wells in.[17] Board field staff also checked salt water disposal practices and ensured, as far as possible, that flares were lit at all times. The products of combustion of gas usually posed a much less serious threat to the environment than did the gas itself. This was particularly true when the gas flared contained hydrogen sulphide (H_2S).

The inspection policy of the Board was comprehensive, and the number of regular inspections almost guaranteed that serious divergence from prescribed policies would be minimal. Special inspections were carried out when complaints were lodged by members of the general public adversely affected by some shortcoming or sudden failure at a wellsite or battery site.

RESOURCE EVALUATION, EVOLUTION OF RESOURCE CONSERVATION, SAFETY AND ENVIRONMENTAL MANAGEMENT IN HANDLING NATURAL GAS PRODUCTION

Evaluation of Reserves

The *Gas Resources Preservation Act,* 1949, assigned to the Board responsibility to ensure "the preservation and conservation" of the province's natural gas resources for "the present and future needs" of Albertans. To be sure that natural gas exported from the province was surplus to Alberta's requirements, it was necessary therefore for the Conservation Board to have current authoritative data on gas reserves. At the same time, there was a reluctance in Alberta, as noted in the 1948 report of the Dinning Commission, to rely on periodic estimates developed by federal authorities such as the Geological Survey of Canada, which had done substantial work and was the principal source of provincial gas reserve estimates throughout the 1930s and 1940s. The Geological Survey had provided good assessments, but there was some discomfort about relying upon a federal agency, that was also perceived as speaking for consumer interests in eastern Canada, to make assessments of Alberta's reserves. Thus, from 1949, the Board had a priority obligation that demanded ongoing careful study and analysis of Alberta's oil and gas reservoirs. Later in the 1950s, as more proposals were brought forward to remove gas from Alberta, the Board's own need for dependable information was reinforced by a strong demand from individuals, cities, towns, interest groups and politicians for good data on reserves, reserve growth trends, and growth in provincial requirements for natural gas. The industry also required consistent authoritative data on all pools from which to develop submissions respecting pipelines to new markets.

The Board's mandated responsibility to monitor the province's natural gas reserves and requirements also lent new moral authority to its traditional pursuit—the conservation of natural gas. The trend of the Board's estimates of proved reserves is shown in Figure APP.X.1. "Proved reserves" (sometimes referred to as "established reserves") is the term used for both oil and gas to refer to quantities that meet prescribed conditions of assurance and in which there is a high degree of confidence. Initial proved reserves in Alberta grew rapidly from 4 trillion cubic feet in 1948 to 33 trillion cubic feet in 1960. The Canadian Petroleum Association (CPA) began publishing its own estimates of proved reserves in 1956. The lower trend of its estimates is explained mainly by the CPA's lack of access to confidential Board data on certain reserves.

Limited markets in the 1940s and early 1950s meant that there was little incentive to drill additional wells to define better many discovered gas pools, particularly those that were remote from existing and proposed pipelines. Thus, there was a large amount of gas between, and too far from, widely spaced wells to qualify as proved reserves. A portion of the gas that did not qualify as proved was classified as "probable reserves" and received some consideration by Board and industry in reviews of proposals for new pipelines.

Part of the uncertainty that existed about reserves related to doubt about the recoveries that would be achieved in many of the pools, particularly those for which there was little or no production history. Board estimates made considerable allowance for this uncertainty; however, the reserves of some pools, such as the Pincher Creek Rundle A pool, which developed severe water production problems soon after production commenced, had to be reduced. On the other hand, reserves of other pools, such as the Waterton Rundle-Wabamun A pool, performed better than projected, and estimates were substantially increased on the basis of production behaviour. Overall, the proved reserves grew substantially beyond the amount classified as proved in 1962. About 47% of the province's 1991 initial established gas reserves are in pools discovered before the 1961 year-end. This compares with 63% of the 1991 initial established crude oil reserves, which are in pools discovered to the same date.

The Board also maintained estimates of the proved reserves that ultimately would be developed in the province. These estimates were based on the average addition to reserves per exploratory well drilled and on an estimate of the total number of exploratory wells expected to be drilled. Estimates were also made by applying the ratio of reserves discovered per volume of sediments favourable for the occurrence of natural gas in extensively explored basins in the United States to the total volume of such sediments identified in Alberta. Since the exploration was directed mainly towards crude oil through this period, the estimates for natural gas proved to be extremely conservative. In the early 1960s, as shown in Figure APP.X.3, the Board estimated the ultimate proved reserves to be in the range of 75 to 90 trillion cubic feet, as compared with 170 trillion cubic feet estimated as the ultimate potential in 1991. The initial established (almost fully proved) reserves alone at the 1991 year-end totalled 120 trillion cubic feet, over 60 trillion cubic feet of which had already been produced.

Flaring Restrictions

All gas produced from oil pools had to be processed to remove the heavier hydrocarbons, and all products, including the residue gas, had to be disposed of in a manner satisfactory to the Board. Where it could be demonstrated to the Board's satisfaction that gas conservation was not feasible, flaring of the solution gas could be continued. There were a considerable number of fields in which flaring was permitted during the 1950s.

Flare pits had to meet special regulations. For example, the Drilling and Production Regulations, 1957, required that all flare pits and ends of flare lines be constructed and safeguarded in such a way that no hazard to property and forest cover would be created. A minimum distance of 330 feet was required from any right-of-way and occupied buildings of any kind. Some leeway from this restriction was granted in special circumstances if, in the opinion of the Board, it was justified. Distances required from other oil well equipment was normally 150-feet, and in the case of treaters the limit was reduced to 75 feet. These regulations applied to both drilling and producing operations, even though the use of flare lines for

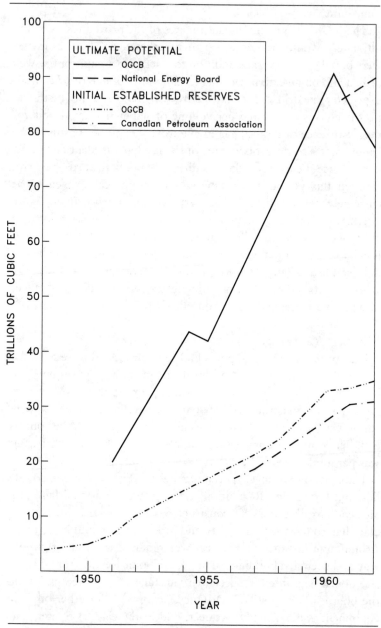

drilling operations was temporary. To make sure that any blow-back of the flare to the production tanks would not result in fires, flame arresters were required on vent lines to the flare pits.

Flares were required to be lit at all times. This was a necessary measure in all areas, but particularly in farming localities or other settled areas, where there were houses in the vicinity. The escape of unburnt gas could present an explosion hazard if it should seep into a farm building with a flame heater or where a spark or flame might be generated. Gas containing hydrogen sulphide further increased the danger and could damage facilities through increased corrosion. In the early 1950s, much of the paint used for external purposes on farm buildings was lead based, and as often as not white. Such houses, if in the path of unburnt hydrogen sulphide, could undergo a rapid colour transformation from white to a dirty, dull grey as the lead oxide in the paint was altered to lead sulphide.

Occasionally, flares went out, but they were usually relit within a short period by the battery operator. A few times, for whatever reason, flares were not relit until discomfort or damage was occasioned to people or property in the area. The Board would sometimes discover such situations on routine inspections, and in others would be promptly notified by the concerned parties. If the problem was severe, or recurred after warning, enquiries were held and penalties assessed. The incidence of unlit flares remained a continuing problem in the 1950s. It led the Board, early in 1960, to require the installation of equipment at the end of the flare line to ensure that the gas was automatically relit whenever the flare went out.

Gas Processing

The recovery of solution gas through oilfield conservation schemes, such as at Leduc and Redwater, and the approval of the removal of large amounts of marketable dry gas to eastern Canada and the United States required that gas processing plants be built. These plants varied in capacity from a few million to over 300 million cubic feet per day. The plants removed natural gas liquids, comprising propane, butanes and pentanes plus, as well as nonhydrocarbons such as hydrogen sulphide. At least a portion of these constituents had to be removed to render the gas suitable for transportation through pipelines and for domestic consumption. These by-products of the marketing of natural gas, including sulphur, served growing markets and added to the revenue of the producing companies, the province and the freehold mineral owners. Seventeen such plants were in operation by 1958 year-end, and others were under construction or in the design stage (see Table APP.X.3).

TABLE APP.X.3 Alberta Gas Processing Plants, 1958

Year on Stream	Area Name	Location
1933	Turner Valley	06-20-02W5
1950	Leduc-Woodbend	34-50-26W4
1951	Jumping Pound	13-25-05W5
1954	Acheson	02-53-26W4
1954	Bonnie Glen	17-47-27W4
1954	Golden Spike	22-51-27W4
1956	Nevis	33-38-22W4
1956	Samson	09-44-24W4
1957	Pincher Creek	23-04-29W4
1957	Provost	19-36-05W4
1958	Alexander	16-56-27W4
1958	Carbon	17-29-22W4
1958	Cessford	08-24-12W4
1958	Cessford	31-22-11W4
1958	Pembina	22-49-10W5
1958	Pembina	24-48-07W5
1958	St. Albert	26-54-25W4

SOURCE: ERCB

Gas gathering and processing programs were an essential part of schemes to market gas from gas reservoirs as well as from oilfields, and such programs were brought forward as markets were established, and in the process added greatly to the Board's evaluation and field responsibilities. Proper design of plants required detailed accurate analyses of the natural gas and proper regard for Alberta's severe winter climatic conditions. Improperly designed plants could lead to unnecessary flaring of gas and products and poor recovery of by-products from the gas stream. Section 38 of the 1957 *Oil and Gas Conservation Act* required, among other things, that "No scheme for the gathering, storage or disposal of gas ... shall be proceeded with unless the Board, by order, has approved the scheme upon such terms and conditions as the Board may prescribe." Some of the conditions specified in the approvals issued by the Board in the 1950s pursuant to section 38 were the following:

(a) All gas not required for lease fuel shall be gathered.

(b) The gas shall be processed for the complete recovery of marketable residue gas, or reinjected to an underground formation.

(c) Liquid hydrocarbons removed from the gas in the preparation of residue gas shall be completely recovered.

(d) Separate recovery of propane and butanes or increased recovery thereof shall be obtained if in the Board's opinion this is warranted by market conditions.

(e) The recovery in the form of elemental sulphur of not less than 'X' per cent of the sulphur contained in the gas shall be achieved ('X' varied from 85 to 95 depending on plant location with respect to centres of population, volume of acid gas and acid gas composition).

(f) No flaring of produced hydrocarbons shall occur except in cases of emergency.

(g) Adequate surface or underground storage for liquids shall be provided.

Operators of gas processing plants were required to submit monthly reports summarizing total disposition of the gas produced. Field inspectors would visit the plant periodically and discuss any problems relating to gas gathering, scheduling of wells, and the distribution of products, with a view to keeping flaring to a minimum. Such a visit was followed by a report to the Board's gas department outlining the plant operations and any problems, including occasional flaring of raw gas, residue gas and products. This report assisted the gas department in assessing the continued suitability of the requirements of the approval, and determining if adjustments to the required level of gas or product recovery should be made.

The opportunities to increase the recovery of natural gas by special depletion methods are not nearly as extensive as for crude oil. Gas cycling, however, in which dry gas is injected into a gas reservoir to displace wet or sour gas, is economical in some instances, and in such cases was required by the Board. These projects increased the portion of the heavier natural gas liquids recovered by 5 to 30%, depending on the nature of the reservoir and the liquids. Improvement in the recovery of hydrogen sulphide, and hence of sulphur by cycling, can also be significant where the hydrogen sulphide content of the gas is high.

The Petroleum and
Natural Gas Industry in Alberta

1882—G.M. Dawson, Director of the Geological Survey, and Dr. Robert Bell investigated the oil sands along the Athabasca River and suggested a development strategy.

1883 (DEC)—Natural gas discovered by Canadian Pacific Railway while drilling for water at Langevin station on the main line southeast of Calgary.

1890—Natural gas discovered while drilling for coal at Medicine Hat.

1897—A federal government well drilled at Pelican Rapids on the Athabasca River discovered the Pelican natural gas field. It blew wild until sealed in 1918.

1902 (SEPT)—Well drilled by John Lineham's Rocky Mountain Development Co. on Cameron Creek near Waterton Lakes in the southwest corner of the province produced the first oil in Alberta.

1905—A.W. Dingman organized the Calgary Natural Gas Company.

1908 (OCT)—Calgary Natural Gas Company drilled and found natural gas on the Colonel Walker estate in southeast Calgary.

1909 (FEB)—The Bow Island gas field was discovered by Eugene Coste.

1910 (AUG)—Eugene Coste formed the Prairie Fuel Gas Company (reorganized in 1911 to become the Canadian Western Natural Gas, Light, Heat and Power Company) and then obtained the natural gas rights to CPR lands at Bow Island.

1912 (JULY)—W.S. Herron and A.W. Dingman formed the Calgary Petroleum Products Company to drill for oil on Sheep Creek in Turner Valley.

1912 (JULY)—A natural gas pipeline was completed to Calgary from Bow Island.

1914 (JAN)—Canada, Department of the Interior, Petroleum and Natural Gas Regulations, section 29, first conservation measure.

1914 (MAY)—Dingman No. 1 struck oil, starting the first Turner Valley oil boom.

1915—Alberta Board of Public Utility Commissioners came into being.

1916—Viking gas field discovered.

1917—Imperial Oil, subsidiary of Standard Oil of New Jersey, began drilling its first wells in western Canada.

1918—First plant for production of helium established in Bow Island for use in dirigibles.

1919—Trial runs undertaken at Bow Island at the helium recovery facility on a program to supply helium for airstrips and balloons.

1920 (DEC)—Calgary Petroleum Products Co. re-formed as the Royalite Oil Co., with Imperial Oil being the majority shareholder.

1921 (JULY)—Alberta: Election of United Farmers of Alberta government under Herbert Greenfield.

1921 (DEC)—Natural gas from the Dingman wells in Turner Valley (then owned by Royalite) was tied into the Canadian Western Natural Gas, Light, Heat and Power Co. Ltd. pipeline system.

1923 (OCT)—Edmonton received its first delivery of natural gas from the Viking gas field.

1923 (NOV)—British Petroleums Ltd. found indications of oil in the Wainwright area that led to the discovery of the Wainwright field in 1926.

1924 (OCT)—Royalite No. 4's spectacular deeper discovery in the Madison Limestone spurred renewed exploration in Turner Valley and opened up the immense gas reserves of that formation.

1925—Alberta's first oil pipeline was constructed from Turner Valley to the Imperial Oil refinery in Calgary.

1926 (APR)—Alberta: The *Oil and Gas Wells Act* gave authority to issue conservation regulations in specified areas, but none were issued.

1926 (JUNE)—Re-election of United Farmers of Alberta under John Brownlee.

1926 (OCT)—Ottawa: Minister of the interior issued the first gas conservation regulation. It restricted gas flow to 40% of the potential capacity of any well on crown lands, but no enforcement procedures were introduced.

1926 (NOV)—F.P. Fisher's "Report on Southern Alberta Gas Situation" outlined the conservation problem in Turner Valley.

1930 (MAY)—Ottawa authorized the first natural gas conservation scheme in Alberta. The Canadian Western Natural Gas, Light, Heat and Power Co. of Calgary was permitted to store excess gas from Turner Valley in the Colorado formation of the Bow Island gas field, which would be available for future use.

1930 (JUNE)—Re-election of Brownlee government.

1930 (OCT)—Alberta, Saskatchewan and Manitoba gained control of lands and resources from the federal government.

1931—Calgary applied to the Alberta Board of Utility Commissioners for a reduction in the rate paid for natural gas and brought the question of natural gas waste to a head.

1931 (MAR)—Alberta: Department of Lands and Mines created.

1931 (MAR)—Alberta: The *Oil and Gas Wells Act, 1931*, authorized the lieutenant-governor-in-council to make regulations respecting the drilling and production operations of oil and natural gas wells.

1931 (MAY)—First regulation under the *Oil and Gas Wells Act* required that the flow of gas or gas and oil from *every* well within the province be restricted to 40% of the current potential capacity as shown by the last monthly gauge.

1931 (JUNE)—Alberta: First royalty regulation on petroleum and natural gas produced on provincial lands was set at 5% of well output or value to January 1935, thereafter 10%.

1931 (JULY)—Regulations respecting drilling and production operations of oil and natural gas wells were put in force. Flow of gas now restricted to 25% of potential as shown by last monthly gauge.

1931 (OCT)—Government and industry committee (headed by F.P. Fisher) submitted "Proposed Turner Valley Agreement" and recommended a co-operative program for producing Turner Valley field. In March 1932, in a subsequent report, Fisher recommended that the field be unitized by legislation.

1932 (JAN)—Petition to Alberta minister of lands and mines from Turner Valley oil operators and leaseholders asking that all production restrictions be removed.

1932 (APR)—*Turner Valley Gas Conservation Act* established a three-man Turner Valley Gas Conservation Board (TVGCB) to solve the "waste" gas problem at Turner Valley.

1932 (MAY)—First TVGCB order. It set an aggregate production maximum of 200 million cubic feet per day and prescribed a daily rate of permitted production from each well.

1932 (JUNE)—Spooner Oils case. Challenge to first TVGCB order.

1933 (OCT)—*Spooner Oils Ltd. and A.G. Spooner v. Turner Valley Gas Conservation Board and the Attorney General of Alberta* (Supreme Court). Decided that the Board lacked authority to enforce conservation measures.

1933–34—A request by the Alberta government for federal legislation to enforce conservation measures was rejected by the federal government.

1934—Texas: Clymore Case lasts into 1935 and establishes legal authority of Texas Railroad Commission.

1934—First export of Alberta natural gas from Red Coulee field on Alberta-Montana border to United States.

1935 (AUG)—Alberta: Election of Social Credit government under William Aberhart.

1935 (SEPT)—Charles C. Ross (engineer and conservation proponent) appointed first Social Credit minister of lands and mines.

1936 (JUNE)—Turner Valley Royalties No. 1, which had commenced drilling in April 1934, discovered crude oil and launched Turner Valley on its third and greatest oil boom.

1937—Turner Valley crude oil production exceeded local market requirements for the first time.

1937—Alberta government officials met with the Oil and Gas Association in Calgary to discuss the waste gas problem.

1937 (JAN)—Alberta: Resignation of Charles C. Ross, minister of lands and mines, succeeded by N.E. Tanner.

1938 (APR)—Alberta: The *Oil and Gas Conservation Act, 1938,* called for the creation of a three-man Petroleum and Natural Gas Conservation Board (PNGCB) to effect the conservation of the province's oil and natural gas resources and to enforce the provisions of the *1931 Oil and Gas Wells Act.*

1938 (JUNE)—Ratification of an amendment to the mineral resources transfer agreement by the federal government enabled Alberta to proceed with a strengthened *Oil and Gas Resources Conservation Act.*

1938 (JULY)—Alberta: PNGCB established. W.F. Knode, a petroleum engineer experienced in conservation matters from Texas, was appointed first Conservation Board chairman.

1938 (AUG)—The new Conservation Board's first Order decreed that, without Board permission to do otherwise, the number of wells in the Turner Valley Field would be restricted to one for every 40 acres.

1938 (OCT)—Alberta: Royal commission appointed to inquire into matters connected with petroleum and petroleum products, chaired by the Honourable Mr. Justice A.A. McGillivray.

1938 (NOV)—Alberta: The *Oil and Gas Resources Conservation Act,* a refined and strengthened version of the earlier (April) Act.

1938 (NOV)—At its first meeting under the new legislation, the Conservation Board issued three orders. No. 1 put forth a daily production limit for each producing oil well in the province; No. 2 told well owners to render all necessary assistance to enable Board personnel to conduct well tests; and No. 3 provided a schedule of monthly allowable production for each gas well in the province.

1939 (JULY)—Board Chairman Knode goes to England with Alberta government delegation to seek British financing for an oil pipeline to Lake Superior. (Included are N.E. Tanner and W.S. Campbell from Alberta and G.S. Hume from Ottawa.)

1939 (AUG)—Conservation Board cuts back allowable production for Royalite wells in favour of smaller producers.

1939 (AUG)—PNGCB Deputy Chairman Charles W. Dingman acted as chairman, upon expiration of W.F. Knode's contract.

1940 (MAR)—Alberta: Re-election of Aberhart government.

1940 (SEPT)—Market demand for Turner Valley oil exceeded production, allowable production set on an engineering basis.

1940 (SEPT)—Conservation Board "Hearings" policy formalized.

1940 (DEC)—Discovery of Princess oilfield by California Standard Co. [wholly owned subsidiary of Standard Oil Company of California].

1941 (JAN)—Ottawa's wartime "oil controller" requested increased Turner Valley crude production against the advice of the PNGCB, which feared permanent damage to the Turner Valley field.

1941 (MAR)—On advice of PNGCB Board chairman and others, Alberta government set aside 15 areas of crown (petroleum and natural gas) reserves.

1941 (MAR)—Alberta: The *Oil and Gas Resources Conservation Act Amendment Act*.

1941 (MAY)—Alberta: Revised royalty system.

1942—Turner Valley oil production peaked.

1942 (JAN)—Dr. G.G. Brown's conservation plan for Turner Valley submitted to Conservation Board.

1942 (MAY)—J.J. Frawley appointed Conservation Board chairman.

1942 (JULY)—Conservation Board adopted Brown Plan for north half of Turner Valley.

1943 (MAR)—Dr. Thomas R. Weymouth conservation plan for Turner Valley related to recovery of isobutane for aviation fuel.

1943 (MAR)—*Rowley v. Petroleum and Natural Gas Conservation Board*.

1943 (AUG)—*Model Oils v. Petroleum and Natural Gas Conservation Board*. Compelled the Conservation Board to enforce its Order against Royalite for exceeding the allowable production limit to meet requirements for strategic reasons.

1943 (SEPT)—Dr. E.H. Boomer appointed Conservation Board chairman.

1943 (NOV)—Various wells shut in by Conservation Board for production in excess of the Board's designated limit.

1943 (NOV)—Conservation Board approved a compensation scheme in accordance with the principle that the Turner Valley field was a common reservoir, and if one operator produced in excess of the allowable for some special reason, owners of other wells in the field should be compensated by sharing the revenue earned by the additional production.

1944 (MAR)—Alberta Natural Gas Utilities Board created, chaired by G.M. Blackstock.

1944 (AUG)—Alberta: Election of Social Credit government headed by E.C. Manning.

1944 (DEC)—The Jumping Pound natural gas field was discovered by Shell Oil of Canada.

1945 (MAR)—Model No. 1 and No. 2 wells shut in by PNGCB for overproduction.

1945 (MAR)—Carleton No. 1 and Pacalta No. 2 purchased by PNGCB to be used as input wells for repressuring of the south end of Turner Valley.

1945 (MAY)—Well shut in in Conrad field after PNGCB inspection revealed various infractions.

1945 (JUNE)—PNGCB given responsibility for regulation of tank farm operations.

1945 (JUNE)—First unitization agreement, Jumping Pound field.

1946—Production of Alberta light crude oil had declined to an amount insufficient to meet regional market demand.

1946 (JAN)—Dr. D.L. Katz brought from University of Michigan to review application of Brown Plan.

1946 (MAR)—First meeting in Board's new office at 514–Eleventh Avenue S.W., Calgary.

1946 (OCT)—A.G. Bailey appointed Conservation Board chairman.

1947 (FEB)—Imperial Leduc No. 1 discovered the Leduc oilfield and drew international attention to Alberta's petroleum potential.

1947 (MAR)—The *Right of Entry Arbitration Act* created a three-member board to mediate disputes arising between different parties holding surface and mineral title to the same piece of property.

1947 (MAY)—PNGCB began consultation with industry to work out development regulations for the Leduc field.

1947 (JUNE)—PNGCB Chairman A.G. Bailey resigned. D.P. Goodall appointed acting chairman.

1947 (AUG)—Alberta: Major revision of petroleum and natural gas regulations. Beginning of bonus bidding system for petroleum and natural gas leaseholds.

1948 (FEB)—I.N. McKinnon appointed Conservation Board chairman.

1948 (FEB)—Leduc hearings with respect to crude oil market sharing commence.

1948 (MAR)—Atlantic No. 3 blowout.

1948 (APR)—G.S. Hume and A. Ignatieff report, "Natural Gas Reserves of the Prairie Provinces."

1948 (AUG)—The first experimental project on enhanced recovery in Alberta undertaken in the Turner Valley Rundle pool by Royalite Oil Company, with the injection of water and some gas into a selected area of the pool.

1948 (AUG)—Alberta: Re-election of Manning government.

1948 (OCT)—Pincher Creek sour gas field discovered by British American Oil Co.

1948 (OCT)—Redwater oilfield discovered by Imperial Oil.

1948 (NOV)—Mercury-Leduc No. 1 blowout.

1948 (NOV)—Dinning Commission appointed by Alberta government to investigate the province's natural gas reserves and requirements. Report submitted March 1949.

1949 (FEB)—Alberta's light crude oil production potential exceeded the requirements of the three prairie provinces.

1949 (MAR)—Alberta: Dinning Commission found that there was insufficient proven natural gas reserves to warrant export from the province.

1949 (MAR)—*An Act to Amend The Oil and Gas Resources Conservation Act,* to define and confirm the authority of the Conservation Board to fix allowable crude oil production in accordance with "reasonable market demand."

1949 (APR)—Canada: *Pipe Lines Act* asserted federal right to regulate oil and gas pipelines crossing provincial or international boundaries.

1949 (APR)—Alberta: Department of Lands and Mines divided to form two new ministries, the Department of Lands and Forests and the Department of Mines and Minerals.

1949 (JULY)—The *Gas Resources Preservation Act* empowered the PNGCB to control the removal of gas from the province through removal permits.

1949 (JULY)—*An Act to Amend the Pipe Line Act*, transferred responsibility for approving pipeline construction proposals from the Department of Public Works to the Conservation Board.

1949 (OCT)—Frank McMahon's Westcoast Transmission Co. Ltd. applied to the Conservation Board to be allowed to export natural gas from the Peace River area to Vancouver and the U.S. Pacific Northwest.

1950 (APR)—PNGCB officially designated areas of the 10 producing oilfields in Alberta. "F" (Field) Orders issued for first time.

1950 (APR)—The *Oil and Gas Resources Conservation Act* updated conservation regulations to keep abreast of the dramatic development that had occurred since 1947.

1950 (MAY)—Leduc-Woodbend field's gas conservation plant opened at Devon, Alberta, by Imperial Oil.

1950 (JUNE)—Gordon Plotke was granted the first gas export permit under the *Gas Resources Preservation Act* (from Red Coulee field to Montana).

1950 (AUG)—Continental Oil Co. of Canada Ltd. applied to the Board to have Imperial Oil declared a common purchaser of oil from the Leduc-Woodbend field.

1950 (SEPT)—Fenn-Big Valley oilfield discovered by British American Oil Co.

1950 (DEC)—Initial proration plan to meet the problem of excess crude oil production capacity and to ensure that market was equitably divided among oil pools and wells. Maximum Permissible Rate (MPR) concept.

1950 (DEC)—Completion of Interprovincial pipeline from Edmonton to Lake Superior.

1951 (JAN)—Conservation Board rejected Westcoast and other export applications on grounds that Alberta's proven natural gas reserves were still inadequate in relation to Alberta's 30-year requirements.

1951 (APR)—Shell Jumping Pound gas plant completed.

1951 (JUNE)—McColl-Frontenac and Union Oil permitted to export 10 billion cubic feet of natural gas per year for five years to the Anaconda Copper Mining Company in Montana. The Conservation Board found that the gas to be exported was not surplus to Alberta's long-term needs but granted the permit as a special measure to assist the Korean War effort.

1951 (SEPT)—Wizard Lake oilfield discovered by Texaco.

1951 (SEPT)—Alberta Petroleum and Natural Gas Conservation Board reopened gas export hearings. Five applicants sought Board approval.

1952 (JAN)—Bonnie Glen oilfield discovered by Texaco.

1952 (MAR)—Westcoast Transmission Company Ltd., pursuant to PNGCB recommendation, granted permission to export natural gas from the Peace River area in northern Alberta. The Conservation Board recommended against proposals to export gas from fields in the southern part of the province.

1952 (APR)—PNGCB formally expressed concern to Leduc-Woodbend operators regarding gas flaring.

1952 (AUG)—Alberta: Re-election of Manning government. The question of natural gas exports was the central issue of the election.

1952 (SEPT)—N.E. Tanner resigned as Alberta's minister of mines and minerals, succeeded by Premier E.C. Manning.

1952 (DEC)—Wells shut in by PNGCB for exceeding allowable oil production.

1953—Interprovincial pipeline extended to Sarnia, Ontario.

1953—Completion of the Trans Mountain pipeline to Vancouver.

1953 (JAN)—*Borys v. CPR and Imperial Oil* (Judicial Committee of the Privy Council) confirmed the "Rule of Capture" and refined the definition of petroleum and natural gas.

1953 (JAN)—Wells shut in by PNGCB for failing to follow drilling and production requirements.

1953 (MARCH)—Sturgeon Lake South oilfield discovered by Amerada Minerals.

1953 (JUNE)—Pembina oilfield discovered by Mobil/Seaboard.

1953 (NOV)—Conservation Board Report on the natural gas export applications of Canadian-Montana Pipeline Co., Trans-Canada Pipe Lines Ltd., Trans-Canada Grid of Alberta Ltd., Canadian Delhi Oil Ltd. and Western Pipe Lines.

1953 (DEC)—PNGCB general order to stop flaring gas—Leduc-Woodbend field.

1954—Portable gas liquid recovery plants completed at Big Valley and Winterburn were the first of seven processing plants completed by 1956, some to meet PNGCB orders to conserve gas.

1954 (APR)—*Alberta Gas Trunk Line Act* provided for the creation of a sole, integrated, privately owned gas-gathering system for the entire province.

1954 (MAY)—Conservation Board Report on the natural gas export application of the merged applicants: Trans-Canada Pipe Lines Ltd., Trans-Canada Grid of Alberta, and Canadian Delhi Oil Ltd. and Western Pipe Lines under the name of Trans-Canada Pipe Lines Ltd.

1954 (JULY)—Wells shut in by PNGCB in Morinville area for improperly disposing of salt water and poor wellsite maintenance.

1955—Trans Mountain pipeline extended to Anacortes, Washington.

1955 (JUNE)—PNGCB launched Peace River Well Abandonment Program.

1955 (JUNE)—Alberta: Social Credit government elected for the fourth time under E.C. Manning.

1955 (JULY)—PNGCB ordered hundreds of wells in Redwater field shut in by November unless an acceptable conservation program was put in place.

1955 (OCT)—PNGCB approved an Imperial Oil scheme for processing "waste" gas from Redwater field.

1956 (MAY-JUNE)—Trans-Canada Pipe Lines debate in Parliament.

1956 (OCT-NOV)—Suez Crisis, curtailed availability of crude oil from Persian Gulf states.

1957—First plant to process natural gas for removal from Alberta by Trans-Canada Pipe Lines completed in the Provost gas field near Consort, Alberta.

1957—Interprovincial pipeline extended to Port Credit, Ontario.

1957—Water injection schemes in Pembina Cardium pool marked the real beginning of large enhanced recovery programs in Alberta.

1957 (APR)—Alberta: The *Oil and Gas Conservation Act* refined conservation legislation and changed the name of the Board to the Oil and Gas Conservation Board (OGCB).

1957 (JULY)—Standard well spacing for oil wells expanded from 40 to 80 acres in that portion of the province west of the Fifth Meridian.

1957 (JULY)—Completion of the first leg of the Alberta Gas Trunk Line gathering system.

1957 (JULY)—Virginia Hills oilfield discovered by Home Oil.

1957 (AUG)—Revised proration plan to discourage excessive drilling in established pools.

1957 (OCT)—Ottawa: Borden Royal Commission on Energy established.

1957 (OCT)—Westcoast Transmission began exporting natural gas to the United States.

1957 (NOV)—Swan Hills oilfield discovered by Home Oil.

1957 (NOV)—Ottawa: Final report of Gordon Commission on Canada's Economic Prospects recommended development of a comprehensive energy policy and the creation of a national energy authority.

1958—Conservation Board submission to the Borden Royal Commission on Energy.

1958 (FEB AND MAY)—Premier E.C. Manning's appearance before the Royal Commission on Energy.

1958 (APR)—Alberta: The *Pipe Line Act* 1958, placed pipelines under the jurisdiction of the Department of Mines and Minerals, and required application for a permit to construct and a licence to operate a pipeline in the province.

1958 (JUNE)—Atomic detonation scheme proposed for oil sands by Richfield Oil Company.

1958 (OCT)—"First Report" of (Borden) Royal Commission on Energy recommended the formation of a national energy board.

1958 (OCT)—Trans-Canada pipeline completed to Toronto, Ottawa and Montreal.

1958 (NOV)—Carson Creek North oilfield discovered by Mobil.

1959—First gas cycling scheme in Canada to recover natural gas liquids (NGL) at Windfall D-3A Pool.

1959 (MAY)—Judy Creek oilfield discovered by Imperial Oil.

1959 (JUNE)—Alberta: Re-election of Manning government.

1959 (JULY)—*National Energy Board Act* gave the NEB regulatory control over the export and import of natural gas and electricity and transportation rates, administrative responsibility for the *Pipe Line Act,* and the responsibility of advising on all energy matters of concern to the federal government.

1959 (JULY)—"Second Report" of the Royal Commission on Energy.

1959 (AUG)—I.N. McKinnon named chairman of National Energy Board (leave of absence from OGCB until 1962).

1959 (AUG)—Second report of (Borden) Royal Commission on Energy recommended against construction of a pipeline to bring western crude oil to Montreal.

1959 (SEPT)—G.W. Govier appointed deputy chairman during McKinnon's leave of absence.

1960—Water injection commenced to Leduc D-3 reservoir to supplement the natural water drive.

1960 (NOV)—Alberta: OGCB recommends approval of the Great Canadian Oil Sands project.

1961 (FEB)—National oil policy announced by Ottawa, western Canadian oil to be limited to markets west of the Ottawa Valley.

Imperial to Metric Units

VOLUME

1 cubic foot of gas (14.65 psia and 60°F)	= 0.028174 cubic metre of gas (101.325 kilopascals and 15°C)
1 Can. barrel of ethane (equil. pressure and 60°F)	= 0.1580 cubic metre of ethane (equil. pressure and 15°C)
1 Can. barrel of butanes (equil. pressure and 60°F)	= 0.1588 cubic metre of butanes (equil. pressure and 15°C)
1 Can. barrel of propane (equil. pressure and 60°F)	= 0.1587 cubic metre of propane (equil. pressure and 15°C)
1 Can. barrel of oil or pentanes plus (equil. pressure and 60°F)	= 0.1589 cubic metre of oil or pentanes plus (equil. pressure and 15°C)
1 Can. barrel of water (equil. pressure and 60°F)	= 0.1590 cubic metre of water (equil. pressure and 15°C)
1 (UK) long ton (2240 lbs)	= 1.016 tonne
1 short ton (2000 lbs)	= 0.907 tonne
1 British thermal unit (BTU) (60°)F	= 1.055 kilojoules

LENGTH

1 mile	= 1.61 kilometres
1 foot	= 30.48 centimetres

AREA

1 acre	= 0.4047 hectares
10 acres	= 4.047 hectares
40 acres	= 16.19 hectares
160 acres	= 64.75 hectares
640 acres or 1 square mile	= 259.1 hectares

Metric to Imperial Units

VOLUME

1 cubic metre of gas (101.325 kilopascals and 15°C)	= 37.49 cubic feet of gas (14.65 psia and 60°F)
1 cubic metre of ethane (equil. pressure and 15°C)	= 6.330 Can. barrels of ethane (equil. pressure and 60°F)
1 cubic metre of butanes (equil. pressure and 15°C)	= 6.2968 Can. barrels of butanes (equil. pressure and 60°F)
1 cubic metre of propane (equil. pressure and 15°C)	= 6.300 Can. barrels of propane (equil. pressure and 60°F)
1 cubic metre of oil or pentanes plus (equil. pressure and 15°C)	= 6.2929 Can. barrels of oil or pentanes plus (equil. pressure and 60°F)
1 cubic metre of water (equil. pressure and 15°C)	= 6.2901 Can. barrels of water (equil. pressure and 60°F)
1 tonne	= 0.984206 (UK) long tons (2240 lbs)
1 tonne	= 1.02311 short tons (2000 lbs)
1 kilojoule	= 0.9482133 British thermal units (BTU) (60°F)

LENGTH

1 kilometre	= 0.62 miles
1 centimetre	= 0.39 inches

AREA

1 hectare	= 2.47 acres

GLOSSARY

ABANDON (A WELL)—Convert a drilled well to a condition where it can be left indefinitely without further attention and will not damage fresh water sands, reservoirs of prospective interest or prospective mining operators, or interfere with normal operations conducted at the surface in the vicinity of the well. This usually requires setting cement plugs at appropriate depths in the well as well as close to the surface, cutting off the casing at a depth below the land surface, and restoring the land surface to a condition close to that existing before drilling operations were undertaken.

ALLOWABLE—The amount of oil or gas a well is permitted to produce in accordance with an order of the Board after application of any applicable penalty factors.

ANNULUS—The space between two concentric lengths of pipe or between pipe and the hole in which it is located.

API GRAVITY SCALE—A scale developed for the American Petroleum Institute and used extensively throughout the world for comparing the relative densities of crude oils. The API gravity is related to relative density by the following formula: Degrees $API = \frac{(141.5)}{(\tau - 131.5)}$ where τ is the relative density of the crude oil to water at 60°F.

AQUIFER—An underground accumulation of water contained in porous rock or sand.

ASSOCIATED GAS—Natural gas that is in contact with or dissolved in crude oil in the reservoir.

BATTERY (TANK BATTERY)—Equipment suitably arranged to process or store crude oil from one or more wells.

B/D or bbl/d—Barrels per day.

BITUMEN or CRUDE—Petroleum that exists in the semi-solid or solid phase in natural deposits.

BLOWOUT—An uncontrolled flow of gas, oil or other fluids from a well.

BLOWOUT PREVENTER (BOP)—Equipment installed or that might be installed at the wellhead to control pressures and fluids during drilling, completion and certain workover operations.

BOARD—Unless otherwise stated, the Oil and Gas Conservation Board, or Petroleum and Natural Gas Conservation Board, as it was previously called.

BOTTOM WATER—Water occurring below oil or gas in an underground reservoir.

BOTTOM-HOLE—The lowest or deepest part of a well. Sometimes used as an adjective to identify equipment or operations conducted in the lower area of a well.

BRING IN A WELL—To complete a well and put it on production.

BUTANES—In addition to its normal scientific meaning, a mixture mainly of butanes that ordinarily might contain some propane or pentanes plus.

CASED HOLE—A wellbore in which casing has been set.

CASING (SURFACE PRODUCTION)—The steel pipe placed in a well to prevent the collapse of the hole, to provide a conduit for fluids during the drilling or producing operations, or to which wellhead equipment may be anchored.

CASING PRESSURE—The pressure in the casing, drill pipe or casing tubing annulus, as measured at the surface.

CASING SHOE—The bottom end of the casing.

CASING STRING—The pipe run in a well, for example, surface string, intermediate string, production string.

CASING-HEAD GAS—Gas produced with oil in oil wells, the gas being taken from the well through the casing head at the top of the well (the annular space between the casing and tubing). Casing-head gas contains liquid hydrocarbons in solution, which can be separated in part by a reduction in pressure at the wellhead, or might be separated more completely in a separator, absorption plant or by other manufacturing process. The term casing-head gas is also sometimes more broadly applied to gas released from oil through a modern separator.

CLOSED-IN—A term applied to a well capable of producing oil, gas or other fluids but temporarily shut in, or to measurements conducted on a well in that circumstance.

CONDENSATE—Hydrocarbons that are in the gaseous state under reservoir conditions but become liquid either in passage up the hole or at the surface.

CORING—A drilling operation to obtain a cylindrical sample or core of rock using a special drilling bit.

CORRELATIVE RIGHTS—The opportunity afforded, so far as it is practicable to do so, to the owner of each property in a pool to produce without waste his just and equitable share of the oil or gas, or both, in a pool.

DEVELOPMENT WELL—A well drilled in the proven part of a field.

DIKE—See fire wall.

DIRECTIONAL SURVEY—A procedure to measure the direction and amount that a wellbore has deviated from the vertical.

DISCOVERY WELL—An exploratory well that encounters a new and previously untapped petroleum or natural gas deposit. A successful wildcat well.

DISPOSABLE GAS—See MARKETABLE GAS.

DISPOSAL WELL—A well through which water (usually salt water) or other fluids are injected into a subsurface formation.

DISSOLVED GAS—Natural gas that is in solution in crude oil (see also SOLUTION GAS and CASING HEAD GAS).

DOGHOUSE—A small utility shack on the drilling rig floor, heated during the winter months. Tour reports are usually kept there, along with the licence to drill and instructions for the crew.

DOLOMITE—A type of sedimentary rock composed essentially of calcium carbonate and magnesium carbonate.

DRAINAGE UNIT—See SPACING UNIT.

DRILL PIPE—The heavy seamless pipe used to rotate the drilling bit and to circulate the drilling fluid.

DRILL STRING—The column or string of drill pipe.

DRILLER—The member of the rig crew who operates the rig controls and supervises the rig and crew for one of the two or three shifts that the rig is operating each day.

DUMP BAILER—A bailing device with a release valve, used to place or spot material (such as cement slurry) at the bottom of a well.

ENHANCED RECOVERY—The increased recovery from a pool achieved by artificial means, or by application of energy extrinsic to the pool, including injection of fluids, chemicals or heat.

ESTABLISHED RESERVES—Those reserves recoverable under current technology and present and anticipated economic conditions specifically proved by drilling, testing or production, plus a judgement portion of contiguous recoverable reserves that are interpreted from geological, geophysical or similar information with reasonable certainty to exist. (See also PROVED RESERVES.)

ETHANE—In addition to its normal meaning, a mixture mainly of ethane that ordinarily might contain some methane or propane.

EVALUATION WELL—A well drilled or operated pursuant to an experimental scheme approved by the Board.

EXPERIMENTAL WELL—A well being drilled or operated pursuant to an experimental scheme approved by the Board.

FIELD—A designated surface area underlain, or appearing to be underlain, by one or more pools, or the subsurface regions vertically beneath such surface area or areas.

FIRE WALL (*or* DIKE)—An earth wall built around oil tanks and other oil handling equipment to contain any oil that might be discharged accidentally. It also serves to block the spread of a fire or to give protection when emergency action is taken.

FISH—The object being sought downhole during fishing operations.

FISHING—Attempting to recover tools, cable, pipe or other objects from the wellbore that have become lost in the well. Many special and ingeniously designed fishing tools are used to recover objects lost downhole.

FLOW LINE—The pipe through which oil or gas travels from the well to the field processing facility. It is usually buried below plough depth.

FLOWING PRESSURE—The pressure at the wellhead of a flowing well.

FLOWING WELL—A well that produces without any means of artificial lift.

FORMATION—A designated subsurface strata that is composed throughout of substantially the same kind of rock.

FORMATION PRESSURE—The pressure exerted by fluids in a formation. Also called reservoir pressure or shut-in bottom-hole pressure.

GAS—Raw gas or marketable gas, or any constituent of raw gas, condensate, crude bitumen or crude oil, that is recovered in processing and is gaseous at the conditions under which its volume is measured or estimated.

GAS CAP—The portion of an oil-producing reservoir occupied by free gas.

GAS-CAP DRIVE—The drive mechanism by which oil is recovered through the expansion of gas in a cap overlying the oil in a reservoir.

GAS INJECTION—Injection of natural gas under high pressure into a producing reservoir through an input or injection well often as part of an enhanced recovery operation.

GAS LIFT—The raising or lifting of liquid from a well by means of injecting gas into the liquid in the well bore.

GAS-OIL RATIO (GOR)—The amount of gas in cubic feet produced with a barrel of oil.

GAS PROCESSING PLANT—A facility designed (1) to recover natural gas liquids or natural gas that might or might not have been processed through lease separators and field facilities and (2) to condition the natural gas to meet market requirements.

GAS WELL—A well producing mainly natural gas.

GATHERING LINES—The flow lines between wells and a central lease or plant facility.

GRADIENT—Rate of change, for example, pressure with depth.

GRADIENT, PRESSURE—Pressure change with depth, expressed in pounds per square inch per foot (psi/ft).

GRADIENT, TEMPERATURE—Temperature change with depth, expressed in degrees Fahrenheit.

HELIUM—A mixture mainly of helium that ordinarily might contain some nitrogen and methane.

HOLE—The wellbore.

HYDROCARBON—A compound consisting of molecules of hydrogen and carbon. Petroleum is a mixture of many hydrocarbons.

HYDROGEN SULPHIDE (H_2S)—A naturally occurring poisonous gas.

HYDROSTATIC PRESSURE—The pressure exerted by a column of fluid, such as drilling mud in the wellbore.

IN SITU COMBUSTION—The setting fire to some portion of the oil in a reservoir (1) to produce combustion and distillation to drive oil ahead of them to producing wells and (2) to heat the oil so it will flow more readily.

INITIAL POTENTIAL (IP)—The initial capacity of a well to produce.

INJECTED GAS—Gas injected into a formation for storage, to maintain or restore reservoir pressure or otherwise to enhance recovery. Also, gas injected for gas-lift.

INJECTION WELL—A well that is used for injecting fluids into an underground stratum.

KILL A WELL—To stop a well from producing, usually by filling it with heavy fluids to restore control or so that surface connections may be safely removed for well servicing or workover.

LEASE—(1) A legal document that conveys to an operator the right to drill for oil and gas, or (2) the tract of land on which a lease has been obtained to explore for or to produce oil or gas or to locate equipment.

LEASE AUTOMATIC CUSTODY TRANSFER (LACT or ACT)—The measurement and transfer of oil from the producer's tanks to the oil purchaser's pipeline on an automatic basis without a representative of either having to be present.

LIMESTONE—A type of sedimentary rock rich in calcium carbonate. Limestone sometimes serves as a reservoir rock for petroleum.

LIQUIFIED PETROLEUM GASES (LPGS)—Hydrocarbon gases comprise mainly propane or butane, or a mixture of them. (See NATURAL GAS LIQUIDS (NGLS).)

LOAD FACTOR (RE NATURAL GAS)—A term used in natural gas contracts indicating the average daily requirement divided by the maximum daily requirement, stated as a percentage. Also the average daily throughput of a pipeline, gas plant or other facility divided by the maximum daily throughput, stated as a percentage.

LOST CIRCULATION—The loss of drilling fluid from the wellbore into a permeable formation.

MARGINAL WELL—A low producing rate well where the revenue from production only slightly exceeds the operating cost.

MARKETABLE GAS—A mixture, mainly of methane originating from raw gas through processing for the removal or partial removal of some consituents, that meets specifications for use as a domestic, commercial or industrial fuel, or as an industrial raw material.

MCF—1,000 cubic feet (usually applied to natural gas).

METHANE—In addition to its normal meaning, a mixture mainly of methane that ordinarily might contain some ethane, nitrogen, helium or carbon dioxide.

MISCIBLE FLOOD—An oil-recovery process that involves the injection of a fluid, which mixes readily with the oil, followed by a displacing fluid.

MUD (DRILLING)—The fluid circulated down the drill pipe and up the annulus during drilling. It removes drill cuttings and keeps the bit cool during drilling operations.

MULTIPLE COMPLETION—A well equipped to produce oil and/or gas separately from more than one reservoir.

NATURAL or RAW GAS—A mixture of hydrocarbons and varying quantities of nonhydrocarbons that exists either in the gaseous phase or in solution with crude oil in natural underground reservoirs.

NATURAL GAS LIQUIDS (NGLs)—Those hydrocarbon portions of reservoir gas that are liquefied at the surface in gas processing plants.

NIPPLE—A short section of pipe with screw-threads at each end for couplings.

NOMINATION—In Alberta, a statement by the purchaser of oil to the Board setting out the volume and type of oil required, the refinery location, and the means of transportation.

NON-ASSOCIATED GAS—Natural gas in reservoirs that do not contain significant quantities of crude oil.

OFFSET WELL—A well drilled near an existing well.

OIL—Condensate or crude oil, or a constituent of raw gas, condensate or crude oil, that is recovered in processing and is liquid at the conditions under which its volume is measured or estimated.

OIL AND GAS SEPARATOR—An item of production equipment used to separate the liquid components of the well stream from the gaseous components.

OIL SANDS—Sands and other rock materials that contain crude bitumen and include all other mineral substances in association therewith. Sometimes referred to as tar sands.

OIL-SANDS DEPOSIT—A natural reservoir containing, or appearing to contain, an accumulation of oil sands separated, or appearing to be separated, from any other such accumulation.

OPEN HOLE—The uncased portion of a well.

ORIFICE METER—An instrument commonly used to measure the flow of fluid (usually gas) in a pipe.

OVERPRODUCED—The status of a well that has produced more than its allowable.

PENTANES PLUS—A mixture mainly of pentanes and heavier hydrocarbons that ordinarily might contain some butanes and which is obtained from the processing of raw gas, condensate or crude oil.

PERMEABILITY—A measure of the transmissibility of fluids through rock.

PINNACLE REEF FORMATION—A geological term referring to a spirelike formation, usually composed of limestone, in which hydrocarbons might be trapped.

POOL—A natural underground reservoir containing, or appearing to contain, an accumulation of oil, gas or both separated from any other such accumulation.

POOLING—A term frequently used interchangeably with unitization but more properly used to describe the bringing together of small tracts sufficient for the granting of a well permit or licence under applicable spacing rules, as distinguished from the term unitization, which is used to describe the joint operation of all, or some portion of, a producing reservoir.

POROSITY—The open or void space within rock.

POROSITY (OF A RESERVOIR ROCK)—The percentage that the volume of the pore space bears to the total bulk volume. The pore space determines the amount of space available for storage of fluids.

POROUS FORMATION—A strata of substantially the same kind of rocks containing openings or small spaces within the rock that are usually filled with some fluid, such as oil, water or gas.

PRESSURE DRAWDOWN—The reduction in a well's bottom-hole pressure.

PRESSURE MAINTENANCE—Maintaining reservoir pressure by injecting fluid, normally water or gas, or both.

PRIMARY RECOVERY—The amount of oil and/or gas produced from a reservoir by the reservoir's natural sources of energy. This includes gas-cap drive, dissolved gas drive, water drive, or any combination of these.

PROCESSING PLANT—See GAS PROCESSING PLANT.

PRODUCTION—The yield of an oil or gas well.

PRODUCTION CASING—Usually the last string of casing set in a well; the casing string set from the surface to the top of or through the producing formation and inside of which is usually suspended the tubing string.

PROJECT—A pool, or part thereof, in which operations in accordance with a scheme for enhanced recovery of oil, approved by the Board, are conducted.

PROPANE—In addition to its normal meaning, it means a mixture mainly of propane that ordinarily might contain some ethane or butanes.

PRORATION—The system of allocation of the provincial demand for crude oil among pools and wells.

PROVED RESERVES—These are reserves delineated by drilling, ditching, running adits, testing or production, plus a judgement portion of those further contiguous reserves that are generally delineated by geological, seismic or similar information and can be reasonably counted upon.

RELIEF WELL—A well drilled near, and deflected into or near, the wellbore of a well that is out of control to bring the wild well under control.

RESERVOIR—A subsurface porous and permeable rock body that contains oil and/or gas.

RESERVOIR PRESSURE—The pressure at the face of the producing formation when the well is shut in.

RESIDUAL GAS—The gas remaining after processing of raw gas for the removal or partial removal of some constituents (see MARKETABLE GAS).

REWORKING A WELL—Restoring production from an existing formation (remedial operations) when it has fallen off substantially or ceased altogether. Also called WORKOVER.

RIG—The derrick, draw-works and attendant surface equipment of a drilling or workover unit.

ROUGHNECK—A member of the rig crew who usually works on the rig floor.

ROYALTY INTEREST—The fraction of the oil and gas retained by the mineral rights owner under the lease agreement.

SANDSTONE (or SAND)—A compacted sedimentary rock composed mainly of the minerals quartz or feldspar. Sandstone is a common rock in which petroleum and water accumulate.

SEPARATOR—A pressure vessel used for the purpose of separating gas from crude oil and water.

SERVICE WELL—A nonproducing well used for injecting liquid or gas into the reservoir for enhanced recovery. Also a salt water disposal well, a water supply well, or a well used to measure reservoir conditions.

SHUT-IN PRESSURE—Pressure as recorded at the wellhead when the valves are closed shutting in the well.

SOLUTION GAS—Natural gas dissolved in crude oil and held under pressure in the oil in a reservoir.

SOLUTION-GAS DRIVE—A natural drive mechanism where an oil reservoir derives its energy for production from the expansion of the natural gas in solution in the oil.

SOUR CRUDE OIL (SOUR CRUDE)—An oil containing free sulphur or other sulphur compounds.

SOUR GAS—Natural gas containing hydrogen sulphide in measurable concentrations.

SPACING UNIT—Sometimes referred to as a Drainage Unit, the term refers to the area allocated to a well pursuant to the Drilling and Production Regulations or a spacing order for a particular geological formation or series of formations. In Alberta, since the early 1960s, a distinction has been made between drilling spacing units and production spacing units. A drilling spacing unit is the area in the formation, in which a well is proposed to be completed, for which the person drilling the well must have the oil or gas rights. A production spacing unit is the area allocated to a well in determining its production allowable. Normally, it comprises the same area as the drilling spacing unit, but in some circumstances might comprise two or more drilling spacing units.

SPUD IN—Commence drilling of a well.

STOCK TANK—A lease tank into which a well's production is run.

STRIPPER WELL—A well nearing depletion that produces small amounts of oil or gas.

SULPHUR DIOXIDE (SO_2)—A poisonous gas formed by burning hydrogen sulphide.

SURFACE CASING—The first string of casing to be set in a well. Its principal purposes are to protect fresh water sands and to provide a means of anchoring well control equipment.

SWEET OIL—Oil containing little or no sulphur and hydrogen sulphide.

SYNTHETIC CRUDE OIL—A mixture, mainly of pentanes and heavier hydrocarbons that might contain sulphur compounds, that is derived from crude bitumen and is liquid at the conditions under which its volume is measured or estimated and includes all other hydrocarbon mixtures so derived.

TOUR REPORTS (Pronounced "Tower")—The daily record of activities at a drilling rig.

TRAP (GEOLOGIC)—An arrangement of rock strata or structures that halts the migration of oil and gas and causes them to accumulate.

TRAVELLING BLOCK—A moveable portion of a pulley system suspended from the rig crown, and under the direct control of the driller. Used along with other equipment to guide, maintain or alter the depth of the drill pipe or other strings of pipe within the wellbore during drilling or reworking operations.

TREATER—A fired apparatus specifically designed and used for separating gas and water from crude oil.

ULTIMATE RESERVES—Those reserves ultimately expected to have been proved after all exploration and production has been completed.

UNIT OPERATOR—The company designated to operate a unitized property.

UNITIZATION—The process whereby the owners of adjoining properties pool their reserves and form a single unit for the operation of the properties by only one of the owners. The production from the unit is then divided on the basis established in the unit agreement. The purpose of such agreement is to produce the reserves more efficiently, increasing the recovery for every participant. It is used frequently where an enhanced recovery process is proposed.

WASTE—In addition to its ordinary meaning, it means wasteful operations.

WASTEFUL OPERATIONS—The locating, spacing, drilling, equipping, completing, operating or producing of a well in a manner that results, or tends to result, in reducing the quantity of oil, gas or crude bitumen recoverable from a pool or oil-sands deposit or tends to cause excessive surface loss or destruction of oil, gas or crude bitumen.

WATER FLOODING—One method of enhanced recovery in which water is injected into an oil reservoir to force additional oil out of the reservoir rock and into the wellbores of producing wells.

WATER WELL—A well drilled to obtain a fresh water supply to support drilling and/or production operations.

WATER-CONING—The upward encroachment of water into a well because of pressure drawdown at the well due to production.

WELL—Any orifice in the ground made by drilling, boring or any other manner to obtain natural gas, petroleum or other resources.

WELL EFFLUENT—The substances that flow from a well.

WELL LICENCE—The authorization to drill a well issued by the Conservation Board.

WELLHEAD—The equipment used to maintain surface control of a well.

WILDCAT WELL—A well drilled in a previously unexplored area.

WORKING INTEREST—The operating interest under an oil and gas lease.

WORKOVER—See REWORKING A WELL.

PREFACE

1.   Gerald D. Nash, *United States Oil Policy 1890–1964* (Pittsburgh: University of Pittsburgh Press, 1968), pp. 20–21 and 34–38.

2.   *The Financial Post,* "The Financial Post 500" (Summer 1986). Ranked by sales and operating revenue, seven of the 20 largest corporations are in the oil and gas sector; by value of exports, seven of the 20; and by level of capital spending, eight of the top 20.

3.   Eric J. Hanson, *Dynamic Decade: The Evolution and Effects of the Oil Industry in Alberta* (Toronto: McClelland and Stewart, 1958), pp. 200–209.

4.   John Richards and Larry Pratt, *Prairie Capitalism: Power and Influence in the New West* (Toronto: McClelland and Stewart, 1979).

5.   Alvin Finkel, *The Social Credit Phenomenon in Alberta* (Toronto: University of Toronto Press, 1989), p. 116. Texas regulatory practice, which is usually cited as the model for Alberta legislation, was in fact geared more towards protecting the interests of the smaller independent western-based oil companies.

6.   See, for example, Philip Smith, *The Treasure-Seekers: The Men Who Built Home Oil* (Toronto: Macmillan, 1978), one of the best company histories. Earle Gray, *Wildcatters: The Story of Pacific Petroleums and Westcoast Transmission* (Toronto: McClelland and Stewart, 1982); Peter Foster, *From Rags to Riches: The Story of Bow Valley Industries Ltd.* (Calgary: Bow Valley Industries Ltd., 1985); Allan Anderson, *Roughnecks and Wildcatters* (Toronto: Macmillan, 1981). An exception is Fred Stenson's *Waste to Wealth: A History of Gas Processing in Canada.* It reflects that none of the industry's component parts has had a closer relationship with the Board than the gas processing sector.

7.   Among the best studies are John N. McDougall, *Fuels and the National Policy* (Toronto: Butterworth, 1982), and G. Bruce Doern and Glen Toner, *The Politics of Energy: The Development and Implementation of the NEP* (Toronto: Methuen Publications, 1985). Studies such as C. Lloyd Brown-John's *Canadian Regulatory Agencies* (Toronto: Butterworth, 1981) do comment on the ERCB, but still just in passing and as one of a host of regulatory bodies being considered.

8.   The limited and hasty consideration of the ERCB's place in the regulatory picture is underlined by the above authors' confusion about when the Alberta Board was formed and the circumstances that surrounded its creation. See Doern and Toner, pp. 69 and 494; McDougall, p. 69.

9. See, for example, Marver H. Bernstein, *Regulating Business by In-dependent Commission* (Princeton: Princeton University Press, 1955); Louis M. Kohlmeier, *The Regulators* (New York: Harper and Row, 1969); Paul M. MacAvoy, *The Crisis of the Regulatory Commissions* (New York: W.W. Norton, 1970).

10. David Crane, *Controlling Interest: The Canadian Gas and Oil Stakes* (Toronto: McClelland and Stewart, 1982), p. 148.

INTRODUCTION

1. Erich W. Zimmerman, *Conservation in the Production of Petro-leum: A Study in Industrial Control* (New Haven: Yale University Press, 1957), p. 29.

2. Wallace F. Lovejoy and Paul T. Homan, *Economic Aspects of Oil Conservation Regulations* (Baltimore: Johns Hopkins University Press, 1967), pp. 16–17, quoting Stephen L. McDonald "The Economics of Con-servation," paper presented to the Rocky Mountain Petroleum Economics Institute, Boulder, Colorado, 18 June 1964, p. 6.

3. Ibid.

4. For a more thorough discussion of this subject, see also Anthony Scott, *Natural Resources: The Economics of Conservation* (Toronto: Uni-versity of Toronto Press, 1955), and Samuel P. Hays, *Conservation and the Gospel of Efficiency* (Cambridge: Harvard University Press, 1959).

5. See, for example, IOCC Governors' Special Study Committee, *A Study of Conservation of Oil and Gas in the United States, 1964* (Okla-homa City: IOCC, 1964). Lovejoy and Homan, pp. 27–28, note that in the above study "the rather amorphous character of industry thinking on con-servation is well illustrated."

6. ERCB, George W. Govier, "Oil and Gas Conservation," paper pre-sented before the Canadian Institute of Mining and Metallurgy, Western Annual Meeting, Vancouver, 1950, pp. 1–2. Conservation authorities dif-fer somewhat in their definitions of waste. Alberta's *Oil and Gas Conser-vation Act* elaborates on the definition in Section 1.(1), subsections w and x.

7. J.W. Amyx, D.M. Bass, Jr., and R.L. Whiting, *Petroleum Reservoir Engineering* (New York: McGraw-Hill, 1960).

8. R.J.W. Douglas et al., "Geotectonic correlation chart for western Canada," Ottawa: Geological Survey of Canada, 1970.

9. Robert E. Hardwicke, "Some Legal and Economic Aspects of Con-servation Regulation," in Wallace F. Lovejoy and I. James Pikl, Jr., eds., *Essays on Petroleum Conservation Regulation* (Dallas: Southern Method-ist University, 1960), p. 95.

10. Govier, p. 4.

11. The discussion on reservoir drives has relied heavily upon Nicholas J. Constant, *Improved Recovery, Oil and Gas Production Series,* ed., Karen I. Stelzner (Austin: Petroleum Extension Service, Division of Contin-uing Education, University of Texas, 1983), pp. 21–35; Lovejoy and Homan, pp. 61–62; and Govier, pp. 8–15.

12. *Westmoreland Natural Gas Company v. DeWitt,* 130 Pa. 225, 18

Atl. 724, 1889. See also Robert E. Hardwicke, "The rule of capture," *Mississippi Law Journal* 13 (1941), pp. 381–416.

13. *Borys v. Canadian Pacific Railway and Imperial Oil Limited* (P.C., 1953, Lord Porter), 7WWR (NS), 550.

14. *Ohio Oil Company v. Indiana*, 177 U.S. 190, 20S.Ct.585, 44L. Ed. 729 (1900), as cited by Zimmerman, p. 100.

15. Lovejoy and Homan, p. 50. Each of these approaches in itself has a long history of legislative action, judicial interpretation and technical development.

16. Ibid.

17. Zimmerman, pp. 328–36. In the early development of an Alberta oil pool, when many of its characteristics must be estimated, a maximum permissible rate per well (MPR) is prescribed using broad criteria. This rate is based upon the maximum rate that would be appropriate considering production efficiency, including the orderly development of facilities to accommodate production. This rate is reviewed periodically to consider its continued suitability, and it might evolve into a pool MER when more is learned about reservoir conditions and the pool's characteristics are better understood. Also note that, for some pools, the MER might be less than the economic rate; hence, a higher rate might be necessary to make continued production economic and thus sustain the pool's operation.

18. Energy Resources Conservation Board (ERCB) Minutes, Petroleum and Natural Gas Conservation Board, Order No. 1, 4 August 1938. This is an order under the first of the two 1938 conservation acts. Unless indicated otherwise, ERCB-identified sources are to be found at the Board's Calgary headquarters.

19. Stephen L. McDonald, *Petroleum Conservation in the United States: An Economic Analysis* (Baltimore: Johns Hopkins University Press, 1971), p. 150.

20. Lovejoy and Homan, pp. 75–79. See also McDonald, pp. 229–52; E.V. Rostow, *A National Policy for the Oil Industry* (New Haven: Yale University Press, 1948); and M.G. DeChazeau and A.E. Kahn, *Integration and Competition in the Petroleum Industry* (New Haven: Yale University Press, 1959).

21. McDonald, p. 198.

22. Alberta, Revised Statutes of Alberta, 1980, *Oil and Gas Conservation Act*, ch. o–5, ss. 70 and 71.

23. M.A. Adelman, *The World Petroleum Market* (Baltimore: Johns Hopkins University Press, 1972); John M. Blair, *The Control of Oil* (New York: Pantheon Books, 1976); and Robert Engler, *The Politics of Oil: A Study of Private Power and Democratic Directions* (New York: Macmillan, 1961).

24. Robert E. Allen, "Control of California oil curtailment," AIME *Transactions*, vol. 92, 1931, 47–66, and "Some Effects of Curtailment on the Potential and Recovery of Petroleum in California," *Mining and Metallurgy* 15 (December 1934), pp. 486–88.

25. Alberta, "The Report of a Royal Commission Appointed by the Government of the Province of Alberta Under The Public Inquiries Act to Inquire Into Matters Connected With Petroleum and Petroleum Products," Edmonton, 1940.

26. J.L. Tiedje, "Pioneers in petroleum technology," *Journal of Canadian Petroleum Technology* 21 (November/December 1982), p. 41, and Dianne Newell, *Technology on the Frontier: Mining in Old Ontario* (Vancouver: University of British Columbia Press, 1986), pp. 121–27. The Petroleum industry in Canada began about 1850 near the exposed petroleum seepages in Enniskillen township. Asphalt produced from this source by C.N. Tripp won honourable mention at the Paris Universal Exhibition in 1855. Tripp sold his business to James M. Williams in 1856, who immediately sank a well into the gum beds. The well's success led Williams to build in 1857 what might have been North America's first oil refinery to produce lamp oil and asphalt.

27. Gerald D. Nash, *United States Oil Policy 1890–1964* (Pittsburgh: University of Pittsburgh Press, 1968), pp. 15–16.

28. Quoted by Zimmerman, p. 136.

29. For the best discussion on early Oklahoma legislation, see W.P.Z. German, "Legal History of Conservation of Oil and Gas in Oklahoma," in *Legal History of Conservation of Oil and Gas, A Symposium* (American Bar Association, 1938).

30. The 1909 Act launched a long legal battle that was not concluded until 1922 when the U.S. Supreme Court, in *Pierce Oil Company v. Phoenix Refining Company*, upheld the constitutionality of the Oklahoma statute.

31. Oklahoma Corporation Commission Report, 1915 (Oklahoma City, 1916).

32. Zimmerman, Table XV, p. 284.

33. Nash, p. 82. Doherty, formerly a consulting engineer, was President of H.L. Doherty and Company, fiscal agent for the Cities Service Companies.

34. For a statement on Doherty's conservation ideas, see *The New York Times*, 20 November 1924. Doherty was not the first to draw attention to the importance of unitization. In 1916, William F. McMurray and James O. Lewis of the U.S. Bureau of Mines, in a technical paper entitled "Underground Wastes in Oil and Gas Fields and Methods of Prevention," recommended unit operation and also expressed the view that compulsion would be required, cited by Zimmerman, p. 122, n. 20.

35. Nash, pp. 83–84.

36. Zimmerman, pp. 124–25; Nash, pp. 86–100.

37. McDonald, pp. 36–37.

38. Nash, p. 145, and *Panama Refining Co. v. Ryan*, 239 U.S. 388 (1935).

39. Interstate Oil Compact Commission, *The Compact's Formative Years 1931–1935* (Oklahoma City: IOCC, 1954), pp. 38–49.

40. Lovejoy and Homan, p. 44.

ONE—TURNER VALLEY: "WASTE" GAS AND THE EARLY
CONSERVATION MOVEMENT

1. The Rocky Mountain Development Co. was one of several seeking oil along Cameron Creek near Waterton Lakes at the turn of the century.

Oil showings had been reported in the area by the Boundary Commission survey in 1874. A.P. Patrick, a dominion land surveyor and director of the Rocky Mountain Development Co., staked his claim along the creek in 1889. Beginning drilling in November 1901, this company did achieve enough production to create a stir of interest from 1902 to 1905. For an account of the activities along Cameron Creek, see George de Mille, *Oil in Canada West, the Early Years* (Calgary: George de Mille, 1969), pp. 91–100.

2. Glenbow Archives (hereafter GA), "The Alberta Oil Fields," Prospectus of the Rocky Mountain Development Co. Ltd. of Alberta, 1902, p. 12.

3. Chester Martin, *"Dominion Lands" Policy*, ed., Lewis H. Thomas (1938; reprint ed., Toronto: McClelland and Stewart, 1973) and Kirk N. Lambrecht, *The Administration of Dominion Lands, 1870–1930* (Regina: Canadian Plains Research Centre, 1991) are the best studies of federal land and resource policy in western Canada.

4. The federal government retained direct control of land and natural resources within the boundaries of Indian reservations, national parks, certain military reserves, and all unsurrendered lands in the Yukon, Northwest Territories and Arctic islands.

5. *Statutes of Canada*, 35 Vic., ch. 23.

6. Ibid., s. 105.

7. Ibid., s. 38(1).

8. Order-in-Council, No. 1070, 1887. In 1889, the geographic limitation was removed and the reservation of mineral title applied to all dominion lands.

9. Order-in-Council, P.C. No. 2774, 18 December 1890.

10. Order-in-Council, 7 March 1884.

11. Order-in-Council, P.C. No. 1822, 6 August 1898.

12. Order-in-Council, P.C. No. 893, 31 May 1901.

13. Order-in-Council, P.C. No. 513, 23 March 1904.

14. Order-in-Council, P.C. No. 2287, 26 December 1906.

15. Order-in-Council, P.C. No. 414, 11 March 1910.

16. Ibid., ss. 13–15.

17. Order-in-Council, P.C. No. 1951, 12 October 1910. For a discussion of the Admiralty's role in the formulation of Canada's initial petroleum policy, see D.H. Breen, "Anglo-American rivalry and the evolution of Canadian petroleum policy to 1930," *Canadian Historical Review* LXII, 3 (September 1981), pp. 283–304.

18. H.H. Somerville, *A History of Crown Rentals, Royalties and Mineral Taxation in Alberta to December 31, 1972*, ENR Report No. 1 (Edmonton: Alberta Department of Energy and Natural Resources, 1977), p. 5–4.

19. W. Kaye Lamb, ed., *Journals and Letters of Alexander Mackenzie*, Hakluyt Society, Extra Series, no. 4 (Toronto: Macmillan, Canada, 1970), p. 129.

20. John Macoun, "Geological and Topological Notes by Professor Macoun, on the Lower Peace and Athabasca Rivers," Geological Survey of Canada, *Report of Progress for 1874–6* (Montreal: Geological Survey of Canada, 1877), as quoted by B.G. Ferguson in *Athabasca Oil Sands* (Edmonton: Alberta Culture and Canadian Plains Research Centre, 1985), p.

14. Ferguson offers an excellent account of the investigation and development of the oil sands from 1875 to 1951.

21. Robert Bell, "Report on Part of the Athabasca River," in Geological Survey of Canada, *Report of Progress for 1882–4* (Montreal: Dawson Brothers, 1884), as discussed in Ferguson, pp. 15–19.

22. The Survey's continued interest in the Athabasca Basin is manifest in the numerous reports that they continued to publish about the region, and more particularly in their unsuccessful drilling endeavour at Athabasca Landing in 1894 and subsequently at the confluence of the Pelican and Athabasca rivers in 1897. The latter well encountered an uncontrollable natural gas flow and was left to run wild for 21 years.

23. Later known as Carlstadt and, after 1915, as Alderson.

24. *Calgary Herald*, 12 December 1883 and 16 January 1884. See also George de Mille, p. 63. A second successful well was drilled in 1884 only eight feet away from the first, and it continued to produce a small volume of gas until 1954.

25. *Calgary Herald*, 29 October 1884.

26. G.M. Dawson "On Certain Borings in Manitoba and the Northwest Territory," *Proceedings and Transactions of the Royal Society of Canada for the Year 1886* (Montreal: Dawson Brothers, 1887), vol. IV, p. 96.

27. GA, CPR Papers, BN.2/C212/1, 8 February 1906, Second Vice-president W. Whyte to Eugene Coste.

28. The company began to withhold coal rights in 1904.

29. GA, CPR Papers, BN.2/C212G/1, 5 November 1908, J.S. Dennis, Assistant to the Second Vice-president, to W. Whyte.

30. For a more detailed discussion of this and related oil and gas activities in Alberta before 1924, see David H. Breen, "The CPR and Western Petroleum, 1904–24," in Hugh A. Dempsey, ed., *The CPR West: The Iron Road and the Making of a Nation* (Vancouver: Douglas & McIntyre, 1984), pp. 229–44.

31. Coste Gas Wells

Location	Depth (ft)	Output (cu ft/24 hrs)	Pressure (lbs)
Medicine Hat:			
*SE/4 31–12–5 W4M	1,000	1,500,000	500
Dunmore Jct:			
NE/4 09–12–5 W4M	1,300	2,000,000	450
Bow Island:			
SW/4 15–11–11 W4M	1,916	5,250,000	500–600
Suffield:			
SE/4 10–13–9 W4M	700	90,000	70
Brooks:			
SW/4 33–18–14 W4M	2,630	280,000	—
Bassano:			
NW/4 17–21–18 W4M	1,240	37,000	—

*LAND DESCRIPTION = Southeast Quarter-Section-Township-Range Meridian

684 –

32. GA, CPR Papers, BN.2/C212C/2, 15 June 1910, Shaughnessy to Whyte.

33. National Archives of Canada (hereafter NAC), RG85, vol. 2002, 157100, "Report on Investigation" of Canadian Western Natural Gas, Light, Heat and Power Company Ltd., pp. 4–8.

34. GA, CPR Papers, BN2/C212G/4, Memorandum of Agreement, 15 February 1911.

35. *Calgary Herald,* 15 January 1914.

36. See the *Alberta Gazette,* May-August 1914.

37. GA, CPR Papers, BN.2/C212G/1923, 10 December 1915, Naismith to D.C. Coleman.

38. Hugh Grant, "Canadian Petroleum Industry: An Economic History, 1900–1960," Ph.D., University of Toronto, 1986, from J.S. Ewing, "The History of Imperial Oil" (Boston: Business History Foundation Inc., Harvard Business School, 1951), ch. XII, sec. B, p. 9.

39. George S. Gibb and Evelyn H. Knowlton, *History of Standard Oil Company (of New Jersey), 'The Resurgent Years' 1911–1927* (New York: Harper and Brothers, 1956), p. 89.

40. Canada, Order-in-Council, P.C. No. 154, 19 January 1914, "Petroleum and Natural Gas Regulations." For a discussion of the origin of Section 40, see D.H. Breen, "Anglo-American rivalry and the evolution of Canadian petroleum policy to 1930," *Canadian Historical Review* LXII, 3 (September 1981), pp. 283–304.

41. GA, CPR Papers, BN.2/C212G/431, 5 October 1916, "Minutes of the seventh meeting of the advisory committee."

42. To check Imperial's enthusiasm for the Viking-Wainwright area, the CPR hired C.A. Fisher, a well-known consulting geologist from Denver, to investigate and report on the area. Fisher's favourable assessment reinforced the railway's inclination towards a cautious approach.

43. GA, CPR papers, BN.2/C212G/547, 6 February 1917, W.C. Teagle to E.W. Beatty. The area was bounded on the north by Township 53; on the south by Township 34; on the west by Range 18, west of the 4th Meridian; and on the east by Range 20, west of the 3rd Meridian.

44. Ibid., 16 February 1917, A.M. Nanton to Beatty.

45. NAC, RG85, vol. 1683, 143207, 8 July 1918, Sir Reginald McLeod, Shell Transport Company Ltd., to Sir Robert Borden. McLeod refers to "certain communications" of April and May 1917.

46. Ibid., 30 July 1917, Lieutenant-colonel A.T. Shillington, attorney for R.N. Benjamin, Shell Transport Company Ltd., to W.S. Roche, Minister of the Interior.

47. Ibid., 8 August 1918, Sir Reginald McLeod to Arthur Meighen, with enclosure "Memorandum With Regard of an Interview at the Savoy Hotel Granted by the Hon. Arthur Meighen to Sir Reginald McLeod and Mr. R.N. Benjamin on the 13 July 1918."

48. Ewing, ch. XII, section C. See endnote 38.

49. NAC, RG85, vol. 1863, 143207, 1 March 1919, W.J. Hanna to Arthur Meighen.

50. GA, CPR Papers, BN.2/C212G/547, 24 November 1919. The block reserved was now confined to railway lands in Townships 38, 39 and 40; Ranges 6, 7 and 8; West of the 4th Meridian; also 29 October 1919, "Draft form of lease for petroleum rights for Imperial Oil Company," lease terms provided for a yearly rental of 50¢ per acre and the right to take 10% of production or accept a cash royalty of 10% of the market value of petroleum or natural gas produced.

51. Ibid., 19 February 1918, J.R. Cowell, Clerk, Legislative Assembly of Alberta, to Arthur Meighen, and 26 December 1918, Charles Stewart to Arthur Meighen.

52. Ibid., p. 470, 18 October 1920, Dennis to Nanton.

53. Ibid., 14 May 1920, "Minutes of the thirty-seventh meeting of the Advisory Committee of the Department of Natural Resources."

54. GA, A.W. Dingman Papers, A. D584, 3, 20 November 1920, Director's Meeting, and 21 December 1920, Special General Meeting of Calgary Petroleum Products Ltd.

55. Gibb and Knowlton, pp. 260, 275.

56. Dominion Bureau of Statistics, *Canadian Mineral Statistics 1886–1956 and Mining Events, 1604 to 1956.* Reference paper no. 68 (Ottawa: Queen's Printer, 1957), p. 106. Active local companies included Canada Southern Oil and Refining Company, McDougall-Segur Exploration Company, Alberta Pacific Consolidated, McLeod Oil Company and United Oils.

57. Order-in-Council, P.C. No. 154, 19 January 1914.

58. Ibid., s. 29.

59. Ibid.

60. See, for example, Alberta, 1938, ch. 15, *An Act for the Conservation of Oil and Gas Resources of the Province of Alberta*, s. 14, and 1950, ch. 46, *An Act to Provide for the Conservation of the Oil and Gas Resources of the Province of Alberta*, ss. 16 and 30.

61. NAC, RG85, vol. 1939, 112938, 31 May 1918, H.H. Rowatt, Mines Branch, to Mitchell, Secretary to the Minister of the Interior.

62. U.S. *Congressional Record*, 1919, 4169.

63. Order-in-Council, P.C. No. 105, 29 January 1920.

64. Order-in-Council, P.C. No. 1230, 29 May 1919.

65. Order-in-Council, P.C. No. 2433, 3 December 1919.

66. Order-in-Council, P.C. No. 1920, 29 January 1920.

67. NAC, RG15, A3, vol. 25, Alberta Resources Commission, "Information Prepared for Dominion Counsel," p. 17.

68. Order-in-Council, P.C. No. 2614, 29 October 1920.

69. NAC, RG85, vol. 1839, 112938, 10 June 1913, Graham Greene, Admiralty, to Undersecretary of State, Colonial Office.

70. Ibid., 14 August 1916, H.H. Rowatt to Mitchell, Secretary to the Minister of the Interior.

71. In unsurveyed lands, the words "crown reserve" had to be written or printed on the stakes marking the area to be reserved. H.H. Somerville, *A History of Crown Rentals, Royalties and Mineral Taxation in Alberta to December 31, 1972*, ENR Report no. 1 (Edmonton: Dept. of Energy and Natural Resources, 1977), pp. (5) 9–10.

72. Order-in-Council, P.C. No. 953, 24 March 1921. The person whose bid was accepted had to deposit the amount of the bonus offered immediately with the Mining Recorder, together with the rental for the first year of the tract applied for. The Department reserved the right to refuse any or all bids.

73. Order-in-Council, P.C. No. 235, 6 February 1901, and P.C. No. 1059, 9 May 1907.

74. In the interest of conservation, after September 1918 no mine could be opened before plans showing the manner in which it was proposed to open up, develop and work the property were approved by the mining inspector.

75. Canada, Commission of Conservation, *Report of the First Annual Meeting* (Ottawa: Mortimer Company Ltd., 1910), pp. 4–5.

76. Canada, *An Act Establishing the Commission for the Conservation of Natural Resources*, 8–9 Edward VII, ch. 27, assented to 19 May 1909.

77. Canada, Commission of Conservation, *Report of the First Annual Meeting* (Ottawa: Mortimer Company Ltd., 1910), p. 5. In accepting the chairmanship, Sifton publicly disassociated himself from further active participation in party political affairs.

78. These were Committee on Fisheries, Game and Fur-bearing Animals; Committee on Forests; Committee on Lands; Committee on Minerals; Committee on Press and Co-operating Organizations; Committee on Public Health; and Committee on Waters and Water-powers.

79. Frank D. Adams, *The National Domain in Canada and Its Proper Conservation*. Presidential Address before the Royal Society of Canada, 1914 (Ottawa: Lowe-Martin Co. Ltd., 1915), p. 47.

80. Ibid., p. 7.

81. Ibid.

82. Ibid., p. 48.

83. Canada, Commission of Conservation, *Report of the Third Annual Meeting* (Montreal: John Lovell and Son Ltd., 1913), p. 29.

84. Ibid.

85. Canada, Commission of Conservation, *Report of the Fifth Annual Meeting* (Toronto: Bryant Press Ltd., 1914), pp. 52–65.

86. Canada, Commission of Conservation, *Report of the Sixth Annual Meeting* (Toronto: Bryant Press Ltd., 1915), p. 63.

87. Ibid., p. 68. Adams calculated that gas was escaping at a daily rate of 2,900,000 cubic feet, and that at an average price for natural gas of 16.4¢ per 1,000 cubic feet in 1913, this represented an accumulated value of $2,951,098 wasted gas.

88. Ibid., pp. 68–69.

89. Ibid., p. 52.

90. NAC, RG15, A3, vol. 25, Alberta Resources Commission, p. 40.

91. Order-in-Council, P.C. No. 418, 26 February 1919.

92. NAC, RG15, A3, vol. 25, Alberta Resources Commission, p. 41.

93. NAC, RG85, vol. 1988, 151163, 23 November 1919, S.E. Slipper to O.S. Finnie. Slipper provided detail of what kind of information should be expected on drillers' daily reports, as well as the specifics on how pressure tests be conducted.

94. W.J. Dick, "Importance of Bore-hole Records and Capping Gas Wells in Canada," Commission of Conservation, *Report of the Fifth Annual Meeting* (Toronto: Bryant Press Ltd., 1914), pp. 52–65.

95. *Calgary Herald*, 23 August 1905.

96. See, for example, *Calgary Herald*, 10 June 1905; *Morning Albertan*, 10 April 1913; *Calgary News Telegram*, 2 April 1913; *Calgary Herald*, 4 July 1921; *Morning Albertan*, 14 July 1921; *Calgary Herald*, 25 May 1926 and 25 June 1926.

97. NAC, RG15, A3, vol. 25, Alberta Resources Commission, p. 42.

98. Ibid.

99. NAC, RG15, A3, vol. 25, Alberta Resources Commission, p. 42.

100. *Edmonton Bulletin*, 14 April 1921.

101. Ibid., 13 April 1921.

102. *Edmonton Journal*, 8 March 1921.

103. *Edmonton Bulletin*, 13 April 1921.

104. *Edmonton Journal*, 18 April 1921.

105. *Edmonton Bulletin*, 19 April 1921.

106. NAC, RG15, vol. 26, Book K, pp. 32–33, 23 November 1921, Victor Ross to Sir James Lougheed.

107. Order-in-Council, P.C. No. 4613, 21 December 1921.

108. NAC, RG85, vol. 2002, 157100.

109. Ibid. The company was asking for a rate of 50¢ per 1,000 cu. ft. during the winter months, and a rate of 75¢ per 1,000 cu. ft. during the summer, with a minimum charge of $2.00 per month for service in place of the existing flat rate of 35¢ per 1,000 cu. ft.

110. "Report of the Natural Gas Conference," Calgary, Canada, 15 August 1921, p. 3.

111. Ibid., p. 6.

112. Ibid., p. 7.

113. Provincial Archives of Alberta (hereafter PAA), 69.289, Premiers' Papers, 321, 1 September 1921, D.F. Patterson, Calgary Citizens' Protective League to Premier Herbert Greenfield.

114. Ibid., 23 September 1921, Medicine Hat United Farmers of Alberta, Local 866, to Premier Greenfield; 19 October 1921, Woolchester United Farmers of Alberta, Local 937, to Premier Greenfield.

115. Ibid., 13 October 1921, Wood Gundy and Company Ltd. to Premier Greenfield.

116. Ibid., 1 June 1922, H.B. Pearson to George Hoadley, Minister of Agriculture and Acting Premier.

117. Ibid., 11 November 1921, S.H. Adams to Premier Greenfield.

118. NAC, RG85, vol. 2002, 157100, 30 December 1920, O.S. Finnie to H.H. Rowatt.

119. For Ontario's concurring view, see ibid., 9 March 1920, T.W. Gibson, Deputy Minister of Mines, to W. Pearce.

120. *Science*, 16 January 1920.

121. NAC, RG85, vol. 2002, 157100, 25 November 1921, S.E. Slipper to O.S. Finnie. C.W. Dingman also pointed out how such a restriction would hamper the search for oil, see 28 November 1921, C.W. Dingman Memorandum, "Regarding the Restriction of the Use of Natural Gas to Domestic Purposes."

122. Ibid., 28 November 1921, S.E. Slipper to O.S. Finnie. Slipper had also suggested the royalty approach earlier in the year, see RG85, vol. 2005, 158697, 29 March 1921, Slipper to Finnie. In 1923, Slipper left the Department to work for Canadian Western, Natural Gas, Light, Heat and Power Co. Ltd.

123. NAC, RG85, vol. 2005, 158697, 12 August 1921, H.H. Rowatt to the Deputy Minister of the Interior.

124. Ibid., 29 April 1927, H.H. Rowatt to the Deputy Minister of the Interior.

125. Ibid., 6 May 1921, H.H. Rowatt to the Deputy Minister of the Interior, and 26 June 1930, Memorandum, C.F. Spence.

126. It is probable that Deputy Minister W.W. Cory's decision to visit Alberta was motivated by Charles Stewart, the former premier of Alberta, who had just become minister of the interior.

127. NAC, RG85, vol. 2016, 165928, 25 February 1922, "Investigation Conducted by W.W. Cory, Deputy Minister, Department of the Interior, Federal Oil Regulations."

128. Ibid., 9 May 1922, Memorandum, H.H. Rowatt.

129. The bounty issue was one that stirred strong emotions in Alberta. One Turner Valley oilman reminded Cory that the bounty paid to eastern Canadians up to 1922 totalled $2,918,760 whereas the rentals paid on oil and gas leases in the province of Alberta amounted to $2,384,000. "Consequently we have practically paid the bounty received by the oil men of Eastern Canada" (Mr. Livingston). The problem was that the bounty was paid

only for crude petroleum having a specific gravity of not less than 0.8235 at 60°F. But for the output of one Turner Valley well, all Turner Valley oil was lighter and therefore not eligible for the bounty of 1½¢ per Imperial gallon.

130. Ibid., 25 February 1922, "Conference with Mr. Cory and Mr. Slipper, and a few of the bona fide petroleum and natural gas operators," pp. 8–9.

131. GA, CPR Papers, BN.2/C212G/547, 12 June 1922, R.B. Bennett to E.W. Beatty. For a more extensive discussion of Imperial-CPR petroleum leasehold negotiations, see D.H. Breen, "The CPR and Western Petroleum, 1904–24," in Hugh A. Dempsey, ed., *The CPR West: The Iron Road and the Making of a Nation* (Vancouver: Douglas & McIntyre Ltd. 1984), pp. 241–44.

132. GA, CPR Papers, BN.2/C212G/547, "Canadian Pacific Railway Company and Royalite Oil Company Ltd.: Lease of Petroleum Rights," 1 August 1922.

133. Site: Legal Subdivision 12, Section 7, Township 20, Range 2, West of the 5th Meridian.

134. Charles C. Ross, "Petroleum and natural gas development in Alberta," *Transactions of the Canadian Institute of Mining and Metallurgy, 1926*, vol. XXIX, p. 321.

135. Ibid., p. 324.

136. The separation occurred in a 3 x 16-foot cylinder through a relatively simple process. As Charles Ross explained to his fellow engineers, "separation starts as soon as the production enters the separator from the well. The gas expands, thereby losing much pressure, and gives up the vapour which it carries. The gas accumulates in the separator and is drawn off under its own pressure, the oil falls to the bottom of the separator and its outlet is controlled by means of a [governor]. Owing to the low temperature of the oil and gas at Royalite No. 4 this controlling device had to be removed." Ibid.

137. *Imperial Oil Review* 9, no. 4 (1925).

138. Charles C. Ross, p. 327. Production in western Ontario during 1924 was 153,766 barrels, total production from the Turner Valley field was 171,500 barrels. See NAC, RG45, vol. 39, 153 m l, Department of Mines, "Petroleum and Natural Gas Wells," July 1926.

139. *Edmonton Journal*, 15 November 1923, 16 November 1923 and 4 November 1924. See also NAC, RG85, vol. 2024, 173554.

140. GA, CPR Papers, BN.2/C212G/546, extract from minutes of the sixty-fifth meeting of the Advisory Committee of the Department of Natural Resources, 15 December 1924; P.L. Naismith to A.M. McQueen, 22 December 1924.

141. *Bulletin of the Canadian Institute of Mining and Metallurgy* 171, July 1926, pp. 760–63.

142. PAA, Premiers' Papers, 69.289, 251, 3 October 1923, Premier Greenfield to A. Chard, Freight Rate Supervisor, Alberta, Department of Railways and Telephones; 15 October 1923, Chard to Greenfield, "Supplementary Report, Gasoline Price Investigation," and 10 January 1924, Chard to Greenfield, "Further Report, Gasoline Investigation."

143. Ibid., 345 and 168B.

144. The anger of the Alberta government was mollified by what seemed promising negotiations then under way with the federal government on the transfer of the control of natural resources to the province. NAC, RG15, vol. 11, 1 August 1925, Memorandum to the Deputy Minister of the Interior,

Re: "Transfer of Albertan Natural Resources" and "Memorandum of Agreement."

145. NAC, RG85, vol. 2030, 181172, 16 July 1925, Herbert Greenfield to Charles Stewart.

146. *Calgary Herald,* 23 June 1925, editorial quoting the *New York World.*

147. NAC, RG85, vol. 2030, 181172, 2 September 1925, Charles Stewart to Herbert Greenfield.

148. *Statutes of Alberta,* 1926, ch. 26, "An Act to Provide for the Regulation of Oil and Gas Wells;" *Edmonton Journal,* 25 March 1926.

149. Ibid., s. 4.

150. Ibid., s. 5.

151. NAC, RG85, vol. 2030, 181172, 2 August 1926, J.E. Brownlee to J.H. Woods.

152. PAA, 65.74, 773a, 9 November 1926, F.P. Fisher to J.E. Brownlee, p. 3.

153. Ibid., pp. 8–10.

154. *Calgary Herald,* 6 October 1926.

155. PAA, Premiers' Papers, 69.289, 320, 28 May 1928, J.E. Brownlee to Charles Stewart, Minister of the Interior.

156. Ibid., 31 May 1928, C. Stewart to J.E. Brownlee, and 7 June 1928, Brownlee to Stewart.

157. Ibid., 18 December 1928, "Memorandum re Wastage of Natural Gas in the Turner Valley Field," p. 1.

158. Ibid., p. 2.

159. Ibid.

160. Ibid., p. 3.

161. Ibid., p. 4.

162. Ibid.

163. Ibid., pp. 6–7. That naphtha was more valuable than the heavier California crude did not offset Turner Valley's inferior production, "assuming that naphtha is worth four times what crude petroleum is worth, the gas produced in California for an amount [of liquid] of equal value to that in Turner Valley, would be only 16,000 cubic feet."

164. Ibid., p. 6.

165. Ibid., 23 February 1929, mayor of Calgary to C. Stewart. For further expression of opposition to gas exports, see also 22 March 1929, Red Deer Board of Trade to Premier J. Brownlee.

166. Ibid., 31 May 1929, Stewart to Brownlee.

167. On 14 December 1929, the federal government and the Province of Alberta signed the "Memorandum of Agreement for the Transfer of the Natural Resources to the Province."

168. Ibid., "Memorandum on the Progress of Investigations with Reference to Utilization and Conservation of Waste Gas in the Turner Valley Field."

169. Ibid., 25 October 1929, "First Progress Report of the Committee on Waste Gas."

170. Ibid., 8 January 1930, "Report of the Committee on the Conservation and Utilization of Waste Gas in Turner Valley, Alberta."

171. Ibid., p. 4.

172. Ibid., p. 5.

173. Ibid., 9 January 1930, Charles Camsell to J.E. Brownlee.

174. Ibid., 24 January 1930, A. Davison (Mayor of Calgary) to J.E. Brownlee.

175.   Ibid., 30 January 1930, J.E. Brownlee to Charles Stewart. See also *Calgary Albertan,* 3 April 1930.

176.   Licensed under the authority of the "Electricity and Fluid Explora-tion Act" of 1907. Between 1928 and 1930, several groups presented pro-posals to remove Alberta gas by pipeline to Regina and Winnipeg. One of the stumbling blocks seems to have been Imperial Oil's strenuous objec-tions to most of the proposals. See NAC, RG85, vol. 1650, 3142; vol. 1651, 3142C; vol. 1652, 3364.

177.   In contemporary industry usage, the term "export" is used to refer to natural gas leaving Canada and the broader term "removal" is applied to all gas transmitted out of the province.

178.   Canada, the *Alberta Natural Resources Act,* 20–21 George V, ch. 3, assented to 30 May 1930, and Order-in-Council, P.C. No. 2453, 1930, Memorandum of Agreement, setting October 1 as the date of transfer.

179.   PAA, Premiers' Papers, 69. 289, 320, 13 March 1930, J.A. Allan to J.E. Brownlee.

180.   Ibid., 10 March 1930, "Draft Copy of the Proposed Petroleum and Natural Gas Regulations, Applicable to all Lands, Crown and Freehold, Within the Province of Alberta."

181.   Alberta, Department of Lands and Mines, *Annual Report,* 1931, p. 24; NAC, RG15.

182.   NAC, RG15, A3, Alberta Resources Commission, pp. 49–51.

183.   Alberta, Department of Lands and Mines, *Annual Report,* 1931, p. 9.

184.   NAC, RG85, vol. 1975, 145058, "Schedule of Oil Producing Wells on Dominion Lands Which Have Operated or May Have Been Operated at Intervals During the Calendar Year, 1929," and 145059, "Schedule of Gas Producing Wells Which have Operated or May Have Been Operated at In-tervals During the Calendar Year, 1929."

185.   PAA, Premiers' Papers, 69.289, 320, 12 February 1931, "Turner Valley Gas Waste," W. Calder.

186.   Ibid., p. 1.

187.   Ibid., pp. 8–9.

188.   *Edmonton Journal,* 18 February 1931. The president of Imperial Oil was informed by telegram on 3 March 1931 of the premier's desire to discuss conservation with industry representatives. See PAA, Premiers' Pa-pers 69.289, 320.

189.   *Edmonton Journal,* 11 March 1931.

190.   Ibid.

191.   Ibid.

192.   Ibid., and *Edmonton Bulletin,* 12 March 1931.

193.   *Statutes of Alberta,* 1931, ch. 42, "An Act Respecting Oil and Gas Wells," assented to 28 March 1931.

194.   Ibid., s. (9). The new legislation also retained the former general clause assigning the right to make tests "for my purpose," s. (p).

195.   Ibid., s. (x).

196.   Alberta, Order-in-Council, OC No. 493–31, 6 May 1931. Potential capacity was determined from the most recent monthly gauge measure-ment.

197.   F.P. Fisher, "General Report of the Proposed Turner Valley Agree-ment," Edmonton: 1932, p. 6.

198.   Ibid.

199.   Alberta, Order-in-Council, OC No. 769–31, 10 July 1931, "Regula-tions Respecting Drilling and Production Operations of Oil and Natural Gas Wells."

200. PAA, Premiers' Papers, 69.289, 320, 25 July 1931. Mayor Andrew Davison to J.E. Brownlee, and 27 July 1931, Brownlee to Davison.

201. Fisher, p. 8.

202. *Calgary Herald,* 5 September 1931, "Draft Report" and related comment. The committee also prepared and mailed to each independent company a document showing the probable earnings of each company producing naphtha under the pooling plan compared with likely earnings under restriction by means of gas proration, both based on an allowance of 100 million cubic feet per day. Fisher, p. 16.

203. *Calgary Herald,* 5 September 1931.

204. PAA, W.S. Herron Papers, 78.230, 4 September 1931, "An Operator's Point of View on Turner Valley."

205. Ibid.

206. *Calgary Herald,* 5 September 1931, copy of the city's letter to the premier.

207. Fisher, pp. 3 and 18.

208. Ibid., p. 23.

209. PAA, W.S. Herron Papers, 78.230, 4, "Memorandum for Submission to the Agriculture Committee of the Legislative Assembly of Alberta on Behalf of the Independent Oil Companies of Turner Valley."

210. Fisher, p. 23.

211. Ibid., p. 24.

212. Ibid.

TWO—THE TURNER VALLEY GAS CONSERVATION BOARD AND
THE FAILURE OF ALBERTA'S FIRST PRORATION PROGRAM

1. *Edmonton Journal,* 12 February 1932.

2. Ibid., 12 March 1932.

3. Ibid., 18 February 1932.

4. Ibid., 27 February 1932.

5. PAA, Premiers' Papers, 69.289, 320, "Turner Valley Problem," and *Edmonton Journal,* 12 and 19 March 1932.

6. *Edmonton Journal,* 17 March 1932.

7. Ibid., 18 March 1932.

8. Ibid., 21 March 1932.

9. Ibid.

10. *Edmonton Bulletin,* 21 and 22 March 1932, see the remarks of Dr. O.B. Hopkins, chief geologist to Imperial Oil Company; also *Edmonton Journal,* 21 March 1932.

11. *Edmonton Journal,* 31 March 1932.

12. Alberta, *Statutes,* 1936, ch. 6, 4 (2).

13. Ibid., s. 18.

14. Ibid., s. 25.

15. Ibid., s. 10. "Public" meant any officer of the Petroleum and Natural Gas department of the provincial government, any officer of a company distributing gas, and an officer of any community in which gas was being distributed and any person having an ownership interest in any well.

16. *Edmonton Journal,* 6 April 1932.

17. Alberta, Order-in-Council, OC No. 362/32, 23 April 1932.

18. Aytenfisu, K.M., "The University of Alberta: Objectives, Structure and Role in the Community, 1908–1928" (unpublished Master's thesis, University of Alberta, 1982), pp. 71–75. Further evidence of Tory's philosophy is his initiative in the 1921 formation of the Scientific and Industrial

Research Council of Alberta, Canada's first provincially sponsored re
search council. In 1928, Tory left the University to head the National
Research Council of Canada.

19.   The Turner Valley Gas Conservation Board (hereafter TVGCB) Calgary office was at 128–7 Avenue SW.

20.   ERCB, TVGCB, Minutes, 4 May 1932, General Order No. 1 "Restricting Gas Production in Turner Valley Field."

21.   ERCB, "Report of the Turner Valley Gas Conservation Board," 22
February 1933, p. 4.

22.   PAA, 65.741, Turner Valley Gas Conservation Board, 773a, 27 August 1932, R.J. Gibb, President, Association of Professional Engineers of
Alberta, to Hon. Mr. Justice Carpenter.

23.   Alberta Supreme Court, Appellate Division, *Spooner Oils Limited
and Spooner v. Turner Valley Conservation Board and Attorney General
for Alberta*, transcript, p. 3.

24.   A similar action launched by the Richfield Oil Corp. Ltd. was
dropped when Spooner was successful in obtaining an interim injunction.

25.   PAA, 74.101, 483c, 7 May 1932, W.C. Fisher, Managing Director,
Model Oils to the TVGCB.

26.   Ibid., 23 August 1932, A.A. Carpenter to the McLeod Oil Company Ltd.

27.   Ibid., 9 July 1932, E. Pearle Miller, Acting Secretary-Treasurer,
New McDougall-Segur Oil Company Ltd., to the TVGCB. Although some
companies eventually surrendered to the Board's threat, Spooner Oils Limited, Widney Oils Ltd., Wellington Oil Company Ltd., Structure Oil and
Gas Company and Okalta Oils Ltd. chose to accept the TVGCB's challenge
in court.

28.   Ibid., 6 July 1932, S.J. Helman, K.C., to A.A. Carpenter.

29.   New drilling and production regulations restricted the flow of gas
to 0.57542 of the volume of flow of gas that would be produced if the well
were operated at a pressure equal to 80% of the closed-in pressure of the
well. This provision was varied on 10 occasions before being discontinued
on 13 April 1933.

30.   PAA, 74.101, 483c, 5 August 1932, J.H. McLeod, Manager, Imperial Oil Ltd., to TVGCB.

31.   *Western Weekly Reports*, 2, pp. 641–50, and ERCB, Alberta Supreme Court, Appellate Division, *Spooner Oils Limited and Spooner v.
Turner Valley Gas Conservation Board and Attorney General of Alberta*,
pp. 1–21.

32.   PAA, 74.101, 483c, "Statement of Claim," *Spooner Oils Ltd. and
Spooner v. the TVGCB and the Attorney General of Alberta*.

33.   PAA, 74.101, 485b, "Report of Public Inquiry Held by Turner Valley Gas Conservation Board, January 10 to January 13, 1933;" see also
PAA, 65.74, 773a, A.J. Scott to A.A. Carpenter, 12 January 1932 [*sic*], with
attached resolution.

34.   Alberta, *Annual Report of the Department of Lands and Mines for
1933* (Edmonton: King's Printer, 1934), pp. 58–60.

35.   Ibid., p. 65. "Under this plan the more efficient the production, the
lower is the gas-oil ratio, in other words, the larger is the production quota.
The operator who is producing under the lowest gas-oil ratio possible for
his wells obtains a higher gas quota in subsequent revisions than the operator who obtains the same amount of naphtha at a greater expenditure of
gas because of the fact that the acreage potential is maintained at a higher
level under efficient production methods when gas is not unduly wasted."

36. Ibid., p. 66.

37. Ibid., p. 54. With wells unmetered, there could be no efficient policing of the field to ensure observance of the Board's orders. See also PAA, 65.74, 774, 6 February 1933, J. McLeish, Director, Mines Branch, to A.A. Carpenter.

38. Alberta, *Annual Report of the Department of Lands and Mines for 1934* (Edmonton: King's Printer, 1935), p. 15.

39. Ibid., 1933, pp. 47–48.

40. Ibid., 1934, p. 50.

41. Cited in George R. Elliott, "Conservation of natural gas: with special reference to Turner Valley, Alberta," *Transactions of the Canadian Institute of Mining and Metallurgy*, XXXVII (1934), pp. 557–59.

42. *Edmonton Journal*, 8 March 1933.

43. Ibid., 9 March 1933.

44. Ibid.

45. *Edmonton Journal*, 7 April 1933. The motion first presented did not include a field production ceiling and drew considerable opposition before being amended to state a maximum production of 240 million cu. ft. a day.

46. Alberta, *Annual Report of the Department of Lands and Mines for 1933* (Edmonton: King's Printer, 1934), p. 41.

47. Canada, Supreme Court, *Spooner Oils Ltd. and Spooner v. Turner Valley Gas Conservation Board and the Attorney General of Alberta.* S.C.R. 1933, pp. 630–49.

48. Regulations made under the authority of the *Dominion Lands Act,* 1908, c. 20.

49. Agreement of 14 December 1929, given "the force of law" by the *BNA Act,* 1930, c. 26.

50. *Edmonton Bulletin,* 3 October 1933.

51. *Calgary Herald,* 4 October 1933.

52. Erich W. Zimmerman, *Conservation in the Production of Petroleum* (New Haven: Yale University Press, 1957), pp. 254–55, and *Clymore Production Co v. Thompson,* 13 F. Supp. 469, W.D. Texas, 1936.

53. Stephen S. McDonald, *Petroleum Conservation in the United States: An Economic Analysis* (Baltimore: Johns Hopkins University Press, 1971), pp. 36–38.

54. *Edmonton Bulletin,* 22 September 1933. On 30 June 1934, a jury found in favour of MacMillan, but the trial judge declared the verdict not consonant with the law or the facts and dismissed the charge. The Alberta Supreme Court upheld the opinion of the lower court judge. A publicly subscribed fund helped the MacMillans take the case to the Supreme Court of Canada where on 1 March 1937 a decision overruling the Alberta courts awarded Vivian MacMillan $10,000.

55. Alberta, *Annual Report of the Department of Lands and Mines for 1934* (Edmonton: King's Printer, 1935), p. 40.

56. Ibid., p. 41.

57. Ibid., p. 42.

58. *Edmonton Journal,* 23 February 1935.

59. *Calgary Albertan,* 14 August 1935.

60. ERCB interview, 19 February 1987, with Hubert H. Somerville, Alberta Deputy Minister, Department of Mines and Minerals, 1952–73.

61. John J. Barr, *The Dynasty: The Rise and Fall of Social Credit in Alberta* (Toronto: McClelland and Stewart Ltd., 1974), p. 84, and *Calgary Herald,* 6 January 1937.

62.   H.H. Somerville, correspondence with author, January 1987. Calder resigned on 13 September 1935.

63.   PAA, 74.101, 5502, 2 December 1935, C.W. Dingman to J. Harvie,
Deputy Minister, Department of Lands and Mines.

64.   PAA, Premiers' Papers, Reel No. 84, 812, 28 November 1935 Ever-
ett Marshall, Editor, *Western Oil Examiner,* to Premier William Aberhart.

65.   PAA, 74.101, 550a, 2 December 1935, C.W. Dingman to J. Harvie.

66.   PAA, Premiers' Papers, Reel 84, see file 812.

67.   Ibid., 3 December 1935, Ernest Moore to William Aberhart.

68.   PAA, 74.101, 550a, January 1936, C.W. Dingman "Turner Valley
Production."

69.   Ibid., p. 5.

70.   Ibid., and Saskatoon *Star-Phoenix,* from 9 April 1936.

71.   *Calgary Herald,* 19 June 1936.

72.   Ibid.

73.   Ibid., 20 June 1936.

74.   *The Financial Post,* 27 June 1936.

75.   NAC, RG21, 161.36.9, 18 September 1937, W.S. Campbell, Chairman, Petroleum Producers Association, to T.S. Crerar, Minister of Mines.

76.   *Calgary Herald,* 17 October 1936.

77.   Ibid., 27 December 1936, quoting the *Toronto Star.*

78.   The limit of 1,920 acres and the 40¢ per acre cash bond were
dropped in favour of prospecting permits, at 10¢ per acre per year or 5¢
per acre for six months, for such an acreage as the minister might decide.

79.   PAA, Premiers' Papers, Reel 83, 802. 30 September 1936, "Speech
made by Hon. C.C. Ross, Minister of Lands and Mines, Alberta Government, to the Canadian Mining and Metallurgical Institute Convention at
Edmonton."

80.   For a hint of this, see *Calgary Herald,* 9 January 1937, editorial
"Explanation in Order."

81.   *Calgary Herald,* 6 January 1937, see also G. Homer Durham, *N.
Eldon Tanner: His Life and Service* (Salt Lake City, Utah: Deseret Book
Company, 1982), pp. 70–71.

THREE—FOUNDING THE PETROLEUM AND NATURAL GAS
   CONSERVATION BOARD, 1937–38

1.   G. Homer Durham, *N. Eldon Tanner: His Life and Service.* (Salt
Lake City, Utah: Deseret Book Company, 1982), p. 70.

2.   *Calgary Herald,* 6 January 1937.

3.   Ibid., 9 January 1937, quoting the *Edmonton Bulletin.*

4.   Durham, pp. 48–49 and 56–57.

5.   Alberta, *Annual Report of the Department of Lands and Mines for
1937* (Edmonton: King's Printer, 1938), p. 9. This surplus was up almost
$350,000 from the previous year.

6.   Ibid., p. 15.

7.   *Calgary Herald,* 30 January 1937.

8.   Ibid.

9.   Ibid.

10.   Ibid.

11.   For a discussion of legislation on crown reserves, see H.H. Somerville, *A History of Crown Rentals, Royalties and Mineral Taxation in Alberta* (Edmonton: Energy and Natural Resources, 1977), pp. 5/18–5/19.

12. Alberta, Order-in-Council, OC No. 225, 3 March 1937. Odd-numbered sections of land situated north of the north boundary of township 52 were also set aside as part of the crown reserve.

13. NAC, RG85, vol. 1839, 112938, 10 June 1913, Graham Greene, Admiralty, to Undersecretary of State, Colonial Office.

14. *Calgary Herald,* 5 April 1937.

15. Ibid.

16. *Calgary Herald,* 7 September 1937.

17. Ibid., 11 September 1937. The study was to be undertaken by the producers, who were to have access to all available statistical and technical information of the Department of Lands and Mines. Although the Department of Lands and Mines annual report for the fiscal year ended 31 March 1938 states that an "order" limiting production to 65% of capacity was issued by the department, press reports make clear that this was a "voluntary" agreement "supervised" by the Imperial and British American refineries in Calgary.

18. *Calgary Herald,* 13 September 1937.

19. Ibid.

20. Alberta, *The Case for Alberta* (Edmonton: King's Printer, 1938), p. 225. Author's italics.

21. NAC, RG21, 161.36.9, pt.1, 18 September 1937, W.S. Campbell, Petroleum Producers Association, to Hon. T.A. Crerar, Minister of Mines, with enclosure "Production, Refining and Distribution of Crude Oil."

22. Ibid. Regina to Calgary 466 miles, to Oklahoma 1,452 miles; Winnipeg to Calgary 822 miles, to Oklahoma 1,198 miles.

23. Ibid. Imperial Oil had justified the reduction in the price offered for Turner Valley crude after 1 September 1937 with the argument that they had to meet Cut Bank, Montana, competition.

24. Ibid.

25. Ibid., 27 September 1937, T.G. Madgwick to W.B. Timm, Chief, Bureau of Mines.

26. *Calgary Herald,* 25 November 1937.

27. Ibid.

28. Ibid. 4 January 1938. See also the *Financial Post,* 8 January 1938. Royalite also reduced its pipeline rate for gathering and delivering Turner Valley crude oil to Calgary from 17¢ to 15¢ per barrel, and the price of gasoline to consumers was dropped by 1½¢, making the price of standard gasoline 27¢ a gallon (7¢ of which was provincial tax).

29. Ibid., 5 January 1938.

30. *Calgary Herald,* 14 January 1938.

31. NAC, RG21, 161.36.9, pt. 1, 27 January 1938, F.M. Steel, Petroleum Engineer, to John McLeish, Director, Mines Branch, Department of Mines and Resources. See also *Calgary Herald,* 8 January 1938, "Will 'Gang Up' On Alberta Oil."

32. Ibid., 10 February 1938, G.G. Brown to George H. Sedgewick, Chairman of Tariff Board.

33. Ibid., 4 March 1938, T.G. Madgwick to W.B. Timm, Chief, Bureau of Mines.

34. Ibid.

35. ERCB, N.E. Tanner File, 17 March 1938, N.E. Tanner to John W. Finch, Director, Bureau of Mines. See Board history files—Tanner, N.E.

36. Ibid., 18 March 1938, John W. Finch to N.E. Tanner. Finch followed with a second telegram on March 23, saying that he "should have included each member of firm Parker, Foran, Knode and Boatright."

37.   William R. Childs, "The Transformation of the Railroad Commission of Texas, 1917–1940: Business-Government Relations and the Importance of Personality, Agency Culture and Regional Differences," *Business History Review* 65 (Summer 1991), pp. 286–87 and 307. See also *Time*, 2 July 1934, pp. 49–50. The issue that precipitated Parker's resignation was his lack of authority to dismiss staff who were appointed to positions on the basis of their political connections rather than professional qualifications.

38.   ERCB, N.E. Tanner File, 21 March 1938, N.E. Tanner to R.D. Parker, and 24 March 1938, N.E. Tanner to W.F. Knode. Knode was available for a fee of $50 a day plus expenses.

39.   Ibid., 24 March 1938. J. Harvie, Deputy Minister of Lands and Mines, to C.W. Jackson, Secretary, Department of Mines and Resources, Ottawa.

40.   Ibid., Alberta Legislature, 1938 (Special Session), "Evidence Taken Before the Agriculture Committee in Connection With Bill No. 1, An Act for the Conservation of the Oil and Gas Resources of the Province of Alberta," p. 2, and GA, transcript of tape-recorded interview with Margaret Knode, 29 July 1981, and Texas Railroad Commission, "Oil and Gas Circular No. 16–B," 15 May 1934, see facing page.

41.   ERCB, N.E. Tanner File, 6 April 1938, W.F. Knode to N.E. Tanner.

42.   Marver H. Bernstein, *Regulating Business by Independent Commission* (Princeton: Princeton University Press, 1955), p. 26.

43.   Rene Dussault and Louis Borgeat, *Administrative Law: A Treatise,* trans. Murray Rankin (Toronto: Carswell, 1986), p. 127.

44.   Joshua Treulmin Smith: *Government by Commissions Illegal and Pernicious* (1849), p. 157.

45.   Dussault and Borgeat, p. 127.

46.   *Statutes of Alberta*, 1907, ch. 12, *The Public Health Act.*

47.   Ibid., 1915, ch. 6, *The Public Utilities Act.*

48.   Commissioners could be removed by the lieutenant-governor-in-council on address to the Legislative Assembly.

49.   The only avenue of appeal open was a question involving the jurisdiction of the Board, and this had to "lie to the court *en banc* from any final decision of the Board, but such appeal can be taken only by permission of the court *en banc*...."

50.   For example, the Alberta Minimum Wage Board 1922, the Liquor Control Board 1924, the Eugenics Board 1928, the Provincial Farm Loan Board 1929, the Provincial Parks Board 1930 and the Poultry Marketing Board 1934.

51.   Bernstein, p. 20.

52.   *Statutes of Alberta*, 1938, ch. 15, "An Act for the Conservation of Oil and Gas Resources of the Province of Alberta," s. 5.

53.   W.R. Childs, p. 309.

54.   Ibid., pp. 331–32.

55.   *Statutes of Alberta*, 1938, ch. 15, s. 8.

56.   Ibid., s. 3 and s. 27.

57.   Indian lands remained under federal jurisdiction and were not subject therefore to Alberta's regulations.

58.   Ibid., s. 14(j). By way of elaboration, this section stipulated that any proration formula adopted by the Board would include acreage and bottom-hole pressure as some of its factors, and that in no formula would a potential factor have more than 25% weight.

59.   Ibid., s. 14(o).

60. Ibid., s. 28.

61. Ibid., s. 20.

62. NAC, RG21, vol. 37, 161.36.9, pt. 1, 10 May 1938, T.G. Madgwick, Technical Assistant, Bureau of Mines, Department of Mines and Resources, memorandum "An Act for the Conservation of Oil and Gas Resources of the Province of Alberta." Madgwick, the senior petroleum engineer in the Department, was well-acquainted with the situation in Alberta and fully informed on conservation practice elsewhere. In his judgement, the Alberta legislation was in "accordance with the aim of the best current oil field practice" and showed an awareness of "the latest ideas from the practice in the United States."

63. *Edmonton Bulletin,* 8 April 1938.

64. *Calgary Herald,* 7 April 1938.

65. *Edmonton Bulletin,* 8 April 1938, regarding the Alberta Legislature's refusal to ratify the unemployment insurance amendment.

66. PAA, Premiers' Papers, Reel 83, 810B, correspondence of L.L. Plotkins, January-November 1938.

67. *Calgary Herald,* 17 May 1938.

68. Ibid., 18 and 19 May 1938.

69. Ibid., 2 June 1938.

70. Ibid., 9 June 1938. Before 1924, only a few wells were operating in Turner Valley, and they were producing low-pressure gas or oil from above the limestone horizon. Calgary gas consumers were then supplied with gas from the Bow Island field near Medicine Hat.

71. Ibid., 2 June 1938.

72. PAA, Premiers' Papers, Reel 83, 810B, 21 June 1938, William Aberhart to William Lyon Mackenzie King. Tanner drafted the telegram.

73. Alberta, Order-in-Council, OC No. 759, 27 June 1938.

74. GA, transcript of tape-recorded interview with Margaret Knode, 29 July 1981, pp. 16 and 24.

75. Alberta, Order-in-Council, OC No. 663, 10 June 1938, and OC No. 1101, 30 August 1938. Alberta cabinet ministers received $6,000 per year, the premier $8,500, and the deputy minister of lands and mines, $4,750.

76. Ibid., OC No. 816, 2 July 1938, OC No. 883, 16 June 1938, and OC No. 887, 19 July 1938. The government had unofficially named Calgary as the location for the new Board on 31 May 1938. See *Calgary Herald,* 31 May 1938.

77. ERCB, "Petroleum and Natural Gas Conservation Board, Minutes to 28 October 1938" (hereafter cited as "Minutes to 28 October 1938"), 1st Meeting, 4 July 1938, pp. 3–4.

78. *Time,* 2 July 1934, p. 50.

79. ERCB, "Minutes to 28 October 1938," 1st Meeting, 4 July 1938, p. 3. The Board considered that it already had sufficient recent data to determine the proper potential of gas wells.

80. *Western Oil Examiner,* 30 July 1938.

81. ERCB, "Minutes to 28 October 1938," 4th Meeting, 4 August 1938, pp. 15–17.

82. The *Oil and Gas Conservation Act* 1938 mentioned the possibility of a hearing only in the context of an appeal from the annual assessment made against producing wells to finance the operation of the Board.

83. *Western Oil Examiner,* 20 August 1938.

84. ERCB, "Minutes to 28 October 1938," 5th Meeting, 13 August 1938, pp. 18–19.

85. Ibid., 6th Meeting, 19 August 1938, pp. 22–23, and 7th Meeting,

26 August 1938, pp. 25–26. The latter increase was partly the result of the revelation that a number of Turner Valley wells had been unable to produce their daily quotas.

86. Ibid., 8th Meeting, 31 August 1938, pp. 30 and 35.

87. Ibid., p. 28.

88. Ibid., p. 32.

89. Ibid. "The allotment to each well in respect of the acreage to which the right to drill and operate the well relates to an extent of not more than One Hundred and Sixty acres, (hereafter referred to as the well acreage) shall be that portion of the total monthly allowable production from the Field which the well acreage bears to the total productive acreage of the Field."

90. *Calgary Herald*, 9 September 1938.

91. Ibid., 14 September 1938, and *Western Oil Examiner*, 10 September 1938.

92. ERCB, "Minutes to 28 October 1938," 10th Meeting, 14 September 1938, p. 40. The extension was intended to give naphtha producers additional time to fill market requirements and to allow time for the Alberta Petroleum Association to complete research on repressuring the Turner Valley field.

93. *Calgary Herald*, 16 September 1938.

94. For information on Mayland and other Turner Valley investor-promoters, see Douglas E. Cass, "Investment in the Alberta Petroleum Industry, 1912–30" (unpublished Master's thesis, Dept. of History, University of Calgary, 1985).

95. ERCB, recorded telephone interview, William Epstein (New York), 4 March 1989. Epstein was the top student (Gold Medallist) of the 1935 graduating law class at the University of Alberta. He articled with Smith, Egbert and Smith, and but for a year (1937–38) studying International Law at the London School of Economics, he remained with his articling firm until June 1942 when he enlisted in the Canadian Army to serve overseas. Throughout this period in Calgary, he worked mainly with A.L. Smith and J.J. Frawley on matters of concern to the Petroleum and Natural Gas Conservation Board, particularly the drafting of the second 1938 conservation Act. Immediately after the war, Epstein began a distinguished career with the United Nations in the field of disarmament and nuclear non-proliferation.

96. Ibid.

97. *Calgary Herald*, 5 October 1938.

98. Ibid., 16 September, 1 and 11 October 1938.

99. ERCB, recorded telephone interview, William Epstein (New York), 4 March 1989.

100. Alberta, Attorney General, Central Records, Edmonton, *Mercury Oils Ltd. v. Attorney General For Alberta*. Judgement of the Honourable Mr. Justice Ives.

101. *Calgary Herald*, 12 and 13 October 1938. Helman and Mahaffy immediately launched an appeal.

102. Ibid., 27 October 1938. See also 17 October 1938.

103. Ibid., 28 October 1938.

104. Ibid., 29 October 1938.

105. ERCB. Epstein speculates "that what happened was that Ives called Art Smith after the case and said, 'look I gave you a break here—I accepted your argument [that] *certiorari* was not the best way to handle this case. But, boy, this law, the Act is no damn good, if you've got any sense, don't come crying now before they start their lawsuit.... I think what you better

do is get a new draft drafted that is not *ultra vires*' and he said, you better do it soon...'" Recorded telephone interview, William Epstein (New York), 4 March 1989.

106. ERCB, "Minutes to 28 October 1938," 12th Meeting, 21 September 1938, p. 3; 13th Meeting, 18 October 1938, p. 3, and 14th Meeting, 26 October 1938, p. 3.

107. *Calgary Herald,* 1 November 1938; *Western Oil Examiner,* 29 October 1938.

108. See, for example, ibid., 28 October 1938, "The Oil Situation."

109. See, for example, *Calgary Herald,* 23 November 1938, "Anxiety in Valley."

110. Alberta, *Report of a Royal Commission,* appointed by the Government of the Province of Alberta under The Public Inquiries Act to inquire into matters connected with Petroleum and Petroleum Products (Edmonton: King's Printer 1940), Commission preamble.

111. *Calgary Herald,* 25 October 1938.

112. Ibid., 28 October 1938.

113. PAA, Premiers' Papers, Reel 84, 812, 9 November 1938, W.C. Fisher to Members of the Legislature, "Summary of Representations Made By the Independent Oil and Gas Producers in Turner Valley," Re *Petroleum and Natural Gas Conservation Act,* 1938.

114. Ibid., p. 3.

115. Recorded telephone interview, William Epstein (New York), 4 March 1989.

116. See *Statutes of Alberta,* 1938, ch.1, *An Act for the Conservation of the Oil and Gas Resources of the Province of Alberta,* s. 44.

117. *Calgary Herald,* 15 November 1938.

118. Ibid., 16 November 1938.

119. ERCB, Alberta Legislative Assembly, 1938 (Special Session), "Evidence Taken Before the Agriculture Committee in Connection With Bill No. 1, An Act for the Conservation of the Oil and Gas Resources of the Province of Alberta," pp. 3–4.

120. Ibid., p. 4.

121. Ibid., p. 5.

122. Ibid., pp. 2, 164 and 166.

123. Ibid., pp. 129, 132 and 139.

124. Ibid., p. 92. At the same time, operators had high praise for Charles Dingman. "We all know and like Mr. Dingman. I have known Mr. Dingman for years, and I appreciate Mr. Dingman. He has worked among us, and we think he has our good will and we have his." p. 92.

125. Ibid., p. 148.

126. Ibid., p. 149.

127. Ibid., p. 150. See also, *Edmonton Journal* and *Edmonton Bulletin,* 21 November 1938.

128. GA, transcript of tape-recorded interview with Margaret Knode, 29 July 1981, p. 17.

129. ERCB, Alberta Legislative Assembly, 1938 (Special Session), "Evidence....," p. 150. (See Agriculture Committee Report 1938 in Board history files.)

130. Ibid., p. 154. For the compensation provision in the *Oil and Gas Conservation Act* of April 1938, see *Statutes of Alberta,* 1938, ch. 15, s. 17, and s. 14, f, h and i.

131. See Alberta Petroleum Association, President's statement of objection, Alberta Legislative Assembly, 1938 (Special Session), "Evidence...." (See Agriculture Committee Report 1938 in Board history files.)

132. *Edmonton Journal,* 23 November 1938.

133. ERCB, Alberta Legislative Assembly, 1938 (Special Session), "Evidence...," p. 161. (See Agriculture Committee Report 1938 in Board history files.)

134. *Edmonton Journal,* 23 November 1938.

135. NAC, RG21, vol. 38, 161.36.9/2, 16 March 1939, T.G. Madgwick, "Memorandum Re: Acts on Conservation of Oil and Gas Reserves of Alberta, etc."

FOUR—ESTABLISHING A REGULATORY FOUNDATION, 1938–1947

1. *Calgary Herald,* 24 November 1939.

2. Alberta, "Alberta's Oil Industry," *The Report of a Royal Commission Appointed by the Government of the Province of Alberta Under The Public Inquiries Act to Inquire Into Matters Connected With Petroleum and Petroleum Products* (Edmonton: Imperial Oil Limited, 1940), pp. 79–80 (hereafter McGillivray Commission *Report*). The British American Oil Company, the second-largest refiner, opened a modern cracking and topping plant in Calgary in the spring of 1939. In addition to the five refineries mentioned above, there was a small refinery operating at Lethbridge using Montana crude and limited production from several tiny plants in the Wainwright area.

3. This is but a rough list, as the balance of ownership in various wells is difficult to determine. For the purposes of this listing, the operating company is assumed to be the majority owner.

4. Alberta, Department of Industries and Labour, Provincial Bureau of Statistics, *Alberta Facts and Figures 1954,* pp. 160, 167 and 245.

5. Alberta, Department of Lands and Mines, *Annual Report* to 31 March 1939, p. 68.

6. Ibid., p. 70. Turner Valley's percentage of the total gas produced in the province was close to 90%.

7. ERCB, transcript of tape-recorded interview with Gordon Connell, 17 June 1986 and ERCB, Minutes, 30 November 1938.

8. Ibid. The stenographers earned $80 and $90 compared to $75 for the three junior engineers. ERCB, transcript of tape-recorded interview with Phyllis Van Aalst and Jean Howells, June 1988.

9. ERCB, Minutes, 30 November 1938, p. 6. Allowables were set for all wells with a depth greater than 3,500 feet and ranged from a low of 15 barrels per day to a high of 415 barrels.

10. Ibid., "Interim Order No. 3," p. 8.

11. See Knode's comments before the Alberta Legislative Assembly just a few days before the Board's first natural gas order. ERCB, Alberta Legislative Assembly, 1938 (Special Session). "Evidence Taken Before the Agriculture Committee in Connection with Bill No. 1, An Act for the Conservation of the Oil and Gas Resources of the Province of Alberta," pp. 143–49. *Calgary Herald,* 30 November 1938. It appears that the naphtha producers were also successful in persuading the government to permit a special well-spacing arrangement for that portion of Turner Valley south of the northern boundary of Township 18.

12. Alberta, Department of Lands and Mines, *Annual Report,* to 31 March 1939, p. 65. Conservation Board estimate of total gas withdrawn.

Gas cap	1,005,198,000 Mcf.
Oil zone	33,600,000 Mcf.
Total	1,038,798,000 Mcf.

13. Alberta, Order-in-Council, OC No. 45–39, 11 January 1939, "Regulations under the Oil and Gas Wells Act, 1931." See also J.L. Irwin, "Alberta, Oil Province of Canada, in 1938," Department of Lands and Mines, January 1939.

14. Alberta, Order-in-Council, OC No. 45–39, 11 January 1939, "Regulations Under the Oil and Gas Wells Act, 1931," s. l(g).

15. Ibid., s. l(g)xii

16. Ibid., s. 7(3), author's italics. This section of the regulations also included the possibility of an exemption, "Where, in the opinion of the Board the surface and subsurface conditions are such that it is not practicable to drill the well so as to be at least 440 yards from any other producing well."

17. Under the infamous Rule 17.

18. The 4° allowance assumed 40-acre spacing; for lesser spacing, only 3° deviation from the vertical was permitted.

19. S.E. Slipper, *Manual for Operators Under Oil and Gas Regulations* (Ottawa: King's Printer, 1922).

20. ERCB, Minutes, 11 April 1939. This was the fifth meeting under the revised Act.

21. Sec. 17.

22. ERCB, Minutes, 18 May 1939, "To His Honour the Lieutenant Governor in Council: Memorandum as to Schemes For Compensation Under the Oil and Gas Resources Conservation Act."

23. NAC, RG27, vol. 38, 161.36.9/3, January 1939, J.L. Irwin, "Alberta, Oil Province of Canada, in 1938," Department of Lands and Mines, pp. 7–8.

24. Great Britain, Public Record Office (hereafter PRO), POWE 33, 287, "Petroleum: Canada-Alberta-Turner Valley, 1937–38," 16 December 1937, Wm. J. Benton to the Rt. Hon. Neville Chamberlain, PM. See other letters in this file. The British Admiralty had monitored oil development in Alberta since 1906.

25. Ibid., 15 September 1938, G.R. Fleming, Lieutenant Colonel ret. to Undersecretary of State, Foreign Office, and, 15 September 1938, "Statement for Press by R.A. Brown, Sr. Chairman, Alberta Petroleum Association." It was determined later that the alleged interest of the German corporation "Tropicorp" was unfounded, but this did little to slow the momentum of interest and concern.

26. Ibid., 20 September 1938, "Canada, Turner Valley Oil Field." Note of meeting.

27. Ibid., 27 October 1938, Confidential memorandum, "Note of talk at Canada House on the Alberta Oil Position." See also, 24 October 1938, *Financial Times* report of British government interest in a pipeline to Fort William or Vancouver.

28. For background on the decision to send a Canadian delegation to the United Kingdom, see NAC, RG27, vol. 38, 161.39.9/3, 9 January 1939, memorandum, G.D. Mallory, Department of Trade and Commerce, to C.W. Jackson, Department of Mines and Resources.

29. Ibid., 8 March 1939, N.E. Tanner to O.D. Skelton, Undersecretary of State for External Affairs. Not all in Ottawa were as supportive as Crerar. The Chief Geologist of the Canadian Geological Survey noted astutely that the producers, refiners and railways held opposing views on the matter of Turner Valley oil, and he wondered if Dr. Hume's presence on the delegation might suggest that the federal government favoured one or other of the opposing groups. 28 February 1939, Dr. G.A. Young to E.C.C. Lynch, Chief Bureau of Geology and Topography.

30. PRO, POWE 33, 289, "Petroleum: Canada-Alberta-Turner Valley,

1939–1947," 14 April 1939, "Note of meeting held in the Dominions office."

31. Ibid., p. 2.

32. Ibid., p. 6. One British official wondered why, "if the Alberta projects were really a good proposition, that more financial backing could not be raised in Canada."

33. Ibid., 20 April 1938, "Meeting With Canadian Delegates Concerning Canadian Oil Resources," p. 1. This was assuming that the field would be extended over 16,000 acres.

34. Ibid., 24 April 1938, "Turner Valley Oil Field: Alberta," p. 2.

35. Ibid., p. 5. Gordon Connell, the Board field engineer in charge at Turner Valley, does not remember police being sent, but he does recall watchmen being hired "to keep an eye on shut-in wells in some cases." Notes received from G. Connell, 21 March 1989.

36. Ibid., 20 April 1938, "Meeting With Canadian Delegates Concerning Canadian Oil Resources," pp. 2–3.

37. Ibid., 24 April 1938, "Turner Valley Oil Field: Alberta," pp. 1–8. See also NAC, RG27, vol. 38, 161.36.9/3, 26 May 1938. O.D. Skelton, Undersecretary of State for External Affairs, to Deputy Minister of Mines and Resources, with enclosure, "Alberta Oil Delegation: Record of Interviews."

38. PRO, POWE, 33, 289, 24 July 1939, Stephen L. Holmes to C.W. Dixon.

39. NAC, RG27, vol. 38, 161.36.9/3, 23 May 1939, Office of the High Commissioner for Canada, London, to the Secretary of State for External Affairs, Ottawa.

40. Ibid., and Calgary Herald, 5 June 1939.

41. NAC, RG27, vol. 38, 161.36.9/3, 6 June 1939, Dr. G.S. Hume to Dr. G.A. Young, Chief Geologist. This estimate was based upon 20,000 barrels an acre, which Hume pointed out was an estimate that the Imperial Oil Company agreed with but was held by most others to be a minimum figure.

42. Ibid., 22 June 1939, "Conference re Turner Valley Oil: Meeting of the Fuel Committee of the Cabinet."

43. Ibid.

44. Ibid.

45. Financial Times, London, 26 July 1939.

46. NAC, RG21, vol. 38, 161.36.9/2, 3 August 1939, Dr. C. Camsell to T.A. Crerar.

47. Ibid.

48. NAC, RG27, vol. 38, 161.36.9/3, 28 August 1938, T.G. Madgwick to Dr. C. Camsell, p. 12. After examining the performance history of producing wells in Turner Valley, Madgwick concluded that to maintain production at 60,000 barrels a day for 12 years would require the completion of 860 wells. At 40 acres to the well, 34,400 acres would be required, and this was considerably in excess of any estimate of Turner Valley's potentially productive area.

49. NAC, RG21, vol. 38, 161.36.9/2, 7 September 1939. Dr. C. Camsell to N.E. Tanner, and, 18 September 1939, Dr. C. Camsell to N.E. Tanner.

50. McGillivray Commission, Report.

51. Ibid., p. 2.

52. Ibid., p. 15.

53. The Connolly Law prohibited the interstate transportation of oil or oil products produced in violation of state law.

54. Ibid., p. 56.

55. Legislative Library of Alberta, transcript of evidence presented before the "Royal Commission appointed by the Government of the Province of Alberta under The Public Inquiries Act to inquire into matters connected with Petroleum and Petroleum Products," vol. 2, pp. 110–12 (hereafter McGillivray Commission, Transcripts).

56. Ibid., vol. 1, pp. 9–11. See also Alberta, *Journals of the Sixth Session of the Eighth Legislative Assembly*, p. 125. The $3\frac{1}{3}¢$ per gallon was derived from the posted field price of $1.26 per barrel for Turner Valley 46-degree gravity crude oil, which was the average gravity of the refinery run. The $16\frac{1}{2}¢$ per gallon was the Calgary per gallon wholesale price exclusive of provincial tax. The resolution in the Legislature claimed that "all reductions in the price of gasoline and distillate since May 1935 had been achieved at the expense of the crude oil producers, that there had been no appreciable reduction in the spread between the field price of crude oil and the wholesale price of refined products," p. 10. Also noted in the resolution was the Dominion Tariff Board finding of May 1935 that "there appeared to be no justification for the price of gasoline in Calgary," p. 9.

57. *Calgary Herald*, 28 October 1939.

58. McGillivray Commission, Transcripts, vol. 134, pp. 14936–37. The ordinance was passed because wood was highly important, being used for building and heating homes and constructing furniture and other essentials, and because the transportation of this vital commodity from any great distance was an avoidable hardship.

59. Ibid., vol. 134, p. 14970.

60. Ibid., vol. 137, p. 15336.

61. Ibid., vol. 137, p. 15335, Dr. Frey, and vol. 135, p. 15112, R.V. LeSueur.

62. McGillivray Commission, *Report*, pp. 26 and 29, and Transcripts, vol. 135, p. 15114.

63. McGillivray Commission, Transcripts, vol. 137, p. 15336, "and when I say 'equitable' I think it follows that this goes back into economics."

64. Ibid.

65. Frey presented the Arkansas Conservation Act (Arkansas Act No. 105 of 1939) as a model to be followed.

66. Ibid., vol. 126, pp. 14028–29 and p. 14157, and vol. 127, pp. 14232–37, Dr. Brown; vol. 131, pp. 14653–60, and vol. 132, pp. 14676, 14766, 14783, Dr. Frey.

67. Mahaffy's argument was ably supported at certain points by Eric Harvie, who attended commission hearings on behalf of the Anglo-Canadian group, and by L. Plotkins, owner of Lion Oils.

68. McGillivray Commission, Transcripts, vol. 135, p. 15007.

69. Ibid., p. 15009.

70. Ibid., p. 15049.

71. Ibid., pp. 15050–51.

72. Ibid., vol. 146, pp. 16245–48. The use of natural gasoline in the refining process enabled the refiner to reduce the amount of expensive tetraethyl lead and still achieve the same high octane value.

73. Ibid., p. 16250.

74. Ibid., pp. 16250–61. See also ERCB, Letters to Operators, 11 October 1939, "Tentative Proposals For the Restriction of Gas Production From the Turner Valley Gas Cap."

75. Board announcements of proposed regulatory change included notice of a date when representations would be received from those af-

fected by the proposal; but Mahaffy argued that these sessions, in addition to being completely voluntary, gave the intervener little satisfaction because the Board did not have to present a formal decision that addressed the evidence presented.

76.   McGillivray Commission, Transcripts, vol. 146, pp. 16277–78. See also pp. 16282 and 16284–85.

77.   Ibid., pp. 16285–86.

78.   Ibid., p. 16271.

79.   Ibid., pp. 16287–88. Dr. Boatright (Knode's Texas colleague) subscribed to the free movement school, whereas Dr. Link, Dr. Silas F. Shaw, Mr. Stanley Gill and Mr. Davies supported the little or no movement theory.

80.   Ibid., p. 16289. Knode's position with the Board had by this time changed from chairman to technical adviser.

81.   Ibid., p. 16291.

82.   Ibid., p. 16308.

83.   Ibid.

84.   Ibid., pp. 16306 and 16309.

85.   McGillivray Commission, Report, p. 264.

86.   Ibid., pp. 258–66. McGillivray Commission, Transcripts, vols. 158 and 159.

87.   McGillivray Commission, Report, p. 261.

88.   Ibid.

89.   Ibid., p. 265. Author's italics.

90.   Ibid., p. 234.

91.   Ibid., p. 235. McGillivray admitted that an "ideal" five-member board might be too costly and that for the time being a three-member board comprising individuals who shared the identified attributes of the ideal five would be acceptable.

92.   Ibid., p. 215.

93.   Ibid., p. 248. For the complete list of McGillivray's conservation recommendations, see pp. 234–38 and 248–52.

94.   Ibid., p. 3.

95.   Ibid., pp. 250–52.

96.   Ibid., p. 252. Particularly section 498 of the Criminal Code.

97.   Ibid., p. 30.

98.   Ibid., p. 31.

99.   Ibid., McGillivray added, "It might be thought that more wells than necessary will not be drilled because it would be uneconomic to do so in the case of proration to a limited market. As this is contrary to the 'oilman mind' and to the whole history of the industry wherever the 'Rule of Capture' has had play, in our opinion the thought cannot be entertained."

100.   Ibid., p. 42.

101.   Ibid., p. 215.

102.   Ibid., p. 234.

103.   Ibid.

104.   Ibid., pp. 256–57. Imperial Oil cut the price of gasoline on 24 July 1939 and again on 25 October 1939.

105.   Carlo Caldarola, ed., Society and Politics in Alberta (Toronto: Methuen, 1979), p. 373. In the 57-seat Legislature (reduced from 63 in 1935), Social Credit won 36 seats, the Liberal-Conservative Independents 19, Labour 1 and the Liberals 1. In 10 of the Social Credit seats, the margin of victory was less than 200 votes.

106.   William Knode chaired his last meeting of the Board on 10 June 1939, the day after he returned from London. In all, Knode chaired only

four meetings of the Board under the revised 1938 *Conservation Act,* ERCB, Minutes, 30 November 1938—5 July 1939.

107. ERCB, Minutes, 21 September 1939. See also *Calgary Herald,* 19 September 1940, where it is stated that Knode resigned when his engineering firm entered business in Alberta. No mention of "resignation" is to be found in the Board minutes, the *Annual Report* of the Department of Lands and Mines, the N.E. Tanner Papers or the E.C. Manning Papers. Gordon Connell, the Board's senior engineer in Turner Valley throughout the period of Knode's chairmanship, remembers that Knode's departure came about as a result of his failure to estimate Turner Valley's producing potential accurately. ERCB, transcript of tape-recorded interview with Gordon Connell, 17 June 1986, pp. 16–17.

108. Brown was one of the experts selected by J.J. Frawley to appear before the McGillivray Commission. See George Granger Brown and Donald L. Katz, "The Active Crude Reserves in Turner Valley Field Computed From Production and Pressure Data," 28 December 1939.

109. Dingman had been in government service since joining the newly formed Petroleum and Natural Gas Division of the federal Department of the Interior in 1920.

110. Alberta, Department of Lands and Mines, *Annual Report* to 31 March 1941, p. 16. It is probable that the advice sought was that of Dr. John Frey.

111. ERCB, Minutes, 31 August 1940. Dingman, who seems to have been less than enthused about Knode's appointment in the first instance, delivered the three-month notice of termination to the Knode household in person.

112. *Western Oil Examiner,* 7 and 14 September 1940 and *Calgary Herald,* 12 September 1940.

113. *Western Oil Examiner,* 21 September 1940. For industry reaction to Frawley's proposed Board, see *Western Oil Examiner,* 6 January 1940, 3 and 10 February 1940.

114. Ibid., 16 November 1940 and Alberta, Department of Lands and Mines, *Annual Report* to 31 March 1941, p. 16.

115. Erich W. Zimmermann, *Conservation in the Production of Petroleum: A Study in Industrial Control* (New Haven: Yale University Press, 1957), pp. 162–67.

116. See, for example, Robert E. Allen, "Control of California Oil Curtailment," AIME Transactions, vol. 92 (1931), pp. 47–66, and "Some Effects of Curtailment on the Potential and Recovery of Petroleum in California," *Mining and Metallurgy* 15, no. 336 (December 1934), pp. 486–89.

117. Robert E. Burke, *Olson's New Deal for California* (Berkeley and Los Angeles: University of California Press, 1953), pp. 116–18.

118. *Calgary Herald,* 1 October 1940. See also ERCB, Minutes, 30 March 1939, 11 April 1939 and 10 May 1939.

119. ERCB, Minutes, 18 October 1940 and 5 December 1940.

120. Ibid., 24 September 1940. "In all cases where operators apply to this Board for variations or exceptions in regard to well spacing, drilling or completion procedure, or other departures from established practice or procedure prescribed by this Board, it shall be the policy of the Board to hold hearings, open to all interested and affected operators in the following specific cases: 1. Where observed conditions indicate that a wider spacing program than one well to forty acres may efficiently recover the reserves of the particular reservoir. 2. Where the exception requested by the operator

is such that the producing formation may be endangered, or the recovery of reservoirs jeopardized by the failure of any of the operations proposed. 3. Where the proposed exception contemplates the spacing of a well on an area less than the approved spacing plan for the particular field or part thereof. 4. Where the proposed exception contemplates spacing a new well closer than the established distance to the boundary of a separately owned or competitively operated well area."

121.   *Calgary Herald,* 15 October 1940. The merger followed the suggestion of J.J. Frawley before the McGillivray Commission and the subsequent advice of the Alberta Petroleum Association. The offices of the two agencies were already in the same building, but on separate floors.

122.   Ibid., 21 September 1940. Cottrelle was from Ontario and was a prominent banker and industrialist. He was chairman of Union Gas as well as a director of the Canadian Bank of Commerce and various companies in the manufacturing sector.

123.   Machine tools and power controllers were added a few months later.

124.   NAC, RG2, vol. 1427, Order-in-Council PC 2818, 28 June 1940. Amended by PC 1195, 19 February 1941, and PC 2448, 8 April 1941.

125.   Cottrelle began by appointing a seven-man advisory committee. Six members were presidents or directors of eastern refineries, the other was a representative of the largest petroleum importer.

126.   *Calgary Herald,* 19 July 1940 and 21 September 1940.

127.   *Western Oil Examiner,* 30 November 1940.

128.   ERCB, Letters to Operators, 12 December 1940.

129.   *Western Oil Examiner,* 14 December 1940.

130.   Ibid. See also *Calgary Herald,* 13 December 1940.

131.   ERCB, Minutes, 20 December 1940, and Allen to "All Turner Valley Oil Producers," 27 December 1940.

132.   Ibid., 16 January 1941.

133.   Ibid.

134.   Ibid., 31 January 1941. Ottawa wanted 26,000 bbls. a day.

135.   Ibid., 22 January 1941. Letter to "All Turner Valley Operators."

136.   Ibid., 17 February 1941, "Summary of Suggestions for Allocation Procedure Revision." Letter to "All Turner Valley Producers," 17 February 1941, and "Production Allocation Procedure," February 1941. Acreage and gas-oil ratios were retained as corrective or penalty factors. Previously a quarter (one-third in February) of the allocation was based specifically on each of these factors. The acreage factor in early conservation practice was used as a device to protect correlative rights. In Turner Valley, it had the effect of reducing allotments to wells on less than 40-acre tracts, but in January 1941 this was applicable to only 13 wells.

137.   Ibid. The formula assigned 5 barrels daily for the first 1,000 feet of producing depth, 5½ barrels for the second 1,000 feet, 6 barrels for the third 1,000 feet, and so on. The corollary of a minimum allowance or a maximum well allocation was also debated but not adopted. The reasoning was that, just as there is some minimum below which a well cannot produce without premature abandonment and loss of reserves, there is similarly a maximum production beyond which a well's ultimate production is compromised.

138.   ERCB, Minutes, 3 March 1941.

139.   Alberta, Order-in-Council 279–41, 6 March 1941, "Regulations under The Oil and Gas Wells Act, 1931, respecting Drilling and Production Operations of Oil and Gas Wells." Henceforth, the surface location of all wells was to be as near as possible to the centre of a legal subdivision. Also,

the "right" of those whose only holding was less than a legal subdivision to drill one well was dropped.

140.   Alberta, Order-in-Council 278/41, 31 March 1941, and 1213/41, 26 August 1941. Alberta, Department of Lands and Mines, *Annual Report* to 31 March 1941, p. 51.

141.   *Western Oil Examiner*, 15 March 1941. In keeping with past practice, change in the regulations was preceded by consultation with representatives of the oil producers, see *Calgary Herald*, 6 March 1941.

142.   *Alberta Gazette*, 31 March 1941, p. 253, provincial reserves as distinguished from crown reserves set aside by Order-in-Council, 225/37, 3 March 1937. See H.H. Somerville, *A History of Crown Rentals, Royalties and Mineral Taxation in Alberta to December 31, 1972* (Edmonton: Alberta Energy and Natural Resources, 1977), pp. 5–18. The 15 areas totalled 14,112 square miles. A short time later, the large Antelope plains area in southeastern Alberta was acquired by the Department of National Defence for chemical warfare research, leaving 14 provincial reserves. The Sand River area later became part of the Primrose Lake Air Weapons Range. ERCB, Virginia Hills, Kaybob and Cynthia, Provincial Reserves, 1941. Contrary to the assertion sometimes made that Allen was assisted by Deputy Minister of Lands and Mines John Harvie and Professor of Geology J.A. Allan at the University of Alberta, the chairman of the Conservation Board made the selection on his own. Virginia Hills and Cynthia he named after his daughters, and Kaybob apparently is a contraction of his own and his wife's first names. See 9 April 1941, J.A. Allan to John Harvie, Deputy Minister of Lands and Mines.

143.   *Calgary Herald*, 31 March 1941.

144.   Ibid., 4 April 1941.

145.   *Edmonton Journal*, 5 April 1941. Alberta *Statutes*, 1941, ch. 68, "An Act to authorize the Granting of Franchises for providing Public Services in Petroleum and Natural Gas Areas," 8 April 1941.

146.   *Edmonton Bulletin*, 5 April 1941 and 8 April 1941. *Calgary Herald*, 5 April 1941.

147.   The industry's concern was accentuated by two additional bills that cleared the Legislature at the same time, "The Unit Operation of Mineral Resources Act" (1941, ch. 69), which provided for provincial participation in unitized development of minerals, and "The Geophysical and Geological Exploration Regulation Act" (1941, ch. 70), which forbade anyone from undertaking any geophysical operations or examinations of subsurface geology without a licence from the Minister of Lands and Mines. Both Acts were to come into effect if and when the government might choose.

148.   ERCB, "Letters to Operators," 9 April 1941. R.E. Allen to "All Absorption Plant Owners, Turner Valley Field." Of this total amount, oil wells averaged 94 million cubic feet daily or 64.4%, whereas gas wells contributed 52 million cubic feet daily or 35.6%. Calgary, which was by far the largest natural gas consumer, averaged 18 million cubic feet daily over the eight-month period.

149.   Ibid., and *Calgary Herald*, 18 April 1941.

150.   ERCB, Minutes, 15 May 1941.

151.   In January 1940, the Petroleum and Natural Gas Conservation Board rated Princess No. 2 at 520 barrels daily, by July the well was averaging 47 barrels daily.

152.   *Calgary Herald*, 2 June 1941.

153.   Ibid. See also ERCB, Minutes, 26 May 1941.

154.   ERCB, Compensation Scheme 1939–45, 7 June 1941, R.E. Allen to Honourable N.E. Tanner.

155. Ibid.

156. Through the first 2½ years of the war, Ottawa did little to stimulate petroleum exploration or to curb nonproductive use of gasoline. Central Canadian refiners expected to be able to rely on cheap U.S. crude oil.

157. ERCB, Compensation Scheme 1939–45, 7 June 1941, R.E. Allen to Honourable N.E. Tanner.

158. Ibid.

159. ERCB, Minutes, 7 June 1941.

160. Alberta, Department of Lands and Mines, *Annual Report* to 31 March 1942, p. 15.

161. ERCB, Minutes, 26 May 1941.

162. Ibid. On 26 May 1941, the Board hired L.A. McLeod, R.T. Helmer and W.C. Carter as junior engineers to replace the departing Gordon Connell, S.S. Cosburn and Joseph Gleddie.

163. ERCB, Minutes, 15 July 1941. *Calgary Herald*, 14 July 1941, 31 July 1941 and 1 August 1941.

164. *Calgary Herald*, 6 August 1941. See also *Calgary Herald*, 25 June 1941 and 5 July 1941, and *Western Oil Examiner*, 12 July 1941.

165. *Calgary Herald*, 7 August 1941.

166. Ibid. 15 October 1941.

167. *Western Oil Examiner*, 1 November 1941.

168. ERCB, Minutes, 17 October 1941.

169. ERCB, "Minutes of Proceedings of a Sitting of the Alberta Petroleum and Natural Gas Conservation Board . . . to hear and discuss the *Report* of Dr. George Granger Brown," 9 January 1942.

170. Ibid.

171. ERCB, Brown Plan, 1941–44, 10 December 1941, Dr. G.G. Brown to J.J. Frawley.

172. Ibid., "Suggested Modifications in the Suggested Procedure for Immediately Regulating the Production of Petroleum and Natural Gas in the Turner Valley Field," 26 January 1942. See also Board Minutes, 16 February 1942.

173. Ibid.

174. See, for example, *Calgary Herald*, 3 February 1942. *Western Oil Examiner*, 12 July 1941.

175. ERCB, Alberta Petroleum Association, Advisory Committee, 27 April 1942, N.E. Tanner to H. Greenfield, President, Alberta Petroleum Association. See also ERCB, Minutes, 16 February 1942.

176. Ibid. The Alberta Petroleum Association subsequently called a meeting (nonmember oilmen were also invited to attend), and the following were elected by secret ballot: R.A. Brown, Sr., R.E. Trammell, K.M. Doze, W.S. Herron and W.H. Jones. The committee appointed initially by the Association to discuss the Brown plan consisted of J.G. Spratt, R.E. Trammell, F.F. Reeve, G.A. Watt and K.M. Doze.

177. *Calgary Herald*, 26 February 1942.

178. Ibid., 3 July 1942. *Western Oil Examiner*, 4 July 1942.

179. ERCB, W. Epstein, interview, 4 March 1989. Epstein, as a young lawyer with the Calgary firm Smith, Egbert and Smith, also assisted in drafting the 1938 conservation legislation.

180. For illuminating certain aspects of Frawley's career and personality beyond what can be gleaned from correspondence at the ERCB, in the Premiers' Papers, and from Commission transcripts, the author is indebted to Frawley's son, Mark P. Frawley, director of legal services for the Ontario Human Rights Commission, and Ronald D. Karoles of Stewart

and Karoles, Edmonton, who, as a young lawyer, was sent to Ottawa to assist J.J. Frawley. See J.J. Frawley (Chairman) file in Board history files.

181.   *Western Oil Examiner,* 27 December 1941. *Calgary Herald,* 18 November 1941.

182.   Alberta, Order-in-Council 737/42, 26 May 1942. Frawley's salary was set at $6,000 per annum, $4,000 less than that paid to Robert Allen.

183.   See, for example, "Crisis in Oil," *Saturday Evening Post,* 20 and 27 November 1943.

184.   NAC, RG21, vol. 36, 161.36, pt. 6., H.L. Keenleyside to the Deputy Minister of Mines and Resources, 16 January 1943, with enclosure, "Suggested informal Memorandum to the Canadian Government on the Need for a Concerted Program of Wildcatting for Oil in Western Canada." The memorandum was passed from J.D. Hickerson of the U.S. State Department to Hugh Keenleyside, Canada's assistant undersecretary of state for external affairs and Hickerson's counterpart on the Permanent Joint Board on Defence.

185.   ERCB, Brown Plan, 1941–44. "Suggested Modification...," 26 January 1942, p. 6.

186.   PAA, Premiers' Papers, Reel 84, 811B, 16 July 1943, J.J. Frawley to N.E. Tanner.

187.   ERCB, Thomas R. Weymouth, "The Conservation of Gas in the Turner Valley Field," p. 3. See also Board history file.

188.   Ibid., p. 33.

189.   Ibid., p. 29. Of the total estimated remaining reserve of 402.2 billion cubic feet of dry gas likely to be recovered from the Turner Valley field, Weymouth projected that only 223.1 billion could be utilized with current facilities, the rest would be dissipated in flares or otherwise.

190.   ERCB, Minutes, 12 April 1943. The March 3 draft agreement was developed from a proposal that Weymouth obtained from Royalite about February 17. See PAA, Premiers' Papers, Reel 84, 811B, 16 July 1943, J.J. Frawley to N.E. Tanner.

191.   ERCB, Minutes, 12 April 1943.

192.   Ibid.

193.   ERCB, Thomas R. Weymouth, "The Conservation of Gas...," p. 38.

194.   PAA, Premiers' Papers, Reel 84, 811B, 16 July 1943, J.J. Frawley to N.E. Tanner.

195.   ERCB, Compensation Scheme, 1939–45, "In the Trial Division of the Supreme Court of Alberta, Affidavit," D.P. McDonald, 23 June 1943.

196.   PAA, Premiers' Papers, Reel 84, 811B, 16 July 1943, J.J. Frawley to N.E. Tanner (the hearing was adjourned at Royalite's request), and 30 June 1943, D.P. McDonald to N.E. Tanner.

197.   ERCB, Compensation Scheme, 1939–45, "Gas Wells Operated by Royalite Oil Co. Ltd." 5 July 1943.

198.   Ibid., 30 June 1943, D.P. McDonald to N.E. Tanner.

199.   PAA, Premiers' Papers, Reel 84, 811B, 21 November 1942, J.J. Frawley to J.H. McLeod, President Royalite Oil, with copies to Andrew Davison, Mayor of Calgary, H.R. Milner, President of Canadian Western Natural Gas, and G.R. Cottrelle. The letter informed Royalite that the allowables had been fixed according to the Brown plan, and as a result it was estimated that Royalite would have produced the entire quota by about 1 May 1943. The company was advised that it must take immediate steps to obtain additional gas supply elsewhere.

200.   ERCB, Compensation Scheme, 1939–45, 16 July 1943, H. Crabtree, President, Allied War Supplies Corporation, to J.J. Frawley.

201. Ibid., 2 July 1943, J.J. Frawley to N.E. Tanner.

202. PAA, Premiers' Papers, Reel 84, 811B, 16 July 1943, J.J. Frawley to N.E. Tanner.

203. Ibid., 17 July 1943, E.C. Manning to C.D. Howe.

204. NAC, C.D. Howe Papers, MG27, 111, B2, vol. 35, f37, 20 July 1943, G.R. Cottrelle to W.S. Bennett, Executive Assistant to the Minister of Munitions and Supply, and 22 July 1943, G.R. Cottrelle to C.D. Howe.

205. Ibid., 26 July 1943, C.D. Howe to E.C. Manning.

206. ERCB, "Impact of Higher Wartime Production Rates on Turner Valley Oil," G.A. Warne and R. Purvis to D.H. Breen, 21 November 1986. See Board history file.

207. PAA, Premiers' Papers, Reel 84, 811B, 5 August 1943, E.C. Manning to C.D. Howe.

208. ERCB, Compensation Scheme, 1939–45, Board Order 300, 5 August 1943.

209. Ibid., Alberta, Order-in-Council, 1289/43, 10 August 1943. OC 1289/43 was largely drafted by J.J. Frawley. The Board subsequently collected from Royalite and distributed $196,603.00 to Turner Valley producers.

210. Ibid. Text of a statement for the press by Premier Manning, see *Albertan*, 13 August 1943.

211. Ibid., 3 September 1943, J.J. Frawley to E.C. Manning.

212. Ibid., "Memorandum in Respect to the Production, Sale, Processing and Distribution of Natural Gas in the Turner Valley Field, Prepared by Model Oils Ltd. for Submission to the Lieutenant Governor in Council, of the Province of Alberta." 4 August 1943; 30 November 1943, T.R. Weymouth to J.J. Frawley; and 4 December 1943, R.V. LeSueur, vice-president, Imperial Oil Ltd., to J.J. Frawley.

213. Alberta *Statutes*, 1944, ch. 4, "An Act to establish the Natural Gas Utilities Board and to prescribe its duties." See sections 68–70 particularly. The Act was drafted by J.J. Frawley.

214. Alberta, In the Matter of "The Natural Gas Utilities Act" and In the Matter of an Enquiry into Scheme to be adopted for Gathering, Processing and Transmission of Natural gas in Turner Valley, Natural Gas Utilities Board, Decision, vol. 1, pp. 1–8 and vol. 2, pp. 175–76.

215. ERCB, Minutes, 1 and 12 March 1945. Board Order No. 660 informed the operator (British American Oil Company Ltd.) that the two wells were being purchased by the Board, as directed by section 16 of the *Oil and Gas Resources Conservation Act,* to be used by British American Gas Utilities Ltd. for the purpose of repressuring the field. The owners of the Carleton Royalties well were unco-operative and the Board proceeded under section 55 of the Act to effect compulsory purchase of the well for $7,500. In June, the Board designated three wells operated by Royalties to be used as input wells by the Madison Natural Gas Company. Minutes, 5 June 1945.

216. The 1932 seismic tests were at the Twin River Structure in southern Alberta.

217. Located in LSD 16–35–3–23 W4M, the well was logged on 8 November 1939 by the Halliburton Oil Well Cementing Co. Ltd.

218. ERCB, J.R. Pow, "The Core Research Facility and its Background," June 1989, including inter-office memorandum, I.M. Cook to J.J. Frawley, 30 September 1942. The Board started out with stringent requirements that "The operator of *every* well . . . shall immediately upon encountering a geological formation . . . from which production . . . may be anticipated, unless otherwise ordered by the Board, core and adequately test such . . .

information. All information as obtained shall be forwarded to the Board . . . and no further drilling shall be conducted until so instructed by the Board." This resulted in constant requests for exemption, and in 1941 the provision was amended to say that cores were to be taken only "when ordered by the Board."

219.   Ibid.

220.   Alberta, Order-in-Council, 148–43, February 1943. The Board's files contain electric logs for 55% of the wells drilled in 1943, compared to only 34% for the previous year.

221.   ERCB, J.R. Pow, "Regulatory Policies Relating to Core and Drill Cuttings and their Impact," 23 June 1989, including letter of Allan R. Stander, Curator, Core Research Center, The University of Texas at Austin, to Bob Pow. See "Core Research Facility and its Background" file in Board history file.

222.   ERCB, G.A. Warne, "Progress in Reservoir Engineering: 1930's and 1940's," 20 June 1989.

223.   "Edward Herbert Boomer (1900–1945)," *Transactions of the Royal Society of Canada,* 1946, pp. 79–81. G.W. [Govier], "Edward Herbert Boomer," *The New Trail,* vol. IV, 1946, p. 34, *Western Oil Examiner,* 3 November 1945. See under "Dr. E.H. Boomer, Chairman," in Board history file.

224.   ERCB, Minutes, 15 March 1945. The Board found that Model-Spooner No. 1 had been opened up and had produced nearly a full year's allowable in a few days. Model was notified that the well was to be shut in until further notice.

225.   Ibid., 2 February 1945. Henceforth, all applications for a licence to drill had to include an official well name that did not duplicate any previously recorded name in the *Official Well Name Register.*

226.   Alberta, Order-in-Council, OC No. 706, 1 May 1945, Drilling and Production Regulations, section 7(3). If incidental work cost more than the deposit, the excess amount was chargeable to the operator.

227.   Ibid., OC No. 1690, 16 October 1945.

228.   ERCB, Minutes, 12 June 1945. The Board insisted that it have complete authority, "with no interference from the Department of Public Works." Board jurisdiction did not cover storage tanks at refineries or absorption plants.

229.   ERCB, "Vermilion Oil Field," 12 May 1943, D.P. Goodall to M.D. Kemp in Board history file.

230.   Manyluk, interview, 29 April 1992. Manyluk was stationed in Vermilion in 1946 and charged with the task of bringing production practice up to an acceptable standard. He recalled that conditions were even worse on the Saskatchewan side of the field where, in one instance, a huge pit was used as a holding pond for 100,000 barrels of oil and water.

231.   ERCB, Minutes, 24 August 1944. See also 13 September 1944.

232.   Ibid., 5 April 1945. See also 26 September 1944. On this latter date, the Board decreed that any new gas wells in the area would only be approved on the basis of one well per section (640 acres), and operators were informed that they must equip their wells with approved meters to enable proper gas measurement.

233.   ERCB, Minutes, 30 May 1945.

234.   Ibid., 18 October 1940. See also 24 September 1940. Under a 40-acre drilling pattern, distances of 1,320 feet between wells and 660 feet from a quarter section line are prescribed.

235.   Ibid., 20 December 1940. A well could be located up to 117 feet radius from the exact centre without special Board approval. The bottom

of the drill hole had to be within the square comprising the central 10 acres of the tract.

236. Ibid., 1 August 1941. The location of Conestoga No. 1 was LSD 1–30–50–5 W4M. Instead of the usual Department of Lands and Mines approval, the well licence has only Frawley's signature.

237. Ibid., 31 July and 7 August 1945. After contacting all the operators in the Princess field, and finding that none were opposed, the Board agreed to sanction 20-acre spacing on sections 14, 15, 16, 21, 22, 23, 26, 27, 28, 33, 34 and 35 of Township 19, Range 12, West of the 4th Meridian.

238. *Western Oil Examiner,* 6 March 1943. Rowley v. Petroleum and Natural Gas Conservation Board, *Western Weekly Reports,* vol. 1, 1943, pp. 470–76. In this action, the Conservation Board once again fought off the allegation that it lacked constitutional authority for certain of its actions.

239. Zimmerman, p. 338. The Texas Railroad Commission's first spacing rule, rule 37, allowed for exceptions "in order to prevent waste and protect vested rights" (this wording is still in use today). Exceptions to "protect the vested rights" of small producers have been granted in thousands of instances. See also Gerald Forbes, *Flush Production* (Norman: University of Oklahoma Press, 1942), pp. 92 and 188–90.

240. The location of the discovery well, Shell No. 4–24–5, was LSD 4–24–25–5 W4M.

241. ERCB, Minutes, 4 May 1945.

242. Ibid., 2 June and 5 July 1945. See Drilling and Production Regulations, ss. 42, 43 and 44.

243. Ibid., 23 October 1945.

244. *Western Oil Examiner,* 3 and 17 November 1945.

245. Alberta, Order-in-Council, No. 1043/44. ERCB, transcript of tape-recorded interview with Dr. G.W. Govier, 8 June 1987, tape 1, pp. 9, 10 and 23.

246. ERCB, Minutes, 16 January 1946, 25 February 1946 and 27 March 1946.

247. Ibid., 3 January 1947.

248. Ibid., 3 June 1946 and 31 May 1946.

249. *Western Oil Examiner,* 5 October 1946.

250. ERCB, Minutes, 13 June 1946. See Regulation No. 11. On farms where gas consumption was less than 500 Mcf per year, a gas well could be drilled with an assigned acreage of 10 acres, provided that the gas produced was not used for pumping water for irrigation purposes or consumed by any person other than the farm owner. See also Regulations No. 12 and No. 13.

251. Ibid., 5 April 1945.

252. Ibid., 18 December 1945. With the schedule came the instructions that it was to be "strictly adhered to in future staff dealings."

253. Ibid., 12 February 1947, $15 was close to a 10% increase for most employees.

254. The low percentage of the oil-in-place recovered at Turner Valley appears to be substantially the result of the poor quality of the reservoir. Application of good engineering practices earlier, however, especially in the control of the production of high gas-oil ratio wells and the maintenance of pressure where gas-cap encroachment could occur, might have permitted recovery of perhaps another 3% of the oil-in-place, or about 30 million barrels. The depletion of the gas cap in the early life of the pool, mainly to recover condensate, wasted a substantial amount of gas, but because the gas was poorly connected to the oil-bearing section of the

pool, oil recovery was not as seriously affected as might have been. I am indebted to Dr. Ross Purvis, Manager of the ERCB Oil Department, for this assessment.

255.   David F. Prindle, *Petroleum Politics and the Texas Railroad Commission* (Austin: University of Texas Press, 1981), p. 63.

256.   Effective conservation of casing-head gas did not happen until after the *Texas Railroad Commission v. Fleur Bluff Oil Co.* case in 1949.

257.   B.M. Murphy, *Conservation of Oil and Gas* (Chicago: American Bar Association, 1949), p. 302, endnote 7.

258.   Ibid., p. 308.

259.   Ibid., p. 309.

260.   Ibid., pp. 304–5.

261.   The great advantage of a superior database upon which to plan recovery schemes in Alberta was noted as early as 1943 by the U.S. expert Dr. Thomas Weymouth. Writing about his gas conservation plan for Turner Valley, he explained that not only was it feasible from an engineering standpoint but also it was "sound from an economic standpoint. The records of operating experience in this field are so full and complete that it is felt that the estimates based upon them contain fewer uncertain factors than is usually the case in problems such as this, and that therefore they may safely be relied upon." ERCB, T. Weymouth, "The Conservation of Gas in the Turner Valley Field," 1943, p. 31.

FIVE—THE LEDUC DISCOVERY AND THE NEW REGULATORY ENVIRONMENT

1.   Vern Hunter, 1957b, letter in Appendix 1 of D.B. Layer, "The Leduc Oil Field Discovery: Summary of Documentation Prepared for Imperial Oil" (Calgary: Esso Resources, 1979).

2.   Imperial-Leduc No. 1 was spudded in on 20 November 1946 in LSD 5 Sec. 22, Twp. 50, Rg. 26, w4 Meridian and brought into production 13 February 1947. The well was abandoned in July 1974 after producing more than 300,000 barrels. Imperial-Leduc No. 2, at LSD 1 Sec. 16, Twp. 50, Rg. 26, w4 Meridian, was spudded in on 12 February 1947 and completed as a producer on 27 May 1947. D.B. Layer, pp. 15-16.

3.   Alberta, Department of Mines and Minerals, *Annual Report* for the year ended 31 March 1950, p. 14.

4.   D.B. Layer, pp. 5-6. Shell Oil, on the other hand, decided to abandon further exploration efforts.

5.   *Western Oil Examiner*, 25 January 1947. Speaking in Toronto, E.V. Murphree, vice-president of Standard Oil Development (Standard Oil of New Jersey's research arm), assured his audience that his company had developed a process for the conversion of natural gas to gasoline that was economically competitive with the production of gasoline from crude oil. Murphree was a high-profile figure, and his words carried great weight. He had been chairman of the planning board responsible for the technical and engineering aspects of the Manhattan Project (the atomic bomb). Related to this, Murphree directed the construction of a heavy-water plant in British Columbia and supervised much of the work on the centrifugal method of separating uranium isotopes. Henrietta M. Larson et al., *History of Standard Oil Company (New Jersey): New Horizons, 1927-1950* (New York: Harper and Row, Publishers, 1971), pp. 168 and 520. Dr. Donald Katz also proclaimed to Calgary oilmen his faith in the eventual production of gasoline from natural gas. *Western Oil Examiner*, 23 February 1946.

6. The Viking-Kinsella field was already being tapped by the local natural gas utility company serving the Edmonton area.

7. Imperial Oil, Toronto Advisory Group, "Comparative Prospects—Western Canada," April 1946, cited by D.B. Layer, pp. 6-7. At the same time, Dr. Ted Link, Imperial's chief geologist, circulated a questionnaire among the company's geologists asking where in Canada they thought oil was most likely to be found and why. The strongly preferred area, the "Central Alberta Plains," was the first choice of 18 of the 32 geologists, and the second choice of six.

8. Imperial's rights covered 46% of the acreage within the 24 townships. The rest was freehold land, with Canadian Pacific Railway being the largest private mineral holder.

9. W.O. Twaits (copy of 1957 letter), in D.B. Layer, p. 14.

10. Imperial-Leduc No. 1, LSD 5, Sec. 22, Twp. 50, Rg. 26, W4 Meridian.

11. Petroleum and Natural Gas Conservation Board (hereafter PNGCB), *Alberta Petroleum Industry*, 1947 (Calgary, 1948), p. 19. Leduc No. 1 produced 42,590 barrels during 1947, but this was over an eleven- rather than an eight-month period, as was the case with Leduc No. 2.

12. ERCB, Minutes, 20 February 1947.

13. Ibid., 16 May 1947.

14. Ibid., 31 May 1947.

15. After he resigned, Bailey became manager of the Land and Exploration Department of Husky Oil and Refining Ltd. In March 1949, he left Husky to start his own company—the A.G. Bailey Company Ltd., offering petroleum consultant and management services for oil exploration and development operations. *Daily Oil Bulletin*, 23 February 1949. Note that there are citations from the *Daily Oil Bulletin* and from the *Oil Bulletin*. Both were edited by Carl Nickle. The former was a daily publication and the latter appeared on Friday afternoons with more extended comment.

16. Durham, p. 100.

17. For an indication of Tanner's development philosophy, see *Oil in Canada*, 7 February 1949, "'Open Door' Alta. Oil Policy Guided by Hon. N.E. Tanner," pp. 12-13 and N.E. Tanner, "Government Policy Regarding Oil-Sand Leases and Royalties," *Proceedings: Athabasca Oil Sands Conference* (Edmonton: King's Printer, 1951).

18. Alberta, *Statutes*, 1947, ch. 24, "An Act to Provide for the Exercise of any Right of Entry and for the Determination of the Compensation to be paid therefor."

19. In determining the amount of compensation, the Board could take into account such factors as the value of the land, the amount of land that could be permanently damaged by the operator, the adverse effect on remaining land, and the nuisance and inconvenience that might arise from drilling operations.

20. Recollected by M. Blackadar regarding a particular incident in Joseph Lake field.

21. See, for example, *Western Oil Examiner*, 28 June 1947.

22. Alberta, Order-in-Council, OC No. 278, 6 March 1941. There was no limit to the number of reservations. The deposit was $750 for each 20,000 acres if the exploration was restricted to surface geological investigation, and $2,000 for each 20,000 acres if geophysical operations on subsurface geology were involved. See also OC No. 60, 14 January 1944.

23. Ibid., OC No. 716, 11 July 1947.

24. *Oil Bulletin*, 11 July 1947.

25. The new regulations built upon the base established in 1944, when the vast territory north of the North Saskatchewan River, along with un-

surveyed forest reserves, was combined with the 15 "provincial reserves" to create crown reserves. In June 1946, the southern boundary of Alberta's northern crown reserve territory was moved northward from the North Saskatchewan River to the north boundary of Township 60. Alberta, Order-in-Council, OC No. 1111, 20 June 1946.

26. Alberta, Order-in-Council, OC No. 860, 19 August 1947. An oil discovery obliged the reservation holder to convert immediately to leasehold tenure.

27. The bonus bidding system originated with the Department of Interior regulations. See Order-in-Council, P.C. No. 53, 24 March 1921.

28. Alberta, Order-in-Council, OC No. 860, 26(a), 19 August 1947. See also Department of the Interior, *Manual for Operators Under Oil and Gas Regulations* (Ottawa: F.A. Acland, 1922), p. 21, re. section 41 of Regulations.

29. Alberta, Order-in-Council, OC No. 860, 13(a), 19 August 1947.

30. *Calgary Herald*, 28 August 1947.

31. *Edmonton Bulletin*, 3 and 4 March 1948.

32. *Edmonton Journal*, 4 March 1948. The petroleum and natural gas rights to some of these sections came from sources other than the Crown.

33. Alberta, Order-in-Council, OC No. 308, 29 March 1948.

34. Ibid. Further, the regulations stated that leased areas had to be square or rectangular in shape, that the maximum length of any tract was four miles, and in no case could the length exceed twice the breadth.

35. Alberta, Department of Mines and Minerals, *Annual Reports*, 1948-1954. The Canadian company, Home Oil was the most frequent purchaser.

36. PNGCB, *Alberta Petroleum Industry* (Calgary, 1948). *Oil Bulletin*, 16 January 1948.

37. Although Goodall's appreciation of the practical problems faced by operators in the field enabled him to know when the literal application of a regulation should not apply and to offer his Board colleagues expert advice about the formulation and application of regulations at the field level, at times this same quality also made him more vulnerable to pressure from the industry or companies seeking relief from one or other of the regulations on "practical" grounds. In May 1947, for example, the Board under Bailey's guidance persuaded the industry that *all* wells should be electrologged before production casing was run or before the well was abandoned. A month later, the Board with Goodall in charge was persuaded by industry that it was *not* necessary to have all wells electrologged, that electrologs should be run only at those wells designated specifically by the Board.

38. ERCB, interview, D.P. (Red) Goodall, 4 April 1981. "When I was there alone with just [as] a temporary Chairman I tended to just leave things the way they were, not get anything new stirred up because I didn't want to try to handle it myself. But after we got a full Board with McKinnon and Govier, then it was a different matter," tape 2, p. 5.

39. ERCB, Minutes, 12 November 1947.

40. Calgary *Herald*, 29 November 1947.

41. ERCB, Advisory Committee to Conservation Board, Alberta Petroleum Association, Minutes, 10 December 1947.

42. Ibid., See also ERCB, interview, D.P. (Red) Goodall, 4 April 1981, tape 2.

43. Ever cautious, Premier Manning was careful to give the Western Canadian Petroleum Association prior notice of McKinnon's appointment. Canadian Petroleum Association, Western Canada Petroleum Association, Minutes, Board of Directors, 5 February 1948.

44. ERCB, R. McKinnon, Draft, "Ian Nicholson McKinnon, 1906-1976: A Biographical Sketch," September 1989. See Board history file, Board Chairman—I.N. McKinnon.

45. ERCB, Minutes, 23 January 1950.

46. ERCB, interviews, G.W. Govier, 6 August 1981, 8 June 1987 and 27 July 1989; Jack W. Patrick, 6 May 1986; D.R. Craig, 29 July 1986; J.G. Stabback, 21 January 1987; H.H. Somerville, 19-20 May 1987; and J.R. Pow, 12 July 1989. *Western Oil Examiner*, 7 February 1948. *Oil in Canada*, 23 November 1961, p. 9.

47. ERCB, transcript of tape-recorded interview with George Govier, 8 June 1987, tape 1, p. 9.

48. GA, transcript of tape-recorded interview with George Govier, 6 August 1981, tape 1, p. 2.

49. By analysing gas meter charts carefully Govier tried to determine what might be achieved by scheduling oil production so as to minimize the peaking of production of solution gas beyond the capacity of the existing gas gathering system. ERCB, transcript of tape-recorded interview with George Govier, 8 June 1987, tape 1, p. 24.

50. At the University of Michigan, Brown directed a chemical engineering department that was oriented towards the petroleum and natural gas industry and considered to be one of the best two or three such schools in the United States.

51. ERCB, transcript of tape-recorded interview with George Govier, 8 June 1987, tape 1, pp. 14-15, and GA, transcript, 6 August 1981, p. 2.

52. ERCB, Minutes, 2 February 1948.

53. *Calgary Herald*, 9 February 1948.

54. *Daily Oil Bulletin*, 9 February 1948.

55. ERCB, Minutes, 11 February 1948. At the meeting with the advisory committee on 11 February 1948, Govier presented his estimate of the optimum initial allowable for each zone based upon assessed minimum and maximum values for porosity, connate water, percent recovery and decline rate. His figure for the D-2 zone was 102 barrels per day and for the D-3 zone 111 per day, "pending the time when it would become necessary to express the allowable in terms of reservoir fluid and take specific cognizance of water production and gas-oil ratios."

56. Ibid., 23 February 1948.

57. See Aubrey Kerr, *Atlantic No. 3, 1948* (Calgary: S.A. Kerr, 1986), p. 78, quoting Jack Pettinger. Kerr offers an excellent assessment of the Atlantic No. 3 disaster.

58. ERCB, Minutes, 12 April 1948. It was understood that the Board would later recover the cost of Kinley's services from Atlantic. Kinley first visited the wellsite on May 9.

59. ERCB, Atlantic No. 3, general file, 10 June 1948, M. Kinley to I.N. McKinnon.

60. Kerr, p. 84.

61. *Calgary Herald*, 13 May 1948.

62. *Calgary Herald*, 19 May 1948. See also Kerr p. 87.

63. Kerr, p. 95. It is unclear whether Imperial approached the Board with a proposed solution or whether the Board asked Imperial for a specific recommendation about what should be done.

64. ERCB, Minutes, 13 May 1948.

65. Kerr, pp. 97-98.

66. ERCB, Minutes, 18 May 1948. See also, *Oil Bulletin*, 14 May 1948.

67. Kerr, p. 104. Moroney also insisted upon the inclusion in his "team" of two of Royalite/Imperial's old Turner Valley hands: driller

Charlie Visser and mud man Jim Tod. The two possessed skills critical to the success of Moroney's plan of operations.

68. ERCB, Minutes, 20 May 1948, including V.J. Moroney to I.N. McKinnon, 19 May 1948.

69. Flooding implies establishing a water-saturated area at the base of the well. The water would then be displaced into the well by natural pressure and its heavier weight relative to oil was intended to result in the "killing" of the well.

70. The *Financial Post* reported Moroney's appointment as supervisor of operations and incorrectly described him as "a temporary member of the Petroleum and Natural Gas Conservation Board." *Financial Post*, 27 May 1948.

71. This effort was complicated because normal deviation surveys run at Atlantic No. 3 made it impossible to determine the precise bottom-hole position.

72. *Calgary Albertan,* 7 September 1948.

73. *Financial Post,* 18 September 1948.

74. For an excellent discussion of the technical difficulties and the creative solutions arrived at on site, see Kerr, pp. 142-55.

75. ERCB, "Oil and Gas Well Blows, Uncontrolled Blows, and Blowouts: 1924-1983," List. From 1948 to the end of 1958, there were 22 blowouts, ranging from relatively minor situations that took from half a day to 2 days to bring under control to several that took a month or more to master, including Hudson Bay Union Liberal No. 7, which required 150 days. See Board history file, "Blowout Wells—Alberta Historical Listing."

76. Kerr, pp. 55-62.

77. *Edmonton Bulletin,* 13 May 1948.

78. Alberta, Order-in-Council, OC No. 860, 19 August 1947. See clause 24(a).

79. Alberta, Department of Mines and Minerals, *Annual Report* for the year ended 31 March 1948, p. 106.

80. Alberta, Order-in-Council, OC No. 308, 29 March 1948. See clause 24(b)(i). See also H.H. Somerville, *A History of Crown Rentals, Royalties and Mineral Taxation in Alberta to December 31, 1972,* ENR Report, No. 1, (Edmonton: Dept. of Energy and Natural Resources, 1977) pp. 5-28 and 5-29. The maximum royalty payable on any renewal was to be "that fixed by the Lieutenant Governor in Council or in force at the time of issue of such renewal."

81. The maximum area of a petroleum and natural gas lease was reduced from 16 to nine sections.

82. *Oil Bulletin,* 14 May 1948.

83. Ibid., 7 and 14 May 1948.

84. Ibid., 7 May 1948.

85. Ibid.

86. Frank Manyluk, note to author, June 1992.

87. *Edmonton Bulletin,* 21 May 1948.

88. *Oil Bulletin,* 21 May 1948.

89. Ibid.

90. Ibid. See also the *Oil and Gas Journal,* Tulsa, 27 May 1948. Along with Tanner's reassurance, the delegates heard Col. Ernest Thompson, chairman of the Texas Railroad Commission, argue the case in favour of free enterprise with a minimum of government interference.

91. In his book, *Atlantic No. 3,* Aubrey Kerr suggests that the Board was dilatory, that belated and effective action came only after the government, under pressure from Imperial Oil, signalled to the Conservation

Board that it was time to take more forceful action. Although it can be argued that the Board might have moved sooner to take direct control of Atlantic No. 3, it should also be noted that, had the Board proposed such action before the first week in May, the Alberta cabinet would almost certainly have turned it down. McKinnon probably understood this. If any aspect of the Board's role is to be questioned, it is the approval given on March 6 to allow Atlantic to continue drilling despite the unquestioned evidence of great risk.

92.   Believing that a state of insurrection existed, Texas Governor Ross S. Sterling declared martial law on 17 August 1931 and shut in the entire East Texas oilfield, but individual properties were not seized and the government did not take over the operation of oil wells. Louisiana was one of the few oil- and gas-producing states before the 1950s to have a comprehensive statute authorizing a government agency to take charge of the capping of a wild or uncontrolled gas well where the owner or operator had failed, refused or neglected to bring the well under control. Recognizing some need for regulation in this area, the Texas Legislature passed a bill in 1947, not to expand the powers of the Texas Railroad Commission but to permit the organization of corporations with special skills and equipment "for the purpose of fighting fires and blowouts in oil wells and gas wells and oil and gas wells." B.M. Murphy, *Conservation of Oil and Gas: A Legal History* (Chicago: American Bar Association, 1948), p. 455, and, n.a., *Conservation of Oil and Gas: A Legal History* (Chicago: American Bar Association, 1938), pp. 223-24.

93.   GA, M1722, f289, "A Challenge to the People of Alberta: Alberta's Oil." See also *Daily Oil Bulletin*, 13 August 1948, for industry response.

94.   *Calgary Herald*, 13 August 1948.

95.   Ibid.

96.   Ibid., 18 August 1948.

97.   Ibid., 13 August 1948.

98.   See, for example, *Calgary Herald*, 14 August 1948, "The CCF and Oil."

99.   Alberta, *A Report on Alberta Elections 1905-1982* (Edmonton: Chief Electoral Officer, 1983), pp. 66-70.

100. *Calgary Herald*, 18-19 August 1948. Ironically, many city votes in favour of private power appear to have been cast by citizens in Edmonton, Lethbridge and Medicine Hat who feared the possible expropriation of their profit-making municipal power systems and a consequent rise in property taxes.

101.   *Financial Post*, 11 September 1948.

102.   *Calgary Herald*, 11 September 1948.

103.   ERCB, Atlantic No. 3, general file, 9 October 1948, Porter, Allen and Millard to Ian McKinnon.

104.   Ibid.

105.   Ibid., "Memorandum For Petroleum and Natural Gas Conservation Board, Re: Atlantic No. 3 Well," 18 December 1948.

106.   Mercury-Leduc No. 1, LSD 8-33-50-25 W4M. ERCB, "Oil and Gas Well Blows, Uncontrolled Blows, and Blowouts." See history file, "Blowout Wells—Alberta Historical Listing."

107.   ERCB, Atlantic No. 3, general file, minutes of meeting 26 January 1949.

108.   The Conservation Board retained funds in the trust account beyond mid-1951, because the lease remained to be cleaned up to the Board's satisfaction.

109.   Well-spacing regulations would normally have permitted four wells. The combined impact of the restrictions, therefore, had the effect of

reducing, until the excess production had been offset, Atlantic's production to 1/3 of what it might normally have expected from the leasehold. The reduction from ½ to 1/3 had come at the request of Imperial Oil at the meeting on 26 January 1949.

110. Alberta, *Statutes, An Act to Determine All Claims Arising From the Atlantic No. 3 Oil Well disaster,* ch. 17, 29 March 1949.

111. Ibid., s. 5

112. Kerr, pp. 161-63 and 172. Kerr observes that, had Atlantic been held responsible for the estimated future oil recovery loss of 2.5 million barrels, at $2.90 per barrel the bill to Atlantic would have been nearly $7.7 million. A less harsh, and perhaps more reasonable, expectation would have been to charge Atlantic the cost of a water injection program to restore pressure lost to the reservoir because of the blowout. Atlantic's original settlement proposal offered the setting aside of a sum for this purpose. It seems that the Conservation Board studied this possibility, but in the end it elected not to proceed with such a plan.

113. The author wishes to thank Ronald J. McKinnon, professor of international economics at Stanford University, for drawing attention to this point.

114. *Calgary Herald,* 9 February 1948. See also 12 April 1948.

115. G.W. Govier, "Alberta Proration Reflects Sound Engineering and Economic Principles," *World Oil,* March 1952, p. 240.

116. ERCB, G.W. Govier, "Oil and Gas Conservation," a paper presented before the Canadian Institute of Mining and Metallurgy, Western Annual Meeting, Vancouver, November 6-8, 1950.

117. ERCB, Alberta Petroleum Association, Advisory Committee to Conservation Board, Minutes, 25 October 1948, p. 3.

118. ERCB, G.W. Govier, "Oil and Gas Conservation," p. 10.

119. Ibid. Reservoir and laboratory tests had shown that excessive rates of production not only resulted in channelling (water bypassing and cutting off pockets of oil or gas) but could also render the water drive virtually ineffective as a displacement process. Experience and theory held that, under ideal conditions, recoveries as high as 80% of the oil in the reservoir might be expected; however, with poor engineering practice, recovery could be as low as 20%.

120. ERCB, G.W. Govier, "Oil and Gas Conservation," p. 12.

121. Ibid., p. 12. Govier explained that, although restricting the rate of production seldom altered the basic displacement mechanism, it could in some circumstances prevent excessive channelling of gas, which could reduce overall recovery. Also, in some cases where the reservoir was characterized by high vertical permeability, permitting easy vertical movement of gas, the evolved solution gas might segregate to form a "derived gas cap." Whenever there was such a possibility, there was much to be gained by restricting the rate of production.

122. Ibid., p. 13.

123. Ibid., p. 14.

124. ERCB, interview, G. Govier, 8 June 1987, tape 2, pp. 3-5.

125. ERCB, Minutes, 11 December 1947 to 4 October 1949, see Board orders 2L through 29L. Some flexibility was allowed, the 100-barrel and 150-barrel figures were meant as average rates, and production during any 24-hour period from the D-2 was allowed to go as high as 125 barrels per day and in the D-3 up to 175 per day.

126. This adjustment came as Imperial Oil called attention to one of its wells in the north part of the field that had begun to produce some salt water. *Calgary Herald,* 12 April 1948, and Board order 4L, 19 April 1948.

127. ERCB, Minutes, 23 December 1948, see Board order 14L.

128. *Calgary Herald,* 12 April 1948. *Daily Oil Bulletin,* 9 February 1948.

129. *Calgary Herald,* 20 March 1950. The Leduc-Woodbend field produced 26,936 barrels from 374 wells.

130. *Western Oil Examiner,* 12 March 1949.

131. Speaking for Royalite, former Board engineer Gordon Connell submitted that this formula would curb the flow of oil from one property to another and permit each operator to recover whatever oil was directly beneath his property. *Western Oil Examiner,* 12 March 1949, and *Calgary Herald,* 9 March 1949.

132. *Calgary Herald,* 16 March 1949. *Daily Oil Bulletin,* 16 March 1949.

133. ERCB, Minutes, 23 April 1949. Maximum recovery permitted in any 24-hour period was set at 550 barrels.

134. *Daily Oil Bulletin,* 28 April 1949. *Calgary Herald,* 22 April 1949 and 15 May 1950.

135. Canada, *Parliamentary Debates,* Session 1949, pp. 2642-54.

136. Hardwicke had made a name for himself in the volatile world of Texas oil in the 1930s. Before coming to Alberta, he had just finished reviewing Texas oil and gas conservation law for the American Bar Association. See Robert E. Hardwicke, "Texas 1938-48," in B.M. Murphy, ed., *Conservation of Oil and Gas: A Legal History* (Chicago: American Bar Association, 1948).

137. Durham, p. 100, quoting from Tanner's diary.

138. The new departments came into existence on 1 April 1949. Before this date, the authority for making petroleum and natural gas regulations emanated from the *Provincial Lands Act.* This Act was replaced by the *Public Lands Act* and the *Mines and Minerals Act* (Alberta, *Statutes,* 1949, ch. 66), both of which came into force on 1 April 1949.

139. ERCB, Robert E. Hardwicke Correspondence, Robert E. Hardwicke to Kenneth A. McKenzie, Acting Legislative Counsel, 6 June 1949. Board history file entitled "Acts and Letters PNGCB about 1949."

140. Many hydrocarbon compounds, collectively known as "condensate," "naphtha" or "distillate," exist in a gaseous state, under the originally prevailing temperature and pressure conditions in the reservoir, but become liquids when the pressure is reduced, either by bringing them to the surface or by reducing the pressure in the reservoir.

141. ERCB, Robert E. Hardwicke Correspondence, Robert E. Hardwicke to Kenneth A. McKenzie, 6 June 1949. Board history file entitled "Acts and Letters PNGCB about 1949."

142. Ibid., Robert E. Hardwicke to G.W. Govier, 22 June 1949.

143. Ibid., "as phoned to McKenzie," 29 June 1949. This memo records approval of the definition by Ian McKinnon, D.P. Goodall, G.W. Govier, lawyer Marsh Porter, and industry representatives, Art Bessemer, Don Wilson and Walker Taylor from Imperial Oil. This definition has to be understood in conjunction with the agreed upon definition for natural gas: "gas means all natural gas both before and after it has been subjected to any treatment or process by absorption, purification, scrubbing or otherwise, and includes all other fluid hydrocarbons not defined as oil." Alberta, *Statutes,* 1949 (Second Session), ch. 5. s. 2(d).

144. Canada, *Parliamentary Debates,* Session 1949, p. 2515. See also *Calgary Herald,* 4 July 1949, regarding the July session of the Alberta Legislature.

145. Canada, *Parliamentary Debates,* pp. 2642-45.

146. Ibid., p. 2652.

147.   ERCB, Robert E. Hardwicke Correspondence, Robert E. Hardwicke to Kenneth A. McKenzie, 6 June 1949. See Board history file entitled "Acts and Letters PNGCB about 1949."

148.   Canadian Petroleum Association, Calgary (hereafter CPA), Western Canada Petroleum Association (hereafter WCPA), Board of Directors Minutes, 23 June 1949, p. 3. The WCPA was reorganized to become the CPA in December 1952.

149.   Ibid., 24 June 1949, pp. 2-3.

150.   Ibid., p. 5.

151.   Alberta, *Statutes*, 1949 (Second Session), ch. 5.

152.   Ibid., s. 3.

153.   Ibid.

154.   Ibid., s. 16c. By order, the Board could also relieve any common purchaser, after due notice and hearing, from the purchasing of oil or gas of inferior or different quality or grade.

155.   Alberta, *Statutes*, 1949 (Second Session), ch. 6, *An Act To Amend the Pipe Line Act.* See also *Revised Statutes of Alberta*, 1942, vol. III, ch. 315, *An Act Relating to the Construction of Pipe Lines.*

156.   Alberta, *First Annual Report of the Department of Mines and Minerals*, 1950, p. 11. See also Alberta, *Statutes*, 1944, ch. 4, *the Natural Gas Utilities Act.* Repealed, 1949 (Second Session), ch. 4, *An Act To Repeal the Natural Gas Utilities Act.* The Natural Gas Utilities Board consisted of two members, the chairman of the Board of Public Utility Commissioners (BPUC) and the chairman of the Petroleum and Natural Gas Conservation Board (PNGCB). With repeal, the duties of the Natural Gas Utilities Board were simply divided between the PNGCB and the BPUC. A Gas Utilities Board was re-established in 1962. Alberta, *Statutes*, 1962, ch. 28, *An Act To Amend the Gas Utilities Act.*

157.   Alberta, *Statutes*, 1949 (Second Session), ch. 2, *An Act To Provide for the Preservation, Conservation and Effective Utilization of the Gas Resources of the Province.*

158.   Ibid., ch. 5, s. 44a. Significant was subsection 11, which stated, "If the order or regulation of the Board is varied or set aside by a judgement of the Appellate Division or other Appellate tribunal, its operation shall *not* be suspended by any such judgement until the time for appeal from that judgement has expired, or if an appeal has been taken, until the appeal has been finally disposed of." Author's italics. The Manning government was not going to allow the courts to be used as a delay strategy.

159.   CPA, WCPA, Board of Directors Minutes, 3 January 1950.

160.   Ibid., 2 and 6 March 1950. Oilmen focused their attention on areas formerly covered by the *Oil and Gas Wells Act.* They expressed concern about the cost of drilling licences, the need to have a licence to operate a drilling rig, and the new idea that drilling employees should also be licensed. The industry also hoped that their portion of the cost of maintaining the Board might be reduced.

161.   Alberta, *Statutes*, 1950, ch. 46. See Part III, s. 30, Regulations established by the Board were subject to cabinet approval. Also, Alberta, *Revised Statutes*, 1942, vol. 1, ch. 67, the *Oil and Gas Wells Act.*

162.   Alberta, *Statutes*, 1950, ch. 46, s. 16, and Alberta, *Revised Statutes*, 1942, vol. 1, ch. 66, s. 16. Another change was the clarification and expansion of the range of records that operators were required to keep and submit to the Board. Notice that any unitization scheme required Board approval and a statement of the Board's mandate to promote such schemes also appeared in the 1950 Act.

163.   CPA, WCPA, Board of Directors Minutes, 28 April 1950.

164. Ibid., 2 June 1950.

165. *Calgary Herald,* 15 May 1950.

166. Ibid., 16 and 31 May 1950.

167. *The Oil Beacon,* 12 March 1948. See also *Oil Bulletin,* 7 May 1948, and *Edmonton Journal,* 4 October 1951. The author is indebted to Aubrey Kerr for information on "Sugar" Schultz.

168. ERCB, Minutes, 29 August 1950. See letter, Fred A. Schultz, Continental Oil Company of Canada Ltd., to Petroleum and Natural Gas Conservation Board, 22 August 1950.

169. Ibid. See also *Calgary Herald,* 25 August 1950.

170. CPA, WCPA, Minutes of an Extraordinary Special Meeting of the Entire Membership 18 September 1950. "The Chairman [J.G. Spratt] explained... The Conservation Board had advised they did not wish to interfere in the matter if industry could handle it, but if formal application were made to the Board to take over proration they would be prepared to do so."

171. Ibid. By this date, there is a notable change in the membership character of the WCPA, the predominant voice was that of the large American companies.

172. Ibid., p. 4.

173. Ibid. The WCPA's solicitor and executive vice-president, G.W. Auxier, KC, had given the association in writing his considered opinion that, although the Board clearly had jurisdiction to prorate within a pool, there was a serious question about its power to prorate among fields.

174. *Calgary Herald,* 19 September 1950.

175. ERCB, Board orders 32L, 2C and 2W. As the *Daily Oil Bulletin,* 2 October 1950, explained,

> The Gas-oil Ratio Penalty applies to wells in fields affected which have gas-oil ratios exceeding 1,000 cubic feet per barrel at zero pounds separator pressure or 800 cubic feet per barrel at 200 pounds separator pressure, for last previous producing month. Under the formula, clean oil allowable would be cut to about 70 per cent of normal if gas-oil ratio is 1,400 cubic feet per barrel, 50 per cent of normal if ratio is 2,000, 20 per cent of normal if ratio is 5,000, 10 per cent of normal if ratio is 10,000, and 5 per cent of normal if ratio is 20,000.

> The Water-oil Ratio Penalty applies to wells producing more than two barrels of water for each 100 barrels of oil. Under the formula, clean oil allowable would be cut to 95 per cent of normal if a well makes 10 barrels of water for each 100 of oil; 90 per cent if water yield is 20 barrels for each 100 of oil; 87 per cent if water yield is 30 barrels for 100 of oil; 83 per cent if water yield is 40 barrels for each 100 of oil; 80 per cent if water yield is 50 barrels for each 100 of oil; 67 per cent if water is 100 barrels for each 100 of oil; 50 per cent if two barrels of water are produced for each barrel of oil; 40 per cent if three barrels of water are produced per barrel of oil and so on.

176. ERCB, Hearing No. 22, 23 October 1950, "Submission of the Western Canada Petroleum Association with Respect to Proration of Oil Production Based on Market Demand." From the initial stage of a pool's development, maximum permissible rates (MPRs) for oil pools may be prescribed to prevent reservoir or surface waste and to protect the interests of owners and to promote orderly development. The maximum efficient rate (MER) of

a pool is the maximum rate at which the pool may be produced without incurring reservoir waste. Considerable production history and extensive knowledge of the characteristics of a pool are required before the pool's MER can be defined accurately.

177. Ibid., "Submission of Pacific Petroleums Ltd., Altantic Oil Company Ltd., Princess Petroleum Ltd., Allied Oil Producers Ltd., Calvan Petroleums Ltd., . . . [et al.]."

178. Ibid., "Submission of Royalite Oil Company, Ltd., with Respect to Proration of Oil Production in the Province of Alberta Based on Market Demand."

179. Ibid., "Submission of Imperial Oil Ltd. With Respect to Proration of Oil Production in the Province of Alberta Based on Market Demand."

180. Ibid., Imperial's italics.

181. *Calgary Herald,* 24 October 1950. The first Board Nomination hearing was held 22 November 1950.

182. *Daily Oil Bulletin,* 19 September, 4 October and 23 November 1950. At a public ceremony in Edmonton on 4 October 1950, Premier Manning opened the 439-mile Edmonton-Regina section of the line. Regina refiners were expected to require 20,000 barrels daily. The Interprovincial Pipeline started out as a plan approved by the Imperial Board of Directors in May 1948 to build a pipeline to Regina. It soon became apparent that there was more oil in the Edmonton area than originally anticipated, and the decision was made to extend the line to Superior, Wisconsin. The Interprovincial Pipeline Co. was incorporated by the Canadian Parliament in April 1949. By October 1949, Interprovincial was a public company, and within a few years Imperial's holding in the company dropped to 33% as British American (Gulf), Canadian Oil Companies Ltd. and Montreal Trust assumed significant equity ownership.

183. ERCB, Letters To All Operators, "The Petroleum and Natural Gas Conservation Board Plan For Proration to Market Demand," 1 December 1950. See also *Calgary Herald,* 29 November 1950, *Daily Oil Bulletin,* 29 November 1950.

184. The need for cabinet approval of MD Orders was eliminated later. For an excellent and more detailed discussion of Alberta's 1950 proration system, see "Alberta Proration to Market Demand," presented by Board secretary Vernon Millard to the Interstate Oil Compact Commission meeting in Banff in September 1952 and subsequently published in *Oil in Canada,* 13 October 1952, pp. 30-33.

SIX—THE NATURAL GAS EXPORT DEBATE

1. G.S. Hume and A. Ignatieff, "Natural Gas Reserves of the Prairie Provinces" (Ottawa: Department of Mines and Resources, April 1948), p. 4. The authors expressed their thanks for the data supplied by the Conservation Board, without which they could not have prepared their report, and their particular gratitude for the help given by senior Board engineer G.E.G. Liesemer.

2. Ibid., "Natural Gas Reserves of Prairie Provinces (Supplement to Main Report)," December 1948.

3. NAC, RG25, B3, 2125-1207 (Washington Embassy file), "Statement of Thomas A. Stone, Minister at the Canadian Embassy, Washington, D.C. before the Federal Power Commission," Washington, 17 May 1948.

4. Alberta, "Natural Gas Commission: Enquiry into Reserves and Consumption of Natural Gas in the Province of Alberta," Report, p. 96.

Evidence presented to the Commission by A.R. Crozier, fuel controller, province of Ontario.

5. Ibid., "The Secretary of State for External Affairs to The Canadian Ambassador to the United States," teletype message, 5 April 1948.

6. Ibid., "Statement of Thomas A. Stone, Minister at the Canadian Embassy, Washington, D.C. before the Federal Power Commission, Washington," 17 May 1948.

7. Ibid., United States of America, Federal Power Commission, "Petition to Intervene," H.H. Wrong, Canadian ambassador.

8. Ibid., Department of State, Washington to The Canadian Embassy, Washington, 10 August 1948.

9. Alberta, "Natural Gas Commission," Report, p. 3.

10. Dinning was president of Burns and Company Ltd. and closely involved with a number of other leading Calgary companies outside the petroleum sector. Marler was president of the Alberta Federation of Agriculture from 1947 to 1956. In 1948, he was the Canadian delegate to the International Federation of Agriculture convention in Paris, and in 1956, part of the Canadian contingent at the United Nations Wheat Conference in Geneva, Switzerland.

11. Alberta, "Natural Gas Commission," Report, pp. 37-45 and 117. Others providing estimates were Robert Pot, engineer, Imperial Oil Ltd.; W.C. Spooner, consulting geologist, Shreveport, Louisiana; Stanley J. Davies, consulting engineer, Calgary; Gordon Connell, Madison Natural Gas Company Ltd.; Dr. J.D. Weir, geologist, California Standard Company; Dr. H.H. Beach, geologist, McColl-Frontenac Oil Co. Ltd.; and Dr. A.D. Brokaw, Northwest Natural Gas Co.

12. Ibid., p. 56.

13. Ibid., p. 84. Net natural gas withdrawals (after deducting gas injected in Turner Valley and Bow Island) in 1948 totalled 57.55 Bcf. Of this, 9.7 Bcf was flared and the remaining difference between the total withdrawn and the amount marketed is accounted for by field use and shrinkage.

14. Ibid., p. 93. From this total, the British Columbia Electric Company expected to purchase 7.3 billion cubic feet per year for distribution in the Vancouver area.

15. Ibid.

16. The existing crown royalty was 15% of the wellhead value of the gas produced or a minimum of 3/4¢ per Mcf.

17. Ibid., Transcript, vol. 23, p. 2223. See also Thomas Brook, p. 2234, and K.M. Doze, Transcript, vol. 21, p. 1978.

18. Ibid., vol. 25, p. 2370.

19. Ibid., vol. 21, p. 1895.

20. Ibid., p. 1899.

21. Ibid., p. 1949.

22. Ibid., Exhibit No. 15, "Brief of Towns of Olds, Innisfail and Didsbury and Villages of Penhold, Bowden, Carstairs and Airdrie."

23. Ibid., Exhibit No. 18, "Brief of Town of Leduc."

24. Ibid.

25. Ibid., Exhibit No. 17, "Brief of the Executive Committee of the Union of Alberta Municipalities." Exhibit No. 16, "Brief of Alberta Branch, Canadian Manufacturers Association." See also Exhibit No. 24, "Brief of Edmonton Chamber of Commerce," pp. 4-5. Edmonton businessmen called attention to the acute power shortage in eastern Canada and asserted that, given the "ample evidence" that eastern manufacturers were al-

ready examining the possibilities of establishing branch plants in Alberta, a premature policy of natural gas export would be a great folly.

26. In 1931, Zinder left the Commonwealth Edison Company to become the chief rate analyst for the Public Service Commission of Wisconsin. He joined the Federal Power Commission in 1937, and became acting chief of its Division of Rates and Statistics until 1945 when he left to become an associate of E. Holley Poe and Associates, a consulting firm based in New York, Washington and Chicago and principally engaged in engineering rate and management work for natural gas companies.

27. Ibid., *Report*, p. 19, and Exhibit No. 31, "Natural Gas as a Factor in the Location of Industry," prepared for the Natural Gas Industry Committee by E. Holley Poe and Associates, New York, 1946, p. 12.

28. Ibid., Exhibit No. 30, "Economic Factors Which Control Gas Field Development: A Case Study of the Hugoton and Amarillo Fields in the Mid-Continent Region," prepared by Petroleum Statistics Bureau, Inc., p. 33.

29. Ibid., *Transcript*, vol. 23, p. 2171.

30. Ibid., p. 2180.

31. Ibid., *Transcript*, vol. 19, p. 1717, and Exhibit No. 67, "Submission of the Research Council of Alberta."

32. Ibid., *Report*, p. 80.

33. Ibid., *Transcript*, vol. 24, pp. 2273-2312.

34. Ibid., *Transcript*, vol. 25, pp. 2393-2408.

35. Ibid., Exhibit No. 98, "Letter, Minister of Mines and Natural Resources, Province of Manitoba," and Exhibit No. 93, "Brief of the Saskatchewan Power Commission."

36. Ibid., *Report*, p. 87. Alberta's actual consumption in 1960 was just over 141 billion cubic feet.

37. Ibid., pp. 121-22.

38. Ibid., p. 122.

39. Ibid., p. 123.

40. Ibid., The Commission studied revenues that would be obtained by the pipeline companies at the prices that they indicated that they could expect to get in markets outside the province. Since, in the Commission's view, most gas was a by-product of the search for and development of oil, and insofar as the costs of petroleum development might be covered in the return from oil sales, it was concluded that the cost-revenue equation was favourable and that it would be possible for the companies "to contribute very substantially to government revenues."

41. Ibid.

42. Ibid., p. 126.

43. Ibid.

44. Ibid., p. 127.

45. Dinning was concerned that the foundation study by Hume and Ignatieff on natural gas reserves be constantly updated. See also Alberta, *The Report of a Royal Commission Appointed by the Government of the Province of Alberta Under The Public Inquiries Act to Inquire Into Matters Connected With Petroleum and Petroleum Products* (Edmonton, 1940), pp. 250-52.

46. H. Ray Milner came to Edmonton to practise law in 1912 shortly after graduating from Dalhousie University. In addition to developing a successful law practice, he also established himself as one of the city's more successful businessmen. As president of the Canadian Western Natural Gas, Light, Heat and Power Company Ltd., Northwestern Utilities Ltd.

and Anglo-Canadian Oil Co., and as director of the Calgary and Edmonton Corporation and Home Oil Company, Milner's interests by the late 1940s were focused in the oil and gas sector. From 1929 to 1939, he was president of the Alberta Conservation Association, and in 1942 he was honoured as Edmonton's "Citizen of the Year."

47. Alberta, "Natural Gas Commission," Exhibit No. 42, "Canadian Western Natural Gas Company Ltd., and Northwestern Utilities, Ltd., Submission to the Natural Gas Commission."

48. Ibid., pp. 7-8.

49. Ibid., *Report*, pp. 123-24.

50. Ibid.

51. Ibid., Exhibit No. 42 argues the advantages of having Canadian Western Natural Gas and Northwestern Utilities build and operate an Alberta trunk line system.

52. Canada, *Parliamentary Debates*, First Session, 1949, pp. 2666-67. Bill 238 to incorporate Interprovincial Pipe Line Co.; Bill 239 to incorporate Alberta Natural Gas Co.; Bill 240 to incorporate Westcoast Transmission Co. Ltd.; Bill 241 to incorporate Trans-Northern Pipeline Co.; Bill 242 to incorporate British American Pipeline Co.; and Bill 243 to incorporate Western Pipelines.

53. Ibid., p. 2776.

54. Canada, *Revised Statutes of Canada*, 1952, ch. 211, "An Act respecting Oil or Gas Pipelines."

55. Canada, *Parliamentary Debates*, Second Session, 1949, pp. 1245-52 and 1805-13.

56. Alberta, *Statutes*, 1949 (Second Session), ch. 2, "An Act to provide for the Preservation, Conservation and Effective Utilization of the Gas Resources of the Province," s. 7(1).

57. Ibid., s. 9.

58. Alberta, *Statutes*, 1949 (Second Session), ch. 6, "An Act to amend *The Pipe Line Act*." The Board, rather than the minister's staff, would evaluate pipeline proposals.

59. Ibid., s. 7.

60. Alberta, *Statutes*, 1949 (Second Session), ch. 8, "An Act to amend the Public Utilities Act," s. 70a.

61. Ibid., s. 70e.

62. *Calgary Herald*, 5 July 1949.

63. *Edmonton Bulletin*, 5 July 1949.

64. *Edmonton Journal*, 6 July 1949.

65. *Calgary Herald*, 6 July 1949.

66. *Edmonton Journal*, 6 July 1949.

67. *Edmonton Bulletin*, 7 July 1949.

68. *Calgary Herald*, 7 July 1949, and *Edmonton Journal*, 7 July 1949.

69. *Calgary Herald*, 7 July 1949.

70. *Edmonton Bulletin*, 7 July 1949.

71. *Calgary Herald*, 7 July 1949.

72. The Opposition was able to persuade the government to make one modest concession. In response to Prowse's and MacDonald's persistence, the *Gas Resources Preservation Act* was amended to make it compulsory for members of the Legislature to be notified when companies seeking to export gas were accorded a hearing by the Conservation Board.

73. PAA, Premiers' Papers, Reel 153, file 1627A, "Press Statement by Hon. E.C. Manning, July 19, 1949." See also *Calgary Herald*, 19 July 1949.

74. Ibid., 20 July 1949.

75.   ERCB, D.P. Goodall, *Gas Export, 1950-1960* (Calgary, Oil and Gas Conservation Board, 1961), pp. 8-9.

76.   Canada, *Parliamentary Debates,* Second Session, 1949, p. 1246.

77.   Ibid.

78.   Ibid., p. 1247.

79.   Ibid., p. 1250.

80.   Ibid., p. 1248.

81.   Ibid., p. 1249.

82.   Ibid., p. 1808.

83.   Ibid., pp. 1251-52 and 1469, quoting Premier Manning's reply to Senator Turgeon's telegram of 17 October 1949. Thomas represented the constituency of Wetaskiwin.

84.   Ibid., pp. 1679-1680. Thomas blamed the federal government for the waste that had occurred in Turner Valley and he claimed that the Alberta government had "immediately put through conservation measures" when the province gained control of natural resources in 1930, p. 1469.

85.   Ibid., First Session, 1950, pp. 328–30.

86.   Ibid., p. 2486.

87.   Ibid., p. 2487. Howe suggested that the Westcoast Transmission Company was a blind for Sunray Oil Corporation of Tulsa, Oklahoma.

88.   Ibid., pp. 2513-16. After passage of the Alberta Natural Gas Bill and a brief debate, the House of Commons also passed a bill to incorporate Prairie Transmission Lines Ltd. A parallel filibuster also marked this bill.

89.   Edward W. Constant II, "Cause or Consequence: Science, Technology, and Regulatory Change in the Oil Business in Texas, 1930-1975," *Technology and Culture,* vol. 30 (1989), p. 440.

90.   Ibid., p. 443.

91.   For a more extensive discussion of this topic, see Richard W. Hooley, *Financing the Natural Gas Industry: The Role of Life Insurance Investment Policies* (New York: Columbia University Press, 1961).

92.   Sunray, followed by McMahon, were the predominant shareholders of Pacific Petroleums. For a more detailed discussion of Westcoast's organization, see Earle Gray, *Wildcatters: The Story of Pacific Petroleums and Westcoast Transmission* (Toronto: McClelland and Stewart, 1982).

93.   *Oil in Canada,* 5 December 1949.

94.   *Calgary Herald,* 30 January 1950. In addition to Tanner, the MLAS present were Social Creditors F.C. Colborne, Calgary, C.R. Wood, Stony Plain and W. Tomyn, Willingdon; Liberals, J. Harper Prowse, Edmonton, and Hugh John MacDonald, Calgary; and the Independent Social Credit member Arthur Wray, Banff-Cochrane.

95.   ERCB, transcript of Westcoast Application, Hearing No. 5, vol. 1, pp. 12-16.

96.   All submissions had to be read by the persons presenting them, unless otherwise permitted by the Board, and entered as exhibits.

97.   These were the main expert witnesses called by Westcoast; a number of other secondary expert witnesses were also called.

98.   Dodge had been brought to Alberta by Shell Oil the previous year to comment on the Jumping Pound gas field before the Dinning Commission.

99.   ERCB, transcript of Westcoast Application, Hearing No. 5, 1 February 1950, vol. 5, pp. 213-14.

100.   Ibid., vol. 5, p. 203.

101.   Ibid., p. 204.

102.   *Calgary Herald,* 1 February 1950. See also *Calgary Herald,* 8, 13, 15, 16 and 17 February 1950.

103. ERCB, transcript of Westcoast Application, Hearing No. 5, 18 December 1949, vol. 1, pp. 14-15; 15 February 1950, vol. 15, pp. 1089-90.

104. Ibid., 17 February 1950, vol. 17, pp. 1262-71.

105. *Calgary Herald,* 3 March 1950, and *Edmonton Journal,* 3 March 1950.

106. *Calgary Herald,* 22 March 1950.

107. *Calgary Albertan,* 28 March 1950, and *Edmonton Bulletin,* 28 March 1950.

108. *Calgary Herald,* 12 April 1950.

109. ERCB, transcript of Westcoast Application, Hearing No. 5, 13 April 1950, vol. 20, pp. 1516-23. Fenerty compelled Dr. Charles Hetherington to admit that Westcoast's claim that Alberta had a 50-year reserve supply did not mean *deliverable* supply.

110. Ibid., vol. 20, pp. 1439-40.

111. The applicants and dates of application were Northwest Natural Gas Company, 19 August 1949; Western Pipe Lines, 10 February 1950; Gordon M. Plotke, 27 March 1950; and Prairie Pipe Lines Limited, 11 April 1950.

112. A permit to export natural gas was required under the provisions of the *Electricity and Fluid Exportation Act* administered by the Department of Trade and Commerce. For a history of the Plotke wells, see ERCB, Vanalta 1931-50.

113. Having participated as an intervening party in the first Westcoast hearing, Northwest was alerted to Calgary's great apprehension and the important role that the city would continue to play at the natural gas export hearings. A. Faison Dixon, New York president of the Northwest Natural Gas Company and its associates, Alberta Natural Gas Company and Alberta Natural Gas Grid Ltd., assured Calgary in his opening remarks that his grid system would make large additional reserves available to the city. He emphasized that market for gas "at fair prices" would encourage exploratory drilling and noted that negotiations were under way with Dominion Bridge Company to supply large diameter piping, which might mean the construction of a pipe mill in Canada, "perhaps in Calgary." *Calgary Herald,* 29 May 1950.

114. *Calgary Herald,* 20 June 1950.

115. For a full discussion, see ERCB, transcript of Westcoast Application, Hearing No. 1A, 20-21 June 1950.

116. *Daily Oil Bulletin,* 22 June 1950.

117. Ibid., Nickle held that the Conservation Board need not concern itself with pipeline routes, construction costs and other economic considerations. These matters should be left to be considered by the federal Board of Transport Commissioners. The only legitimate question for the Board was whether there was a surplus of natural gas available for export.

118. Western Pipe Lines was sponsored by a Winnipeg group, headed by the brokerage firm Osler, Hammond and Nanton, and incorporated by Parliament on 30 April 1949.

119. ERCB, the Petroleum and Natural Gas Conservation Board, "Interim Report with respect to applications now before the Board for permission to remove gas or cause it to be removed from the Province under the provisions of the Gas Resources Preservation Act," January 1951, Appendix 2, 29 August 1950, Milner, Steer, Dyde, Poirier, Martland and Layton to Chairman, PNGCB (hereafter PNGCB, Interim Report, January 1951).

120. Ibid.

121. Ibid., Appendix 3, 20 September 1950, G.W. Auxier, Executive Vice-President WCPA, to Chairman, PNGCB.

122. Ibid., Appendix 5, 16 September 1950, C.D. Howe, Minister of Trade and Commerce, to N.E. Tanner, Minister of Mines and Minerals.

123. Ibid.

124. Ibid., Appendix 4, 23 September 1950. Tanner also told McKinnon that he had reminded Howe of Alberta's policy not to give favourable consideration to natural gas export applications until the government was assured that there was sufficient gas to meet the province's present and future domestic and industrial requirements. Tanner did not inform McKinnon, however, that he had also told Howe that the Alberta government believed the province's present proven reserves were already sufficient.

125. PAA, Premiers' Papers, Reel 153, 1627B, 8 August 1950, "Memorandum of Alberta Natural Gas Grid Ltd., Alberta Natural Gas Company, Northwest Natural Gas Company, for Hon. E.C. Manning, Premier of Alberta, and Hon. N.E. Tanner, Minister of Mines and Minerals.

126. NAC, C.D. Howe Papers, MG27, 111, B20, vol. 31, F8-2-1, 28 September 1950, T.H. Jenkins to C.D. Howe.

127. Ibid.

128. Ibid., 22 November 1950, T.H. Jenkins to C.D. Howe.

129. Ibid., 23 September 1950, N.E. Tanner to C.D. Howe. Author's italics.

130. Ibid., 28 September 1950, C.D. Howe to N.E. Tanner.

131. Ibid., 22 November 1950, T.H. Jenkins to C.D. Howe.

132. Ibid., 25 November 1950, C.D. Howe to T.H. Jenkins.

133. PAA, Premiers' Papers, Reel 155, 1637, 7 December 1950, C.D. Howe to E.C. Manning containing teletype 1, December 1950, Canadian Ambassador to the United States to The Secretary of State for External Affairs.

134. Ibid., 13 January 1951, E.C. Manning to C.D. Howe.

135. Those appearing or submitting briefs were Westcoast Transmission Company Ltd.; Northwest Natural Gas Company and Alberta Natural Gas Grid Ltd.; Western Pipelines; Prairie Pipelines Ltd. (Pacific Northwest Pipeline Corporation); McColl-Frontenac Oil Company Ltd. and Union Oil Company of California; Alberta Research Council; Alberta Power Commission; Canadian Western Natural Gas Co. and Northwestern Utilities Ltd.; Imperial Oil Ltd.; Canadian Gulf Oil Co.; Alberta Inter-Field Gas Lines Ltd.; and Seaboard Oil Co. of Delaware.

136. PNGCB, Interim Report, 20 January 1951, p. 15.

137. Ibid., p. 59. A further 219 billion cubic feet was estimated to exist beyond economic reach.

138. Ibid., p. 59.

139. Ibid., pp. 59-60. The recommendation to suspend hearings was qualified with the provision that the Board would reopen the hearings sooner if compelling evidence warranted.

140. Ibid.

141. Calgary Herald, 25 January 1951.

142. The Board's report also advised that Alberta's existing oil and gas regulations were more appropriate to the development of oil lands and recommended that the government consider offering "gas only" rights in certain areas so that larger blocks of reserve and lease lands more suitable for natural gas exploration could be made available.

143. Calgary Herald, 25 January 1951.

144. Ibid., 27 January 1951.

145. Daily Oil Bulletin, 26 January 1951.

146. NAC, C.D. Howe Papers, MG27, 111, B20, vol. 31, F8-2-1, 26 January 1951, John J. Connolly to C.D. Howe.

147. Ibid., 21 February 1951, C.D. Howe to Premier Manning, with enclosure 16 February 1951, Charles E. Wilson to C.D. Howe.

148. Ibid.

149. Ibid.

150. PAA, Premiers' Papers, Reel 155, 1637, 1 March 1951, E.C. Manning to C.D. Howe.

151. Alberta, *Statutes,* 1951, ch. 36. Assented to 7 April 1951. The Act was necessary to circumvent the provisions of the *Gas Resources Preservation Act,* particularly the requirement that an export permit could only be granted after a hearing by the Conservation Board.

152. *Edmonton Journal,* 27 February 1951.

153. Ibid., 10 March 1951.

154. Ibid., 31 March and 4 April 1951.

155. *Calgary Herald,* 1 and 2 March 1951.

156. Ibid., 6 March 1951; see also 3 March 1951.

157. *Edmonton Journal,* 14 March 1951.

158. *Calgary Herald,* 22 March 1951.

159. ERCB, D.P. Goodall, pp. 81-89. The volume of natural gas authorized in the proposed permit (10 billion cubic feet per year at a maximum rate of 40 million cubic feet per day) met the production schedule submitted by McColl-Frontenac and Union Oil in their original August 1950 application. See transcript, Hearing No. 4, p. 285.

160. *Calgary Herald,* 24 March 1951. NAC, C.D. Howe Papers, MG27, 111, B20, vol. 31, F8-2-1, 21 February 1951, C.D. Howe to Premier Manning.

161. *Edmonton Journal,* 3 April 1951.

162. Ibid.

163. *Calgary Herald,* 3 April 1951.

164. *Edmonton Journal,* 3 April 1951.

165. See, for example, *Calgary Herald* and *Edmonton Journal,* 3 April 1951.

166. *Edmonton Journal,* 6 April 1951.

167. PAA, Premiers' Papers, Reel 155, 1637, 25 May 1951, H.R. Milner to E.C. Manning.

168. Ibid.

169. Ibid., 1643, 16 February 1951, C.D. Howe to N.E. Tanner; see also 13 February 1951, N.E. Tanner to C.D. Howe. The duty to be imposed upon Alberta oil was 11 to 22¢ per barrel.

170. University of Alberta, Archives (hereafter UAA), E.C. Manning Papers, interview transcript, 3 July 1981, pp. 36-38. With reference to the *Act to Permit the Temporary Export of Gas to Montana for Essential Defence Production,* Manning recalled that his government "had the Bill passed in three days and the thing was law" only to find that U.S. regulatory authorities took months to give their approval. He sees this as a baneful and "classic example of government bureaucracy," not as further evidence that the emergency was hardly as great as his government alleged.

171. Directors included Cortelyou Ladd Simonson, New York, investment banker; John J. Connolly, Ottawa lawyer and well-known Liberal; William J. Dick, Edmonton consulting engineer; Harvey R. MacMillan, Vancouver industrialist; John W. Moyer, Calgary oilman; and Austin C. Taylor, Vancouver mining executive.

172. Fish's Prairie Pipelines was originally intended to bring Texas gas to Canada, whereas its corporate twin, Pacific Northwest Pipelines, carried Alberta gas to Washington and Oregon.

173. NAC, C.D. Howe Papers, MG27, 111, B20, vol. 31, F8-2-1, 15 December 1951, T.H. Jenkins to C.D. Howe. The Fish Engineering Corporation placed its Texas Illinois Pipeline to Chicago in service on 5 December 1951. Jenkins also informed Howe that by trading western Canadian gas for Texas gas in Ontario, Canada would in effect be substituting Alberta gas for U.S. coal, and thereby reducing the country's U.S. dollar requirements. See also, Jenkins to Howe, 25 June, 3 August, 20 November and 26 December 1951.

174. R. Bothwell and W. Kilbourn, *C.D. Howe: A Biography* (Toronto: McClelland and Stewart Ltd., 1979), pp. 287-88.

175. PAA, Premiers' Papers, Reel 155, 1640, 20 September 1951, H.R. Milner to E.C. Manning, quoting Arthur Logan, General Counsel of Northwest Natural Gas Company, to Paul H. Graves of Graves, Kizer [sic] and Graves, Spokane, Washington, 31 March 1950.

176. Ibid., 12 September 1951, Lionel Baxter to E.C. Manning.

177. *Daily Mail*, London, 18 July 1951.

178. *Time*, 24 September 1951, pp. 42-48.

179. Ibid., p. 42.

180. Ibid., p. 48.

181. PAA, Premiers' Papers, Reel 155, 1638, 22 October 1951, "Radio Address—N.E. Tanner."

182. Ibid.

183. Alberta, *Statutes*, 1949 (Second Session), ch. 2, "The Gas Resources Preservation Act," sec. 9.

184. ERCB, transcript, Hearing No. 6. "Validity of Section Nine of the Gas Resources Preservation Act. ch. 2. Statutes of Alberta, 1949 (Second Session)" Exhibit 124. See the transcript of Hearing No. 6 for a full discussion of this matter; for a summary, see D.P. Goodall, *Gas Export, 1950-1960* (Calgary: Oil and Gas Conservation Board, 1961), pp. 173-76.

185. *Edmonton Bulletin* and *Calgary Herald*, 7 July 1949.

186. NAC, C.D. Howe Papers, MG27, 111, B20, vol. 31, F22. Charles E. Wilson to C.D. Howe, 4 March 1952. See also the numerous letters in this file from others regarding natural gas export.

187. PAA, Premiers' Papers, Reel 155, 1640, 1 and 2 April 1952, L.E. Drummond, Alberta and North West Chamber of Mines and Resources, to N.E. Tanner. The resolution was moved by W.A. MacDonald, publisher of the *Edmonton Journal* and seconded by H.O. Patriquin, an Edmonton chartered accountant.

188. Ibid., Federal Power Commission Docket G-580.

189. Ibid., p. 4.

190. Ibid., p. 5.

191. ERCB, PNGCB, *Report*, 29 March 1952. For example, DeGolyer and McNaughton, consultants for Canadian Delhi Oil Ltd. and Trans-Canada Pipe Lines, estimated the province's current disposable reserve to be 8.6 trillion cubic feet; the estimate of Westcoast Transmission was 7.8 trillion cubic feet, p. 11. The Board acknowledged that wide differences of opinion existed as to the reserves of individual fields, "since estimates of partially developed reserves are contingent upon a number of judgement factors (for acreage, thickness, etc.) often differently assessed by competent persons." As a matter of unstated policy, the Board prudently chose to err on the side of underestimation rather than overestimation.

192. Ibid., p. 31. On behalf of Canadian Delhi and Trans-Canada Pipe Lines, consultant H. Harries submitted a comprehensive study of Alberta's natural gas needs showing a 30-year need of 3.358 trillion cubic feet. The

other applicants "more or less intimated" that they accepted the Board figure of 3.1 trillion cubic feet given in the Interim Report.

193. Ibid., pp. 8-10.

194. Ibid., p. 10. Although the Board estimated disposable Peace River
gas reserves to be 524 billion cubic feet, Westcoast, on the basis of several
recent discoveries, but single-well assessments, was confident that it had a
disposable gas reserve of 1,192 billion cubic feet. The Board advised it
would specify annual rates applicable beyond the fifth year later, allowing
time for Westcoast to better define the gas reserves in the supply area.
Communities that the Board expected to gain natural gas service were
Spirit River, Rycroft, Falher, Woking, Webster, Sexsmith, Clairmont and
Grande Prairie.

195. Ibid., p. 8.

196. *Edmonton Journal,* 19 March 1952.

197. Ibid.

198. PAA, Premiers' Papers, Reel 155, 1638, 31 March 1952. Lucien
Maynard to C.D. Howe and, 2 April 1952, C.D. Howe to Lucien
Maynard. Alberta's concern centred on the Canada Pipelines Act, which
purported to give the federal Board of Transport Commissioners jurisdiction over any pipeline gathering gas for transmission beyond Alberta's borders.

199. *Edmonton Journal,* 2 April 1952; *Calgary Herald,* 2 April 1952.

200. PAA, Premiers' Papers, Reel 155, 1638, telegram, 4 April 1952, S.A.
Swensrud, President, Gulf Oil Corporation, Pittsburgh, Pennsylvania, to
The Hon. E.C. Manning.

201. Ibid., telegram, 7 April, Swensrud to Manning, and telegram, 5
April 1952, Manning to Swensrud.

202. Ibid., 1640, 7 April 1952, P.H. Bohart, Vice President, Canadian
Gulf Oil Co., Tulsa, Oklahoma, to N.E. Tanner, with attached press
release.

203. Ibid., See also *Daily Oil Bulletin,* 7 April 1952, and *Calgary
Herald,* 7 April 1952.

204. *Calgary Herald,* 5 April 1952.

205. Ibid., 4 April 1952.

206. Ibid., 7 April 1952.

207. *Daily Oil Bulletin,* 12 February 1952. The *Bulletin* predicted that
development of the Pincher Creek field would "bring an industrial revolu-
tion to southern Alberta."

208. PAA, Premiers' Papers, Reel 155, 1638, 3 April 1952, C.J. Bundy,
President, Pincher Creek and District Chamber of Commerce, to E.C.
Manning and, 4 April 1952, telegram, Cardston and District Chamber of
Commerce to E.C. Manning.

209. Manning had promised some "off-the-cuff" remarks (*Calgary
Herald,* 8 April 1952), yet his papers reveal a degree of preparation not
usually apparent in his presentations in the Legislature.

210. *Calgary Herald,* 9 April 1952.

211. Ibid., 10 April 1952.

212. PAA, Premiers' Papers, Reel 155, 1638, 9 April 1952, Henry Young,
President UFA, to E.C. Manning, and 1640, 10 May 1952, Henry Young to
Members of the Legislative Assembly.

213. Ibid.

214. Ibid., 1640, 17 April 1952, A. Faison Dixon to N.E. Tanner.

215. Ibid., 1638, 28 April 1952, "Radio Address—N.E. Tanner."

216. Ibid.

217. Ibid.
218. Ibid.
219. Ibid., 1640, 5 May 1952, "Provincial Affairs," J. Harper Prowse.
220. Ibid.
221. Ibid.
222. Ibid.
223. Ibid., 1638, 10 May 1952, E.C. Manning to Frank McMahon, and 22 May 1952, Frank McMahon to E.C. Manning. McMahon anticipated that there might be some delay before the FPC, and he expressed his hope that there would be some flexibility regarding the November 1 deadline.
224. ERCB, Goodall, pp. 24-25. The specified area of withdrawal was bounded on the west by the B.C.-Alberta border, on the south by the north boundary of Township 68, on the east by Range 15, West of the 5th Meridian, and on the north by the southern boundary of Township 90.
225. UAA, E.C. Manning Papers, Interview No. 25, 14 July 1981. According to Tanner's biographer, Manning told Tanner that he too was beginning to feel the burden of many years in office and had been thinking of leaving the political arena. He asked Tanner if he would consider assuming the premier's mantle rather than leaving. Durham, pp. 105 and 140. Manning makes no mention of this.
226. Social Credit's approach to rural electrification underlines the party's adherence to free enterprise philosophy. Unlike the CCF, who saw a government-owned power system as the vehicle to bring power to Alberta farms, Social Credit proposed low-interest loans to "rural electrification associations" for the construction of co-operatively owned transmission facilities.
227. *Calgary Herald*, 5 August 1952.
228. See, for example, ibid., 4 August 1952.
229. Caldarola, p. 373. See Chapter 4, Note 105.
230. *Calgary Herald*, 6 August 1952.
231. *Daily Oil Bulletin*, 6 August 1952.
232. *Time*, 24 September 1951, p. 47.
233. Charles Merrill had met Tanner in Barbados in the spring of 1950 when the Alberta politician was helping the colonial administration, the British Oil and Petroleum Company, Trinidad Leaseholds Ltd. and Gulf Oil draft a mutually acceptable petroleum conservation policy. Durham, p. 141.
234. For an example of Tanner's continentalist view, see: "'Open Door' Alta. Oil Policy Guided by Hon. N.E. Tanner," *Oil in Canada*, 7 February 1949, and *Calgary Herald*, 14 September 1954.
235. PAA, Premiers' Papers, Reel 156, f1654, Frank McMahon to N.E. Tanner, 25 August 1952.
236. The definition of what might constitute an emergency was left to the Conservation Board.
237. PAA, Premiers' Papers, Reel 156, f1654, Frank McMahon to N.E. Tanner, 25 August 1952.
238. Ibid., N.E. Tanner to Frank McMahon, 5 September 1952.
239. Ibid., C.D. Howe to E.C. Manning, 12 September 1952.
240. Earle Gray, *Wildcatters*, p. 154.
241. Quoted in Gray, *Wildcatters*, p. 157.
242. Ibid.
243. *Calgary Herald*, 18 August 1953, see also 7 August 1953.
244. Ibid., 15 May 1954. U.S. coal interests also lobbied intensively against natural gas imports into the Pacific Northwest.

245. *Calgary Herald*, 3 June 1954.

246. Quoted in NAC, RG21, vol. 38, 161.362/2, Pipelines General File, 17 June 1954, John Davies, Economics Branch, Trade and Commerce, to Marc Boyer, Deputy Minister of Mines and Technical Surveys, enclosure, "Notes on Final Oral Hearings of The U.S. Federal Power Commission Concerning Natural Gas For the Pacific Northwest."

247. Ibid., 18 June 1954.

248. Quoted in Gray, *Wildcatters*, pp. 168-69.

249. *Calgary Herald*, 10 June 1954.

250. Ibid., 20 and 23 July 1954.

251. Pacific Northwest also acquired a 25% interest in Westcoast. For a more detailed discussion of the relations between the two groups, see Gray, *Wildcatters*, pp. 172-81.

252. Alberta, *Statutes*, 1955 (Second Session), ch. 1. Author's italics. Among those campaigning for the deletion of section 9 was the Canadian Petroleum Association, which had been prompted to act by Clint Murchison's Delhi Oil. See Canadian Petroleum Association, Minutes, Board of Directors Meeting, 12 January 1954.

253. NAC, RG21, vol. 38, 161.362/2, 7 June 1954, "Notes on Final Oral Hearings of the U.S. Federal Power Commission Concerning Natural Gas For the Pacific Northwest." John Davis noted that Westcoast's lawyers tried to demonstrate to the FPC that it could have indirect but still effective jurisdiction, and that this drew considerable interest and comment from the commissioners. "They explored [the idea] thoroughly...." p. 7.

254. PAA, Premiers' Papers, Reel 187, 2027A, 31 March 1931, Lucien Maynard, Attorney General, to C.D. Howe, and 2 April 1952, C.D. Howe to Lucien Maynard. See also 29 December 1956, J.J. Frawley to E.C. Manning.

255. Also being considered at this hearing was the application of the Canadian-Montana Pipeline Company to remove additional gas from the Pakowki Lake area just north of the Alberta-Montana boundary.

256. PAA, Premiers' Papers, Reel 154, 1634, 29 January 1953, Alan Williamson to E.C. Manning. The most influential of the independent studies was that commissioned by Imperial Oil and jointly prepared by the Stanford Research Institute and the University of Western Ontario (School of Business Administration). Imperial sent a copy of the report to Manning on April 6. The covering letter explained that, as a producer of gas, the opening up of new markets was of great interest to the company, although it had no financial interest in any of the current projects. Hence, Imperial sought independent "objective data" with which to judge the relative merits of the various proposals. Imperial wanted Manning to note that "the report indicates important differences in the price that can be paid for natural gas at main gathering points when following different routes to the main market." G.L. Stewart, Imperial Oil, to E.C. Manning, 6 April 1953. The report showed that the Western Pipe Lines proposal would yield a better price for field producers.

257. Ibid., 11 February 1953, E.C. Manning to A.H. Williamson.

258. Ibid., quoting 2 January 1953, C.D. Howe to E.C. Manning. This letter reaffirmed what Howe had said to Manning in the presence of Premier Frost of Ontario the previous October. The Canadian government would not permit gas exports to the United States, other than to the Pacific Northwest, unless the needs of eastern Canada were met first.

259. Canada, *Commons Debates*, 13 March 1953, p. 2929.

260. ERCB, transcript, Hearing No. 13, 11 May–21 September 1953, vol. 27, p. 2049.

261. Ibid., see, for example, examination of D.G. Hawthorn, vol. 16, pp. 1151-53.

262. ERCB, The Petroleum and Natural Gas Conservation Board, "Report to the Lieutenant Governor in Council With Respect to the Applications under the Gas Resources Preservation Act of: Canadian-Montana Pipeline Co.; Trans-Canada Pipe Lines Ltd., Trans-Canada Grid of Alberta Ltd., and Canadian Delhi Oil Ltd.; and Western Pipe Lines," 24 November 1953, pp. 8–9. The pattern of reserves growth, together with evidence presented at the hearings, convinced the Board that the province could "safely anticipate" the development of further established reserves at an average rate of at least 1¼ to 1½ trillion cubic feet per year for the next eight to 10 years. The report explained that about 35% of the increase in reserves was due to the expansion of previously known reserves (of which 73% came from development of the Pincher Creek, Medicine Hat and Cessford fields) and about 65% from new discoveries.

263. Ibid., pp. 13-14. The Board did approve the application of the Canadian-Montana Pipeline Company to remove all the additional gas that could be produced in the Pakowki Lake area (estimated to equal 334 billion cubic feet), just north of the Alberta-Montana border.

264. Ibid., p. 14.

265. Ibid., p. 11.

266. Ibid., pp. 11 and 99-106.

267. For a discussion of the Manning-St. Laurent October talks, see *Calgary Herald*, 27, 28 and 30 October 1953.

268. *Calgary Herald*, 30 July 1953.

269. This was also the position of Ontario as well as of the federal Conservative and CCF parties.

270. This was an implicit conclusion of the Stanford study prepared for Imperial Oil and of the report released in November by one of the United States' largest pipeline construction contractors, Bechtel Corporation. (For discussion, see *Calgary Herald*, 16 November 1953.) In the *Daily Oil Bulletin* and in the Canadian Parliament, Carl Nickle had argued vigorously in favour of "common sense" economic rather than political factors deciding the route of a pipeline to eastern Canada. See, for example, *Daily Oil Bulletin*, 30 October 1953.

271. PAA, Premiers' Papers, Reel 154, 1634, "Press Statement—December 3, 1953." *Calgary Herald*, 4 December 1953.

272. Ibid. On 3 December 1953, Manning also sent a letter to Trans-Canada and Western Pipe Lines stating the advantage of a single application, and saying that he was "quite certain" that if such an application were received the Conservation Board would quickly recommend and the government approve the required export permit.

273. *Globe and Mail* and *Calgary Herald*, 4 December 1953.

274. Robert Bothwell and William Kilbourn, p. 285.

275. UAA, E.C. Manning Papers, Interview No. 24, 3 July 1981.

276. *Calgary Herald*, 9 January 1954.

277. UAA, E.C. Manning Papers, Interview No. 24, 3 July 1981, pp. 42–43.

278. Ibid., Interview No. 25, 14 July 1981, pp. 38–39. Howe raised the question of what financial inducements would have to be made, only to be informed by Manning that this would not be a deciding factor in Tanner's case, serving Canada's national interest would count for much more.

279. ERCB, The Petroleum and Natural Gas Conservation Board, "Report to the Lieutenant Governor in Council With Respect to the Merged Applications under the Gas Resources Preservation Act of: (a)

Trans-Canada Pipe Lines Limited, Trans-Canada Grid of Alberta Ltd., and Canadian Delhi Oil Ltd., and (b) Western Pipe Lines, under the Name of Trans-Canada Pipe Lines Ltd.," 10 May 1954.

280. PAA, Premiers' Papers, Reel 155, 1641, Arnold R. La Force, Second Vice President, Securities, Metropolitan Life Insurance Co., to Lehman Brothers, 15 January 1954.

281. Canada, *Commons Debates*, 15 March 1956, pp. 2165-66.

282. See William Kilbourn, *Pipeline* (Toronto: Clark Irwin and Co., 1970); Earle Gray, *The Great Canadian Oil Patch* (Toronto: Maclean-Hunter Ltd., 1970); Robert Bothwell and William Kilbourn, *C.D. Howe, A Biography* (Toronto: McClelland and Stewart Ltd., 1979).

283. Perhaps Manning took into account the warning he received from Walter Campbell, an Edmonton oilman and a past president of the old Alberta Petroleum Producers Association, that Milner and his group were really just a front for U.S. interests, particularly International Utilities, the owners of the Edmonton and Calgary utilities companies over which Milner presided. For reference, see NAC, RG21, vol. 38, 161.362/2, 15 November 1955, W.S. Campbell to Dr. G.S. Hume.

284. PAA, Premiers' Papers, Reel 154, 1634, 10 December 1953. Carl Nickle to E.C. Manning.

285. To legitimize their preserve, two class B shares were set aside for the government; for the other groups, a total of 2,000 class B shares were available on application, with the number to be determined on the basis of the amount of capital invested in the province and other factors, see Alberta, *Statutes*, 1954, ch. 37, *An Act to Incorporate a Gas Trunk Pipe Line Company to Gather and Transmit Gas Within the Province.*

286. For Manning's recollections on AGTL's formation, see UAA, E.C. Manning Papers, Interview No. 27, 21 July 1981, pp. 3-8.

287. PAA, Premiers' Papers, Reel 155, 1641, for various written responses from those who attended the sessions with Manning on 19 and 26 February 1954.

288. Ibid., Reel 154, 1634, 10 December 1953, Carl Nickle to E.C. Manning.

289. Ibid., Reel 155, 1641, 17 February 1954, "Further Considerations Regarding the Trunk Line."

290. *Edmonton Journal*, 1 April 1954.

291. Ibid.

292. PAA, Premiers' Papers, Reel 155, 1641, 21 January 1954, H.R. Milner to E.C. Manning. There was speculation that Ontario Hydro might be asked to assume gas transmission responsibility.

293. Nova Corporation, AGTL, Minutes, 5 June 1954.

294. See, for example, Ibid., 29 November and 8 December 1954 and 27 May 1955.

295. The Alberta *Pipe Line Act* of 1952 assigned to the Conservation Board the responsibility to hear all pipeline construction proposals and to make a recommendation to the minister of highways for approval or rejection.

296. Ibid., 29 September 1954.

297. Gas production from oilfields was related to the prorated crude oil demand, which peaked in the summer rather than in the winter, when the damend for natural gas peaked.

298. ERCB, transcript of Alberta Gas Trunk Line Company Pipe Line Application, Hearing No. 25, 23 November 1954.

299. Nova Corporation, AGTL, Minutes, 2 November 1954, and letter of 4 November 1954, CPA to AGTL. Concern about how extensively the Conservation Board might use the trunk line system as a vehicle to market

"waste" gas led to a meeting on 17 December 1954 of TCPL, AGTL producers' representatives and the Conservation Board called by Premier Manning to discuss "whether or not some assurance could be given by the Conservation Board that when field facilities and feeder pipeline facilities were constructed that the gas production allowables from such a field would not be reduced to the point where such facilities would become uneconomic." AGTL President Will subsequently informed his fellow directors that "the Conservation Board would shortly be writing a letter to this Company on that subject and [he] felt that the terms of the letter would prove to be satisfactory to this company." Minutes, 17 December 1954.

300. Ibid., AGTL, Minutes, 12 October 1955, reporting on meeting held Saturday, October 8.

301. ERCB, transcript, Alberta Gas Trunk Line Company Pipe Line Application, Hearing No. 56, 22 November 1955, p. 10.

302. Ibid., p. 11.

303. Ibid., Exhibit No. 2, "Engineering Report on Proposed Gas Pipeline System," 5 November 1955, Introduction, p. 1.

304. Ibid., transcript, pp. 216-32.

305. Ibid., p. 227.

306. Ibid., Report, pp. 4 and 11. Nova Corporation, AGTL, Minutes, 29 November 1955.

307. Ibid., p. 4. The report implied that this seemed odd, when it was well known that TCPL had just concluded an agreement with Tennessee Gas Transmission giving Tennessee the first call on up to 200 million cubic feet of gas per day when it became available west of the Great Lakes.

308. Ibid., p. 3.

309. Ibid., pp. 4 and 9-10.

310. Ibid., p. 9.

311. Ibid., p. 8.

312. Ibid., p. 12.

313. Alberta, *Statutes,* 1952, ch. 67, s. 6.

314. NAC, RG21, vol. 38, 161.362/2, 12 January 1956, George Hume to Walter S. Campbell. Campbell was president of one of the oldest and still active Turner Valley companies, the Dalhousie Oil Co. Ltd.

315. *Oil in Canada,* 3 January 1955, for full text of speech.

316. *Calgary Herald,* 2 December 1954, see also *Edmonton Journal,* 3 December 1954.

317. PAA, Premiers' Papers, Reel 155, 1635, 3 December 1954, E.C. Manning to Louis St. Laurent.

318. Ibid., 8 December 1954, George Prudham to E.C. Manning.

319. ERCB, Permit, Trans-Canada Pipe Lines Limited, 14 May 1954.

320. PAA, Premiers' Papers, Reel 155, 1635, 21 December 1954, I.N. McKinnon to N.E. Tanner.

321. Ibid.

322. Ibid., 22 December 1954, E.C. Manning to N.E. Tanner.

323. ERCB, Permit Number TC 59-2, "Form of New Permit." Section 8 of the new permit provided that where agreement on the price to be paid for the gas could not be reached, the price would be determined by the Board of Utility Commissioners.

324. Alberta, *Statutes,* 1952, ch. 67, s. 5.

325. As it turned out, Gulf's Pincher Creek gas field did not live up to expectation. This added to the amount of natural gas that TCPL had to draw from other fields. It suggests that the originally proposed AGTL route would have proved more economical; however, a detailed engineering analysis would be necessary to confirm this.

326. PAA, Premiers' Papers, Reel 155, 1641, 9 November 1956, W.F. Knode to E.C. Manning.

327. Canadian Petroleum Association, Board of Directors Minutes. "Notes of the Meeting With Premier Manning, November 25th, 1954." At this meeting, a five-person delegation from the Association met with Manning, McKinnon and Govier to discuss various gas export concerns, one of which was the question of "priority of residue gas from any condensate or distillate fields . . . over dry gas which in periods of over-supply would have to be cut back." Although the industry expressed recognition that there would be an overall recognition of conservation, it cautioned that the continued development of the industry "should not be prejudiced by subsidizing short-term conservation requirements or local consumption situations."

SEVEN—CONSERVATION AND THE STRUGGLE TO EXPAND THE MARKET FOR ALBERTA CRUDE OIL

1. ERCB, OGCB, "Outlook for Alberta Oil," R.J. Cooper and W.D. Keller, 14 November 1958.

2. ERCB, Hearing No. 85, 27–28 May 1957, "Submissions to Proration Hearing."

3. Ibid. Almost all exploratory drilling in western Canada was done by drilling contractors. Contracts were usually arranged on a combination footage and daywork basis.

4. Ibid.

5. Ibid.

6. Ibid.

7. Ibid., Mobil's submission provided detailed information on investment and operating costs for a 5,000-foot well, which indicated that with oil at $2.50 per barrel and an allowable of 42 b/d (the existing economic allowance rate) a four-year pay-out would result. Its analysis excluded income taxes and finding costs, but made allowance for the crown royalty of 11.4%.

8. Ibid.

9. Ibid.

10. ERCB, "Letters to all Operators," Re: Proration Plan and the Economic Allowance, 30 August 1957.

11. Industry opinion concerning the relative weight that should be assigned to the economic allowance and the MPR in the proration plan fell into three groups. Those who favoured (1) the present plan, wherein the prorated allowable for a well was calculated as the economic allowance plus some fraction of the MPR; (2) the "residual MPR plan," wherein the prorated allowable was equal to an economic allowance plus some fraction of the difference between the MPR and the economic allowance; and (3) the "floor allowance plan," wherein the prorated allowable was equal to some fraction of the MPR, but not less than a "floor allowance."

12. *Oil in Canada*, 7 October 1957.

13. Ibid., 4 November 1957.

14. *Daily Oil Bulletin*, 15 November 1957.

15. PAA, Premiers' Papers, Reel 187, 2030A, 17 October 1957, I.N. McKinnon to E.C. Manning.

16. Ibid.

17. Ibid., 8 November 1957, E.C. Manning to J.G. Diefenbaker.

18. Ibid., 8 November 1957, E.C. Manning to Donald Fleming.

19. Ibid., 9 December 1957, E.C. Manning to W.M.V. Ash, president, Shell Oil of Canada Ltd., Toronto. Others at the meeting were A.F. Campo, president, Canadian Petrofina Ltd., Montreal; T.C. Twyman, president, McColl-Frontenac Oil Co. Ltd., Montreal; M.S. Beringer, president, British American Oil Co. Ltd., Toronto; R.J. Hull, president, Cities Service Oil Co. Ltd., Toronto; W.T. Askew, president, Sun Oil Co. of Canada, Toronto; J.G. Case, vice-president and director, and F.W. Bartlett, director, Socony Mobil Oil Co. Inc., New York; Harold Rea, president, Canadian Oil Companies, Toronto; J.R. White, president, Imperial Oil, Toronto; R.G. Follis, chairman of the board, and T.S. Peterson, president, Standard Oil of California, San Francisco; and E.R. Filley, senior vice-president, The Texas Company, New York.

20. Ibid., 31 December 1957, W.O. Twaits to E.C. Manning.

21. Ibid.

22. Refinery capacity had expanded much faster in Quebec than in Ontario between 1946 and 1955 (by 159,000 b/d in Quebec and by 80,000 b/d in Ontario), and a large volume of the petroleum products consumed in Ontario came from Montreal refineries and offshore crude oil. See Ibid., 31 December 1955, W. Harold Rea, Canadian Oil Companies, Ltd., to E.C. Manning; and G.M. Furnival, "Canada's Petroleum Industry," *The Engineering Journal* 38 (August 1955), pp. 1045–46.

23. PAA, Premiers' Papers, Reel 187, 2030A, 31 December 1957, W.O. Twaits to E.C. Manning.

24. Canada, Royal Commission on Energy, *Second Report,* July 1959, pp. 94–96.

25. Canada, *Parliamentary Debates,* 15 October 1957, pp. 12–13.

26. Henry Borden, CMB, QC, was president of Brazilian Traction, Light and Power Co.

27. The Independent Petroleum Association of America and others had persuaded the U.S. government to place quotas upon various categories of imported oil, and through 1957 pressure was mounting to have those quotas made more stringent.

28. *Calgary Herald,* 27 December 1957.

29. Ibid.

30. M.L. Natland, "Project Oilsand," in the *K.A. Clark Volume* (Edmonton: Research Council of Alberta, 1963), p. 143.

31. Ibid., p. 144.

32. For a discussion of Karl Clark's contribution to oil-sands research, see Karl A. Clark, *Oil Sands Scientist: The Letters of Karl A. Clark 1920–1949,* edited by Mary Clark Sheppard (Edmonton: University of Alberta Press, 1989).

33. Barry Glen Ferguson, *Athabasca Oil Sands: Northern Resource Exploration 1875–1951* (Regina: Canadian Plains Studies Centre and Alberta Culture, 1985), p. 91. Ferguson's is the best of the several accounts that deal with oil-sands development. See also J. Joseph Fitzgerald, *Black Gold With Grit. The Alberta Oil Sands* (Sidney, B.C.: Gray's Publishing Ltd., 1978).

34. Ferguson, p. 118.

35. Ibid., pp. 125–28.

36. S.M. Blair, "Report on the Alberta Bituminous Sands," Alberta, *Sessional Paper,* No. 49, 1951, p. 1.

37. In 1950, Edmonton prices for Redwater, Leduc and Golden Spike crude oil averaged $2.73, $3.05 and $3.03 respectively.

38. Board of Trustees Oil Sands Project, *Proceedings, Athabasca Oil Sands Conference* (Edmonton: King's Printer, 1951).

39.   H.H. Somerville, *A History of Crown Rentals, Royalties and Mineral Taxation in Alberta to December 31, 1972* (Edmonton: Alberta Energy and Natural Resources, 1977), pp. 3–3 to 3–6. Formerly, an applicant could obtain a small tract of land upon which to install production equipment. If he was able to demonstrate his process to the satisfaction of the minister after two years, he could *then* apply for a bituminous lease of up to 3,840 acres. The new bituminous sands policy also distinguished the bituminous sands area from conventional oil lands in that it did not provide for the setting aside of crown reserves.

40.   Ferguson, p. 145.

41.   PAA, Premiers' Papers, Reel 154, f1630, "Oil Sands (Bituminous Sands)." Reference is to ss. 3 (c) and 16 of the *Oil and Gas Resources Conservation Act, 1950.*

42.   Ibid.

43.   CPA, Board of Directors Minutes, 17 February 1955.

44.   *Calgary Herald* and *Edmonton Journal,* 24 March 1955. In addition to Great Canadian Oil Sands Ltd., Can-Amera Oil Sands Development Ltd. of Calgary was believed to have completed designs for a full-scale separation plant. Eleven other companies had active exploration programs under way in the Fort McMurray region. See also *Edmonton Journal,* 2 March 1955, "Oil Sands Lease Terms Outlined."

45.   Alberta, *Statutes,* 1955, ch. 57, "An Act Relating to Statutes Affecting Bituminous Sands Operations."

46.   PAA, Premiers' Papers, Reel 187, 2033, 22 July 1958, G.W. Govier to E.C. Manning.

47.   Ibid.

48.   Ibid., 22 October 1958, G.W. Govier to E.C. Manning. The meeting was attended by 18 persons from the Departments of Agriculture, Economic Affairs, Health, Highways, Lands and Forests, Municipal Affairs, Mines and Minerals, Public Welfare, and from the Alberta Research Council and the Oil and Gas Conservation Board.

49.   Ibid.

50.   *Calgary Herald,* 2 February 1959.

51.   Ibid., 3 February 1959.

52.   ERCB, Oil Sands—Nuclear Test, Press Conference, 13 February 1959, transcript.

53.   Ibid., p. 6.

54.   Ibid., p. 10.

55.   Ibid.

56.   Ibid., p. 16.

57.   Ibid., p. 17.

58.   Ibid., p. 20.

59.   Ibid., pp. 24–25.

60.   Ibid., p. 27.

61.   Ibid., p. 29.

62.   Ibid.

63.   Ibid., p. 33.

64.   Ibid., p. 30. The heat retention presumed detonation in a "dry medium" (where there was no flow through of fluid), which Natland and his colleagues believed was characteristic in the Beaverhill Lake limestone and McMurray oil-sand formations.

65.   Ibid., p. 32. The exposure was 6,000 milliroentgens for 44 hours.

66.   See, for example, *Edmonton Journal* and *Calgary Herald,* 14 February 1959.

67.   PAA, Premiers' Papers, Reel 187, 2033, 17 February 1959, G.W. Govier to E.C. Manning. The others on Govier's list were Dr. D.A.L. Dick,

a therapeutic radiologist from the Department of Public Health; Dr. G. Garland, a geophysicist at the University of Alberta; Dr. C.P. Gravenor, a geologist at the Alberta Research Council; and Dr. H.E. Gunning, a chemist at the University of Alberta.

68. ERCB, Oil Sands—Nuclear Test. See, for example, "Evaluation of Ground-Water Conditions in the Grand Rapids Formations, Bohn Lake—Fort McMurray Area," prepared by Richfield and Imperial Oil's committee on groundwater studies.

69. Ibid., 24 April 1959, F. Manyluk to G. Govier, with enclosure, "Project Cauldron: Operational Concept."

70. Ibid., Alberta Technical Committee, Minutes, 27 April 1959.

71. Ibid., Alberta Technical Committee, Minutes of a Special Joint Meeting, 9 May 1959. Curiously, it was Dr. A. Booth of the Department of National Health and Welfare who expressed the concern that "meaningful" test results should not be prejudiced by unnecessarily restrictive safety requirements. Dr. Govier expressed the contrary opinion that safety was the primary concern.

72. Ibid., Minutes, 15 June 1959, p. 5.

73. Natland, "Project Oilsand," p. 155.

74. ERCB, Oil Sands—Nuclear Test, 6 July 1959, "Some Aspects of Public Health and Safety," and 17 August 1959, "Some Aspects of Public Health and Safety."

75. Ibid., 30 December 1959, and 19 May 1960, D.A.L. Dick to D.R. Craig, Secretary, Alberta Technical Committee.

76. PAA, Premiers' Papers, Reel 187, 2033. See, for example, 17 September 1959, S. Weijer, radiation geneticist, University of Alberta, to E.C. Manning; 29 September 1959, Rev. G.G. Pybus, Edmonton Presbytery United Church of Canada, to E.C. Manning.

77. UAA, E.C. Manning Papers, Interview No. 28, 12 August 1921, p. 34.

78. Arthur M. Johnson, *The Challenge of Change: The Sun Oil Company, 1945–1977* (Columbus: Ohio State University Press, 1983), p. 128. Canadian Oil Companies (later Shell Canada) agreed to take the remaining production.

79. Ibid., pp. 8–19.

80. Ibid., p. 130.

81. Ibid., p. 142, quoting J. Howard Pew's remarks at the GCOS plant opening, 30 September 1967.

82. Ibid., p. 129. J. Howard Pew told the Presbyterian Men's Council in March 1965 that Premier Manning preached "one of the greatest sermons I have ever heard."

83. UAA, E.C. Manning Papers, Interview No. 27, 21 July 1960, pp. 29–30.

84. Alberta, *Statutes*, 1960, ch. 74, *An Act to Amend The Oil and Gas Conservation Act*. The amendment also gave the Board authority to make whatever regulations governing the recovery of oil or the abandonment of operations in the oil sands it saw fit. Exemption of oil sands synthetic crude from the provincial crude oil allocation system also was confirmed.

85. ERCB, Oil and Gas Conservation Board, "Report to the Lieutenant Governor in Council with respect to the Application of Great Canadian Oil Sands Ltd.," November 1960, p. 32.

86. Ibid., pp. 38 and 45. The presence of Dr. K.A. Clark should not be confused with that of T.P. Clarke, managing director of GCOS, who also appeared at the hearings to give evidence and respond to certain of the interveners' queries.

87. Ibid., p. 77.

88. Ibid., p. 80.

89. Ibid.

90. Canada, Royal Commission on Energy, *First Report* (October 1958), pp. 90–91. In addition to the Chairman Henry Borden, the other appointed commissioners were J.-Louis Lévesque, an investment dealer in Montreal; George Edwin Britnell, an economist at the University of Saskatchewan; Gordon G. Cushing, vice-president of the Canadian Congress of Labour in Ottawa; Dr. Robert D. Howland, an economist and adviser to the Nova Scotia government; and Leon J. Ladner, a prominent Conservative and lawyer from Vancouver. Cushing, who soon resigned to become assistant deputy minister of labour, was replaced by Dr. R.M. Hardy, Dean of the Faculty of Engineering at the University of Alberta.

91. PAA, Premiers' Papers, Reel 187, 2031, 21 January 1958, E.C. Manning to J.G. Diefenbaker.

92. D.P. Goodall, pp. 146–49.

93. Ibid., pp. 158–61. What opponents found particularly alarming was that Alberta and Southern had contracts or letters of agreement for over 80% of the gas that they proposed to purchase. In addition, these agreements covered not only known gas reserves but also options from major mineral rights holders for their gas "when" it was discovered.

94. *Calgary Herald,* 3 February 1958.

95. Canada, Royal Commission on Energy, Transcript of Hearings, 3 February 1958, p. 9.

96. Ibid., p. 10.

97. *Calgary Herald,* 5 February 1958. See also Govier's testimony regarding the submission, *Calgary Herald,* 6 February 1958.

98. *Calgary Herald,* 7–11 February 1958.

99. Ibid., 12 February 1958. In a subsequent submission to the Commission, B.C. Electric expressed strong objection at having to pay 32¢ when the export price at the border was 22¢ per 1,000 cubic feet. See also *Calgary Herald,* 19 February and 24 April 1958. The revelation caused a furore in Vancouver. See Vancouver *Province* editorial, 22 April 1958.

100. *Calgary Herald,* 15 February 1958.

101. Ibid., 17 February 1958.

102. Ibid., 19 February 1958.

103. Ibid., 19 and 25 February 1958. The seven large volumes that comprised the TCPL brief provided detailed information on the company's organization and details of employment and stock option agreements between the company and some of its officials, including Board Chairman N.E. Tanner and President Charles S. Coates.

104. Ibid., 1 March 1958.

105. Ibid., 27 February 1958.

106. Ibid., 6 February 1958.

107. Ibid., 18 March 1958.

108. Ibid., 19 March 1958. Manning explained that the delay was caused by the government's decision to wait until the most recent round of gas export hearings was completed before the Conservation Board and the latest information was available.

109. Canada, Royal Commission on Energy, Transcript of Hearings, 29 April 1958, p. 3990.

110. Ibid., pp. 3991–92.

111. Canada, *Royal Commission on Canada's Economic Prospects, Final Report,* November 1957, pp. 146–47.

112. Canada, Royal Commission on Energy, Transcript of Hearings, 29 April 1958, p. 3996.

113. Ibid., pp. 3996–97.

114. *Calgary Herald,* 8 April 1958.

115. Philip Smith, *The Treasure-Seekers: The Men Who Built Home Oil* (Toronto: Macmillan, 1978), p. 192.

116. Associated with Home Oil and Western Decalta were Canadian Devonian Petroleums Ltd., Canadian Homestead Oils Ltd., Canpet Exploration Ltd., Colorado Oil and Gas Ltd., Consolidated East Crest Oil Company Ltd., Consolidated Mic Mac Oils Ltd., Medallion Petroleums Ltd., Merrill Petroleums Ltd., Okalta Oils Ltd., and Westburne Oil Company Ltd.

117. Canada, Royal Commission on Energy, "Submission to the Royal Commission on Energy by Canadian Devonian Petroleums Ltd., et al.," Calgary, February 1958, I-9.

118. Ibid., II-15.

119. Ibid., II-10.

120. Ibid., I-18.

121. Ibid., I-20.

122. Ibid., II-18.

123. Ibid., II-20–22. Levy's examination of the pipeline tariff structure and wellhead pricing suggested that there was room for a downward movement of about 20¢ per barrel.

124. Ibid., III-16. The duty was 10.5¢ per barrel for light oils down to 25° API, and 5.25¢ for heavier crudes.

125. Ibid., III-19. Levy pointed out that owned-supply and long-term purchase contracts "may provide the refiners with foreign crudes substantially below the delivered prices assumed in calculations based on posted prices," II-25. He also noted that the Montreal market was of special importance to the parent companies of the Montreal refineries because it provided an important source of hard currency sales, II-26.

126. Ibid., III-19.

127. Ibid., III-15. The $271 million does not include the cost of imported refined products, which totalled almost 100,000 barrels per day.

128. Ibid., Citing "Board of Transport Commissioners, Judgement in the matter of the application of Trans Mountain Oil Pipe Line Company," 13 December 1951, p. 5.

129. Ibid., Introduction, p. 10. Of the 390,000 barrels of petroleum and refined products, crude oil accounted for 290,000 barrels.

130. For Walter J. Levy's examination before the Commission, see *Calgary Herald,* 12–13 May 1958. See also Canada, Royal Commission on Energy, "Submission to the Royal Commission on Energy by Amurex Oil Co., et al," Calgary, February 1958. This submission by Amurex and eight other independent oil companies took a position between that of the Home group and the majors. They urged that strong representation be made to the U.S. government. If expanded U.S. markets did not materialize "in a reasonable period of time," *then* "immediate consideration should be given to the most practical means of commencing movement of western Canadian crude oil into the Montreal market," p. 17. In addition, they pointed out the tax advantages enjoyed by U.S. petroleum companies and urged that Canadian taxation policy be brought into line and thus end this disadvantage faced by Canadian companies.

131. *Calgary Herald,* 2 May 1958. Shell estimated that the cost per barrel to get Alberta crude to Montreal by pipeline would be $3.34, 18¢ higher than the figure given in the Levy report.

132. Ibid., 3 May 1958.

133. Canada, Royal Commission on Energy, "Submission to The Royal Commission on Energy, Imperial Oil Limited," May 1958, p. 4.

134. Ibid., pp. 39–41.

135. Ibid., p. 43.

136. Ibid., p. 45.

137. Ibid., p. 8.

138. *Calgary Herald,* 6 May 1958.

139. Ibid., 6 May 1958. See also 9 May 1958 for an account of Frawley's admonition of the British American Oil Company for failing to be "Canadian minded." For an example of the Conservation Board's role in the preparation of Manning's submission, see PAA, 72.278, 392, Department of Federal and Intergovernmental Affairs, 15 May 1958, Vernon Millard to E.C. Manning, Memorandum "Summary of Main Points of Testimony Regarding Proration to Market Demand."

140. Ibid., 8 April 1958, J.J. Frawley to E.C. Manning.

141. Canada, Royal Commission on Energy, Transcript of Hearings, 16 May 1958, pp. 5989 and 5995–96.

142. Ibid., p. 5999.

143. Ibid., p. 6004.

144. Ibid., pp. 6003–4.

145. Ibid., p. 6004.

146. *Calgary Herald,* 11 June 1958. See also, PAA, Premiers' Papers, Reel 187, 2026C, 11 June 1958, L.H. Fenerty to E.C. Manning.

147. *Calgary Herald,* 21 June 1958.

148. Ibid., 25 and 26 June 1958.

149. PAA, Premiers' Papers, Reel 187, 2030B, 26 June 1958, I.N. McKinnon to E.C. Manning.

150. Ibid., Reel 187, 2031, 27 June 1958, E.C. Manning to J. Diefenbaker.

151. Canada, Royal Commission on Energy, "Supplementary Submission to The Royal Commission on Energy, Imperial Oil Limited," July 1958, I-1. Ontario accounted for approximately 40% of the total Canadian demand for petroleum products.

152. Ibid., I-2.

153. Ibid., II-6.

154. Ibid., II-2.

155. Ibid., II-9.

156. Ibid., I-4.

157. NAC, RG99, vol. 83, M/18/2, 15 July 1958, E.C. Manning to H. Borden.

158. Ibid., p. 5. Manning also pointed out that Venezuela's "exploration concession" was similar to Alberta's Petroleum and Natural Gas Reservation in that, after a three-year period, the company could convert up to 50% of this average into an "exploitation concession," with the rest of the land returning to the Venezuelan government. There is evidence to suggest that the Venezuelan 50% concession reversion was modelled upon that in Alberta.

159. Ibid., p. 7.

160. PAA, Premiers' Papers, Reel 187, 2026C, 7 August 1958, J.J. Frawley to E.C. Manning, regarding Toronto *Globe and Mail,* 7 August 1958, "Alberta Oil Seeks Wider Markets."

161. *Edmonton Journal,* 6 August 1958.

162. PAA, 72.278/33, 392, "Suggested Statement By Premier Re: Vernon Taylor's Speech to Edmonton Kiwanis Club, August 5th, 1958."

163. Ibid., 11 August 1958, J.J. Frawley to I.N. McKinnon.

164. PAA, Premiers' Papers, Reel 187, 2026D, 30 September 1958, H.J. Wilson, Deputy Attorney General, to E.C. Manning.

165. ERCB, Chairmen's Correspondence, 29 August 1958, I.N. McKinnon to H.J. Borden.

166. Canada, Royal Commission on Energy, *First Report,* October 1958, p. x.

167. Ibid., p. xi.

168. Ibid., p. xiii. The NEB would nonetheless be required to submit an annual report to the minister of trade and commerce.

169. See, for example, *Edmonton Journal,* 14 February 1959, editorial, "Another 'Killing' In Alberta Gas."

170. Canada, Royal Commission on Energy, *First Report,* p. viii. Author's italics.

171. Ibid., p. 54.

172. Ibid., pp. viii and 59–65.

173. Ibid., pp. 78–83. The *Calgary Herald,* 27 October 1958, noted that Tanner's option of 60,000 shares at $8 each would have yielded a return of $1,457,000, in mid-September 1958, when the stock was selling for $34.50 a share. On the Commission's findings regarding Tanner's stock options, Manning reminisced, "the criticism was 90% political, and I think it was very, very unfair." UAA, E.C. Manning Papers, Interview No. 25, 14 July 1981, p. 42.

174. *Calgary Herald,* 27 October 1958.

175. Ibid.

176. Ibid., 30 October 1958.

177. *Oil in Canada,* 3 November 1958.

178. Ibid., 10 November 1958. M.J. Coldwell was the federal leader of the "socialist" CCF party and Solon Low was the federal leader of the Social Credit party.

179. Ibid.

180. Ibid., 24 November 1958.

181. Ibid., 8 December 1958.

182. Ibid.

183. Ibid., pp. 10–11. Nordegg, formerly a coal-mining town, in the Alberta foothills.

184. *Calgary Herald,* 15 December 1958.

185. Ibid., 30 October 1958.

186. ERCB, Goodall, p. 150. See also ERCB, Transcript of Hearing, No. 120, 6–20 January 1959.

187. Ibid., p. 173.

188. Ibid., pp. 110–12.

189. *Calgary Herald,* 8 January 1959. See also ERCB, Transcript of Hearing, No. 120.

190. *Calgary Herald,* 14 January 1959.

191. Ibid., 15 January 1959.

192. Ibid., 20 January 1959.

193. PAA, Premiers' Papers, Reel 187, 2026E, 21 January 1959, "Special Gas Committee to His Worship the Mayor and Members of Council."

194. Ibid.

195. Ibid.

196. Ibid. The Calgary delegation consisted of Mayor D. MacKay, Alderman J.J. Hanna, S.J. Helman, QC, and S.J. Davies. See also *Calgary Herald,* 2 and 7 February 1959.

197. PAA, Premiers' Papers, Reel 187, 2026E, "City of Calgary Submission to the Rt. Hon. John Diefenbaker, Prime Minister of Canada," January 1959, p. 2.

198. Ibid., p. 5.

199. Ibid., 4 February 1959, "Mayor MacKay's Address Over All Calgary Radio Stations."

200. Ibid., See "Statement of Bruce F. Willson, Executive Vice President Canadian Western Natural Gas Company Ltd. In reply to Mayor D.H. MacKay," "Radio Broadcast by Carl Nickle." "Radio Script for A.G. Bailey."

201. Ibid., 10 February 1959, John W. Proctor, General Manager, CPA, to E.C. Manning.

202. Ibid., 27 February 1959, E.C. Manning to J.G. Diefenbaker. See also in this file "Canadian Petroleum Association Submission to the Prime Minister of Canada."

203. Ibid. Among the provisions drawn to attention were s. 46 of the *Oil and Gas Conservation Act,* which authorized the Conservation Board to order the owner or operator of a well, gas pipeline or processing plant to deliver gas to such person or distribution system as it might name, and s. 58 of the *Public Utilities Act,* which empowered the Board of Public Utility Commissioners to change the rates or charges levied under contract by a utility.

204. *Calgary Herald,* 7 February 1959.

205. In February, the monthly production allowable peaked at 412,859 barrels per day; by May, it stood at 301,674 barrels.

206. See, for example, Bruce Phillips, "Tories Face Energy Board Dilemma," *Calgary Herald,* 4 February 1959, and PAA, Premiers' Papers, Reel 187, 2026E, 6 March 1959, J.J. Frawley to E.C. Manning.

207. *Edmonton Journal* and *Calgary Herald,* 27 May 1959. It was also known that Ottawa had also approached Hubert Somerville, Alberta's deputy minister of mines and minerals on the possibility of assuming the energy board chairmanship. See interview with Somerville in Aubrey Kerr, *Leduc* (Calgary: by author, 1991), pp. 257–64.

208. *Edmonton Journal* and *Calgary Herald,* 27 May 1959.

209. See, for example, PAA, 72.464/65, 26 April 1960, I.N. McKinnon to E.C. Manning. Manning and Frawley were being consulted on amendments to the *National Energy Board Act.*

210. Ibid., s. 22.

211. Ibid., s. 22.

212. Both the federal *Pipelines Act* and the *Exportation of Power and Fluids and Importation of Gas Act* were repealed.

213. Canada, *Statutes,* 1959, ch. 46, s. 87.

214. Also elected in 1955 were 2 CCF, 3 Conservatives, 1 Independent, 1 Independent Social Credit, 1 Liberal-Conservative and 1 Coalition.

215. Caldarola, p. 373.

216. PAA, Premiers' Papers, Reel 187, 2026F, 22 June 1959, J.J. Frawley to E.C. Manning.

217. The Canadian-American Committee was established in 1957 by a group of senior Canadian and U.S. businessmen "to study problems arising from growing interdependence between Canada and the United States." Co-chairmen of the governing executive committee in 1959 were Robert M. Fowler, president of the Canadian Pulp and Paper Association, and R. Douglas Stuart, chairman of the board of the Quaker Oats Company. John Davis was the author of the important volume on Canada's energy resources prepared for the Gordon Commission.

218. NAC, RG99, vol. 98.92, 17 February 1959, "Notes on Visit to Western Oil Industry," D.H. Fullerton.

219. Ibid.

220. Canada, Royal Commission on Energy, *Second Report,* July 1959, p. 144.

221. Ibid., pp. 76 and 144.

222. Ibid., p. 76.

223. Ibid., p. 144.

224. The commissioners had difficulty coming to an agreement on the text of the "Final Report." In the end, University of Saskatchewan economist G.E. Britnell insisted on attaching a strong dissenting memorandum, and R.D. Howland and R.M. Hardy added personal addenda.

225. Doern and Toner, pp. 80–81.

226. For an example of the Alberta government's controlled disappointment, see *Daily Oil Bulletin*, 9 December 1959, E.C. Manning speech to the Canadian Association of Oilwell Drilling Contractors, 8 December 1959.

227. UAA, E.C. Manning Papers, Interview No. 28, 12 August 1981, p. 23.

EIGHT—THE PETROLEUM AND NATURAL GAS CONSERVATION
BOARD: ORGANIZATION AND REGULATION OF FIELD
DEVELOPMENT, 1948–1959

1. ERCB, Oil and Gas Conservation Act, 1956–62, "CPA Conservation Report 1956," p. 5, CPA, Alberta Division, Board of Directors, Minutes, Special Meeting, 27–28 August 1956. N.A. Macleod file.

2. ERCB, CPA, "Conservation Report," p. 3.

3. Ibid., p. 5.

4. Ibid.

5. Ibid., p. 6.

6. Ibid., pp. 9 and 17.

7. Ibid., pp. 18 and 20.

8. Ibid., p. 17. According to the report, "payouts in the order of 5 years were usually considered reasonable" as a measure of whether or not a conservation proposal was "economic."

9. Ibid.

10. CPA, Alberta Division, Board of Directors, Minutes, Special Meeting, 27–28 August 1956. Companies represented at the meeting were Sun Oil, Royalite, Hudson's Bay Oil and Gas, Bailey, Selburn Oil and Gas, Amerada Petroleum, California Standard, Mobil Oil, Texaco, British American Oil, Union Oil of California, Dome Exploration, Stanolind Oil and Gas, McColl-Frontenac, Canadian Homestead Oils, Husky Oil and Refining, General Petroleums, Richfield, Home Oil, Shell, Great Plains Development and Gulf Oil.

11. ERCB, N. Macleod, file, Oil and Gas Conservation Act, CPA Submission, 31 August 1956, p. 4.

12. Ibid., p. 25, the reference is to s. 83(4).

13. Ibid., pp. 6–8. The reference is to s. 34(1)(2). Section 42, authorizing the Board to order the construction of pipelines for the purpose of conserving gas, and section 44, giving the Board authority to direct the purchase of residue gas, were also amended to conform to the prudent operator standard.

14. It was proposed that the court should have authority not only to confirm or set aside the appealed order but also, if it chose, to vary the order as it deemed appropriate. In such a case, lawyers rather than engineers would become the final arbiters of conservation practice.

15. Ibid., p. 24, the reference is to ss. 77 and 82.

16. CPA, Alberta Division, Board of Directors, Minutes, 11 and 30 October and 30 November 1956.

17. Ibid., Minutes, 23 January 1957.

18. Ibid., Minutes, 29 January 1957.

19. Ibid., Minutes, 1 February 1957.

20. Ibid.

21. Alberta, *Statutes,* 1957, ch. 63, s. 73.

22. Ibid., s. 75. In special situations, the Board had discretion to hear an application made by owners of less than 50% of the interests; however, no unitization order could come into effect without the written approval of 75% of the holders of drilling and production rights. See s. 76.

23. *Edmonton Journal,* 4 April 1957.

24. Ibid.

25. See, for example, PAA, Premiers' Papers, Reel 186, 2026B, 1 May 1957, "Revised Draft Drilling and Production Regulations." Manning's papers reveal a personal and remarkably detailed interest in the drilling and production regulations created and administered by the Board under the authority of the Conservation Act.

26. See, for example, Texas Railroad Commission Annual Reports.

27. Saskatchewan, *Revised Statutes,* 1953, ch. 327, and 1956, ch. 62.

28. ERCB, interview Govier, 8 June 1987, tape 1, p. 27.

29. Ibid.

30. ERCB. See, for example, "A Syllabus on a Proposed Training Course for Technicians and Engineers in the Technical and Engineering Aspects of Gas Work in the Field," 17 April 1957. Locate under "Gas Measurement Field Courses, 1957," in Board history files.

31. The evidence indicates that Premier Manning had complete confidence in McKinnon's management, and that he regularly approved the Board's budget as it was submitted. From 1950, the government contributed 40% of the Board's annual expenditure and the industry was assessed for the remaining portion.

32. The Texas Railroad Commission was unable to supply this information.

33. Texas, *Annual Report of the Oil and Gas Division of the Texas Railroad Commission 1958* (Austin), p. 461.

34. The OGCB's salary costs totalled $995,284.

35. U.S. salary levels were generally higher than Canadian equivalents during the 1950s, and it is possible that the Oil and Gas Division's staff in Austin and in the field might have been less than that of their Alberta counterparts. For a comparative summary discussion of U.S. and Canadian wage rates in the 1950s, see Paul Wonnacott and Ronald Wonnacott, *U.S.-Canadian Free Trade: The Potential Impact on the Canadian Economy* (Washington: National Planning Association, 1968), pp. 15–19.

36. ERCB, M.R. Blackadar, "The Devon Office—1949," August 1949. See Board history files under Field Duties.

37. Occasionally, drilling was discovered to have commenced without a licence, but rumours aside there is no record of the same owner having done it a second time.

38. ERCB, Minutes, 9 December 1953, Gas Conservation Order No. GC1.

39. Ibid., 18 February 1955.

40. ERCB, Gas Conservation Leduc Field, "McLeod, Kavanagh, Oil City, Gas Conservation Scheme," E.M. Foo and A. Warke, 7 March 1955. The Board's study indicated that "the overall average gross rate of return for the eight years of operation to be 1.64 per cent, out of which interest on bonds must be paid. The plant during its time of operation would not be depreciated to its salvage value." See also, ERCB, Minutes, 14 March 1955.

41. Ibid., 10 February 1954.

42. Ibid., 8 July 1955, Order No. C 88.

43. Ibid., 5 October 1955. The approved scheme called for gathering and processing "not less than ninety (90) percent of the gas, produced from wells in the Redwater Field" identified in Board Order GC 3.

44. ERCB, see Order No. GC5, June 1959 (Harmattan-Elkton field), and Order No. GC6, July 1959 (Innisfail field).

NINE—THE PETROLEUM AND NATURAL GAS CONSERVATION BOARD IN RETROSPECT

1. David F. Prindle, *Petroleum Politics and the Texas Railroad Commission* (Austin: University of Texas Press, 1981), p. 151. See also William R. Childs, "The Transformation of the Railroad Commission of Texas, 1917–1940: Business-Government Relations and the Importance of Personality, Agency Culture and Regional Differences," *Business History Review*, vol. 65 (Summer 1991), pp. 285–344, and Jones A. Clark, *Three Stars for the Colonel: The Biography of Ernest O. Thompson* (New York: Random House, 1954).

APPENDIX X

1. ERCB, Minutes, 15 August 1944.
2. ERCB, "Letter to All Operators," 9 April 1952.
3. ERCB, Drilling and Production Regulations under the *Oil and Gas Conservation Act* (O.C. 958/57), s. 58.
4. This evolved into the much larger and well-equipped Core Research Centre in 1983, which is expected to meet space requirements well into the next century. The earliest drill cuttings stored in the Board's facility appear to be from the Peace River Petroleum No. 1 well located in 10–31–83–21 W5M, which was completed on 20 May 1920.
5. ERCB, Board history file, entitled "Core Research Facility and Background," Subject "Chronological Account of the Core Drill Cuttings Facilities, and Storage and Examination Facilities," J.R. Pow, August 1989, p. 4. This essay offers an excellent review of Board policy on a centralized core research facility.
6. Ibid., p. 10.
7. Ibid., pp. 1–2.
8. Lyn Brown, "Core Research Centre geologists 'gusher' of raw data," *The Pegg* (Association of Professional Engineers, Geologists and Geophysicists of Alberta), 13 (September 1985), pp. 1 and 5.
9. ERCB, "Letter to All Operators," 10 March 1952.
10. Ibid., 4 April 1960.
11. ERCB, transcript of tape of Bill Unrau. First Univac, Model 120, arrived in 1958 and was put into operation in January 1959.
12. Alberta, Order-in-Council, OC No. 1193/52.
13. Alberta, the *Oil and Gas Conservation Act*, ss. 73 and 74.
14. ERCB, Pooling order no. P1. Among its provisions was clause 7, which provided that sums payable to the owner of Lot 10, Block 1, who was deceased, shall be paid to the public trustee. See Board history file entitled "Orders Issued by the Board."
15. ERCB, Hearing No. 85, transcript.
16. ERCB, Dual Completion File History Project. A review of Dual Completion Practices and Regulations, July 1957, by the Dual Comple-

tions Working Committee of the Interprovincial Petroleum and Natural Gas Committee Note recommendations.

17. ERCB, Miscellaneous Order 5813, 27 August 1958. This order was rescinded on 20 November 1958 and replaced by Order No. MU1 initiating a specific system of orders for multizone pools.

18. The Board bottom-hole pressure units used an Amerada R.P.G. bottom-hole pressure gauge, together with the equipment necessary to lower the gauge into the hole and to measure the run depth. A special bottom-hole truck was used to transport the equipment and to provide the power to raise and lower the gauge. The bottom-hole technicians, who worked alone, required considerable strength and agility as they maintained their balance on the wellhead and single-handedly attached the lubricator (housing the gauge) and bolted it into place.

19. Alberta, *Statutes,* 1950, ch. 46, s. 34 (e,f).

20. Ibid., 1957, ch. 63, the *Oil and Gas Conservation Act,* s. 37.

21. PNGCB, Order No. 1348, 26 April 1957, respecting the Joffre Viking pool.

22. PNGCB, Hearing No. 000014, 7 November 1953, respecting Leduc-Woodbend D-2A pool.

23. ERCB, report ST 91–18, p. 2–66.

24. PNGCB, Order No. Misc. 5913, 5 May 1959, respecting Pembina Cardium A pool.

25. ERCB, report ST 91–18, p. 2–88.

26. These and other more advanced recovery methods introduced in Alberta crude oil pools by 1991 had increased oil recovery from an average of 18% by natural means to 27%. Recoveries for light and medium crude oil pools were substantially better, and those from heavy crude oil pools were considerably less favourable. See ERCB, report ST 92–18, pp. 2–5.

27. ERCB, interview with H.H. Somerville, 18 July 1984, by Aubrey Kerr, p. 17 of transcript of Tape 2.

28. ERCB, "Drilling and Production Regulations," 1959. These regulations specified the pressure rating requirement for blowout prevention equipment (BOPs); the bleed-off and kill line diameters and tie-in points, as well as tie-down requirements; the two types of BOPs to be installed on wells drilled below 4,000 feet, including one hydraulic (bag) type and one double-gate type; the distances that BOP controls had to be located from the rig for a well drilling below 4,000 feet; and that the drilling crew had to be able to demonstrate evidence of having received training in blowout prevention and well control.

29. ERCB, "Drilling and Production Regulations," 1950, amended, s. 22.

30. ERCB, History Project File, Casing, Surface Vent, article by R.W. Edgecombe.

31. ERCB, Schedule of Wells 1954, Canadian Prospect Big Valley No. 2 well.

32. ERCB, "Drilling and Production Regulations," 1957, s. 51.

33. One of the more serious cases was the false reporting of production in the Stettler field by Gulf Oil. See ERCB, Minutes, 3 June 1955, with letter 3 June 1955, I.N. McKinnon to E.D. Loughney, Vice-president, Canadian Gulf Oil Company.

34. Alberta, the *Oil and Gas Wells Act,* 1931, s. 38(5).

35. ERCB, "Vermilion Oil Field," 12 May 1943, D.P. Goodall to M.D. Kemp. See Board history file, "Vermilion Field Inspection, 1943."

36. ERCB, Minutes, 3 September 1954, Order No. W2, "An order relating to the disposal of salt water in the Joarcam Field."

37.   See, for example, ERCB, Minutes, 2 July 1954; 28 February, 3 March and 12 April 1955; 9 August and 7 November 1957; 6 October and 2 December 1958.

38.   IOCC, Research Committee, *Production and Disposal of Oilfield Brines in the United States and Canada* (Oklahoma City: IOCC, 1960), p. 59. It is important to note that in these older producing regions the volume of salt water produced was also much higher than in Alberta. Total brine production was two to 16 times total oil production and represented an immense disposal problem.

39.   Ibid., pp. 62–65. In April 1967, the Texas Railroad Commission issued a state-wide rule prohibiting disposal in surface pits as of 1 January 1969. TRC, "Special Order Amending Rule 8 of the General Conservation Rules of Statewide Application as recompiled April 16, 1964." (3 April 1967).

40.   D. Schmeekle, "Application of automatic control to oil production in Western Canada," *Oil in Canada,* 16 July 1956.

41.   ERCB, Board history file, "Automatic Custody Transfer."

42.   ERCB, Development Department "Operations Summary," 1958.

43.   ERCB, Letters to Operators, November 1957, "Lease Clean-ups."

44.   See, for example, ERCB, "Report of Examiners," 8 November 1956, regarding operations at Hargal No. 1 Battery.

45.   ERCB, Minutes, 4 June 1957.

SELECTED BIBLIOGRAPHY

The Petroleum Industry in Alberta
to 1961

The following selected bibliography is intended to assist those primarily interested in the history of the petroleum industry in Alberta to 1961. Included are primary and secondary sources that focus directly upon, or at least shed important light on some aspects of the oil and gas sector in Alberta, and which proved to be helpful for this study. The order in which the bibliography is arranged is as follows:

MANUSCRIPT SOURCES

National Archives of Canada
Provincial Archives of Alberta
Glenbow Alberta Archives
University of Alberta Archives
Library of the Alberta Legislature
Energy Resources Conservation Board
Public Record Office, London, England
National Archives, Washington, D.C.
Other Manuscript Sources

PUBLISHED SOURCES

Canada
Alberta
Energy Resources Conservation Board
Newspapers
Periodicals
Articles
Books

UNPUBLISHED STUDIES

MANUSCRIPT SOURCES

National Archives of Canada (NAC), Ottawa

C.D. Howe Papers
Department of Energy, Mines and Resources
Department of the Interior
     Dominion Lands Branch
Department of Mines and Resources
Department of Mines and Technical Surveys
External Affairs
     Washington Embassy
Geological Survey of Canada
Munitions and Supply
     Wartime Industries Control Board
National Energy Board
Northern Affairs Program
     Mining Lands and Yukon Branch
Northern Ontario Pipeline Corporation
Privy Council Office
     Minutes of the Cabinet War Committee and its predecessor

Provincial Archives of Alberta (PAA), Edmonton

Alberta, Legislative Assembly, *Alberta Scrapbook Hansard*
Department of Energy and Natural Resources
Department of Federal and Intergovernmental Affairs
Premiers' Papers (microfilm)
     H. Greenfield
     J.E. Brownlee
     W. Aberhart
     E.C. Manning
W.S. Herron Papers
Turner Valley Gas Conservation Board
Public Utilities Board

Glenbow Alberta Institute (GA), Calgary

The holdings at the Glenbow Archives in Calgary include the papers of numerous small oil companies and many individuals involved in the petroleum industry. Interviews prepared for the Petroleum Industry Oral History Project are also available. The collections most helpful to this study were the following:

Canadian Pacific Railway
Canadian Petroleum Association
Royalite Oil Company
A.W. Dingman Papers
Interviews
     Elmer Berlie
     Margaret Knode

D.P. McDonald
Robert Welch

University of Alberta Archives (UAA), Edmonton

Karl A. Clark Papers
E.C. Manning (interview transcript)
William Pearce Papers
Research Council of Alberta

Library of the Alberta Legislature (Edmonton)

Transcript of Hearings, and Submissions to the *Royal Commission appointed by the Government of the Province of Alberta Under the Public Inquiries Act to Inquire Into Matters Connected with Petroleum and Petroleum Products* (McGillivray Commission, 1939).

Energy Resources Conservation Board, Calgary (ERCB)

Annual Report of Field Operations 1957–1962
Alberta Petroleum Association—Conservation Board
    Advisory Committee 1942–1949
Alberta Legislative Assembly, 1938 (Special Session), "Evidence Taken Before the Agriculture Committee in Connection with Bill No. 1, An Act for the Conservation of the Oil and Gas Resources of the Province of Alberta."
Atlantic No. 3, General File
Chairmens' Correspondence
Decision Reports
Drilling and Production Regulations, 1930–1956
Govier, G.W., "Addresses and Publications"
Hardwicke, Robert E., Correspondence 1949
Hearings: Transcripts and Submissions
History Project Files
Letters to Operators, 1931–1960
Minutes
    Turner Valley Gas Conservation Board, April-November 1932
    Petroleum and Natural Gas Conservation Board, 1938–1957
    Oil and Gas Conservation Board, 1957–1971
    Energy Resources Conservation Board from 1971
Report, Transcript of Hearings, and Submissions to *The Province of Alberta Natural Gas Commission Enquiry into Reserves and Consumption of Natural Gas in the Province of Alberta* (Dinning Commission, 1949).
Tanner, N.E., Correspondence 1938

Interviews

A large number of oral histories have been collected at the ercb. Those used for this study were the following:

Bohme, Vic. E.
Connell, Gordon

Craig, Doug
Edgecombe, Rod
Epstein, William
Goodall, D.P.
Govier, George
Manyluk, A.F.
Millard, V.
Pow, J.R.
Somerville, H.H.
Stabback, J.

Public Record Office (PRO), London, England

POWE 33,287, "Petroleum: Canada-Alberta-Turner Valley, 1937–1938"
POWE 33,289, "Petroleum: Canada-Alberta-Turner Valley, 1939–1947"

National Archives (NA), Washington, D.C.

General Records of the Department of State
    John Hickerson files 1941–47
    Office of Intelligence and Research
    Petroleum Division
National Security Council
National Security Resources Board
Office of Strategic Services
Petroleum Administration for War
Records of Foreign Service Posts
    Records of Diplomatic Posts
    Records of Consular Posts
Records of the U.S. Joint Chiefs of Staff

Other Manuscript Sources

Alberta Gas Trunk Line (AGTL), Nova Corporation, Calgary
    Minutes, from 1954
Canadian Petroleum Association, Calgary (CPA)
    Minutes, Western Canada Petroleum Association 1948–1952
    Minutes, CPA, from 1952
Imperial Oil Company Ltd. Corporate Archives, Toronto

PUBLISHED SOURCES

Canada

Parliamentary Debates.
Royal Commission on Canada's Economic Prospects, Final Report,
    1957.
Royal Commission on Energy, First Report, 1958; Second Report,
    1959.
Slipper, S.E. *Manual for Operators Under Oil and Gas Regulations.* De-
    partment of the Interior, 1922.

Alberta

Blair, S.M. Report on the Alberta Bituminous Sands, Alberta, *Sessional Paper*, No. 49, 1951.
The Case for Alberta. Edmonton: King's Printer, 1938.
Clark, K.A., ed. *Proceedings*, Athabasca Oil Sands Conference, September 1951. Edmonton: Board of Trustees, Oil Sands Project, 1951.
Department of Lands and Mines. *Annual Report, 1930–1949*.
Department of Mines and Minerals. *Annual Report* from 1950.
Irwin, J.L. *Alberta, Oil Province of Canada 1938*. Edmonton: Department of Lands and Mines, 1939.
The Report of a Royal Commission Appointed by the Government of the Province of Alberta Under the Public Inquiries Act To Inquire Into Matters Connected With Petroleum and Petroleum Products. Edmonton: Imperial Oil Limited, 1940.
Statutes of Alberta from 1926.

Energy Resources Conservation Board

Alberta Petroleum Industry, Annual Report 1940 to 1950.
Alberta Oil and Gas Industry, Annual Report, 1951–1964.
Conservation in Alberta, Annual Report, 1966–1981.
Goodall, D.P., *Gas Export 1950–1960*, 1961.
PNGCB, *Schedule of Wells Drilled For Oil and Gas*, from 1938.
PNGCB, *Interim Report With Respect to the Applications now before the Board for Permission to Remove Gas or cause it to be Removed from the Province under the Provisions of the Gas Resources Preservation Act*, 1951.
PNGCB, *Report With Respect to the Applications Under the Gas Resources Preservation Act of: Westcoast Transmission Company Ltd. and Westcoast Transmission Company Ltd. (Alberta Incorporation); Alberta Natural Gas Company, Alberta Natural Gas Grid, Ltd. and Northwest Natural Gas Company; Western Pipe Lines, Prairie Pipe Lines Ltd. and Prairie Transmission Lines Ltd.; McColl-Frontenac Oil Company Ltd. and Union Oil Company of California; Canadian Delhi Oil Ltd. and Trans-Canada Pipe Lines Ltd.*, 1952.
PNGCB, *Report to the Lieutenant Governor in Council With Respect to the Applications under the Gas Resources Preservation Act of: Canadian-Montana Pipe Line Co.; Trans-Canada Pipe Lines Ltd.; Trans-Canada Grid of Alberta Ltd. and Canadian Delhi Oil Ltd.; and Western Pipe Lines*, 1953.
PNGCB, *Report to the Lieutenant Governor in Council With Respect to the Merged Applications under the Gas Resources Preservation Act of: (a) Trans-Canada Pipe Lines Ltd.; Trans-Canada Grid of Alberta Ltd. and Canadian Delhi Oil Ltd.; (b) Western Pipe Lines, under the Name of Trans-Canada Pipe Lines Ltd.*, 1954.
PNGCB, *Report on an Application of Great Canadian Oil Sands Limited Under Part VI A of the Oil and Gas Conservation Act*, 1964.
Report of the Operations of the Energy Resources Conservation Board. Annual Report, 1963 to 1978.

Submission by the Alberta Energy Resources Conservation Board to the Restrictive Trade Practices Commission, 1983.

Newspapers

Calgary Albertan
Calgary Herald
Daily Oil Bulletin
Edmonton Bulletin
Edmonton Journal
Turner Valley Observer

Periodicals

Imperial Oil Review
Oil in Canada
Weekly Oil Bulletin
Western Oil Examiner

Articles

Ballem, John B. "Pipelines and the Federal Transportation Power." *Alberta Law Review* 29, no. 3 (1991): 617–32.

Beach, Floyd K. "History of Land Regulations and Land Grants in Western Canada." *Canadian Oil and Gas Industries* 7, nos. 5 and 6 (May and June 1954).

Breen, D.H. "Anglo-American Rivalry and the Evolution of Canadian Petroleum Policy to 1930." *Canadian Historical Review* 62, no. 3 (1981): 283–304.

————. "The CPR and Western Petroleum, 1904–24." In *The CPR West: The Iron Road and the Making of a Nation,* edited by Hugh A. Dempsey. Vancouver: Douglas & McIntyre, 1984.

Craig, E.H. Cunningham. "The Empire's Oil: Alberta will be an Important Producer Soon." *Canadian Mining Journal* LI, no. 29 (July 18, 1930): 696–98.

Dawson, G.M. "On Certain Borings in Manitoba and the Northwest Territory." *Proceedings and Transactions of the Royal Society of Canada for the Year 1886,* Vol. IV. Montreal: Dawson Brothers, Publishers, 1887, pp. 85–99.

Debanné, J.G. "Evolution of Canadian Oil Policy and Canadian-U.S. Energy Relations." In Edward W. Erickson and Leonard Waverman, *The Energy Question,* Vol. 2, Toronto: University of Toronto Press, 1974.

Elliott, George R. "Conservation of Natural Gas: With Special Reference to Turner Valley, Alberta." *Transactions of the Canadian Institute of Mining and Metallurgy* XXXVII (1934) pp. 557–59.

"Early History of Alberta Gas and Oilfields." *Canadian Mining Journal* LI, no. 45 (Nov. 7, 1930), 107–9.

Furnival, G.M. "Canadian Petroleum Industry." *The Engineering Journal* 38 (August 1955): 1035–46.

Grommelin, Michael. "Government Management of Oil and Gas in Alberta." *Alberta Law Review* 13, no. 1, (1975): 146–211.

Khoury, Nabil. "Prorationing and the Economic Efficiency of Crude Oil Production." *Canadian Journal of Economics* 2, no. 3 (August 1969): 443–48.

McKinnon, I.N. "The Administration of Oil and Natural Gas Resources in Alberta." Proceedings of the *Institute of Public Administration of Canada* (1952): 143–66.

MacDonald, R.D. "Canadian Oil Province." *Quarterly Review of Commerce* IX, no. 4 (Autumn 1942): 270–74.

Natland, M.L. "Project Oilsand." In *The K.A. Clark Volume.* Edmonton: Research Council of Alberta, 1963.

Ross, Charles C. "Petroleum and Natural Gas Development in Alberta." *Transactions of the Canadian Institute of Mining and Metallurgy,* Vol. 29 (1926), pp. 317–46.

Thompson, Andrew R. "Sovereignty and National Resources, a Study of Canadian Petroleum Legislation." *University of British Columbia Law Review* IV, no. 2 (1969): 161–93.

————. "Basic Contrasts Between Petroleum Land Policies of Canada and the United States." *University of Colorado Law Review* 36, no. 1 (1964): 187–221.

Watkins, G.C. and J.T. Fong. "Comparisons in Economic Development: Alberta and Texas." *Journal of Energy and Development* 6, no. 1 (1980): 124.

Watkins, G.C. and L. Waverman. "Canadian Natural Gas Export Pricing Behavior." *Canadian Public Policy* XI (July 1985): 415–26.

Books

American Petroleum Association. *Petroleum Facts and Figures.* New York: American Petroleum Institute, 1959 to 1971.

Anderson, Allan. *Roughnecks and Wildcatters.* Toronto: Macmillan of Canada, 1981.

Ballem, John Bishop. *The Oil and Gas Lease in Canada.* Toronto: University of Toronto Press, 1973.

Beach, F.K. and J.L. Irwin. *The History of Alberta Oil.* Edmonton: Department of Lands and Mines, 1940.

Breen, David H., ed. *William Stewart Herron: Father of the Petroleum Industry in Alberta.* Calgary: Historical Society of Alberta, 1984.

Canadian Petroleum Association. *Statistical Yearbook.* Calgary: Canadian Petroleum Association, 1955–1971.

Clark, Karl A. *Oil Sands Scientist: The Letters of Karl A. Clark 1920–1949,* edited by Mary Clark Sheppard, Edmonton: University of Alberta Press, 1989.

de Mille, George. *Oil in Canada West: The Early Years.* Calgary: by author, 1969.

Doern, G. Bruce and Glen Toner. *The Politics of Energy: The Development and Implementation of the NEP.* Toronto: Methuen, 1985.

Durham, G. Homer. *N. Eldon Tanner: His Life and Service.* Salt Lake City, Utah: Deseret Book Company, 1982.

Embry, Ashton F., ed. *Fifty Years of Canadian Petroleum Geology.* Tulsa, Oklahoma: American Association of Petroleum Geologists, 1978.

Ferguson, Barry G. *Athabasca Oil Sands: Northern Resource Exploration 1875–1951*. Regina: Alberta Culture and Canadian Plains Research Center, 1985.

Finch, David. *Traces Through Time: The History of Geophysical Exploration for Petroleum in Canada*. Calgary: Canadian Society of Exploration Geophysicists, 1985.

Finkel, Alvin. *The Social Credit Phenomenon in Alberta*. Toronto: University of Toronto Press, 1989.

Fitzgerald, J. Joseph. *Black Gold With Grit: The Alberta Oil Sands*. Sidney, British Columbia: Gray's Publishing Ltd., 1978.

Foran, Max. *Earning Our Stripes: Fifty Years in Canada*. Calgary: Chevron Canada Resources Publishers, 1988.

Foster, Peter. *From Rigs to Riches: The Story of Bow Valley Industries Ltd*. Calgary: Bow Valley Industries Ltd., 1985.

Freeman, J.M. *Biggest Sellout in History: Foreign Ownership in Alberta's Oil and Gas Industry and the Oil Sands*. Edmonton: Alberta New Democratic Party, 1966.

Gibb, George S. and Evelyn B. Knowlton. *History of Standard Oil Company (New Jersey) The Resurgent Years 1911–1927*. New York: Harper and Brothers, 1956.

Gray, Earle. *The Great Canadian Oil Patch*. Toronto: Maclean-Hunter Ltd., 1970.

————. *Wildcatters: The Story of Pacific Petroleums and Westcoast Transmission*. Toronto: McClelland and Stewart, 1982.

Hanson, Eric J. *Dynamic Decade*. Toronto: McClelland and Stewart Ltd., 1958.

Hillborn, James D., ed. *Dusters and Gushers: The Canadian Oil and Gas Industry*. Toronto: Pitt Publishing, 1968.

House, J.D. *The Last of the Free Enterprisers: The Oilmen of Calgary*. Toronto: Macmillan, 1980.

James Richardson and Sons. *Western Canada Oils*. Winnipeg: James Richardson and Sons, 1947–1954.

————. *Western Canadian Oils: Including Natural Gas Review*. Winnipeg: James Richardson and Sons, 1955–1960.

Johnson, Arthur M. *The Challenge of Change: The Sun Oil Company 1945–1977*. Columbus: Ohio State University Press, 1983.

Kennedy, J. de N. *History of the Department of Munitions and Supply*. 1. Ottawa: E. Cloutier, 1950.

Kerr, Aubrey. *Atlantic No. 3*. Calgary: by author, 1986.

————. *Corridors of Time*. Calgary: by author, 1988.

————. *Leduc*. Calgary: by author, 1991.

Kilbourn, William. *Pipeline*. Toronto: Clark Irwin and Co., 1970.

Lambrecht, Kirk N. *The Administration of Dominion Lands, 1870–1930*. Regina: Canadian Plains Research Centre, 1991.

Larson, Henrietta M. et al. *History of Standard Oil Company (New Jersey) New Horizons, 1927–1950*. New York: Harper and Row Publishers, 1971.

Lyon, J. *Dome: The Rise and Fall of the House That Jack Built*. Toronto: Macmillan of Canada, 1983.

McDougall, John. *Fuels and the National Policy*. Toronto: Butterworths, 1982.

Nickle, C.O. *The Valley of Wonders, the Story of Turner Valley*. Calgary: The Oil Bulletin, 1942.

Pratt, L. *The Tar Sands: Syncrude and the Politics of Oil*. Edmonton: Hurtig Publishers, 1976.

Richards, John and Larry Pratt. *Prairie Capitalism: Power and Influence in the New West*. Toronto: McClelland and Stewart, 1979.

Schmidt, John. *Growing up in the Oil Patch*. Toronto: Natural Heritage/Natural History Inc., 1989.

Shaffer, Ed. *Canada's Oil and the American Empire*. Edmonton: Hurtig Publishers, 1983.

Sheep River Historical Society. *In the Light of the Flares: The History of the Turner Valley Oil Fields*. Turner Valley, 1979.

Skinner, Walter E., ed. *The Oil and Petroleum Year Book*. London: Walter E. Skinner, 1910–1961.

Smith, Philip. *The Treasure-Seekers: The Men Who Built Home Oil*. Toronto: Macmillan of Canada, 1978.

Somerville, H.H. *A History of Crown Rentals, Royalties and Mineral Taxation in Alberta to December 31, 1972*. Edmonton: Department of Energy and Natural Resources, 1977.

Stenson, Fred. *Waste to Wealth: A History of Gas Processing in Canada*. Calgary: Canadian Gas Processors Association, 1985.

Tanner, James N. *Reserves of Hydrocarbons in Alberta: A Review of Canadian Petroleum Association and Alberta Energy Resources Conservation Board Estimates and Methodology*. Calgary: Canadian Energy Research Institute, 1986.

UNPUBLISHED STUDIES

Bronson, Harold E. "Review of Legislation Pertaining to Petroleum Resources, Government of Alberta, 1930–1957" (MA thesis, University of Alberta, 1958).

Cass, Douglas E. "Investment in the Alberta Petroleum Industry, 1912–1930" (MA thesis, University of Calgary, 1985).

Conder, John E.H. "The Disposition of Crown Petroleum and Natural Gas Rights in Alberta" (MA thesis, University of Alberta, 1963).

Dagher, J.H. "Effect of the National Oil Policy on the Ontario Petroleum Refining Industry" (PhD dissertation, McGill University, 1968).

Ewing, J.S. "The History of Imperial Oil Ltd." Business History Foundation Inc., Harvard University Business School, Boston, 1951.

Finch, David A. "Turner Valley Oilfield Development, 1914–1945" (MA thesis, University of Calgary, 1985).

Grant, Hugh. "The Canadian Petroleum Industry: An Economic History, 1900–1960" (PhD dissertation, University of Toronto, 1986).

Istvanffy, Daniel I. "Turner Valley: Its Relationship to the Development of Alberta's Oil Industry" (MA thesis, University of Alberta, 1950).

Jones, Llewellyn May. "The Search for the Hydrocarbons: Petroleum and Natural Gas in Western Canada, 1883–1947" (MA thesis, University of Calgary, 1978).

Levy, W.J. "Market Outlets for Canadian Crude Oil: Problems and Prospects," presented by R.A. Brown, President of Home Oil Co. Ltd. et al. to the Royal Commission on Energy. February 1958.

Shaffer, E.H. "Employment Impact of Oil and Natural Gas in Alberta: 1961–70." [Department of Economics] University of Alberta, 1970.

Watkins, G.C. "Canadian Public Policy and the Export of Natural Gas." Discussion paper series No. 75, 1982. Department of Economics, University of Calgary.